THE MECHANICS OF SOLIDS AND STRUCTURES

Dr D. W. A. Rees

*Department of Manufacturing
and Engineering Systems
Brunel University*

McGRAW-HILL BOOK COMPANY

London · New York · St Louis · San Francisco · Auckland
Bogotá · Guatemala · Hamburg · Lisbon · Madrid · Mexico
Montreal · New Delhi · Panama · Paris · San Juan · São Paulo
Singapore · Sydney · Tokyo · Toronto

Published by
McGRAW-HILL Book Company (UK) Limited
Shoppenhangers Road, Maidenhead, Berkshire SL6 2QL
Telephone 0628   23432
Fax 0628   35895

British Library Cataloguing in Publication Data
Rees, David, 1947–
   The mechanics of solids and structures.
   1. Materials. Mechanics
   I. Title
   620.1'12

   ISBN 0-07-707222-7

Library of Congress Cataloging-in-Publication Data

Rees, David, 1947–
   The mechanics of solids and structures / David Rees.
      p.   cm.
   Includes bibliographical references.
   ISBN 0-07-707222-7
     1. Structural analysis (Engineering)   2. Strength of materials.
   I. Title.
   TA645.R39   1990   89-13684
   624.1'71—dc20   CIP

1234   93210
Illustrated by the author.
Typeset by Thomson Press (India) Limited, New Delhi.
Printed and bound in Page Bros (Norwich) Ltd.

In memory of my father
A. L. REES (1906–1984)

CONTENTS

PREFACE

This book has been developed from subject matter and examples that I have used in my teaching of solid mechanics, strength of materials and structures in universities and polytechnics over the last fifteen years. It is intended as an undergraduate degree and National Diploma text at all levels on engineering courses in which solid mechanics and structures form a part. The recently revised syllabus of the Council of Engineering Institution's part 2 examination in the mechanics of solids has also been covered. Since the contents have been selected to illustrate where overlapping topics in civil, aeronautical and materials engineering employ common principles, they should serve engineering students of all disciplines. This broadening of the subject base is also aligned to those enhanced engineering courses that have now been run successfully by academia in collaboration with UK industry for more than a decade.

A concise approach has been employed for the theoretical developments in order to provide the space for many illustrative examples. These are all worked in SI units. The solutions to selected examples are also given in Imperial units. It should become obvious that these calculations are all related to the load-carrying capacity of materials used in engineering design. Among the requirements are the choice of material, its physical shape, the assessment of the nature of imposed loading and its effect on life expectancy. The text illustrates where and how the necessary techniques are to be employed in each case. The reader will soon recognize, for example, that under elastic loading the solution to the stress and strain suffered by a material invariably becomes that of satisfying three requirements: equilibrium, compatibility and the boundary conditions.

The aim has been to provide mostly self-contained chapters with a logical and clear presentation of the subject matter. Reference to definitions given in other chapters is occasionally made in order to avoid distraction from the main argument. Here, introductory sections are included to explain the interrelation between chapters. The text is arranged with shorter, earlier chapters containing the necessary fresher material for the more advanced sophister material appearing in later chapters. Wherever possible, the use of general chapter titles has been employed in preference to chapters on specific topics, so that the more wide-ranging principles of the subject can be emphasized. This text also contains the necessary prerequisite material for specialist subject matter that has not been included, namely, the experimental measurement of stress and strain, the theoretical treatments of stress in concentration and between bodies in contact, and three-dimensional elasticity. Some recent developments in the field are reflected with an introduction to plane finite elements and by the inclusion of the final two introductory chapters on inelastic deformation, fatigue and fracture. Research into the mechanics of deformation and fracture is presently pursued world-wide. The motivation lies in the knowledge that inherent crack-like defects in the structure of a material may drastically reduce its load-bearing capacity under plasticity, creep and fatigue conditions.

Acknowledgements are made to Imperial College, London; Trinity College, Dublin; Kingston Polytechnic and to the C.E.I. for granting permission to include questions from their past examination papers as worked examples and exercises. However, all the worked solutions are those of the author and not of these consenting institutions. The author also thanks his past teachers, colleagues and students, now scattered far and wide, who have all helped to shape this work.

D. W. A. Rees

LIST OF NOTATIONS

$1, 2, u, v$	Principal directions
$a, b, B, c, d, D, h, l, L, t$	Linear dimensions
A, Δ	Areas
$[B], [D], [H], [K], [S]$	Finite-element matrices
C, T	Couple or torque
$C(t)$	Compliance
$C_c, k, n, \mu, \sigma, E$	Strut constants
D	Flexural rigidity
D_x, D_y, i_x, i_y	First moments of area
e_u, e_v, e_x, e_y, h, k	Eccentricities
e_{xy}, e_{xz}, e_{yz}	Angular distortions
E, G, K, v	Elastic constants
E_t	Tangent modulus
$\mathbf{f}, \mathbf{s}, \mathbf{t}$	Force vectors
f, k	Flexibility and stiffness
F, Q	Shear forces
$F_u, F_v, F_x, F_y, S_x, S_y, S_z$	Cartesian forces
$\hat{F}_x, \hat{F}_y$	Equivalent forces
$h, \bar{x}, \bar{y}, \bar{z}, \bar{X}, \bar{Y}$	Centroid positions
I_u, I_v	Principal second moments of area
I_x, I_y, I_{xy}	Second moments of area
$\mathbf{I}$	Moment-of-area matrix
I_z, J	Polar second moments of area
I_1, I_2, I_3	Strain invariants
J_1, J_2, J_3	Stress invariants

k, m, p	Unit loadings
K_I, K_{II}, K_{III}	Stress intensities
K_{Ic}, G_{Ic}	Fracture toughnesses
K_f	Fatigue strength reduction factor
K_r, K_s	Buckling coefficients
K_t	Stress concentration
l, m, n	Direction cosines
$l_{ij}, \mathbf{L}$	Transformation matrices
L, Q, S	Load, shape and safety factors
M	Bending moment
M_r, M_θ	Polar moment components
M_x, M_y	Cartesian moment components
$\hat{M}_x, \hat{M}_y$	Equivalent bending moments
n, N	Cyclic life
N, ω	Rotational and angular speed
$\mathbf{N}$	Normal force
P, P', P'', W	Concentrated loading
q	Shear flow
Q	Heat transfer and activation energy
$r_i, r_o, R_r, R_\theta, R_x, R_y$	Radii
R	Characteristic gas constant
R_A, R_B	Force reactions
s	Perimeter dimension
S_E, S_L	Endurance and fatigue lives
S_o	Surface tensions
S_P	Plastic work
t, T	Temperature
$\mathbf{u}_x, \mathbf{u}_y, \mathbf{u}_z$	Unit vectors
U, U^*	Strain and complementary energy
V, V^*	Potential and complementary energy
w	Distributed loading
W, W^*	External and complementary work
x, y, X, Y	Cartesian coordinates
Y, k, σ_c	Yield stresses
α	Coefficient of linear expansion
β, λ	Angular measures
$\delta, \Delta, u, v, w, x$	Displacements
ϵ, σ	Engineering direct strain and stress
γ, τ	Engineering shear strain and stress
$\bar{\epsilon}^P, \bar{\sigma}$	Equivalent plastic strain and stress
$\epsilon_{ij}, \mathbf{E}$	Strain tensor and matrix respectively
$\epsilon_x, \epsilon_y, \epsilon_z, \gamma_{xy}$	Cartesian tensor strain components
$\epsilon_r, \epsilon_\theta, \epsilon_z, \gamma_{r\theta}$	Polar tensor strain components
$\epsilon_1, \epsilon_2, \epsilon_3$	Principal strains

θ, ϕ	Angular twist and unit twist
ϕ, ψ	Torsion functions in x, y
ρ	Density
$\sigma_\theta, \tau_\theta$	Stress components on an inclined plane
$\sigma_{ij}, \mathbf{S}$	Stress tensor and matrix respectively
$\sigma_x, \sigma_y, \sigma_z, \tau_{xy}$	Cartesian tensor stress components
$\sigma_\theta, \sigma_r, \tau_{r\theta}$	Polar stress components
$\sigma_1, \sigma_2, \sigma_3$	Principal stresses
∇	Biharmonic operator
ω, D	Damage parameters
$\omega_x, \omega_y, \omega_{xy}$	Rigid-body rotation components

PROPERTIES OF AREAS

1.1 MOMENTS OF AREA

The properties of a cross-section that resist externally applied loading are the area A of that section and the first (i) and second (I) moments of area about its centroidal axes. The following definitions apply to an element of area, $dA = dX \times dY$, in the section with respect to (w.r.t.) the general Cartesian coordinates X, Y in Fig. 1.1, for any arbitrary origin O.

First moments: $\qquad i_X = \int_A Y \, dA, \qquad i_Y = \int_A X \, dA.$

Centroid: $\qquad \bar{X} = i_Y/A, \qquad \bar{Y} = i_X/A.$ $\qquad\qquad$ (1.1)

Second moments: $\quad I_X = \int_A Y^2 \, dA, \quad I_Y = \int_A X^2 \, dA, \quad I_{XY} = \int_A X Y \, dA.$

It should be emphasized that these properties refer to area and not mass. The first moment of mass, the centre of gravity and the mass moment of inertia are defined in a similar manner when mass replaces area in the respective definitions given in Eq. (1.1). However, mass properties appear more in the study of dynamics than in statics. It will be seen later that area properties are essential for the statical analyses of the stresses arising in loaded bodies.

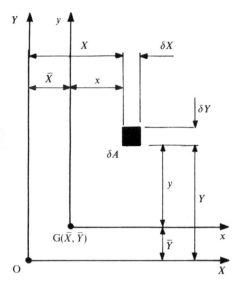

Figure 1.1 Parallel axes.

1.2 PARALLEL AND PERPENDICULAR AXES

When the axes X and Y originate from the centroid G, then $\bar{X} = \bar{Y} = 0$ and $i_X = i_Y = 0$. I_X may otherwise be transferred from axes X, Y to centroidal axes x, y (Fig. 1.1) using the *parallel-axis theorem*. From Eq. (1.1),

$$I_X = \int Y^2 \, dA = \int (y + \bar{Y})^2 \, dA = \int (y^2 + 2y\bar{Y} + \bar{Y}^2) \, dA.$$

Substituting for the centroidal axes:

$$I_x = \int y^2 \, dA, \qquad I_y = \int x^2 \, dA, \qquad I_{xy} = \int xy \, dA \quad \text{and} \quad i_x = \int y \, dA = 0$$

leads to the parallel-axis theorem;

$$I_X = I_x + A\bar{Y}^2.$$

Similarly it can be shown that

$$I_Y = I_y + A\bar{X}^2, \qquad I_{XY} = I_{xy} + A\bar{X}\bar{Y},$$

in which the signs of $\bar{X}$ and $\bar{Y}$ are determined from the positive x, y coordinate directions defining the first quadrant. The *perpendicular-axis theorem* enables the polar second moment of area I_z about a centroidal axis z, to be found from I_x and I_y.

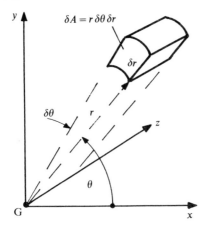

Figure 1.2 Perpendicular axes.

Representing dA in polar coordinates (Fig. 1.2) it follows that

$$I_z = \int r^2 \, \mathrm{d}A = \int (x^2 + y^2) \, \mathrm{d}A$$

and with similar substitutions for I_x and I_y

$$I_z = I_x + I_y$$

where, in this theorem, the subscripts x, y and z now denote centroidal values.

1.3 PRINCIPAL SECOND MOMENTS OF AREA

These are equal to the centroidal values I_x and I_y when the product moment $I_{xy} = 0$, indicating that x and y are axes of symmetry. When, for the centroid G,

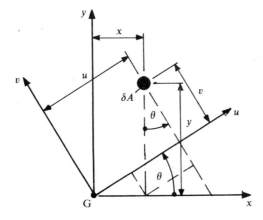

Figure 1.3 Principal axes.

$I_{xy} \neq 0$ it is necessary to identify the principal I values with orthogonal symmetry axes u and v (Fig. 1.3) such that

$$I_u = \int_A v^2 \, dA, \qquad I_v = \int_A u^2 \, dA, \qquad I_{uv} = \int_A uv \, dA = 0,$$

where, from the geometry given in Fig. 1.3,

$$u = x \cos \theta + y \sin \theta, \tag{1.2a}$$

$$v = y \cos \theta - x \sin \theta. \tag{1.2b}$$

Substituting from Eqs (1.2) leads to:

$$I_u = I_x \cos^2 \theta + I_y \sin^2 \theta - I_{xy} \sin 2\theta \tag{1.3a}$$

$$I_v = I_y \cos^2 \theta + I_x \sin^2 \theta + I_{xy} \sin 2\theta \tag{1.3b}$$

$$I_{uv} = \tfrac{1}{2}(I_x - I_y) \sin 2\theta + I_{xy} \cos 2\theta = 0. \tag{1.3c}$$

The inclination θ of u, w.r.t. x, follows from putting $I_{uv} = 0$ in Eq. (1.3c):

$$\tan 2\theta = 2I_{xy}/(I_y - I_x). \tag{1.4}$$

Now θ may be eliminated between Eqs (1.3) and (1.4) to give the principal values I_u, I_v in terms of I_x, I_y and I_{xy}:

$$I_{u,v} = \tfrac{1}{2}(I_x + I_y) \pm \tfrac{1}{2}\sqrt{[(I_x - I_y)^2 + 4I_{xy}^2]} \tag{1.5}$$

noting that $I_u + I_v = I_x + I_y$.

1.4 MATRIX REPRESENTATION

The reader should also note that Eqs (1.3a, b and c) conform to the transformation law;

$$\mathbf{I'} = \mathbf{L} \mathbf{I} \mathbf{L}^{\mathrm{T}} \tag{1.6}$$

where $\mathbf{I}$ is the matrix containing I_x, I_y, and I_{xy}. $\mathbf{I'}$ is the matrix of transformed I values for any inclined axes, including u and v, whose directions are defined in $\mathbf{L}$, a 2×2 matrix of direction cosines:

$$\mathbf{L} = \begin{bmatrix} l_{ux} & l_{uy} \\ l_{vx} & l_{vy} \end{bmatrix} = \begin{bmatrix} \cos \theta & \cos (90 - \theta) \\ \cos (90 + \theta) & \cos \theta \end{bmatrix} = \begin{bmatrix} \cos \theta & \sin \theta \\ -\sin \theta & \cos \theta \end{bmatrix}$$

in which l_{ux} is the cosine of the angle u makes with x, l_{vx} is the cosine of the angle v makes with x, etc. (Fig. 1.3). Then from Eq. (1.6), for the u, v axes,

$$\begin{bmatrix} I_u & I_{uv} \\ I_{uv} & I_v \end{bmatrix} = \begin{bmatrix} \cos \theta & \sin \theta \\ -\sin \theta & \cos \theta \end{bmatrix} \begin{bmatrix} I_x & I_{xy} \\ I_{xy} & I_y \end{bmatrix} \begin{bmatrix} \cos \theta & -\sin \theta \\ \sin \theta & \cos \theta \end{bmatrix}.$$

With $I_{uv} = 0$ the principal values in Eq. (1.5) follow from the expansion of the determinant:

$$\det \begin{vmatrix} I_x - I & I_{xy} \\ I_{xy} & I_y - I \end{vmatrix} = 0.$$

1.5 GRAPHICAL SOLUTION

In coordinates $[(I_x, I_y), I_{xy}]$, Eq. (1.5) describes a circle, with centre $[(I_x + I_y)/2, 0]$ and the square of its radius equal to the discriminant. The principal values I_u and I_v lie on opposite ends of the horizontal diameter, where $I_{xy} = I_{uv} = 0$, as shown in Fig. 1.4. Note that the calculated value of I_{xy}, together with its sign, is plotted along with I_x to fix point A. The sign of I_{xy} is changed when accompanying I_y to fix point B. The circle is then drawn with AB as its diameter. In Fig. 1.4(a), with I_{xy} positive and $I_x > I_y$, then $I_u < I_v$ in order that the u-axis makes an acute angle θ with the x-axis. Figure 1.4(b) gives the construction corresponding to $I_x > I_y$ with a negative I_{xy}. Here $I_u > I_v$ for the acute angle θ between the axes of u and x. These directions are established about the focus point F, a singular point on the circle through which any pair of orthogonal axes may be projected. Where these cut the circle again will determine the associated I_x, I_y and I_{xy}, or I_u and I_v for the principal axes. It follows that F is found by projecting the x-direction through A or the y-direction through B.

The following worked examples illustrate the application of the foregoing theory in respect of some elementary plane areas and also for typical cross-sections found in structures appearing later in this book.

Example 1.1 Find all the properties of the rectangular area $b \times d$ given in Fig. 1.5.

$$i_x = \int_{-d/2}^{d/2} y(b \times dy) = (b/2)|y^2|_{-d/2}^{d/2} = 0;$$

$$i_y = \int_{-b/2}^{b/2} x(d \times dx) = (d/2)|x^2|_{-b/2}^{b/2} = 0;$$

$$i_X = \int_0^d Y(b \times dY) = (b/2)|Y^2|_0^d = (b/2)d^2;$$

$$i_Y = \int_0^b X(d \times dX) = (d/2)|X^2|_0^b = (d/2)b^2.$$

$$\bar{X} = i_Y/A = b/2;$$
$$\bar{Y} = i_X/A = d/2.$$

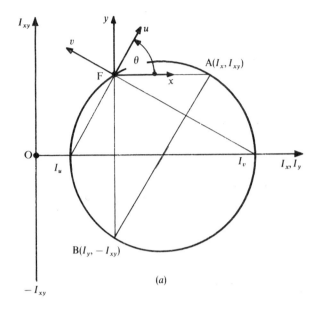

(a)

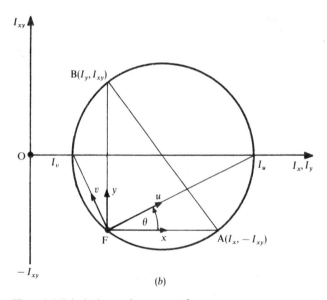

(b)

Figure 1.4 Principal second moments of area.

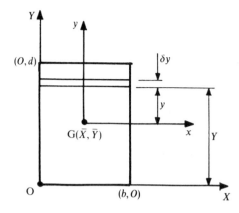

Figure 1.5 Rectangle.

$$I_x = \int_{-d/2}^{d/2} y^2(b \times dy) = (b/3)|y^3|_{-d/2}^{d/2} = bd^3/12;$$

$$I_y = \int_{-b/2}^{b/2} x^2(d \times dx) = (d/3)|x^3|_{-b/2}^{b/2} = db^3/12;$$

$$I_z = I_x + I_y = bd(b^2 + d^2)/12;$$

$$I_{xy} = \int_{-b/2}^{b/2} \int_{-d/2}^{d/2} xy(dx\,dy)$$

$$= \int_{-b/2}^{b/2} x\,dx \left|\frac{y^2}{2}\right|_{-d/2}^{d/2} = 0 \qquad (x \text{ and } y \text{ are axes of symmetry});$$

$$I_X = \int_0^d Y^2(b \times dY) = (b/3)|Y^3|_0^d = bd^3/3;$$

$$I_Y = \int_0^b X^2(d \times dX) = (d/3)|X^3|_0^b = db^3/3;$$

$$I_{XY} = \int_0^b \int_0^d XY(dX\,dY) = (b^2/2)\int_0^d Y\,dY = b^2d^2/4$$

(or from $I_{XY} = A\bar{X}\bar{Y}$).

Note that I_X, I_Y and I_{XY} can be checked from parallel axes:

$$I_X = I_x + A\bar{X}^2 = bd^3/12 + (bd)(d/2)^2 = bd^3/3$$
$$I_Y = I_y + A\bar{Y}^2 = db^3/12 + (db)(b/2)^2 = db^3/3$$
$$I_{XY} = I_{xy} + A\bar{X}\bar{Y} = \quad 0 \quad + (bd)(b/2)(d/2) = b^2d^2/4.$$

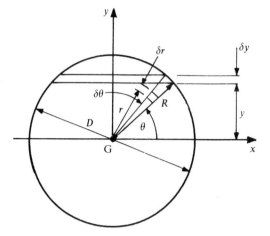

Figure 1.6 Circle.

Example 1.2 Find I_x, I_y and I_z for the circular section in Fig. 1.6.

$$I_x = \int_{-R}^{R} y^2 [2(R^2 - y^2)^{1/2} \, dy]$$

and, putting $y = R \sin \theta$,

$$I_x = 2 \times 2R^4 \int_{0}^{\pi/2} \sin^2 \theta \cos^2 \theta \, d\theta$$
$$= (R^4/2)|\theta - (1/4) \sin 4\theta|_0^{\pi/2}$$
$$= \pi R^4/4 = \pi D^4/64,$$

but $I_x = I_y$ and, therefore,

$$I_z = I_x + I_y = \pi R^4/2 = \pi D^4/32.$$

Alternatively, from Fig. 1.2

$$I_z = \int_0^R \int_0^{2\pi} r^2 (dr \times r \, d\theta)$$
$$= 2\pi \int_0^R r^3 \, dr = \pi R^4/2 = \pi D^4/32.$$

Example 1.3 Find the position of the centroid for the area enclosed between the positive X- and Y-axes, the parabola $Y^2 = 4aX$ and

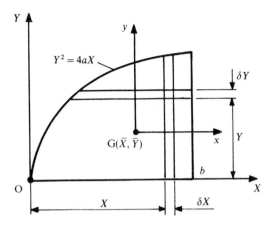

Figure 1.7 Parabola.

$X = b$ in Fig. 1.7.

$$A = \int_0^b Y \, dX = 2a^{1/2} \int_0^b X^{1/2} \, dX = (4/3)(ab^3)^{1/2};$$

$$i_X = \int_0^{2\sqrt{(ab)}} Y(b - X) \, dY = \int_0^{2\sqrt{(ab)}} Y(b - Y^2/4a) \, dY = ab^2;$$

$$i_Y = \int_0^b X(Y \, dX) = 2a^{1/2} \int_0^b X^{3/2} \, dX = (4/5)(ab^5)^{1/2};$$

$$\bar{X} = i_Y/A = 3b/5;$$

$$\bar{Y} = i_X/A = (3/4)(ab)^{1/2}.$$

Example 1.4 Find $\bar{Y}, I_x$ and I_X for the isoceles triangle, base b and height a, in Fig. 1.8.

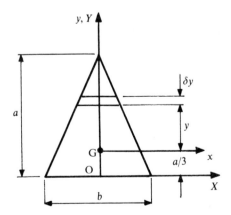

Figure 1.8 Isosceles triangle.

In the coordinates X, Y the equation of the right sloping side is

$$Y = a - (2a/b)X$$

from which it follows that;

$$i_X = \int_0^a Y(2X \, dY) = (b/a) \int_0^a (aY - Y^2) \, dY = ba^2/6;$$

$$\bar{Y} = i_X/A = (ba^2/6)/(ab/2) = a/3.$$

In the centroidal coordinates (x, y) the equation of the right sloping side is $y = -(2a/b)x + 2a/3$, which gives $x = b(2/3 - y/a)/2$ and thus

$$I_x = \int y^2 \, dA = \int y^2(2x \, dy) = \int_{-a/3}^{2a/3} y^2[b(2/3 - y/a)] \, dy$$

$$= b \int_{-a/3}^{2a/3} (2y^2/3 - y^3/a) \, dy = a^3b/36$$

and $I_X = I_x + A\bar{Y}^2 = a^3b/36 + (ab/2)(a/3)^2 = a^3b/12$.

Example 1.5 Find I_x, I_y and I_z for an ellipse with major axis $2a$ and minor axis $2b$, in Fig. 1.9.

The equation of an ellipse is

$$x^2/a^2 + y^2/b^2 = 1,$$

from which it follows that the second moment of area about the centroidal x-axis is

$$I_x = \int_{-b}^{b} y^2(2x \, dy) = 2 \int_{-b}^{b} a(1 - y^2/b^2)^{1/2} y^2 \, dy$$

$$= (2a/b)|[b^2 y(b^2 - y^2)^{1/2}/8] - [y(b^2 - y^2)^{3/2}/4] + b^2 \sin^{-1}(y/b)|_{-b}^b$$
$$= \pi ab^3/4.$$

Similarly, $I_y = \pi ba^3/4$ and therefore $I_z = I_x + I_y = (\pi ab/4)(a^2 + b^2)$.

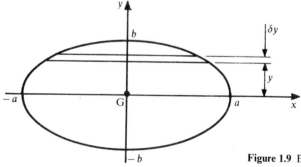

Figure 1.9 Ellipse.

Example 1.6 Find $I_x, I_y, I_{xy}, I_u, I_v$ and θ for the right-angled triangle of base b and height a, lying in the positive quadrant of X, Y in Fig. 1.10.

The equation of the sloping side is

$$Y = a - (a/b)X$$

and hence the first moments and centroids are given by

$$i_X = \int Y(X\,dY) = (b/a)\int_0^a (a - Y)Y\,dY;$$

$$= (b/a)|(aY^2/2) - (Y^3/3)|_0^a = a^2b/6;$$

$$\bar{Y} = i_X/A = (a^2b/6)(2/ab) = a/3.$$

Similarly, $i_Y = b^2a/6$ and $\bar{X} = i_Y/A = b/3$.

The second moments for the axes X, Y are

$$I_X = \int Y^2(X\,dY) = (b/a)\int_0^a (a - Y)Y^2\,dY$$

$$= (b/a)|aY^3/3 - Y^4/4|_0^a = a^3b/12;$$

$$I_Y = \int_0^b X^2(Y\,dX) = b^3a/12;$$

$$I_{XY} = \int\int XY\,dX\,dY = \int_0^a \left(\int_0^X XY\,dX \right)dY = \int_0^a |(1/2)X^2 Y|_0^{b(a-Y)/a}\,dY$$

$$= (b^2/2a^2)\int_0^a (a^2Y - 2aY^2 + Y^3)\,dY = a^2b^2/24.$$

I_{XY} may be checked from integrating the product moment of the

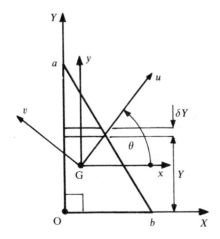

Figure 1.10 Right-angled triangle.

elemental area $Y \, dA$ about its centroid $(X/2, Y)$. That is,

$$I_{XY} = \int XY \, dA = \int_0^a (b/2a)(a-Y)Y[(b/a)(a-Y) \, dY] = a^2b^2/24.$$

Then, for the centroidal axes (x, y), by parallel axes,

$$I_x = I_X - A\bar{Y}^2 = a^3b/12 - (ab/2)(a/3)^2 = a^3b/36;$$
$$I_y = I_Y - A\bar{X}^2 = b^3a/12 - (ab/2)(b/3)^2 = b^3a/36;$$
$$I_{xy} = I_{XY} - A\bar{X}\,\bar{Y} = a^2b^2/24 - (ab/2)(a/3)(b/3) = -a^2b^2/72;$$

and, finally, from Eqs (1.4 and 1.5),

$$\tan 2\theta = -ab/(b^2 - a^2)$$
$$I_{u,v} = (ab/72)[(a^2 + b^2) \pm (a^4 + b^4 - a^2b^2)^{1/2}].$$

Note that if $a = b$ then $\theta = 45°$, $I_u = a^4/24$ and $I_v = a^4/72$.

Example 1.7 Find the position of the centroid and I_x, I_y, I_{xy}, I_u and I_v for a quadrant of a circle, radius r, lying in the positive quadrant of X, Y in Fig. 1.11.

Because of symmetry it is only necessary to find the centroid about one axis (say X):

$$i_X = \int_0^r YX \, dY = \int_0^r Y(r^2 - Y^2)^{1/2} \, dY = r^3/3(=i_Y);$$

$$\bar{Y} = i_X/A = (r^3/3)(4/\pi r^2) = 4r/3\pi(= \bar{X});$$

$$I_X = \int_0^r Y^2(r^2 - Y^2)^{1/2} \, dY = (r^4/4)\int_0^{\pi/2} \sin^2 2\beta \, d\beta = \pi r^4/16(= I_Y);$$

$$I_{XY} = \int_0^r (X \, dY)YX/2 = \tfrac{1}{2}\int_0^r (r^2 - Y^2)Y \, dY = r^4/8.$$

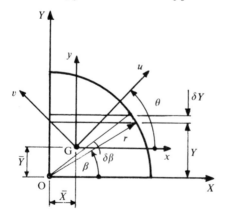

Figure 1.11 Quadrant.

Then, from parallel axes,

$$I_x = I_X - A\bar{Y}^2 = (\pi r^4/16) - (\pi r^2/4)(4r/3\pi)^2 = (\pi/16 - 4/9\pi)r^4(=I_y),$$
$$I_{xy} = I_{XY} - A\bar{X}\bar{Y} = r^4/8 - (\pi r^2/4)(4r/3\pi)^2 = r^4(1/8 - 4/9\pi),$$

when, from Eqs (1.4) and (1.5), $\theta = 45°$ and

$$I_u = r^4(\pi/16 + 1/8 - 8/9\pi),$$
$$I_v = r^4(\pi/16 - 1/8).$$

Example 1.8 Find the centroidal position and I_x for a closed equilateral tubular section of mean side length a and thickness t, with one side lying parallel to the y-direction (Fig. 1.12).

For simplicity, a third variable, s, is now introduced to define the distance along the sloping side as shown. Then, since X is an axis of

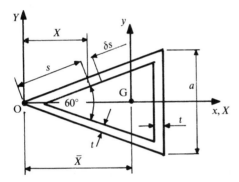

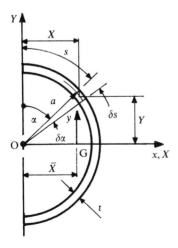

Figure 1.12 Closed tube.

Figure 1.13 Open tube.

symmetry, $\bar{Y} = 0$ and

$$i_Y = \int_A X \, dA = (at)(\sqrt{3}\,a/2) + 2\int_0^a (s\cos 30°)(t\,ds) = \sqrt{3}\,a^2 t;$$

$$\bar{X} = i_Y/A = \sqrt{3}\,a^2 t/3at = a/\sqrt{3} \qquad \text{(i.e. one-third of the axis length);}$$

$$I_x = I_X = \int_A Y^2 \, dA = ta^3/12 + 2\int_0^a (s\sin 30°)^2(t\,ds) = a^3 t/4.$$

Example 1.9 Find I_X and I_Y for an open semicircular tubular section (Fig. 1.13) of mean radius a and thickness t, when the diameter coincides with the Y-axis.

$$I_X = \int_A Y^2 \, dA = 2\int (a\cos\alpha)^2(t\,ds)$$

$$= 2\int_0^{\pi/2} (a\cos\alpha)^2(ta\,d\alpha) = \pi a^3 t/2;$$

$$I_Y = \int X^2 \, dA = 2\int (a\sin\alpha)^2(t\,ds)$$

$$= 2\int_0^{\pi/2} (a\sin\alpha)^2(ta\,d\alpha) = \pi a^3 t/2.$$

Example 1.10 Find I_X for the thin-walled trapezoidal section shown in Fig. 1.14.

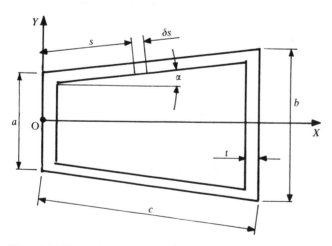

Figure 1.14 Trapezoidal tube.

Taking the vertical webs and sloping sides separately,

$$I_X = (tb^3/12 + ta^3/12)_{webs} + \int_{sides} Y^2 \, dA;$$

where, for the sides,

$$\int_A Y^2 \, dA = 2 \int_0^c (a/2 + s \sin \alpha)^2 (t \, ds)$$

$$= tc[a^2/2 + ac \sin \alpha + (2c^2/3) \sin^2 \alpha],$$

therefore,

$$I_X = t[b^3/12 + a^3/12 + a^2 c/2 + ac^2 \sin \alpha + (2c^3/3) \sin^2 \alpha].$$

Example 1.11 Determine analytically I_u, I_v and θ for the solid section given in Fig. 1.15. (K.P. 1976)

The centroidal position $(\bar{X}, \bar{Y})$ is found by taking first moments of area about axes X, Y. This gives

$$(20 \times 20 \times 10) + (\pi/2)(20)^2 20 = [(20 \times 20) + (\pi/2)(20)^2] \bar{X},$$

$$\therefore \quad \bar{X} = -16.11 \, \text{mm}.$$

$$(20 \times 20 \times 10) + (\pi/2)(20)^2 [20 + 4(20)/3\pi]$$

$$= [(20 \times 20) + (\pi/2)(20)^2] \bar{Y},$$

$$\therefore \quad \bar{Y} = 21.3 \, \text{mm}.$$

Noting that G for a semicircle is $4r/3\pi$ above its horizontal diameter, it follows from the parallel axes theorem that

$$I_x = 20(20)^3/12 + (21.32 - 10)^2 20^2 + \pi(40)^4/128$$

$$- [\pi(40)^2/8]\{(80/3\pi)^2 - [(80/3\pi) - 1.3]^2\}$$

$$I_x = 11.44 \times 10^4 \, \text{mm}^4;$$

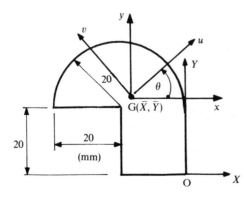

Figure 1.15 Complex section.

$$I_y = 20(20)^3/12 + (16.11 - 10)^2(20)^2 + \pi(40)^4/128$$
$$\quad + (\pi/8)(40)^2(20 - 16.11)^2,$$
$$I_y = 10.06 \times 10^4 \, \text{mm}^4;$$
$$I_{xy} = (\pi/8)(40)^2(16.11 - 20)[(80/3\pi) - 1.3]$$
$$\quad + (20)^2(16.11 - 10)(10 - 21.3),$$
$$I_{xy} = -4.52 \times 10^4 \, \text{mm}^4;$$

and substitution into Eqs (1.4 and 1.5) gives

$$\theta = 41°, \qquad I_u = 15.3 \times 10^4 \, \text{mm}^4 \qquad \text{and} \quad I_v = 6.1 \times 10^4 \, \text{mm}^4.$$

Example 1.12 Determine for the section given in Fig. 1.16(a) the position of the centroid $(\bar{X}, \bar{Y})$ and I_x, I_y, I_z for the centroidal axes x, y and z. Using both analytical and graphical methods determine I_u, I_v and the inclination, θ, of the principal axes u and v, w.r.t. x and y.

SOLUTION TO SI UNITS GIVEN IN FIG. 1.16(a) First moments about axes X, Y in Fig. 1.16(a) give

$$(30 \times 10 \times 5) + (40 \times 10 \times 30) + (20 \times 10 \times 55) = 900\bar{Y}$$
$$\therefore \quad \bar{Y} = 27.22 \, \text{mm};$$
$$(60 \times 10 \times 5) + (10 \times 10 \times 15) + (20 \times 10 \times 20) = 900\bar{X}$$
$$\therefore \quad \bar{X} = 9.45 \, \text{mm};$$
$$I_x = 20(32.78)^3/3 - 10(22.78)^3/3 + 30(27.22)^3/3 - 20(17.22)^3/3$$
$$\quad = 36.35 \times 10^4 \, \text{mm}^4;$$
$$I_y = 60(9.45)^3/3 + 10(20.55)^3/3 + 10(10.55)^3/3 + 40(0.55)^3/3$$
$$\quad = 4.98 \times 10^4 \, \text{mm}^4;$$
$$I_{xy} = 600(+2.78)(+4.45) + 100(+27.78)(-5.55)$$
$$\quad + 200(+22.22)(+10.55)$$
$$\quad = 3.89 \times 10^4 \, \text{mm};$$
$$I_z = I_x + I_y = 41.33 \times 10^4 \, \text{mm}^4.$$

Analytically, from Eqs (1.4) and (1.5),

$$\tan 2\theta = 2(+3.89)/(4.98 - 36.35)$$
$$\therefore \quad \theta = 83°;$$
$$I_u, I_v = \{(36.35 + 4.98)/2 \pm \tfrac{1}{2}\sqrt{[(36.35 - 4.98)^2 + 4(3.89)^2]}\} \times 10^4;$$
$$\therefore \quad I_u = 4.47 \times 10^4 \, \text{mm}^4;$$
$$I_v = 36.87 \times 10^4 \, \text{mm}^4.$$

The corresponding graphical solution is given in Fig. 1.16(b). It is seen

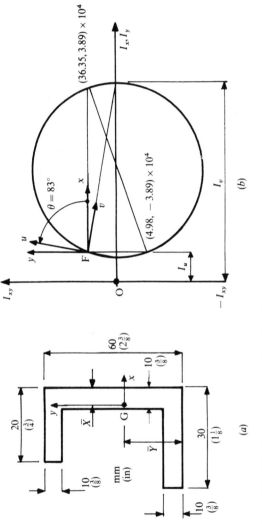

Figure 1.16 Principal I values for a channel section.

17

that since $I_x > I_y$ and I_{xy} is positive, then $I_u < I_v$ in accordance with Fig. 1.4(a). A scaled construction confirms the analytical I_u, I_v and θ values.

ANALYTICAL SOLUTION TO IMPERIAL UNITS GIVEN IN FIG. 1.16(a)

$$(1\tfrac{1}{8} \times \tfrac{3}{8})\tfrac{3}{16} + (1\tfrac{5}{8} \times \tfrac{3}{8})(\tfrac{3}{8} + \tfrac{13}{16}) + (\tfrac{3}{4} \times \tfrac{3}{8})2\tfrac{3}{16} = (\tfrac{21}{16})\bar{Y}$$
$$\therefore \quad \bar{Y} = 1.08 \text{ in};$$

$$(2\tfrac{3}{8} \times \tfrac{3}{8})\tfrac{3}{16} + (\tfrac{3}{4} \times \tfrac{3}{8})(\tfrac{3}{8} + \tfrac{3}{8}) + (\tfrac{3}{8} \times \tfrac{3}{8})(\tfrac{3}{8} + \tfrac{3}{16}) = (\tfrac{21}{16})\bar{X}$$
$$\therefore \quad \bar{X} = 0.348 \text{ in};$$

$$I_x = (1.125 \times 1.08^3/3) - (0.75 \times 0.705^3/3) + (0.75 \times 1.295^3/3)$$
$$- (0.375 \times 0.92^3/3)$$
$$= 0.8304 \text{ in}^4;$$

$$I_y = (2.375 \times 0.348^3/3) + (0.375 \times 0.402^3/3) + (0.375 \times 0.777^3/3)$$
$$+ (1.625 \times 0.027^3/3)$$
$$= 0.100 \text{ in}^4;$$

$$I_{xy} = (2.375 \times 0.375)(+0.1075)(+0.1605)$$
$$+ (0.375 \times 0.375)(+1.1075)(-0.2145)$$
$$+ (0.375 \times 0.75)(-1.08)(-0.402)$$
$$= 0.1041 \text{ in}^4;$$

$$I_z = I_x + I_y = 0.9304 \text{ in}^4.$$

Analytically, from Eqs (1.4) and (1.5);

$$\tan 2\theta = 2 \times 0.1041/(0.100 - 0.8304),$$
$$\therefore \quad \theta = 82.05°.$$

$$I_u, I_v = (0.8304 + 0.100)/2 \pm \tfrac{1}{2}\sqrt{[(0.8304 - 0.100)^2 + 4(0.1041)^2]}.$$
$$\therefore \quad I_u = 0.0854 \text{ in}^4;$$
$$I_v = 0.845 \text{ in}^4.$$

EXERCISES

1.1 Determine I_X, I_Y, I_{XY} and I_x, I_y, I_{xy} for the parabola in Fig. 1.7.

1.2 Find I_X directly for the triangular section in Fig. 1.8, then, given the position of the centroid, find I_x and I_z from the appropriate theorems.

1.3 Write down the expression for the polar second moment of area (I_z) for the ellipse in Fig. 1.9.

1.4 Examine the effect on I_{XY}, I_{xy}, I_u, I_v and θ when the areas in Figs 1.10 and 1.11 appear in the second quadrant of X, Y, i.e. the areas are rotated about Y leaving the positions of the x, y and X, Y axes unchanged.

1.5 Find I_Y, I_y and I_z for the triangular tube in Fig. 1.12.

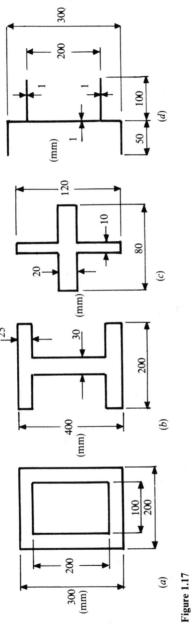

Figure 1.17

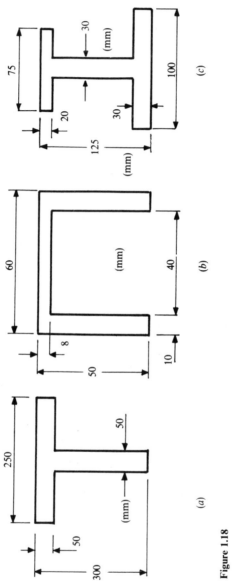

Figure 1.18

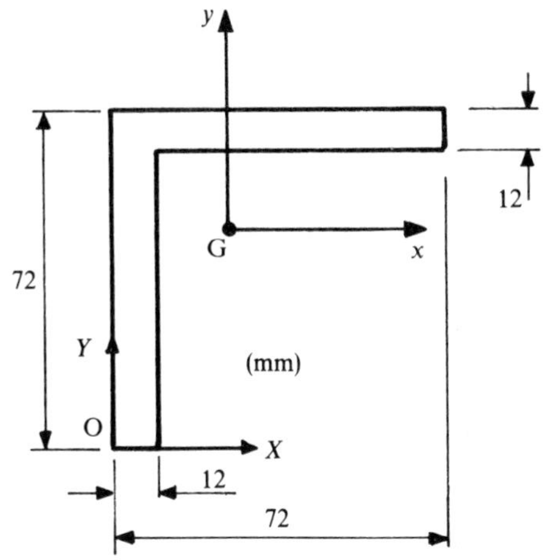

Figure 1.19

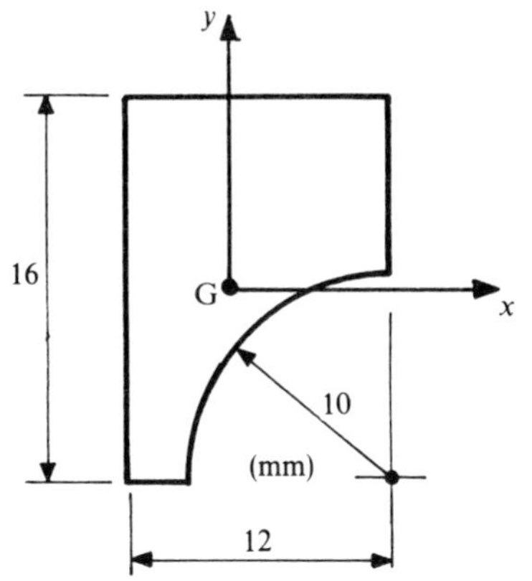

Figure 1.20

1.6 Show that the horizontal position of the centroid for the open tube in Fig. 1.13 given by $\bar{X} = 2a/\pi$ and that $I_y = a^3 t(\pi/2 - 4/\pi)$.

1.7 Find I_x for the centroidal axis when the structure in Fig. 1.14 is modified with a semicircular tube of the same thickness that (i) replaces the right vertical web and (ii) is added to this web to form a two-cell tube. The diameter equals the length of web.

1.8 Show, for a thin-walled circular tube of mean radius a and wall thickness t, that the second moment of area about any diameter in the cross-section is $\pi a^3 t$. Determine I_z from the perpendicular axis theorem and check directly using polar coordinates.

1.9 Find I_x and I_y for each of the four symmetrical sections in Fig. 1.17(a)–(d) and check that $I_{xy} = 0$. (All dimensions in millimetres.)

1.10 Find the position of the centroid and I_x for each of the sections shown in Fig. 1.18(a)–(c). (All dimensions in millimetres.)

1.11 Determine, for the equal-angle section shown in Fig. 1.19, the position of the centroid and I_x, I_y and I_{xy}. Find by both analytical and graphical methods the principal second moments of area and the inclination of their axes w.r.t. x and y.

Answer: $\bar{X} = -22.4$ mm; $\bar{Y} = 49.6$ mm; $I_x = I_y = 73.5 \times 10^4$ mm⁴; $I_{xy} = 42.4 \times 10^4$ mm⁴; $I_u = 31.1 \times 10^4$ mm⁴; $I_v = 115.9 \times 10^4$ mm⁴, $\theta = 45°$.

1.12 Find I_u, I_v and θ for the section shown in Fig. 1.20.

Answer: $I_u = 791.1$ mm⁴; $I_v = 2241.9$ mm⁴; $\theta = 51.35°$ (K.P. 1976).

TWO

ANALYSIS OF FORCES

2.1 COPLANAR, CONCURRENT FORCES

A system of forces **A**, **B**, **C**, etc., acting through a common point O, in the same plane (Fig. 2.1(a)), is in equilibrium when the algebraic sum of their force components $\sum F$, for the x and y directions, is zero. The polygon of force vectors (Fig. 2.1(b)) will then close. It follows from the force components shown in Fig. 2.1 that

$$A_x + B_x + C_x + D_x + E_x = 0, \qquad \sum F_x = 0, \qquad (2.1a)$$

$$A_y + B_y + C_y + D_y + E_y = 0, \qquad \sum F_y = 0, \qquad (2.1b)$$

in which the sign convention is implied that upward forces are positive. When a system of coplanar, concurrent forces is not in equilibrium (Fig. 2.2(a)) their effect may be replaced by a resultant force **R**, whose x and y components ($\sum R_x$ and $\sum R_y$ in Fig. 2.2(b)) are the algebraic sum of the force components for these directions. The magnitude of the resultant vector **R** equals that of the opposing equilibrant vector that would close the force polygon in Fig. 2.2(b). Then from Fig. 2.2, the x, y components for the magnitude $|\mathbf{R}|$ are

$$\sum R_x = A_x + B_x + C_x + D_x + E_x, \qquad \sum R_y = A_y + B_y + C_y + D_y + E_y,$$

where the magnitude and direction of **R** (w.r.t. x) are given by

$$|\mathbf{R}| = \sqrt{[(\sum R_x)^2 + (\sum R_y)^2]}, \qquad \theta_x = \tan^{-1}[(\sum R_y)/(\sum R_x)]. \qquad (2.2a, b)$$

Example 2.1 Four forces **A**, **B**, **C**, and **D** with respective magnitudes of 20, 30, 40 and 15 N act at a point in a plane as shown in Fig. 2.3(a).

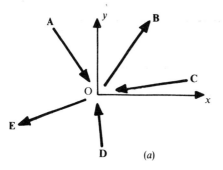

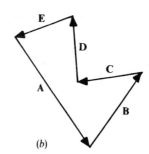

Figure 2.1 Coplanar forces in equilibrium.

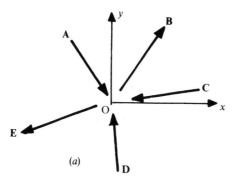

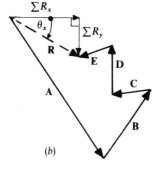

Figure 2.2 Non-equilibrium coplanar, forces.

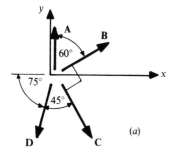

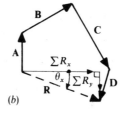

Figure 2.3 Non-equilibrium coplanar, concurrent forces.

Find the magnitude, direction and sense of their resultant.

$$\sum R_y = A_y + B_y + C_y + D_y$$
$$= 20 + 30 \sin 30° - 40 \sin 60° - 15 \sin 75° = -14.13 \text{ N};$$
$$\sum R_x = A_x + B_x + C_x + D_x$$
$$= 0 + 30 \cos 30° + 40 \cos 60° - 15 \cos 75° = 42.10 \text{ N};$$
$$|\mathbf{R}| = [(-14.13)^2 + (42.10)^2]^{1/2} = 44.4 \text{ N};$$
$$\theta_x = \tan^{-1}(-14.13/42.10) = -18.56° \text{ anticlockwise from } x.$$

The sense follows from the directions of R_x and R_y, or directly from the alternative polygon solution given in Fig. 2.3(b).

2.2 COPLANAR, NON-CONCURRENT FORCES

Two systems of coplanar forces may be identified when they do not pass through a common point.

1 Parallel Forces

When a parallel force system, $\mathbf{F}_i$ with magnitudes $|\mathbf{F}_i| \equiv F_1, F_2, F_3, \dots$, is in equilibrium (Fig. 2.4) both these forces and their moments about any arbitrary point A will balance. Thus, accounting for the force directions shown in Fig. 2.4:

$$-F_1 + F_2 - F_3 + F_4 + F_5 = 0, \quad \text{i.e. } \sum |F_i| = 0,$$

and, with positive clockwise moments about A,

$$-F_1 a_1 + F_2 a_2 + F_3 a_3 - F_4 a_4 - F_5 a_5 = 0, \quad \text{i.e. } M_A = \sum |F_i| a_i = 0.$$

When a parallel force system is not in equilibrium (Fig. 2.5), it will exert a force and a moment that are its resultant values.

Taking the resultant force $\mathbf{R}$ to act in the direction shown at a

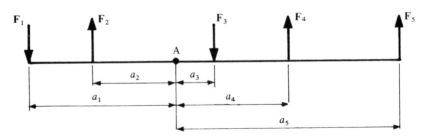

Figure 2.4 Coplanar, non-concurrent parallel forces.

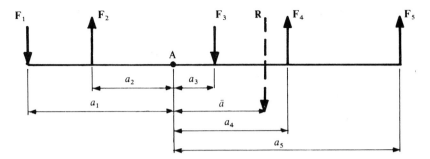

Figure 2.5 Non-equilibrium parallel forces.

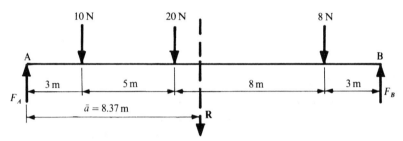

Figure 2.6 Resultant of parallel forces.

perpendicular distance $\bar{a}$ from A, it follows that

$$|\mathbf{R}| = \sum |\mathbf{F}_i| = -F_1 + F_2 - F_3 + F_4 + F_5;$$
$$M_A = |\mathbf{R}|\bar{a} = \sum |\mathbf{F}_i|a_i = -F_1 a_1 + F_2 a_2 + F_3 a_3 - F_4 a_4 - F_5 a_5.$$

Example 2.2 Determine the resultant of the three forces shown in Fig. 2.6 and the position, $\bar{a}$, at which this acts from the point A. If these forces are equilibrated by two further forces acting at A and B, find their magnitudes.

$$|\mathbf{R}| = \sum |\mathbf{F}_i| = -10 - 20 - 8 = -38\,\text{N}\downarrow;$$
$$M_A = 38\bar{a} = (3 \times 10) + (8 \times 20) + (16 \times 8);$$
$$\bar{a} = 8.37\,\text{m}.$$

For complete equilibrium:

$$\sum M_A = 0 = (8.37 \times 38) - 19F_B, \Rightarrow F_B = 16.74\,\text{N}\uparrow;$$
$$\sum F_i = 0 = -38 + 16.74 + F_A, \Rightarrow F_A = 21.26\,\text{N}\uparrow.$$

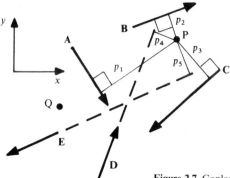

Figure 2.7 Coplanar, non-concurrent forces in equilibrium.

2 Non-parallel Forces

Here the conditions for equilibrium are the balance of x and y force components and moment balance about any two points P and Q. That is,

$$\sum R_x = 0, \qquad \sum R_y = 0, \tag{2.3}$$

$$M_P = 0 \qquad \text{and} \qquad M_Q = 0. \tag{2.4}$$

Then, for Fig. 2.7,

$$\sum R_x = A_x + B_x + C_x + D_x + E_x = 0,$$
$$\sum R_y = A_y + B_y + C_y + D_y + E_y = 0,$$
$$\sum M_P = -Ap_1 + Bp_2 + Cp_3 + Dp_4 + Ep_5,$$

and $\sum M_Q = 0$ for any other point Q.

When this system of forces is not in equilibrium, then the resultant force $|\mathbf{R}|$ and its inclination w.r.t. x are again given by Eqs (2.2(a and b)). In addition, there will be a resultant moment $M_P = |\mathbf{R}|p$ that is entirely dependent upon the position of P. These resultants may be found analytically or in a graphical construction known as the funicular polygon. The following two examples illustrate each method.

Example 2.3 A horizontal beam, loaded as shown in Fig. 2.8, is supported on rollers at A and hinged at B. Find the magnitude and direction of the support reactions analytically and graphically.

Applying the four conditions in Eqs (2.3) and (2.4) and noting that the roller reaction is vertical:

$$M_B = 0 = (0.75 \times 2 \sin 60°) + (1.75 \times 1.25) - (3 \times 1.5 \sin 60°)$$
$$- (3.75 \times 2 \sin 45°) + 5R_A = 0.$$

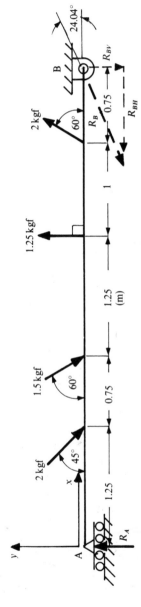

Figure 2.8 Beam in equilibrium under non-parallel forces.

27

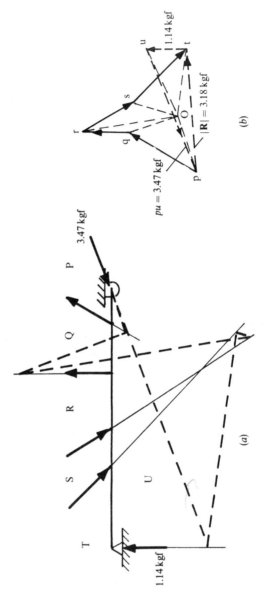

Figure 2.9 Non-parallel force resultant by graphical method.

From which $R_A = 1.143$ kgf (an equilibrant). Force equilibrium yields:

$(+ \rightarrow) \sum R_x = 0 = 2 \cos 45° + 1.5 \cos 60° + 2 \cos 60° + R_{BH}$,

$\therefore \quad R_{BH} = -3.164$ kgf;

$(+ \uparrow) \sum R_y = 0 = 1.143 - 2 \sin 45° - 1.5 \sin 60°$
$\qquad\qquad + 1.25 + 2 \sin 60° + R_{BV}$,

$\therefore \quad R_{BV} = -1.412$ kgf,

$\therefore \quad |R_B| = \sqrt{[(-3.164)^2 + (1.412)^2]} = 3.465$ kgf (an equilibrant);

$\qquad \theta_x = \tan^{-1}(R_{BV}/R_{BH}) = \tan^{-1}[(-1.412)/(-3.164)]$

$\qquad\quad = 24.04°, \qquad \text{as shown};$

$\qquad M_A = (1.25 \times 2 \sin 45°) + (2 \times 1.5 \sin 60°) - (1.25 \times 3.25)$
$\qquad\qquad - (4.25 \times 2 \sin 60°) + (5 \times 1.412) = 0.$

The final condition is unnecessary but provides a useful check.

In the *funicular polygon* construction (Fig. 2.9) the spaces between all forces in a scaled diagram (a) are lettered according to *Bow's notation*. Starting from the space P, draw that part of the force polygon (b) from p − t. Choose any point O and draw in the lines Op, Oq, Or, etc. Lines parallel to these (shown broken) are then drawn to link the lines of action of the corresponding forces in (a), starting at the right-hand support point, where the direction of the reaction is unknown. This construction terminates at a point on the left-hand vertical reaction. Join this point to the starting point in (a) and draw a parallel line in (b) through O to fix point u in the polygon. Vectors **tu** = 1.14 kgf and **pu** = 3.47 kgf provide the magnitude and direction of the support reactions. Note that the polygon also supplies the resultant of the applied forces $|R| = 3.18$ kgf in the vector **pt**.

Example 2.4 A square plate ABCD weights 6 kgf (13 lbf) with side length 12 m (40 ft) (Fig. 2.10). If it is in equilibrium under the action of six forces,

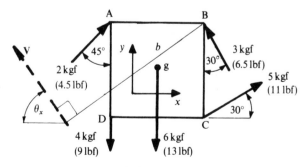

Figure 2.10 Plate in equilibrium.

including self-weight, find the magnitude and direction of the sixth force and the perpendicular distance at which this force acts from B.

Let the unknown force be **V** acting at b from B in direction shown.

SOLUTION TO SI UNITS

$$\sum F_x = 0 = 2\cos 45° - |\mathbf{V}|\cos\theta_x - 3\sin 30° + 5\cos 30°;$$
$$|\mathbf{V}|\cos\theta_x = 4.2442;$$
$$\sum F_y = 0 = 2\sin 45° + |\mathbf{V}|\sin\theta_x + 3\cos 30° + 5\sin 30° - 6 - 4 = 0;$$
$$|\mathbf{V}|\sin\theta_x = 3.4887;$$
$$\therefore \quad |\mathbf{V}| = \sqrt{[(4.2442)^2 + (3.4887)^2]} = 5.4934\,\text{kgf}.$$
$$\theta_x = \tan^{-1}(3.4887/4.2442) = 39.41°;$$
$$\sum M_B = 0 = 5.4934b - (4 \times 12) - (6 \times 6) - (12 \times 5\cos 30°)$$
$$+ (12 \times 2\cos 45°);$$
$$\therefore \quad b = 21.66\,\text{m}.$$

Check by taking moments about any other point, say A, where $\sum M_A = 0$:

$$\sum M_A = -(12 \times 3\cos 30°) - 12(5\cos 30° + 5\sin 30°)$$
$$+ 5.4934(21.66 - 12\sin 39.41°) + (6 \times 6) = 0.$$

Figure 2.11(a) and (b) give the corresponding graphical solution. Spaces P, Q, R, S, T and U between the forces in (a) define the funicular polygon pqrstu in (b). Vector **up** has the magnitude $|\mathbf{V}| = 5.49$ kgf required to equilibrate the five given forces. The position N, through which V

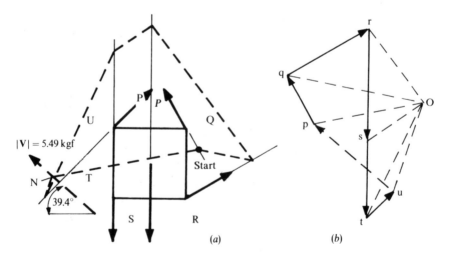

Figure 2.11 Funicular polygon construction.

acts, is found from linking the forces with lines parallel to Op, Oq, Or, Os, Ot and Ou as shown.

ANALYTICAL SOLUTION TO IMPERIAL UNITS GIVEN IN EXAMPLE 2.4

$$\sum F_x = 0 = 4.5\cos 45° - |\mathbf{V}|\cos\theta_x - 6.5\sin 30° + 11\cos 30°;$$
$$|\mathbf{V}|\cos\theta_x = 10.173;$$
$$\sum F_y = 0 = 4.5\sin 45° + |\mathbf{V}|\sin\theta_x + 6.5\cos 30° + 11\sin 30° - 13 - 9;$$
$$|\mathbf{V}|\sin\theta_x = 7.689;$$
$$|\mathbf{V}| = \sqrt{[(10.173)^2 + (7.689)^2]} = 12.752\,\text{lbf.}$$
$$\theta_x = \tan^{-1}(7.689/10.173) = 37.08°;$$
$$\sum M_B = 0 = 12.752b - (9 \times 40) - (13 \times 20) - (40 \times 11\cos 30°)$$
$$+ (40 \times 4.5\cos 45°);$$
$$\therefore \quad b = 68.52\,\text{ft.}$$

Check by taking moments about A ($\sum M_A = 0$):

$$\sum M_A = -(40 \times 6.5\cos 30°) - 40(11\cos 30° + 11\sin 30°)$$
$$+ 12.752(68.52 - 40\sin 37.08°) + (13 \times 20) = 0.$$

2.3 THE FREE BODY DIAGRAM

When a device consists of more than one load-bearing component, it is often convenient to isolate each component in a free body diagram (f.b.d.) for the analyses of its forces and moments under equilibrium conditions.

Example 2.5 If a normal force of 500 N is required at the jaws of the adjustable pliers in Fig. 2.12 determine, for the position given, the force **Q** at the grips.

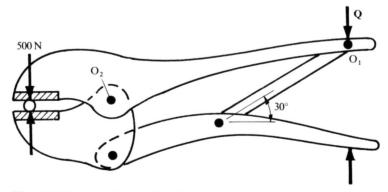

Figure 2.12 Forces applied to plier grip and jaw.

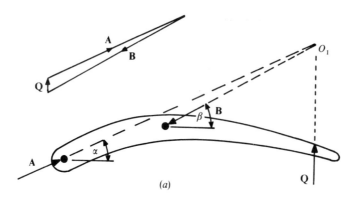

(a)

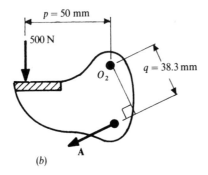

(b)

Figure 2.13 Free body diagrams for grip and jaw.

Taking the f.b.d. for the lower grip and jaw in Fig. 2.13(a) and (b) respectively, each then appears as an example of coplanar forces in equilibrium. Since the three forces **A**, **B** and **Q** in (a) are concurrent at O_1, their directions are known. From moment equilibrium in Fig. 2.13(b),

$$\sum M_{O_2} = 0 = -(500 \times p) + (|\mathbf{A}| \times q), \Rightarrow |\mathbf{A}| = 653 \text{ N}.$$

Returning to Fig. 2.13(a):

$$\sum F_y = 0 = +|\mathbf{Q}| - B_y + A_y, \quad \sum F_x = 0 = -B_x + A_x, \Rightarrow |\mathbf{Q}| = 30 \text{ N},$$

where $A_x = |\mathbf{A}| \cos \alpha$, $A_y = |\mathbf{A}| \sin \alpha$, $B_x = |\mathbf{B}| \cos \beta$, and $B_y = |\mathbf{B}| \sin \beta$.

Alternatively, the force polygon shown for (a) supplies **Q** directly once **A** is known. Note that α and β must also be known in each method.

Example 2.6 Find the moment M required to lift the nose wheel assembly in Fig. 2.14 when the link BC is vertical. Take the weight of AO to be 800 N acting at the centroid g but neglect the weights of links BC and DC. (K.P. 1973.)

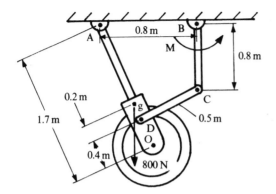

Figure 2.14 Nose wheel assembly.

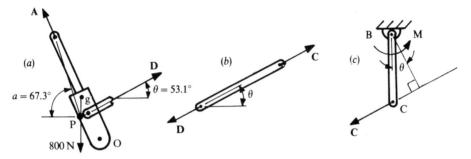

Figure 2.15 Free body diagrams for wheel links.

The f.b.d. for each link is given in Fig. 2.15(a), (b) and (c). In (a) the three forces must be concurrent at point P and in equilibrium when

$$\sum F_x = 0 = |\mathbf{D}| \cos \theta - |\mathbf{A}| \cos \alpha,$$
$$\sum F_y = 0 = -800 + |\mathbf{D}| \sin \theta + |\mathbf{A}| \sin \alpha,$$

where, from the geometry, $\theta = 53.1°$ and $\alpha = 67.3°$. The simultaneous solution to these equations then gives $|\mathbf{D}| = 358$ N. Since this magnitude of force is exerted at C in the direction of link DC then from Fig. 2.15(c):

$$\mathbf{M}_B = 0 = -\mathbf{M} + \mathbf{C}(BC \cos \theta);$$
$$\therefore \quad M = 358 \times 0.6 \times 0.8 = 172 \, \text{N m}.$$

2.4 FORCES IN FRAMEWORKS

In two-dimensional frameworks the externally applied loading and the equi-librating reactions at support points form a coplanar, non-concurrent force system for which Eqs (2.3) and (2.4) apply. In addition, the externally applied loading and the bar forces at each joint are concurrent and in equilibrium, satisfying Eqs (2.1). The bar forces may be determined when this latter condition is embodied in a composite force polygon for all joints. This is sometimes referred to as the *Maxwell–Cremona diagram*. *Bow's notation* is again used to label the spaces between all forces in the frame. Alternatively, the direct application of Eqs (2.1) results in a system of simultaneous force equations that are soluble provided the frame is statically determinate. This approach is readily extended to three-dimensional (3D) frames by the method of *tension coefficients*. A direct application of the non-concurrent equilibrium condition (Eqs (2.3) and (2.4)) is that of the *method of sections*, which enables isolated bar forces to be found. The following examples illustrate each of these approaches.

Example 2.7 Determine the forces HQ, QK, PQ and GP in the plane frame of Fig. 2.16 by both graphical and analytical methods, $(1 t \equiv 1000\,\text{kgf})$.

In the graphical approach the exterior force polygon abcdefghjka in Fig. 2.17 can be constructed when one reaction is known. Since the right-hand reaction R_r is normal to the roller, the magnitude follows by taking moments about the hinge (whose reaction direction is unknown). That is,

$$(2.5 \times 10/\sqrt{3}) + (1.5 \times 5) + (1 \times 20/\sqrt{3}) + (2.5 \times 10)$$
$$+ (3.5 \times 15) + 2(20 \times \sqrt{3}/2) = 20R_r$$

from which $R_r = 7.28\,\text{t}$. The closing vector **ka** supplies the magnitude and inclination of the hinge reaction (R_h). Starting at the right-hand joint,

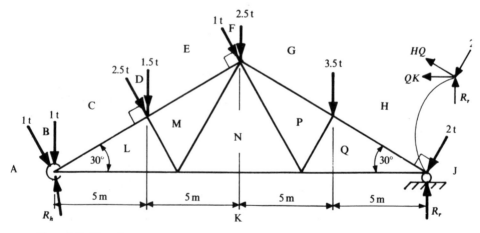

Figure 2.16 Plane frame.

where there are the least number of unknown bar forces (two), add vectors **hq** and **kq** to represent the directions of the forces in bars HQ and KQ so that a closed polygon results from working in a clockwise direction around this joint. The intersection point q fixes their magnitudes. In a similar manner, forces PQ and GP are supplied from the vectors **pq** and **gp** in the force polygon ghqp pertaining to the next joint, for which there are two unknowns. The force vector directions in these closed polygons also supply the action exerted by each bar at the joint, as shown inset in Fig. 2.16. Thus, a pulling action is associated with a tensile tie bar and a pushing action with a compressive strut. The complete diagram in Fig. 2.17 maps all the bar forces when equilibrium at each joint is considered in turn.

Alternatively, by the analytical approach, Eqs (2.1) supply the following horizontal and vertical equilibrium equations for the same two joints when bar tension (tie) is assumed throughout with positive forces

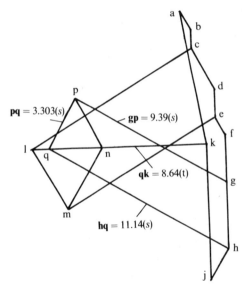

Figure 2.17 Maxwell–Cremona diagram for the frame in Fig. 2.16.

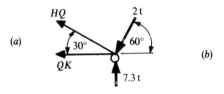

(a)

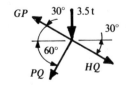

(b)

Figure 2.18 Joint equilibrium.

in the $\rightarrow$ and $\uparrow$ directions (Fig. 2.18(a) and (b))

$$- HQ \cos 30° - QK - 2 \cos 60° = 0;$$
$$7.28 + HQ \sin 30° - 2 \sin 60° = 0; \tag{a}$$
$$\Rightarrow HQ = -11.14\,\text{t}, \qquad QK = 8.64\,\text{t}.$$
$$- GP \cos 30° - PQ \cos 60° + HQ \cos 30° = 0;$$
$$- 3.5 + GP \sin 30° - HQ \sin 30° - PQ \sin 60° = 0; \tag{b}$$
$$\Rightarrow PQ = -3.03\,\text{t}, \qquad GP = -9.39\,\text{t}.$$

The $-$ sign indicates that the bar is pushing on the joint i.e. the bar itself acts as a strut.

Example 2.8 Determine the forces in tonnes for the bars in the Warren girder in Fig. 2.19 by the methods of sections and tension coefficients.

Horizontal force and moment equilibrium to the left of the sections in Fig. 2.20(a, b and c) and to the right in (d) gives

(a) $\sum F_y = 2.5 - F_{AE} \sin 60° = 0;$
  $M_E = (5 \times 2.5) + (5\sqrt{3})F_{AB} = 0;$
  $\Rightarrow F_{AE} = 5/\sqrt{3}, \qquad F_{AB} = -5/(2\sqrt{3}).$
(b) $\sum F_x = -5/(2\sqrt{3}) + (5/\sqrt{3}) \cos 60° - (5/\sqrt{3}) \cos 60°$
  $\qquad + F_{ED} + F_{BE} \cos 60° = 0;$
  $M_B = (2.5 \times 10) - (5 \times 2) - (5\sqrt{3})F_{ED} - (5/3)10 \cos 30°$
  $\qquad + (5/3)10 \cos 30° = 0;$
  $\Rightarrow F_{ED} = \sqrt{3}, \qquad F_{BE} = -1/\sqrt{3}.$
(c) $\sum F_y = 2.5 - 2 + (5/\sqrt{3}) \cos 30° - (5/\sqrt{3}) \cos 30°$
  $\qquad - (1/\sqrt{3}) \cos 30° + (1/\sqrt{3}) \cos 30° - F_{BD} \cos 30°;$

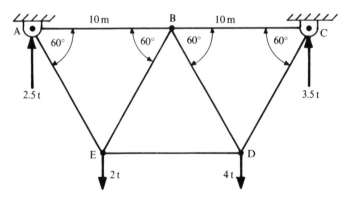

Figure 2.19 Warren girder under load.

$$M_D = (2.5 \times 15) - (2 \times 10) + (5\sqrt{3})F_{BC} = 0;$$
$$\Rightarrow F_{BD} = 1/\sqrt{3}, \qquad F_{BC} = -7/(2\sqrt{3}).$$
(d) $\sum F_y = 3.5 - \sqrt{3}F_{CD}/2 = 0, \qquad \Rightarrow F_{CD} = 7/\sqrt{3}.$

In the method of tension coefficients, the tensile force/unit length of bar (t) in Fig. 2.21 appears in the force component expressions as

$$F_x = |F|\cos\theta = |F|x/l = tx, \qquad F_y = |F|\sin\theta = |F|y/l = ty,$$

where $t = F/l$ and x and y are the projections of the bar length l in the x–y directions. Then, from assuming tension at each joint in Fig. 2.19

joint A: $\uparrow \; 2.5 - (5\sqrt{3})t_{AE} = 0, \qquad \Rightarrow t_{AE} = 1/(2\sqrt{3}), \qquad F_{AE} = 5/\sqrt{3};$

$\rightarrow 10t_{AB} + 5t_{AE} = 0, \qquad \Rightarrow t_{AB} = -1/(4\sqrt{3}), \quad F_{AB} = -5/(2\sqrt{3});$

joint E: $\uparrow \; -2 + (5/\sqrt{3})t_{AE} \qquad \Rightarrow t_{BE} = -1/(10\sqrt{3}), \quad F_{BE} = -1/\sqrt{3};$

$\qquad + (5\sqrt{3})t_{BE} = 0,$

$\rightarrow -5t_{AE} + 5t_{BE} + 10t_{ED} = 0, \quad \Rightarrow t_{ED} = 3/(10\sqrt{3}), \qquad F_{ED} = \sqrt{3};$

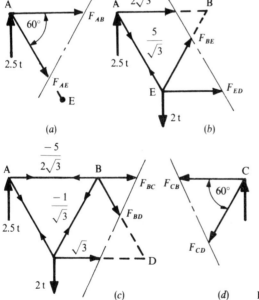

(a)

(b)

(c)

(d)   **Figure 2.20** Method of sections.

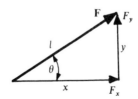

Figure 2.21 Tension coefficients.

joint B: $\uparrow -(5/3)t_{BE} - (5/3)t_{BD} = 0$, $\Rightarrow t_{BD} = 1/(10\sqrt{3})$, $F_{BD} = 1/\sqrt{3}$;

        $\rightarrow -10t_{AB} - 5t_{BE} + 5t_{BD}$ $\Rightarrow t_{BC} = -7/(20\sqrt{3})$, $F_{BC} = -7/(2\sqrt{3})$;

         $+ 10t_{BC} = 0$,

joint C: $\uparrow 3.5 - (5\sqrt{3})t_{CD} = 0$, $\Rightarrow t_{CD} = (7\sqrt{3})/10$, $F_{CD} = 7/\sqrt{3}$.

Example 2.9 Determine the forces in the three bars of the space-frame in Fig. 2.22 using the method of tension coefficients.

     Orthogonal force equilibrium at B:

$$(x\rightarrow) - 3 - 6t_{BA} + 4t_{BD} + 13t_{BC} = 0;$$

$$(y\uparrow) - 5 - 12t_{BA} - 12t_{BC} - 12t_{BD} = 0;$$

$$(z\uparrow) + 6t_{BA} - 8t_{BD} = 0;$$

from which the solutions are

$$t_{AB} = -0.327, \quad t_{BC} = +0.155, \quad t_{BD} = -0.245;$$

$$F_{AB} = ABt_{AB} = \{12^2 + 6^2 + 6^2\}^{1/2}(-0.327) = -4.80\,t;$$

$$F_{BC} = BCt_{BC} = \{12^2 + 13^2\}^{1/2}(+0.155) = 2.78\,t;$$

$$F_{BD} = BDt_{BD} = \{12^2 + 8^2 + 4^2\}^{1/2}(-0.245) = -3.67\,t.$$

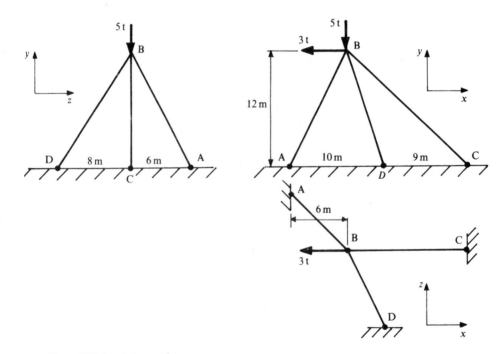

Figure 2.22 Loaded space frame.

EXERCISES

Concurrent Forces

2.1 Calculate the force F required to balance the lever ABCD in Fig. 2.23 and find the magnitude and direction of the support reaction at B.

    Answer: $F = 12.97\,\text{kN}$; $Re = 15\,\text{kN}$ at $30°$ clockwise from vertical.

2.2 The action of the control lever in Fig. 2.24 is such that a force of 400 N, when applied at the lever end, is resisted by a clockwise couple of 600 N m at A. Determine the necessary dimension l for the lever.

    Answer: $l = 1.53\,\text{m}$.

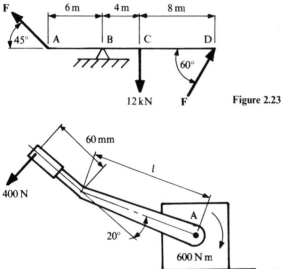

Figure 2.23

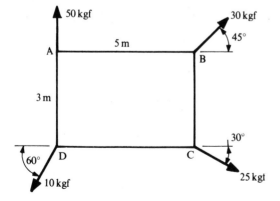

Figure 2.24

Figure 2.25

2.3 Four forces act on the rectangular plate ABCD in Fig. 2.25. Find the magnitude, direction and position from B of the resultant force.

Answer: 63 kgf, 38° clockwise from vertical, 2.16 m perpendicular from B.

2.4 The rectangular lamina ABCD in Fig. 2.26 is hinged at A and supported by the string SC. Determine the tension in the string and the reaction components at the hinge when the lamina supports a vertical load of 3 kgf at B in addition to the plate weight of 2 kgf.

Answer: 6.48 kgf; $A_x = 3.24$ kgf; $A_y = 0.61$ kgf.

2.5 The light, horizontal, 20 m long beam in Fig. 2.27 is hinged at the left end and is simply supported at its right end. Determine the magnitude and direction of the reactions and the resultant of the applied forces shown.

Answer: 36 kgf↑, 35 kgf at 18° to the vertical, 71 kgf at 10° to the vertical, 11.25 m from left end.

2.6 Show that the gusset plate in Fig. 2.28 is in equilibrium under the active forces shown.

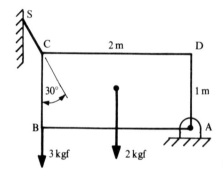

Figure 2.26

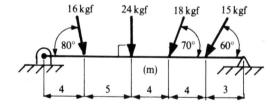

Figure 2.27

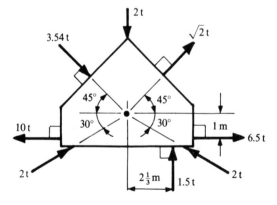

Figure 2.28

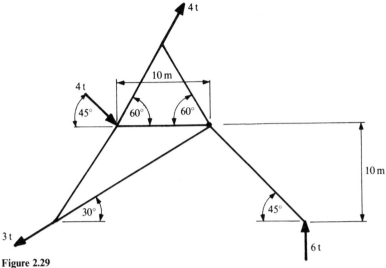

Figure 2.29

2.7 Determine the magnitude and direction of the line of action of the resultant of the forces acting on the frame in Fig. 2.29.

Answer: 6 t at 71° to the horizontal, and 5 m from the 6 t force.

Free Body Diagrams

2.8 Determine, for the jib crane shown in Fig. 2.30, the wall reactions at A and B and the force in the rod AC when the horizontal beam supports a force of 50 kN at mid-span.

2.9 If the jib-crane in Fig. 2.31 supports a vertical load of 5 kN at 2 m from B, determine (i) the forces acting at joints A, B and at the contact support E and (ii) the force in rod DC.

Answer: 12.2 kN at 65.75°; 11.11 kN at 3.1°; 11.1 kN at 0°; 12.4 kN at 26.6°.

2.10 The principle of the lifting mechanism shown in Fig. 2.32 is that a horizontal force H, applied at the slider S, will raise a weight W through the pin joints at B, C and E and a further slider at D. Find the force, H, necessary to support $W = 150$ kN and the total vertical shear force at the mid-point E. Take $CS = BD = x = 1$ m.

Answer: 260 kN; 103.6 kN.

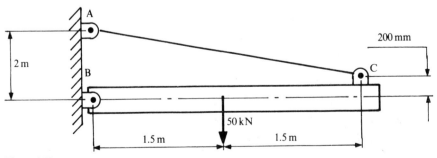

Figure 2.30

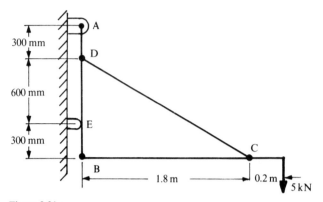

Figure 2.31

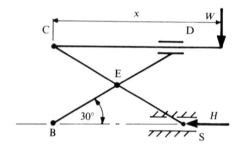

Figure 2.32

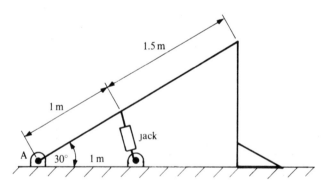

Figure 2.33

2.11 Find the force exerted by the jack in Fig. 2.33 as it initially raises a triangular block of 5 kN while the apex remains in contact with the ground. What is the corresponding shear force in the pin at A? (K.P. 1973)

2.12 During taxiing a lateral force of 400 N is developed at the wheel of the retractable aircraft undercarriage shown in Fig. 2.34. If the internal diameter of the hydraulic jack is 50 mm find the necessary internal pressure to maintain equilibrium.

Answer: 4.12 bar. (K.P. 1973)

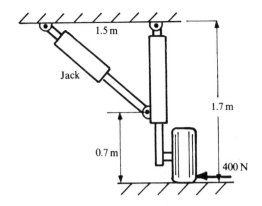

1.5 m

Jack

1.7 m

0.7 m

400 N

Figure 2.34

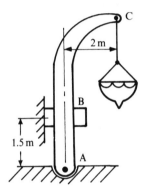

C

2 m

B

1.5 m

A

Figure 2.35

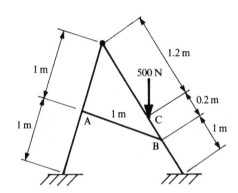

1.2 m

1 m

500 N

0.2 m

1 m

A

1 m

C

1 m

B

Figure 2.36

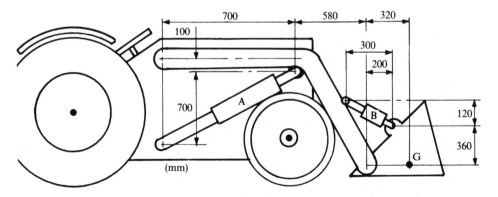

700

580

320

100

300

200

700

A

120

B

360

G

(mm)

Figure 2.37

2.13 The boat in Fig. 2.35 is supported by two identical davits ABC. Find the reactions at the pin A and guide B when a vertical load of 3.84 kN is transmitted to each davit at C.

Answer: $R_A = 6.4$ kN at $36.9°$ anticlockwise from horizontal; $R_B = 5.12$ kN horizontal.

2.14 A string, AB, connects the unequal legs of the step ladder in Fig. 2.36. What is the force in AB and what are the ground reactions when a man of weight 500 N rests at C?

Answer: 73 N, 200 N, 300 N.

2.15 Find the forces in the hydraulic cylinders A and B in Fig. 2.37 when the shovel carries a total vertical load of 8 kN at the centroid G.

Answer: 22.65 kN; 3.145 kN. (K.P. 1974)

Frameworks

2.16 The simply-supported frames in Figs 2.38–2.41 are subjected to the vertical loading shown. Determine graphically the forces in the bars. Check these values analytically.

Answer: (t = tie; s = strut)

CK	KD	KF	FE	FJ	JB	JA	AB	BC	
0.35(t)	0.65(s)	4.10(t)	2.40(s)	5.80(t)	0.55(s)	1.05(t)	1.00	0.85	(kgf)
LQ	QN	QR	RM	RP	OP	PL			
2(s)	0	5(s)	5(s)	4(t)	6	5			(t)
AE	BF	CG	EF	FG	DE	DG	AD	DC	
1660(s)	1400(s)	1660(s)	460(s)	400(t)	1440(t)	1200(t)	833.33	1166.67	(kgf)
AC	DC	DJ	FE	BC	ED	EH	AB	FH	
14(t)	10.4(t)	6.9(t)	16.9(s)	16.9(s)	10.4(t)	14(t)	5	5	(t)

2.17 Using the method of sections, determine the forces in the lettered bars for the frames in Figs 2.42–2.44 when each frame is hinged at A and rests on rollers at B.

Answer: (t = tie; s = strut) $XY = 0.89$ kN(t), $R_B = 0.81$ kN; $XY = 2.1$ t(s), $WX = 4.3$ t(s), $YZ = 2.75$ t(t), $R_A = 5.17$ t at $41.5°$, $R_B = 0.62$ t; $XY = 4.2$ t(t), $R_A = 10.8$ t at $20°$ anticlockwise, from vertical, $R_B = 8.4$ t.

2.18 The bars of the bracket shown in Fig. 2.45 are pinned at A and B and fixed to a vertical wall at C, D, E and F. Calculate, using the method of tension coefficients, the axial force induced in each bar when a 10 tonne weight is applied vertically at B.

2.19 Determine the forces in each bar of the frame in Fig. 2.46 when it is fixed to a wall at points A, B, C and D and carries the loading in kilograms-force shown.

	BE	EC	DF	BF	AF	FE
Answer: (t = tie; s = strut)	504(t)	−284(s)	−284(s)	−365(s)	178.3(t)	237.6(t)

2.20 The three bars OA, OB and OC in Fig. 2.47 support forces of 2 kgf, 1.50 kgf and 2 kgf at the pinned joint O. Find the force in bar OC.

Answer: 1.26 kgf(s).

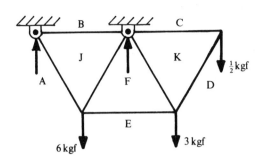

Figure 2.38

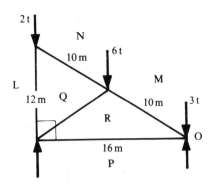

Figure 2.39

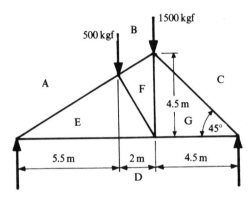

Figure 2.40

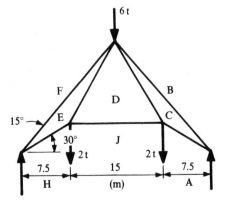

Figure 2.41

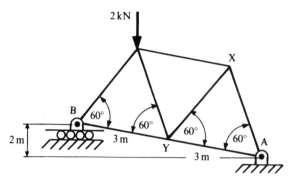

Figure 2.42

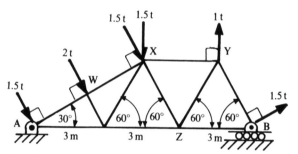

Figure 2.43 ⁓

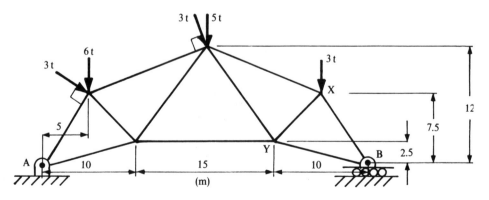

Figure 2.44

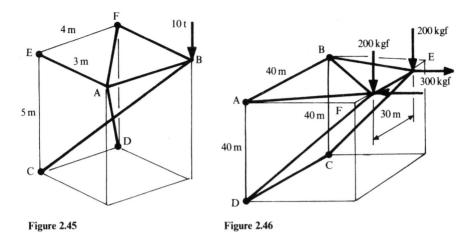

Figure 2.45

Figure 2.46

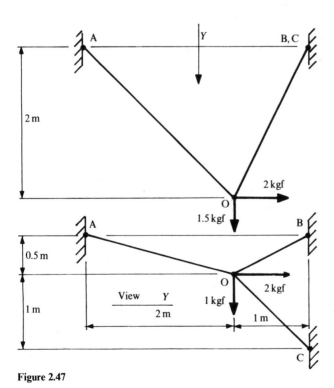

Figure 2.47

THREE

SIMPLE STRESS AND STRAIN

3.1 DIRECT STRESS AND STRAIN

Direct stress is the internal normal force a material exerts per unit of its area A resulting from externally applied loading W. When the latter is uniaxial of either a tensile $(+)$ or compressive $(-)$ nature, the stress is simply;

$$\sigma = \pm W/A. \tag{3.1}$$

The corresponding direct strain ϵ, is the amount by which the material deforms per unit of its length l. Thus, for a displacement $\pm x$:

$$\epsilon = \pm x/l. \tag{3.2}$$

Provided the loading is elastic, the displacement is fully recoverable on unloading. Since metallic materials are usually linear in their load–displacement behaviour, it follows from Eqs (3.1) and (3.2) that the ratio between σ and ϵ is then a constant for a particular material. This is the *modulus of elasticity* (*Young's modulus*), defined as

$$E = \sigma/\epsilon = Wl/Ax. \tag{3.3}$$

When A varies with l in the function $A = A(l)$, the displacement follows from Eq. (3.3) as

$$x = (W/E) \int_0^l dl/A(l). \tag{3.4}$$

Example 3.1 A tie bar on a pressing machine is 2 m long and 40 mm diameter. What is the stress and the extension under a tensile load of 100 kN? Take $E = 205\,000\,\text{MN/m}^2$.

Direct substitution into Eqs (3.1) and (3.3) gives

$$\sigma = W/A = +4 \times 100 \times 10^3/\pi(40)^2 = 79.6 \, \text{MPa};$$
$$x = Wl/AE = +4 \times 100 \times 10^3 \times 2 \times 10^3/[\pi(40)^2 \times 205000] = 0.78 \, \text{mm}.$$

Where, in design, it is required that the working stress σ_w, must remain elastic, a *safety factor*, S, is used in conjunction with some maximum allowable stress value σ_{max}. Then, by definition,

$$S = \sigma_{max}/\sigma_w \qquad (S > 1). \qquad (3.5)$$

Figures 3.1(a) and (b) show that, depending upon the design criterion for a given material, σ_{max} may be equated to the *yield stress Y*, the *ultimate tensile strength* (UTS), the stress at the *limit of elastic proportionality* (LP) or a particular *proof stress* value. The proof stress for 0.1% offset strain is shown (0.1 per cent PS) in Fig. 3.1(b). It follows from Eq. (3.5) that σ_w is respectively: Y/S, UTS/S, LP/S, or (0.1 per cent PS)/S. The latter is used for materials that do not display a sharp yield point (Fig. 3.1(b)) at which the transition from linear-elastic behaviour to non-linear, elastic-plastic behaviour is so gradual that the LP becomes ill defined. To determine a proof stress value a line is drawn parallel to the initial elastic line offset by the given amount of strain.

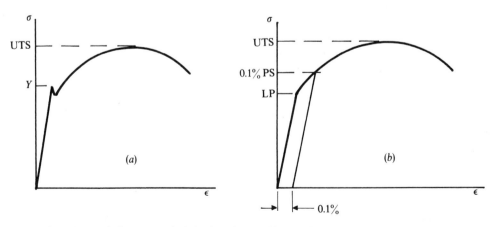

Figure 3.1 Typical stress–strain behaviour for metallic materials.

Example 3.2 A solid circular bar is to carry a force of 220 kN (22 tonf). Find a suitable bar diameter (a) when the strain in the bar is not to exceed 0.05 per cent and (b) by employing a factor of safety of 8 with a UTS value of 465 MPa (30 tonf/in²). What is the strain in (b)? Take $E = 207$ GPa (13 400 tonf/in²).

SOLUTION TO SI UNITS
(a) It follows from Eq. (3.3) that

$$E = W/A\epsilon = 4W/\pi d^2 \epsilon$$
$$\therefore \quad d = \sqrt{(4W/\pi E\epsilon)}$$
$$= \sqrt{(4 \times 220 \times 10^3 \times 10^2/\pi \times 207\,000 \times 0.05)} = 52\,\text{mm}.$$

(b) From Eq. (3.5),

$$\sigma_w = \sigma_{max}/S = 465/8 = 58.13\,\text{MPa},$$

where $\sigma_w = W/A = 4W/\pi d^2$.

$$\therefore \quad d = \sqrt{(4W/\pi\sigma_w)} = \sqrt{(4 \times 220 \times 10^3/\pi \times 58.13)} = 69.42\,\text{mm},$$
$$\epsilon = \sigma_w/E = 58.13/207\,000 = 0.000\,28 = 0.028\ \text{per cent}.$$

SOLUTION TO IMPERIAL UNITS
(a) From Eq. (3.3),

$$d = \sqrt{(4W/\pi E\epsilon)}$$
$$= \sqrt{(4 \times 22 \times 10^2/\pi \times 13\,400 \times 0.05)} = 2.045\,\text{in}.$$

(b) From Eq. (3.5),

$$d = \sqrt{(4W/\pi\sigma_w)} = \sqrt{(4 \times 22 \times 8/\pi \times 30)} = 2.733\,\text{in},$$
$$\epsilon = \sigma_w/E = 30/(8 \times 13\,400) = 0.028\ \text{per cent}.$$

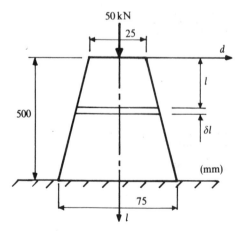

Figure 3.2 Tapered column.

Example 3.3 The tapered solid alloy column in Fig. 3.2 has a diameter which varies uniformly from 25 mm to 75 mm over a length of 500 mm. Find the change in length and the maximum stress in the column under an axial compressive force of 50 kN. Take $E = 100$ GPa.

In the d vs. l axes shown, the equation of the sloping side is $d/2 = 12.5 + (0.05)l$. Then from Eq. (3.4),

$$A(l) = \pi(d/2)^2 = \pi(12.5 + 0.05l)^2;$$

$$x = (W/E\pi) \int_0^{500} (12.5 + 0.05l)^{-2}\, dl$$

$$= -(W/0.05E\pi)|(12.5 + 0.05l)^{-1}|_0^{500}$$

$$= (-50 \times 10^3/\pi \times 0.05 \times 100 \times 10^3)(12.5^{-1} - 37.5^{-1})$$

$$= -0.17\,\text{mm};$$

$$\therefore \quad \sigma_{max} = W/A_{min} = -4 \times 50 \times 10^3/\pi(25)^2 = -101.86\,\text{MPa}.$$

3.2 SHEAR STRESS AND STRAIN

In common with Eq. (3.1), the *shear stress* τ sustained by a material in maintaining equilibrium with an applied shear force F that acts over an area A is given by

$$\tau = F/A \tag{3.6}$$

where A depends upon the mode of shear. In the riveted joints of Fig. 3.3, for

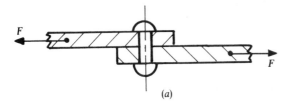

(a)

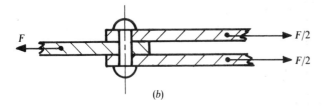

(b)

Figure 3.3 Riveted joints.

example, the force F per rivet (of area, a) in (a) is carried by a single rivet, i.e. *single shear*, while in the double lap joint in (b) the effective resisting area is doubled, i.e. *double shear*. Thus, from Eq. (3.6), the rivet stresses are respectively $\tau = F/a$ and $\tau = F/2a$.

Example 3.4 The shear stress in an axle of solid circular section is limited to 37.5 MPa. What is its smallest permissible diameter when it acts in double shear under a load of 20 kN?

From Eq. (3.6), with the section in double shear,

$$\tau = F/(2a) = 4F/2\pi d^2,$$
$$\therefore \quad d = \sqrt{(2F/\pi\tau)} = \sqrt{(2 \times 20 \times 10^3/\pi \times 37.5)} = 18.43 \text{ mm}.$$

Shear strain γ is a dimensionless measure of angular distortion. That is, in Fig. 3.4, it is defined as

$$\gamma = \tan \phi \simeq \phi, \tag{3.7}$$

where ϕ is the angular change in the right angle measured in radians. Since the elastic shear displacement is small, it follows from Fig. 3.4 that $\phi \simeq x/l$. To complete the comparison with direct strain in Eqs (3.1) to (3.3), when x increases linearly with F in a material's elastic region, a *modulus of rigidity* (*shear modulus*) is defined from Eqs (3.6) and (3.7) as the ratio between shear stress and strain. That is,

$$G = \tau/\gamma = Fl/Ax. \tag{3.8}$$

A safety factor S is also applied with shear design for the calculation of a working shear stress τ_w, from some maximum shear stress value τ_{max} for the material. Thus,

$$S = \tau_{max}/\tau_w, \tag{3.9}$$

in which τ_{max} may be equated to the shear yield stress k, or the ultimate shear strength USS. These are related to the maximum direct tensile stresses in Eqs (3.5). That is, $k = Y \times Q$ and $USS = UTS \times Q$ where, from yielding criteria (Chapter 16), $1/2 \leq Q \leq 1/\sqrt{3}$. Riveted joints may be formed with

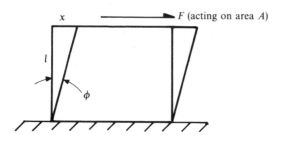

Figure 3.4 Shear distortion.

single or multiple rows of rivets. The following examples illustrate that the necessary pitch, diameter and joint efficiency are determined from the possible failure modes: tearing of the plate, shearing and crushing of the rivets. The problem is complicated with the additional effect of torsion for eccentrically loaded groups of rivets, bolts or studs when each rivet, etc., will then sustain different total shear loads.

Example 3.5 The fork-end in Fig. 3.5 is to carry a force F of 35.5 kN. Determine, using a safety factor of 2 throughout, (a) the diameter d of the pin when the USS is 386 MPa; (b) the necessary thickness t of the mild steel rocker arm to avoid bearing failure under an ultimate bearing stress (UBS) of 620 MPa; (c) dimension A of the rocker to avoid shear failure from the pin under an ultimate stress of 250 MPa; and (d) dimension B of the rocker to prevent tensile failure under a UTS of 310 MPa.

(a) $\tau_w = \text{USS}/S = F/2a,$

$\therefore \quad d = \sqrt{(2F/\pi\tau_w)} = \sqrt{(4F/\pi \times \text{USS})} = \sqrt{(4 \times 35.5 \times 10^3/\pi \times 386)}$
$= 10.82 \text{ mm}.$

(b) $\sigma_w = \text{UBS}/S = F/dt$

where $d \times t$ is the projected contact area. Hence, t is found from

$t = FS/d \times \text{UBS} = 35.5 \times 10^3 \times 2/10.82 \times 620 = 10.58 \text{ mm}$

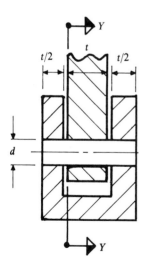
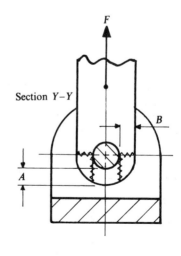

Figure 3.5 Loaded fork-end assembly.

and the fork thicknesses are $t/2 = 5.29$ mm, assuming that the rocker and fork are made of the same material.

(c) $\tau_w = \text{USS}/S = (F/2)/[(A + d/2)t]$;

      $250/2 = 17.85 \times 10^3/(A + 5.41)10.58$;

      $A = 8.09$ mm.

(d) $\sigma_w = \text{UTS}/S = (F/2)/(B \times t)$;

      $310/2 = 17.83 \times 10^3/(B \times 10.58)$;

      $B = 10.88$ mm.

Example 3.6 The double riveted butt joint in Fig. 3.6 connects two plates 25 mm thick with 30 mm diameter rivets. Determine the necessary pitch p of the rivets if their ultimate shear and compressive strengths are 370 and 695 MPA respectively. The UTS for the plates is 465 MPa: what is the efficiency of the joint?

The joint is made equally strong in tension and shear when the tearing force in the central plate equals the shear force in the rivets. With two rivets per pitch, each in double shear, it follows that

$$(p - d)t \times \text{UTS} = 2 \times 2 \times (\pi \times d^2/4) \times \text{USS};$$

$$(p - 30)25 \times 465 = \pi \times 30^2 \times 370, \Rightarrow p = 120 \text{ mm}.$$

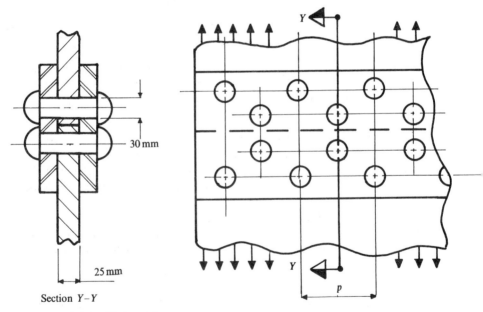

Section $Y-Y$

Figure 3.6 Double butt joint.

Note that safety factors would only be influential here if they differed between tension and shear. Now, by definition,

$$\text{Joint efficiency} = \frac{\text{least value of failure load}}{\text{tensile strength of original plate}} \qquad (3.10)$$

where the numerator is the least of the following.

Tearing strength of plate per pitch

$$= (p - d)t \times \text{UTS}$$
$$= (120 - 30)25 \times 465 = 1.046\,\text{MN}.$$

Resistance of rivets to shearing/pitch

$$= \text{no. of rivets/pitch} \times \text{shear mode} \times \text{area of rivet} \times \text{USS}$$
$$= 2 \times 2 \times (\pi d^2/4) \times \text{USS} = \pi \times 30^2 \times 370 = 1.046\,\text{MN}.$$

Crushing resistance of rivets/pitch

$$= \text{no. of rivets/pitch} \times \text{projected area} \times \text{UCS}$$
$$= 2 \times (d \times t) \times \text{UCS} = 2 \times 30 \times 25 \times 695 = 1.043\,\text{MN}.$$

It appears that the rivets will crush first when, from Eq. (3.10),

$$\text{Per cent joint efficiency} = 1.043 \times 10^6 \times 100/(120 \times 465 \times 25)$$
$$= 74.8 \text{ per cent}.$$

In practice, however, as the shear mode factor associated with double shear is nearer 1.75 than 2, then either the plate tearing strength or the rivet shear strength would dictate joint efficiency for this design.

Example 3.7 Find the rivet with the greatest load for the single-lap eccentrically loaded bolt group in Fig. 3.7(a). If ultimate strengths of the bolt material are 350 MPa in shear and 600 MPa in compression, find the minimum values of rivet diameter and plate thickness using a safety factor of 3.

The analysis of a single bolt in Fig. 3.7(b) shows that the net force f/bolt is the vector sum of the forces **s** and **t** due to shear and torsion respectively. When a group of n bolts sustains a shear force **F**, then,

$$|\mathbf{s}| = |\mathbf{F}|/n \qquad (3.11)$$

which each bolt reacts in a direction parallel to **F** but in an opposite sense. If O is the centroid of the shear areas in the group, then the applied torque

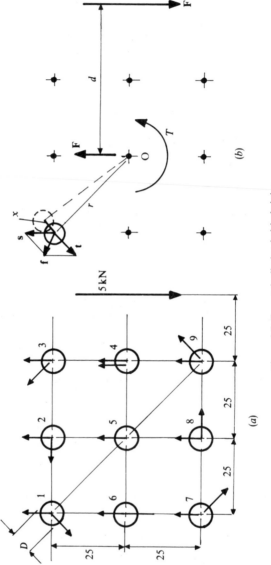

Figure 3.7 Eccentrically loaded bolt joint.

T is equilibrated by the reacting bolt torques:

$$T = |\mathbf{F}|d = \sum_i (|\mathbf{t}|r)_i \qquad (3.12)$$

where $i = 1, 2, 3, \ldots, n$. Now with a shear displacement x, at the bolt under torque we find equal shear strains. That is, $x_1/r_1 = x_2/r_2$ for any two adjacent bolts. Since x is proportional to the respective shear stresses in the bolts, then $\tau_i/r_i = K$ where K is a constant. Thus, with the same area A for each bolt,

$$|\mathbf{t}_i| = A\tau_i = (AK)r_i$$

and substitution into Eq. (3.12) gives the torque equilibrium equation,

$$T = |\mathbf{F}|d = \sum (AK)r_i^2 = (AK)\sum r_i^2 = (|\mathbf{t}|/r)_i \sum r_i^2,$$

from which the individual bolt forces under torques are

$$|\mathbf{t}_i| = (|\mathbf{F}|d)r_i / \sum r_i^2 \qquad (3.13)$$

These act in a direction perpendicular to the radial line joining the rivet centre to O in an opposite sense to the applied torque $T = |\mathbf{F}|d$. Finally, the net force for any bolt in the group is the vector sum

$$\mathbf{f}_i = \mathbf{s} + \mathbf{t}_i.$$

For a rivet working shear stress $\tau_w = USS/S$ and a plate working crushing strength $\sigma_w = UCS/S$, the rivet diameter D, and plate thickness t, are determined from

$$|\mathbf{f}| = \pi D^2 \tau_w / 4 \qquad \text{and} \qquad |\mathbf{f}| = Dt\sigma_w, \qquad (3.14a, b)$$

where $|\mathbf{f}|$ is then the maximum value for the group. In Fig. 3.7(a), Eq. (3.11) gives $|\mathbf{s}| = 5/9 = 0.556\,\text{kN}$. Then, from Eq. (3.13),

$$|\mathbf{F}|d = 5 \times 50 = 250\,\text{kN mm}, \qquad \sum r_i^2 = 4(35.36)^2 + 4(25)^2 = 7500\,\text{mm}^2;$$
$$|\mathbf{t}_5| = 0, \qquad |\mathbf{t}_1| = |\mathbf{t}_3| = |\mathbf{t}_7| = |\mathbf{t}_9| = 250 \times 35.36/7500 = 1.1787\,\text{kN};$$
$$|\mathbf{t}_2| = |\mathbf{t}_4| = |\mathbf{t}_6| = |\mathbf{t}_8| = 250 \times 25/7500 = 0.833\,\text{kN}.$$

The vector directions given show that $\mathbf{f}$ will be greatest in the rivet line 3–4–9. They are

$$|\mathbf{f}_4| = |\mathbf{s}| + |\mathbf{t}_4| = 0.556 + 0.833 = 1.389\,\text{kN}.$$

Then, from the cosine rule and Eqs (3.14(a) and (b):

$$|\mathbf{f}_3| = |\mathbf{f}_9| = \sqrt{[(0.556)^2 + (1.1787)^2 - (2 \times 0.556 \times 1.1787)\cos 135°]}$$
$$= 1.62\,\text{kN (greatest)};$$

$$D = \sqrt{(4|\mathbf{f}_3|/\pi\tau_w)} = \sqrt{(12|\mathbf{f}_3|/\pi \times \text{USS})}$$
$$= \sqrt{(12 \times 1.62 \times 10^3/\pi \times 350)} = 4.2\,\text{mm};$$
$$t = |\mathbf{f}_3|/D\sigma_w = 3 \times |\mathbf{f}_3|/D \times \text{UCS}$$
$$= 3 \times 1.62 \times 10^3/4.2 \times 600 = 1.93\,\text{mm}.$$

3.3 COMPOUND BARS

Stresses arise in compound bars from direct and indirect mechanical modes of loading and from constrained thermal expansion.

Mechanical Effects

Mechanical strain may be induced in compound bars by the direct application of a force or by bolt-tightening forces. In each case the principles of force equilibrium and strain compatibility enable the component stresses and deformations to be found.

Composite bars When a composite bar or reinforced column made from two different materials A and B (Fig. 3.8) of the same length l, supports an axial tensile or compressive force W, then each component contributes to the force while suffering the same amount of strain ϵ.

The equilibrium and compatibility condition are expressed in

$$W = W_A + W_B \tag{3.15}$$

$$\epsilon = (\sigma/E)_A = (\sigma/E)_B \tag{3.16}$$

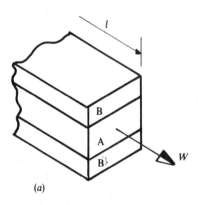

(a)

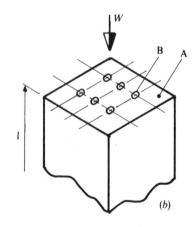

(b)

Figure 3.8 Composite bars.

Now, since $\sigma_A = (W/A)_A$ and $\sigma_B = (W/A)_B$, Eqs (3.15) and (3.16) give

$$W_A = WE_A A_A/(E_A A_A + E_B A_B) \qquad (3.17a)$$

$$W_B = WE_B A_B/(E_A A_A + E_B A_B) \qquad (3.17b)$$

and the axial displacement x for both bars is

$$x_A = x_B = W_A 1/E_A A_A = W_B 1/E_B A_B. \qquad (3.18)$$

Example 3.8 The composite bar in Fig. 3.8(a) consists of a $25\,\mathrm{mm} \times 6\,\mathrm{mm}\,(1\,\mathrm{in} \times \frac{1}{4}\,\mathrm{in})$ strip of steel bonded between two $25\,\mathrm{mm} \times 10\,\mathrm{mm}\,(1\,\mathrm{in} \times \frac{1}{2}\,\mathrm{in})$ strips of brass for a common length of $130\,\mathrm{mm}\,(5\,\mathrm{in})$. Calculate the stresses in each material and the extension of the bar under a tensile force of $100\,\mathrm{kN}\,(10\,\mathrm{tonf})$. Take the elastic constants as: $E = 84.8\,\mathrm{GPa}$ $(5490\,\mathrm{tonf/in^2})$ for brass and $E = 207\,\mathrm{GPa}\,(13\,400\,\mathrm{tonf/in^2})$ for steel.

SOLUTION TO SI UNITS Identifying A with steel and B with brass, Eqs (3.16)–(3.18) give

$$\sigma_A = WE_A/(E_A A_A + E_B A_B)$$
$$= 100 \times 10^3 \times 207 \times 10^3/[(207\,000 \times 150) + (84\,800 \times 500)]$$
$$= 281.82\,\mathrm{MPa};$$

$$\sigma_B = E_B \sigma_A/E_A$$
$$= 84\,800 \times 281.82/207\,000 = 115.45\,\mathrm{MPa};$$

$$x = (\sigma 1/E)_A$$
$$= 281.82 \times 130/207\,000 = 0.177\,\mathrm{mm}.$$

SOLUTION TO IMPERIAL UNITS

$$\sigma_A = WE_A/(E_A A_A + E_B A_B)$$
$$= 10 \times 13\,400/[13\,400 \times (\tfrac{1}{4}) + 5490 \times 2(\tfrac{1}{2})] = 15.16\,\mathrm{tonf/in^2};$$

$$\sigma_B = E_B \sigma_A/E_A$$
$$= 5490 \times 15.16/13\,400 = 6.21\,\mathrm{tonf/in^2};$$

$$x = (\sigma l/E)_A$$
$$= 15.16 \times 5/13\,400 = 0.0057\,\mathrm{in}.$$

Example 3.9 The reinforced concrete column in Fig. 3.8(b) is $375\,\mathrm{mm}$ square with eight steel rods each of area $625\,\mathrm{mm^2}$ embedded into it for a length of $1.8\,\mathrm{m}$. If the compressive stress for concrete is not to exceed $4.1\,\mathrm{MPa}$, compare the total compressive loads that the column can carry with and without reinforcement. Find the stress in the steel and at the steel–concrete interface. $E = 207\,\mathrm{GPa}$ for steel and $E = 13.75\,\mathrm{GPa}$ for concrete.

Without reinforcement, the maximum concrete load is

$$W_{max} = A_c \sigma_c = (375)^2 \times 4.1 = 576.56 \text{ kN}.$$

With reinforcement, the areas of steel and concrete are

$$A_s = 8 \times 625 = 5000 \text{ mm}^2;$$
$$A_c = (375)^2 - 5000 = 135\,625 \text{ mm}^2.$$

Then from Eqs (3.15)–(3.17), with subscripts c and s replacing A and B:

$$W = W_c + W_s = \sigma_c A_c + \sigma_s A_s = \sigma_c(A_c + E_s A_s/E_c)$$
$$= 4.1[135\,625 + (207 \times 5000/13.75)] = 864.68 \text{ kN};$$
$$\sigma_s = E_s \sigma_c/E_c = 207 \times 4.1/13.75 = 61.72 \text{ MPa}.$$

The stress, σ_s may be checked from Eq. (3.17b):

$$\sigma_s = \frac{WE_s}{(A_s E_s + A_c E_c)} = \frac{864.68 \times 10^3 \times 207\,000}{(5000 \times 207\,000) + (135\,625 \times 13\,750)} = 61.72 \text{ MPa}.$$

The interface stress σ_i is found from $W_c - W_s$. From Eqs (3.17a):

$$W_c = 864.68 \times 10^3 \times 135\,625/[135\,625 + (207 \times 5000/13.75)]$$
$$= 556.06 \text{ kN};$$
$$W_s = W - W_c = 864.68 - 556.06 = 308.62 \text{ kN}.$$

The difference $W_c - W_s$ is the load carried by the total peripheral interface area. Then, for n rods each of diameter $d = \sqrt{(625 \times 4/\pi)} = 28.21$ mm,

$$\sigma_i = (W_c - W_s)/n\pi dl$$
$$= (556.06 - 308.62) \times 10^3/(8\pi \times 28.21 \times 1800) = 0.194 \text{ MPa}.$$

Bolted assembly In the bolted assembly in Fig. 3.9, both the bolt b, and the cylinder c, carry the same magnitude of load P, but in opposite senses, while undergoing the same strain.

Given that the cylinder of initial length l_c contracts by an amount x_c after tightening the nuts on rigid end plates, P is found from

$$x_c = Pl_c/A_c E_c, \quad \Rightarrow P = A_c E_c x_c/l_c,$$

Alternatively, if the amount δ is given by which the nut is tightened, i.e. $\delta = $ pitch $\times$ no. of turns, and x_b and x_c are the corresponding displacements of the bolt and cylinder respectively, then with the bolt in tension and the cylinder in compression the compatibility requirement is that final length of bolt = final length of tube. That is,

$$l_b + x_b - \delta = (l_c + 2t) - x_c,$$
$$\therefore \quad x_b + x_c = \delta, \tag{3.19}$$

where the initial bolt length l_b, is greater than the initial cylinder length l_b,

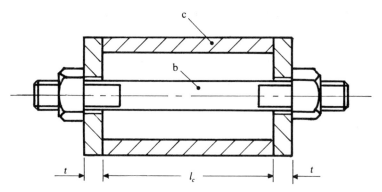

Figure 3.9 Bolt-loaded compound assembly.

by two end-plate thicknesses, i.e. $l_b = l_c + 2t$. Now, as the displacements are given by

$$x_b = Pl_b/A_b E_b \quad \text{and} \quad x_c = Pl_c/A_c E_c,$$

then P is found by substituting these in to Eq. (3.19). This gives

$$P(l_c + 2t)/A_b E_b + Pl_c/A_c E_c = \delta. \tag{3.20}$$

If t is small then $l = l_c \simeq l_b$ and Eq. (3.20) becomes

$$P = A_b E_b A_c E_c \delta / l(A_b E_b + A_c E_c). \tag{3.21}$$

In each case the stresses are $\sigma_b = P/A_b$ (tension), $\sigma_c = P/A_c$ (compression). With a subsequent tensile force W applied to the assembly, Eqs (3.15)–(3.17) again apply, where $W_A(=W_b)$ and $W_B(=W_c)$ provide the change in forces from this pre-stressed state.

Temperature Effects

A compressive internal stress $\sigma = E\epsilon = E\alpha \Delta t$, will be induced in a bar of length l under a temperature rise Δt when the free axial extension $x = \alpha \Delta t\, l$, is prevented (α is the coefficient of linear expansion for the bar material). In a similar manner, preventing or constraining the free expansion of a compound bar assembly will result in thermally induced stress and strain. The following cases arise.

Two bars of different materials 1 and 2 connected end-to-end between fixed platens (Fig. 3.10) With a temperature rise of Δt, the free displacements in each bar would be additive, to give a total unconstrained displacement of

$$x = x_1 + x_2 = \Delta t(\alpha_1 l_1 + \alpha_2 l_2). \tag{3.22}$$

Since displacement is prevented, each bar experiences the same induced force.

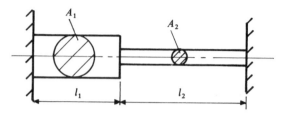

Figure 3.10 Composite bar with expansion prevented.

That is,

$$\sigma_1 A_1 = \sigma_2 A_2, \tag{3.23}$$

where compressive stresses σ_1 and σ_2 depend upon Eq. (3.22) in the following form to ensure zero axial displacement:

$$\alpha_1 l_1/E_1 + \sigma_2 l_2/E_2 + \Delta t(\alpha_1 l_1 + \alpha_2 l_2) = 0. \tag{3.24}$$

Combining Eqs (3.23) and (3.24):

$$\sigma_1 = -\Delta t(\alpha_1 l_1 + \alpha_2 l_2)/[(l_1/E_1) + (A_1 l_2/A_2 E_2)]. \tag{3.25}$$

Example 3.10 Equal lengths of copper and steel bar with the same diameters are assembled as in Fig. 3.10. Determine the stress in each material after a temperature rise of $10\,°C$. For steel, $E = 207\,\text{GPa}$, $\alpha = 12.5 \times 10^{-6}/°C$. For Cu, $E = 105\,\text{GPa}$, $\alpha = 18 \times 10^{-6}/°C$.

Since $A_1 = A_2$ and $l_1 = l_2$ substitution into Eqs (3.23) and (3.25) gives

$$\begin{aligned}
\sigma_1 &= -\Delta t(\alpha_1 + \alpha_2)E_1 E_2/(E_1 + E_2) = \sigma_2 \\
&= -10(12.5 + 18) \times 207 \times 105/[(207 + 105) \times 10^3] \\
&= -21.25\,\text{MPa}.
\end{aligned}$$

Concentric tubes or balanced strips of different materials A and B held between rigid platens (Fig. 3.11(a) and (b)) Assuming $\alpha_A < \alpha_B$ the free expansion caused by Δt in these assemblies is shown in Fig. 3.11(c). Since the actual diplacements are equal and of an intermediate value, when they are bonded at their ends by platens, it follows that A will be put in tension and B in compression. The induced stresses must satisfy two conditions:

(i) Equilibrium—the net load is zero, which gives

$$\sigma_A A_A + \sigma_B A_B = 0. \tag{3.26}$$

(ii) Compatibility—the displacements are equal, giving

$$\alpha_A l \Delta t + \sigma_A l/E_A = \alpha_B l \Delta t + \sigma_B l/E_B. \tag{3.27}$$

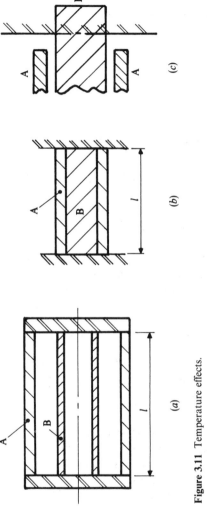

Figure 3.11 Temperature effects.

Then, from Eqs (3.26) and (3.27),

$$\sigma_A = \Delta t(\alpha_B - \alpha_A)(A_B E_A E_B)/(A_B E_B + A_A E_A). \tag{3.28}$$

When there is an initial difference in the lengths of A and B, l in Eq. (3.27) cannot be cancelled. Note that Eqs (3.26) and (3.27) may be used to find the amount of stress by which pre-tensioned bolts in a bolted flange assembly will relax following a temperature rise.

3.4 MECHANICAL AND TEMPERATURE EFFECTS

Other compound-bar problems may involve a combination of mechanical and thermal effects. The solutions are found from superimposing their separate effects, as the following example illustrates.

Example 3.11 Find, for the assembly shown in Fig. 3.9, the stresses in the steel (s) bolt and copper (c) tube when the nut is tightened by one revolution on a thread of pitch 0.5 mm. What are the net stresses when the pre-stressed assembly is subsequently subjected to (i) an axial tensile force of 5 kN and (ii) a temperature rise of 10 °C? Neglect the thickness of the end-plates. For the cylinder: $l_c = 2.5$ m, i.d. $= 38$ mm, o.d. $= 50$ mm, $E_c = 104$ GPa and $\alpha_c = 18 \times 10^{-6}/°C$. For the bolt: $E_s = 207$ GPa, $d_s = 12.5$ mm and $\alpha_s = 12.2 \times 10^{-6}/°C$.

The steel and copper areas are

$$A_s = (\pi/4)(12.5)^2 = 122.72 \, \text{mm}^2,$$
$$A_c = (\pi/4)[(50)^2 - (38)^2] = 829.38 \, \text{mm}^2.$$

Then from Eq. (3.21),

$$P = \delta(A_s E_s A_c E_c)/l(A_c E_c + A_s E_s)$$

$$= \frac{0.5 \times 122.72 \times 207 \times 829.38 \times 104 \times 10^3}{2.5 \times 10^3[(829.38 \times 104) + (122.72 \times 207)]} = 3924.74 \, \text{N};$$

$$\therefore \quad \sigma_c = -3924.74/829.38 = -4.732 \, \text{MPa};$$
$$\sigma_s = 3924.74/122.72 = 31.98 \, \text{MPa}.$$

(i) From Eqs (3.17), the bolt stress due to the axial force W is

$$\sigma_s = W E_s/[E_s A_s + E_c A_c]$$

$$= \frac{5 \times 10^3 \times 207 \times 10^3}{10^3[(207 \times 122.72) + (104 \times 829.38)]} = 9.27 \, \text{MPa}$$

and from Eq. (3.16) the corresponding cylinder stress is

$$\sigma_c = E_c\sigma_s/E_s = 104 \times 9.27/207 = 4.657\,\text{MPa}.$$

The net stresses become

$$\sigma_c = -4.732 + 4.657 = -0.075\,\text{MPa},$$
$$\sigma_s = 31.98 + 9.27 = 41.25\,\text{MPa}.$$

(ii) From Eq. (3.28), the cylinder stress due to a temperature rise Δt is

$$\sigma_c = (\alpha_s - \alpha_c)\,\Delta t\, E_c/(1 + A_c E_c/A_s E_s)$$
$$= \frac{(12.2 - 18)10^{-6} \times 10 \times 104 \times 10^3}{1 + (829.38 \times 104)/(122.72 \times 207)} = -1.37\,\text{MPa},$$

and from Eq. (3.26) the corresponding bolt stress is

$$\sigma_s = 1.37 \times 829.38/122.72 = 9.275\,\text{MPa}.$$

The net stresses are then

$$\sigma_c = -4.732 - 1.37 = -6.102\,\text{MPa},$$
$$\sigma_s = 31.98 + 9.275 = 41.26\,\text{MPa}.$$

EXERCISES

Direct Stress and Strain

3.1 A 180 m length of pipe run is subjected to a longitudinal tensile force of 50 kN. If the outside and inside diameters are 110 and 100 mm respectively, calculate the increase in the pipe length given $E = 193\,\text{GPa}$.

Answer: 28.3 mm.

3.2 A steel bar 300 mm long is 6.5 mm square for 100 mm of its length, the remaining length being 50 mm in diameter. Determine the extension of the bar when it is subjected to a tensile load of 220 kN. Take $E = 207\,\text{GPa}$.

Answer: 0.13 mm.

3.3 A metal tube of o.d. 76 mm and 1.65 m long is to carry a compressive load of 60 kN. If the allowable tensile stress is 77 MPa, calculate the internal diameter of the tube. Given $E = 92.6\,\text{MPa}$, find by how much the tube will shorten.

Answer: 69.3 mm; 1.4 mm.

3.4 A steel bolt 25 mm in diameter with an effective length of 250 mm and a thread pitch of 2.5 mm passes through a rigid block. Calculate the load imparted to the bolt when the clamping nut is initially tightened 1/4 turn with a spanner. If the UTS for the bolt material is 925 MPa what is the bolt safety factor? Take $E = 208\,\text{GPa}$.

Answer: 265 kN; 1.8.

3.5 A stepped shaft is 50, 35 and 20 mm in diameter for respective lengths of 125, 75 and 100 mm. Calculate the axial extension and the diametral contractions under a load of 7 MN given $E = 205\,\text{GPa}$, $v = 0.27$. (*Hint*, see Chapter 4.)

Answer: 1.57 mm; 0.0235 mm; 0.0335 mm; 0.0587 mm.

3.6 A tapered alloy column with end diameters of 25 and 50 mm and a length of 300 mm is subjected to an axial tensile load of 100 kN. Calculate the axial displacement and the maximum stress. Take $E = 80$ GPa.

   Answer: 0.38 mm; 204 MPa.

3.7 A solid steel bar 0.4 m long has a square cross-section of side length increasing linearly from 20 mm at one end of the bar to 30 mm at the other. What is the greatest axial tensile load that this bar can carry when the maximum permissible stress is 300 MPa and by how much does the bar extend under this load if $E = 200$ GPa?

   Answer: 120 kN; 0.4 mm.

3.8 To facilitate design calculations, a screwed thread may be simulated by a geometrical form shown in Fig. 3.12. Show that the tensile elongation in each simulated pitch, based upon simple elastic theory, is given by $4PF/\pi EDd$, where E is Young's modulus and F is the applied tensile force. Stress concentrations are not to be considered. (C.E.I., Pt. 2, 1972)

3.9 A tensile test on a 15 mm diameter steel testpiece with a 50 mm gauge length produced the recording of axial force vs. displacement shown in Fig. 3.13. Estimate the upper and lower yield points, the UTS, the percentage elongation at fracture, and Young's modulus.

   Answer: 237, 220, 285 MPa; 15 per cent; 205 GPa.

3.10 The stress–strain curve in Fig. 3.14 consists of two regions. Select from the following terms those which apply to any part the deformation in each region: elastic, recoverable, linear, elastic-plastic, partly recoverable, non-linear, permanent, hardening, unstable.

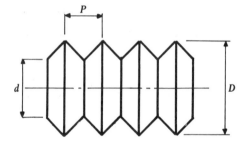

Figure 3.12

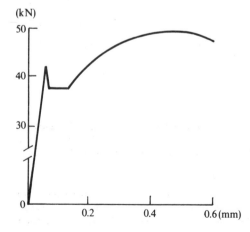

Figure 3.13

Figure 3.14

3.11 Calculate the modulus of elasticity from the load-vs.-deflection readings given, when the testpiece gauge diameter and length were 15 and 50 mm respectively.

Load (kN)	0	10	20	30	40	50	60	70	80	90	100
Deflection (mm)	0	0.013	0.027	0.040	0.054	0.067	0.081	0.094	0.108	0.121	0.135

Answer: $E = 210$ GPa.

3.12 In a tensile test on a 12.5 mm diameter test-piece, with a 50 mm gauge length, the following load and extension readings were obtained. Determine the 0.1 per cent proof stress and the stress at the limit of elastic proportionality.

Load (kN)	0	5.56	11.13	16.45	22.25	27.80	33.38	38.94	44.50	53.40
Ext. (mm)	0	0.01	0.020	0.031	0.041	0.051	0.061	0.071	0.081	0.102

Load (kN)	58.30	64.97	70.09	76.10	81.44	85.00	88.33	91.67	93.90
Ext. (mm)	0.112	0.127	0.142	0.163	0.183	0.203	0.224	0.254	0.279

How would the British Standard relate the test-piece diameter and gauge length? (See BS 18 (1987) *Method for Tensile Testing of Metals.*)

Shear Force, Stress and Strain

3.13 Estimate the forces required to punch (i) a 75 mm diameter hole in a 5 mm thick aluminium plate with an ultimate shear strength (USS) of 95 MPa and (ii) a 20 mm diameter hole in a 2.5 mm thick boiler plate with a USS of 415 MPa. What are the compressive stresses in the punch in each case?

Answer: 112 kN; 65.2 kN; 25.4 MPa; 207.5 MPa.

3.14 A tie bar in a steel structure (Fig. 3.15) carries a load of 100 kN through a 25 mm diameter pin in double shear. Calculate, using a safety factor of 3, suitable dimensions t, B and D of the bar end if tensile failure of the bar and bursting of the lug are to be avoided. The respective ultimate

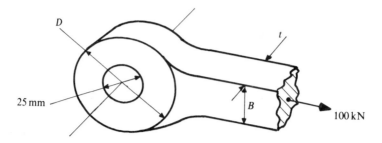

Figure 3.15

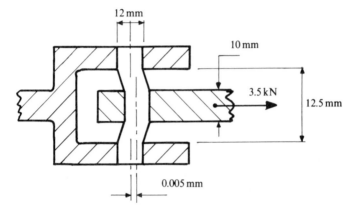

Figure 3.16

stresses are 260 and 320 MPa. What safety factor has been used for the pin, given that its ultimate strength in single shear is 400 MPa?

Answer: 4.

3.15 The central portion of a circular pin, 12 mm diameter, is displaced 0.005 mm from its unstrained centre line as shown in Fig. 3.16, owing to a 3.5 kN force. The ends of the pin pass through fork ends that maintain the original centre line position. Find the shear stress and strain in the pin and the value of its shear modulus.

3.16 A shaft is required to transmit 220 kW of power (P) at a speed (N) of 600 rev/min through two flanges fastened with 6 bolts. If these bolts all lie on a pitch circle diameter of 200 mm, determine the necessary bolt diameters using a safety factor of 2. The ultimate shear strength of the bolt material is 300 MPa. Note that torque $T = P/2\pi N$.

Answer: 5 mm.

3.17 A press tool produces 100 washers/stroke in a 1.5 mm thick material whose ultimate shear strength is 110 MPa. If the dimensions of the washers are to be 12 mm o.d. and 6 mm i.d., calculate the power required to drive a crankshaft when rotating at 30 rev/min given that the rod, which connects the crankshaft to the press, has a 25 mm stroke.

3.18 A lap joint in two plates 12.5 mm thick is formed with a single row of 20 mm diameter rivets. Determine the maximum possible pitch of the rivets and the efficiency of the joint with

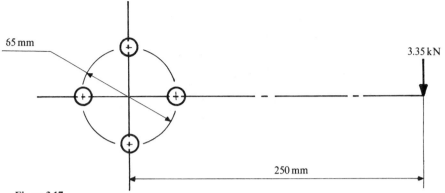

Figure 3.17

possible failure from tearing of the plate and shearing and crushing of the rivets. The respective ultimate strengths are 430, 355 and 710 MPa.

Answer: 48 mm; 53.3 per cent; 52.8 per cent; 76.5 per cent.

3.19 Three rows of 20 mm diameter rivets are used to form a lap joint between two 15 mm thick plates. Find the maximum allowable pitch of the rivets and the efficiency of the joint in tearing of the plates and shearing of the rivets given that the respective ultimate strengths are 90 and 70 MPa. What is the bearing stress?

Answer: 80 mm; 72 per cent, 74 per cent.

3.20 A butt joint consists of a single row of rivets in 20 mm thick plate. Using a safety factor of 3, with ultimate strengths of 380 and 700 MPa in shear and compression, determine the rivet pitch and diameter.

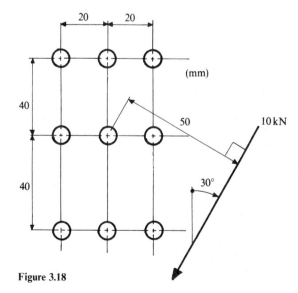

Figure 3.18

3.21 A riveted lap joint consists of nine rivets arranged in three rows of three with 25 mm spacing. Determine the load on each rivet when the joint carries 2 kN applied parallel to, but at a distance of 50 mm away from, the vertical centre line.

3.22 In Fig. 3.17 a 3.35 kN vertical force acts at a distance of 250 mm from the centre of a 65 mm pitch circle diameter (p.c.d.) on which lie four equi-spaced studs arranged as shown. If the studs act in single shear, calculate the net shear force on each. Hence determine a suitable stud diameter using a safety factor of 3 with an ultimate shear strength of 525 MPa.

Answer: 7.28 kN max; 8 mm.

3.23 For the bolted assembly given in Fig. 3.18, determine all the bolt forces when they act in single shear. Using a factor of safety of 2.5, find a suitable bolt diameter and plate thickness given that their ultimate shear and compressive strengths are 375 and 690 MPa respectively.

Compound Bars—Mechanical Effects

3.24 Two bronze strips 8 mm thick and 250 mm wide are securely fixed to the sides of a rectangular steel bar of the same width and length. Find the thickness of steel required to limit its stress to 155 MPa when the compound bar is loaded in tension to 200 kN. Find also the stress in the bronze and the extension of the bar over a 1 m length. $E = 110$ GPa for bronze and $E = 207$ GPa for steel.

Answer: 17 mm; 83 MPa; 0.75 mm.

3.25 A circular mild steel bar 37.5 mm diameter is firmly enclosed by a bronze tube of outer diameter 62.5 mm for a common length of 150 mm. Find the stress in each material and the change in length under an axial compressive force of 200 kN. Take $E = 200$ and 100 GPa for steel and bronze respectively.

Answer: 92.6 MPa; 46.3 MPa; 0.07 mm.

3.26 A bar of steel and two bars of brass each with a 25×15 mm cross-section are to carry jointly a load of 20 kN over a common length of 180 mm. If the bars are bonded in parallel so that each extends by the same amount, determine the load carried by each bar. Take $E = 207$ and 83 GPa for steel and brass respectively.

Answer: 8.9 kN brass; 11.1 kN steel.

3.27 A 25 mm composite square section is formed from a central brass strip 5 mm thick bonded between two 10 mm strips of epoxy resin for a length of 750 mm. Calculate the maximum axial load and the extension given that the respective allowable stresses for brass and epoxy are 155 MPa and 60 MPa. $E_{brass} = 86$ GPa, $E_{resin} = 3100$ MPa.

Answer: 27.5 kN; 1.4 mm.

3.28 A short steel tube of 100 mm o.d. and 10 mm thick is firmly surrounded by a brass tube of the same thickness for a common length of 100 mm. If the compound tube carries a total compressive load of 5 kN, calculate by how much the tube will shorten and the load carried by each material. $E_{steel} = 201$ GPa, $E_{brass} = 85$ GPa.

Answer: 3.2 kN; 1.8 kN; 0.000 46 mm.

3.29 Determine the maximum axial tensile load that may be applied to a 250 mm long bonded composite tube without causing yielding in either of its components. Brass inner ring: 50 mm i.d., 75 mm o.d., yield stress 160 MPa, $E = 105$ GPa. Steel outer ring: 75 mm i.d., 100 mm o.d., yield stress 210 MPa and $E = 210$ GPa.

3.30 A concrete column ($E = 20.7$ GPa) is reinforced by two steel bars ($E = 207$ GPa) of 25 mm diameter. Calculate the dimensions of a square-section column that has to support 400 kN in compression if the stress in the concrete is not to exceed 6.9 MPa.

3.31 A concrete column, 3.05 m high with 380 mm square section, is reinforced by four steel rods, 25 mm in diameter, evenly spaced within the cross-section. If the concrete carries an axial load of

600 kN, determine the stresses in the steel and concrete and the change in length of the column. $E_s = 207$ GPa, $E_c = 13.8$ GPa.

Answer: 52.4 MPa; 3.49 MPa; 0.772 mm.

3.32 A concrete column is reinforced by a 90° equal-angle 0.1 m long by 0.01 m thick. Calculate the stress in each material when the length of reinforcement is shorter by 0.15 mm than the length of the concrete. It is known that the concrete stress is 20 MPa for an unreinforced column of equal 1.25 m length. Take $E_s = 20E_c = 200$ GPa.

Answer: 188 MPa; 21.2 MPa.

3.33 A 12.5 mm diameter steel rod passes centrally through a 2.5 m length of copper tube 50 mm o.d. and 38 mm i.d. The tube is closed at each end by 25 mm thick rigid steel plates secured by nuts on the threaded ends of the rod. Calculate the stresses set up in the steel and the copper (a) when the tube contracts 0.5 mm after the nuts are tightened, (b) when the nuts, of 0.5 mm pitch, are tightened by two revolutions. Note that the rod and cylinder are of unequal lengths. $E_s = 207$ GPa, $E_c = 105$ GPa.

Answer: (a) 140.6 MPa; −20.8 MPa.

Compound Bars—Temperature Effects

3.34 A tie bar 50 mm diameter and 6 m long realigns a bulging vessel by alternate heating and cooling to a maximum of 300 °C. If the vessel offers a maximum resistance of 200 kN, determine the greatest alignment for each heating cycle when the nuts at the end of the bar are finger tight at 300 °C. Take $E = 207$ GPa, $\alpha = 12 \times 10^{-6}/°C$ for the bar.

Answer: 19.2 mm.

3.35 A solid bronze and a steel bar of equal lengths with diameters of 50 mm and 30 mm respectively are placed together end to end with their axes horizontal between close-fitting rigid walls. Calculate the stresses induced in each material when the temperature is raised by 55.55 °C (100 °F). Take $E = 208$ GPa, $\alpha = 11.7 \times 10^{-6}/°C$ ($6.5 \times 10^{-6}/°F$) for steel and $E = 108$ GPa, $\alpha = 18 \times 10^{-6}/°C$ ($10 \times 10^{-6}/°F$) for bronze.

Answer: 72.9 MPa; 206 MPa.

3.36 A phosphor bronze tube 40 mm long, 40 mm i.d. and 50 mm o.d., forms a pipeline with a second plastic tube of the same i.d. but with a 55 mm o.d. and a 60 mm length. The assembly is held together by rigid plates that prevent axial expansion of the tubes but allow free fluid flow. Calculate the stresses induced in each tube when the fluid temperature is raised by 75 °C from the unstressed state at room temperature. Take $E = 80$ GPa, $\alpha = 20 \times 10^{-6}/°C$ for the smaller tube, $E = 12$ GPa, $\alpha = 18 \times 10^{-6}/°C$ for the larger tube.

Answer: 179 MPa; 112 MPa. (K.P. 1975)

3.37 A compound tube is formed from an outer stainless steel tube, 50 mm o.d., 47.5 mm i.d. and an inner mild steel tube of 6 mm wall thickness. The two tubes are welded together at their ends to give a 1.5 mm radial clearance. Assuming that the tubes are free to expand between the ends when heated, calculate the tube stresses for $\Delta t = 55.55$ °C. $E = 172$ GPa, $\alpha = 18 \times 10^{-6}/°C$ for stainless steel; $E = 207$ GPa, $\alpha = 12 \times 10^{-6}/°C$ for mild steel.

Answer: 44.8 MPa; 14.5 MPa.

3.38 A compound bar is formed from a stainless steel rod placed inside a mild steel tube, welded together at their ends, over a length of 600 mm. The cross-sectional area of the rod is 750 mm² and that of the tube is 1625 mm². If an extension of the compound bar of 0.375 mm results from a temperature rise of 38.88 °C, determine the load induced in the rod and the stress in the tube. What is α for the tube if $E = 207$ GPa? Take $E = 165.4$ GPa, $\alpha = 18 \times 10^{-6}/°C$ for stainless steel.

Answer: 9.6 kN; 5.72 MPa; $\alpha = 15.35 \times 10^{-6}/°C$.

3.39 A cast iron cylinder cover is bolted to the flanges of a cast iron cylinder by wrought iron bolts. The latter occupy 1/10 of the area of the mating cast iron surfaces. If, after the bolts are tightened, the temperature is raised by 166.67 °C, find the amount by which the bolt stresses will relax. Take $E = 92.6\,\text{MPa}$, $\alpha = 10.8 \times 10^{-6}/°C$ for cast iron; $E = 185.2\,\text{GPa}$, $\alpha = 12.1 \times 10^{-6}/°C$ for wrought iron.

3.40 A long brass rod just fits in a steel tube of external diameter 60 mm and internal diameter 30 mm when the materials are at 20 °C. If the rod and tube are heated to a temperature of 100 °C, calculate the maximum direct stresses induced in the steel and the brass. (*Hint:* see Chapter 4.) For steel, take $E = 200\,\text{GPa}$, $v = 0.3$ and $\alpha = 10 \times 10^{-6}/°C$; for brass, take $E = 80\,\text{GPa}$, $v = 0.3$ and $\alpha = 16 \times 10^{-6}/°C$.

Compound Brass—Mechanical and Temperature Effects

3.41 A compound rectangular bar is made from a layer of steel plate 50 mm wide and 20 mm thick lying between copper strips 50 mm wide by 5 mm thick. The strips are fixed firmly together at the ends of their common 900 mm room-temperature lengths. Determine the stresses in the steel and the copper when the temperature is raised by 55.56 °C. What are the stresses and the final extension of the assembly when a tensile force of 4.5 kN is applied at the increased temperature? $E = 207\,\text{GPa}$, $\alpha = 12.6 \times 10^{-6}/°C$ for steel and $E = 110\,\text{GPa}$, $\alpha = 18 \times 10^{-6}/°C$ for copper.

Answer: 26 MPa; 13 MPa; 3.4 MPa; 1.81 MPa; 0.762 mm.

3.42 A flat steel bar 25 mm wide and 5 mm thick is placed between two aluminium alloy bars each 25 mm wide by 10 mm thick to form a composite bar with 25 mm square section. The bars are fixed together at their ends at a temperature of 10 °C. Find the stress in each bar when the temperature is then raised to 48.88 °C and, also, when an additional force of 20 kN is applied at this temperature. Take $E = 207\,\text{GPa}$, $\alpha = 11.5 \times 10^{-6}/°C$ for steel; $E = 69\,\text{GPa}$, $\alpha = 23 \times 10^{-6}/°C$ for Al alloy.

Answer: 52.89 MPa(St); -13.22 MPa(Al); 121.46 MPa(St); 9.64 MPa(Al).

3.43 If, after tightening the bolted assembly in Exercise 3.33 by two revolutions of the nut, it is subsequently subjected to (i) an axial tensile force of 4 kN and (ii) a temperature rise of 20 °C, determine, for each case, the net stresses arising in the steel bolt and the copper tube. Take $\alpha = 12 \times 10^{-6}/°C$ and $\alpha = 18 \times 10^{-6}/°C$ for steel and copper respectively.

3.44 A horizontal brass tube is clamped between rigid end covers by six equally spaced, 6 mm diameter steel bolts as shown in Fig. 3.19. The tube is precompressed at room temperature,

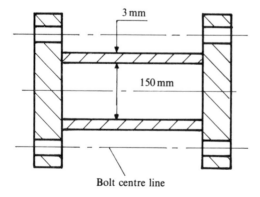

Bolt centre line   **Figure 3.19**

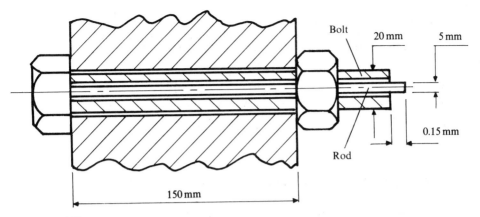

Figure 3.20

(15 °C) by tensioning the bolts to $40 \, \text{N/mm}^2$. Determine the stresses in both the bolts and the tube when the temperature of the bolts is 0 °C and the temperature of the tube is 60 °C. Assume the following values for material constants: $E = 20 \times 10^4 \, \text{N/mm}^2$, $\alpha = 12 \times 10^{-6}/°C$ for steel; $E = 10 \times 10^4 \, \text{N/mm}^2$, $\alpha = 18 \times 10^{-6}/°C$ for brass. (C.E.I. Pt. 2, 1972)

3.45 The steel bolt shown in Fig. 3.20 is pre-tensioned over a fixed length of 150 mm. A hole drilled through the centre of the bolt contains a loose-fitting aluminium alloy rod that is attached to the left-hand bolt head. In this way, tightening the right-hand nut strains the bolt but not the pin. Determine the design load for the bolt when the right-hand ends of the rod and bolt shank are level. By how much are this load and the end alignment altered with the introduction of fluid to the bolt hole that raises the temperature of the whole bolt assembly by 60 °C? Assume that the 150 mm length is unaffected. Take $E = 207 \, \text{GPa}$, $\alpha = 11. \, 5 \times 10^{-6}/°C$ for steel and $E = 70 \, \text{GPa}$, $\alpha = 24 \times 10^{-6}/°C$ for aluminium alloy.

ELASTIC CONSTITUTIVE RELATIONS

4.1 THE FOUR ELASTIC CONSTANTS

Both E and G were defined in Chapter 3. Two further elastic constants can be defined for materials operating under elastic systems of stress.

Poisson's ratio, v Poisson's ratio is defined (with E) for a uniaxial system of elastic stress, as the ratio between the lateral ϵ_y and axial Λ_x strains. This is a dimensionless elastic constant, lying in the range 1/4–1/3 depending upon the material. In order to make the ratio positive, a minus sign must appear with ϵ_y in the definition, since the strains will have opposite sense for both tension and compression. Thus, in Fig. 4.1, if δx is the increase in length l under tension and δy is the contraction in a lateral dimension w, then

$$v = -\frac{\epsilon_y}{\epsilon_x} = -\frac{l\delta y}{w\delta x}. \tag{4.1}$$

Equation (4.1) is also valid for compression, when δx and δy are the respective decrease in length and increase in width. The lateral strain may also be found from E and v for a given axial stress σ. That is, from Eq. (4.1),

$$\epsilon_y = -v\epsilon_x = -v\sigma/E. \tag{4.2}$$

Bulk modulus K Bulk modulus is the ratio between the mean or hydrostatic stress σ_m and volumetric strain $\delta V/V$. That is,

$$K = \frac{\sigma_m}{(\delta V/V)}. \tag{4.3}$$

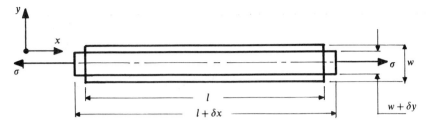

Figure 4.1 Axial and lateral displacements under tension.

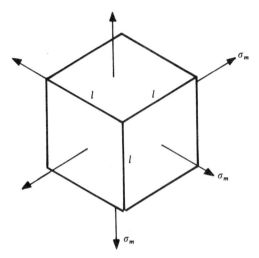

Figure 4.2 Hydrostatic pressure.

This system arises directly when a solid is subjected to mutually perpendicular equal stresses typical of fluid pressure (Fig. 4.2). The ratio remains positive irrespective of whether σ_m is tensile or compressive because δV will change sign accordingly. Assuming tension for a unit cube ($l = 1$), the increase in length of each side is $\delta x = \epsilon \times l = \epsilon$. Then

$$\delta V = \text{strained volume} - \text{initial volume},$$

$$\delta V / V = (1 + \epsilon)^3 - 1 \simeq 3\epsilon,$$

ignoring ϵ^2 and ϵ^3 terms. From Eq. (4.3),

$$K = \sigma_m / 3\epsilon. \tag{4.4}$$

The hydrostatic system will arise indirectly when normal stresses are unequal. When σ_1, σ_2 and σ_3 act as in Fig. 4.3, the mean of these, σ_m, the hydrostatic component, will cause elastic compressibility only. The remaining stresses, $\sigma_1 - \sigma_m$, $\sigma_2 - \sigma_m$ and $\sigma_3 - \sigma_m$, the deviatoric components, are those

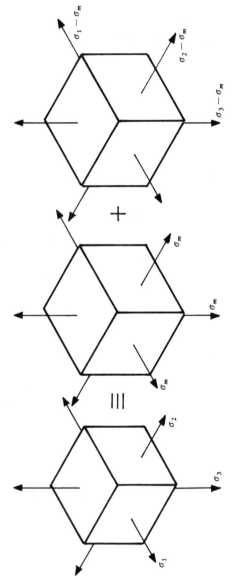

Figure 4.3 Hydrostatic and deviatoric components of applied stress.

responsible for distortion without volume change. It follows that σ_m in Eq. (4.3) becomes

$$\sigma_m = (\sigma_1 + \sigma_2 + \sigma_3)/3 \qquad (4.5)$$

and, with associated length changes (strains) of ϵ_1, ϵ_2 and ϵ_3 in the sides of a unit cube $(V = 1)$,

$$\delta V/V = (1 + \epsilon_1)(1 + \epsilon_2)(1 + \epsilon_3) - 1 \simeq \epsilon_1 + \epsilon_2 + \epsilon_3.$$

Equation (4.3) now becomes

$$K = (\sigma_1 + \sigma_2 + \sigma_3)/3(\epsilon_1 + \epsilon_2 + \epsilon_3) \qquad (4.6)$$

4.2 RELATIONSHIP BETWEEN E, G AND v

This relationship is established from a consideration of the stress state in pure shear (Fig. 4.4(a)). Firstly, it must be realized that the shear stresses τ_{AD} and τ_{BC} on opposite faces resulting from the shear force cannot exist without an equilibrating or complementary action on each adjacent face. The complementary shear stresses τ_{AB} and τ_{CD} are found from moment equilibrium, say about point A where, for thickness t,

$$\tau_{BC} \times (BC \times t) \times AB = \tau_{CD} \times (CD \times t) \times AD$$

$$\therefore \quad \tau_{CD} = \tau_{BC} \quad \text{and} \quad \tau_{AB} = \tau_{AD} \quad \text{(since } \tau_{AD} = \tau_{BC}\text{)}.$$

That is, the shear stresses are all the same, simply labelled τ hereafter.

Now consider the induced state of normal σ_θ and shear stress τ_θ, acting along an oblique plane BT inclined at θ as shown in Fig. 4.5(a). The conversion of stresses to forces and their resolution into directions parallel and perpendicular to BT is given in Fig. 4.5(b). The following equilibrium

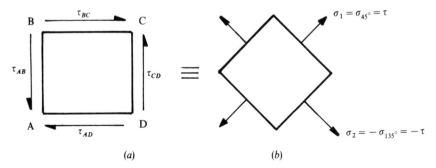

Figure 4.4 Analysis of stress in pure shear.

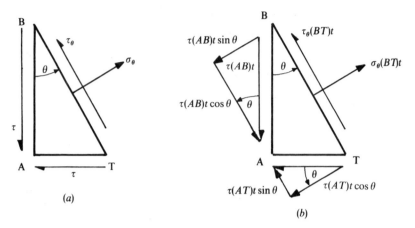

(a)

(b)

Figure 4.5 State of stress on an oblique plane in pure shear.

equations apply to these directions noting that $AB = BT \cos \theta$, $AT = BT \sin \theta$:

$$\sigma_\theta(BT)t = \tau t\, AB \sin \theta + \tau t\, AT \cos \theta,$$
$$= \tau t(BT \cos \theta) \sin \theta + \tau t(BT \sin \theta) \cos \theta$$
$$= 2(BT)\tau t \sin \theta \cos \theta,$$
$$\therefore \quad \sigma_\theta = \tau \sin 2\theta.$$
$$\tau_\theta(BT)t = \tau t\, AB \cos \theta - \tau t\, AT \sin \theta$$
$$= \tau t(BT \cos \theta) \cos \theta - \tau t(BT \sin \theta) \sin \theta$$
$$= (BT)\tau t(\cos^2 \theta - \sin^2 \theta),$$
$$\therefore \quad \tau_\theta = \tau \cos 2\theta.$$

These equations show that, when $\theta = 45°$, $\sigma_{45°} = \tau$, $\tau_{45°} = 0$ and that when $\theta = 135°$, $\sigma_{135°} = -\tau$, $\tau_{135°} = 0$. Pure shear is therefore equivalent to the stress system shown in Fig. 4.4(b), where tensile and compressive stresses act on the 45° planes in the absence of shear stress ($\sigma_2 = -\tau$ implies compression). This is the principal element for the system to which we shall return in Chapter 13. An examination of the shear distortion in Fig. 4.6 shows that

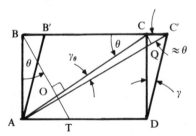

Figure 4.6 Shear distortion.

the direct strain normal to BT is, by definition,

$$\epsilon_\theta = (AC' - AC)/AC \simeq (CC' \cos \theta)/AC.$$

Now, as elastic distortions are small $\theta \simeq$ angle(AC'B) and $\gamma \simeq$ angle(CDC') (rads). Thus $CC' = AB\gamma$ and $AB = AC \sin \theta$, when ϵ_θ becomes

$$\epsilon_\theta = (AC \sin \theta)\gamma \cos \theta/AC,$$

$$\epsilon_\theta = (\gamma/2) \sin 2\theta.$$

The rotation angle CAC' for the direction AC, in radians, is given by

$$CAC' = CQ/AC \simeq (CC'/AC) \sin \theta \simeq \gamma \sin^2 \theta.$$

Replacing θ with $\theta + \pi/2$, the rotation angle for the perpendicular direction BT is $\gamma \sin^2 (\theta + \pi/2) = -\gamma \cos^2 \theta$. The sum of these rotations defines the change in the right angle BOC, i.e. the shear strain γ_θ, as

$$\gamma_\theta = \gamma(\sin^2 \theta - \cos^2 \theta) = -\gamma \cos 2\theta$$

The ϵ_θ and γ_θ equations show that, when $\theta = 45°, \epsilon_{45°} = \gamma/2, \gamma_{45°} = 0$ and that when $\theta = 135° \epsilon_{135°} = -\gamma/2, \gamma_{135°} = 0$. i.e. the normal strains in the 45° and 135° directions for the equivalent system in Fig. 4.4(b) are respectively tensile and compressive, both equal in magnitude to half the applied shear strain, there being no angular change between adjacent sides. Recognition of this equivalent biaxial stress system enables the normal stress and strain to be related. The total normal strain ϵ_1, for the 1-direction is composed of the sum of two parts: (i) a direct tensile strain σ_1/E and (ii) a lateral strain under σ_2. For the direction of σ_2 shown, the lateral component is found from Eq. (4.2) to be $-\nu\sigma_2/E$. Then,

$$\epsilon_1 = (1/E)[\sigma_1 - \nu\sigma_2]. \tag{4.7a}$$

Substituting $\epsilon_1 = \epsilon_{45°} = \gamma/2, \sigma_1 = \sigma_{45°} = \tau$ and $\sigma_2 = \sigma_{135°} = -\tau$ into Eq. (4.7a) relates the equivalent system to the system of pure shear that produced it:

$$\gamma/2 = (1/E)[\tau - \nu(-\tau)] = (1 + \nu)\tau/E. \tag{4.7b}$$

Now, as $G = \tau/\gamma$ (Eq. (3.8)) it follows that Eq. (4.7b) becomes

$$E = 2G(1 + \nu). \tag{4.8}$$

Equation (4.8) can also be derived from the $\sigma - \epsilon$ relation for the 2-direction.

4.3 RELATIONSHIP BETWEEN E, K AND ν

Consider the total strain, ϵ, produced by hydrostatic stress for any one stress direction in Fig. 4.2. It is composed of the sum of three parts: (i) the direct tensile strain $\epsilon_{(i)} = \sigma_m/E$ and (ii) two lateral compressive strains, produced by the remaining stresses, which from Eq. (4.2) are $\epsilon_{(ii)} = -\nu\epsilon_{(i)} = -\nu\sigma_m/E$.

The total strain for any direction is then

$$\epsilon = \sigma_m/E - v\sigma_m/E - v\sigma_m/E = (\sigma_m/E)(1 - 2v). \tag{4.9}$$

Substituting Eq. (4.9) into Eq. (4.4),

$$K = \sigma_m/3[(\sigma_m/E)(1 - 2v)]$$

from which,

$$E = 3K(1 - 2v). \tag{4.10}$$

Further relationships may be found between G, K and v by eliminating E and v between Eqs (4.8) and (4.10):

$$G = 3KE/(9K - E) = 3K(1 - 2v)/2(1 + v). \tag{4.11}$$

Example 4.1 A metal bar 50 mm in diameter and 150 mm long is subjected to a tensile force of 75 kN. If the change in length and diameter are 0.08 and -0.0065 mm respectively, find the four elastic constants.

Using Eqs (3.3), (4.1), (4.8) and (4.10):

$E = \sigma/\epsilon_x = Wl/Ax$
$\quad = 75 \times 10^3 \times 150/[(\pi/4)(50)^2 \times 0.08] = 71\,619.72\,\text{kN/mm}^2 = 71.62\,\text{GPa};$
$v = -\epsilon_y/\epsilon_x = -(\Delta d)l/d \times x$
$\quad = 0.0065 \times 150/(50 \times 0.08) = 0.244;$
$G = E/2(1 + v) = 71.62/2(1 + 0.244) = 28.80\,\text{GPa};$
$K = E/3(1 - 2v) = 71.62/3(1 - 0.488) = 46.63\,\text{GPa}.$

Example 4.2 A cylindrical steel bar 50 mm (2 in) in diameter and 305 mm (1 ft) long is subjected to a fluid pressure of 7.75 bar (112.5 psi). Find the change in volume and the modulus of rigidity, given that $E = 204\,\text{GPa}$ ($29 \times 10^6\,\text{lbf/in}^2$) and $v = 0.294$. (10 bar = 1 MPa.)

SOLUTION TO SI UNITS Using Eq. (4.10) to find the bulk modulus,

$$K = E/3(1 - 2v) = 204/3(1 - 0.588) = 165.05\,\text{GPa}.$$

Since the pressure is compressive, $\sigma_m = -p$ and Eq. (4.3) becomes

$K = -p/(\delta V/V)$
$\therefore \quad \delta V = -pV/K = -0.775 \times (\pi/4)(50)^2 \times 305/(165.05 \times 10^3)$
$\quad = -2.812\,\text{mm}^3.$

SOLUTION TO IMPERIAL UNITS Again, from Eq. (4.10),

$$K = E/3(1 - 2v) = 29 \times 10^6/3(1 - 0.588) = 23.46 \times 10^6\,\text{lbf/in}^2,$$

and from Eq. (4.3),

$$\delta V = -pV/K = -11.25 \times 10^3 \times (\pi/4)(2)^2 \times 12/(23.46 \times 10^6)$$
$$= -0.0181 \, \text{in}^3.$$

Example 4.3 Find the change in volume of a 1 m long steel bar with 75 mm square section under an axial compressive force of 20 kN, given $E = 207$ GPa and $G = 82$ GPa.

The hydrostatic stress component in Eq. (4.5) is simply

$$\sigma_m = \sigma/3 = -20 \times 10^3/3(75)^2 = -1.185 \, \text{MPa}$$

and, from Eqs (4.3) and (4.11),

$$K = GE/3(3G - E) = 82 \times 207/3[(3 \times 82) - 207] = 145.08 \, \text{GPa},$$
$$\delta V = \sigma_m V/K = -1.185 \times (75)^2 \times 1000/(145.08 \times 10^3) = -45.95 \, \text{mm}^3.$$

4.4 GENERAL RELATIONS BETWEEN NORMAL STRESS AND STRAIN

The stress–strain relations employed in Eqs (4.7) and (4.9) are particular forms of a more general relation that applies when unequal mutually perpendicular stresses σ_1, σ_2 and σ_3 act. Thus, in the first diagram of Fig. 4.3 the total normal strain for the 1-direction (ϵ_1) is composed of the sum of a direct strain σ_1/E due to σ_1 and lateral strains $-v\sigma_2/E$ and $-v\sigma_3/E$ due to σ_2 and σ_3 respectively. The normal strains for the 2- and 3-directions are similarly composed of their direct and lateral strain components. The full stress–strain relations are then

$$\epsilon_1 = (1/E)[\sigma_1 - v(\sigma_2 + \sigma_3)] \qquad (4.12a)$$

$$\epsilon_2 = (1/E)[\sigma_2 - v(\sigma_1 + \sigma_3)] \qquad (4.12b)$$

$$\epsilon_3 = (1/E)[\sigma_3 - v(\sigma_1 + \sigma_2)] \qquad (4.12c)$$

Also, $x_1 = \epsilon_1 l_1, x_2 = \epsilon_2 l_2$ and $x_3 = \epsilon_3 l_3$, where x are the changes in length of side lengths l. Equations (4.12b and c) follow from (a) with a rotation in the subscripts. Adding Eqs (4.12a–c) leads to

$$(\epsilon_1 + \epsilon_2 + \epsilon_3) = (1 - 2v)(\sigma_1 + \sigma_2 + \sigma_3)/E \qquad (4.13)$$

from which the comparison with Eq. (4.6) again confirms Eq. (4.10). Note that as v approaches $1/2$ in Eq. (4.13), $K \to \infty$ as a material becomes elastically incompressible, but this applies only to certain non-metallic materials, e.g. rubber. In the case of plane, or two-dimensional stressing in Fig. 4.7, where

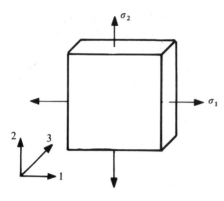

Figure 4.7 Plane stress.

$\sigma_3 = 0$, Eqs (4.12) become

$$\epsilon_1 = (1/E)(\sigma_1 - v\sigma_2), \tag{4.14a}$$

$$\epsilon_2 = (1/E)(\sigma_2 - v\sigma_1), \tag{4.14b}$$

$$\epsilon_3 = -(v/E)(\sigma_1 + \sigma_2). \tag{4.14c}$$

Hence a direct, through-thickness strain is induced in the 3-direction from the lateral strains produced by σ_1 and σ_2. This Poisson effect is evident, for example, in the thinning of the plate in Fig. 4.7 under the in-plane tensile stresses. Solving Eqs (4.14a and b) simultaneously leads to the inverted forms

$$\sigma_1 = E(\epsilon_1 + v\epsilon_2)/(1 - v^2), \tag{4.15a}$$

$$\sigma_2 = E(\epsilon_2 + v\epsilon_1)/(1 - v^2), \tag{4.15b}$$

where stress may be calculated from measured strain knowing the elastic constants. Typical room-temperature values for the elastic constants of some of the more important metallic materials are given in Table 4.1.

Table 4.1 Elastic constants

Material	E(GPa)	G(GPa)	K(GPa)	v
Carbon and alloy steels	207	81	157	0.28
Cast iron	96.5	41	50.3	0.18
18/8 stainless steel	200	77	167	0.30
Aluminium alloys	75	28.5	65.8	0.31
Copper	110	41.3	107.5	0.33
Brass	103.5	39	101.5	0.33
Bronze	117	44.8	102.6	0.31
Titanium	118	45	104	0.31
Tungsten	400	157	290	0.27
Nimonic alloys	210	80	184	0.31

Example 4.4 The reduction in volume of a solid block is to be restricted to $1 \, \text{mm}^3$ under the action of three mutually perpendicular forces: $5 \, \text{kN}$ in tension on opposite $75 \times 50 \, \text{mm}$ faces and $10 \, \text{kN}$ in compression on $100 \times 50 \, \text{mm}$ faces. Find the magnitude and nature of the maximum force that may act on the $100 \times 75 \, \text{mm}$ faces and the changes in length of each side. Take $K = 150 \, \text{GPa}$, $v = 0.27$ (see Fig. 4.3) and

$$\sigma_2 = -10 \times 10^3/50 \times 100 = -2 \, \text{MPa},$$
$$\sigma_3 = 5 \times 10^3/75 \times 50 = 4/3 \, \text{MPa}.$$

Let the unknown stress be σ_1, then from Eqs (4.3) and (4.5),

$$\sigma_m = (\sigma_1 + \sigma_2 + \sigma_3)/3 = (\sigma_1 - 2 + 4/3)/3 = (3\sigma_1 - 2)/9,$$
$$\therefore \quad \delta V = \sigma_m V/K = (3\sigma_1 - 2)V/9K,$$

and hence,

$$\sigma_1 = 3K \, \delta V/V + 2/3$$
$$= 3 \times 150 \times 10^3(-1)/(100 \times 50 \times 75) + 2/3$$
$$= -1.2 + 0.667 = -0.533 \, \text{MPa},$$
$$W_{\text{max}} = \sigma_1 A = (100 \times 75)(-0.533) \, \text{N} = -4 \, \text{kN}.$$

Now, $E = 3K(1 - 2v) = 3 \times 150 \times 10^3(1 - 0.54) = 207\,000 \, \text{MPa}$, when, from Eq. ((4.12a–c),

$$\epsilon_1 = (1/E)[\sigma_1 - v(\sigma_2 + \sigma_3)] = (10^{-3}/207)[-0.533 - 0.27(-2 + 1.333)]$$
$$= -1.71 \times 10^{-6},$$
$$\epsilon_2 = (1/E)[\sigma_2 - v(\sigma_1 + \sigma_3)] = (10^{-3}/207)[-2 - 0.27(-0.533 + 1.333)]$$
$$= -10.71 \times 10^{-6},$$
$$\epsilon_3 = (1/E)[\sigma_3 - v(\sigma_1 + \sigma_2)] = (10^{-3}/207)[1.333 - 0.27(-0.533 - 2)]$$
$$= +9.74 \times 10^{-6},$$
$$\therefore \quad x_1 = -1.71 \times 0.050 = -0.0535 \, \mu\text{m},$$
$$x_2 = -10.71 \times 0.075 = -0.803 \, \mu\text{m}$$

and

$$x_3 = 9.74 \times 0.10 = 0.974 \, \mu\text{m}.$$

4.5 CYLINDRICAL AND SPHERICAL VOLUMETRIC STRAINS

When the dimensions of a cylindrical and spherical vessel change by the action of external loading, e.g. pressure of a fluid and axial forces, then the volumetric strain $\delta V/V$ expresses the ratio between the change in internal volume to the original volume.

Cylinder

Let the initial length l and internal diameter d increase by the small amounts δl and δd respectively. The volumetric strain is

$$\delta V/V = (\pi/4)[(d + \delta d)^2(l + \delta l) - d^2 l]/(\pi/4)d^2 l$$
$$= [d^2 l + 2dl(\delta d) + l(\delta d)^2 + d^2(\delta l) + 2d(\delta d)(\delta l) + (\delta d)^2(\delta l) - d^2 l]/(d^2 l)$$
$$\simeq [2dl(\delta d) + d^2(\delta l)]/(d^2 l)$$
$$= (2\delta d/d) + (\delta l/l).$$

Now, by definition, the axial and circumferential strains (at the inner diameter) are, in respective polar coordinates z and θ:

$$\epsilon_z = \delta l/l,$$
$$\epsilon_\theta = [\pi(d + \delta d) - \pi d]/\pi d = \delta d/d,$$
$$\therefore \quad \delta V/V = 2\epsilon_\theta + \epsilon_z. \tag{4.16}$$

The strains ϵ_z and ϵ_θ depend upon the stress state existing in the wall of the vessel. For *thick-walled* vessels (Chapter 15), a triaxial stress state exists and the corresponding polar coordinate forms of Eqs (14.12), where θ, r and z replace 1, 2 and 3 respectively, apply to all points in the wall. When the vessel is *thin-walled*, in which case the ratio of diameter to thickness is greater than 20, it becomes reasonable to assume a biaxial stress state in the plane of the wall, as the radial stress through the wall is negligibly small by comparison. Equations (4.14) then apply. Moreover, as a closed thin cylindrical vessel containing fluid under pressure is a statically determinate problem, the resulting tensile axial (σ_z) and circumferential (or hoop) (σ_θ) stresses in the cylinder wall may be found directly from force equilibrium in Fig. 4.8.

Vertical equilibrium across the diametral plane in Fig. 4.8(a) gives

$$2\sigma_\theta t l = pdl, \quad \Rightarrow \sigma_\theta = pd/2t, \tag{4.17a}$$

where p acts on the projected area $d \times l$. Horizontal equilibrium across the vertical section in Fig. 4.8(b) gives

$$\pi dt \sigma_z = (\pi/4)d^2 p, \quad \Rightarrow \sigma_z = pd/4t. \tag{4.17b}$$

Substituting Eqs (4.14) and (4.17) into Eq. (4.16) leads to the volumetric strain for a thin cylinder:

$$\delta V/V = (2/E)(\sigma_\theta - v\sigma_z) + (1/E)(\sigma_z - v\sigma_\theta)$$
$$= (1/E)[\sigma_\theta(2 - v) + \sigma_z(1 - 2v)]$$
$$= (pd/4tE)(5 - 4v). \tag{4.18}$$

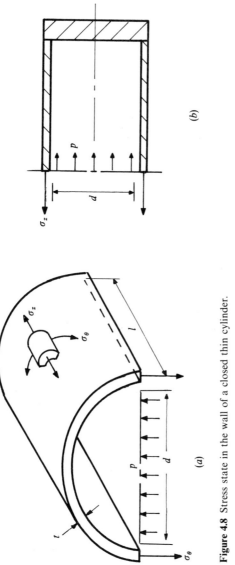

(b)

(a)

Figure 4.8 Stress state in the wall of a closed thin cylinder.

Sphere

When the internal diameter d increases by the small amount δd, then

$$
\begin{aligned}
\delta V/V &= (\pi/6)[(d + \delta d)^3 - d^3]/(\pi/6)d^3 \\
&= [d^3 + 3d^2(\delta d) + 3d(\delta d)^2 + (\delta d)^3 - d^3]/d^3 \\
&\simeq 3d^2(\delta d)/d^3 \\
&= 3(\delta d)/d,
\end{aligned}
$$

$$
\therefore \quad \delta V/V = 3\epsilon_\theta, \tag{4.19}
$$

in which ϵ_θ depends upon the stress state existing in the wall at the inner diameter. When the radial stress in a thin-walled sphere $(d/t > 20)$ is considered negligible, horizontal equilibrium across the vertical section in Fig. 4.9 gives

$$
\pi dt\sigma_\theta = (\pi/4)d^2 p, \quad \Rightarrow \sigma_\theta = pd/4t, \tag{4.20}
$$

which defines the uniform in-plane biaxial stress state. Then, from Eqs (4.14), (4.19) and (4.20), the volumetric strain for a thin sphere is

$$
\begin{aligned}
\delta V/V &= (3/E)(\sigma_\theta - v\sigma_\theta) \\
&= (3pd/4tE)(1 - v). \tag{4.21}
\end{aligned}
$$

When a cylindrical or spherical pressure vessel is fabricated from riveted plate, the design of the joint is based upon the working stresses calculated from Eqs (4.17) and (4.20). Safety factors and joint efficiencies, as defined in Eqs (3.5), (3.9) and (3.10), are employed to avoid tearing of the plate and shearing of the rivets as outlined in the previous chapter. Note that in a thin-walled fabricated vessel, where the least value of the failure load is determined from tearing of the plate along the joint, Eq. (3.10) becomes

$$
\text{Joint efficiency} = \text{tearing strength of joint/UTS of plate} \tag{4.22}
$$

Example 4.5 A thin-walled closed cylinder with inner diameter 500 mm, thickness 10 mm and internal volume 0.5 m³ is filled with a fluid at a gauge pressure of 20 bar (2 MPa). What volume of fluid will be ejected from the

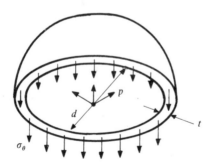

Figure 4.9 Thin-walled sphere.

cylinder when the pressure is reduced to atmospheric, assuming that the fluid is (a) incompressible and (b) compressible? Take $E = 210\,\text{GPa}$, $v = 0.25$ for the cylinder, and $K = 2500\,\text{MPa}$ for the compressible fluid.

Equation (4.18) gives the amount δV by which the cylinder recovers its internal volume on release of the pressure. This is equal to the amount of incompressible fluid ejected. That is,

$$\delta V = (pdV/4tE)(5 - 4v)$$
$$= 2 \times 500(5 - 1) \times 0.5 \times 10^9/(4 \times 10 \times 210 \times 10^3)$$
$$= 238.1 \times 10^3\,\text{mm}^3.$$

The amount δV by which a compressible fluid is compressed under pressure is found from Eq. (4.3) with $\sigma_m = -p$:

$$\delta V = -pV/K$$
$$= -2 \times 0.5 \times 10^9/2500 = -400 \times 10^3\,\text{mm}^3.$$

On release of the pressure, this additional volume is ejected to give a total spillage of $638.1 \times 10^3\,\text{mm}^3$.

Example 4.6 A thin-walled cylinder has two hemispherical caps of the same material welded to its ends. Determine the sphere-to-cylinder thickness ratio that will ensure the same radial displacement at the weld when the vessel is pressurized. If the 760 mm (30 in) inner diameter cylindrical portion is made from 37.5 mm (1.5 in) thick steel in a length of 2 m (6.5 ft), what additional volume of water must be pumped in when the vessel is full before the greatest stress in the vessel exceeds a limiting value of 125 MPa (8 tonf/in²)? Take $E = 207\,\text{GPa}$ (13 400 tonf/in²), $v = 0.25$ for steel, and $K = 690\,\text{MPa}$ (45 tonf/in²) for water.

SOLUTION TO SI UNITS Equality of radial displacement implies equality of hoop strain. With subscripts s and c for sphere and cylinder respectively, Eq. (4.14a) gives

$$\epsilon_\theta = [(1/E)(\sigma_\theta - v\sigma_\theta)]_s = [(1/E)(\sigma_\theta - v\sigma_z)]_c.$$

Substituting from Eqs (4.17 and 4.20), where p, d and E are common,

$$(pd/4tE)_s(1 - v)_s = (pd/4tE)_c(2 - v)_c,$$
$$\therefore \quad t_s/t_c = (1 - v)/(2 - v) = (3/4)/(7/4) = 3/7.$$

Then for $t_c = 37.5\,\text{mm}$ and $t_s = 16.07\,\text{mm}$, the limiting stress must correspond to the lesser of the two water pressures:

$$p = (2t\sigma/d)_c = 2 \times 37.5 \times 125/760 = 12.34\,\text{MPa},$$
$$p = (4t\sigma/d)_s = 4 \times 16.07 \times 125/760 = 10.57\,\text{MPa}.$$

The total additional volume δV_t to be pumped in is given by

$$\delta V_t = \delta V_c + \delta V_s + \delta V_w.$$

Substituting from Eqs (4.3), (4.18), (4.20) and (4.21):

$$\delta V_t = (pdV_c/4t_cE)(5 - 4v) + (3pdV_s/4t_sE)(1 - v) + pV_t/K$$
$$= (pd/4Et_c)[(5 - 4v)V_c + 7(1 - v)V_s] + pV_t/K, \qquad (i)$$

where

$$V_c = (\pi/4)760^2 \times 2000 = 907.29 \times 10^6 \, \text{mm}^3,$$
$$V_s = \pi d^3/6 = \pi \times 760^3/6 = 229.85 \times 10^6 \, \text{mm}^3,$$
$$V_t = V_c + V_s = 1137.14 \times 10^6 \, \text{mm}^3,$$
$$\therefore \quad \delta V_t = (10.57 \times 760 \times 10^6/4 \times 37.5 \times 207 \times 10^3)$$
$$\times [907.29 + (7 \times 0.75 \times 229.85)]$$
$$+ (10.57 \times 1137.14 \times 10^6/690)$$
$$= 258.72(907.29 + 1206.71) + (17.42 \times 10^6)$$
$$= (0.547 + 17.42) \times 10^6 \, \text{mm}^3 = 17.97 \times 10^6 \, \text{mm}^3.$$

Clearly the water compressibility is the dominant contributor.

SOLUTION TO IMPERIAL UNITS Here, since $t_c = 1.5 \, \text{in}$, then $t_s = (3/7)$ $1.5 = 0.643 \, \text{in}$. The lesser water pressure is found from

$$p = (4t\sigma/d)_s = 4 \times 0.643 \times 8 \times 2240/30 = 1536 \, \text{psi}.$$

And in Eq. (i):

$$V_c = (\pi/4)30^2 \times 6.5 \times 12 = 55.135 \times 10^3 \, \text{in}^3,$$
$$V_s = \pi \times 30^3/6 = 14.137 \times 10^3 \, \text{in}^3,$$
$$V_t = V_s + V_c = 69.272 \times 10^3 \, \text{in}^3.$$
$$\therefore \quad \delta V_t = (1536 \times 30/4 \times 13\,400 \times 2240 \times 1.5)$$
$$\times [55.135 + (7 \times 0.75 \times 14.137)] \times 10^3$$
$$+ (1536 \times 69.272 \times 10^3/45 \times 2240)$$
$$= 33.10 + 1055.6 = 1088.7 \, \text{in}^3.$$

Example 4.7 A cylindrical boiler, welded from plates 15 mm thick, is to withstand a gauge pressure of 15 bar. If the efficiencies of the longitudinal and circumferential joints are 85% and 50% respectively, calculate the boiler diameter using a factor of safety $S = 4$, when the UTS of the plate material is 340 MPa. (1 MPa = 10 bar.)

The tearing strength (TS) of the joint is, from Eq. (4.22),

$$\text{TS} = \text{joint efficiency} \times \text{UTS}. \qquad (i)$$

Now, from Eq. (3.5), the working stress becomes;

$$\sigma_w = TS/S \qquad (ii)$$

Combining Eqs (i) and (ii):

$$\sigma_w = \text{joint efficiency} \times UTS/S. \qquad (iii)$$

For the circumferential direction, Eq. (iii) gives

$$(\sigma_\theta)_w = 0.85 \times 340/4 = 72.25 \, \text{MPa},$$

and from Eq. (4.17a)

$$d = 2t(\sigma_\theta)_w/p = 2 \times 15 \times 72.25/1.5 = 1445 \, \text{mm}.$$

For the axial direction, Eq. (iii) gives

$$(\sigma_z)_w = 0.5 \times 340/4 = 42.5 \, \text{MPa}$$

and from Eq. (4.17b)

$$d = 4t(\sigma_z)_w/p = 4 \times 15 \times 42.5/1.5 = 1700 \, \text{mm}.$$

The lesser diameter of 1.445 m governs the design.

EXERCISES

Elastic Constants

4.1 A bar of metal originally 38 mm in diameter and 102 mm long deforms elastically by the respective amounts -0.0025 mm and 0.022 mm under a 50 kN tensile force. Find E, G, K and v for the metal.

4.2 A 25 mm diameter steel bar, 200 mm long, increases by 0.01 mm under an axial tensile force of 50 kN and twists $0.82°$ under a torque of 220 N m. Determine the four elastic constants, given that $G = Tl/J\theta$ from Chapter 9.
 Answer: $E = 203.7$ GPa; $G = 80.2$ GPa; $v = 0.27$; $K = 147.62$ GPa.

4.3 The elastic constants for a piece of steel are assumed to be $E = 207$ GPa, $G = 80.6$ GPa. Find the apparent value of v and assess the percentage error in v if the assumed values of E and G are respectively 2 per cent greater and 3 per cent less than the true values.
 Answer: 0.28; 0.29 per cent.

4.4 An electrical strain gauge of resistance 1000 Ω is bonded axially to a tensile testpiece of cross-sectional area 645 mm². When a tensile load of 190 kN is applied, the resistance of the gauge changes by 3 Ω. Given that the gauge factor is 2.1 and $v = 0.3$, find the moduli of elasticity and rigidity for the material. *Hint*: gauge factor = resistance charge/strain.

4.5 Strain gauges are bonded in axial and lateral directions to the top surfaces of a beam 25 mm wide and 6.5 mm thick. The gauges lie between 400 mm rigid supports. When equal unit loads are applied at an overhang of 200 mm beyond each support, the strains in the gauges are respectively $9.89 \times 10^{-6}/N$ and $-2.61 \times 10^{-6}/N$. Find E and v for the beam material given that the axial stress at the gauge point is found from; $\sigma = \text{overhang} \times \text{thickness}/2I$ for a unit applied load (see Chapter 5).
 Answer: $E = 207$ GPa; $v = 0.27$.

4.6 A bar of steel 38 mm in diameter is subjected to a tensile force of 100 kN. The measured extension of a 200 mm gauge length is 0.0864 mm and the corresponding reduction in diameter is 0.0048 mm. Calculate E, v, G and K.

Answer: 205 GPa; 0.29; 79 GPa; 169.5 GPa.

4.7 A titanium alloy bar 38 mm in diameter twists through an angle of 1.1° over a length of 0.6 m under a torque of 254 N m. The same bar, when pulled with an axial force of 200 kN, resulted in a resistance change of $\delta R/R = 0.365\%$ for an axially bonded strain gauge with a gauge factor $\lambda = 2.2$. Calculate G, E and v for the alloy given that $G = Tl/J\theta$ (see Chapter 9) and $\epsilon = (1/\lambda)(\delta R/R)$.

Answer: 38.8 GPa; 105.5 GPa; 0.34.

Constitutive Relations

4.8 A block of steel is subjected to orthogonal stresses 77 and 61.75 MPa in tension and 92.5 MPa in compression. Find the strains in orthogonal directions given that $E = 208$ GPa and $v = 0.286$.

Answer: 412, -635, 318, ($\times 10^{-6}$).

4.9 A bar of metal 50 mm $\times$ 75 mm in section is 250 mm long. Determine the changes in dimensions that take place when it is subjected to an axial tensile force of 300 kN, a compressive force of 3 MN on its 75 $\times$ 250 mm faces and a tensile force of 2 MN on its remaining 50 $\times$ 250 mm faces. Take $E = 210$ GPa, $v = 0.26$.

Answer: -0.053; 0.0646; 0.0952 mm.

4.10 A rectangular section bar is to be prevented from lateral contraction under an axial tensile stress of 15.5 MPa. What is the intensity of the lateral stress required and by how much is the axial strain reduced? Take $E = 207$ GPa and $v = 0.27$.

Answer: 33.3%; 28.7 MPa.

4.11 An axial compressive force of 200 kN is applied to a 75 mm square-section bar 200 mm long. Lateral stresses are applied to restrict lateral strains to 1/4 of what they would be if unrestricted. Given $E = 70$ GPa and $v = 1/3$, find the lateral stresses and the restricted and unrestricted axial displacements.

Answer: -13.31 MPa; -0.075 mm; and -0.1016 mm.

4.12 If, for a bar with the dimensions in Exercise 4.11, restricted lateral strains are to be one-third of their unrestricted values under an axial compressive force of 225 kN, determine the necessary lateral stresses and the change in length of the bar. Take $E = 208$ GPa, $v = 0.25$.

Answer: 8.59 MPa; -0.0335 mm.

4.13 A square-section ($a \times a$) rubber block of thickness t is prevented from changing its section dimensions when a compressive force F is applied normally to the square faces. Show that the change in the thickness is $\delta t = (Wt/a^2 E)[1 - 2v^2/(1 - v)]$.

4.14 Determine the thickness ratio between the spherical ends and cylindrical body of a pressurized vessel when, across the weld, (a) the axial strain is the same and (b) the hoop stress is the same. Take $v = 1/4$.

Answer: 3/2; 1/2.

4.15 A thin-walled cylinder of internal diameter 75 mm and thickness 3 mm is 3 m long. Determine the changes in length, diameter and thickness for an applied internal pressure of 1.5 bar. Take $E = 210$ GPa, $v = 0.27$.

Answer: $\delta l = 0.615$ mm; $\delta d = 0.0579$ mm; $\delta t = -1.083 \times 10^{-3}$ mm.

4.16 Find the change in length of a thin-walled cylinder of 25 mm i.d., 1.5 mm thick and 380 mm long, when it is subjected to an internal pressure of 2.2 MPa and a simultaneous axial force of 1.75 kN. Take $E = 205$ GPa.

Answer: 1.66 mm.

4.17 A solid cylindrical shaft, 250 mm long and 50 mm in diameter, is made from aluminium alloy. The periphery of the shaft is constrained in such a way as to prevent lateral strain. Calculate the axial force that will compress the shaft by 0.50 mm. Determine the change in length of the shaft when the lateral constraint is removed but the axial force remains unaltered. Calculate the required reduction in axial force for the non-constrained shaft if the axial strain is not to exceed 0.2 per cent. Assume the elastic constants $E := 70 \times 10^3$ N/mm^2 and $v = 0.3$. (C.E.I. Pt.2, 1973.)

Volumetric Strain

4.18 A steel bar 25 mm in diameter and 3 m long carries an axial tensile force of 60 kN. Find the extension, reduction in diameter and change in volume, given $E = 201$ GPa and $v = 0.3$.

4.19 A bar 30 mm long and 20 mm in diameter is subjected to an axial compressive load of 30 kN. If the length and diameter change by -0.015 and 0.0027 mm respectively, determine the change in volume when the bar is immersed to a depth of 3 km in sea water with a density of 815.5 kg/m^3.
Answer: -163 mm^3.

4.20 A 50 mm steel cube is subjected to a tensile stress of 185 MPa and a compressive stress of 247 MPa acting in perpendicular directions on its faces. Find the change in volume. Take $E = 208$ GPa.
Answer: -15.73 mm^3.

4.21 A rectangular block is subjected to tensile forces of 120 kN on its 75×50 mm faces, 800 kN on its 50×250 mm faces and a compressive force of 900 kN on its 75×250 mm faces. Find the change in volume and the dimensional changes of the sides, given that the elastic constants are $E = 105$ GPa and $v = 0.35$.

4.22 A steel block with volume 1.64×10^6 mm^3 is subjected to three mutually perpendicular stresses, of which two are tensile of magnitudes 155 and 125 MPa and the third is to be compressive. If the change in volume is restricted to 685 mm^3, find the magnitude of the compressive stress and the magnitudes of three principal strains. Take $E = 207$ GPa, $v = 0.3$.
Answer: -61.75 MPa; -702×10^{-6}; 463×10^{-6}; 657×10^{-6}.

4.23 A closed steel tube of 900 mm inside diameter, 3 m long and 10 mm thick, is subjected to a fluid test pressure of 1.5 MPa. Find the volume of fluid ejected when a valve is opened for both compressible ($K = 3500$ MPa) and incompressible fluids. Take $E = 207$ GPa and $v = 0.3$ for steel.
Answer: 1.19×10^6 mm^3; 0.373×10^6 mm^3.

4.24 A closed copper tube of 50 mm i.d., 1.25 m long and 1.25 mm thick, is filled with water under pressure. Determine the change in pressure when an additional 3275 mm^3 of water is pumped in. Take $E = 104$ GPa and $v = 0.3$ for copper, and $K = 2070$ MPa for water.
Answer: 1.56 MPa.

4.25 A 1 m long thin cylinder of diameter 600 mm is pressurized to 1.375 MPa. Find the necessary thickness if the maximum tensile stress is not to exceed 82.5 MPa. What are the axial and circumferential strains and the increase in capacity under this pressure? Take $E = 207$ GPa, $v = 0.3$.
Answer: 5 mm; 3.4×10^{-4}; 20.32×10^4 mm^3.

4.26 A thin-walled cylinder 4.5 m long, 510 mm in diameter and 5 mm thick is prevented from axial displacement under an internal pressure of 82.5 bar. Determine the maximum tensile stress and the increase in the internal volume at this pressure. Take $E = 165$ GPa, $v = 0.24$.
Answer: 413 MPa; 4.6×10^6 mm^3.

4.27 A thin-walled steel sphere with 600 mm mean radius and 25 mm thickness contains oil at atmospheric pressure. To what depth may the sealed sphere be immersed in sea water of density

1040 kg/m³ if the maximum compressive stress in the sphere wall is restricted to 92.5 MPa? Take $E = 207\,\text{GPa}$, $v = 0.25$ for steel, and $K = 3445\,\text{MPa}$ for oil. (*Hint*: $\delta V/V$ for oil and sphere are the same.)

Answer: 1097 m.

4.28 A spherical vessel 1525 mm in diameter with 75 mm thick wall is made from a material whose ultimate compressive strength is 77 MPa. If the internal pressure of the sealed vessel remains constant as it is submerged in the sea ($\rho = 1040\,\text{kg/m}^3$), determine the safe depth and the corresponding reduction in the internal volume. Take $E = 166\,\text{GPa}$ and $v = 0.25$.

Answer: 1513 m; $1947 \times 10^3\,\text{mm}^3$.

Pressure Vessels

4.29 A thin-walled steel sphere is subjected to an internal pressure of 5 bar. If the inner diameter is 2 m, determine the necessary thickness using a factor of safety of 5 with a UTS of 450 MPa for the steel used.

4.30 A thin-walled cylindrical bottle with hemispherical ends is 150 mm in diameter with a parallel length of 760 mm and constant thickness. It is to contain oil at a gauge pressure of 35 MPa. Determine the required thickness for a safety factor of 2 if the vessel material can withstand a maximum tensile stress of 345 MPa. What additional volume of oil ($K = 3445\,\text{MPa}$) must be pumped in to double the pressure? Take $E = 207\,\text{GPa}$ and $v = 0.28$ for the bottle material.

Answer: 15.22 mm; $182 \times 10^3\,\text{mm}^3$.

4.31 If a working pressure of 1.4 MPa exists in a 2.4 m diameter cylinder, find the necessary thickness for a safety factor of 6 when the UTS of the steel is 465 MPa and the efficiency of the longitudinal joint is 75 per cent.

Answer: 29 mm.

4.32 The plates of a boiler are 12.5 mm thick and the internal pressure to which they are subjected is 14 bar. If the tensile stress in the material is limited to 137.5 MPa and the efficiency of the longitudinal and circumferential joints are 75 and 30 per cent respectively, determine the maximum permissible diameter of the boiler.

Answer: 1.5 m.

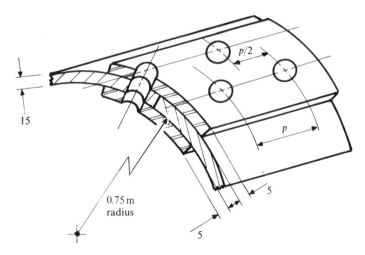

Figure 4.10

4.33 A cylindrical boiler 1.2 m diameter is to be made from steel plates 12.5 mm thick with a safe working tensile stress of 82.5 MPa. Given that the efficiencies of the longitudinal and circumferential joints are 75 and 45 per cent respectively, determine the safe working pressure.

Answer: 13 bar.

4.34 The longitudinal butt joint in a cylinder consists of 5 mm thick cover plates held together on either side of the joint with two longitudinal rows of staggered rivets with equal pitch *p* (Fig. 4.10). Using a factor of safety of 4, estimate the diameter and spacing of the rivets when the cylinder, of diameter 1.5 m and thickness 15 mm, is pressurized to 10 bar. Take UTS = 470 MPa for the plate, material and the USS = 200 MPa for the rivets.

FIVE

SHEAR FORCE AND BENDING MOMENT DISTRIBUTIONS

5.1 INTRODUCTION

In the analysis of the stresses in loaded beams and structures, it is necessary to know the manner in which the shear force and bending moment vary with the length. For a single-bay beam with two supports, the application of equilibrium principles alone is sufficient to determine these distributions, since the beam is statically determinate. However, because beams with two or more bays and many structures are statically indeterminate, the analysis becomes more complex. Here a trial-type solution, known as the moment distribution method, is employed. The alternative theorem of three moments (see Chapter 7), commonly used with continuous beams, employs the relationships that exist between moments, slopes and deflections.

5.2 SINGLE-SPAN BEAMS

The distributions in shear force F, and bending moment M, may be presented graphically by the following method:

1. Find the support reactions from force and moment equilibrium.
2. To find F at any section, calculate the net force to the left of that section taking downward forces as positive. The result is equivalent to plotting either F ($\downarrow + \text{ve}$) from left to right, or F ($\uparrow + \text{ve}$) from right to left along the length.
3. To find M at any section, take moments either to the right or left of that

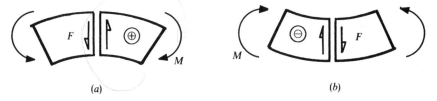

Figure 5.1 Convention for shear forces and bending moments.

section, depending upon which is easier, according to the convention: hogging moments are positive and sagging moments are negative as respectively illustrated in Fig. 5.1(a) and (b).

Some texts make use of the opposite convention, in which sagging is positive. It will be seen that the present convention, which is used consistently throughout this book, avoids the necessity for the introduction of minus signs for the stress in the engineer's theory of bending (Chapter 6) and for the curvature in the buckling of struts (Chapter 8).

Relationships Between w, F and M

Consider an elemental length δz of a beam carrying a uniformly distributed load, w/unit length. Let the force and moments vary in the manner shown in Fig. 5.2. Applying vertical force equilibrium:

$$(F + w\,\delta z) = (F + \delta F), \quad \Rightarrow w = \delta F/\delta z, \tag{5.1}$$

i.e. rate of loading $\equiv$ slope of F diagram and, from taking moments about O:

$$(M + \delta M) + \delta z(w\,\delta z)/2 = M + (F + \delta F)\,\delta z, \quad \Rightarrow F = \delta M/\delta z, \tag{5.2}$$

i.e. F is the derivative of M, or M is the area beneath the F diagram when the products of infinitesimals are ignored. Note also from Eq. (5.2) that the maximum M will occur where F is zero.

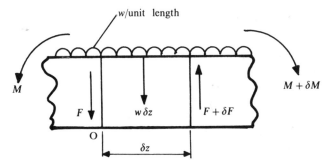

Figure 5.2 Beam element.

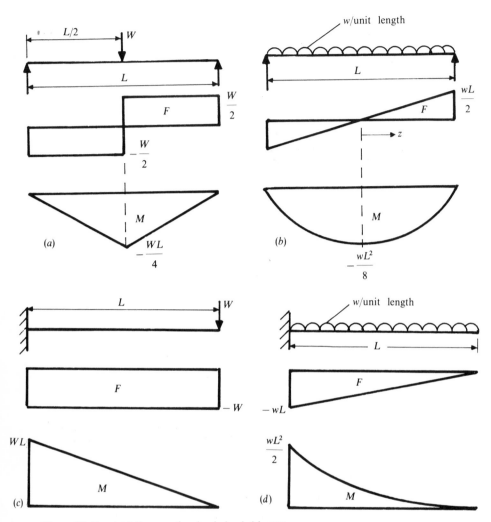

Figure 5.3 *F* and *M* diagrams for simply loaded beams.

The reader should now verify the *F* and *M* diagrams given for the four common beams in Fig. 5.3(a–d). The expression for the maximum bending moment in each case is worth remembering.

Example 5.1 Draw the *F* and *M* diagrams for the beam in Fig. 5.4.

R.H. reaction—moments about L.H. end:

$$(3 \times 400) + (8 \times 300) = 10 \text{ R.H.},$$
$$\therefore \quad \text{R.H.} = 360 \text{ N}.$$

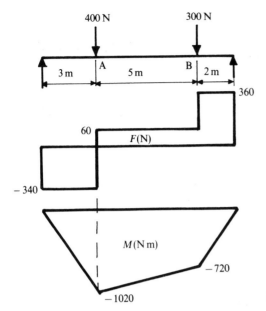

Figure 5.4 F and M diagrams.

L.H. reaction—vertical equilibrium:

$$400 + 300 = \text{R.H.} + \text{L.H.},$$

$$\therefore \quad \text{L.H.} = 340 \text{ N.}$$

F diagram: Start at L.H. end with $+\text{ve}\, F(\downarrow)$ from a horizontal datum. Diagram must close on same datum at R.H. end. Or, start at R.H. end with $+\text{ve}\, F(\uparrow)$.

M diagram:

At R.H. and L.H. ends, $\qquad M = 0;$

At A to left, $\qquad M_A = -(3 \times 340) = -1020 \text{ N m};$

At B to right, $\qquad M_B = -(2 \times 360) = -720 \text{ N m.}$

Both M_A and M_B sag the beam. Then, $M_{\max} = 1020 \text{ Nm}$, where the F diagram crosses the horizontal datum.

Example 5.2 Draw the shear force and bending moment diagrams for the beam with mixed loading in Fig. 5.5.

Reactions:

$$(3 \times 400) + (5.5 \times 500) = 10R_B, \qquad \Rightarrow R_B = 395 \text{ N};$$

$$R_A + R_B = 900, \qquad \Rightarrow R_A = 505 \text{ N.}$$

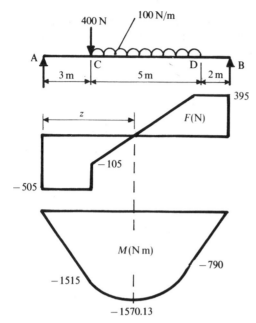

Figure 5.5 F and M diagrams.

F diagram:

$$F_A = -R_A = -505\,\text{N};$$
$$F_C = -505 + 400 = -105\,\text{N};$$
$$F_D = -505 + 400 + (100 \times 5) = 395\,\text{N} = F_B;$$
$$F = 0 = -505 + 400 + 100(z-3), \qquad \Rightarrow z = 4.05\,\text{m}.$$

M diagram:

$$M_A = M_B = 0;$$
$$M_C(\text{to left}) = -(3 \times 505) = -1515\,\text{N m (sag)};$$
$$M_D(\text{to right}) = -(2 \times 395) = -790\,\text{N m (sag)};$$
$$M_{\max} = -(4.05 \times 505) + (400 \times 1.05) + 100(1.05)^2/2$$
$$= -1570.13\,\text{N m} \quad \text{where} \quad F = 0.$$

Points of Contraflexure

In the previous examples, the bending moment diagrams are all of the same sign. With other forms of loading, the bending moment diagrams may contain both positive and negative portions. The points of contraflexure (or inflection) are points of zero bending moment, i.e. points where the beam changes its curvature from hogging to sagging. The determination of the position of these points in the length is illustrated in the following three examples.

Example 5.3 Construct the F and M diagrams for the loaded beam in Fig. 5.6. Find the position and magnitude of the maximum bending moment and the points of contraflexure.

Reactions—moments about A:

$$(5 \times 4) + (8 \times 8) - (2 \times 3) = 10R_B;$$
$$R_B = 7.8\,\text{kN}, \qquad R_A = 2 + 4 + 8 - 7.8 = 6.2\,\text{kN}.$$

F diagram is constructed from R.H. end where $\downarrow$ forces are $-$ve. M diagram to left or right:

$$M_A = +(2 \times 3) = 6\,\text{kN m};$$
$$M_B = +(2 \times 1 \times 1) = 2\,\text{kN m};$$
$$M_C = (1 \times 1 \times 0.5) + (8 \times 2) - (5 \times 6.2)$$
$$= -14.5\,\text{kN m} = M_{\text{max}} \qquad (F = 0);$$
$$M_D = (7 \times 2) - (4 \times 6.2) = -10.8\,\text{kN m};$$
$$M_E = (4 \times 1 \times 2) - (2 \times 7.8) = 7.6\,\text{kN m}.$$

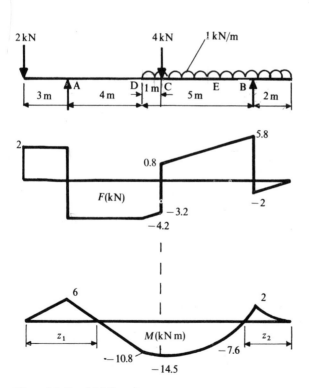

Figure 5.6 F and M diagrams.

Points of contraflexure:

At z_1 from L.H. end:

$$(2z_1) - (z_1 - 3)6.2 = 0, \qquad \Rightarrow z_1 = 4.45 \, \text{m}.$$

At z_2 from R.H. end:

$$-7.8(z_2 - 2) + (z_2 \times 1 \times z_2/2) = 0,$$
$$z_2^2 - 15.6z_2 + 31.2 = 0, \qquad \Rightarrow z_2 = 2.36 \, \text{m}.$$

Example 5.4 Find the maximum bending moment and point of contraflexure for the cantilever beam in Fig. 5.7.

F diagram: Note that with our convention, it is more convenient to start at the R.H. free-end with $\downarrow F(-\text{ve})$, since it is then unnecessary to calculate the fixing reaction. The latter is the force (0.2 kN) required to close the F diagram at the L.H. end.

M diagram to left or right:

$$M_A = 5 \times 0.1 \times 2.5 = 1.25 \, \text{kN m} = M_{\text{max}};$$
$$M_B = (8 \times 0.1 \times 4) - (3 \times 1) = 0.2 \, \text{kN m};$$
$$M_C = (8 \times 0.1 \times 8) - (7 \times 1) = -0.6 \, \text{kN m}.$$

Contraflexure—z from R.H. end:

$$M = [8 \times 0.1(z - 8 + 4)] - 1(z - 5) = 0, \qquad \Rightarrow z = 9 \, \text{m}.$$

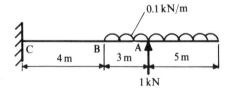

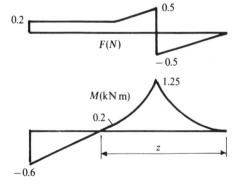

Figure 5.7 F and M diagrams.

Example 5.5 Sketch the F and M diagrams for the beam in Fig. 5.8 and determine the position and magnitude of the maximum bending moment and the position of the point of contraflexure.

SOLUTION TO SI UNITS
Reactions:

$$(2 \times 20) + (6 \times 20) + (10 \times 15) + (30 \times 10 \times 5) = 8R_B,$$
$$\therefore \quad R_B = 226.3 \text{ kN}.$$
$$R_A = [20 + 20 + 15 + (30 \times 10)] - 226.3 = 128.7 \text{ kN}.$$

F diagram:

$$F = 0 = -128.7 + 20 + 30z_0, \qquad \Rightarrow z_0 = 3.62 \text{ m}.$$

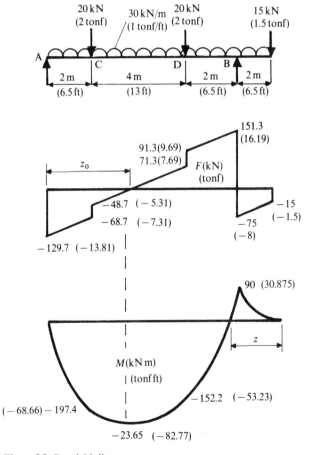

Figure 5.8 F and M diagrams.

M diagram to left or right:

$$M_B = +(2 \times 15) + (2 \times 30 \times 1) = 90 \, \text{kN m};$$
$$M_C = -(128.7 \times 2) + (30 \times 2 \times 1) = -197.4 \, \text{kN m};$$
$$M_D = -(128.7 \times 6) + (20 \times 4) + (6 \times 30 \times 3) = -152.2 \, \text{kN m};$$
$$M_{max} = -(128.7 \times 3.62) + (1.62 \times 20) + 15(3.62)^2$$
$$= -236.9 \, \text{kN m} \quad \text{where} \quad F = 0.$$

Contraflexure: z from R.H. end:

$$M = 15z - 226.3(z - 2) + 30z^2/2 = 0;$$
$$15z^2 - 211.3z + 452.6 = 0;$$
$$z = 2.64 \, \text{m}.$$

SOLUTION TO IMPERIAL UNITS GIVEN IN FIG. 5.8.
Reactions:

$$(6.5 \times 2) + (19.5 \times 2) + (32.5 \times 1.5) + (32.5 \times 1 \times 16.25) = 26R_B,$$
$$\therefore \quad R_B = 24.188 \, \text{tonf}.$$
$$R_A = [2 + 2 + 1.5 + (32.5 \times 1)] - 24.188 = 13.813 \, \text{tonf}.$$

F diagram:

$$F = 0 = -13.813 + 2 + (1 \times z_0), \quad \Rightarrow z_0 = 11.813 \, \text{ft}.$$

M diagram:

$$M_B = (6.5 \times 1.5) + (6.5 \times 1 \times 3.25) = 30.875 \, \text{tonf ft};$$
$$M_C = -(13.813 \times 6.5) + (1 \times 6.5 \times 3.25) = -68.66 \, \text{tonf ft};$$
$$M_D = -(13.813 \times 19.5) + (13 \times 2) + (1 \times 19.5 \times 9.75) = -53.23 \, \text{tonf ft};$$
$$M_{max} = -(13.813 \times 11.813) + (5.313 \times 2) + (1 \times 11.813 \times 5.907)$$
$$= -82.77 \, \text{tonf ft}.$$

Contraflexure:

$$M = 1.5z - 24.188(z - 6.5) + z^2/2 = 0;$$
$$z^2 - 45.376z + 314.44 = 0;$$
$$z = 8.535 \, \text{ft}.$$

5.3 THE MOMENT DISTRIBUTION METHOD

Application to Continuous Beams

Where a beam has two or more spans, a convenient method for determining the F and M diagrams employs the technique in which the moments at a

Table 5.1 Fixed-end moments (FEM)

Beam	R.H. FEM	L.H. FEM
	Wa^2b/L^2	Wab^2/L^2
	$wL^2/12$	$wL^2/12$
	$wa^2(6L^2 - 8aL + 3a^2)/12L^2$	$wa^2(4aL - 3a^2)/12L^2$
	$(Mb/L)(1 - 3a/L)$	$(Ma/L)(1 - 3b/L)$
	$(1/L^2)\int wz(L-z)^2\,dz$	$(1/L^2)\int wz^2(L-z)\,dz$
	0	$3EI\delta/L^2$
	$6EI\delta/L^2$	$6EI\delta/L^2$

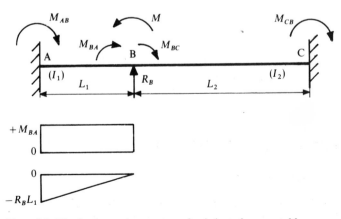

Figure 5.9 Distribution and carry-over of unbalanced moment M.

given support are distributed from the left and right sides until they become equal. The method is as follows.

1. Fix the beam at all simple supports and calculate the left- and right-hand fixed-end moments from Table 5.1.
2. Release at simply supported ends, distribute unbalanced moments and carry-over (Eqs (5.5)) to construct the fixing moment diagram.
3. Isolate each bay and construct the free M diagram (Section 5.2).
4. The net M diagram is then found from the difference between ordinates in the fixing and free moment diagrams.

Fixed-end moments (FEM) Derivations of some of the expressions appearing in Table 5.1 are given in Chapter 7. The convention here is that clockwise FEMs are positive and therefore the signs will normally alternate within each bay and on either side of each support in moving from left to right along the beam.

Carry-over moments (COM) In Fig. 5.9, since the FEMs on either side of the support B will not usually balance, there is introduced an out of balance moment M at B, which is distributed into bays BA and BC. That is,

$$M = M_{BA} + M_{BC},\qquad(5.3)$$

where the directions of M_{BA} and M_{BC} must oppose the directions of M in order to maintain moment equilibrium at the support point. Moreover, it is an equilibrium requirement that a proportion of each distributed moment will be carried over to the ends A and C.

M_{AB} and M_{CB} are carry-over moments, found from the condition that the deflection v, at B, is zero. Applying Mohr's deflection theorem (see Chapter 7) to bay AB where the M diagram is composed of two parts the uniform distributed moment M_{BA} and that due to the reaction R_B at B:

$v_B = (1/EI)$(moment of area of M diagram from A to B about B) $= 0$;
$$M_{BA}L_1(L_1/2) - (R_BL_1^2/2)(2L_1/3) = 0;$$
$$R_B = 3M_{BA}/2L_1\qquad(5.4)$$

The net moment applied at A is then

$$M_{BA} - (3M_{BA}/2L_1)L_1 = -M_{BA}/2.$$

The reacting the COM at A and, similarly, the COM at C, take the opposite sense to the moments applied at A and C. It follows that each COM in Fig. 5.9 has the same sense with magnitude:

$$M_{AB} = M_{BA}/2,\qquad M_{CB} = M_{BC}/2.\qquad(5.5a, b)$$

Distribution factors (DF) Furthermore, the slope i, at B, must be constant

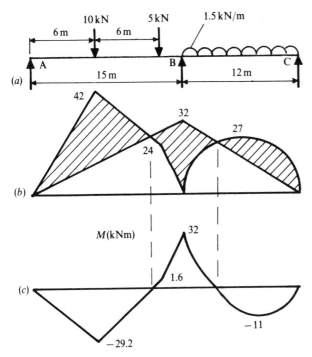

Figure 5.10 Free, fixing and net moment diagrams for a two-bay simply supported beam.

in bays AB and BC. If L_1, I_1 are the length and second moment of area for bay AB, the application of Mohr's slope theorem (see Chapter 7) to Fig. 5.9 gives

$$i_B = (1/EI)(\text{area of } M \text{ diagram from A to B})$$
$$= (1/EI_1)(M_{BA}L_1 - R_BL_1^2/2) = M_{BA}L_1/4EI, \quad (5.6)$$

in which R_B is supplied from Eq. (5.4). If L_2 and I_2 are the respective length and second moment of area for bay BC, then, similarly

$$i_B = M_{BC}L_2/4EI_2. \quad (5.7)$$

Equating (5.6) and (5.7) leads to

$$M_{BA}/M_{BC} = I_1L_2/I_2L_1. \quad (5.8)$$

Equations (5.3) and (5.8) supply the *distribution factors*:

$$\frac{M_{BA}}{M} = \frac{(I_1/L_1)}{(I_1/L_1) + (I_2/L_2)}, \qquad \frac{M_{BC}}{M} = \frac{(I_2/L_2)}{(I_1/L_1) + (I_2/L_2)}. \quad (5.9a, b)$$

That is, M is divided between the two bays in the ratio of their stiffnesses, I/L. It should be noted that Eqs (5.9) are based upon encastre supports. In distributing moments to a simply supported bay, the stiffness becomes $(3/4)(I/L)$. The following examples illustrate the application of Table 5.1 and Eqs (5.5) and (5.9) to continuous beams.

Example 5.6 Construct the net bending moment diagram for the two-bay beam in Fig. 5.10(a) showing the maximum values.

Fixed-end moments from Table 5.1:

Bay AB   R.H. FEM $= (10 \times 6^2 \times 9/15^2) + (5 \times 12^2 \times 3/15^2) = 24\,kN\,m$;
Bay AB   L.H. FEM $= (10 \times 9^2 \times 6/15^2) + (5 \times 3^2 \times 12/15^2) = 24\,kN\,m$;
Bay BC   R.H. FEM $=$ L.H. FEM $= (1.5 \times 12^2/12) = 18\,kN\,m$.

DFs at joint B from Eqs (5.9):

$$\frac{M_{BA}}{M} = \frac{(3/4)(I/15)}{(3/4)(I/12 + I/15)} = \frac{4}{9};$$

$$\frac{M_{BC}}{M} = \frac{(3/4)(I/12)}{(3/4)(I/12 + I/15)} = \frac{5}{9}.$$

Note that the 3/4 factor cancels with simple supports throughout.

Fixing moments: the distribution and carry-over moment table (Table 5.2) ensures that the moments at B are equal and that there can be no moments at the ends A and C. The fixing moment diagram is superimposed upon the free moment diagrams in Fig. 5.10(b). Finally, the difference between corresponding ordinates in Fig. 5.10(b) establishes the net moment diagram in Fig. 5.10(c).

Example 5.7 Find the fixing moments for the beam in Fig. 5.11.

SOLUTION TO SI UNITS Fixed-end moments from Table 5.1:

Bay AB   L.H. FEM $=$ R.H. FEM $= 1 \times 15^2/12 = 18.75\,kN\,m$;
Bay BC   R.H. FEM $= 10 \times 6^2 \times 4/10^2 = 14.4\,kN\,m$;
        L.H. FEM $= 10 \times 4^2 \times 6/10^2 = 9.6\,kN\,m$.

Table 5.2

Joint	A	B		C
Member	AB	BA	BC	CA
DF		4/9	5/9	
Initial FEM	-24	$+24$	-18	$+18$
Release at A and C and COM	$+24$	$+12$	-9	-18
Modified FEM	0	$+36$	-27	0
Unbalanced moment			$+9$	
First distribution		-4	-5	
Final fixing moments	0	$+32$	-32	0

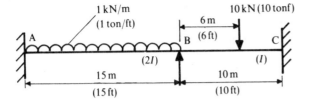

Figure 5.11 Two-bay encastre beam.

DFs at B from Eqs (5.9):

$$\frac{M_{BA}}{M} = \frac{2I/15}{(2I/15)+(I/10)} = \frac{4}{7};$$

$$\frac{M_{BC}}{M} = \frac{I/10}{(2I/15)+(I/10)} = \frac{3}{7}.$$

Fixing moments: it is only necessary to balance the moments at B and carry over to establish the moments at A and C (Table 5.3).

SOLUTION TO IMPERIAL UNITS GIVEN IN FIG. 5.11 Fixed-end moments:

Bay AB   L.H. FEM = R.H. FEM = $1 \times 15^2/12 = 18.75$ tonf ft;

Bay BC   R.H. FEM = $10 \times 6^2 \times 4/10^2 = 14.4$ tonf ft;

L.H. FEM = $10 \times 4^2 \times 6/10^2 = 9.6$ tonf ft.

The DFs are again 4/7 and 3/7. It follows that the moment distribution Table 5.3 again applies, where now the Imperial units of moment, tonf ft, replace the SI units, kN m.

Example 5.8 Construct the moment diagram for the beam in Fig. 5.12(a).

Fixed-end moments from Table 5.1:

Bay AB   L.H. FEM = $5 \times 3^2 \times 2/5^2 = 3.6$ kN m;

R.H. FEM = $5 \times 2^2 \times 3/5^2 = 2.4$ kN m;

Table 5.3

Joint	A	B		C
Member	AB	BA	BC	CB
DF		4/7	3/7	
Initial FEM	−18.75	+18.75	−9.6	+14.40
Unbalanced moment		+9.1		
Distribution and COM	−2.6	−5.2	−3.9	−1.95
Final fixing moments	−21.35	+13.55	−13.5	+12.45

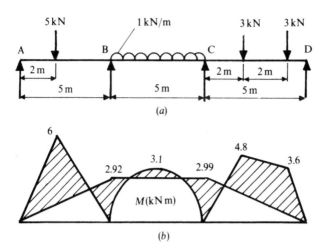

Figure 5.12 Free, fixing and net moment diagrams for a three-bay simply supported beam.

Bay BC   L.H. FEM = R.H. FEM = $1 \times 5^2/12 = 2.08$ kN m;

Bay CD   L.H. FEM = $(3 \times 3^2 \times 2/5^2) + (3 \times 1^2 \times 4/5^2) = 2.64$ kN m;

R.H. FEM = $(3 \times 2^2 \times 3/5^2) + (3 \times 4^2 \times 1/5^2) = 3.36$ kN m.

DFs at B and C from Eqs (5.9):

$$\frac{M_{BA}}{M} = \frac{(3/4)(I/5)}{(3/4)(I/5) + I/5} = \frac{3}{7};$$

$$\frac{M_{BC}}{M} = \frac{(I/5)}{(3/4)(I/5) + I/5} = \frac{4}{7};$$

$$\frac{M_{CB}}{M} = \frac{(I/5)}{(3/4)(I/5) + I/5} = \frac{4}{7};$$

$$\frac{M_{CD}}{M} = \frac{(3/4)(I/5)}{(3/4)(I/5) + I/5} = \frac{3}{7}.$$

Fixing moments: these must balance at B and C and equal zero at A and D (Table 5.4). The free and fixing moments are superimposed in Fig. 5.12(b). The net moment diagram is then given by the ordinates in the shaded region.

Example 5.9 Draw the bending moment diagram for the continuous beam in Fig. 5.13(a). Calculate the support reactions and draw the F diagram.

Table 5.4

Joint	A	B		C		D
Member	AB	BA	BC	CB	CD	DC
DF		3/7	4/7	4/7	3/7	
Initial FEM	−3.6	+2.4	−2.08	+2.08	−2.64	+3.36
Release A and D and COM	+3.6	+1.8			−1.68	−3.36
Net fixing moments	0	+4.2	−2.08	+2.08	−4.32	0
Unbalanced moment			+2.12		−2.24	
First distribution		−0.91	−1.21	+1.28	+0.96	
COM		0	+0.64	−0.61	0	
Unbalanced moment			+0.64		−0.61	
Second distribution		−0.27	−0.37	+0.34	+0.26	
COM		0	+0.17	−0.19	0	
Unbalanced moment			+0.17		−0.19	
Third distribution		−0.07	−0.10	+0.11	+0.08	
COM		0	+0.06	−0.05	0	
Unbalanced moment			+0.06		−0.05	
Fourth distribution		−0.025	−0.035	+0.03	+0.02	
COM		0	+0.015	−0.017		
Final fixing moments	0	+2.925	−2.91	+2.98	−3.0	0

Fixed-end moments from Table 5.1:

Bay AB FEMs = 0;

Bay BC L.H. FEM = R.H. FEM = $6 \times 1^3/2^2 = 1.5\,\text{kN m}$;

Bay CD L.H. FEM = R.H. FEM = $3 \times 2^2/12 = 1.0\,\text{kN m}$;

Bay DE $M_D = 3 \times 1^2/2 = 1.5\,\text{kN m (hogging)}, \quad M_E = 0$.

Table 5.5

Joint	A	B		C		D		E
Member	AB	BA	BC	CB	CD	DC	DE	ED
DF	0	0.5	0.5	0.73	0.27			
Initial FEM	0	0	−1.5	+1.50	−1.0	+1.0	−1.5	0
Balance at D						+0.5		
COM					+0.025			
Net fixing moments	0	0	−1.5	+1.5	−0.75	+1.5	−1.5	0
Unbalanced moment			−1.5		+0.75			
First distribution		+0.75	+0.75	−0.55	−0.20			
COM	+0.37		−0.28	+0.37				
Second distribution		+0.14	+0.14	−0.27	−0.10			
COM	+0.07		−0.14	+0.07				
Third distribution		+0.07	+0.07	−0.05	−0.02			
COM	+0.035		−0.025	+0.035				
First distribution		+0.012	+0.012	−0.026	−0.01			
COM	+0.006		−0.013	+0.006				
Final FMs	+0.48	+0.972	−0.986	+1.085	−1.080	+1.5	−1.5	0

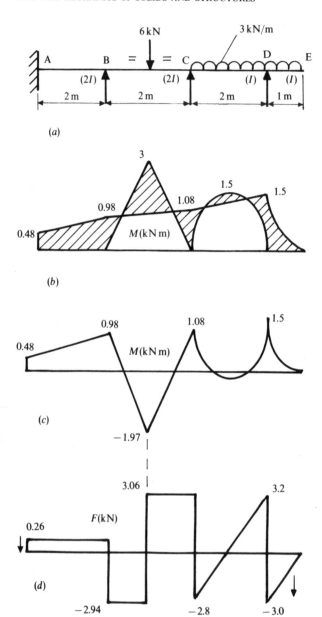

Figure 5.13 M and F diagrams for a continuous beam.

DFs at B and C from Eqs (5.9):

$$\frac{M_{BA}}{M} = \frac{(2I/2)}{(2I/2)+(2I/2)} = 0.5 = \frac{M_{BC}}{M};$$

$$\frac{M_{CB}}{M} = \frac{(2I/2)}{(2I/2)+(3/4)(I/2)} = 0.73;$$

$$\frac{M_{CD}}{M} = \frac{(3/4)(I/2)}{(2I/2)+(3/4)(I/2)} = 0.27.$$

Here it is necessary to balance moments at B, C and D (Table 5.5). The latter is achieved first, since $M_D = 1.5\,\text{kN m}$ is obviously the net moment at D.

Free moments:

Bay BC $M_{max} = WL/4 = 6 \times 2/4 = 3\,\text{kN m}$,

Bay CD $M_{max} = wL^2/8 = 3 \times 2^2/8 = 1.5\,\text{kN m}$.

These calculations enable the net bending moment diagram to be constructed in Figs 5.13(b) and (c) consistent with the convention that hogging moments are positive.

Reactions:

$$\sum M_A = 0 = -0.48 + (6 \times 3) + (3 \times 3 \times 5.5) - 2R_B - (4 \times 5.86) - (6 \times 6.2), \quad \Rightarrow R_B = 3.2\,\text{kN};$$

$$\sum M_B = 0 = -0.98 + (6 \times 1) + (3 \times 3 \times 3.5) - 2R_C - (4 \times 6.2), \quad \Rightarrow R_C = 5.86\,\text{kN};$$

$$\sum M_C = 0 = -1.082 + (3 \times 3 \times 1.5) - 2R_D, \quad \Rightarrow R_D = 6.2\,\text{kN};$$

$$\sum F = 0 = 6 + (3 \times 3) - 3.2 - 5.86 - 6.2 - R_A, \quad \Rightarrow R_A = -0.26\,\text{kN}.$$

Thus the shear force diagram in Fig. 5.13(d) is constructed from the right-hand side with downward forces negative consistent with $F = \delta M/\delta z$.

Beams with Misaligned Supports

If, for a continuous beam, one or more supports are at a different level from the remainder, then moments will be induced at all interior supports and at the ends if these are encastre.

Example 5.10 Find the moments at B and C for the beam in Fig. 5.14 when the support at B lies 40 mm below the level of A, C and D. Take $E = 200\,\text{GPa}$, $I = 4 \times 10^6\,\text{mm}^4$.

Fixed-end moments: all are found from $M = 6EI\delta/L^2$ (see Table 5.1)

Bay AB L.H. FEM $= 6 \times 200 \times 10^3 \times 3 \times 4 \times 10^6 \times 40/12^2 \times 10^6$

$\qquad\qquad\qquad = 4 \times 10^6\,\text{N mm} = 4\,\text{kN m} = \text{R.H. FEM};$

Bay BC L.H. FEM $= 6 \times 200 \times 10^3 \times 10 \times 4 \times 10^6 \times 40/24^2 \times 10^6$

$\qquad\qquad\qquad = 3.333 \times 10^6\,\text{N mm} = 3.333\,\text{kN m} = \text{R.H. FEM};$

Bay CD FEMs $= 0$.

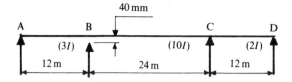

Figure 5.14 Continuous beam with misaligned support.

DFs at B and C from Eqs (5.9):

$$\frac{M_{BA}}{M} = \frac{(3/4)(3I/12)}{(3/4)(3I/12) + 10I/24} = 0.31;$$

$$\frac{M_{BC}}{M} = \frac{(10I/24)}{(3/4)(3I/12) + 10I/24} = 0.69;$$

$$\frac{M_{CB}}{M} = \frac{(10I/24)}{(3/4)(2I/12) + 10I/24} = 0.77;$$

$$\frac{M_{CD}}{M} = \frac{(3/4)(2I/12)}{(3/4)(2I/12) + 10I/24} = 0.23.$$

Because the end A is simply supported, it is first necessary to release the FEM at this point and carry over (Table 5.6). Note from Table 5.1 that in this case, without external loading, the initial FEMs will have the same sense within each bay but that the sense alternates from bay to bay.

Table 5.6

Joint	A	B		C		D
Member	AB	BA	BC	CB	CD	DC
DF		0.31	0.69	0.77	0.23	
Initial FEM	−4	−4	+3.333	+3.333	0	0
Release at A and COM	+4	+2				
Net FEM	0	−2	+3.333	+3.333	0	0
Unbalanced moment		+1.333		+3.333		
First distribution	0	−0.413	−0.920	−2.566	−0.767	
COM			−1.283	−0.460		
Second distribution	0	+0.398	+0.885	+0.354	+0.106	
COM			+0.177	+0.443		
Third distribution	0	−0.055	−0.122	−0.341	−0.102	
COM			−0.171	−0.061		
Fourth distribution	0	+0.053	+0.118	+0.047	+0.014	
COM			+0.024	+0.059		
Fifth distribution	0	−0.007	−0.017	−0.045	−0.014	
COM			−0.023	−0.009		
Sixth distribution	0	+0.007	+0.016	+0.007	+0.002	
COM			+0.004	+0.008		
Final FEM	0	−2.017	+2.021	+0.769	−0.761	0

5.4 MOMENT DISTRIBUTION FOR STRUCTURES

The following three examples illustrate how the moment distribution method may be adapted to loaded structures.

Example 5.11 Find the vertical reaction at A for the structure in Fig. 5.15(a) when the ends A and B are encastre, C is pinned and O is a rigid joint.

Fixed-end moment from Table 5.1:

$$\text{Bay OA} \quad \text{L.H. FEM} = 2 \times 2^2 \times 1/3^2 = 0.89 \, \text{kN};$$
$$\text{R.H. FEM} = 2 \times 1^2 \times 2/3^2 = 0.44 \, \text{kN};$$

Struts OC and OB FEMs are zero.

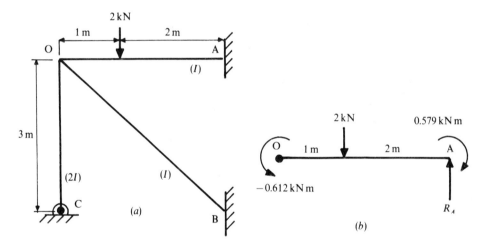

Figure 5.15 Three-bar plane structure.

Table 5.7

Joint	C		O			A	B
Member	CO	OC	OB	OA	AO	BO
DF		0.467	0.220	0.313		
Initial FEM	+0	−0	+0	−0.89	+0.44	−0
Unbalanced moment			−0.89			
First distribution		+0.416	+0.196	+0.278		
COM	+0				+0.139	+0.098
Final FMs	+0	+0.416	+0.196	−0.612	+0.579	+0.098

DFs at O from Eqs (5.9):

$$\frac{M_{OA}}{M} = \frac{(I/3)}{(I/3) + (I/3/2) + (3/4)(2I/3)} = 0.313;$$

$$\frac{M_{OB}}{M} = \frac{(I/3/2)}{(I/3) + (I/3\sqrt{2}) + (3/4)(2I/3)} = 0.220;$$

$$\frac{M_{OC}}{M} = \frac{(3/4)(2I/3)}{(I/3) + (I/3\sqrt{2}) + (3/4)(2I/3)} = 0.467.$$

It follows, in this case, that the L.H. FEM is the unbalanced moment for the joint O, which is distributed within Table 5.7.

Note that with three bars meeting at O, the moments in each bar are different. Taking moments about O with OA as the free body diagram (f.b.d.) in Fig. 5.15(b)

$$\sum M_O = 0 = -0.612 + (2 \times 1) + 0.579 - 3R_A, \quad \Rightarrow R_A = 0.657 \, \text{kN}.$$

Example 5.12 The structure in Fig. 5.16 is encastre at points A, D and F and is hinged at E. Find the bending moments at A, F and D.

Fixed-end moments from Table 5.1:

Bay AB L.H. FEM = R.H. FEM = $1 \times (1/2) \times (1/2)^2/1^2 = 0.125 \, \text{kN m}$;

Bay BC L.H. FEM = $3 \times 2^2 \times 1/3^2 = 1.33 \, \text{kN m}$,

R.H. FEM = $3 \times 1^2 \times 2/3^2 = 0.67 \, \text{kN m}$;

Bay CD L.H. FEM = R.H. FEM = $1 \times 2^2/12 = 0.33 \, \text{kN m}$;

Bays BE and CF FEMs are zero.

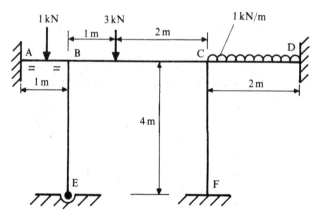

Figure 5.16 Encastre structure.

DFs at B and C from Eqs (5.8):

$$\frac{M_{BA}}{M} = \frac{(I/1)}{(I/1) + (3/4)(I/4) + (I/3)} = 0.658;$$

$$\frac{M_{BE}}{M} = \frac{(3/4)(I/4)}{(I/1) + (3/4)(I/4) + (I/3)} = 0.123;$$

$$\frac{M_{BC}}{M} = \frac{(I/3)}{(I/1) + (3/4)(I/4) + (I/3)} = 0.219;$$

$$\frac{M_{CB}}{M} = \frac{(I/3)}{(I/3) + (I/4) + (I/2)} = 0.308;$$

$$\frac{M_{CF}}{M} = \frac{(I/4)}{(I/3) + (I/4) + (I/2)} = 0.231;$$

$$\frac{M_{CD}}{M} = \frac{(I/2)}{(I/3) + (I/4) + (I/2)} = 0.461.$$

Since E is pinned, it carries no moment and therefore need not enter into the Table 5.8. However, if a horizontal force were applied to

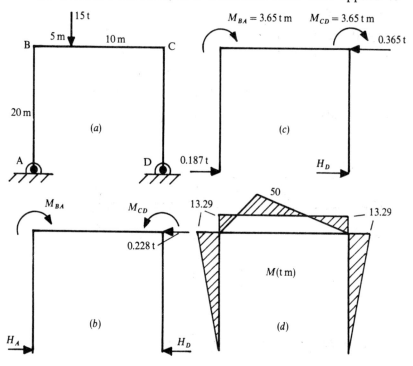

Figure 5.17 Moment analysis of a portal frame.

BE then the FEM at E would need to be released and carried over to B. The distribution is continued until the unbalanced carry-over moments at B and C become negligibly small. The moments at A, F and D then appear in the first and final two columns respectively.

Example 5.13 Construct the bending moment diagram for the portal frame in Fig. 5.17(a) when EI is constant (1 tonne (t) $\equiv 10^3$ kg).

Fixed-end moments:

$$\text{Bay BC} \quad \text{R.H. FEM} = 10 \times 5^2 \times 15/15^2 = 16.67 \, \text{t m};$$
$$\text{L.H. FEM} = 5 \times 10^2 \times 15/15^2 = 33.33 \, \text{t m};$$

Struts AB and CD   all FMs = 0.

DFs:

$$\frac{M_{BA}}{M} = \frac{(3/4)(I/20)}{(3/4)(I/20) + (I/15)} = 0.36 = \frac{M_{CD}}{M};$$

$$\frac{M_{BC}}{M} = \frac{(I/15)}{(3/4)(I/20) + (I/15)} = 0.64 = \frac{M_{CB}}{M}.$$

The FEMs are first distributed in the usual way, but these are later corrected to account for the effect of horizontal side-way. The non-sway FEMs enable the horizontal forces at A and D to be found by treating AB and DC as f.b.d.'s. Then, from Fig. 5.17(b),

$$\sum M_B = 0 = M_{BA} - 20H_A, \quad \Rightarrow H_A = 15.57/20 = 0.778 \, \text{t};$$
$$\sum M_C = 0 = M_{CD} - 20H_D, \quad \Rightarrow H_D = 11.0/20 = 0.550 \, \text{t};$$

and, for horizontal force equilibrium of the frame,

$$\sum H = 0 = H_A - H_D - H_C, \quad \Rightarrow H_C = 0.778 - 0.55 = 0.228 \, \text{t}.$$

The non-sway FEMs in Table 5.9 are valid only when a horizontal equilibrating force of 0.228 t is applied at B. Without this, the frame sways sideways by an amount δ under H_C. Table 5.1 shows, from the FEM expression $M = 3EI\delta/L^2$, that the moments M_{BA} and M_{CD}, induced at B and C as a consequence of side-sway, are inversely proportional to the squares of the respective lengths AB and CD. Thus $M_{CD} = M_{BA}$, with each moment having the same sense. This relationship is used for a second-moment distribution in Table 5.9 with an assumed value for M_{BA} (say 5 t m) to find the induced sway moments with rigid joints at B and C. The sway moments produce equal horizontal reactions $H_A = H_D = 3.65/20 = 0.187$ t requiring an equilibrating force of 0.365 t at C (Fig. 5.17(c)). The true corrected sway moments are found from the

Table 5.8

Joint	A	B			C			F	D
Member	AB	BA	BE	BC	CB	CF	CD	FC	DC
DFs		0.658	0.123	0.219	0.308	0.231	0.461		
Initial FEM	−0.125	+0.125	0	−1.33	+0.67	0	−0.33	0	+0.33
Imbalance			−1.205			+0.340			
First distribution		+0.793	+0.148	+0.264	−0.105	−0.079	−0.157		
COM	+0.397			−0.053	+0.132			−0.040	−0.0785
Second distribution		+0.033	+0.006	+0.011	−0.040	−0.030	−0.060		
COM	+0.017			−0.020	+0.006			−0.015	−0.030
Third distribution		+0.013	+0.003	+0.004	−0.0019	−0.0014	−0.0028		
COM	+0.007			−0.001	+0.002			−0.0007	−0.0014
Fourth distribution		+0.0007	+0.00012	+0.00022	−0.00062	−0.00046	−0.00092		
COM	+0.0003			−0.0003	+0.00011			−0.0002	−0.0005
Final FM	+0.296	+0.965	+0.157	−1.127	+0.663	−0.111	−0.551	−0.056	+0.220

117

Table 5.9

Joint	A		B		C	D
Member	AB	BA	BC	CB	CD	DC
DF		0.36	0.64	0.64	0.36	
Initial FEM	0	0	−33.33	+16.67	0	0
First distribution		+12	+21.3	−10.7	−6	
COM			−5.4	+10.7		
Second distribution		+1.95	+3.45	−6.85	−3.85	
COM			−3.43	+1.73		
Third distribution		+1.24	+2.19	−1.10	−0.63	
COM			−0.55	+1.10		
Fourth distribution		+0.20	+0.35	−0.70	−0.40	
COM			−0.40	+0.20		
Fifth distribution		+0.15	+0.25	−0.13	−0.07	
COM			−0.07	+0.13		
Sixth distribution		+0.03	+0.04	−0.08	−0.05	
Non-sway FEM		+15.57	−15.57	+11.0	−11.0	0
Initial sway M	0	+5		+5		0
First distribution		−1.8	−3.20	−3.20	−1.80	
COM			−1.60	−1.60		
Second distribution		+0.60	+1.00	+1.00	+0.60	
COM			+0.50	+0.50		
Third distribution		−0.20	−0.30	−0.30	−0.20	
COM			−0.15	−0.15		
Fourth distribution		+0.05	+0.10	+0.10	+0.05	
Sway M	0	+3.65	−3.65	−3.65	+3.65	0
Corrected sway M	0	−2.28	+2.28	+2.28	−2.28	0
Final FEM	0	+13.29	−13.29	+13.28	+13.28	0

condition that there must be no horizontal force at C, since the horizontal reactions at A and D will balance. That is, between Figs 5.17(b and c):

$$0.228 + 0.365C = 0,$$

where $C = -0.625$ is the multiplication factor used to correct the sway moments found from the second distribution. The sign of C depends upon the sense assumed for the initial sway moments M_{BA} and M_{CD}. The final FEMs are then the sums of the corrected sway and non-sway moments in the above table. Finally, the bending moment diagram in Fig. 5.17(d) is constructed, where, for bay BC, the net moment at any point is the difference between the free ($M_{max} = -50\,\text{t m}$) and fixing moment ordinates as shown by the shaded region.

EXERCISES

Shear Force and Bending Moment Diagrams For Single-span Beams

5.1 A beam 10 m long is simply supported at its ends and carries two concentrated loads of 30 kN and 50 kN at a distance of 3 m from each end. Draw the F and M diagrams.

5.2 A beam of 12 m length carries a uniformly distributed load of 200 kg/m and rests on two simple supports distance l apart with an equal overhang at each end. If $M = 0$ at midspan, find l. What is l if M_{max} is to be kept as small as possible? Sketch the F and M diagrams for the latter case and determine the points of contraflexure.

Answer: 6 m; 7.1 m; 0.99 m.

5.3 Draw F and M diagrams for the beam in Fig. 5.18 and find the position of the point of contraflexure.

Answer: $M_{max} = 22$ t m; 8 m from L.H. end; 9.667 m from R.H. end.

5.4 Find the magnitude and position of M_{max} and the points of inflection for the beam in Fig. 5.19.

Answer: $M_{max} = 7.25$ t m; 7.85 m from R.H. end; 4.34 m from R.H. end; 3.71 m from L.H. end.

5.5 Sketch, for the beam in Fig. 5.20, the F and M diagrams. Determine the position and magnitude of M_{max} and the point of contraflexure.

Answer: $M_{max} = 20.1$ t m; 6 m from L.H. end; 7.82 m from R.H. end.

5.6 Find w (t/m) for the simply supported beam in Fig. 5.21 in order that $F = 0$ at 5 m from the L.H. end. Then draw the F and M diagrams and find the points of inflection.

Answer: 1/13 t/m; 0.75 m; 2.66 m.

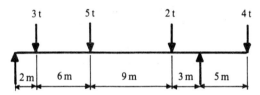

Figure 5.18

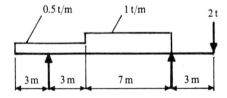

Figure 5.19

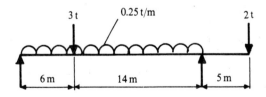

Figure 5.20

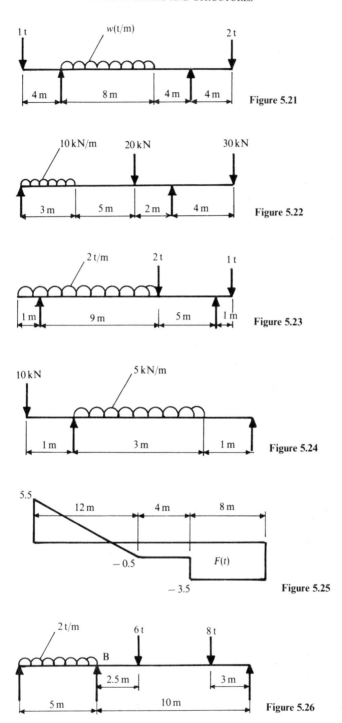

Figure 5.21

Figure 5.22

Figure 5.23

Figure 5.24

Figure 5.25

Figure 5.26

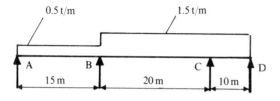

Figure 5.27

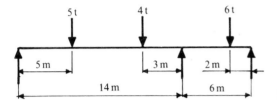

Figure 5.28

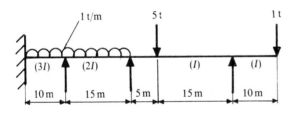

Figure 5.29

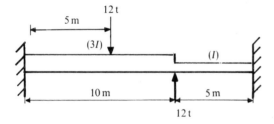

Figure 5.30

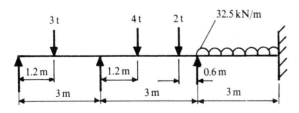

Figure 5.31

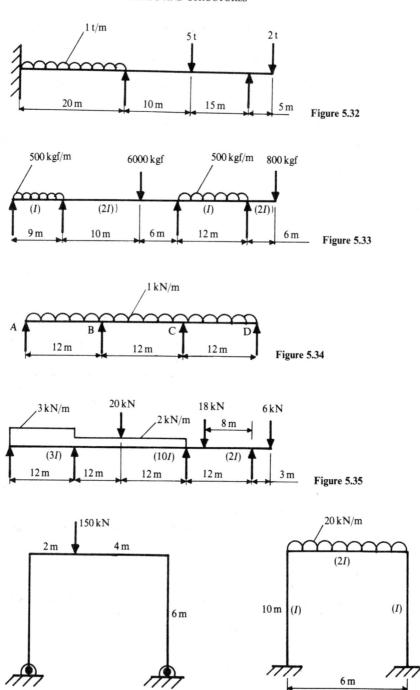

Figure 5.32

Figure 5.33

Figure 5.34

Figure 5.35

Figure 5.36

Figure 5.37

5.7 Calculate, for the beam in Fig. 5.22, the position and magnitude of the maximum hogging and sagging bending moments.

5.8 Find the points of contraflexure and the magnitude and position of F_{max} and M_{max} for the beam if Fig. 5.23.

Answer: 1.142 m from R.H. end, 1.075 m from L.H. end; 12.94 t at left support, 40.66 t m at 7.47 m from L.H. end.

5.9 Find the maximum hogging and sagging bending moments for the beam in Fig. 5.24.

Answer: 10 kN m, 6.7 kN m at 1 m and 3.4 m respectively from L.H. end.

5.10 Figure 5.25 shows the F diagram for a simply supported beam. Find the beam loading and M_{max}.

Answer: 30.25 t m at 11 m from L.H. end.

Moment Distribution–Continuous Beams

5.11 A simply supported beam of constant cross-section is loaded as shown in Fig. 5.26. Draw the F and M diagrams showing the maximum values.

Answer: 16.69 t at B; 16.52 t m at 8 t.

5.12 The continuous beam in Fig. 5.27 rests on four level supports. Determine the bending moments and the reaction at each support.

Answer: at A, 1.33 t; at B, 20.77 t, 36.3 t m; at C, 27.32 t, 44.15 t m; at D, 3.08 t.

5.13 The beam in Fig. 5.28 is of uniform section. Draw the shear force and bending moment diagrams indicating principal values.

5.14 Calculate the fixed-end moments at each end of the continuous beam given in Fig. 5.29.

Answer: 19.5 t m, 3 t m.

5.15 The continuous beam in Fig. 5.30 is built-in at one end and propped at the three positions shown. If the section varies as shown, draw the M and F diagrams indicating salient values.

5.16 Draw the F and M diagrams for the propped cantilever in Fig. 5.31. What is the maximum moment?

Answer: 26.6 kN m at the fixed end.

5.17 The beam in Fig. 5.32 has a uniform section throughout. Draw the F and M diagrams showing the maximum values.

Answer: $M_{max} = 37.75$ t m at the fixed end.

5.18 Determine the magnitude and position of the maximum bending moment for the beam in Fig. 5.33.

Answer: 13.561 kgf m at the 6000 kgf force.

5.19 Calculate the bending moments at points B and C for the beam in Fig. 5.34 and hence construct the M diagram showing the maximum value.

Answer: 14.4 kN m.

5.20 Determine all the fixing moments for the continuous beam in Fig. 5.35.

Answer: 107.6 kN m; 73.6 kN m; 18 kN m.

5.21 A continuous beam rests on three simple supports A, B and C. If the level of B is 20 mm below that of A and C, determine the fixing moment at B given that AB = 5 m, BC = 10 m, $I_{AB} = 12 \times 10^8$ mm^4, $I_{BC} = 40 \times 10^8$ mm^4 and $E = 210$ GPa.

Moment Distribution—Structures

5.22 Construct the bending moment diagrams for the equal-legged portal frame in Fig. 5.36 where EI is constant and the legs are pinned to the foundations. ($M_{max} = 111.5$ kN m at 50 kN.)

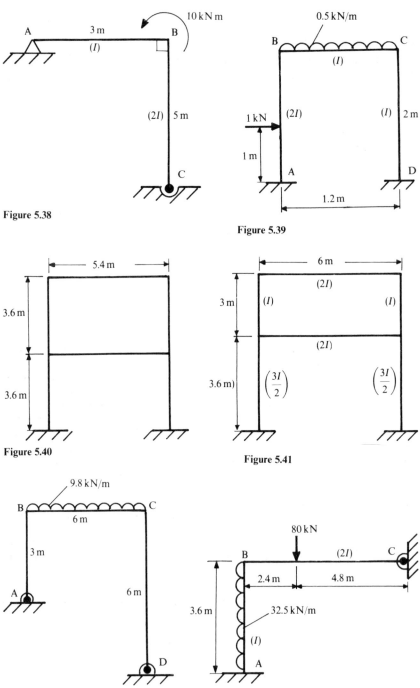

Figure 5.38

Figure 5.39

Figure 5.40

Figure 5.41

Figure 5.42

Figure 5.43

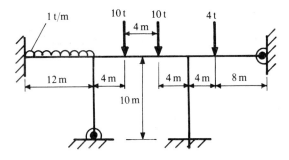

Figure 5.44

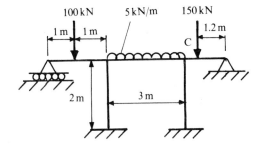

Figure 5.45

5.23 The symmetrical portal frame in Fig. 5.37 is rigidly fixed into its foundations. Establish the moment distribution diagram for the frame.

5.24 The frame ABC in Fig. 5.38 is simply supported at A and pinned at C. Find the horizontal and vertical reactions at A and C for a concentrated moment of 10 kN m at B.

5.25 The frame in Fig. 5.39 carries a concentrated horizontal force of 1 kN in addition to uniformly distributed loading of 0.5 kN/m as shown. Determine the reactions at the rigid foundations A and D and the horizontal displacement at C. Take $E = 75$ GPa, $I = 2.5 \times 10^3$ mm^4.

5.26 The two-storey frame in Fig. 5.40 is manufactured with a central horizontal member that is 4 mm too short. Determine the resulting moment distribution for the frame when this bar is elastically stretched into position. The cross-section of each bar is rectangular 100 mm × 25 mm and $E = 200$ GPa.

5.27 The two-storey frame shown in Fig. 5.41 is subjected to a temperature increase of 25 °C. Determine the resulting induced moment distribution given that $\alpha = 15 \times 10^{-6}/°C$, $E = 210$ GPa and $I = 20.80 \times 10^6$ mm^4.

5.28 Construct the bending moment diagram for the unsymmetrical portal frame in Fig. 5.42, where EI is constant, taking account of side-sway. Show the position and magnitude of the maximum bending moment.

Answer: 23.7 kN m at C; 77.4 kN m at B.

5.29 Find the position and magnitude of the maximum bending moment for the structure in Fig. 5.43 when it is fixed at A and hinged at C.

Answer: 77.4 kN m at B.

5.30 Determine the fixing moments and construct the bending moment diagram for the structure in Fig. 5.44 given that I is constant.

5.31 Determine the maximum of the true fixing moments for the structure in Fig. 5.45 when the effect of side-sway is accounted for.

Answer: 51 kN m at C.

THEORIES OF BENDING

6.1 BENDING OF STRAIGHT BEAMS

The engineer's theory of bending (ETB) enables the longitudinal bending stress σ, to be found anywhere in the cross-section of an initially straight beam when the bending moment M at that section is known. Let the radius of curvature R in Fig. 6.1(a) define the unstrained longitudinal fibre OO, which is referred to as the neutral axis (NA) for the cross-section. With the angle θ (radians) subtended by a portion of the beam at the centre of curvature, C the strain in a fibre AB, at a positive distance y, from the NA on the tensile side is

$$\epsilon = [(R + y)\theta - R\theta]/R\theta = y/R$$

and, therefore, the longitudinal bending stress is expressed as

$$\epsilon = \sigma/E = y/R. \tag{6.1}$$

Now consider the equilibrium of moments for any cross-section in Fig. 6.1(b). The hogging moment M is equilibrated by the sum of all elemental moments $\delta M = (\sigma\, dA)\, y$, that are produced by the action of σ on the area, δA, shown. That is, for a cross-section of area A,

$$M = \int_A (\sigma\, dA)y. \tag{6.2}$$

Substituting from Eq. (6.1) into Eq. (6.2):

$$M = (E/R) \int_A y^2\, dA = EI/R, \tag{6.3}$$

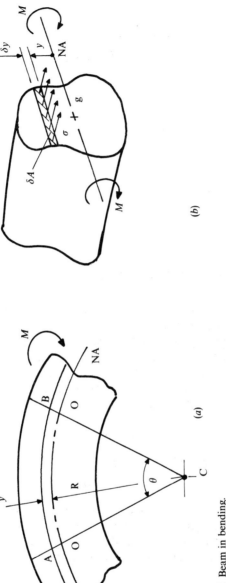

Figure 6.1 Beam in bending.

where the integral defines I, the second moment of area about the NA. Combining Eqs (6.1) and (6.3) leads to the well-known, ETB formula.

$$M/I = E/R = \sigma/y. \qquad (6.4)$$

Now, as there is no applied axial force the sum of all elemental forces $\delta F = \sigma \, dA$, over the whole cross-section, must be zero. Then from Eq. (6.1),

$$F = \int_A \sigma \, dA = (E/R) \int_A y \, dA = 0.$$

Since this condition is only satisfied when the first moment of area, defined by the integral, is zero, it follows that the neutral axis of bending passes through the centroid g, of the cross-section.

Example 6.1 A cast iron girder of unsymmetrical I-section (Fig. 6.2) is simply supported at the ends of its 4.5 m (15 ft) length. Calculate the distributed load per metre length that can be carried without the tensile stress exceeding 15 MPa (1 tonf/in²). What is the maximum compressive stress in the beam?

SOLUTION TO SI UNITS. Take first moments of area about the base where centroid g lies at height $\bar{y}$:

$$(375 \times 62.5 \times 31.25) + (312.5 \times 30 \times 218.75) + (125 \times 30 \times 390)$$
$$= 36562.5\bar{y},$$
$$\therefore \quad \bar{y} = 116.12 \text{ mm}.$$

Second moment of area—using $bd^3/3$ for constituent rectangles:

$$I = [375(116.12)^3/3 - 345(53.62)^3/3] + [125(288.88)^3/3 - 95(258.88)^3/3]$$
$$= (177 + 455.07)10^6 = 633.06 \times 10^6 \text{ mm}^4.$$

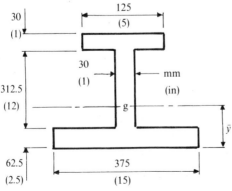

Figure 6.2 Unsymmetrical I-section.

The maximum sagging bending moment for the given loading is

$$M_{max} = -wl^2/8 = -w(4500)^2/8 = -2531.25 \times 10^3 w \text{ N mm}.$$

Then from Eq. (6.4), the maximum tensile stress occurs at the bottom edge of the lower flange where $y = -116.12$ mm. That is,

$$\sigma = My/I = (-2531.25 \times 10^3 w)(-116.12)/633.06 \times 10^6$$
$$= 15 \text{ N/mm}^2 \text{ (MPa)},$$

from which $w = 32.3$ N/mm (kN/m). The maximum compressive stress occurs at the top flange surface, where $y = 405 - 116.12 = 288.88$ mm. Again from Eq. (6.4),

$$\sigma = (-2531.25 \times 10^3 \times 32.3)288.88/(633.06 \times 10^6) = -37.32 \text{ MPa}.$$

SOLUTION TO IMPERIAL UNITS. The centroid position $\bar{y}$ is found from

$$(15 \times 2.5 \times 1.25) + (12 \times 1 \times 8.5) + (5 \times 1 \times 15) = 54.5\bar{y},$$
$$\therefore \quad \bar{y} = 4.108 \text{ in.}$$

The respective calculations become

$$I = (15 \times 4.108^3/3) - (14 \times 1.608^3/3) + (5 \times 11.392^3/3)$$
$$- (4 \times 10.392^3/3)$$
$$= 1294.92 \text{ in}^4.$$
$$M_{max} = -wl^2/8 = -w(15 \times 12)^2/8 = -4050w \text{ lbf in.}$$
$$\sigma = M_{max}\bar{y}/I = (-4050w)(-4.108)/1294.92 = 1 \times 2240 \text{ lbf/in}^2,$$
$$\therefore \quad w = 174.34 \text{ lbf/in} = 0.934 \text{ tonf/ft},$$

and the maximum compressive stress occurs for $y = (15.5 - 4.108) = 11.392$ in.

$$\sigma = My/I = (-4050 \times 174.34)11.392/1294.92$$
$$= -6211.7 \text{ lbf/in}^2 = -2.77 \text{ tonf/in}^2.$$

Example 6.2 A 6 m long, simply-supported beam carries a uniformly distributed load of 30 kN/m together with concentrated loads to 20 kN at 1.5 m from each end (Fig. 6.3(a)). The beam cross-section is fabricated from a symmetrical I-section 0.3 m deep ($I = 198 \times 10^6$ mm^4) with top plates 0.3 m wide × 18 mm thick riveted to the top and bottom flanges (see Fig. 6.3(b)). Calculate the maximum bending stress and find the percentage increase in strength owing to the addition of the plates.

The maximum moment occurs at the centre, where

$$M_{max} = (1.5 \times 20) + (3 \times 1.5 \times 30) - (3 \times 110) = -165 \text{ kN m (sagging).}$$

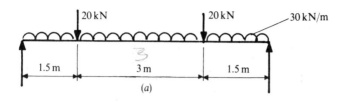

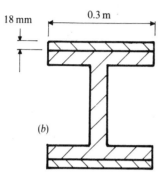

Figure 6.3 Fabricated I-section.

For the I-section alone, Eq. (6.4) gives the maximum tensile stress as

$$\sigma = My/I = (-165)(-150) \times 10^6/198 \times 10^6 = 125 \text{ N/mm}^2.$$

For the strengthened section, by parallel axes,

$$I = (198 \times 10^6) + 2[(300 \times 18^3)/12 + (300 \times 18 \times 159^2)]$$
$$= 471.33 \times 10^6 \text{ mm}^4.$$

The bending stress corresponding to $y = 150 + 18 = 168$ mm in Eq. (6.4) is

$$\sigma = My/I = (-165)(150 + 18) \times 10^6/471.33 \times 10^6 = 58.81 \text{ N/mm}^2.$$

The measure of strength is the maximum moment the beam can carry for a given allowable bending stress, but as $M/\sigma = I/y$ it is only necessary to employ the section modulus, $z = I/y$, for a strength comparison. That is,

$$z(\text{unstrengthened}) = I/y = 198 \times 10^6/150 = 1.32 \times 10^6 \text{ mm}^3,$$
$$z(\text{strengthened}) = I/y = 471.33 \times 10^6/168 = 2.806 \times 10^6 \text{ mm}^3,$$
$$\therefore \quad \text{Strength increase} = (2.806 - 1.32)100/1.32 = 112.54 \text{ per cent.}$$

Example 6.3 Find the maximum tensile and compressive stresses and the radius of curvature when a sagging moment of 30 kN m produces compression in the flange of the T-section given in Fig. 6.4(a). Plot the bending stress distribution. Take E = 208 GPa.

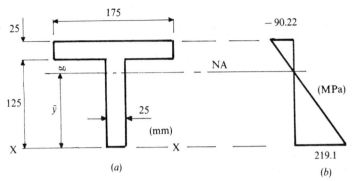

Figure 6.4 T-section.

For the position $\bar{y}$, of the centroid g, take moments about base $X - X$;

$$(125 \times 25 \times 62.5) + (175 \times 25 \times 137.5) = 750\bar{y}, \Rightarrow \bar{y} = 106.25\,\text{mm}.$$

And for rectangles with their bases on the NA, using $bd^3/3$,

$$I = 25(106.25)^3/3 + 175(43.75)^3/3 - 150(18.75)^3/3$$
$$= 14.55 \times 10^6\,\text{mm}^4.$$

Maximum tensile stress at web bottom, $y = -106.25\,\text{mm}$, where

$$\sigma = My/I = (-30 \times 10^6)(-106.25)/(14.55 \times 10^6) = 219.1\,\text{MPa}.$$

Maximum compressive stress at flange top, $y = 150 - 106.25 = 43.75\,\text{mm}$,

$$\sigma = My/I = (30 \times 10^6)(43.75)/(14.55 \times 10^6) = -90.22\,\text{MPa}.$$

As σ is proportional to y for a given M and I, it follows that the stress will vary linearly with the depth between these maxima, passing through zero at the NA (see Fig. 6.4(b)). Finally, from Eq. (6.4).

$$R = EI/M = 208 \times 10^3 \times 14.55 \times 10^6/(30 \times 10^6 \times 10^3) = 100.8\,\text{m}.$$

6.2 ELASTIC BENDING OF INITIALLY CURVED BEAMS

The theory depends upon the relative magnitude of the radius of curvature and the dimensions of the cross-section. Beams with small and large curvatures (Fig. 6.5) may be treated separately.

Large Initial Radius (Small Curvature)

When $R \gg$ dimensions of the cross-section, as is the case with thin rings, then the theory is similar to ETB for straight beams. The same analysis applies,

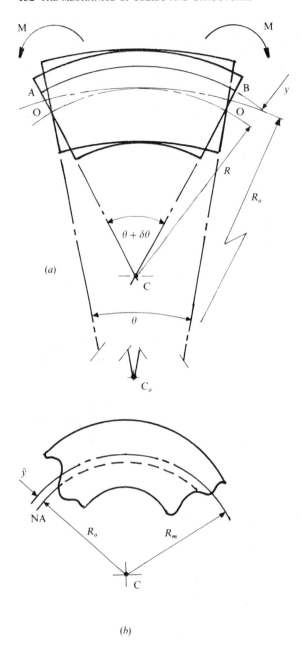

(a)

(b)

Figure 6.5 Bending of beams with initial curvature.

provided the initial curvature R_o, is accounted for in the longitudinal strain expression. Referring to Fig. 6.5(a), the centre of initial curvature C_o, does not generally coincide with the centre C induced by applies bending moments M. The tensile strain in fibre AB at distance y from the unstrained neutral axis OO is

$$\epsilon = \frac{(R + y)(\theta + \delta\theta) - (R_o + y)\theta}{(R_o + y)\theta}, \tag{6.5}$$

where the denominator is the initial length of AB. Since the length of OO remains unchanged,

$$R(\theta + \delta\theta) = R_o\theta. \tag{6.6}$$

Combining Eqs (6.5) and (6.6) gives

$$\epsilon = y(R_o - R)/R(R_o + y) \simeq y[(1/R) - (1/R_o)] \tag{6.7}$$

provided $y \ll R_o$. Since the remaining analysis is identical to that for ETB, it follows that with Eq. (6.7) the three-part ETB formula in Eq. (6.4) is now modified to

$$M/I = E[(1/R) - (1/R_o)] = \sigma/y. \tag{6.8}$$

It is seen from Eq. (6.8) that the bending stress σ again varies linearly with dimension y and that the NA, from which y is measured, must pass through the centroid of the cross-section.

Example 6.4 The semicircular steel arch in Fig. 6.6 rests on rollers at each end support and is 100 mm in width. Find the section depth for a safety factor of 2.5 with an ultimate stress of ± 300 MPa and the radius of

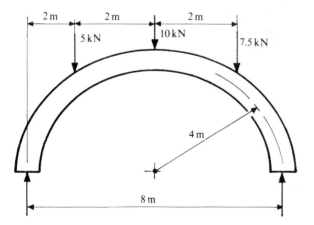

Figure 6.6 Steel arch.

curvature under the maximum bending moment. Take $E = 207\,\text{GPa}$. The support reactions are

$$8R_B = (0.5 \times 2) + (1 \times 4) + (0.75 \times 6), \Rightarrow R_B = 1.188\,\text{kN},$$
$$R_A + R_B = 0.5 + 1 + 0.75, \Rightarrow R_A = 1.063\,\text{kN}.$$

Maximum moment—at centre:

$$M = -(4 \times 1.063) + (2 \times 0.5) = -3.25\,\text{kN m}.$$

Working stress:

$$\sigma = \text{UTS (or UCS)}/S = 300/2.5 = 120\,\text{MPa}.$$

Then from Eq. (6.8),

$$M/(bd^3/12) = \sigma/(d/2),$$
$$\therefore \quad d = \sqrt{6M/b\sigma} = \sqrt{[(6 \times 3.25 \times 10^6)/(100 \times 120)]} = 40.31\,\text{mm}.$$

R at M_{max} (sagging—increasing the radius of curvature): from Eq. (6.8),

$$1/R = M/EI + 1/R_o = -3.25 \times 10^6/(207\,000 \times 100 \times 40.31^3/12)$$
$$+ 1/(4 \times 10^3),$$
$$R = 4.52\,\text{m}.$$

Small Initial Radius (Large Curvature)

When R and the cross-section dimensions are of the same order (Fig. 6.5(b)), the approximation made in Eq. (6.7) is invalid, so that the stress becomes

$$\sigma = yE(R_o - R)/R(R_o + y). \tag{6.9}$$

Thus, as the total axial force is zero,

$$F = \int \sigma \, dA = \frac{E(R_o - R)}{R} \int \frac{y \, dA}{(R_o + y)} = 0 \tag{6.10}$$

where R and R_o are constants at a given section. Furthermore, the resisting moment M for the section is given by

$$M = \int \sigma y \, dA = \frac{E(R_o - R)}{R} \int \frac{y^2 \, dA}{(R_o + y)}$$
$$= \frac{E(R_o - R)}{R} \left(\int y \, dA - R_o \int \frac{y}{(R_o + y)} dA \right). \tag{6.11}$$

Combining Eqs (6.10) and (6.11), M becomes

$$M = E[(R_o - R)/R] \int y \, dA. \tag{6.12}$$

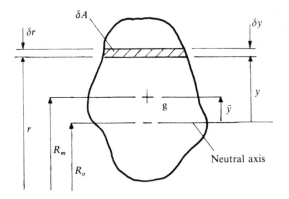

Figure 6.7 NA shift.

However, though y is still measured from the NA, Eq. (6.10) reveals that this axis will now pass through the centroid, g, of the section but shifts by an amount $\bar{y}$ towards the centre of curvature C (Fig. 6.5(b)). Taking first moments of the cross-sectional area A about the NA in Fig. 6.7 it follows that $A\bar{y} = \int_A y\,dA$. Equation (6.12) then becomes

$$M = E[(R_o - R)/R]A\bar{y}. \tag{6.13}$$

The corresponding three-part formula is found from Eqs (6.9) and (6.13) to complete the comparison with Eqs (6.4) and (6.8).

$$\frac{M}{[A\bar{y}(R_o + y)]} = \frac{E(R_o - R)}{R(R_o + y)} = \frac{\sigma}{y}. \tag{6.14}$$

Note in Eq. (6.14) that, when $R < R_o$, a positive hogging moment will produce tension above the NA (y is positive) and compression below the NA (y is negative). Conversely, when $R > R_o$ a negative sagging moment produces compression above and tension below the NA. Since Eq. (6.14) shows that σ is no longer proportional to y, the variation in σ across the section is non-linear. In applying Eq. (6.14), the problem is that of finding $\bar{y}$ for a section of known centroidal or mean radius $R_m = R_o + \bar{y}$.

Rectangular section With the rectangular section in Fig. 6.8, Eq. (6.10) gives

$$\int y\,dA/(R_o + y) = \int y(b\,dy)/[(R_m - \bar{y}) + y]$$

$$= b\int_{\bar{y}-d/2}^{\bar{y}+d/2} [1 - (R_m - \bar{y})/(R_m + y - \bar{y})]\,dy = 0;$$

$$= b|y - (R_m - \bar{y})\ln(y + R_m - \bar{y})|_{\bar{y}-d/2}^{\bar{y}+d/2} = 0;$$

$$\therefore \quad \bar{y} = R_m - \frac{d}{\ln[(R_m + d/2)/(R_m - d/2)]}. \tag{6.15}$$

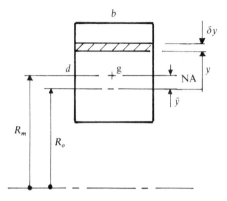

Figure 6.8 Rectangular section.

I-section In the I-section of Fig. 6.9, the web and two flanges are taken separately. It is convenient to introduce a parameter, $z = y - \bar{y}$, measured from the centroidal axis so that with $R_o = R_m - \bar{y}$ the integral in Eq. (6.10) becomes

$$i = \int (z + \bar{y}) \, dA/(R_m + z) = \int [1 - (R_m - \bar{y})/(R_m + z)] \, dA. \qquad (6.16)$$

Top flange ①, $dA = B \, dz$:

$$i_{f1} = B|z - (R_m - \bar{y}) \ln (R_m + z)|_{d/2}^{D/2}$$

$$= B \left\{ (D/2 - d/2) - (R_m - \bar{y}) \ln \left[\frac{(R_m + D/2)}{(R_m + d/2)} \right] \right\}. \qquad (6.17a)$$

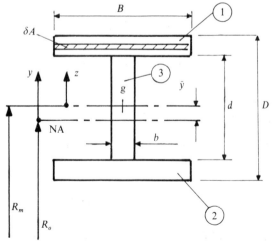

Figure 6.9 I-section.

Web ③, $dA = b\,dz$:

$$i_w = b|z - (R_m - \bar{y})\ln(R_m + z)|_{-d/2}^{d/2}$$

$$= b\left\{d - (R_m - \bar{y})\ln\left[\frac{(R_m + d/2)}{(R_m - d/2)}\right]\right\}.\qquad(6.17b)$$

Bottom flange ②, $dA = B\,dz$. Since the area now lies wholly beneath the centroidal axis, it is simplest to change the sign of z in Eq. (6.16) and then integrate between positive limits for z of $D/2$ and $d/2$. Note that this is not equivalent to integrating Eq. (6.16) between negative limits:

$$i_{f2} = B|z + (R_m - \bar{y})\ln(R_m - z)|_{d/2}^{D/2}$$

$$= B\left\{(D/2 - d/2) + (R_m - \bar{y})\ln\left[\frac{(R_m - D/2)}{(R_m - d/2)}\right]\right\}.\qquad(6.17c)$$

The NA($\bar{y}$) is then found by equating the sum of Eqs (6.17a–c) to zero.

Example 6.5 Compare the maximum tensile and compressive stresses and the radii of curvature of the neutral axes for the initially curved beam in Fig. 6.10(a) with the two sections given in Figs. 6.10(b) and (c) under $M = +10\,\text{kN}\,\text{m}$ shown. Take $E = 207\,\text{GPa}$.

For the rectangular section, $A = 20 \times 10^3\,\text{mm}^2$, $b = 100\,\text{mm}$, $d = 200\,\text{mm}$, $R_m = 200\,\text{mm}$. Then from Eqs (6.14) and (6.15),

$$\bar{y} = 200 - 200/\ln(300/100) = 17.95\,\text{mm},$$

$$\sigma = M y/[A\bar{y}(R_o + y)].\qquad\text{(i)}$$

For maximum tension, $y = +(100 + 17.95)\,\text{mm}$ at the outer radius. Then by Eq. (i),

$$\sigma = \frac{10 \times 10^6 \times 117.95}{20\,000 \times 17.95 \times 300} = +10.95\,\text{MPa}.$$

For maximum compression, $y = -(100 - 17.95) = -82.05\,\text{mm}$, when from Eq. (i),

$$\sigma = \frac{10 \times 10^6 \times (-82.05)}{20\,000 \times 17.95 \times 100} = -22.85\,\text{MPa}.$$

Re-arrange Eq. (6.14) to give the 'strained' radius of the NA as

$$R = R_o/[1 + (M/AE\bar{y})]\qquad\text{(ii)}$$

$$= \frac{200 - 17.95}{1 + [10 \times 10^6/(20\,000 \times 207\,000 \times 17.95)]} = 182.03\,\text{mm}.$$

For the I-section, $A = 7200\,\text{mm}^2$, $B = 100\,\text{mm}$, $D = 200\,\text{mm}$,

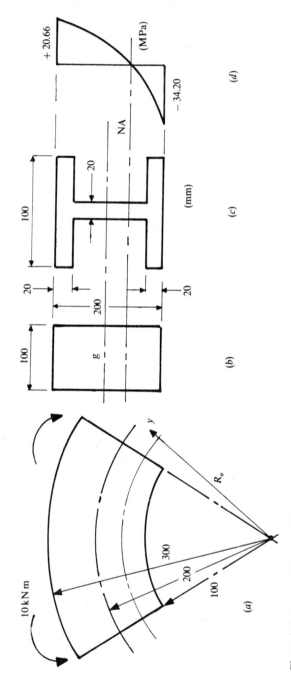

Figure 6.10 Beam with large initial curvature.

$b = 20$ mm, $d = 160$ mm, and $R_m = 200$ mm. Then from Eq. (6.17a–c):

$$i_{f1} = 100[(100 - 80) - (200 - \bar{y})\ln(300/280)]$$
$$= 100[20 - 0.069(200 - \bar{y})];$$
$$i_{f2} = 100[(100 - 80) + (200 - \bar{y})\ln(100/120)]$$
$$= 100[20 - 0.1823(200 - \bar{y})];$$
$$i_w = 20[160 - (200 - \bar{y})\ln(280/120)]$$
$$= 20[160 - 0.8473(200 - \bar{y})].$$

Since $i_{f1} + i_{f2} + i_w = 0$, these equations give $\bar{y} = 28.88$ mm. Then, for maximum tension with $y = (100 + 28.88)$ mm in Eq. (i),

$$\sigma = \frac{10 \times 10^6 \times 128.88}{7200 \times 28.88 \times 300} = +20.66 \text{ MPa};$$

and for maximum compression with $y = -(100 - 28.88) = -71.12$ mm, in Eq. (i),

$$\sigma = \frac{10 \times 10^6 \times (-71.12)}{7200 \times 28.88 \times 100} = -34.20 \text{ MPa}.$$

The variation in the stress between these limits across the section is shown in Fig. 6.10(d). Equation (ii) supplies the strained NA radius as

$$R = \frac{200 - 28.88}{1 + [10 \times 10^6/(7200 \times 207\,000 \times 28.88)]} = 171.08 \text{ mm}.$$

It is seen that for each section the radius of the strained NA differs little from its unstrained value of R_o.

Complex sections The shift in neutral axis $\bar{y}$ from the centroid g for a large-curvature beam with any given cross-section is more conveniently found from the introduction of a new variable $r = R_o + y$ in Fig. 6.7. Since $R_m = R_o + \bar{y}$ the integral in Eq. (6.10) then becomes

$$\int_A \frac{y\,dA}{R_o + y} = \int_A \frac{(r - R_m + \bar{y})}{r}\,dA = 0,$$

$$\therefore \quad A - (R_m - \bar{y})\int_A dA/r = 0,$$

from which $\bar{y}$ can be found once the integral $\int_A dA/r$ is known. (This integral has been derived for a large number of sections including circular, triangular and trapezoidal, by O. Sidebottom and J. Boresi, *Advanced Mechanics of Materials*, Wiley, 1984.) The reader should confirm Eq. (6.15) by applying this approach to the rectangular section in Fig. 6.8 where $A = b \times d$ and $dA = b\,dr$ with limits for r of $R_m + d/2$ and $R_m - d/2$.

6.3 COMBINED ELASTIC BENDING AND DIRECT STRESS

The particular combination of direct stresses produced by bending and tension (or compression) may arise in one of two basic ways.

A Beam Carrying an Axial Force F, in Addition to a Bending Moment M

Since the stresses produced by each action are longitudinal they may be added to give the resultant stress,

$$\sigma = \pm F/A \pm My/I \qquad (6.18)$$

in which the signs account for the sense of the stress above and below the NA, due to bending, in combination with either a tensile or compressive stress due to the axial force. Figure 6.11(a) illustrates the case for positive M and F when Eq. (6.18) becomes

$$\sigma = + F/A + My/I \quad \text{for the top surface, and}$$
$$\sigma = + F/A - My/I \quad \text{for the bottom surface.}$$

The resultant stress in Fig. 6.11(d) is the sum of the distributions due to bending in Fig. 6.11(b) and tension in Fig. 6.11(c). Clearly, the presence of F modifies the position of the NA, which is found by equating (6.18) to zero.

Example 6.6 A simply supported I-section beam 2 m long carries a central concentrated force of $W = 15$ kN in addition to a longitudinal tensile force $F = 60$ kN. Given that the depth of section $= 100$ mm, area $= 1800$ mm^2 and $I = 3.15 \times 10^6$ mm^4, determine the maximum tensile and compressive stresses and the position of the neutral axis.

$$M_{\max} = -WL/4 = -15 \times 10^3 \times 2 \times 10^3/4 = -7.5 \times 10^6 \text{ N mm (sagging)},$$

when from Eq. (6.18), with $y = +50$ mm, for the top surface,

$$\sigma = F/A + My/I = (60 \times 10^3/1800) + (-7.5 \times 10^6)(+50)/(3.15 \times 10^6)$$
$$= 33.33 - 119.1 = -85.77 \text{ MPa}$$

and for the bottom surface with $y = -50$ mm,

$$\sigma = F/A - My/I = (60 \times 10^3/1800) - (-7.5 \times 10^6)(-50)/(3.15 \times 10^6)$$
$$= 33.33 + 119.1 = 152.43 \text{ MPa}.$$

As the NA lies on the side of $(+ \text{ve})$ y,

$$0 = F/A + My/I = 33.33 + (-7.5 \times 10^6)y/(3.15 \times 10^6),$$
$$\therefore \quad y = +14 \text{ mm above the centroid.}$$

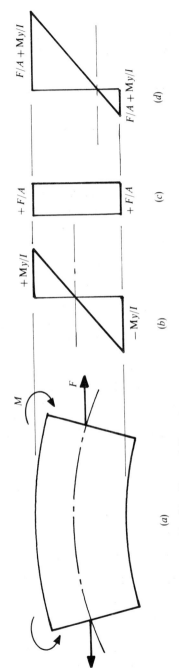

Figure 6.11 Beam under combined bending and direct stress.

A Short Column Carrying an Eccentric Axial Force F

Take the general case of a normal tensile force F, acting in the positive x, y quadrant in Fig. 6.12. The moments, $M_x = Fk$ and $M_y = Fh$ produced by the eccentricity are both hogging about the centroidal axes x and y, w.r.t. this quadrant. The resultant tensile stress at any point $P(x, y)$ is the sum of a direct stress due to F and bending stresses due to M_x and M_y. That is,

$$\sigma = F/A + M_y x/I_y + M_x y/I_x. \tag{6.19a}$$

The stresses in other quadrants are also determined from Eq. (6.19a) with the appropriate signs for x and y. Indeed, Eq. (6.19a) supplies the stress for compressive loadings provided the correct signs are applied. For this purpose it is convenient to arrange the column so that the load always lies in the first quadrant and then substitute for F with the appropriate sign. Where F is applied along an axis of symmetry (a principal axis) only one bending stress term will appear. For example, when F is eccentric to y only, then $M_x = 0$, $M_y = Fh$ and Eq. (6.19a) becomes

$$\sigma = F/A + (Fh)x/I_y. \tag{6.19b}$$

If the stress is not to change sign within a given cross-section, then the extreme load position $F(h, 0)$ follows from putting $\sigma = 0$ in Eq. (6.19b). With the circular section in Fig. 6.13, for example, this condition is ensured when

$$\sigma_A = F/A + M_y x/I_y = 0,$$
$$4F/\pi d^2 + Fh(-d/2)64/\pi d^4 = 0.$$

Thus, the stress may be wholly tensile or compressive, depending upon the sense of F, provided h lies within a circle of radius $h = d/8$.

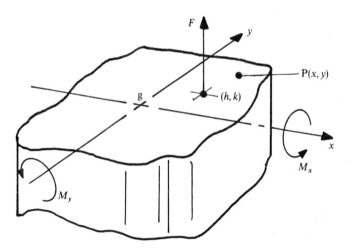

Figure 6.12 Eccentrically loaded column.

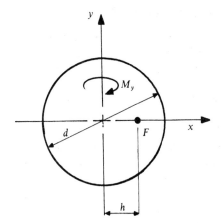

Figure 6.13 Circular section.

Example 6.7 Figure 6.14 gives the plan view of a short vertical channel. Determine the maximum compressive force F if the stress is not to exceed 77 MPa. Where must F lie along x if point Q is unstressed?

The centroid $(\bar{x})$ is found from taking first moments of area about the longest vertical side. That is, (AA)

$$(150 \times 25 \times 12.5) + (2 \times 100 \times 25 \times 50) = 8750\bar{x},$$
$$\bar{x} = 33.93 \text{ mm}, \therefore h = 100 - 33.93 = 66.07 \text{ mm}.$$

Using $bd^3/3$ for rectangles all with one side lying on the y-axis,

$$I_y = (200 \times 33.93^3/3) - (150 \times 8.93^3/3) + 2(25 \times 66.07^3/3)$$
$$= 7.3754 \times 10^6 \text{ mm}^4.$$

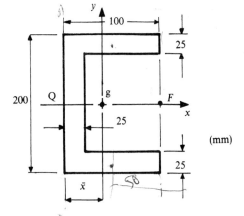

(mm)

$100 \times 25 \times 50 \times 50.$

$25 \times 150 \times 12.5. = \bar{x}$

$(30 \times 10 \times 2 \times 15) +$

$(10 \times 4 \times 5)$

Figure 6.14 Vertical channel.

$100 +$

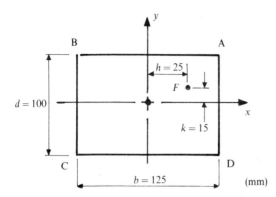

(mm)   **Figure 6.15** Short channel.

The greatest (compressive) stress $\sigma = -77\,\text{MPa}$ occurs all along the flange tips where $x = 66.07\,\text{mm}$. Then from Eq. (6.19b),

$$F = \sigma/(1/A + hx/I_y)$$
$$= -77/[1/8750 + (66.07 \times 66.07)/(7.3754 \times 10^6)] = -109\,\text{kN}.$$

To determine h, with $x_Q = -33.93\,\text{mm}$, Eq. (6.19b) gives

$$\sigma_Q = 0 = 1/A + hx/I_y$$
$$\therefore \quad h = -I_y/Ax = -7.3754 \times 10^6/[8750 \times (-33.93)] = 24.84\,\text{mm}.$$

Example 6.8 The short column of rectangular section in Fig. 6.15 has a 200 kN vertical compressive force applied at F. Calculate the resultant stress at each corner.

$$A = bd = 125 \times 100 = 12.5 \times 10^3\,\text{mm}^2;$$
$$I_x = bd^3/12 = 125(100)^3/12 = 10.4 \times 10^6\,\text{mm}^4;$$
$$I_y = db^3/12 = 100(125)^3/12 = 16.28 \times 10^6\,\text{mm}^4;$$
$$h = 25\,\text{mm}, k = 15\,\text{mm}, F = -200 \times 10^3\,\text{N}.$$

Then for corners $A(62.5, 50)$, $B(-62.5, 50)$, $C(-62.5, -50)$, $D(62.5, -50)$ the direct application of Eq. (6.19a) gives

$$\sigma_A = -200 \times 10^3[(1/12.5 \times 10^3) + (25 \times 62.5/16.28 \times 10^6)$$
$$+ (15 \times 50/10.4 \times 10^6)]$$
$$= -200(0.08 + 0.09\,598 + 0.0721) = -49.62\,\text{MPa};$$
$$\sigma_B = -200(0.08 - 0.09\,598 + 0.0721) = -11.22\,\text{MPa};$$
$$\sigma_C = -200(0.08 - 0.09\,598 - 0.0721) = +17.64\,\text{MPa};$$
$$\sigma_D = -200(0.08 + 0.09\,598 - 0.0721) = -20.78\,\text{MPa}.$$

6.4 ELASTIC BENDING OF COMPOSITE BEAMS

When a beam is fabricated from more than one material the ETB is adapted to the particular cross-section by ensuring that the strains between materials are compatible and that force and moment equilibrium is obeyed. Four common types of cross-section arise.

Balanced Vertical Layers

Assuming materials A and B (Fig. 6.16) to be firmly secured at their common interfaces, they will attain the same radius of curvature under an applied moment M. This approach is also valid for composite beam sections made up of two thin horizontal layers, either bonded or unbonded, in which their individual neutral axes approximate to a common radius of curvature (see Example 6.11. From Eq. (6.4),

$$R = (EI/M)_A = (EI/M)_B, \qquad \Rightarrow M_A = M_B(EI)_A/(EI)_B, \qquad (6.20)$$

$$\sigma_A = (My/I)_A, \qquad \sigma_B = (My/I)_B. \qquad (6.21a, b)$$

The total moment carried by the section (the moment of resistance) is

$$M = M_A + M_B = M_A[1 + (EI)_B/(EI)_A] = M_B[1 + (EI)_A/(EI)_B. \qquad (6.22a, b)$$

Example 6.9 Given that the allowable stresses for steel (A) and wood (B) are 140 and 8.3 MPa respectively, determine the maximum moment for the section in Fig. 6.17. $E_A = 200$ GPa and $E_B = 8.3$ GPa.

$$I_B = (bd^3/12)_B = 200 \times 300^3/12 = 450 \times 10^6 \, \text{mm}^4,$$
$$I_A = 2(bd^3/12)_A = 2 \times 10 \times 300^3/12 = 45 \times 10^6 \, \text{mm}^4.$$

When A and B are fully stressed, Eq. (6.21a, b) gives

$$M_A = (\sigma I/y)_A = 140 \times 45 \times 10^6/150 = 42 \, \text{kN m},$$
$$M_B = (\sigma I/y)_B = 8.3 \times 450 \times 10^6/150 = 24.9 \, \text{kN m}.$$

Then, from Eq. (6.20), when the wood is fully stressed the steel

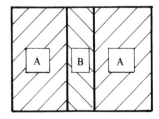

Figure 6.16 Composite beam.

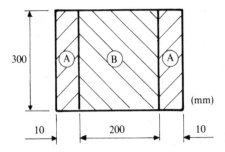

300

10   200   10

(mm)

Figure 6.17 Composite section.

moment is

$$M_A = 24.9(200 \times 10^3 \times 45 \times 10^6)/(8.3 \times 10^3 \times 450 \times 10^6) = 60\,\text{kN m},$$

which indicates that the steel is overstressed. When steel is fully stressed the wood moment is, from Eq. (6.20),

$$M_B = 42(8.3 \times 10^3 \times 450 \times 10^6)/(200 \times 10^3 \times 45 \times 10^6) = 17.43\,\text{kN m},$$

indicating that wood is understressed. That is, for this condition;

$$\sigma_B = (My/I)_B = 17.43 \times 10^6 \times 150/450 \times 10^6 = 5.81\,\text{MPa}.$$

Then, from Eq. (6.22), the maximum moment is,

$$M = M_A + M_B = 42 + 17.43 = 59.43\,\text{kN m}.$$

Horizontal Layers

It is convenient here to work in a section of one material. When A is the stiffer of the two materials, i.e. $E_A > E_B$ in Fig. 6.18(a), the equivalent section in material A becomes that of a tee (Fig. 6.18(b)). One requirement is that when B is replaced by A, all fibres at the same depth suffer the same strain. That is, when σ_A and σ_B act on an elemental area $\delta A = a\,\delta y$ and $\delta B = b\,\delta y$

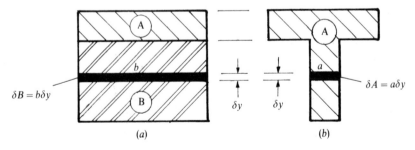

Figure 6.18 Composite beam in A and B with the equivalent section in A.

as shown,

$$\epsilon = (\sigma/E)_A = (\sigma/E)_B, \quad \Rightarrow \sigma_B = \sigma_A E_B/E_A. \tag{6.23}$$

Furthermore, the moments produced about the NA by the forces $\sigma_A \delta A$ and $\sigma_B \delta B$ must be the same between the original and equivalent part of the section. That is,

$$\sigma_A(a\,\delta y)y = \sigma_B(b\,\delta y)y = \sigma_A(E_B/E_A)(b\,\delta y)y.$$

Since the depth limits for y are identical, the equivalent width becomes

$$a = b(E_B/E_A). \tag{6.24}$$

ETB then supplies the stress anywhere in the equivalent section A as

$$\sigma_A = (My/I)_A, \tag{6.25}$$

while the actual stress in material B follows from Eqs (6.23) and (6.25) as

$$\sigma_B = (E_B/E_A)(My/I)_A. \tag{6.26}$$

Dividing Eqs (6.25) and (6.26), the maximum σ_B is found for a given M as

$$\sigma_B/\sigma_A = (E_B/E_A)(My/I)_A(I/My)_A \equiv (E_B y_{AB}/E_A y_A), \tag{6.27}$$

where y_{AB} is the position in A at which the stress in B is a maximum.

Example 6.10 Find the maximum allowable bending moment for the section in Fig. 6.19(a), given maximum permissible stresses of 150 MPa for steel (A) and 8.2 MPa for wood (B), with respective elastic moduli 210 and 8.5 GPa.

Using Eq. (6.24), replace the wood in Fig. 6.19(a) by the equivalent width in steel, $a = 200 \times 8.5/210 = 8.1$ mm (Fig. 6.19(b)). Then, using

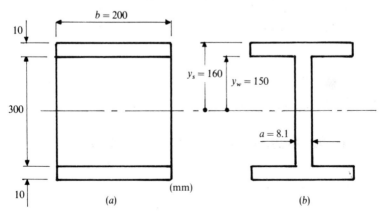

Figure 6.19 Equivalent section.

$$I = bd^3/12,$$

$$I = 200 \times 320^3/12 - 191.9 \times 300^3/12 = 114.35 \times 10^6 \, \text{mm}^4.$$

For fully stressed steel (A) at the outer fibres, Eq. (6.25) gives

$$M = (\sigma I/y)_A = 150 \times 114.35 \times 10^6/160 = 107.2 \, \text{kN m}.$$

If the wood (B) is to be fully stressed at the interface, then Eq. (6.26) gives the corresponding moment that would need to be applied to the equivalent section in Fig. 6.19(b). That is,

$$M = (E_A/E_B)(I/y)_A \sigma_B = (210/8.5)(114.35/150)8.2 = 154.44 \, \text{kN m}.$$

Selecting the lesser M value ensures that the steel is fully stressed while the wood is understressed to a maximum value (Eq. (6.27)) of

$$\sigma_B = 150[(8.5 \times 150)/(210 \times 160)] = 5.69 \, \text{MPa}.$$

Example 6.11 A rectangular $38 \times 10 \, \text{mm}$ bar of steel and one of brass form a composite $38 \times 20 \, \text{mm}$ rectangular section of a beam (Fig. 6.20(a)), which is simply supported over a length of 750 mm. Find the maximum central concentrated load (W) when the brass lies below the steel, if, with allowable stresses of 70 and 105 MPa respectively, (i) each beam can bend independently and (ii) they are bonded along the interface. Take $E_B = 84 \, \text{GPa}$ for brass, $E_A = 207 \, \text{GPa}$ for steel.

In the unbonded case (i), as each beam can bend separately, the maximum moments for the given allowable stresses are, from Eqs (6.21a and b),

$$M_A = (\sigma I/y)_A = 105 \times (38 \times 10^3/12)/5 = 66.5 \, \text{N m};$$
$$M_B = (\sigma I/y)_B = 70 \times (38 \times 10^3/12)/5 = 44.3 \, \text{N m}.$$

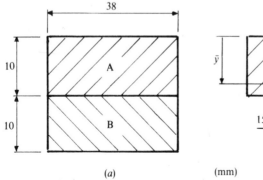
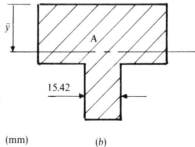

Figure 6.20 Rectangular composite section.

For the same interface radius of curvature and $I_A = I_B$, Eq. (6.20) gives

$$M_A = M_B(E_A/E_B)$$
$$M_A = 44.3(207/84) = 109.3 \, \text{N m}.$$

This indicates that when the brass is fully stressed the steel will be overstressed. Therefore, the requirement is that the steel be fully stressed and the brass understressed. The moment for the brass is

$$M_B = M_A(E_B/E_A) = 66.5(84/207) = 27 \, \text{N m}.$$

Now, from Eq (6.22), the total moment carried by the section at the central load is given as

$$M = 66.5 + 27 = 93.5 \, \text{N m},$$
$$\therefore \quad W = 4M/L = 4 \times 93.5 \times 10^3/750 = 498.5 \, \text{N}.$$

The maximum stress reached in the brass is

$$\sigma_B = (My/I)_B = (27 \times 10^3 \times 5 \times 12)/(38 \times 10^3) = 42.62 \, \text{MPa}.$$

In the bonded case (ii), the equivalent width of steel is, from Eq. (6.24), $a = 38 \times 84/207 = 15.42$ mm. Then, for the equivalent steel section (Fig. 6.20(b)) the centroidal position is found from

$$(38 \times 10 \times 5) + (15.42 \times 10 \times 15) = (380 + 154.2)\bar{y}, \quad \Rightarrow \bar{y} = 7.89 \, \text{mm}.$$

Then using $I = bd^3/3$ for the second moment of area:

$$I = (38 \times 7.89^3/3) + (22.58 \times 2.11^3/3) + (15.42 \times 12.11^3/3)$$
$$= 15.42 \times 10^3 \, \text{mm}^4.$$

With the steel fully stressed along its upper fibres, Eq. (6.25) gives the associated moment:

$$M = (\sigma I/y)_A = 105 \times 15.42 \times 10^3/7.89 = 205.2 \, \text{N m}.$$

From Eq. (6.26) the maximum stress in the brass will then be

$$\sigma_B = (E_B/E_A)(My/I)_A = (84/207)[(205.2 \times 10^3 \times 12.11)/(15.42 \times 10^3)]$$
$$= 65.33 \, \text{MPa}.$$

As the brass is understressed, this condition governs the design when

$$W = 4M/L = 4 \times 205.2 \times 10^3/750 = 1094.4 \, \text{N}.$$

Bimetallic Strip

When two thin, straight strips of dissimilar metals 1 and 2 are joined along their longer sides to form a beam, a temperature change, ΔT, will result in bending and deflection. This principle is employed in certain thermostats.

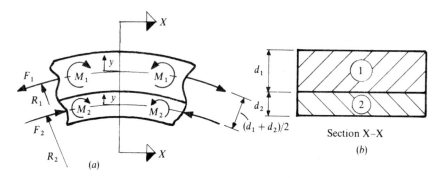

Figure 6.21 Bimetallic strip.

Taking the elastic moduli $E_1 > E_2$, the linear expansion coefficients will obey $\alpha_1 < \alpha_2$ and the beam will bend as shown in Fig. 6.21(a). Longitudinal forces F_1 and F_2 arise owing to restrained thermal expansion. These are tensile and compressive, respectively, because the beam fibres in 1 and 2 are stretched by greater and lesser amounts than their free expansions. The consideration of horizontal force equilibrium normal to a cross-section in Fig. 6.21(b) requires that

$$F_1 = F_2 = F.$$

Now, since there are no external moments, these forces produce a couple that must be resisted by internal moments M_1 and M_2 acting in the opposite sense. That is,

$$M_1 + M_2 = F(d_1 + d_2)/2. \tag{6.28}$$

In any fibre, the total longitudinal strain is composed of a direct strain (F/AE) due to F, a bending strain (My/EI) due to M and a temperature strain $(\alpha l\,\Delta T/l = \alpha\,\Delta T)$ due to ΔT. Since the total strain in 1 and 2 must be compatible at the interface, it follows from the F and M directions shown that,

$$F_1/A_1 E_1 + [M_1(d_1/2)/E_1 I_1] + \alpha_1\,\Delta T$$
$$= -F_2/A_2 E_2 - [M_2(d_2/2)/E_2 I_2] + \alpha_2\,\Delta T. \tag{6.29}$$

Moreover, assuming that the radii of curvature of the central neutral axes of each strip are the same, Eq. (6.20) supplies the relationship

$$M_2 = M_1 E_2 I_2/E_1 I_1. \tag{6.30}$$

Then, combining Eqs (6.28) and (6.30)

$$M_1 = FE_1 I_1(d_1 + d_2)/2(E_1 I_1 + E_2 I_2). \tag{6.31}$$

The behaviour of the strip is thus reduced to the simultaneous solution of F from Eqs (6.29)–(6.31). The axial stress at distance y from the NA in

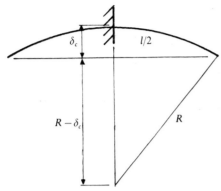

Figure 6.22 Strip deflection.

either material is

$$\sigma = \pm F/A \pm My/I + \alpha \Delta TE. \tag{6.32}$$

The corresponding central deflection δ_c, of the strip when it is simply supported at its ends (Fig. 6.22) is found from

$$R^2 = (R - \delta_c)^2 + (l/2)^2.$$

Neglecting the small quantity δ_c^2

$$\delta_c \simeq l^2/8R$$

where

$$R = R_1 = R_2 = (EI/M)_1 = (EI/M)_2.$$

The deflection δ_e, at the free end of a bimetallic cantilever of length L is found from δ_c by putting $l = 2L$, i.e. the deflection at the end relative to the centre in a simply supported beam (Fig. 6.22). Then

$$\delta_e = L^2/2R.$$

Example 6.12 Calculate the stresses induced at the free and interface surfaces when the copper–steel bimetallic strip in Fig. 6.23(a) is subjected

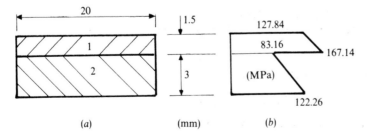

(a)   (mm)   (b)

Figure 6.23 Bimetallic strip.

to a temperature rise of 60 °C. For Cu, $E = 103$ GPa and $\alpha = 17.5 \times 10^{-6}/°C$. For steel, $E = 207$ GPa and $\alpha = 11 \times 10^{-6}/°C$.

Identifying 1 with steel and 2 with copper ($E_1 > E_2$):

$$A_1 = 30\,\text{mm}^2, \qquad I_1 = b_1 d_1^3/12 = 20 \times 1.5^3/12 = 5.625\,\text{mm}^4;$$
$$A_2 = 60\,\text{mm}^2, \qquad I_2 = b_2 d_2^3/12 = 20 \times 3^3/12 = 45\,\text{mm}^4.$$

Equations (6.30) and (6.31) give the moments:

$$M_1 = F \times 207 \times 5.625(1.5 + 3)/2[(207 \times 5.625) + (103 \times 45)]$$
$$= 0.4518\,F,$$
$$M_2 = 0.4518F \times 103 \times 45/207 \times 5.625 = 1.7983F.$$

Since the directions of F and M correspond with those given in Fig. 6.21(a), Eq. (6.29) may be applied directly to find F and M:

$$F(1/A_1E_1 + 1/A_2E_2) + F(0.4518d_1/2E_1I_1 + 1.7983d_2/2E_2I_2)$$
$$= (\alpha_1 - \alpha_2)\,\Delta T;$$
$$\therefore \quad F = \frac{(17.5 - 11)60 \times 10^{-6}}{(0.3228 \times 10^{-6}) + (0.873 \times 10^{-6})} = 326.13\,\text{N};$$
$$M_1 = 147.35\,\text{N mm}, \qquad M_2 = 586.48\,\text{N mm}.$$

The stresses then follow Eq. (6.32). At the steel interface, where M_1 produces tension,

$$\sigma_1 = E_1\epsilon_1 = F/A_1 + M_1d_1/2I_1 + \alpha_1\,\Delta T\,E_1$$
$$= (326.13/30) + (147.35 \times 1.5/2 \times 5.625) + (11 \times 10^{-6} \times 60 \times 207 \times 10^3)$$
$$= 10.87 + 19.65 + 136.62 = 167.14\,\text{MPa}.$$

At the free steel surface (M_1 produces compression),

$$\sigma_1 = F/A_1 - M_1d_1/2I_1 + \alpha_1\,\Delta T\,E_1$$
$$= 10.87 - 19.65 + 136.62 = 127.84\,\text{MPa}.$$

At the copper interface (M_2 produces compression),

$$\sigma_2 = E_2\epsilon_2 = -F/A_2 - M_2d_2/2I_2 + \alpha_2\,\Delta T\,E_2$$
$$= -(326.13/60) - (586.48 \times 3)/(2 \times 45) + (17.5 \times 10^{-6} \times 60 \times 103 \times 10^3)$$
$$= -5.44 - 19.55 + 108.15 = 83.16\,\text{MPa}.$$

At the free copper surface (M_2 produces tension),

$$\sigma_2 = -F/A_2 + M_2d_2/2I_2 + \alpha_2\,\Delta T\,E_2$$
$$= -5.44 + 19.55 + 108.15 = 122.26\,\text{MPa}.$$

The resulting distribution in stress through the section is shown in

Fig. 6.23(b), where strain compatibility at the interface is accommodated by a stress discontinuity at this position.

Reinforced Sections (Steel in Concrete)

Though the following theory is concerned with reinforced concrete, it should also apply anywhere a line of reinforcing rods are inserted into a section of parent material that is inherently weak in bending either on the tensile or compressive sides of the neutral axis. Tensile weakness applies to concrete in each of the following sections.

Rectangular section with a single steel line The presence of steel of total area A_s on the tensile side of the section of breadth B, and steel line depth D, will modify the NA to a non-central position h in Fig. 6.24(a). Since the steel strain ϵ_S, and the tensile strain in the concrete ϵ_{ct}, are compatible at the interface (Fig. 6.24(b)) the corresponding stresses σ_S and σ_{ct} (Fig. 6.24(c)) are found from

$$\epsilon_S = \epsilon_{ct},$$
$$\therefore \quad \sigma_S = (E_S/E_c)\sigma_{ct} = m\sigma_{ct}, \tag{6.33}$$

where $m = E_S/E_c$ is the modular ratio. Now, from Fig. 6.24(c),

$$\sigma_{ct}/(D - h) = \sigma_c/h, \tag{6.34}$$

when, from Eqs (6.33) and (6.34), h is found from

$$\sigma_S = m(D - h)\sigma_c/h, \tag{6.35}$$

where σ_S and σ_c are the maximum allowable stresses for the steel and concrete respectively. If the section is to be designed economically, then A_S is found

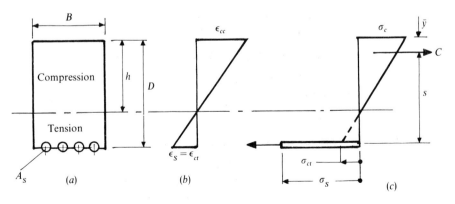

Figure 6.24 Single-line steel reinforcement.

from the balance between the average compressive force, $C = \sigma_c Bh/2$, acting at the centroid ($\bar{y} = h/3$) of the stress distribution for the concrete area above the NA, and the total tensile force $T = \sigma_s A_s$, in the steel. This neglects any tensile stress in the concrete (broken line in Fig. 6.24(c)) to give simply

$$\sigma_c Bh/2 = \sigma_s A_s,$$
$$\therefore \quad A_s = (Bh/2)(\sigma_c/\sigma_s). \tag{6.36}$$

If $s = D - \bar{y}$ is the distance between C and T then because $C = T$, the moment of resistance M, is given by either one of the expressions

$$M_c = Cs = \sigma_c(Bh/2)(D - h/3), \tag{6.37a}$$
$$M_t = Ts = A_s \sigma_s(D - h/3). \tag{6.37b}$$

With an uneconomic section, A_s will not obey Eq. (6.36). When A_s is specified, then h is again found from the zero axial force condition $C = T$ where Eq. (6.35) applies with the concrete understressed. That is,

$$\sigma_c Bh/2 = A_s[m(D - h)\sigma_c/h]$$

from which the following quadratic in h results:

$$Bh^2 + 2A_s mh - 2A_s mD = 0. \tag{6.38}$$

Example 6.13 Determine the moment of resistance for the section in Fig. 6.25 when the steel reinforcement (i) consists of four bars each of 10 mm (1/2 in) diameter and (ii) is chosen economically based upon allowable stresses in the steel and concrete of 125 MPa (18 000 lbf/in²) and 4.5 MPa (600 lbf/in²) respectively. Find the maximum uniformly distributed load (UDL) an 8 m (30 ft) length simply supported beam in each section could carry. Take $m = 15$.

SOLUTION TO SI UNITS
   (i) Uneconomic design

$$A_s = 4 \times \pi(10)^2/4 = 314.14 \text{ mm}^2.$$

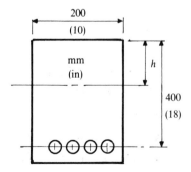

Figure 6.25 Reinforced section.

Then, from Eq. (6.38) the NA position (h) is

$$200h^2 + (2 \times 314.14 \times 15)h - (2 \times 314.14 \times 15 \times 400) = 0,$$
$$h^2 + 47.12h - 18\,848.4 = 0, \quad \Rightarrow h = 115.74\,\text{mm}.$$

M is then the lesser of Eqs 6.37(a) and (b):

$$M_c = Cs = \sigma_c(Bh/2)(D - h/3)$$
$$= 4.5(200 \times 115.74/2)(400 - 115.74/3) \times 10^{-6} = 18.82\,\text{kN m},$$
$$M_t = Ts = A_S\sigma_S(D - h/3)$$
$$= 314.14 \times 125(400 - 115.74/3) \times 10^{-6} = 14.19\,\text{kN m}.$$

This ensures that the steel is fully stressed while the concrete is under-stressed. That is, since $M_c = M_t = 14.19\,\text{kN m}$, Eq. 6.37(a) gives

$$\sigma_c = M_t/(Bh/2)(D - h/3)$$
$$= (2 \times 14.19 \times 10^6)/\{200 \times 115.74[400 - (115.74/3)]\} = 3.39\,\text{MPa}.$$

Then UDL $= 8M_t/l^2 = 8 \times 14.19 \times 10^6/(8000)^2 = 1.77\,\text{N/mm (kN/m)}$.

(ii) Economic design. With both steel and concrete fully stressed, Eq. (6.35) defines the NA position h, as

$$125 = 15(200 - h)4.5/h, \quad \Rightarrow h = 140.26\,\text{mm}.$$

Then the steel area is supplied by Eq. (6.36):

$$A_S = 4.5 \times 200 \times 140.26/(2 \times 125) = 504.94\,\text{mm}^2,$$

and the moment of resistance by Eq. (6.37b):

$$M = 504.94 \times 125(400 - 140.26/3) = 22.3\,\text{kN m}.$$

$$\text{Then UDL} = 8 \times 22.3 \times 10^6/(8000)^2 = 2.79\,\text{k N/m}.$$

SOLUTION TO IMPERIAL UNITS
(i) Uneconomic design

$$A_S = (\pi/4)(\tfrac{1}{2})^2 \times 4 = 0.785\,\text{in}^2.$$

Substituting into Eq. (6.38) gives the quadratic

$$h^2 + 2.36h - 42.4 = 0,$$

from which $h = 5.43\,\text{in}$. M is then the lesser of

$$M_c = (600 \times 10 \times 5.43/2)(18 - 5.43/3) = 263.5 \times 10^3\,\text{lbf in}.$$
$$M_t = (18\,000 \times 0.785)(18 - 5.43/3) = 229 \times 10^3\,\text{lbf in}.$$

Thus, the steel is fully stressed. The concrete stress is

$$\sigma_c = 2 \times 229 \times 10^3/\{10 \times 5.43[18 - (5.43/3)]\} = 522\,\text{lbf/in}^2,$$
$$\text{UDL} = 8M_t/l^2 = 8 \times 229 \times 10^3/(30 \times 12)^2 = 14.14\,\text{lbf/in} = 169.63\,\text{lbf/ft}.$$

(ii) Economic design, Eq. (6.35) gives the NA position for this case as

$$h = 15 \times 600 \times 18/[18\,000 + (15 \times 600)] = 6\,\text{in.}$$

The steel area is, from Eq. (6.36),

$$A_S = 600 \times 10 \times 6/(2 + 18\,000) = 1\,\text{in}^2,$$

and finally from Eq. (6.37):

$$M = M_c = M_t = 1 \times 18\,000(18 - 6/3) = 288 \times 10^3\,\text{lbf in,}$$
$$\text{UDL} = 8 \times 288 \times 10^3/(30 \times 12)^2 = 17.78\,\text{lbf/in} = 213.33\,\text{lbf/ft.}$$

T-section With a single line of reinforcement at the web bottom (Fig. 6.26(a)) it is seen that the same strain and stress distributions (Fig. 6.26(c) and (d)) apply as with the rectangular section.

Here, however, two positions of the NA are possible.

1. NA in the flange—Fig. 6.26(a). All the previous relationships in Eqs (6.33)–(6.38) apply, since $\bar{y} = h/3$.
2. NA in the web—Fig. 6.26(b). Equation (6.37a) is invalid, since C now acts upon the area above the NA. That is, with $C = T$,

$$(\sigma_c/2)[Bd + b(h - d)] = \sigma_S A_S = m A_S \sigma_c (D - h)/h,$$
$$bh^2 + h(Bd - bd + 2m A_S) - 2m A_S D = 0. \qquad (6.39)$$

The moments of resistance expressions are also modified to

$$M_c = Cs = (\sigma_c/2)[Bd + b(h - d)](D - \bar{y}), \qquad (6.40a)$$
$$M_t = Ts = \sigma_S A_S(D - \bar{y}), \qquad (6.40b)$$

where $\bar{y}$ is the centroid of the compressive area, and is found from

$$[Bd + (h - d)b]\bar{y} = Bd(d/2) + b(h - d)[d + (h - d)/2]. \qquad (6.41)$$

Again, Eqs (6.40a and b) may be equated for economical design. Otherwise, selecting the lesser value avoids overstressing the concrete.

Example 6.14 Find the moment of resistance for the T-section in Fig. 6.27 and establish the magnitude of the stresses in the steel and concrete given respective allowable values of 125 and 7 MPa. Take $m = 15$.

First the position of the NA must be determined. Normally, when A_S is given, this implies an uneconomical design. Then, if the NA lies in the

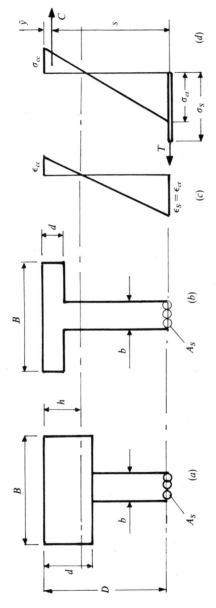

Figure 6.26 Reinforced T-section.

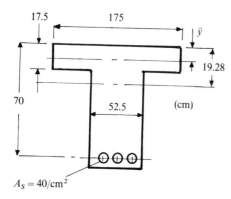

$A_S = 40/\text{cm}^2$

Figure 6.27 Reinforced T-section.

flange, Eq. (6.38) gives

$$175h^2 + (2 \times 40 \times 15)h - (2 \times 40 \times 15 \times 70) = 0,$$
$$h^2 + 6.857h - 480 = 0, \quad \Rightarrow h = 18.75\,\text{cm}.$$

Hence the NA must lie in the web when Eq (6.39) gives

$$52.5h^2 + [(175 \times 17.5) - (52.5 \times 17.5)$$
$$+ (2 \times 15 \times 40)]h - (2 \times 15 \times 40 \times 70) = 0,$$
$$h^2 + 63.69h - 1600 = 0, \quad \Rightarrow h = 19.28\,\text{cm}.$$

Note from Eq. (6.35) that an economic design would require $h \simeq 32\,\text{cm}$ for the given allowable stresses. The centroid $\bar{y}$, is found from Eq. (6.41):

$$(175 \times 17.5 \times 8.75) + (1.78 \times 52.5 \times 18.39) = 3155.95\bar{y}, \quad \Rightarrow \bar{y} = 9.026\,\text{cm}.$$

The moment of resistance is the lesser of Eqs (6.40a and b):

$$M_c = (7/2)[(175 \times 17.5) + 52.5(19.28 - 17.5)](70 - 9.026) \times 10^3/10^6$$
$$= 673.51\,\text{kN m},$$
$$M_t = 125 \times 40(70 - 9.026)10^{-3} = 304.87\,\text{kN m}.$$

By selecting the latter the steel is fully stressed while the concrete stress is found either from Eq. (6.40a) with $M_c = M_t$,

$$\sigma_c = \frac{2M_t}{[Bd + b(h-d)](D - \bar{y})} = \frac{2 \times 304.87 \times 10^6}{192\,431.43 \times 10^3}$$
$$= 3.17\,\text{MPa},$$

or by the proportion of the allowable concrete stress, σ_{ca},

$$\sigma_c = (M_t/M_c)\sigma_{ca} = (304.87/673.51)7 = 3.17\,\text{MPa}.$$

Rectangular section with double reinforcement The stress and strain distributions in the steel (S) and concrete (c) with a line of reinforcement on

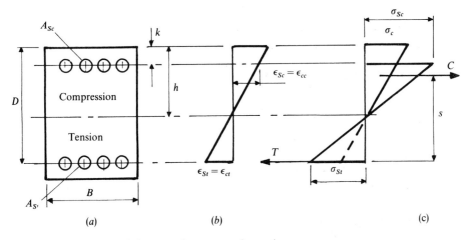

Figure 6.28 Double reinforcement in a rectangular section.

the tensile (t) and compressive (c) sides are shown in Fig. 6.28(a)–(c). Let σ_{St} and σ_{Sc} be the respective tensile and compressive stresses in the steel lines and σ_c be the maximum concrete stress at the compressive surface. The strains in the steel and concrete at the tensile interface are denoted by ϵ_{St} and ϵ_{ct} and at the compressive interface by ϵ_{Sc} and ϵ_{cc} respectively. (S (steel) and c (concrete) are the first subscript; t (tensile) and c (compressive) are the second subscripts.)

Strain compatibility at each interface provides the relationships

$$\epsilon_{St} = \epsilon_{ct}, \quad \Rightarrow \sigma_{St} = m\sigma_c(D - h)/h, \tag{6.42a}$$

$$\epsilon_{Sc} = \epsilon_{cc}, \quad \Rightarrow \sigma_{Sc} = m\sigma_c(h - k)/h. \tag{6.42b}$$

Now, for an economic section with the same maximum allowable tensile and compressive steel stresses ($\sigma_{St} = \sigma_{Sc} = \sigma_S$), it follows from Eqs (6.42a and b) that $D - h = h - k$. That is, the NA passes through the centroid. Assuming the same steel area A_S on the tensile and compressive sides, this area is found from equating horizontal forces C and T, where these are

$$C = \sigma_c Bh/2 + m(h - k)\sigma_c A_S/h - (h - k)\sigma_c A_S/h$$
$$= \sigma_c[Bh/2 + (m - 1)A_S(h - k)/h], \tag{6.43}$$

$$T = A_S\sigma_S = A_S m\sigma_c(D - h)/h. \tag{6.44}$$

Since $M_t = M_c$ for an economic section, the moment of resistance is found from either the concrete or the steel. That is,

$$M_t = Ts = A_S\sigma_S s, \tag{6.45a}$$

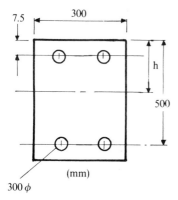

Figure 6.29 Reinforced section.

and by taking moments about the tensile steel line:

$$M_c = Cs = (D - k)A_S\sigma_S - (D - k)A_S(h - k)\sigma_c/h + (\sigma_c Bh/2)(D - h + 2h/3)$$
$$= (D - k)A_S m(h - k)\sigma_c/h - (D - k)A_S(h - k)\sigma_c/h + (D - h/3)Bh\sigma_c/2$$
$$= \sigma_c(D - k)[(Bh/2)(D - h/3)/(D - k) + (m - 1)A_S(h - k)/h]. \qquad (6.45b)$$

The moment arm s, is found from substituting Eq. (6.43) into Eq. (6.45b). If the steel areas on each side are different, it becomes necessary to solve Eqs (6.43)–(6.45) simultaneously for h and for each area A_{Sc} and A_{St} using $C = T$ and $M_t = M_c$ if the section is to remain economic. When the same steel areas are employed with an uneconomic section, the position h of the NA follows from equating Eqs (6.43) and (6.44). This gives the following quadratic in h, from which the moment of resistance is the lesser of Eqs (6.45a and b):

$$BH^2 + 2(h - k)(m - 1)A_S - 2mA_S(D - h) = 0. \qquad (6.46)$$

Example 6.15 Determine the moment of resistance for the reinforced concrete section in Fig. 6.29 given $\sigma_c = 6$ MPa, $\sigma_S = 125$ MPa and $m = 15$. Find the maximum stress in the understressed material.

With $B = 300$ mm, $D = 500$ mm, $k = 75$ mm and $A_S = 2(\pi 30^2)/4 = 1413.72$ mm^2 the NA position is found from Eq. (6.46):

$$h^2 + 273.32h - 80\,582.1 = 0, \quad \Rightarrow h = 178.39 \text{ mm}.$$

For the lever arm s, Eq. (6.43) gives

$$C = 6[(300 \times 178.39/2) + (15 - 1)1413.72(178.39 - 75)/178.39]$$
$$= 229.38 \text{ kN},$$

and from Eq. (6.45b) the concrete moment is

$$M_c = Cs = 6(500 - 75)[(300 \times 178.39/2)(500 - 178.39/3)/(300 - 75)$$
$$+ (14 \times 1413.73)(178.39 - 75)/178.39] = 99.98 \text{ kN m}.$$

$$\therefore \quad s = M_c/C = 99.98 \times 10^3/229.38 = 435.88 \, \text{mm}.$$

Note that as σ_c cancels in equating (6.43) to (6.45b) s is independent of the concrete stress actually achieved. The lesser M is, from Eq. (6.45a),

$$M_t = 1413.72 \times 125 \times 435.88 = 77.03 \, \text{kN m},$$

which indicates that the steel is fully stressed in tension. In compression the steel stress is $(h - k)\sigma_S/(D - h) = 40.19 \, \text{MPa}$ and the maximum compressive concrete stress $= 6(77.03/99.89) = 4.62 \, \text{MPa}$. The latter may also be checked from Eq. (6.42a), where

$$\sigma_c = \sigma_S h/m(D - h) = 125 \times 178.39/15(500 - 178.39) = 4.62 \, \text{MPa}.$$

The reader should confirm that the removal of the compressive steel line results in $M = 76.3 \, \text{kN m}$ ($h = 204.4 \, \text{mm}$), which only slightly impairs the bending strength of this section. With an economic design, however, the area of an additional compressive steel line may be chosen to fully stress the steel on both sides and so improve the resistive moment.

6.5 ASYMMETRIC BENDING

The ETB has so far been applied where the moment axes for a given section have coincided with its principal axes. In beam bending, for example, this arises when the NA is a horizontal axis of symmetry about which the bending moment has been applied. In the more general case (see Fig. 6.30) when the moment axis x in beam bending, or axes x and y in eccentric column bending, do not align with the principal axes (u and v) for the section, it becomes necessary to resolve the applied moments M_x and M_y into M_u and M_v. Let M_x and M_y be hogging w.r.t. the positive x, y quadrant, then in the positive u, v quadrant the right-hand screw rule gives the net hogging bending moments:

$$M_u = M_x \cos \theta - M_y \sin \theta, \tag{6.47a}$$

$$M_y = M_x \sin \theta + M_y \cos \theta, \tag{6.47b}$$

where θ is the inclination between positive u and x. The bending stress σ, at any point $P(u, v)$ is then

$$\sigma = M_u v/I_u + M_v u/I_v, \tag{6.48}$$

in which the determination of the principal quantities u, v, θ, I_u and I_v from x, y, I_x, I_y and I_{xy} have been outlined in Chapter 1. The signs all follow from those positive directions given in Fig. 6.30. Note that M_x and M_y in Eq. (6.47) are negative when sagging w.r.t. the first quadrant.

Example 6.16 Calculate the greatest stresses induced at points A, B and C when the equal-angle section in Fig. 6.31 is mounted as a cantilever 1 m long carrying a vertical and load of 1 kN through its centroid g.

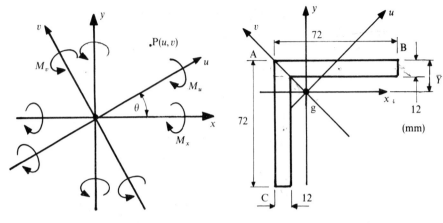

Figure 6.30 Axes of bending.

Figure 6.31 Equal-angle section.

First moments of area about AB give

$$(72 \times 12 \times 6) + (60 \times 12 \times 42) = [(72 \times 12) + (60 \times 12)]\bar{Y}$$
$$\therefore \quad \bar{Y} = 22.4 \text{ mm} = \bar{X}$$

and the second moments of area are (using $bd^3/3$)

$$I_x = I_y = 72(22.4)^3/3 - 60(10.4)^3/3 + 12(49.6)^3/3$$
$$= 73.46 \times 10^4 \text{ mm}^4,$$
$$I_{xy} = (72 \times 12)(+13.6)(+16.4) + (60 \times 12)(-19.6)(-16.4)$$
$$= 42.41 \times 10^4 \text{ mm}^4.$$

Then from Eqs (1.4) and (1.5), $\tan 2\theta = \infty$, $\theta = 45°$ and

$$I_{u,v} = \tfrac{1}{2}(2 \times 73.46 \times 10^4) \pm \tfrac{1}{2}\sqrt{[0 + (4 \times 42.41 \times 10^4)]}$$
$$\therefore \quad I_u = 31.05 \times 10^4 \text{ mm}^4, \qquad I_v = 115.87 \times 10^4 \text{ mm}^4.$$

At the fixed end, $M_x = (1 \times 1) = +1 \text{ kN m}$, $M_y = 0$. Equations (6.47) and (6.48) give

$$M_u = 1 \times \cos 45° - 0 = 0.7071 \text{ kN m},$$
$$M_v = 1 \times \sin 45° + 0 = 0.7071 \text{ kN m}.$$
$$\sigma = (0.7071 \times 10^6)v/(31.05 \times 10^4) + (0.7071 \times 10^6)u/(115.87 \times 10^4)$$
$$= 2.277v + 0.6103u.$$

The stresses follow from the u, v coordinates of points A, B and C. They are, from Eqs (1.2a and b):

A: $x = -22.4 \text{ mm}$, $y = 22.4 \text{ mm}$,
$$u = -22.4 \cos 45° + 22.4 \sin 45° = 0,$$

$$v = 22.4 \cos 45° + 22.4 \sin 45° = 31.68 \,\text{mm},$$
$$\sigma_A = 2.277(31.68) + 0.6103(0) = 72.13 \,\text{MPa}.$$

B: $\qquad x = 49.6 \,\text{mm}, \quad y = 22.4 \,\text{mm},$

$\qquad \therefore \quad u = 50.92 \,\text{mm}, \quad v = -19.24 \,\text{mm},$

$\qquad \sigma_B = 2.277(-19.24) + 0.6103(50.92) = -12.73 \,\text{MPa}.$

C: $\qquad x = -22.4 \,\text{mm}, \quad y = -49.6 \,\text{mm},$

$\qquad \therefore \quad u = -50.92 \,\text{mm}, \quad v = -19.24 \,\text{mm},$

$\qquad \sigma_C = 2.77(-19.24) + 0.6103(-50.92) = -74.87 \,\text{MPa}.$

Example 6.17 If the cross-section in Fig. 6.31 supports a compressive force of 50 kN at the centre of the vertical web, determine the stresses at points A, B and C.

The x, y coordinates of the force point are $(-16.4, -13.6)$. Then the corresponding eccentricities (e_u, e_v) follow from Eq. (1.2):

$$e_u = -16.4/\sqrt{2} - 13.6/\sqrt{2} = -21.22 \,\text{mm},$$
$$e_v = -13.6/\sqrt{2} + 16.4/\sqrt{2} = 1.98 \,\text{mm}.$$

The bending moments about u and v are simply the product of the force and its respective eccentricity. They are, w.r.t. the first u, v quadrant:

$$M_u = -50 \times 1.98 = -99 \,\text{N m (sagging)},$$
$$M_v = -50 \times (-21.22) = 1061 \,\text{N m (hogging)}.$$

The stresses are then, from Eq. (6.48):

$$\sigma = (-99 \times 10^3)v/(31.05 \times 10^4) + (1061 \times 10^3)u/(115.87 \times 10^4)$$
$$= -0.319v + 0.916u \quad (\text{MPa}).$$

$$
\begin{array}{lll}
\text{At A } (0, 31.68), & \qquad & \sigma_A = -10.11 \,\text{MPa}, \\
\text{At B } (50.92, -19.24), & & \sigma_B = 52.78 \,\text{MPa}, \\
\text{At C } (-50.92, -19.24), & & \sigma_C = -40.51 \,\text{MPa}.
\end{array}
$$

Equivalent Moments

In the above examples it was necessary to evaluate I_u, I_v, θ, u and v separately. This may, however, be avoided if θ is first eliminated between Eqs (1.2) and (1.4) and the resulting expressions for u and v are substituted together with Eqs (1.5) and (6.47) into Eq. (6.48). This results in the bending stress expressed in terms of the x, y quantities:

$$\sigma = \hat{M}_x y/I_x + \hat{M}_y x/I_y \tag{6.49a}$$

where the equivalent moments are

$$\hat{M}_x = \frac{M_x - M_y(I_{xy}/I_y)}{1 - I_{xy}^2/I_x I_y},$$ (6.49b)

$$\hat{M}_y = \frac{M_y - M_x(I_{xy}/I_x)}{1 - I_{xy}^2/I_x I_y}.$$ (6.49c)

Example 6.18 A beam with the asymmetric channel section in Fig. 1.16 has bending moments of $M_x = 600\,\text{N m}$ (hogging), $M_y = 400\,\text{N m}$ (sagging) applied w.r.t. positive x, y shown in Fig. 6.32. Determine the maximum tensile and compressive stresses and the inclination α, of the true NA.

It has been shown on page 16 that the c.g. lies at $\bar{Y} = 27.22\,\text{mm}$, $\bar{X} = 9.45\,\text{mm}$, $I_x = 36.35 \times 10^4\,\text{mm}^4$, $I_y = 4.98 \times 10^4\,\text{mm}^4$ and $I_{xy} = 3.89 \times 10^4\,\text{mm}^4$. The equivalent moments are from Eq. (6.49b and c):

$$\hat{M}_x = \frac{600 - (-400)(3.89/4.98)}{1 - 3.89^2/(36.35 \times 4.98)} = 995.60\,\text{N m},$$

$$\hat{M}_y = \frac{-400 - 600(3.89/36.35)}{1 - 3.89^2/(36.35 \times 4.98)} = -506.55\,\text{N m}.$$

The maximum tensile stress will occur at the corner, $P(-10.55, 32.78)$ of the second x, y quadrant where the actions of M_x and M_y are both hogging. Then, from Eq. (6.49a),

$$\sigma_P = (995.6 \times 10^3)(32.78)/(36.35 \times 10^4)$$
$$+ (-506.55 \times 10^3)(-10.55)/(4.98 \times 10^4)$$
$$= 89.79 + 107.3 = 197.09\,\text{MPa}.$$

The maximum compressive stress occurs at the corner, $Q(9.45, -27.22)$

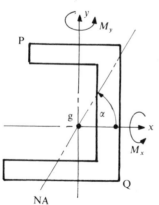

Figure 6.32 Asymmetric channel.

of the fourth x, y quadrant where the actions of M_x and M_y are both sagging.

$$\sigma_Q = [(995.6 \times 10^3)(-27.22)/(36.35 \times 10^4)]$$
$$+ [(-506.55 \times 10^3)(9.45)/(4.98 \times 10^4)]$$
$$= -74.55 - 96.12 = -170.67 \,\text{MPa}.$$

Along the NA it follows from Eq. (6.49a) that

$$\hat{M}_x y / I_x + \hat{M}_y x / I_y = 0,$$
$$[(995.6 \times 10^3)y/(36.35 \times 10^4)]$$
$$+ [(-506.55 \times 10^3)x/(4.98 \times 10^4)] = 0,$$
$$2.739y - 10.172x = 0,$$
$$\therefore \quad \alpha = \tan^{-1}(y/x) = \tan^{-1}(10.172/2.739)$$
$$= 74.93° \text{ (anticlockwise from } x\text{)}.$$

Idealized Sections

Thin-walled beam sections may be idealized into concentrated areas, known as booms, which carry the bending stresses, interconnected with thin webs carrying shear stress only (see Chapter 10). The longitudinal bending stresses in the booms are found from the ETB. In particular, Eq. (6.49) again applies to the case of an asymmetric section.

Example 6.19 The idealized section in Fig. 6.33 is subjected to bending moments $M_x = 10\,\text{kN m}$ (sagging), $M_y = 5\,\text{kN m}$ (hogging) w.r.t. the first x, y quadrant. If the areas of booms A and C are 500 mm² and the areas

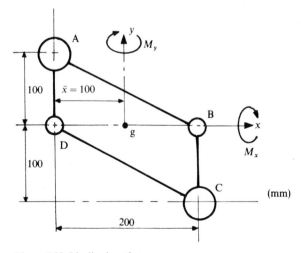

Figure 6.33 Idealized section.

of booms B and D are $200\,mm^2$, find the bending stress in each boom when the webs do not contribute to the total x, y moments of area. (K.P. 1975).

Take first moments of area about a horizontal through C:

$$2(200 \times 100) + (200 \times 500) = 2(200 + 500)\bar{Y}, \quad \Rightarrow \bar{Y} = 100 \text{ mm},$$

and about the vertical passing through A and D:

$$200(500 + 200) = 2(200 + 500)\bar{X}, \quad \Rightarrow \bar{X} = 100\,\text{mm}.$$

For second moment use the form $I = Ah^2$. Then,

$$I_x = 2 \times 500 \times 100^2 = 10 \times 10^6 \text{ mm}^4,$$
$$I_y = 2[(500 \times 100^2) + (200 \times 100^2)] = 14 \times 10^6 \text{ mm}^4,$$
$$I_{xy} = 500(-100)(100) + 500(100)(-100) = -10 \times 10^6 \text{ mm}^4.$$

The equivalent moments are, from Eq. (6.49b and c),

$$M_x = \frac{-10 - 5(-10/14)}{1 - (-10)^2/(10 \times 14)} = -22.5\,\text{kN m},$$

$$M_y = \frac{5 - (-10)(-10/10)}{1 - (-10)^2/(10 \times 14)} = -17.5\,\text{kN m}.$$

The stresses are then given by Eq. (6.49a), i.e. $\sigma = -2.25y - 1.25x$ and

Boom A $(-100, 100)$	$\sigma_A = -225 + 125 = -100\,\text{MPa},$
Boom B $(100, 0)$	$\sigma_B = -125\,\text{MPa},$
Boom C $(100, -100)$	$\sigma_C = 225 - 125 = 100\,\text{MPa},$
Boom D $(-100, 0)$	$\sigma_D = 125\,\text{MPa}.$

6.6 ELASTIC-PLASTIC BENDING OF BEAMS

When an applied moment reaches a particular value (M_Y) the most highly stressed fibres furthest from the NA reach the yield stress (Y) while all other fibres in the cross-section remain elastic. As the moment is increased beyond M_Y, the cross-section becomes partially plastic as the interior fibres approaching the NA successively reach the yield point. This state constitutes an elastic-plastic beam under an applied moment M_{ep}. When the plastic zone has penetrated the whole cross-section on the tensile and compressive sides of the NA, this condition determines the ultimate moment (M_{ult}) a given beam can withstand. Assuming that a beam of any section collapses under M_{ult} and that the beam material behaves in an elastic-perfect plastic manner, in which no increase in stress beyond Y occurs during plastic penetration, the following terms apply.

Shape factor: $Q = M_{ult}/M_Y > 1$, determined solely by the cross-section, i.e. independently of the applied loading.

Load factor: $L = W_{ult}/W$, the ratio between the corresponding collapse load W_{ult}, and the safe elastic working load W, which depends upon the section, the applied loading and the manner of support. When the beam collapses from the formation of a single plastic hinge, the following example shows that Q may then be combined with the definition of the safety factor S in Eq. (3.5) where $\sigma_{max} = Y$, to give $L = QS$.

Axisymmetric Section

Consider a beam with a rectangular section $b \times d$ in Fig. 6.34(a) under a hogging moment. The stress distributions for the transition from elastic to fully plastic behaviour are shown in Figs 6.34(b–d).

Under M_Y in Fig. 6.34(b) the cross-section remains elastic and therefore the ETB applies to the outer fibres, where $\sigma = Y$, $y = d/2$. This gives

$$M_Y = YI/y = Y(bd^3/12)/(d/2) = bd^2 Y/6,$$
$$R = Ey/\sigma = Ed/2Y.$$

Under M_{ep} (Fig. 6.34c), where the plastic zone has penetrated by the amount h from top and bottom edges, the resistive moment exerted by the section is composed of elastic and plastic components such that

$$M_{ep} = M_e + M_p = YI/y + Fs,$$

where $s = d - h$ is the distance between the net force $F = Ybh$ in the plastic regions and $y = (d - 2h)/2$ defines the elastic-plastic boundary. Then

$$M_{ep} = 2Yb(d - 2h)^3/12(d - 2h) + (Ybh)(d - h)$$
$$= Y(bd^2/6)[1 + (2h/d)(1 - h/d)]. \tag{6.50}$$

The radius of curvature is found from applying ETB to the elastic core where $\sigma = Y$ when $y = (d - 2h)/2$. Then,

$$R_{ep} = Ey/\sigma = E(d - 2h)/2Y.$$

Under M_{ult} in Fig. 6.34(d), the resistive moment of the fully plastic section is simply

$$M_{ult} = (Fs)_{ult} = (bdY/2)(d/2) = bd^2 Y/4. \tag{6.51}$$

Hence, the shape factor and elastic safety factor are

$$Q = M_{ult}M_Y = (bd^2 Y/4)/(bd^2 Y/6) = 3/2, \tag{6.52}$$
$$S = Y/\sigma_w, \quad \Rightarrow \sigma_w = Y/S,$$

from which the safe moment is

$$M/I = \sigma_w/y, \quad \Rightarrow M = \sigma_w I/y = YI/Sy.$$

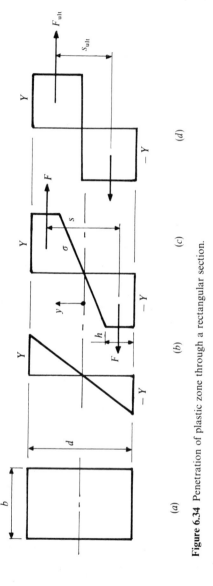

Figure 6.34 Penetration of plastic zone through a rectangular section.

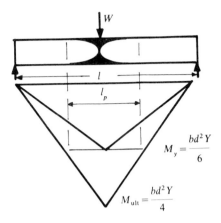

Figure 6.35 Plastic hinge.

In Fig. 6.35, the maximum bending moment is $M = Wl/4$. Hence the working load is

$$W = 4M/l = 4YI/lSy = 2Ybd^2/3lS.$$

The ultimate load and the load factor are defined from Eq. (6.51) as

$$W_{\text{ult}} = 4M_{\text{ult}}/l = bd^2Y/l,$$
$$\therefore \quad L = W_{\text{ult}}/W = (bd^2Y/l)/(2Ybd^2/3lS) = QS.$$

The plastic hinge length l_p, is determined from the M diagram (Fig. 6.35) where, beneath the load $M = M_{\text{ult}}$ and at the extremities of the hinge, $M = M_Y$. It then follows from the geometry of the M diagram that

$$2(M_{\text{ult}} - M_Y)/l_p = 2M_{\text{ult}}/l,$$
$$l_p = l(1 - M_Y/M_{\text{ult}}) = l(1 - 1/S) = l/3.$$

That is, the central hinge extends over one-third of the length for a simply supported beam with rectangular cross-section.

Non-axisymmetric Sections

When the elastic NA ($\bar{y}_e$) does not lie at the central depth of the section, yielding under M_Y will begin first on the most severeley stressed side (see Fig. 6.36a). As the bending moment (M_{ep}) is increased, the plastic zone will penetrate inwards from this side to a depth h, as shown in Fig. 6.36(c). In order that the horizontal tensile (T) and compressive (C) forces remain balanced, the stress distribution in Fig. 6.36(c) is accompanied by a shift in the NA to the new position $\bar{y}_{ep}$ such that $M_{ep} = Cs = Ts$. As M_{ep} is further increased, the stress continues to redistribute, maintaining $C = T$ until fully plasticity is reached under M_{ult} in Fig. 6.36(d) when the NA divides the cross-sectional area equally.

170

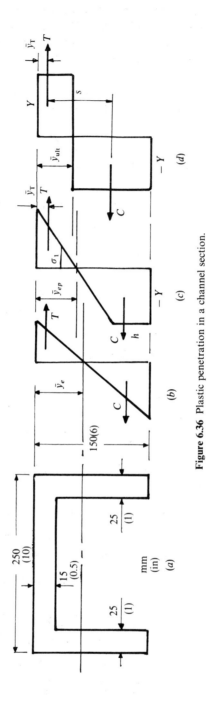

Figure 6.36 Plastic penetration in a channel section.

Example 6.20 Calculate M_Y and M_{ult} for the inverted channel section in Fig. 6.36(a) taking $Y = 235.5\,\text{MPa}$ ($15\,\text{tonf/in}^2$). What is the depth h to which the plastic zone has penetrated the vertical webs when the flange surface first yields? Find the corresponding moment in this case.

SOLUTION TO SI UNITS For initial yielding (Fig. 6.36b), the NA position ($\bar{y}_e$) passes through the centroid. This is found from applying $A\bar{y}_e = \sum(Ay)$ at the flange top:

$$[(250 \times 15) + 2(135 \times 25)]\bar{y}_e$$
$$= (250 \times 15 \times 7.5) + 2(135 \times 25 \times 82.5),$$
$$\bar{y}_e = 55.7\,\text{mm},$$
$$\therefore \quad I = (1/3)[250(55.7)^3 + 2 \times 25(94.3)^3 - 200(40.7)^3]$$
$$= 23.882 \times 10^6\,\text{mm}^4.$$

Yielding commences at the web bottoms where the corresponding moment is

$$M_Y = YI/y = 235.5 \times 23.882 \times 10^6/(150 - 55.7) = 59.64\,\text{kN m}.$$

For full plasticity in Fig. 6.36(d), the NA ($\bar{y}_{ult}$) divides the area in order that there is no net axial force (i.e. $T = C$):

$$[2 \times 25(150 - \bar{y}_{ult})]Y = [(250 \times 15) + 2(y_{ult} - 15)25]Y,$$
$$\bar{y}_{ult} = 45\,\text{mm}.$$

In order to calculate $M_{ult} = Cs = Ts$, it must be recognized that the equivalent force T acts at the centroid ($\bar{y}_T$) of the tensile area in Fig. 6.36(d). This is found from taking moments of that area about the top surface:

$$(250 \times 15)7.5 + 2(30 \times 25)30 = [(250 \times 15) + (2 \times 30 \times 25)]\bar{y}_T$$
$$\bar{y}_T = 13.93\,\text{mm}.$$

C acts at the centroid of the lower web area, giving the moment arm

$$s = (150 - 45)/2 + (45 - 13.93) = 83.57\,\text{mm},$$
$$\therefore \quad M_{ult} = Cs = 235.5 \times 25 \times 105 \times 2 \times 83.57 = 103.33\,\text{kN m}.$$

In the elastic-plastic case (Fig. 6.36c) the NA position ($\bar{y}_{ep}$) is again found from $T = C$. This gives

$$(Y + \sigma_1)(250 \times 15)/2 + 2\sigma_1(\bar{y}_{ep} - 15)25/2$$
$$= (2 \times 25 \times Yh) + 2(150 - \bar{y}_{ep} - h)25Y/2.$$

Substituting $\sigma_1 = Y(1 - 15/\bar{y}_{ep})$ and $h = 150 - 2\bar{y}_{ep}$:

$$1875(2\bar{y}_{ep} - 15) + 25(\bar{y}_{ep}^2 - 30\bar{y}_{ep} + 225)$$
$$= 50\bar{y}_{ep}(150 - 2\bar{y}_{ep}) + 25y_{ep}^2,$$

$$\bar{y}_{ep}^2 - 45\bar{y}_{ep} - 225 = 0,$$
$$\therefore \quad \bar{y}_{ep} = 49.54 \, \text{mm}.$$

From this the depth of penetration into each web is

$$h = 150 - 99.08 = 50.92 \, \text{mm}.$$

In the calculation of M_{ep} for this case, the centroid position $\bar{y}_T$, from the top surface (Fig. 6.36c), at which T acts, is found as

$$(250 \times 15)7.5 + 2(34.54 \times 25)(17.27 + 15)$$
$$= [(250 \times 15) + (2 \times 34.54 \times 25)]\bar{y}_T,$$
$$\bar{y}_T = 15.31 \, \text{mm}.$$

Taking moments about the force line T, M_{ep} is composed of the sum of two parts:

1. that due to plastic stress distribution:

$$2[235.5 \times 25 \times 50.92(150 - 15.31 - 50.92/2)] = 65.49 \times 10^6 \, \text{N mm};$$

2. that due to the elastic stress distribution:

$$2[235.5(150 - 49.54 - 50.92)25/2]$$
$$\times [2(150 - 49.54 - 50.92)/3 + (49.5 - 15.31)] = 19.61 \times 10^6 \, \text{M mm},$$
$$\therefore \quad M_{ep} = 65.49 + 19.61 = 85.1 \, \text{kN m}.$$

SOLUTION TO IMPERIAL UNITS For initial yielding (Fig. 6.36b), apply $A\bar{y}_e = \Sigma(Ay)$ about the flange top:

$$16.5\bar{y}_e = (5 \times 0.25) + 2(5.5 \times 3.25),$$
$$\bar{y}_e = 2.242 \, \text{in},$$
$$I = (10 \times 2.242^3/3) - (8 \times 1.742^3/3) + (2 \times 1 \times 3.758^3/3) = 58.85 \, \text{in}^4.$$
$$M_Y = YI/y = 15 \times 58.85/3.758 = 234.9 \, \text{tonf in}.$$

For full plasticity: $T = C$ (Fig. 6.36d):

$$(10 \times 0.5) + 2(\bar{y}_{ult} - 0.5)1 = 2(6 - \bar{y}_{ult})1$$
$$\therefore \quad \bar{y}_{ult} = 2 \, \text{in}.$$

Now T acts at the centroid ($\bar{y}_T$) of the tensile area. That is,

$$(10 \times 0.5 \times 0.25) + 2(1.5 \times 1 \times 1.25) = 8\bar{y}_T,$$
$$\therefore \quad \bar{y}_T = 0.625 \, \text{in},$$
$$s = (2 - 0.625) + 2 = 3.375 \, \text{in},$$
$$\therefore \quad M_{ult} = Cs = Ts = 15 \times 4 \times 1 \times 2 \times 3.375 = 405 \, \text{tonf in}.$$

For the elastic-plastic section $T = C$ (Fig. 6.36c):

$$(Y + \sigma_1)(10 \times 0.5)/2 + 2\sigma_1(\bar{y}_{ep} - 0.5)1/2 = (2 \times 1 \times Yh) + 2(6 - \bar{y}_{ep} - h)Y/2,$$

where $\sigma_1 = Y(1 - 0.5/\bar{y}_{ep})$ and $h = 6 - 2\bar{y}_{ep}$. The solution is $\bar{y}_{ep} = 2.12$ in, from which the penetration depth is

$$h = 6 - (2 \times 2.12) = 1.76 \text{ in.}$$

The centroid $(\bar{y}_T)$ for T is found from

$$(10 \times 0.5 \times 0.25) + 2(2.12 - 0.5)1(0.5 + 0.81) = 8.24\bar{y}_T,$$

$$\therefore \quad \bar{y}_T = 0.667 \text{ in,}$$

$$\therefore \quad M_{ep} = (2 \times 1.76 \times 1 \times 15 \times 4.453)$$
$$+ (2 \times 2.12 \times 1 \times 15/2)(1.4133 + 1.453)$$
$$= 326.27 \text{ tonf in.}$$

Collapse Under Three Plastic Hinges

In the case of the collapse of an encastre beam from more than one plastic hinge, the bending moment diagram is no longer similar in shape to that for elastic working conditions. This is because when M_{ult} is first reached in the most severely stressed section to form one hinge other hinges necessary for collapse form instantaneously as the stress redistributes to attain M_{ult} in other highly stressed sections.

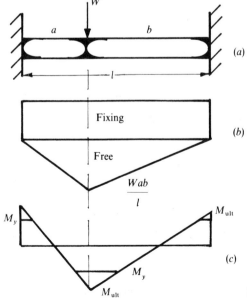

Figure 6.37 Collapse of encastre beam.

Example 6.21 Determine the collapse load and the length of the plastic hinges when the encastre beam in Fig. 6.37(a) carries a concentrated load W, that divides the length l into lengths a and b. Hence deduce the hinge lengths when the load is central and the section rectangular.

In Chapter 7 it is shown how the elastic moment diagram given in Fig. 6.37(b) is constructed from Mohr's theorems by equating the areas of the free and fixing moment diagrams. This results in net moments at the ends and centre that differ. The mechanism necessary for plastic collapse depends upon M_{ult} being reached simultaneously in each of these positions by stress redistribution. Equating moments at the end to the net moment at the load point in Fig. 6.37(b) gives

$$M_{ult}(\text{ends}) = [M(\text{free}) - M(\text{fixing})] \text{ at } W,$$
$$\therefore \quad M_{ult} = Wab/l - M_{ult}.$$

From this the collapse load is $W = 2M_{ult}l/ab$, where M_{ult} is determined separately for the given section. The lengths of the plastic hinges are found from marking off M_Y at the ends and centre in Fig. 6.37(c). Then,

At the L.H. end: $(M_{ult} - M_Y)/l_P = 2M_{ult}/a$,
$$l_P = (a/2)(1 - M_Y/M_{ult}) = (a/2)(1 - 1/Q).$$

At the R.H. end: $(M_{ult} - M_Y)/l_P = 2M_{ult}/b$,
$$l_P = (b/2)(1 - M_Y/M_{ult}) = (b/2)(1 - 1/Q).$$

At the centre: $l_P = (a + b)(1 - M_Y/M_{ult})/2 = (l/2)(1 - 1/Q)$.

With $a = b = l/2$ for a central load and $Q = 3/2$ for a rectangular section (Eq. (6.52)), the hinge lengths become $l/12$ and $l/6$ at the ends and centre.

Example 6.22 Find the uniformly distributed loading w, corresponding to M_Y and M_{ult} for the encastre rectangular section beam in Fig. 6.38(a).

M_Y is determined from the point in the beam carrying the greatest elastic moment. The application of Eq. (7.16) to the free and fixing elastic moment diagrams shows that this is $wl^2/12$ at the fixed ends. Then, from Fig. 6.34(b),

$$M_Y = bd^2Y/6 = wl^2/12,$$
$$\therefore \quad w = 2bd^2Y/l^2.$$

With collapse, M_{ult} is achieved simultaneously at the ends and centre. From the free and fixing moment diagrams in Fig. 6.38(b),

$$M(\text{ends}) = \text{net } M \text{ at centre} = \text{free moment} - \text{fixing moment},$$
$$M_{ult} = wl^2/8 - M_{ult}, \quad \Rightarrow M_{ult} = wl^2/16,$$

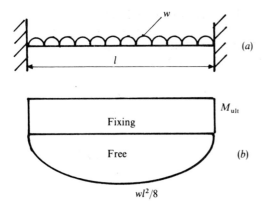

Figure 6.38 Encastre uniformly distributed load beam.

when, from Eq. (6.51),

$$M_{\text{ult}} = bd^2 Y/4 = wl^2/16, \quad \Rightarrow w = 4bd^2 Y/l^2.$$

Residual Bending Stresses

When an elastic-plastic beam is fully unloaded from M_{ep}, a state of residual stress σ_R, will remain in the cross-section after the elastic stresses σ_e, have recovered. Assuming purely elastic unloading from the elastic-plastic stress state σ corresponding to M_{ep}, it follows that

$$\sigma_R = \sigma - \sigma_e. \tag{6.53}$$

For example, with the elastic-plastic rectangular section in Fig. 6.34(c), elastic stress recovers over the whole depth $-d/2 \le y \le d/2$ in unloading from M_{ep} in Eq. (6.50). ETB gives this as

$$\sigma_e = M_{ep}y/I = Y(bd^2/6)[1 + (2h/d)(1 - h/d)](12y/bd^3)$$
$$= (2Y/d)[1 + (2h/d)(1 - h/d)]y. \tag{6.54}$$

Within the plastic zone, $\sigma = Y$ for y in the region $\pm(d/2 - h) \le y \le \pm d/2$. The residual stresses follow from Eqs (6.53) and (6.54),

$$\sigma_R = Y - M_{ep}y/I = Y - (2Y/d)[1 + (2h/d)(1 - h/d)]y. \tag{6.55}$$

Within the elastic zone for y in the region $0 \le y \le \pm(d/2 - h)$,

$$\sigma = M_e y/I = (Yb/6)(d - 2h)^2 y \times 12/b(d - 2h)^3 = 2Yy/(d - 2h), \tag{6.56}$$

which can readily be checked from similar triangles. Then, from Eqs (6.53), (6.54) and (6.56), the residuals are

$$\tau_R = 2Yy/(d - 2h) - M_{ep}y/I$$
$$= 2Yy/(d - 2h) - (2Y/d)[1 + (2h/d)(1 - h/d)]y. \tag{6.57}$$

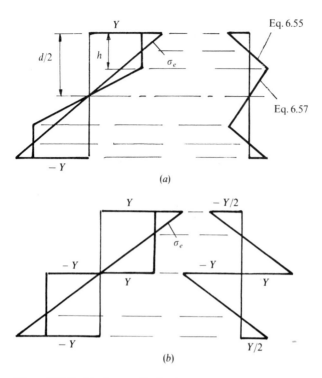

Figure 6.39 Residual stress distributions.

Equations (6.55) and (6.57) show that σ_R varies linearly with y within each zone. Figure 6.39(a) illustrates this variation for the elastic-plastic case, while in Fig. 6.39(b), σ_R for the fully plastic condition is given. That is, $\sigma_R = Y(1 - 3y/d)$ is found by putting $h = d/2$ in Eq. (6.55). This shows that the residual stress is equal to the tensile yield stress Y at the centre ($y = 0$), and one-half of Y at the edges ($y = d/2$).

EXERCISES

Bending of Straight Beams

6.1 A T-section beam 6 m long is simply supported at its ends with the flange uppermost. The width of the flange is 150 mm and its depth is 25 mm. The web thickness is 25 mm. Calculate the greatest distributed loading the beam can support when the maximum tensile and compressive stresses are limited to 55 and 31 MPa respectively.

Answer: 1.16 kN/m.

6.2 A light wooden bridge is supported by six parallel longitudinal timber beams, each being 300 mm deep and 200 mm wide. Each beam may be considered as simply supported over a

4.5 m span. If the allowable bending stress in the timber is 5.5 MPa, calculate the greatest distributed loading the bridge can support.

Answer: 4057 kg/m.

6.3 Calculate the greatest loading that can be carried by a 4 × 4 m square floor when it is simply supported on parallel timber joists 170 mm deep × 45 mm wide spaced at 400 mm intervals.

Answer: 1.79 kN/m².

6.4 A 1.2 m long circular shaft is supported in bearings at its ends and carries concentrated loads of 20 and 30 kN at respective distances of 0.45 and 0.75 m from the left-hand end. If the maximum stress in the material is to be 60 MPa at all points in the length, determine the shaft profile.

6.5 Derive expressions for the position and magnitude of the maximum bending stress in a tapered cantilever with solid circular section carrying a concentrated load at its end.

6.6 A steel pipeline, mean diameter $d = 1$ m, wall thickness $t = 20$ mm and density 5600 kg/m³ is rigidly supported at 10 m intervals. Calculate the maximum stress in the pipe from bending and pressure effects when it carries liquid at a pressure and density of 7 bar and 800 kg/m³ respectively. Take the maximum central moment to be $wl^2/24$ and the hoop stress due to pressure to be $pd/2t$.

Answer: 17.5 MPa.

6.7 The section in Fig. 6.40 is simply supported as a beam over 900 mm and carries a central concentrated load of 2 t. Find the maximum bending moment and the greatest stress in the material.

Answer: 0.45 t m; 86.2 MPa.

6.8 The channel section in Fig. 6.41 is 3 m long and simply supported at its ends with the legs uppermost. Calculate the greatest distributed loading the beam can carry if the tensile and

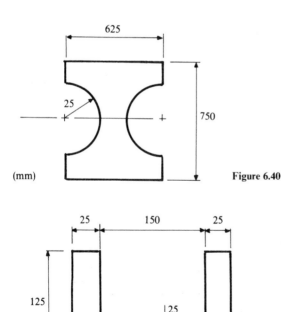

Figure 6.40

Figure 6.41

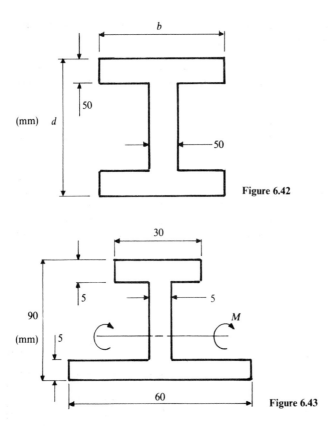

Figure 6.42

Figure 6.43

compressive stresses are limited to 31 and 55 MPa respectively. Sketch the distribution of bending stress across the section.

Answer: 89.2 kg/m.

6.9 The I-section in Fig. 6.42 is to carry a total load of 20 t uniformly distributed over its 7 m simply supported length. If the allowable bending stress is 150 MPa, find the dimensions b and d, where $b = d/2$, using a safety factor of 2.

6.10 The beam section in Fig. 6.43 is subjected to a bending moment M, acting in the sense

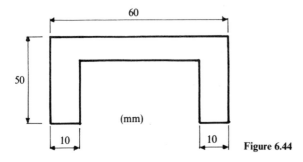

Figure 6.44

shown. If the maximum tensile and compressive stresses are limited to 70 and 95 MPa respectively, calculate the greatest value of M.

Answer: 1772 N m.

6.11 The channel section in Fig. 6.44 is mounted as a cantilever of length 4 m carrying uniformly distributed loading of 20 N/m. Calculate the maximum tensile and compressive stresses in the beam.

Answer: 9.5 MPa; 15.1 MPa.

6.12 An I-beam (Fig. 6.45) is simply supported over a 3.6 m length. If the tensile and compressive stresses are limited to 31 and 23 MPa respectively, what is the greatest distributed loading the beam can support.

Answer: 1.1 kN/m.

6.13 A cantilever beam, 1.8 m long, with the inverted channel section in Fig. 6.46, carries a uniformly distributed loading of 290 N/m together with a concentrated load of 220 N at a point 0.6 m from the free end. Determine the maximum tensile and compressive stresses due to bending.

6.14 Calculate for the T-section shown in Fig. 6.47 the maximum bending stress and the radius of curvature when an applied moment of 30 kN m causes compression in the flange. Take $E = 208$ GPa.

Answer: 219.1 MPa; 100.8 m.

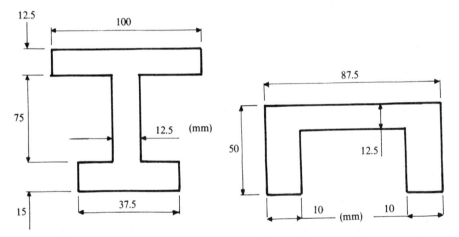

Figure 6.45   **Figure 6.46**

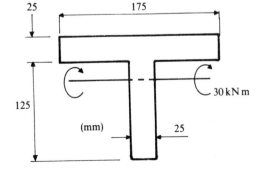

Figure 6.47

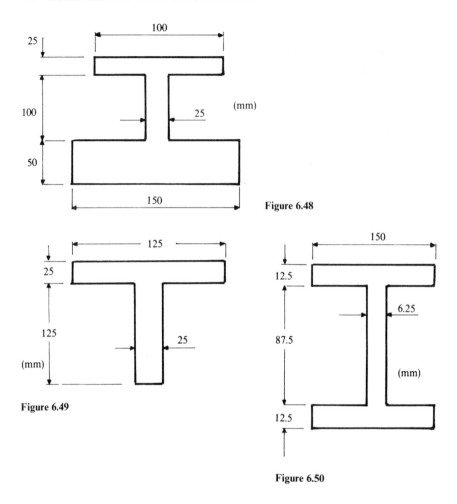

Figure 6.48

Figure 6.49

Figure 6.50

6.15 A cast iron beam 9 m long with the section given in Fig. 6.48 is simply supported at its ends. Given that the density of cast iron is 6900 kg/m³, calculate the maximum tensile and compressive stresses in the beam due to its own weight.

Answer: 13.6; 21.7 MPa.

6.16 The T-section in Fig. 6.49 is the cross-section of the beam loaded as shown in Fig. 5.19 (page 119). Determine the maximum tensile and compressive stresses at the section where the greatest bending moment occurs.

Answer: 7.37 kN/m²; 14.75 kN/m².

6.17 Find the maximum moment for the I-section in Fig. 6.50 when the working stress is limited to 77 MPa. Find the radius of curvature under this moment.

Answer: 15 kN m; 83.3 m.

6.18 Given that $I = 2.04 \times 10^{-4}$ m⁴ for a single 350×150 mm joist in Fig. 6.51 and that the plate dimensions are 375×25 mm, determine the maximum stress in the material when the section is used as a 7 m long simply supported, beam carrying a total load of 20 t distributed uniformly. Propose an improved cross-section for an allowable stress of 150 MPa.

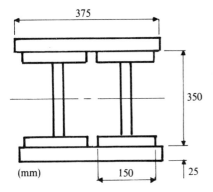

Figure 6.51

Bending of Initially Curved Bars

6.19 The semi-circular cantilever in Fig. 6.52 has a circular section 100 mm in diameter. If the maximum tensile stress in the material is limited to 250 MPa, find the greatest end force P that can be carried.

6.20 The arched beam in Fig. 6.53 carries the concentrated vertical forces shown. Determine, for the point of greatest bending moment, the maximum stress in the material and the radius of curvature for a uniform cross-section 100 mm wide × 20 mm deep. Take $E = 200\,\text{GPa}$.

6.21 Compare the maximum stresses from the large and small initial curvature theories in the

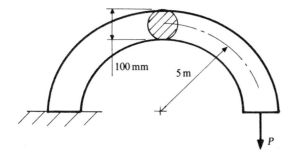

Figure 6.52

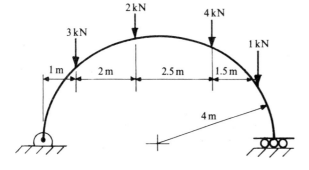

Figure 6.53

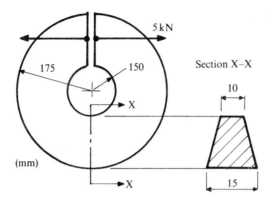

Section X–X

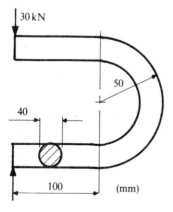

Figure 6.54

case of a split-ring of trapezoidal section when a 5 kN force is applied as shown in Fig. 6.54. Do not neglect the additive effect of the direct stress acting on the section X − X where the moment is a maximum.

6.22 The U-beam in Fig. 6.55 has a 40 mm diameter cross-section. Determine the maximum tensile and compressive stresses in the beam when a compressive force of 30 kN is applied at the free ends.

6.23 Determine the maximum tensile and compressive stresses for a curved beam 150 mm mean radius with rectangular cross-section 50 mm wide × 150 mm deep when it is subjected to a decreasing curvature moment of 15 kN m.

6.24 Plot the variation in stress across the depth of a curved bar with rectangular section; 75 mm wide × 50 mm deep with mean radius 125 mm when it is subjected to a sagging moment of 15 kN m tending to straighten it.

6.25 Show, for a solid circular section of radius r, in a beam with large initial curvature, of mean radius R_m, that the neutral axis shifts by $y = R_m - (1/2)r^2/[R_m - \sqrt{(R_m^2 - r^2)}]$.

6.26 Find the necessary wall thickness for a tube with mean section radius 100 mm when it is to be subjected to a pure bending moment of 10 kN m that decreases the initial 200 mm mean radius without the maximum compressive stress exceeding 65 MPa. What would be the greatest compressive stress in a straight beam with the same cross-sectional dimensions?

Figure 6.55

Combined Bending and Direct Stress

6.27 A cantilever 125 mm long supports a force of 10 kN inclined at 15° to its horizontal axis at the free end. If the section is rectangular, 50 mm wide × 25 mm deep, determine the net distribution across the vertical fixed-end section.

6.28 A simply supported beam, of rectangular section, 75 mm deep with cross-sectional area = 1850 mm^2 and $I = 300 \times 10^4$ mm^4 carries a uniformly distributed load of 45 kN/m over a 1.5 m length. Determine the distribution of stress at mid-span when the beam further supports an axial compressive force of 50 kN.

6.29 A cast iron column of 150 mm o.d. and 112.5 mm i.d. carries a vertical load of 20 t. Find the maximum allowable eccentricity of the load if the maximum tensile stress is not to exceed 30 MPa? What is the maximum compressive stress?

    Answer: 65 mm; 80.7 MPa.

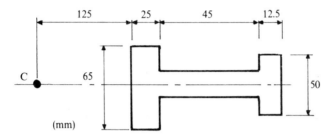

(mm) Figure 6.56

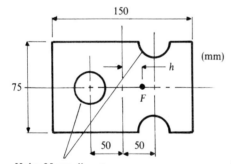

Holes 25 mm diameter   **Figure 6.57**

6.30 In a tensile test on a round bar of 30 mm diameter it was found that points originally 150 mm apart separated by 0.75 mm along one side but decreased by 0.25 mm on the opposite side. Find the axial stress distribution, the axial elongation and the eccentricity of the load. Take $E = 207$ GPa.

    Answer: 33.45 MPa; 0.25 mm; 7.9 mm.

6.31 Figure 6.56 shows the section of a hydraulic riveter. Determine the maximum tensile and compressive stresses in the section when a compressive load of 30 t acts at the anvil centre C.

    Answer: 829.44 MPa; −929.5 MPa.

6.32 The cross-section of a short column is shown in Fig. 6.57. For what value of h, defining the compressive point force F, will the tensile stress at the section edge be zero? Find also the maximum compressive stress when $F = 24$ t.

    Answer: 24.2 mm; 44.97 MPa.

6.33 Derive an expression for the maximum eccentricity h (w.r.t. the centroid) for a rectangular section $b \times d$ such that when a compressive load is applied along a centroidal axis parallel to d, the stress nowhere becomes tensile. Repeat for the case of solid and tubular circular sections.

   Answer: $d/6$; $d/8$; $(D^2 + d^2)/2D$.

6.34 A cast iron column 200 mm external diameter and 165 mm internal diameter carries an eccentric compressive load P at 330 mm from the axis of the column together with an axial compressive load of 800 kN. Calculate the maximum value of P if the limiting tensile and compressive stresses in the material are 31 and 123.5 MPa respectively.

   Answer: 62.78 kN.

6.35 A short tube of o.d. = 200 mm, i.d. = 165 mm is subjected to an eccentric compressive load of 60 kN applied at a distance of 330 mm from the cylinder axis. If the maximum allowable compressive and tensile stresses for the material are 120 and 30 MPa respectively, determine the maximum additional axial compressive load that may be applied. Draw the net stress distribution for the tube cross-section.

6.36 A compressive force F acts normally to the section given in Fig. 6.58. Determine the maximum eccentricity h if there is to be no tensile stress across the section. What is F if the maximum compressive stress is 92.5 MPa at this h value? Establish the area within which F must lie to avoid tensile stress in the section.

   Answer: 25 mm; 976 kN.

6.37 The short column of rectangular cross-section in Fig. 6.59 has a normal vertical compressive force $F = 50$ kN, applied eccentrically to both principal axes. Calculate the resultant stress at each corner of the section.

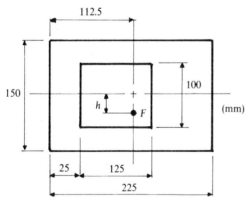

Figure 6.58

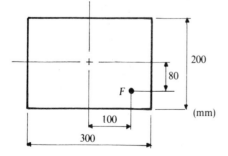

Figure 6.59

Bending of Composite Beams

6.38 A timber beam 200 mm wide and 300 mm deep is reinforced along its bottom edge by a steel plate 200 mm wide and 12.5 mm thick to give a composite beam 312.5 mm deep and 200 mm wide. Calculate the maximum stresses in the steel and timber when the beam carries a uniformly distributed load of 14.6 kN/m over a simply supported span of 6 m. Take E(steel) = 20E(timber).

Answer: 106 MPa, 12.8 MPa.

6.39 A compound beam is formed by brazing a steel strip 25 mm × 6.4 mm on to a brass strip 25 mm × 12.8 mm. When the beam is bent into a circular arc about its longer side, find (a) the position of the neutral axis, (b) the ratio of the maximum stress in the steel to that in the brass, and (c) the radius of curvature of the neutral axis when the steel stress reaches its maximum allowable value of 69 MPa. Take E = 207 GPa for steel, E = 82.8 GPa for brass.

Answer: 7.4 mm; 1.58; 21.75 m.

6.40 A steel strip 50 × 12.5 mm in cross-section is brazed onto a brass strip 50 × 25 mm in section to form a compound beam 50 mm wide × 37.5 mm deep. If the stresses in the steel and brass are not to exceed 138 and 69 MPa respectively, determine the maximum bending moment the beam can carry. What then are the maximum bending stresses in the steel and brass? Take E = 207 GPa for steel, E = 82.8 GPa for brass.

Answer: 1034 Nm; 110 MPa; 68.9 MPa.

6.41 The composite beam in Fig. 6.60 is fabricated from a central steel strip with outer brass strips. If the allowable stresses for each material are 110 and 80 MPa respectively, determine the maximum load that could be applied to this beam at the centre of its 2 m simply supported length. What are the maximum stresses actually achieved in each material? For steel E = 210 GPa and for brass E = 85 GPa.

6.42 The composite beam in Fig. 6.61 is made from timber joists reinforced with a central steel strip. Calculate the moment of resistance for the section when the maximum bending stress in the timber is 8.25 MPa. What then is the maximum bending stress in the steel? Take E = 207 GPa for steel, E = 12.4 for timber.

Answer: 21 kN m; 91.3 MPa.

6.43 A beam of length L, width W and depth D is composed of n bonded layers of different materials. If all layers are $L × W$ and the neutral plane lies within one layer, derive an expression for the bending stress at any point on the cross-section of the beam. The composite beam in

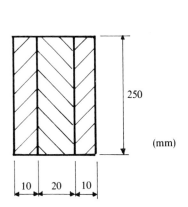

Figure 6.60

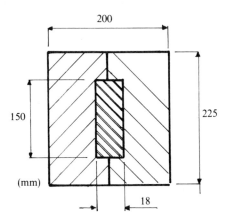

Figure 6.61

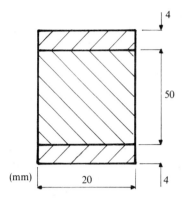

Figure 6.62

Fig. 6.62 is composed of a top plate of duralumin ($E = 70\,\text{GPa}$), a core of expanded plastic ($E = 5\,\text{MPa}$) and a bottom plate of steel ($E = 200\,\text{GPa}$). The beam length is 1000 mm, simply supported at its ends. Calculate the position of the neutral axis and determine safe value of uniformly distributed load when the stress in the core is restricted to 0.02 MPa and the maximum deflection is 10 mm. (C.E.I. Pt. 2, 1975)

6.44 A timber beam 75 mm wide × 150 mm deep is reinforced by aluminium strips 75 mm wide on top and bottom faces of the timber for the whole of its length. The moment of resistance of the composite beam is to be four times that of the timber alone whilst maintaining the same value of maximum bending stress in the timber. Determine the thickness of the aluminium strips and the ratio of the maximum bending stresses in each material. Take $E(\text{Al}) = 7.15\ E(\text{timber})$.
   Answer: 9.3 mm; 8.03.

6.45 A lightweight glass-fibre beam 10 mm wide is reinforced with carbon fibre of the same width along its outer edges to occupy 20 per cent of the total beam volume. If the simply supported length of the beam is 20 times its total depth, determine the maximum central concentrated load that the beam can carry when the allowable bending stresses for each material are 208 MPa for glass fibre and 1040 MPa for carbon fibre. The respective moduli are 16 GPa and 124 GPa.
   Answer: 3.84 kN.

6.46 A bimetallic strip in a temperature controller consists of brass bonded to steel, each material having the same rectangular cross-section and length. Calculate the stresses set up at the outer and interface surfaces in the longitudinal direction for an increase in temperature of 50 °C in the strip. Neglect distortion and transverse stress. For steel, $\alpha = 11 \times 10^{-6}/°\text{C}$, $E = 207\,\text{GPa}$; for brass, $\alpha = 20 \times 10^{-6}/°\text{C}$, $E = 103\,\text{GPa}$. (C.E.I. Pt. 2, 1967)

6.47 A thermostatic control-unit employs a bimetallic cantilevered strip to operate a switch at its free end. If the metal strips are each 100 mm long and 15 mm × 1 mm in cross-section, calculate the common radius of curvature and deflection at the free end for a temperature rise of 50 °C. Take $E = 200\,\text{GPa}$, $\alpha = 10 \times 10^{-6}/°\text{C}$ for one material and $E = 150\,\text{GPa}$, $\alpha = 20 \times 10^{-6}/°\text{C}$ for the other.
   Answer: 2.68 m; 1.87 mm.

Reinforced Concrete

6.48 A reinforced concrete beam is 300 mm wide and 500 mm deep to the centre of the single steel line. If stresses in the concrete and steel of 4.13 and 124 MPa are to be reached under economic design conditions, find the steel area and the distributed weight that can be simply supported over a span of 9 m. Ignore self-weight of the beam and take $m = 15$. Explain why there should be several steel rods each of small cross-sectional area rather than one of large area.

Answer: 1014 mm²; 4.94 kN/m.

6.49 A reinforced concrete beam section is 250 mm wide, 500 mm deep and 50 mm deep to the centre of the steel line where there are four rods each 12.5 mm diameter. If the maximum permissible stresses for steel and concrete are 124 and 4.13 MPa respectively, and the modular ratio is 15, calculate the moment of resistance for the section and the stress in the concrete. Where in the cross-section lies the neutral axis?

Answer: 26 kN m; 3.67 MPa.

6.50 A concrete beam with steel reinforcement is 150 mm wide and 375 mm deep to the steel line, where there are three rods each 12.5 mm diameter. If the maximum permissible stresses for the steel and concrete are 124 and 4.1 MPa respectively, and the modular ratio is 15, calculate the moment of resistance for the section. If the section were to be economically designed, what would then be the steel area and the moment of resistance?

6.51 A reinforced concrete beam covers a span of 12 m and carries a uniformly distributed load of 1.46 kN/m in addition to the beam's self-weight of 1.17 kN/m. The maximum allowable stresses for the steel and concrete are 124 and 4.1 MPa respectively. Assuming economic design, calculate the section dimensions and the steel area when the breadth is 0.6 times the depth to the steel line. Take $m = 15$.

Answer: $d = 500$ mm; $b = 310$ mm; $A_S = 841$ mm².

6.52 A rectangular section concrete beam is 300 mm wide and 200 mm deep to the steel line. The limiting stresses are 6.9 MPa for the concrete and 124 MPa for the steel. If both materials are to be fully stressed, calculate the steel area and the moment of resistance for the section when the modular ratio is 15.

Answer: 760 mm²; 16.75 kN m.

6.53 A reinforced concrete beam with steel-line depth d and breadth $b = d/2$ carries a uniformly distributed load of 1.46 kN/m over a span of 6 m in addition to its self-weight of 1.17 kN/m. Find d and the steel area given that the allowable stresses for steel and concrete are 124 and 4.1 MPa respectively and $m = 15$.

Answer: 335 mm; 314 mm².

6.54 A loading bay 18 m × 6 m is to be supported by 12 evenly spaced tension reinforced concrete cantilevers. Each beam is 6 m long, 300 mm wide with limiting stresses in the steel and concrete of 138 and 10 MPa respectively. If the maximum uniformly distributed load plus self-weight is 9.6 kN/m², find the remaining dimensions of the rectangular cross-section that will support the bay. Take $m = 15$. (C.E.I. Pt. 2, 1969)

6.55 A 500 mm × 250 mm rectangular concrete section is doubly reinforced with two pairs of 28 mm diameter steel rods set 400 mm apart. Determine the moment of resistance for the section with permissible stresses of 124 and 5.5 MPa and $m = 15$.

Answer: 56.75 kN m.

6.56 A concrete beam 300 mm wide with 450 mm steel depth spans 9 m and carries a uniformly distributed load of 8 kN/m. The steel reinforcement consists of four 25 mm bars. Find the maximum stresses in the steel and concrete, given that $m = 15$.

Answer: 106.8 MPa; 6.5 MPa.

6.57 A reinforced T-beam has flange and web widths of 750 and 150 mm respectively. The web steel line is 300 mm from the flange top. If the neutral axis is to lie within the flange width find its position when the maximum stresses in the steel and concrete are limited to 110 and 4.1 MPa respectively. What then is the area of steel reinforcement and the moment of resistance for the section?

Answer: 108 mm; 1520 mm²; 46.3 kN m.

6.58 A reinforced concrete T-beam has a flange 1.5 m wide × 150 mm deep. The web width is 450 mm and the depth to the steel line is 600 mm. If the allowable stresses for the steel and concrete are 124 and 6.9 MPa respectively, and the modular ratio is 15, find the stresses that are actually sustained by each material given that the area of tensile reinforcement is 3750 mm².

Answer: 3.83 MPa; 124 MPa.

6.59 The flange of a reinforced concrete T-beam is 1.5 m wide and 100 mm deep and the steel line in the web lies 375 mm from the flange top. If the limiting stresses for steel and concrete are 110 and 4.1 MPa respectively, and the neutral axis is to coincide with the lower edge of the flange, calculate the moment of resistance for the beam cross-section and the area of steel reinforcement. What are the stresses attained in each material? Take $m = 15$.

Answer: 71.9 kN m; 1820 mm²; 110 MPa; 2.67 MPa.

6.60 A doubly reinforced concrete beam is to sustain a bending moment of 135.6 kN m. The width of the beam is half the effective depth and the compressive steel line lies at 1/10 of this depth. The tensile reinforcement is to be 2 per cent of the effective section area. If the permissible stresses in the steel and concrete are 124 and 6.9 MPa respectively, find the concrete section dimensions and the steel areas. Take $m = 15$.

Answer: $b = 250$ mm; $d = 500$ mm; $A_{st} = 2500$ mm²; $A_{sc} = 1525$ mm².

6.61 The short column of reinforced concrete in Fig. 6.63 supports a compressive load F, applied at the centroid O. What is the safe value of F if the maximum allowable stresses for steel and concrete are 124 and 4.1 MPa respectively. If F is moved distance h away from O as shown, find the value of h and the limiting F if tension in the concrete is to be avoided. Take $m = E_s/E_c = 15$.

Answer: 776.5 kN; 79 mm.

Asymmetric Bending

6.62 A 100 × 75 × 12.5 mm angle section rests on two supports 4.35 m apart with its longer side vertical and the shorter side at the top. Calculate the maximum stress in the angle when a concentrated force of 10 kN is applied at mid-span.

Answer: −96.7 MPa; 85.1 MPa.

6.63 A 1.25 m long steel bar of rectangular section 100 × 37.5 mm is supported in bearings and carries a load of 9 kN at mid-span. If the beam is rotated slowly, find the slope of the longer side w.r.t. x when the bending stress reaches a maximum.

Answer: 20.5°; 117.8 MPa.

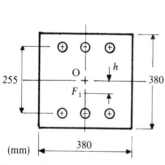

(mm)

Figure 6.63

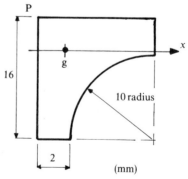

(mm)

Figure 6.64

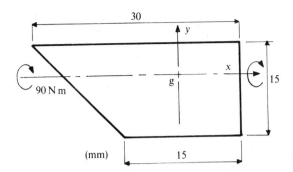

Figure 6.65

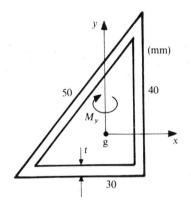

Figure 6.66

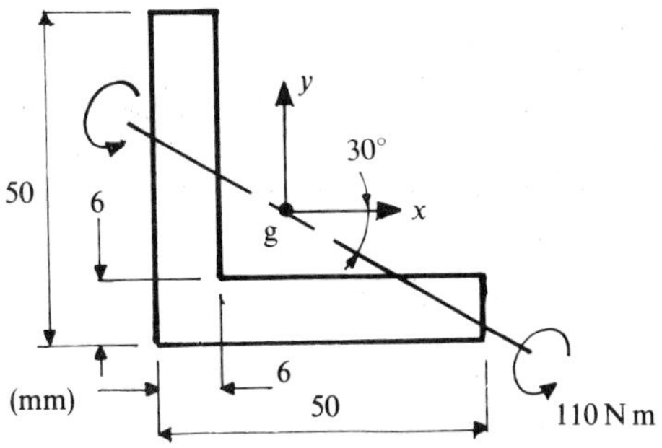

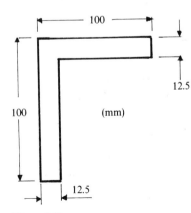

Figure 6.67

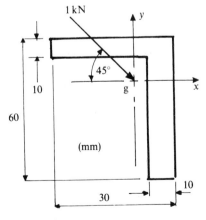

Figure 6.68

Figure 6.69

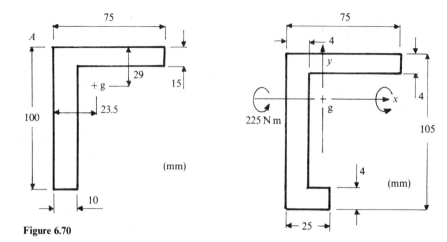

Figure 6.70

Figure 6.71

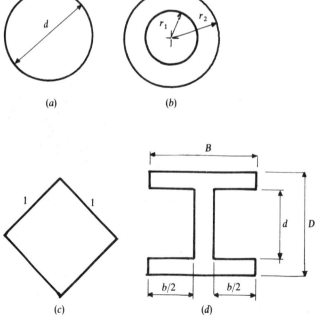

(a)

(b)

(c)

(d)

Figure 6.72

6.64 The stress at point P for the beam cross-section in Fig. 6.64 is restricted to 30 MPa. Find the maximum moment which may be applied about the x-axis.

Answer: 4.97 N m.

6.65 The section in Fig. 6.65 is subjected to a bending moment of 90 N m about the x-axis. Find the greatest bending stress and the inclination of the NA to x.

Answer: 90 MPa; 61°.

6.66 The thin-walled triangular tube in Fig. 6.66 is subjected to a bending moment M_y as shown. Find the inclination of the neutral axis w.r.t. the x-axis.

6.67 The angle section in Fig. 6.67 is subjected to a bending moment of 110 N m about a centroidal axis inclined at 30° to the x-axis as shown. Calculate the principal second moments of area, the position of the neutral plane, and the maximum tensile and compressive stresses. (C.E.I. Pt. 2, 1968)

6.68 A cantilever 500 mm long is loaded at its free end as shown in Fig. 6.68. Calculate the maximum tensile and compressive bending stresses.

6.69 The equal angle section in Fig. 6.69 is used as a cantilever of length 0.9 m carrying a vertical end load passing through the centroid. Determine the maximum value for this load if the bending stress is not to exceed 116 MPa.

Answer: 3.1 kN.

6.70 A simply supported beam of the section given in Fig. 6.70 supports a vertical downward force passing through the centroid that results in a bending moment of 1 kN m at mid-span. Find the bending stress at the corner A, stating whether it is tensile or compressive. (C.E.I. Pt. 2, 1986)

6.71 Calculate the maximum tensile stress and the position of the neutral axis for the section in Fig. 6.71 when a bending moment of 225 N m is applied about the x-axis in the sense shown.

Answer: 14.4 MPa; $-27.5°$.

Elastic-plastic Bending of Beams

6.72 Derive expressions for the moments M_Y, M_{ult} and hence determine $Q = M_{ult}/M_Y$ for each of the axisymmetric beam cross-sections in Fig. 6.72(a–d).

Answer: $Q = 1.7$; $16r_2(r_2^3 - r_1^3)/3\pi(r_2^4 - r_1^4)$; 2; $3(1 - bd^2/BD^2)/2(1 - bd^3/BD^3)$.

6.73 Find M_{ult} for the inverted T-section in Fig. 6.73.

Answer: $M_{ult} = Yah(a + h)/2$.

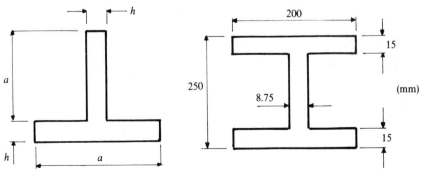

Figure 6.73   **Figure 6.74**

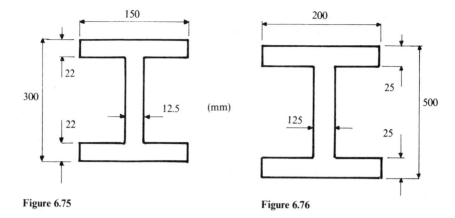

Figure 6.75   **Figure 6.76**

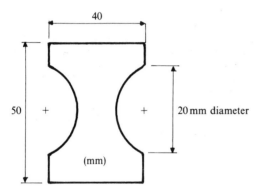

Figure 6.77

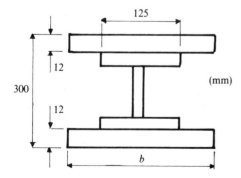

Figure 6.78

6.74 For the I-section in Fig. 6.74 calculate M_Y, the shape factor, and the load factor, with $Y = 324$ MPa, a single plastic hinge, and a safety factor of 2.

Answer: 247, 1.11, 2.22 (k N m).

6.75 Find the bending moment that will just initiate yielding in a circular steel rod 25 mm diameter, given that its tensile yield stress is 300 MPa.

6.76 A simply supported beam 1.25 m long carries a central concentrated load W. If the rectangular cross-section is 25 mm wide $\times$ 75 mm deep, with a yield stress $Y = 278$ MPa, find W when (a) yielding first occurs and (b) yielding penetrates to a depth of 15 mm from top and bottom surfaces at mid-span. Find the length over which yielding has occurred, the radius of curvature for (b) and the residual stress distribution for the unloaded beam. Take $E = 207$ GPa.

Answer: 2126 kgf; 2807 kgf; 303 mm; 89 MPa; 57.8 MPa; 16.75 m.

6.77 Find the moment of resistance in the fully elastic and plastic conditions for the I-section beam in Fig. 6.75 when $Y = 278$ MPa.

6.78 Express in terms of Y, for the I-section beam in Fig. 6.76, the bending moment to cause yielding throughout the depth of the flanges and throughout the whole section. Also establish the plastic penetration depth for an applied moment of 850 kN m with $Y = 278$ MPa.

Answer: $45.3Y$ kN m; $48.6Y$ kN m; 100 mm.

6.79 Determine, for the section in Fig. 6.77, the moment to produce 15 mm of plastic penetration from each edge and the corresponding radius of curvature. Establish the residual stress distribution and radius of curvature upon removal of this moment.

6.80 The steel beam I-section in Fig. 6.78 is simply supported over a length of 5 m and carries a uniformly distributed load of 114 kN/m. Steel reinforcing plates 12 mm thick are welded to each flange. Calculate the plate width b such that the plates are fully plastic at mid-span and the extent of yielding in their length. Find the horizontal shearing stress at the interface for sections where the outer surfaces have just reached the yield point (see Chapter 10). Take the yield stress = 300 MPa and I_g for the unreinforced section = 80×10^{-6} m^4. (C.E.I. Pt. 2, 1971)

6.81 Yielding occurs over the bottom 50 mm of the web for the T-section in Fig. 6.79 under a particular bending moment. If the yield stress remains constant at 278 MPa, find the applied moment, the stress at the flange top and the position of the neutral axis. Repeat for the fully plastic case.

Answer: 71 kN m; 210 MPa; 135.5 mm; 155 mm.

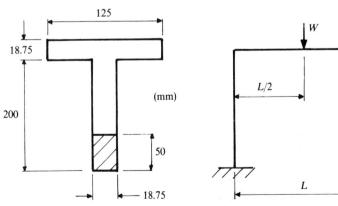

Figure 6.79   **Figure 6.80**

6.82 Derive, from energy considerations, expressions for the collapse load of the fixed base portal frame in Fig. 6.80 in the case of (a) a central vertical load W, (b) a horizontal load P and (c) combined W and P when $P = W/3$.

6.83 A 9 m long beam with 75 mm square section is to carry a uniformly distributed load w. Compare values of w that would cause the beam to collapse corresponding to support over two and three equal-length bays when $Y = 300$ MPa.

6.84 A uniform beam of length l, with rectangular cross-section $b \times d$ is built-in at both ends and carries a vertical load W, at mid-span. Find the extent to which the plasticity spreads in the length when M_{ult} is first achieved.

   Answer: $l/12$; $l/6$; $l/12$.

6.85 A span of 6 m is covered by a beam built-in at both ends. The beam carries three equal concentrated loads W(kN), 1.5 m apart with equal 1.5 m lengths remaining at each end. The central 3 m of the beam is reinforced top and bottom to increase the moment of resistance by 80 per cent. Given that $M_{ult} = 7.5$ kN m for the unreinforced section, find the loading to cause collapse at the centre and at 1.5 m from the ends.

   Answer: 5 kN; 9 kN.

6.86 A 3 m long beam carrying a central concentrated load W is built-in at one end and simply supported on the same level at the other end. If the elastic section modulus, $z = I/y = 15.63 \times 10^3$ mm^3 and $Y = 235$ MPa, find the maximum value of W, using a safety factor of 2 under elastic conditions. Also find the value of W for a load factor of 2 against full plastic collapse.

   Answer: 47 kN; 61 kN.

SEVEN

DEFLECTION AND SLOPES OF BEAMS

7.1 INTRODUCTION

This chapter considers the traditional methods for calculating the slopes and deflection of straight beams due to bending effects only. The deflection that is produced owing to associated shear forces in a long beam is normally negligible by comparison (see Chapter 11). In Chapters 13 and 14 it is shown how the shear deflection and the deflection in curved beams, under any loading combination, may be calculated from the application of energy methods and the principle of virtual work.

7.2 DIFFERENTIAL EQUATION OF FLEXURE

Consider a portion of a beam's neutral axis AB under a positive hogging moment M, in Fig. 7.1. Let the deformed beam lie in axes having z aligned with the undeflected neutral axis such that the *positive dowward* deflections of A and B are v and $v + \delta v$ respectively (Fig. 7.1(a)) where AB subtends an angle $\delta\theta$ at its centre of *positive* curvature O, defined by radius R. The approximation is made that, within the triangle ABC, the slope θ, at point A is given by

$$\tan\theta \simeq \theta(\text{rads}) \simeq \delta v/\delta z, \qquad (7.1)$$

and also from the segment AOB that the change in slope $\delta\theta$ between A and B is approximated by

$$\delta\theta = AB/R \simeq \delta z/R. \qquad (7.2)$$

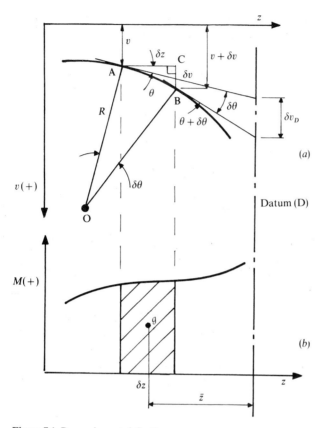

Figure 7.1 Beam element deflection.

Provided the displacements and slopes are small, Eqs (7.1) and (7.2) may be combined in the limit to give

$$\frac{1}{R} = \frac{d\theta}{dz} = \left(\frac{d}{dz}\right)\left(\frac{dv}{dz}\right) = \frac{d^2v}{dz^2}, \tag{7.3}$$

but from the ETB, $1/R = M/EI$, when Eq. (7.3) becomes

$$EI \, d^2v/dz^2 = M. \tag{7.4}$$

Upon successive integration, Eq. (7.4) will yield the slope and deflection at all points in the length, with single expressions, provided M can be expressed as a continuous function in z for the whole length of the beam. In cases of symmetry, it is only necessary to express M in terms of z for the appropriate region, e.g. over half the length of a simply supported beam with central concentrated load. In general, it is only possible to apply the direct integration method to beams with simple loading.

Example 7.1 Find the maximum slope and deflection for a simply supported beam of length L carrying a central concentrated load W (Fig. 7.2).

Because of symmetry, Eq. (7.4) need only be applied for $(0 \leq z \leq L/2)$. Then, by taking moments to the left of the section lying within this region,

$$EI \, d^2v/dz^2 = M = -Wz/2 \quad \text{(sagging)},$$
$$EI \, dv/dz = -Wz^2/4 + C_1,$$
$$EIv = -Wz^3/12 + C_1 z + C_2.$$

The following boundary conditions determine the integration constants:

$$dv/dz = 0 \quad \text{when} \quad z = L/2, \quad \Rightarrow C_1 = WL^2/16,$$
$$v = 0 \quad \text{when} \quad z = 0, \quad \Rightarrow C_2 = 0,$$

and, by inspection, the maximum slope and deflection are

$$dv/dz = C_1/EI = WL^2/16EI, \quad \text{for } z = 0 \qquad \text{(at the ends)},$$
$$v = (-WL^3/96 + WL^3/32)/EI = WL^3/48EI, \quad \text{for } z = L/2 \quad \text{(at the centre)}.$$

Example 7.2 Derive the maximum slope and deflection expression for a cantilever of length L in Fig. 7.3 carrying a concentrated force F, at its end.

The choice of origin is unimportant, though selecting the fixed end ensures that the integration constants are zero. Thus, to the right of the section at z:

$$EI \, d^2v/dz^2 = M = F(L-z) \quad \text{(hogging for } 0 \leq z \leq L),$$
$$EI \, dv/dz = F(Lz - z^2/2) + C_1,$$
$$EIv = F(Lz^2/2 - z^3/6) + C_1 z + C_2.$$

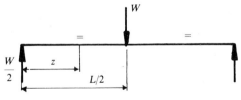

Figure 7.2 Simply supported beam.

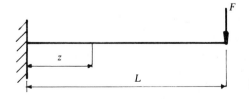

Figure 7.3 Cantilever beam.

Clearly, $C_1 = 0$ and $C_2 = 0$ as $dv/dz = v = 0$ for $z = 0$. The maximum values of slope and deflection are then

$$dv/dz = WL^2/2EI, \qquad\qquad \text{for } z = L \quad \text{(at the free end),}$$
$$v = W(L^3/2 - L^3/6)/EI = WL^3/3EI, \quad \text{for } z = L \quad \text{(at the free end).}$$

Example 7.3 Working from general expressions, determine the maximum slope and deflection and also the slope and deflection for a point 5 m from the fixed end of a cantilever beam in Fig. 7.4 when it carries a total weight of 20 kN evenly distributed over its 10 m length. Take $E = 207$ GPa, $I = 10^9$ mm^4.

$$EI\, d^2v/dz^2 = (w/2)(L - z)^2 \quad \text{(hogging for } 0 \leq z \leq L),$$
$$EI\, dv/dz = (w/2)(L^2z - Lz^2 + z^3/3) + C_1,$$
$$EIv = (w/2)(L^2z^2/2 - Lz^3/3 + z^4/12) + C_1 z + C_2,$$

where $C_1 = C_2 = 0$ as $dv/dz = v = 0$ for $z = 0$. Then for $z = 5$ m,

$$dv/dz = (w/2EI)(L^2 z - Lz^2 + z^3/3)$$
$$= \frac{2(10^2 \times 5 - 10 \times 5^2 + 5^3/3)10^9}{2 \times 207 \times 10^3 \times 10^9} = 0.001\,41 \text{ rad.}$$

$$v = (w/2EI)(L^2 z^2/2 - Lz^3/3 + z^4/12)$$
$$= \frac{2(10^2 \times 5^2/2 - 10 \times 5^3/3 + 5^4/12)10^{12}}{2 \times 207 \times 10^3 \times 10^9} = 4.28 \text{ mm.}$$

The maximum values at $z = L$ are

$$\frac{dv}{dz} = \frac{wL^3}{6EI} = \frac{2 \times 10^3 \times 10^9}{6 \times 207 \times 10^3 \times 10^9} = 0.001\,61 \text{ rad,}$$

$$v = \frac{wL^4}{8EI} = \frac{2 \times 10^4 \times 10^{12}}{8 \times 207 \times 10^3 \times 10^9} = 12.08 \text{ mm.}$$

Example 7.4 Find the position and magnitude of the maximum deflection for the simply supported beam with the linearly varying distributed loading in Fig. 7.5.

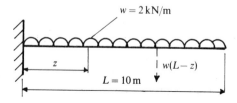

$w = 2$ kN/m

z

$w(L - z)$

$L = 10$ m

Figure 7.4 Cantilever beam.

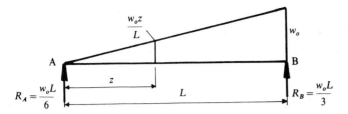

Figure 7.5 Simply supported beam with varying load.

The support reactions R_A and R_B are

$$(2L/3)(w_oL/2) = R_BL, \quad \Rightarrow R_B = w_oL/3,$$
$$R_A = w_oL/2 - w_oL/3 = w_oL/6.$$

Apply Eq. (7.4) with moments to the left of section z:

$$EI\,d^2v/dz^2 = M = -w_oLz/6 + (w_oz/L)(z/2)(z/3) = w_oz^3/6L - w_oLz/6,$$
$$EI\,dv/dz = w_oz^4/24L - w_oLz^2/12 + C_1,$$
$$EIv = w_oz^5/120\,L - w_oLz^3/36 + C_1z + C_2.$$

Applying the boundary conditions:

$$v = 0 \text{ at } z = 0 \text{ gives } C_2 = 0;$$
$$v = 0 \text{ at } z = L \text{ gives } C_1 = w_oL^3/36 - w_oL^3/120 = 7w_oL^3/360.$$

The general deflection equation is then

$$EIv = w_oz^5/120L - w_oLz^3/36 + 7w_oL^3z/360, \tag{i}$$

which has a maximum value at the zero-slope position. That is,

$$dv/dz = w_oz^4/24L - w_oLz^2/12 + 7w_oL^3/360 = 0.$$

This results in the quadratic that determines the position of v_{max}:

$$(z/L)^4 - 2(z/L)^2 + 0.467 = 0,$$
$$\therefore \quad z = 0.519L.$$

Substituting into Eq. (i) leads to

$$EIv_{max} = w_o[(0.519L)^5/120L - L(0.519L)^3/36 + 7L^3(0.519L)/360],$$
$$\therefore \quad v_{max} = w_o(0.002\,247L^4 - 0.003\,89L^4 + 0.010\,098L^4)/EI$$
$$= 0.006\,52\,w_oL^4/EI.$$

7.3 MOMENT AREA METHOD—MOHR'S THEOREMS

It further follows from Eqs. (7.1) and (7.2) that the change in slope and deflection w.r.t. the given datum for the beam in Fig. 7.1(a) are

$$\delta\theta = \delta z/R = (M\,\delta z)/EI$$
$$\delta v = \delta\theta\bar{z} = (\delta z/R)\bar{z} = (1/EI)(M\,\delta z)\bar{z}.$$

Since $M\,dz$ is the elemental shaded area of the M diagram in Fig. 7.1(b), and with $\bar{z}$ measured from the chosen datum, these expressions become

$$\delta\theta(1/EI)(\text{area of the } M \text{ diagram over } dz) \qquad \text{(when } I \text{ is constant)} \qquad (7.5a)$$
$$= (1/E)(\text{area of the } M/I \text{ diagram over } dz) \qquad \text{(when } I \text{ is variable),} \qquad (7.5b)$$
$$\delta v = (1/EI)(\text{moment of area of the } M \text{ diagram}) \qquad \text{(when } I \text{ is constant)} \qquad (7.6a)$$
$$= (1/E)(\text{moment of area of the } M/I \text{ diagram}) \qquad \text{(when } I \text{ is variable).} \qquad (7.6b)$$

Equations (7.5) and (7.6) are Mohr's first and second theorems, though they are more commonly referred to as the moment–area equations. When these equations are employed to find the slope and deflection at a given point in a beam, their application depends upon the method of support. Note that the simply supported and cantilever beams in Examples 7.1–7.4 are statically determinate, requiring only the two equations of equilibrium, i.e. force and moment balance. The propped cantilever, the encastre and the continuous beams in Examples 7.6, 7.9, 7.10, 7.11 and 7.12 are all statically indeterminate in that additional slope and deflections equations are then required to solve for the number of excess or redundant reactions. The latter arise from propping and fixing a beam that could also support the applied loading but with greater deflection on simple supports.

1. Cantilevers

Because both the slope and deflection at the fixed end are zero, by taking this point as the origin (0), Eqs (7.5) and (7.6) may be integrated directly to give absolute values at any point z in the length. That is, since hogging moments produce positive change in slope with v measured downwards, the areas of the M diagram are positive and

$$\theta = (1/EI)(\text{area of } M \text{ diagram from 0 to } z) \qquad (7.7)$$

$$v = (1/EI)(\text{moment of area of } M \text{ diagram from 0 to } z \text{ at } z). \qquad (7.8)$$

Example 7.5 Determine the maximum slope and deflection expressions for the cantilever beam in Fig. 7.6(a) using the moment area method.

With the corresponding hogging bending moment diagram in Fig. 7.6(b), Eqs (7.7) and (7.8) are applied with the origin at the fixed end and, since maximum values are required, with the datum at the free end. Then,

$$\theta = (1/EI)(\text{area of } M \text{ diagram between } z = 0 \text{ and } z = L)$$
$$= (1/EI)(M_{max}L/3) = (1/EI)[(wL^2/2)(L/3)] = wL^3/6EI,$$

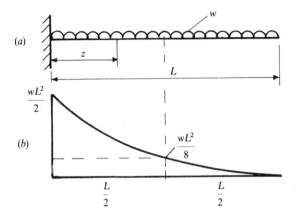

Figure 7.6 Moment area for cantilever.

$v = (1/EI)$(moment of area of M diagram between 0 and

L about the datum)

$= (1/EI)[(LM_{max}/3)(3L/4)] = (1/EI)[wL^2/2)(L/3)(3L/4)] = wL^4/8EI.$

Example 7.6 The cantilever in Fig. 7.7(a) is propped at point P to prevent deflection at that point. Find the free-end slope and deflection given $I = 50 \times 10^6 \, \text{mm}^4$ (2.5 in^4) and $E = 200 \, \text{GPa}$ (13 000 tonf/in^2).

SOLUTION TO SI UNITS The separated bending moment diagrams given in Fig. 7.7(b) correspond to the action of the loading (hogging) and that

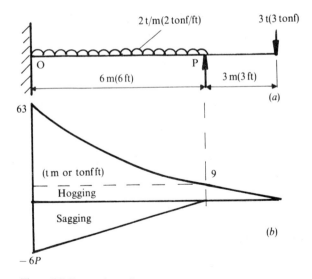

Figure 7.7 Propped cantilever.

of the prop (sagging). Then, from Eq. (7.8) since $v = 0$ at P

$$v = (1/EI)(\text{moment of area of } M\text{-diagrams } 0 - P \text{ about } P) = 0,$$
$$= (1/EI)[(6 \times 54/3)(3 \times 6/4) + (9 \times 6 \times 3) - (6P \times 6/2)(2 \times 6/3)] = 0,$$

from which $P = 9$ t. Equations (7.7) and (7.8) supply the free-end values:

$$\theta = (1/EI)(\text{area of } M \text{ diagram between } z = 0 \text{ and } z = 9 \text{ m})$$
$$= (1/EI)[(54 \times 6/3) + (9 \times 6) + (9 \times 3/2) - (6 \times 9 \times 6/2)]$$
$$= 13.5 (\text{t m}^2)/EI$$
$$= (13.5 \times 10^3 \times 9.81 \times 10^6)/(200 \times 10^3 \times 50 \times 10^6) = 0.0132 \text{ rad}.$$
$$v = (1/EI)(\text{moment of area of } M \text{ diagram between } z = 0 \text{ and } z = 9 \text{ m})$$
$$= (1/EI)\{9 \times 3 \times 2 \times 3)/(2 \times 3) + (9 \times 6 \times 6) + (54 \times 6/3)[3 + (3 \times 6/4)]$$
$$- (6 \times 9 \times 6/2)[3 + (2 \times 6/3)]\}$$
$$= (1/EI)(27 + 324 + 810 - 1134) = 27(\text{t m}^3)/EI$$
$$= (27 \times 10^3 \times 9.81 \times 10^9)/(200 \times 10^3 \times 50 \times 10^6) = 26.49 \text{ mm}.$$

SOLUTION TO IMPERIAL UNITS Note that the imperial units are not equivalent to the SI units. However, since the same numbers for length and loading have been retained, all the previous expressions apply. This gives $P = 9$ tonf and a free-end slope and deflection of

$$\theta = 13.5(\text{tonf ft}^2)/EI$$
$$= (13.5 \times 12^2)/(13\,000 \times 2.5) = 0.0598 \text{ rad},$$
$$v = 27(\text{tonf ft}^3)/EI$$
$$= (27 \times 12^3)/(13\,000 \times 2.5) = 1.436 \text{ in}.$$

Simply Supported Beams

In the application of Eqs (7.5) and (7.6) to find the slope and deflection at C, for the beam in Fig. 7.8, the origin is set at A and the datum at C. Then, under sagging moments the negative change in slope from A and C is, from Eq. (7.5):

$$\theta_C - \theta_A = -(1/EI)(\text{area of } M \text{ diagram between A and C}), \qquad (7.9)$$

where, provided the deflections are small,

$$\theta_A \simeq \Delta v_B/L. \qquad (7.10)$$

Now, with the datum at B the intercept Δv_B, made by the tangents at A and B, is directly supplied from Eq. (7.6) as

$$\Delta v_B = (1/EI)(\text{moment of area of } M \text{ diagram A–B about B}). \qquad (7.11)$$

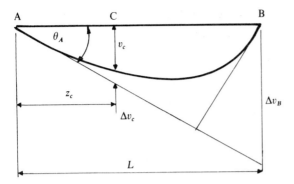

Figure 7.8 Deflection of a simply supported beam.

Combining Eqs (7.9)–(7.11) leads to

$$\theta_C = (1/L)(1/EI)(\text{moment of area of } M \text{ diagram A–B about B})$$
$$- (1/EI)(\text{area of } M \text{ diagram between A and C}). \qquad (7.12)$$

The intercept between the tangents drawn at A and C with datum at C is

$$\Delta v_C = (1/EI)(\text{moment of area of } M \text{ diagram A–C about C}). \qquad (7.13)$$

Now, the true deflection at C is

$$v_C = (z_C/L) \Delta v_B - \Delta v_C. \qquad (7.14)$$

Combining Eqs (7.11), (7.13) and (7.14) leads to

$$v_C = (z_C/L)(1/EI)(\text{moment of area of } M \text{ diagram A–B about B})$$
$$- (1/EI)(\text{moment of area of } M \text{ diagram A–C about C}). \qquad (7.15)$$

Example 7.7 Find expressions for the slope and deflection beneath the load W, for the beam in Fig. 7.9(a).

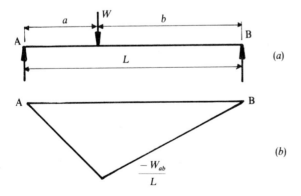

Figure 7.9 Offset load.

From the bending moment diagram in Fig. 7.9(b), the direct application of Eqs (7.12) and (7.15) gives

$$EI\theta = (1/L)(\text{moment of } M \text{ diagram A–B at B})$$
$$- (\text{area of } M \text{ diagram A–W})$$
$$= (1/L)[(Wa^2b/2L)(b + a/3) + (Wab^2/2L)(2b/3)] - Wab^2/2L$$
$$= (Wab/2L^2)(ab + a^2/3 + 2b^2/3 - bL) = (Wab/2L^2)(a^2 - b^2)/3$$
$$= Wab(a - b)/6L.$$
$$EIv = (a/L)(\text{moment of } M \text{ diagram A–B at B})$$
$$- (\text{moment of } M \text{ diagram A–W at W})$$
$$= (a/L)[(Waba/2L)(b + a/3) + (Wabb/2L)(2b/3)] = Wa^2b^2/3EIL.$$

Example 7.8 Calculate the slope and deflection at point C, 40 mm from B, for the centrally loaded, simply supported beam in Fig. 7.10(a). Take $EI = 9 \times 10^6$ Nmm^2, constant throughout the length.

Figures 7.10(b) and (c) are the bending moment and deflection diagrams respectively. The direct application of Eq. (7.12) then gives

$$\theta_C = (1/300)(1/EI)[(3750 \times 150/2)(150 + 50 + 100)]$$
$$- (1/EI)[(150 \times 3750/2) + (110 \times 1000) + (2750 \times 110/2)]$$
$$= (0.2813 - 0.5425)10^6/(9 \times 10^6) = -0.029 \text{ rad.}$$

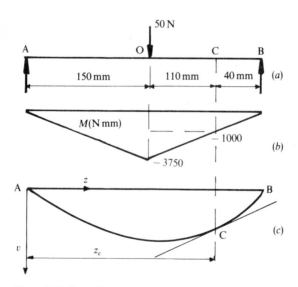

Figure 7.10 Centrally loaded beam.

The minus sign arises because v is measured downwards in $\theta = \mathrm{d}v/\mathrm{d}z$ and thus the slope is negative in our z, v coordinates. Note that, since the slope at the centre O is zero, Eq. (7.5) may also be applied with O as the origin. Then the negative slope change O–C is

$$\theta_C - \theta_O = -(1/EI)(\text{area of } M \text{ diagram between C and O})$$
$$\theta_C = -(1/EI)[(110 \times 1000) + (2750 \times 110/2)]$$
$$= -261.25 \times 10^3 \,(\mathrm{N\,mm^2})/EI$$
$$= -(261.25 \times 10^3)/(90 \times 10^5) = -0.029 \,\mathrm{rad}.$$

The direct application of Eq. (7.15) gives

$$v_C = (260/300)(1/EI)[(3750 \times 150/2)(150 + 50 + 100)]$$
$$- (1/EI)[(3750 \times 150/2)(110 \times 50)$$
$$+ (1000 \times 110 \times 55) + (2750 \times 110/2)(2 \times 110/3)]$$
$$= (73.125 - 62.142)10^6/(9 \times 10^6) = 1.22 \,\mathrm{mm}.$$

Encastre Beams

When both the ends of the beam in Fig. 7.11(a) are built-in, the slope and deflection at each end is zero. In Fig. 7.11(b) the net bending moment diagram, shown in (c), is decomposed into its free sagging components, with the applies

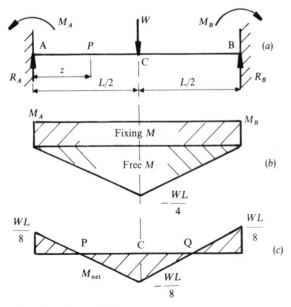

Figure 7.11 Free and fixing moments.

loading simply supported and its fixing hogging component determined by the fixed-end moments.

It follows from Eqs (7.12) and (7.15) that, with the origin at A and the datum at B (or vice versa) in Fig. 7.11(a)

$\theta_B = (1/EI)(\text{area of fixing } M \text{ diagram} - \text{area of free } M \text{ diagram}) = 0,$

$v_B = (1/EI)(\text{moment of area of fixing } M \text{ diagram at B}$
$- \text{moment of area of free } M \text{ diagram at B}) = 0.$

From which it follows that for an encastre beam;

$$\text{Area of free moment diagram} = \text{area of fixing moment diagram.} \quad (7.16)$$

$$\text{Moment of free } M \text{ diagram} = \text{moment of fixing } M \text{ diagram.} \quad (7.17)$$

In general, the net support reactions, R_A and R_B are composed of the free reactions R_1 and R_2 plus the end reaction R due to the couple $M_A - M_B$ that arises when the fixing moments are unequal ($M_A \neq M_B$). Assuming $M_A > M_B$ (as in Fig. 7.12) and noting the corresponding direction of each end reaction, these become

$$R_A = (R_1 + R) = R_1 + (M_A - M_B)/L, \quad (7.18)$$

$$R_B = (R_2 - R) = R_2 - (M_A - M_B)/L. \quad (7.19)$$

Example 7.9 Construct the bending moment diagrams for the beam with a central load in Fig. 7.11(a), and find the fixed-end moments, the maximum deflection, the points of contraflexure and the slopes at these points.

In Fig. 7.11(b) the maximum free moment at the centre is

$$R_A L/2 = R_B L/2 = -(W/2)(L/2) = -WL/4 \quad \text{(sagging).}$$

Equation (7.16) gives the fixing moments:

$$(WL/4)(L/2) = M_A L, \quad \Rightarrow M_A = WL/8 = M_B \quad \text{(hogging).}$$

And at the centre, the net moment in Fig. 7.11(c) is

$$WL/8 - WL/4 = -WL/8.$$

The maximum deflection at centre C is given by

$v_{\max} = (1/EI)(\text{moment of net } M \text{ diagram from A to C about C})$
$= (1/EI)(\text{moment of fixing–free } M \text{ diagrams from A to C about C})$
$= (1/EI)[(L/2)(WL/8)(L/4) - (WL/4)(L/4)(1/3)(L/2)] = WL^3/192EI.$

Contraflexure, points P and Q:

$$M = 0 = M_A - R_A z, \quad \Rightarrow z = M_A/R_A = (WL/8)(2/W) = L/4.$$

Slope at $z = L/4$:

$$\theta = (1/EI)(\text{area of net } M \text{ diagram from A to P})$$
$$= (1/EI)[(1/2)(WL/8)(L/4)] = WL^2/64EI,$$

or,

$$\theta = (1/EI)(\text{area of fixing–free } M \text{ diagrams from A to P})$$
$$= (1/EI)[(L/4)(WL/8) - (L/8)(WL/8)] = WL^2/64EI.$$

Example 7.10 Establish the net bending moment diagram and the points of contraflexure for the beam in Fig. 7.12(a).

For the free M diagram, the support reactions are

$$aW = LR_2, \qquad \Rightarrow R_2 = aW/L,$$
$$R_1 + R_2 = W, \quad \Rightarrow R_1 = bW/L,$$
$$\therefore \quad M_{max} = aR_1 = bR_2 = Wab/L.$$

Then, from applying Eqs (7.16) and (7.17) to Fig. 7.12(b),

$$(L/2)(Wab/L) = M_B L + (M_A - M_B)L/2,$$
$$\therefore \quad M_A + M_B = Wab/L; \tag{i}$$

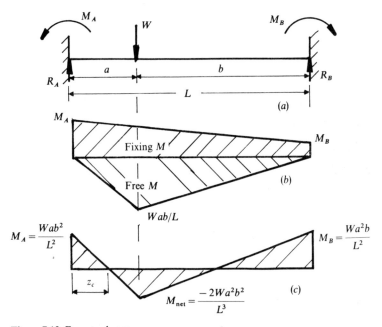

Figure 7.12 Encastre beam.

$$(Wab/L)(b/2)(2b/3) + (Wab/L)(a/2)(b+a/3)$$
$$= M_B L(L/2) + (M_A - M_B)(L/2)(2L/3)$$
$$= 2M_A + M_B = (Wab/L^3)(a^2 + 3ab + 2b^2). \tag{ii}$$

Subtract Eq. (i) from Eq. (ii):

$$M_A = (Wab/L^3)(a^2 + 3ab + 2b^2) - Wab/L = Wab^2/L^2$$
$$M_B = Wab/L - Wab^2/L^2 = Wa^2b/L^2.$$

Then, substituting into Eqs. (7.18) and (7.19):

$$R_A = bW/L + (Wab^2/L^3 - Wa^2b/L^3)/L = W(3ab^2 + b^3)/L^3$$
$$R_B = aW/L - (Wab^2/L^3 - Wa^2b/L^3)/L = W(3a^2b + a^3)/L^3.$$

Check:

$$R_A + R_B = W(a^3 + 3a^2b + 3ab^2 + b^3)/L^3 = W(a+b)^3/L^3 = W.$$

The net moment M_{net} beneath the load W is found from (Fixing moment − free moment at W position), that is,

$$M_{net} = M_B + (M_A - M_B)(b/L) - Wab/L$$
$$= Wa^2b/L^2 + (Wab^2/L^2 - Wa^2b/L^2)(b/L) - Wab/L$$
$$= (Wab/L^3)(aL + b^2 - ab - L^2) = -2Wa^2b^2/L^3.$$

The left contraflexure point in the net moment diagram (Fig. 7.12c) is

$$M = 0 = M_A - R_A z,$$
$$z = M_A/R_z = (Wab^2/L^2)[L^3/W(3ab^2 + b^3)] = aL/(b + 3a).$$

Continuous Beams

Mohr's theorems may be applied to a beam that rests on more than two supports. Consider any two successive spans L_1 and L_2 in Fig. 7.13(a) under arbitrary loading where the moments existing at the ends A, B and C are M_A, M_B and M_C respectively. The corresponding free and fixing moment diagrams in Fig. 7.13(b) take the opposite sense, such that the shaded areas A_1 and A_2 in each bay, which are not common to both, form the net moment diagram given in Fig. 7.13(c). In the general case, let the beam rest on non-level supports at positive vertical distances v_A, v_B and v_C from any horizontal axis, OO, as shown in Fig. 7.13(d). Furthermore, let the second moments of area I_1 and I_2 for each bay differ.

The centroidal positions $\bar{z}_{1,2}$ and $z_{1,2}$ refer to the free and net moment areas S and A in each bay respectively. The slope of the tangent at B is θ_B and the intercepts made by this tangent with vertical datums passing through the tangents at the ends A and C are Δv_A and Δv_C. With the origin at B, Eq. (7.6) yields

$$\Delta v_A = A_1 z_1/EI_1 = \theta_B L_1 + (v_B - v_A), \tag{7.20a}$$

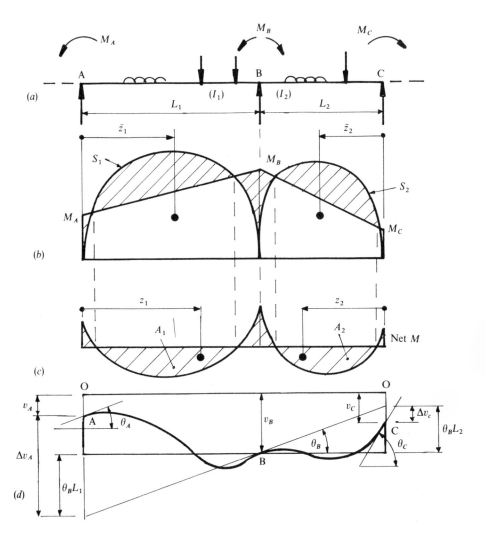

Figure 7.13 Continuous beam.

$$\Delta v_C = A_2 z_2 / EI_2 = -[\theta_B L_2 - (v_B - v_C)], \qquad (7.20b)$$

in which Δv_C is negative upwards. Adding Eqs. (7.20a and b) leads to

$$A_1 z_1 / EI_1 L_1 + A_2 z_2 / EI_2 L_2 = (v_B - v_A)/L_1 + (v_B - v_C)/L_2, \qquad (7.21)$$

but from Fig. 7.13(b) and (c), with hogging positive,

$$A_1 z_1 = \{[M_A L_1^2/2 + (M_B - M_A)L_1^2/3] - S_1 \bar{z}_1\}, \qquad (7.22a)$$

$$A_2 z_2 = \{[M_C L_2^2/2 + (M_B - M_C)L_2^2/3] - S_2 \bar{z}_2\}. \qquad (7.22b)$$

Substituting Eqs (7.22) into Eq. (7.21) leads to a theorem of three moments, due to Clapeyron:

$$M_A L_1 / I_1 + 2 M_B (L_1/I_1 + L_2/I_2) + M_C L_2 / I_2$$
$$= 6[S_1 \bar{z}_1 / I_1 L_1 + S_2 \bar{z}_2 / I_2 L_2] + 6E[(v_B - v_A)/L_1 + (v_B - v_C)/L_2]. \quad (7.23a)$$

By taking the spans in pairs, sufficient equations are obtained to solve simultaneously for all the fixing moments. The method is considerably simplified when, for a uniform beam $(I_1 = I_2)$ resting on level supports $(v_A = v_B = v_C)$, Eq. (7.23a) becomes

$$M_A L_1 + 2 M_B (L_1 + L_2) + M_C L_2 = 6(S_1 \bar{z}_1 / L_1 + S_2 \bar{z}_2 / L_2). \quad (7.23b)$$

Consider now the application of Mohr's first theorem to Fig. 7.13. From Eq. (7.5), with the origin at B for each bay:

$$\theta_A - \theta_B = A_1 / EI_1 = -(1/EI_1)[(M_A + M_B) L_1/2 - S_1], \quad (7.24a)$$

$$\theta_C - \theta_B = A_2 / EI_2 = -(1/EI_2)[(M_B + M_C) L_2/2 - S_2]. \quad (7.24b)$$

These changes in slope are both shown to be negtive for v positive downwards with positive z emanating from B within each bay. Within the first bay,

$$\theta_B = \Delta v_A / L_1 - (v_B - v_A)/L_1 = A_1 \bar{z}_1 / EI_1 L_1 - (v_B - v_A)/L_1.$$

Substituting from Eq. (7.22a),

$$\theta_B = -S_1 \bar{z}_1 / EI_1 L_1 + (L_1/6EI_1)(M_A + 2 M_B) - (v_B - v_A)/L_1. \quad (7.25a)$$

Then from Eqs (7.24a) and (7.25a),

$$\theta_A = (S_1/EI_1)(1 - \bar{z}_1/L_1) - (L_1/6EI_1)(M_B + 2 M_A) - (v_B - v_A)/L_1. \quad (7.26a)$$

Within the second bay,

$$\theta_B = \Delta v_C / L_2 + (v_B - v_C)/L_2 = A_2 z_2 / EI_2 L_2 + (v_B - v_C)/L_2.$$

Substituting from Eq. (7.22b),

$$\theta_B = -S_2 \bar{z}_2 / EI_2 L_2 + (L_2/6EI_2)(M_C + 2 M_B) + (v_B - v_C)/L_2. \quad (7.25b)$$

Then, from Eqs (7.24b) and (7.25b),

$$\theta_C = (S_2/EI_2)(1 - \bar{z}_2/L_2) - (L_2/6EI_2)(M_B + 2 M_C) + (v_B - v_C)/L_2. \quad (7.27a)$$

Equations (7.25)–(7.27) are again simplified for a uniform beam resting on level supports, when they become

$$\theta_B = -S_1 \bar{z}_1 / EIL_1 + (L_1/6EI)(M_A + 2 M_B)$$
$$= -S_2 \bar{z}_2 / EIL_2 + (L_2/6EI)(M_C + 2 M_B), \quad (7.25c)$$

$$\theta_A = (S_1/EI)(1 - \bar{z}_1/L_1) - (L_1/6EI)(M_B + 2 M_A), \quad (7.26b)$$

$$\theta_C = (S_2/EI)(1 - \bar{z}_2/L_2) - (L_2/6EI)(M_B + 2 M_C). \quad (7.27b)$$

Thus, when the moments M_A, M_B and M_C have been established from

Clapeyron's theorem, Eqs (7.25)–(7.27) enable the slopes θ_A, θ_B and θ_C at the support points to be found. The deflection and slope at any other points in a continuous beam may be found from integrating Mohr's theorems with the origin O located at a support point of known slope. Hence, at a given datum within the first bay, distance z from the origin at O at A in Fig. 7.13(c), Eqs (7.5) and (7.6) become

$\theta = \theta_A + (1/EI)$(area of net M diagram from O to z),

$v = \theta_A z + (1/EI)$(moment of net M diagram from O to z about z),

where the areas and moment of areas of the net M diagrams refer to the difference between the areas and moments of areas for the free and fixing moment diagrams. Normally these are respectively composed of negative sagging and positive hogging components. The shaded area in Fig. 7.13(b) and (c) is the area of the net bending moment diagram according to this convention.

Example 7.11 Confirm the bending moment diagram, previously found from moment distribution (see Fig. 5.10(a), page 105) for the two-bay beam in Fig. 7.14(a) (SI units) by using the three-moment theorem. Find the support reactions, the slopes and deflections beneath the 10 kN force in the first bay and also at the centre of the second bay. Take $EI = 105 \, \text{MNm}^2$.

Since $M_A = M_C = 0$ for a two-span beam, Eq. (7.23b) becomes

$$2M_B(L_1 + L_2) = 6(S_1 \bar{z}_1 / L_1 + S_2 \bar{z}_2 / L_2). \qquad \text{(i)}$$

In the application to the corresponding free moment diagrams in Fig. 7.14(b), this equation becomes

$2M_B(15 + 12) = 6[(1/15)$(moment of area of free L.H. M diagram about A)

$\qquad\qquad + (1/12)$(moment of area of free R.H. M diagram about C)$]$.

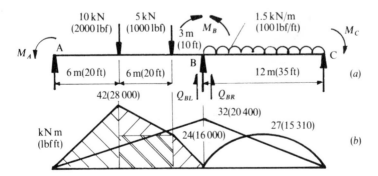

Figure 7.14 Two-bay beam.

Taking L.H. triangular and rectangular constituent areas (kN/m^2) and noting that the R.H. area of a parabola is $2Lh/3$, it follows that Eq. (i) becomes

$$(54/6)M_B = (1/15)[(42 \times 6 \times 2 \times 6/2 \times 3) + (18 \times 6 \times 8/2) + (24 \times 6 \times 9)$$
$$+ (24 \times 3 \times 13/2)] + (1/12)(2 \times 12 \times 27 \times 6/3)$$
$$9M_B = 180 + 108, \quad \Rightarrow M_B = 32\,kN\,m.$$

Let the support reactions and fixing moment M_B act as shown in Fig. 7.14(a). The central reaction R_B is composed of its left- and right-hand bay components Q_{BL} and Q_{BR}. Although M_A and M_C are both zero here, they are included to extend the generality to any continuous beam. Moments about B in bay AB:

$$(5 \times 3) + (10 \times 9) - 15R_A + (M_A - M_B) = 0, \quad \Rightarrow R_A = 4.867\,kN.$$

Moments about A in bay AB:

$$(6 \times 10) + (12 \times 5) + (M_B - M_A) - 15Q_{BL} = 0, \quad \Rightarrow Q_{BL} = 10.13\,kN.$$

Moments about B in bay BC:

$$(1.5 \times 12 \times 6) + (M_C - M_B) - 12R_C = 0, \quad \Rightarrow R_c = 6.333\,kN.$$

Moments about C in bay BC:

$$(1.5 \times 12 \times 6) + (M_B - M_C) - 12Q_{BR} = 0, \quad \Rightarrow Q_{BR} = 11.67\,kN.$$
$$\therefore \quad R_B = Q_{BL} + Q_{BR} = 10.13 + 11.67 = 21.80\,kN.$$

Check:

$$\sum F = [10 + 5 + (12 \times 1.5)] - [4.867 + 6.333 + 21.80] = 0.$$

The slopes at A and C follow from Eqs (7.26b) and (7.27b) with $M_A = M_C = 0$:

$$\theta_A EI = S_1(1 - \bar{z}_1/L_1) - L_1 M_B/6, \tag{i}$$

$$\theta_C EI = S_2(1 - \bar{z}_2/L_2) - L_2 M_B/6, \tag{ii}$$

where $S_1 \bar{z}_1/L_1 = 180\,kNm^2$ and $S_2 \bar{z}_2/L_2 = 108\,kN\,m^2$ were previously calculated in the three moment expression. Referring again to Fig. 7.14(b), the free moment diagram areas are;

$$S_1 = (42 \times 6/2) + (18 \times 6/2) + (24 \times 6) + (24 \times 3/2) = 360\,kN\,m^2,$$
$$S_2 = 2Lh/3 = 2 \times 12 \times 27/3 = 216\,kN\,m^2.$$

Note that fixing moments of 12.8 and 16 kN m exist at each point within bays 1 and 2 respectively in proportion to $M_B = 32\,kN\,m$. Then, from Eqs (i) and (ii), the slopes are

$$\theta_A EI = (360 - 180) - (15 \times 32/6) = 100\,kN\,m^2$$
$$\theta_C EI = (216 - 108) - (12 \times 32/6) = 44\,kN\,m^2.$$

Now from Eqs (7.5c) and (7.6c) at the 10 kN force point, where $z = 6$ m from A,

$$\theta EI = 100 + [(12.8 \times 6/2) - (42 \times 6/2)] = 12.4 \text{ kN m}^2,$$
$$\therefore \quad \theta = 12.4 \times 10^{-5} \text{ rad};$$
$$vEI = (100 \times 6) + [(12.8 \times 6 \times 6)/(2 \times 3) - (42 \times 6 \times 6)/(2 \times 3)]$$
$$= 424.8 \text{ kN m}^3,$$
$$v = 4.25 \text{ mm};$$

and at the centre of the second bay, where $z = 6$ m from C, Eq. (7.5c) gives

$$\theta EI = 44 + [(16 \times 6/2) - (2 \times 6 \times 27/3)] = -16 \text{ kN m}^2,$$
$$\theta = -16 \times 10^{-5} \text{ rad}.$$

Using the centroid position $(\bar{x})$ for the parabola derived on page 9, Eq. (7.6c) gives

$$vEI = (44 \times 6) + [(16 \times 6 \times 6)/(2 \times 3) - (2 \times 6 \times 27 \times 3 \times 6)/(3 \times 8)]$$
$$= 117 \text{ kN m}^3,$$
$$v = 17 \text{ mm}.$$

SOLUTION TO IMPERIAL UNITS Applying Eq. (i) to the free moment diagrams (lbf ft) for each bay:

$$2M_B(50 + 35) = 6\{[(28\,000 \times 20 \times 2 \times 20)/(2 \times 3 \times 50)]$$
$$+ [(40 + 10/3)(16\,000 \times 10)/(2 \times 50)]$$
$$+ [(20 + 20/3)(12\,000 \times 20)/(2 \times 50)]\}$$
$$+ 6[(2 \times 35 \times 15\,310 \times 35)/(3 \times 35 \times 2)],$$

from which $M_B = 20\,400$ lbf ft.

Example 7.12 Confirm the fixing moments prevously found for the beam in Fig. 7.15(a) (see Fig. 5.11) using Clapeyron's theorem.

The free-moment diagrams are given in Fig. 7.15(b). The fixing moments M_A and M_C, together with the central support moment M_B, are unknowns for this continuous beam. M_B is assumed to be the greatest of these to correspond with Fig. 7.13(b) and the resulting form of the three-moment theorem. Since $S_1\bar{z}_1$ and $S_2\bar{z}_2$ are the moments of area of the free M diagrams about the left- and right-hand ends respectively, Eq. (7.23a) becomes

$$15M_A/2I + 2M_B(15/2I + 10/I) + 10M_C/I$$
$$= 6\{(2 \times 15 \times 28.125 \times 7.5/3 \times 2I \times 15)$$
$$+ [(24 \times 6 \times 6/2) + (24 \times 4 \times 2 \times 4/2 \times 3)]/I \times 10\},$$
$$\therefore \quad M_A + 4.67M_B + 1.33M_C = 101.05. \qquad \text{(i)}$$

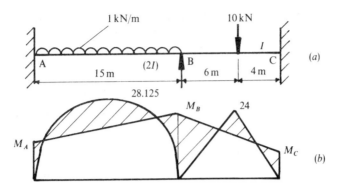

Figure 7.15 Two-bay beam.

Two further relationships are found from the application of Eq. (7.6c) to each bay. Since the deflection intercept between the tangents at A and B is zero with the origin at A and the datum at B, then

$$\Delta v_B = (1/2EI)(\text{moment of } M \text{ diagram from A to B about B}) = 0,$$
$$(15M_A \times 7.5) + (M_B - M_A)(15 \times 15/2 \times 3) - (2 \times 15 \times 28.125 \times 7.5/3) = 0$$
$$2M_A + M_B = 56.25. \tag{ii}$$

Furthermore, the deflection intercept between the tangents at C and B is zero with the origin at C and the datum at B. Thus,

$$\Delta v_B = (1/EI)(\text{moment of } M \text{ diagram from C to B about B}),$$
$$(10M_C \times 5) + (M_B - M_C)(10 \times 10/2 \times 3) - [(24 \times 4 \times 7.333/2)$$
$$+ (24 \times 6 \times 2 \times 6/2 \times 3)]/10 = 0,$$
$$M_B + 2M_C = 38.4. \tag{iii}$$

The simultaneous solution to Eqs (i)–(iii) gives $M_A = 21.365$, $M_B = 13.52$ and $M_C = 12.44$ k N m, in good agreement with the moment distribution solutions in Table 5.3 on page 107. Note that these analyses show that M_C is the least moment.

7.4 STEP FUNCTION APPROACH—MACAULAY'S METHOD

It can be seen with the successive integration of the flexure, Eq. (7.4), that discontinuities will arise in the bending moment expression when z passes points of concentrated forces, moments and abrupt changes in distributed loading. Consequently, as the complexity of the loading increases, the greater becomes the number of differential equations and integration constants to be solved. Fortunately, the amount of work involved may be lessened considerably when the moment expression $M[z]$ in Eq. (7.4) is replaced with the step function $M[z-a]$, in which a defines the points at which discontinuities would arise. Macaulay showed in 1919 that his versatile method permitted successive integration of a single step function. This may readily be applied to simply-supported and single-span encastre beams by following the general rules.

1. Take the origin at the left-hand end.
2. When necessary extend and counterbalance uniformly distributed loading to the right-hand extremity.
3. Establish the function, $M[z-a]$, working to the left of a datum in the last right portion of the beam to the right. Hogging moments are positive.
4. Concentrated moments appear in the form $M[z-a]^0$.
5. Integrate such terms as $[z-a]$ in the form $(1/2)[z-a]^2$. These are to be ignored when in the application of the resulting expressions the value within the brackets becomes negative.
6. Apply the known slope and deflection values at the boundary to find the two constants of integration.

Application to Simply Supported Beams

Following the integration of the step function, the constants are found from applying the boundary condition that the displacement are zero at the support points. The maximum deflection position, which corresponds to that of zero slope, is assumed to lie within an estimated range for z.

Example 7.13 Find the position and magnitude of the maximum deflection and the deflection beneath each force for the beam in Fig. 7.16. Take $E = 207\,\text{GPa}$, $I = 10^{10}\,\text{mm}^4$.

The L.H. reaction R_A, is first found from the R.H. moment equilibrium:

$$(8 \times 10) + (12 \times 4) = 20R_A, \quad \Rightarrow R_A = 6.4\,\text{t} \qquad (R_B = 7.6\,\text{t is unnecessary}).$$

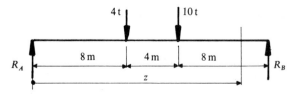

Figure 7.16 Simply supported beam with concentrated loading.

With the datum for M distance z from the L.H. end as shown,

$$EI\,d^2v/dz^2 = -6.4z + 10[z-12] + 4[z-8],$$
$$EI\,dv/dz = -6.4z^2/2 + 10[z-12]^2/2 + 4[z-8]^2/2 + C_1,$$
$$EIv = -3.2z^3/3 + 5[z-12]^3/3 + 2[z-8]^3/3 + C_1z + C_2.$$

The boundary conditions are $v = 0$ when $z = 0$, giving $C_2 = 0$ (ignore $[-]$), and $v = 0$ when $z = 20$ m, which gives $C_1 = 326.25\,(\text{t m}^2)$,

$$\therefore \quad EIv = -1.066z^3 + 5[z-12]^3/3 + 2[z-8]^3/3 + 326.25z.$$

At $z = 8$ m.

$$EIv = -1.066(8)^3 + 326.25(8) = 2064.21\,\text{t m}^3,$$
$$v = (2064.21 \times 10^3 \times 9.81 \times 10^9)/(207 \times 10^3 \times 10^{10}) = 9.783\,\text{mm}.$$

At $z = 12$ m,

$$EIv = -1.066(12)^3 + 2[4]^3/3 + 326.25(12) = 2115.62\,\text{t m}^3,$$
$$v = (2115.62/2064.21)9.783 = 10.03\,\text{mm}.$$

Assuming that v_{max} lies between the loads, then for $dv/dz = 0$,

$$-3.2z^2 + 2[z-8]^2 + 326.25 = 0 \qquad \text{for } 8 \le z \le 12\,\text{m}.$$

The solution $z = 10.3$ m confirms that the assumed range for z is correct.

$$EIv_{max} = -1.066(10.3)^3 + 2[2.3]^3/3 + 326.25(10.3) = 2203.64\,\text{t m}^3,$$
$$v_{max} = (2203.63/2064.21)9.783 = 10.44\,\text{mm}.$$

Example 7.14 Find the slope at the left support and the position and magnitude of the maximum deflection for the beam in Fig. 7.17. Take $EI = 150\,\text{MN m}^2$.

Firstly, the L.H. reaction (R_1) is calculated from

$$(200 \times 1) + (150 \times 1.5) + 100 = 4R_1,$$
$$\Rightarrow R_1 = 131.25\,\text{kN} \quad (R_2 = 218.75\,\text{kN}).$$

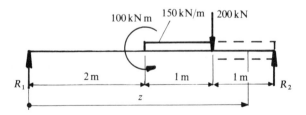

Figure 7.17 Simply supported beam with mixed loading.

Then, in following rules (1) and (2) (page 215) for distributed loading and concentrated moments in particular,

$$EI\, d^2v/dz^2 = M = -13.125z + 100[z-2]^0 + 200[z-3]$$
$$+ 150[z-2]^2/2 - 150[z-3]^2/2,$$
$$EI\, dv/dz = -131.25z^2/2 + 100[z-2] + 200[z-3]^2/2$$
$$+ 75[z-2]^3/3 - 75[z-3]^3/3 + C_1,$$
$$EIv = -131.25z^3/6 + 100[z-2]^2/2 + 100[z-3]^3/3$$
$$+ 75[z-2]^4/12 - 75[z-3]^4/12 + C_1 z + C_2.$$

The boundary conditions are $v = 0$ when $z = 0$, which gives $C_2 = 0$, and $v = 0$ when $z = 4\,$m. This gives $C_1 = 268.23\,\text{kN m}^2$ from

$$0 = -131.25(4)^3/6 + 100[2]^2/2 + 100[1]^3/3$$
$$+ 75[2]^4/12 - 75[1]^4/12 + 4C_1.$$

The slope at the L.H. support, where $z = 0$, is

$$EI\, dv/dz = C_1 = 268.23,$$
$$dv/dz = 268.23/(150 \times 10^3) = 1.788 \times 10^{-3}\,\text{rad}.$$

Assuming that the condition, $dv/dz = 0$, for maximum deflection, occurs in the region $0 \le z \le 2$, the slope equation gives

$$EI\, dv/dz = 0 = -131.25z^2/2 + 268.23, \quad \Rightarrow z = 2.022\,\text{m}.$$

Since this solution refutes the assumed position, a second assumption, that v_{max} lies in the region $2 \le z \le 3$, leads to

$$EI\, dv/dz = 0 = -131.25z^2/2 + 100[z-2] + 75[z-2]^3/3 + 268.23,$$
$$z^3 - 8.625z^2 + 16z - 5.271 = 0.$$

The solution, $z = 2.035\,$m, agrees with the assumed position. Then, from the deflection equation, the maximum value is

$$EIv = -131.25(2.035)^3/6 + 100[0.035]^2/2 + 75[0.035]^4/12$$
$$+ (268.23 \times 2.035) = 361.56,$$
$$\therefore \quad v = 361.56/(150 \times 10^3) = 2.41 \times 10^{-3}\,\text{m} = 2.41\,\text{mm}.$$

Application To Encastre Beams

The Macaulay method may be readily extended to find the slopes and deflections of encastre beams. A complete solution to the support reactions and fixing moments involves the simultaneous use of Eqs (7.18) and (7.19), as the following example shows.

Example 7.15 Determine the support reactions, the fixing moments and

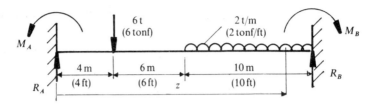

Figure 7.18 Encastre beam.

the maximum deflection for the encastre beam in Fig. 7.18. Take $EI = 150\,\text{MN}\,\text{m}^2$ ($13.5 \times 10^4\,\text{tonf in}^2$).

SOLUTION TO SI UNITS The L.H. reaction R_A, and fixing moment, M_A, now appear in the flexure equation as additional constants. They are determined from the boundary condition of zero slope and deflection at the R.H. end.

$$EI\,\mathrm{d}^2v/\mathrm{d}z^2 = M = M_A - R_A z + 6[z - 4] + 2[z - 10]^2/2,$$

$$EI\,\mathrm{d}v/\mathrm{d}z = M_A z - R_A z^2/2 + 6[z - 4]^2/2 + [z - 10]^3/3 + C_1, \quad \text{(i)}$$

$$EIv = M_A z^2/2 - R_A z^3/6 + 3[z - 4]^3/3 + [z - 10]^4/12$$
$$+ C_1 z + C_2. \quad \text{(ii)}$$

The L.H. boundary conditions are $\mathrm{d}v/\mathrm{d}z = 0$ and $v = 0$ for $z = 0$. It follows from Eqs (i) and (ii) that $C_1 = C_2 = 0$, as is always the case for an encastre beam. Also, $\mathrm{d}v/\mathrm{d}z = 0$ and $v = 0$ when $z = 20\,\text{m}$. From Eqs (i) and (ii):

$$0 = 20M_A - R_A(20)^2/2 + 3[16]^2 + [10]^3/3,$$

$$\therefore \quad 10R_A - M_A = 55.065; \quad \text{(iii)}$$

$$0 = (20)^2 M_A/2 - (20)^3 R_A/6 + [16]^3 + [10]^4/12,$$

$$\therefore \quad 6.667R_A - M_A = 24.617. \quad \text{(iv)}$$

Subtracting Eq. (iii) from (iv) gives $R_A = 9.13\,\text{t}$ and $M_A = 36.2\,\text{t m}$. Assume that v_{max} lies in $4 \le z \le 10$. Then with $\mathrm{d}v/\mathrm{d}z = 0$ in Eq. (i),

$$36.2z - 9.13z^2/2 + 3[z - 4]^2 = 0,$$

$$z^2 - 7.796z - 30.67 = 0, \quad \Rightarrow z = 10.67\,\text{m},$$

which indicates that the assumed position was strictly incorrect. Assuming now that v_{max} occurs in the region $10 \le z \le 20$, then, from Eq. (i),

$$z^3 - 34.695z^2 + 336.6z - 856 = 0.$$

The solution $z = 10.7\,\text{m}$ in this case is, however, only slightly different from that previously found. Equation (ii) supplies the maximum deflection:

$$EIv_{max} = 36.2(10.7)^2/2 - 9.13(10.7)^3/6 + [10.7 - 4]^3 + [10.7 - 10]^4/12$$
$$= 508.95 \, t \, m^3,$$
$$v_{max} = (508.95 \times 9.81)/(150 \times 10^3) = 0.0333 \, m = 33.33 \, mm.$$

The R.H. support reaction and fixing moment are found from Eqs (7.18) and (7.19). The free reactions R_1 and R_2 are supplied from equilibrium conditions with simple supports. Then,

$$(4 \times 6) + (10 \times 2 \times 15) = 20R_2, \quad \Rightarrow R_2 = 16.2 \, t,$$
$$R_1 = (6 + 20) - 16.2 = 9.8 \, t,$$
$$M_B = M_A - L(R_A - R_1) = 36.2 - 20(9.13 - 9.8) = 49.6 \, t \, m,$$
$$R_B = R_2 - (M_A - M_B)/L = 16.2 - (36.2 - 49.6)/20 = 16.87 \, t.$$

SOLUTION TO IMPERIAL UNITS Since the same numbers for forces and lengths have been retained, Eqs (i)–(iv) again apply. It follows that the maximum deflection is found from

$$EIv_{max} = 508.95 \, tonf \, ft^3$$
$$\therefore \quad v_{max} = (508.95 \times 12^3)/(13.5 \times 10^4) = 6.52 \, in.$$

7.5 ASYMMETRIC DEFLECTION

In all the forgoing examples it has been assumed that the principal axes of area, u and v, are coincident with respective horizontal and vertical axes x and y in the cross-section. This applies to symmetrical sections where, with loading parallel to y and $I_x = I_u$, the calculated deflections are the true vertical values. The most straightforward method for calculating deflections of asymmetric cross-sections is to apply any one of the usual methods— successive integration, moment area or Macaulay—separately to the two principal axes u and v. The deflection δ_v in the v-direction is associated with the use of I_u in the chosen method, while for the deflection δ_u in the u-direction, I_v must be employed. Since neither δ_u nor δ_v is aligned with the vertical direction, the vertical (δ_y) and horizontal (δ_x) deflection components may then be found by resolving the δ_u and δ_v deflections. This method relies on the calculation of I_u and I_v from I_x, I_y and I_{xy} and the inclination of the u, v axes to x, y by the methods outlined in Chapter 1. Consider the corresponding *positive* deflections δ, produced along the *negative* x-, y-, u- and v-directions (Fig. 17.19), either by the action of a vertical force F, applied in the negative y-direction (Fig. 7.19a), or by a sagging moment M, applied about the x-axis (Fig. 7.19b).

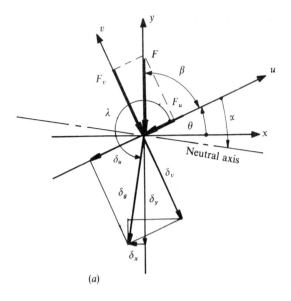

(a)

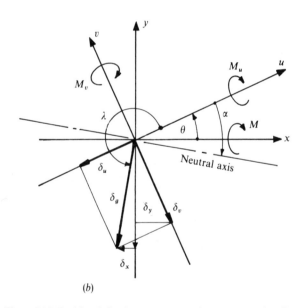

(b)

Figure 7.19 Positive deflection components in u-, v-, x- and y-directions.

It follows from the geometry in each figure that, with α negative:

$$\delta_x = \delta_u \cos\theta - \delta_v \sin\theta = -\delta_g \sin(\theta + \alpha), \tag{7.28a}$$

$$\delta_y = \delta_u \sin\theta + \delta_v \cos\theta = \delta_g \cos(\theta + \alpha), \tag{7.29a}$$

where the movement of the centroid δ_g, is defined in magnitude by

$$\delta_g = \sqrt{[(\delta_x)^2 + (\delta_y)^2]} = \sqrt{[(\delta_u)^2 + (\delta_v)^2]}, \tag{7.30a}$$

and in direction, w.r.t. the u-axis by

$$\lambda = \tan^{-1}(\delta_v/\delta_u). \tag{7.31a}$$

Now the inclination α, of the NA w.r.t. the u-axis may be found from Eq. (6.48):

$$M_u v/I_u + M_v u/I_v = 0, \tag{7.32}$$

where from Fig. 7.19(b) $M_u = M\cos\theta$, $M_v = M\sin\theta$. Their sagging actions ensure the given positive directions for δ_u and δ_v. Equation (7.32) supplies

$$\alpha = \tan^{-1}(v/u) = \tan^{-1}[-(I_u \tan\theta)/I_v]. \tag{7.33}$$

Assume, for example, that Figs 7.19(a) and (b) both apply to a simply supported beam with a central concentrated load. At the span centre,

$$\delta_u = M_v L^2/12EI_v = (M\sin\theta)L^2/12EI_v$$
$$= F_u L^3/48EI_v = (F\sin\theta)L^3/48EI_v, \tag{7.34}$$

$$\delta_v = M_u L^2/12EI_u = (M\cos\theta)L^2/12EI_u$$
$$= F_v L^3/48EI_u = (F\cos\theta)L^3/48EI_u. \tag{7.35}$$

Then, from Eqs (7.31), (7.34) and (7.35), the inclination of displacement δ_g to the u-axis is

$$\lambda = \tan^{-1}(\delta_v/\delta_u) = \tan^{-1}(I_v/I_u \tan\theta). \tag{7.31b}$$

Note from Eqs (7.31b) and (7.33) that the product of the gradients of the NA and the centroidal displacement δ_g is $(\tan\alpha) \times (\tan\lambda) = -1$. That is, their directions are at right angles, $\lambda - \alpha = 270°$ (α is negative). Substituting Eqs (7.34) and (7.35) into Eqs (7.28a) and (7.29a),

$$\delta_x = -\delta_g \sin(\theta + \alpha), \tag{7.28b}$$

$$\delta_y = \delta_g \cos(\theta + \alpha), \tag{7.29b}$$

where, from Eq. (7.30a), (7.34) and (7.35), for the beam in question:

$$\delta_g = (FL^3/48E)\sqrt{[(\sin\theta/I_v)^2 + (\cos\theta/I_u)^2]}. \tag{7.30b}$$

These may further be expressed in terms of the corresponding x, y quantities when θ is eliminated between Eqs (1.3)–(1.5). Assuming $I_y \geq I_x$ and that I_{xy} is positive, it follows from these equations that θ is positive (Fig. 7.19) and

$I_u < I_v$. From Eqs (1.4) and (1.5), with $h = \sqrt{[(I_y - I_x)^2 + 4I_{xy}^2]}$:

$$\tfrac{1}{2}\sin 2\theta(1/I_v - 1/I_u) = \tfrac{1}{2}(2I_{xy}/h)(-h)/(I_x I_y - I_{xy}^2) = -I_{xy}/(I_x I_y - I_{xy}^2), \quad (7.36)$$

$$(\sin^2 \theta)/I_v + (\cos^2 \theta)/I_u = [(I_u + I_v) + (I_v - I_u)\cos 2\theta]/2I_u I_v$$
$$= [(I_x + I_y) + h(I_y - I_x)/h]/2(I_x I_y - I_{xy}^2) = I_y/(I_x I_y - I_{xy}^2). \quad (7.37)$$

Substituting Eqs (1.3), (7.33), (7.36) and (7.37) into Eqs (7.28b) and (7.29b) leads to

$$\delta_x = \hat{F}_x L^3/48EI_y, \quad (7.28c)$$

$$\delta_y = \hat{F}_y L^3/48EI_x, \quad (7.29c)$$

where $\hat{F}_x = -F/(I_x/I_{xy} - I_{xy}/I_y)$ and $\hat{F}_y = F/(1 - I_{xy}^2/I_x I_y)$ are the equivalent forces. Equations (7.28c) and (7.29c) are identical in form to the vertical deflection expression for a symmetrical section. Indeed, this will apply to all the standard deflection expressions for simple loading of asymmetric sections where the equivalent force expressions can be derived. However, the deflection of asymmetric beams under more complex loading is best dealt with by applying the foregoing basic principles. It can then be deduced from Eq. (7.28b), (7.29b) and (7.30b) that with vertical loading of any asymmetric cross-section,

$$\delta_g = \sqrt{[(\sin \theta/I_v)^2 + (\cos \theta/I_u)^2]}(yI_x)$$
$$\delta_y = \sqrt{[(\sin \theta/I_v)^2 + (\cos \theta/I_u)^2]}(yI_x)\cos(\theta + \alpha)$$
$$\delta_x = -\sqrt{[(\sin \theta/I_v)^2 + (\cos \theta/I_u)^2]}(yI_x)\sin(\theta + \alpha)$$

where y is the deflection supplied from the application of any method about the x-axis, as for a symmetrical section; e.g. in the Macaulay method:

$$y = (1/EI_x) \iint (M_x \, dz) \, dz$$

Example 7.16 A 2 m long, aluminium-alloy cantilever has the cross-section shown in Fig. 1.15. Given that the density of this material is $\rho = 2700 \, kg/m^3$ and $E = 70 \, GPa$, find the vertical deflection at the free end due to self-weight. Find also the magnitude and inclination of the maximum deflection at this position.

The vertical distributed loading is given by

$$w = \rho A = 2700 \times 9.81 \times 10^{-9}[(20 \times 20) + (\pi/2)(20)^2] = 0.02724 \, N/mn$$

and it has previously been shown for this section (p. 16) that the principal second moments of area are; $I_u = 15.3 \times 10^4 \, mm^4$, $I_v = 6.1 \times 10^4 \, mm^4$, with an inclination of $\theta = 41°$ for the u-axis. The components of w in the u- and v- directions are then

$$w_u = w \sin\theta = 0.027\,24 \sin 41° = 0.017\,87\,\text{N/mm},$$
$$w_v = w \cos\theta = 0.027\,24 \cos 41° = 0.020\,56\,\text{N/mm}.$$

The moment–area method gave the deflection at the free end of a uniformly-loaded cantilever (Fig. 7.4) as $wL^4/8EI$. This must now be referred to the u-, v- directions, where the corresponding deflections are

$$\delta_u = w_u L^4/8EI_v = 0.017\,87 \times (2000)^4/(8 \times 70 \times 10^3 \times 6.1 \times 10^4) = 8.37\,\text{mm},$$
$$\delta_v = w_v L^4/8EI_u = 0.020\,56 \times (2000)^4/(8 \times 70 \times 10^3 \times 15.3 \times 10^4) = 3.84\,\text{mm}$$

Then from Eq. (7.30a), and w.r.t. Fig. 7.19, the maximum deflection is

$$\delta_g = \sqrt{(\delta_u^2 + \delta_v^2)} = \sqrt{[(8.37)^2 + (3.84)^2]} = 9.21\,\text{mm}$$

with an inclination λ to u (Eq. (7.31a)) of

$$\lambda = \tan^{-1}(\delta_v/\delta_u) = \tan^{-1}(3.84/8.37) = 24.64°.$$

Equations (7.29a) supplies the vertical deflection component:

$$\delta_y = 3.84 \cos 41° + 8.37 \sin 41° = 8.39\,\text{mm}.$$

Alternatively,

$$\delta_y = \delta_g \cos(\theta + \lambda) = 9.21 \sin(41° + 24.64°) = 8.39\,\text{mm}.$$

Example 7.17 A cantilever of length $L = 1\,\text{m}$ carries an end-force of $F = 1\,\text{kN}$ applied through the centroid but inclined at 30° to the vertical of its equal-angle section as shown in Fig. 7.20. Determine the inclination

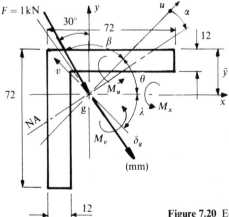

Figure 7.20 Equal-angle section.

of the neutral axis and the vertical and horizontal end displacements. Take $E = 207\,\text{GPa}$.

For the centroid position $\bar{y}$, apply $\sum(Ay) = A\bar{y}$ about the flange top:

$$(72 \times 12 \times 6) + (60 \times 12 \times 42) = [(72 \times 12) + (60 \times 12)]\bar{y},$$
$$\therefore \quad \bar{y} = 22.4\,\text{mm} \ (= \bar{x}).$$

Using the form $I = bd^3/12$,

$$I_x = 72 \times (22.4)^3/3 - 60 \times (10.4)^3/3 + 12 \times (49.6)^3/3$$
$$= 73.5 \times 10^4\,\text{mm}^4 \ (= I_y),$$
$$I_{xy} = \sum Axy = [(72 \times 12)(+13.6)(+16.4)] + [(60 \times 12)(-19.6)(-16.4)]$$
$$= 42.4 \times 10^4\,\text{mm}^4.$$

Then, from Eqs (1.4) and (1.5),

$$\tan 2\theta = 2I_{xy}/(I_y - I_x) = \infty, \quad \Rightarrow \theta = 45°,$$
$$I_{u,v} = (I_x + I_y)/2 \pm (1/2)\sqrt{[(I_x - I_y)^2 + 4I_{xy}^2]}$$
$$= \{(73.5 + 73.5)/2 \pm (1/2)\sqrt{[4(42.4)^2]}\} \times 10^4,$$
$$\therefore \quad I_u = 31.1 \times 10^4\,\text{mm}^4, \quad I_v = 115.9 \times 10^4\,\text{mm}^4,$$

where it should be noted that either Eq. (1.3) or the graphical method confirms that $I_u < I_v$ in this case. Since the force F is now inclined ($\beta = 75°$ to the u-axis in Fig. 7.20), then at the free end the hogging moments about u and v are

$$M_u = F_v L = (F \sin \beta)L, \qquad M_v = F_u L = (F \cos \beta)L. \qquad \text{(i)}$$

Equations (i), (7.32) and (7.33) give the inclination α, of the NA w.r.t. u:

$$\alpha = \tan^{-1}(v/u) = \tan^{-1}(-I_u/I_v \tan \beta) = \tan^{-1}(-0.0719) = -4.11°.$$

The δ_u, δ_v end-displacements take the following form (Example 7.2):

$$\delta_u = F_u L^3/3EI_v, \qquad \delta_v = F_v L^3/3EI_u. \qquad \text{(ii)}$$

From Eqs (ii) and (7.30a) the maximum end-displacement is

$$\delta_g = \sqrt{(\delta_u^2 + \delta_v^2)} = (FL^3/3E)\sqrt{\{[(\cos \beta)/I_v]^2 + [(\sin \beta)/I_u]^2\}}$$
$$= (1 \times 10^3 \times 10^9/3 \times 210 \times 10^3)$$
$$\times \sqrt{[(\cos 75°/115.9 \times 10^4)^2 + (\sin 75°/31.1 \times 10^4)^2]}$$
$$= 4.943\,\text{mm}$$

and from Eqs (ii) and (7.31a) the inclination λ of δ_{max} to the u-axis is

$$\lambda = \tan^{-1}(\delta_v/\delta_u) = \tan^{-1}(I_v \tan \beta)/I_u] \qquad \text{(iii)}$$
$$= \tan^{-1}[115.9(\tan 75°)/31.1] = 85.89°.$$

Thus, the displacement δ_g is inclined at $\theta + \lambda = 45 + 85.89 = 130.89°$ or $310.89°$ to x and at $\lambda - \alpha = 85.99 - (-4.11) = 90°$ to the neutral axis. From Eqs (7.28a) and (7.29a),

$$\delta_x = -\delta_g \sin(\theta + \alpha) = -4.943 \sin(45° - 4.11°) = -3.236 \, \text{mm},$$
$$\delta_y = \delta_g \cos(\theta + \alpha) = 4.943 \cos(45° - 4.11°) = 3.737 \, \text{mm}.$$

Check:

$$\delta_g = \sqrt{(\delta_x^2 + \delta_y^2)} = \sqrt{[(-3.236)^2 + (3.737)^2]} = 4.943 \, \text{mm}.$$

Example 7.18 If the section given in Fig. 1.16, with all dimensions increased by a factor of 10, applies to the beam loaded as shown in Fig. 7.17, determine the maximum and vertical deflections at the point of zero slope. Take $E = 207 \, \text{GPa}$.

It follows from the solution given in Example 1.12 that for this section

$$I_u = 36.87 \times 10^8 \, \text{mm}^4, \qquad I_v = 4.47 \times 10^8 \, \text{mm}^4, \qquad \theta = -7°.$$

In the application of Macaulay's method to the beam in Example 7.14 it was shown that at the point of zero slope $EIv_{\text{max}} = 361.56 \, \text{kNm}^3$ with bending about the x-axis for a symmetrical section. It is deduced that in the present case, where u does not coincide with x, the result of

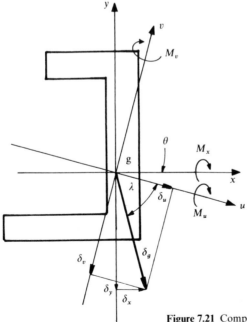

Figure 7.21 Components of deflection.

applying the method separately to the u and v axes, where $M_u = M_x \cos \theta$ and $M_v = M_x \sin \theta$, would give the δu and δv deflections as

$$EIv\,\delta u = \iint (M_v\,dz)\,dz = \sin \theta \iint (M_x\,dz)\,dz$$

$$= 361.56 \sin (-7°) \quad (\text{kN m}^3),$$

$$\therefore \quad \delta_u = 361.56 \times 10^{12} \times (-0.1219)/(207\,000 \times 4.47 \times 10^8) = -0.476\,\text{mm}.$$

$$EI_u\,\delta_v = \iint (M_u\,dz)\,dz = \cos \theta \iint (M_x\,dz)\,dz$$

$$= 361.56 \cos (-7°) \quad (\text{kN m}^3),$$

$$\therefore \quad \delta v = 361.56 \times 10^{12} \times (0.9925)/(207\,000 \times 36.87 \times 10^8) = 0.47\,\text{mm}.$$

The minus sign for δ_u indicates a deflection in the positive u-direction as shown in Fig. 7.21. Then from Eq. (7.30a),

$$\delta_g = \sqrt{(\delta_u^2 + \delta_v^2)} = 0.669\,\text{mm}$$

with an inclination, from Eq. (7.31a), of

$$\lambda = \tan^{-1}(\delta_v/\delta_u) = \tan^{-1}(0.47/-0.476) = -44.64°$$

to the u-axis in Fig. 7.21. That is, $\theta + \lambda = -7° - 44.64° = -51.64°$ to x. From Fig. 7.21, the vertical deflection component is

$$\delta_y = \delta_g \sin (\theta + \lambda) = 0.669 \sin (51.64°) = 0.524\,\text{mm}.$$

7.6 PRINCIPLE OF SUPERPOSITION AND RECIPROCAL LAWS

The superposition principle states that the total elastic displacement at a point in a structure under a given combination of externally applied loading may be obtained by summing the displacements at that point when each load is applied independently to its respective position and in any sequence. Thus, for the cantilever beam in Fig. 7.22 it follows that

$$\delta = \delta_{P1} + \delta_{P2} + \delta_{P3}.$$

The principle further applies to other causes and effects: load–stress,

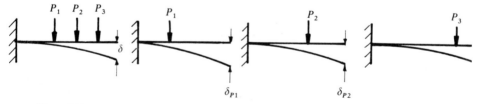

Figure 7.22 Superposition principle.

load–strain, stress–displacement and strain–displacement, provided the two quantities are linearly dependent, as with a material that obeys Hooke's law in its small-strain elastic range.

Example 7.19 Establish the total strain ϵ and displacement δ for a bar of area A and length L when axial forces F_1 and F_2 are applied together.

The independent, linear load–strain relationships are

$$\epsilon_1 = F_1/AE, \qquad \epsilon_2 = F_2/AE,$$

and the independent, linear load–displacement relationships are

$$\delta_1 = F_1L/AE, \qquad \delta_2 = F_2L/AE.$$

Then, from the principle of superposition,

$$\epsilon = \epsilon_1 + \epsilon_2 = \epsilon_2 + \epsilon_1 = (F_1 + F_2)/AE,$$
$$\delta = \delta_1 + \delta_2 = \delta_2 + \delta_1 = (F_1 + F_2)L/AE.$$

Example 7.20 Show that the deflection at a point in the length of a beam under the lateral loading in Fig. 7.17 (Example 7.14) conforms to the principle of superposition.

The necessary generality is achieved by putting $R_1 = 131.25\,\text{kN}$, $M = 100\,\text{kN m}$, $F = 200\,\text{kN}$ and $w = 150\,\text{kN/m}$. Then, from the previously derived deflection expression (see Example 7.14) it may be deduced that

$$EIv = - R_1z^3/6 + (M/2)[z - 2]^2 + (F/6)[z - 3]^3 + (w/24)[z - 2]^4$$
$$- (w/24)[z - 3]^4 + C_1z,$$

where the reaction and integration constant are

$$R_1 = (F + M + 3w/2)/4 \quad (\text{kN})$$
$$C_1 = 8R_1/3 - M/2 - F/24 - 15w/96.$$

Thus, for a given point z in the length it is seen that v depends upon linear terms in the loads R_1, M, F and w. Since this is the condition upon which superposition depends, it follows that v may be found by summing the v at the given point due to each independently applied load. The respective displacement terms are those appearing in the general expression. Under the force F, for example,

$$v = - (F/4)z^3/6 + (F/6)[z - 3]^3 - (F/24)z.$$

Loading Equivalents by Superposition

The previous examples show how any effect may be established by summing the separate effects due to each cause. This further permits the simplifications

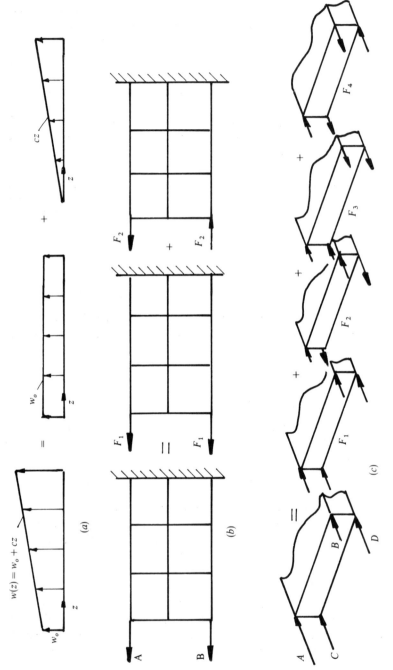

Figure 7.23 Superposition of symmetric and skew-symmetric loadings.

228

to be made in the asymmetrical loadings, given in Fig. 7.23(a–c), when stresses, strains and displacements are required. In each case the equivalent loading systems are composed of symmetric and skew-symmetric components.

The loading must, however, satisfy the following conditions:

(a) $w(z) = w_o + cz$ (c is a constant).
(b) $A = F_1 + F_2$, $B = F_1 - F_2$, from which the equivalent loads are $F_1 = (A + B)/2$, $F_2 = (A - B)/2$.
(c) $A = F_1 + F_2 + F_3 + F_4$, $B = F_1 + F_2 - F_3 - F_4$, $C = F_1 - F_2 + F_3 - F_4$, $D = F_1 - F_2 - F_3 + F_4$, where the equivalent loads, F_i, follow from the simultaneous solution to these equations.

One notable exception to the principle of superposition between load and displacement applies to a long thin strut under an axial compressive force P. In Chapter 8 it is shown that the governing equation between P and the lateral deflection v, (see Fig. 8.1) is

$$EI\, d^2v/dz^2 + Pv = 0,$$

of which the solution is

$$v = A \sin[\sqrt{(P/EI)}z] + B \cos[\sqrt{(P/EI)}z].$$

Clearly, at any point z in the length, v is not linearly dependent upon P as the increasingly large displacements produced up to the onset of instability continuously increase the bending moment (Pz) at z.

The General Linear Load–Displacement Law

It is seen from Example 7.20 that if the applied loadings F, M and w are more generally represented by $P_j (j = 1, 2, 3, \ldots, n)$ then the displacement (Δ) at some point p, has the form

$$\Delta_p = f_{p1}P_1 + f_{p2}P_2 + f_{p3}P_3 + f_{p4}P_4 + ,\ldots, f_{pn}P_n,$$

where, for a given P_j, the flexibility coefficients $f_{11}, f_{12}, f_{13} \cdots f_{1n}$ depend upon the positions of point p and P_j, i.e. functions of z in Example 7.20. Since for a second point q these coefficients will be different, the displacement Δ, at q under the same P_j now has the form

$$\Delta_q = f_{q1}P_1 + f_{q2}P_2 + f_{q3}P_3 + f_{q4}P_4 + ,\ldots, f_{qn}P_n,$$

in which the reason for employing a double subscript notation for f becomes clear. Thus, with a linear dependence of displacement upon the loading, the superposition principle gives the displacement for any point i as the sum of the displacements under each isolated load:

$$\Delta_i = f_{i1}P_1 + f_{i2}P_2 + f_{i3}P_3 + f_{i4}P_4 + ,\ldots, f_{in}P_n. \tag{7.38}$$

When only the displacements at the load points are considered, the restricted form of Eq. (7.38) generates the following system of equations:

$$\Delta_1 = f_{11}P_1 + f_{12}P_2 + f_{13}P_3 + f_{14}P_4 + ,\ldots, f_{1n}P_n,$$
$$\Delta_2 = f_{21}P_1 + f_{22}P_2 + f_{23}P_3 + f_{24}P_4 + ,\ldots, f_{2n}P_n,$$
$$\cdots\cdots\cdots\cdots\cdots\cdots\cdots\cdots\cdots\cdots\cdots,$$
$$\Delta_i = f_{i1}P_1 + f_{i2}P_2 + f_{i3}P_3 + f_{i4}P_4 + ,\ldots, f_{in}P_n,$$
$$\cdots\cdots\cdots\cdots\cdots\cdots\cdots\cdots\cdots\cdots\cdots,$$
$$\Delta_n = f_{n1}P_1 + f_{n2}P_2 + f_{n3}P_3 + f_{n4}P_4 + ,\ldots, f_{nn}P_n,$$

which may be written in the indicial and matrix notations as

$$\Delta_i = f_{ij}P_j, \qquad \{\Delta\} = [f]\{P\} \qquad (7.39a, b)$$

in which the $(n \times n)$ *matrix* $f_{ij} \equiv [f]$ contains the flexibility coefficients and the $(n \times 1)$ *column matrices* $P_j \equiv \{P\} = \{P_1\,P_2\,P_3\,P_4 \cdots P_n\}^T$ and $\Delta_i \equiv \{\Delta\} = \{\Delta_1\,\Delta_2\,\Delta_3\,\Delta_4 \cdots \Delta_n\}^T$ contain the respective loads and load-point displacements. The inverse forms of Eqs (7.39a) and (b) expresses the dependence of the loads upon their displacements:

$$P_j = k_{ji}\Delta_i, \qquad \{P\} = [f]^{-1}\{\Delta\} = [k]\{\Delta\},$$

where $[k]$ is an $(n \times n)$ *matrix* of stiffness coefficients appearing in the equivalent linear law:

$$P_j = k_{j1}\Delta_1 + k_{j2}\Delta_2 + k_{j3}\Delta_3 + k_{j4}\Delta_4 + ,\ldots, k_{jn}\Delta_n \qquad (j = 1, 2, 3, \ldots, n).$$

Maxwell's Reciprocal Law

The particular property of symmetry about the leading diagonal in each matrix $[f]$ and $[k]$ follows from the reciprocal theorem of Maxwell, Referring to Fig. 7.24, the theorem states that the deflection Δ_1 at point 1 caused by the application of a load P_2 to point 2 is the same as the deflection Δ_2, at point 2 with a load P_1, acting at point 1, provided $P_1 = P_2$. Then, from Eq. (7.38),

$$\Delta_1 = f_{12}P_2, \qquad \Delta_2 = f_{21}P_1,$$

and therefore, $f_{12} = f_{21}$ when $P_1 = P_2$. When the theorem is applied to pairs of

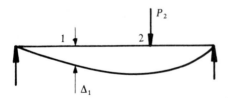
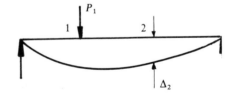

Figure 7.24 Maxwell's reciprocal law.

points in turn under P_n loads, the matrices $[f]$ and $[k]$ connecting the load-point displacements will each possess the symmetry property;

$$f_{ij} = f_{ji}, \qquad k_{ij} = k_{ji}.$$

Example 7.21 Use the reciprocal law and the principle of superposition to find an expression for the end-load that would give the same end-displacement for the cantilever beam in Fig. 7.25.

Under the given loading the principle of superposition gives the end-displacement as

$$\Delta_1 = f_{12} P_2 + f_{13} P_3. \qquad (i)$$

With an end-load P_1 only,

$$\Delta_1 = f_{11} P_1, \qquad \Delta_2 = f_{21} P_1 \quad \text{and} \quad \Delta_3 = f_{31} P_1,$$

where, from the integration method (see Example 7.2), the displacements at point 1, 2 and 3 under this end-load are found by putting $z = 3a$, $2a$ and a respectively in the general deflection equation. This gives

$$\Delta_1 = 27 P_1 a^3 / 3EI = f_{11} P_1, \quad \Rightarrow f_{11} = 27a^3 / 3EI,$$
$$\Delta_2 = 14 P_1 a^3 / 3EI = f_{21} P_1, \quad \Rightarrow f_{21} = 14a^3 / 3EI,$$
$$\Delta_3 = 4 P_1 a^3 / 3EI = f_{31} P_1, \quad \Rightarrow f_{31} = 4a^3 / 3EI.$$

Using the reciprocal law $f_{12} = f_{21}$ and $f_{13} = f_{31}$ Eq. (i) becomes

$$\Delta_1 = (14a^3 / 3EI) P_2 + (4a^3 / 3EI) P_3 = (27a^3 / 3EI) P_1,$$

from which the equivalent end-load is $P_1 = (2/27)(7 P_2 + 2 P_3)$.

Example 7.22 The displacements at three points 1, 2 and 3 in a plate of Hookean material were measured when different loads were applied individually to each point. If initially, when $P_1 = 800$ N, the deflections are $\Delta_1 = 1.5$ mm, $\Delta_2 = -0.5$ mm and $\Delta_3 = -2.5$ mm, find:

(a) Δ_1 given $\Delta_2 = 2$ mm and $\Delta_3 = 1.75$ mm for $P_2 = 600$ N;
(b) Δ_1 and Δ_2 given $\Delta_3 = 5$ mm for $P_3 = 1000$ N;
(c) Δ_1, Δ_2 and Δ_3 when $P_1 = P_2 = P_3 = 1200$ N are applied simultaneously. (K.P. 1977)

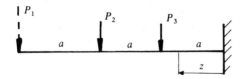

Figure 7.25 Cantilever beam.

Equation (7.39) applies to a Hookean material, where

$$\Delta_1 = f_{11}P_1 + f_{12}P_2 + f_{13}P_3,$$
$$\Delta_2 = f_{21}P_1 + f_{22}P_2 + f_{23}P_3, \qquad \text{(i)}$$
$$\Delta_3 = f_{31}P_1 + f_{32}P_2 + f_{33}P_3.$$

Substituting from the given initial condition in Eq. (i),

$$1.5 = 800f_{11}, \quad \Rightarrow f_{11} = (1.5/800),$$
$$-0.5 = 800f_{21}, \quad \Rightarrow f_{21} = -(0.5/800), \qquad \text{(ii)}$$
$$-2.5 = 800f_{31}, \quad \Rightarrow f_{31} = -(2.5/800).$$

Substituting the information given in (a) into Eq. (i) and using the reciprocal law where appropriate,

$$\Delta_1 = f_{12}P_2 = f_{21}P_2 = -(0.5/800)600 = -0.375 \, \text{mm},$$
$$2 = 600f_{22}, \quad \Rightarrow f_{22} = (2/600), \qquad \text{(iii)}$$
$$1.75 = 600f_{32}, \quad \Rightarrow f_{32} = (1.75/600).$$

Similar substitutions from the information in (b) into Eq. (i), together with the reciprocal law from Eqs (ii) and (iii), gives

$$\Delta_1 = f_{13}P_3 = f_{31}P_3 = -(2.5/800)1000 = -3.125 \, \text{mm},$$
$$\Delta_2 = f_{23}P_3 = f_{32}P_3 = (1.75/600)1000 = 2.917 \, \text{mm}, \qquad \text{(iv)}$$
$$5 = 1000f_{33}, \quad \Rightarrow f_{33} = (5/1000).$$

Finally, Eqs (ii)–(iv) enable the displacements to be solved under condition (c). Then from Eq. (i),

$$\Delta_1 = 1200[(1.5/800) - (0.5/800) - (2.5/800)] = -2.25 \, \text{mm},$$
$$\Delta_2 = 1200[-(0.5/800) + (2/600) + (1.75/600)] = 6.75 \, \text{mm},$$
$$\Delta_3 = 1200[-(2.5/800) + (1.75/600) + (5/1000)] = 5.75 \, \text{mm}.$$

Maxwell–Betti Reciprocal Theorem

A more general reciprocal law deals with the displacements of a structure under a system of loads. Thus, for the cantilever beam in Fig. 7.26(a) let the loads $A_i \equiv \{A\}$ be further displaced at their load-points by amounts $a_i^b \equiv \{a^b\}$ $(i = 1, 2, 3, \ldots, n)$ owing to the addition of a second set of loads $B_j \equiv \{B\}$ $(j = 1, 2, 3, \ldots, m)$ in Fig. 7.26(b) that are not coincident with, or equal in number to, $\{A\}$ in general. When the loads $\{A\}$ are removed and then reapplied, they result in corresponding load-point displacements $b_j^a \equiv \{b^a\}$ of $B_j \equiv \{B\}$ in Fig. 7.26(b).

Let the respective load-point displacements be Δ_i^A and Δ_j^B when the loads A_i and B_j act alone. The total external work that is done by the application

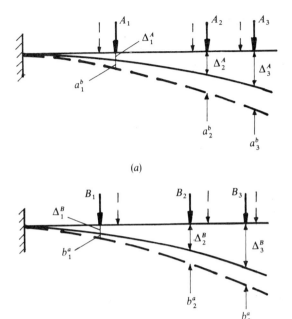

Figure 7.26 Maxwell–Betti reciprocal theorem.

B_j after A_i is

$$W = \tfrac{1}{2} A_i \Delta_i^A + \tfrac{1}{2} B_j \Delta_j^B + \tfrac{1}{2} A_i a_i^b. \tag{7.40}$$

Similarly, when A_i is applied after B_j the total external work done is

$$W = \tfrac{1}{2} B_j \Delta_j^B + \tfrac{1}{2} A_i \Delta_i^A + \tfrac{1}{2} B_j b_j^a. \tag{7.41}$$

In a linear elastic structure the total W is independent of the loading sequence. Equating (7.40) and (7.41), it follows that

$$A_i a_i^b = B_j b_j^a, \tag{7.42a}$$

$$A_1 a_1^b + A_2 a_2^b + A_3 a_3^b + , \ldots, + A_n a_n^b = B_1 b_1^a + B_2 b_2^a + B_3 b_3^a + , \ldots, + B_m b_m^a.$$

The equivalent matrix form is $\{A\}[a^b] = \{B\}[b^a]$, and since $\{A\} = [A]^{\mathrm{T}}$, $\{B\} = [B]^{\mathrm{T}}$ where $[\ \ ]$ denotes a *row matrix*, the usual form is

$$[A]^{\mathrm{T}}[a^b] = [B]^{\mathrm{T}}[b^a]. \tag{7.42b}$$

Either Eqs (7.42a) or (7.42b) is a statement of the reciprocity relation existing between the displacements produced by any two systems of loads when applied in sequence to a give structure.

Example 7.23 A structure is to support a number of loads of equal magnitude A. The deflection beneath each load is to be doubled by the addition of a further set of equal loads of magnitude B. Moreover, the deflection beneath loads B when these act alone in their same positions is to be doubled by the return of loads A to their normal positions. What is the ratio of the load magnitudes in terms of the corresponding flexibility coefficients if these conditions are to be achieved?

Using superscript A and B for each load system, Eq. (7.39a) becomes

$$\Delta_1^A = f_{11}^A A + f_{12}^A A + f_{13}^A A + \cdots, \qquad \Delta_1^B = f_{11}^B B + f_{12}^B B + f_{13}^B B + \cdots$$
$$\Delta_2^A = f_{21}^A A + f_{22}^A A + f_{23}^A A + \cdots, \qquad \Delta_2^B = f_{21}^B B + f_{22}^B B + f_{23}^B B + \cdots$$
$$\Delta_3^A = f_{31}^A A + f_{32}^A A + f_{33}^A A + \cdots, \qquad \Delta_3^B = f_{31}^B B + f_{32}^B B + f_{33}^B B + \cdots$$
$$\cdots\cdots\cdots\cdots\cdots\cdots\cdots, \qquad \cdots\cdots\cdots\cdots\cdots\cdots\cdots$$

$$\text{(i)}$$

With reference to Figs 7.26(a) and (b) the design requirements are that when B is superimposed upon A,

$$a_1^b = \Delta_1^A, \qquad a_2^b = \Delta_2^A, \qquad a_3^b = \Delta_3^A, \qquad \text{(ii)}$$

and when A is superimposed upon B,

$$b_1^a = \Delta_1^B, \qquad b_2^a = \Delta_2^B, \qquad b_3^a = \Delta_3^B, \qquad \text{(iii)}$$

Equation (7.42) also gives, with equal load magnitudes in each set,

$$A(a_1^b + a_2^b + a_3^b + \cdots) = B(b_1^a + b_2^a + b_3^a \cdots). \qquad \text{(iv)}$$

Substituting Eqs (i)–(iii) into (iv) leads to

$$A^2(f_{11}^A + f_{12}^A + \cdots + f_{33}^A + \cdots) = B^2(f_{11}^B + f_{12}^B + \cdots + f_{33}^B + \cdots)$$

and therefore

$$A/B = \sqrt{(\sum f^B / \sum f^A)}.$$

In a beam, for example, the flexibility coefficients f_{ij} can be established from Macaulay's method given the positions in which each load system acts.

EXERCISES

Integration Method

7.1 Derive from the flexure equation general expressions for the slope and deflection of the following beams each of length L. State in each case the position and magnitude of the maximum values.

    (a) Simply supported beam carrying a central concentrated load W. (*Answer*: $WL^2/16EI$; $WL^3/48EI$)

    (b) Simply supported beam carrying a uniformly distributed load, w. (*Answer*: $wL^2/24EI$; $5wL^4/384EI$)

    (c) Cantilever carrying a concentrated end load W. (*Answer*: $WL^2/2EI$; $WL^3/3EI$)

    (d) Cantilever carrying a uniformly distributed load, w. (*Answer*: $wL^3/6EI$; $wL^4/8EI$)

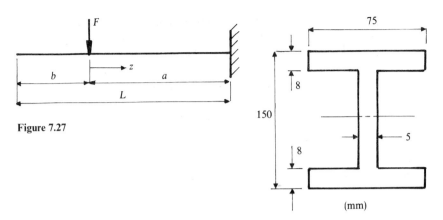

Figure 7.27

Figure 7.28

(e) Cantilever carrying a uniformly distributed load, w, prevented from deflection by a prop at its free end. (*Answer*: $wL^3/48EI$; $wL^4/184.6EI$)

7.2 A steel beam 15 m long is simply supported at its ends and carries a total distributed load of 4 kN/m. Calculate the slope and deflection at a point 6 m from the support. Take $I = 400 \times 10^6$ mm^4, $E = 200$ GPa.

Answer: $0.12°$; 31.4 mm.

7.3 If a cantilever instead of the simply supported beam in Exercise 7.2 is used to carry the distributed load over the same length, calculate the slope and deflection at a point 12 m from the fixed end. What is the maximum deflection?

Answer: $1.6°$; 232.5 mm; 316.4 mm.

7.4 Derive expressions for the slope and deflection at the tip of a cantilever when carrying the linearly varying distributed loading from zero at its tip to a maximum value of w_0 at the fixed end.

Answer: $\theta = w_0L^3/24EI$; $y = w_0L^4/30EI$.

7.5 Find the slope and deflection at the free end of a cantilever when loaded as shown in Fig. 7.27.

Answer: $\theta = Fa^2/2EI$; $y = F(3a^2L - a^3)/6EI$.

7.6 A steel bin of weight 2 kN is to be supported along two parallel sides of its 1.5 m square base with steel cantilevers of the I-section given in Fig. 7.28. If the maximum stress in the steel is not to exceed 85 MPa, calculate the safe load that can be carried in the bin and the maximum deflection under this load.

7.7 A simply supported beam carries a concentrated load W, at distance a from the left-hand support and b from the right-hand support where $a + b = L$ and $a < b$. Determine the slope and deflection at the load point.

Answer: $Wab(a - b)/6L$; $Wa^2b^2/3EIL$.

7.8 The box-section of a 6 m long cantilever is 125 mm wide $\times$ 350 mm deep and 10 mm thick. If it carries a concentrated load of 3 kN at the free end together with a total uniformly distributed load of 5.5 kN, calculate the free-end deflection and the maximum bending stress. Take $E = 210$ GPa.

Moment Area Method

7.9 Use Mohr's theorems to confirm the maximum slope and deflection expressions given as answers to Exercise 7.1.

7.10 The cross-section of a uniform cantilever, 250 mm deep and 3 m long, is symmetrical with $I = 120 \times 10^6$ mm⁴. Calculate, from the moment area method, the magnitude of the total loading and the deflection at the free end in each of the following cases when the maximum bending stress in the material is limited to 75 MPa and $E = 207$ GPa:

    (a) Load concentrated at the free end. (*Answer:* 24 kN; 8.7 mm)
    (b) Load uniformly distributed throughout. (*Answer:* 8 kN/m; 3.25 mm)
    (c) Load uniformly distributed between the mid-point and the free end.
    (d) Load concentrated at a point 1 m from the free end. (*Answer:* 24 kN; 8.4 mm)

7.11 Find the maximum slope and deflection for a 4 m long cantilever that carries a distributed load of 15 kN/m from mid-span to its free end, given $I = 85 \times 10^6$ mm⁴ and $E = 210$ GPa. *Answer:* 0.45°, 23 mm.

7.12 A 4.5 m long cantilever carries a total distributed load of 100 kN. A prop at the free end maintains the level there to 10 mm above the level of the fixed end. Calculate the reactions at the ends and find the maximum bending stress if the section is circular of 500 mm diameter. Take $EI = 625$ MN m².

7.13 Show that a prop that eliminates the deflection at the centre of a cantilever carries $2\frac{1}{2}$ times the concentrated end load W. Show that the free-end deflection of the propped cantilever is $7WL^3/96EI$.

7.14 A steel cantilever 4.5 m long carries a uniformly distributed load of 1.65 t/m over 3.65 m from the built-in end. A free-end prop is to support a load of 2 t when the level of the free end is 7.5 mm above the level of the fixed-end. Find I for the cross-section and the deflection at the free end when $E = 208$ GPa.

    Answer: $I = 93 \times 10^6$ mm⁴; 0.94 mm upwards.

7.15 Calculate the prop load P, the maximum slope and deflection for the cantilevers in Fig. 7.29(a), (b) and (c). Take $E = 200$ GPa, $I = 50 \times 10^3$ mm⁴.

7.16 If the prop in Fig. 7.29(c) is raised to 20 mm above the level of the fixed end, find the weight supported by the prop and the deflection at a point 8 m from the fixed end.

7.17 It is required to design a cantilever seat that is to carry a total distributed load of 3 kN over a length of 4 m from the fixed end together with a concentrated force of 2 kN at the free end of the total 4.5 m length. If the free-end deflection should not exceed 25 mm, determine the minimum value of the stiffness product EI.

    Answer: 3015 kN m².

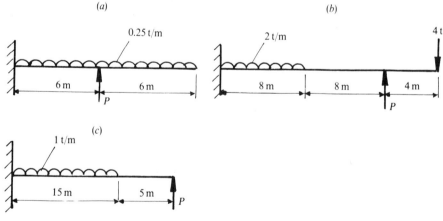

Figure 7.29

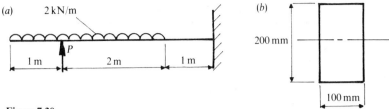

Figure 7.30

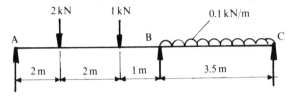

Figure 7.31

7.18 Determine the magnitude of the prop reaction and the slope at the prop for the beam loaded as shown in Fig. 7.30(a) with the cross-section given in Fig. 7.30(b).

7.19 An encastre beam of length L carries a uniformly distributed load w. Find the position and magnitude of the maximum bending moment and the deflection and the points of contraflexure.

Answer: $wL^2/12$, $wL^4/384EI$, $[1 \pm (1/\sqrt{3})](L/2)$.

7.20 An encastre beam carries a uniformly distributed load w in addition to a concentrated load W at its centre. Determine the greatest slope and deflection.

Answer: $(2W + wL)L^2/128EI$; $(2W + wL)L^3/384EI$.

7.21 Verify the solutions previously found from moment distribution for the continuous beams in Figs 5.12 and 5.13, using the three-moment theorem.

7.22 Draw the shear force and bending moment diagrams showing maximum values for the continuous beam in Fig. 7.31. Determine also the slope and deflection beneath the 2 kN force. Take $E = 200\,\text{GPa}$, $I = 80 \times 10^6\,\text{mm}^4$.

Answer: $M_B = 20.4\,\text{kN m}$; $R_A = 0.992\,\text{kN}$; $R_B = 4.34\,\text{kN}$; $R_C = 1.17\,\text{kN}$.

7.23 Find the central fixing moment and the position and magnitude of (i) the maximum bending moment and (ii) the maximum deflection for the continuous beam in Fig. 7.32. Take $EI = 15\,\text{MN m}^2$.

Answer: $M_B = 15.93\,\text{t m}$.

7.24 A beam of flexural stiffness 30 000 N m^2 is fixed horizontally at the left end A and is simply supported at the same level at distances of 4 m and 6 m from that end as shown in Fig. 7.33. A uniformly distributed load of 2500 N/m is carrried between supports B and C. Determine the deflection at a distance of 3 m from A. (C.E.I. Pt. 2, 1970)

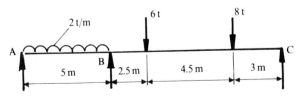

Figure 7.32

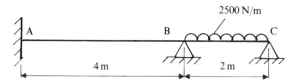

Figure 7.33

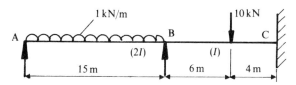

Figure 7.34

7.25 Use the three-moment theorem to construct the bending moment diagram for the beam in Fig. 7.34. Derive from it the shear force diagram.

Answer: $M_B = 18.85$ kN m; $M_c = 9.65$ kN m.

Macaulay's Method

7.26 A simply supported beam 7.5 m long carries a load of 4 t at 1.8 m from the left-hand support and another of 8 t at 1.22 m from the right-hand support. Calculate the position and magnitude of the maximum deflection. Take $I = 500 \times 10^6$ mm⁴, $E = 208$ GPa.

Answer: 4 m from L.H. end; 5.64 mm.

7.27 A uniform steel beam is simply supported at its ends over a span of 8 m. It carries a uniformly distributed load of 6.5 t/m for a length of 3 m starting at a point 2.75 m from the left-hand support and a concentrated load of 8 t acting at a point 1.5 m from the left-hand support. If $I = 500 \times 10^6$ mm⁴, calculate the slope at the left-hand support and the deflection at mid-span. Take $E = 207$ GPa.

Answer: 0.0062 rad; 30.2 mm.

7.28 A horizontal beam of uniform section and length L rests on supports at its ends. It carries a uniformly distributed load w per unit length over length l, from the right-hand support. Determine the value of l for the maximum deflection to occur at the left-hand end of the uniformly distributed load. If the maximum deflection is wL^4/kEI, find the value of k.

Answer: $l = 0.453L$; $k = 179$.

7.29 A steel beam is simply supported at its ends over a span of 6 m. It is loaded with concentrated weights of 5 t and 9 t at 1.8 m and 3.6 m respectively from the left-hand support. Find the position and magnitude of the maximum deflection if $I = 450 \times 10^6$ mm⁴, and $E = 207$ GPa.

Answer: 6.3 mm at 3.1 m from the L.H. end.

7.30 A simply supported beam carries a uniformly distributed load of 8.2 kN/m over its 7.5 m span. At 4.5 m from the left-hand end a further downward force of 80 kN acts together with a clockwise couple of 61 kN m acting in the plane of bending. Determine the position and magnitude of the maximum deflection and construct the bending moment diagram. Take $I = 479 \times 10^6$ mm⁴, $E = 208$ GPa.

Answer: 8.2 mm at 3.96 m from L.H. end, $M_{max} = 210.7$ kN m sagging.

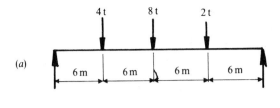

(a)

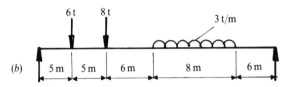

(b)

Figure 7.35

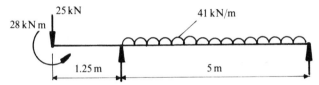

Figure 7.36

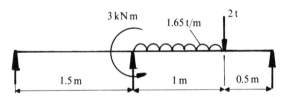

Figure 7.37

7.31 Find the deflection under each concentrated load and the position and magnitude of the maximum deflection for the two beams in Fig. 7.35(a) and (b). Take $E = 207$ GPa, $I = 10^{10}$ mm^4.

7.32 A beam of uniform section is loaded as shown in Fig. 7.36. Determine the slope and deflection at the L.H. end. Sketch the deflected shape and the bending moment diagram inserting principal values. Take $EI = 16\,650$ kN m^2.

Answer: 0.047 rad; 69 mm; 59.25 kN m.

7.33 Find the deflection at a point 1.22 m from the left-hand end of the beam in Fig. 7.37. Take $I = 415 \times 10^6$ mm^4, $E = 206$ GPa.

Answer: 0.15 mm.

7.34 Derive expressions for the slope and deflection at the free end of the cantilever loaded as shown in Fig. 7.38.

Answer: $5Wa^2/2EI$; $6Wa^3/EI$.

7.35 Derive general expressions for the slope and deflection at any point in the beam of Fig. 7.39. Then employ these to find the slope at point A and the deflection at point B.

Answer: $17wL^3/1944EI$; $0.003\,09wL^4/EI$.

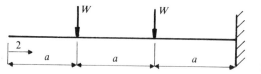

Figure 7.38

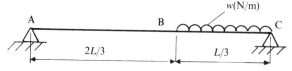

Figure 7.39

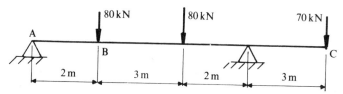

Figure 7.40

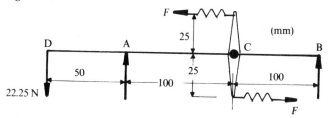

Figure 7.41

7.36 Find the deflection at B and C and the slope, in degrees, at A for the beam in Fig. 7.40.
Answer: 0.608 mm; 1.623 mm; 0.022°.

7.37 Part of a control mechanism consists of a steel tube of 6.35 mm i.d. with wall thickness 0.51 mm. It is simply supported in gimbals at A and B as shown in Fig. 7.41 and a device is attached to the tube at C consisting of a pair of arms and springs which may exert a couple about C. Determine the force F, required in each spring to maintain zero deflection at D when a concentrated force of 22.25 N is applied at that point.
Answer: 200 N. (C.E.I. Pt. 2, 1969)

7.38 The beam in Fig. 7.42 is loaded as shown in (a) and has the I-section given in (b). The left- and right-hand supports are non-rigid with respective stiffnesses of 10^4 and 2×10^4 kN/m. If $E = 200$ GPa, find the maximum deflection of the beam.
Answer: 12.7 mm.

7.39 An encastre beam carries a total load of 20 t uniformly distributed over 3 m of the beam from a point 1.75 m from the left-hand support. If the span of the beam is 6 m, calculate the deflection at mid-span and the value of the maximum slope between the left-hand end and mid-span. Take $E = 207$ GPa, $I = 85 \times 10^6$ mm⁴.
Answer: 10.9 mm; 1.47 m from L.H. end; 0.31°.

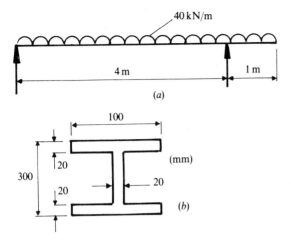

Figure 7.42

7.40 A built-in beam of length L has a clockwise couple M applied at a point on the beam distance a, from the left-hand end. Show that the left-hand fixing reaction and moment are given by $6M(L-a)a/L^3$ and $M(L-a)(3a-L)/L^2$, and that the maximum deflection is given by $Ma(2L-3a)^3/54EI(L-a)^2$ at a distance of $L^2/3(L-a)$ from the left-hand end.

7.41 A horizontal steel beam of uniform section is built-in at each end over a span of 9 m. It carries a uniformly distributed load of 3.25 t/m over 5.5 m measured from the right-hand support. If $I = 330 \times 10^6$ mm⁴ and $E = 207$ GPa, calculate the deflection at mid-span.

Answer: 1 mm.

7.42 A steel beam with both ends fixed has an effective span of 7.5 m and carries a concentrated force of 60 kN at a point 4.5 m from the left-hand end. Calculate the position and magnitude of the maximum deflection and slope due to this load, given that the latter occurs at a point between the left-hand end and the concentrated force. Take $I = 85 \times 10^6$ mm⁴, $E = 208$ GPa.

Answer: 7.25 mm at 4.16 m from the L.H. end; 2.08 m from L.H. end.

7.43 Find the magnitude and position of the maximum deflection for the encastre beam in Fig. 7.43. Take $I = 333 \times 10^6$ mm⁴, $E = 208$ GPa.

Answer: 2.6 mm; 4.86 m from L.H. end.

7.44 Determine R_A, M_A and the deflection at mid-span for the encastre beam in Fig. 7.44. Take $I = 415 \times 10^6$ mm⁴, $E = 208$ GPa.

Answer: $R_A = 91$ kN; $M_A = 110$ kN m; 1.65 mm.

7.45 Find the values of the fixing moments, the reaction at A and the maximum deflection for an encastre beam when it is loaded as shown in Fig. 7.45. Take $I = 75 \times 10^6$ mm⁴, $E = 207$ GPa.

Answer: $R_A = 18.4$ kN; $M_A = 30.5$ kN m; 4.42 mm at 4.1 m from the L.H. end.

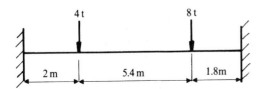

Figure 7.43

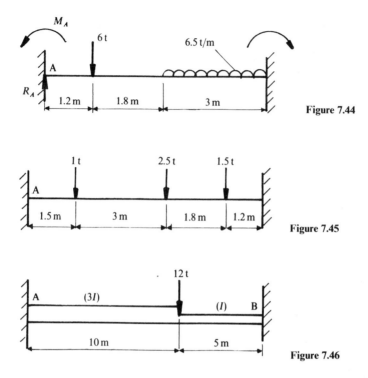

Figure 7.44

Figure 7.45

Figure 7.46

7.46 Determine the fixed-end moments at A and B and the deflection beneath the 12 t load for the encastre beam with variable section in Fig. 7.46. Take $EI = 24.2 \, \text{MN m}^2$.

Answer: 13.33 t m, 26.67 t m, 50 mm.

Asymmetric Deflection

7.47 If the cantilever beam in Exercise 7.11 is made from an angle section with equal arms that are vertical and horizontal, calculate the magnitude and direction of the free-end deflection given that the principal second moments of area are $36 \times 10^6 \, \text{mm}^4$ about a 45° axis passing through the apex and $9.4 \times 10^6 \, \text{mm}^4$ about a perpendicular axis.

7.48 If the simply supported beam in Exercise 7.26 is made from equal-angle section with one horizontal side supporting the loading, calculate the magnitude and direction of the maximum deflection given that the principal I-values are $366 \times 10^6 \, \text{mm}^4$ about an axis passing through the apex and $94.2 \times 10^6 \, \text{mm}^4$ about a perpendicular axis.

7.49 An aluminium beam with the cross-section in Fig. 1.20 is used as a 2 m long cantilever with the y-axis vertical. If the material is of density $2700 \, \text{kg/m}^3$ and $E = 70 \, \text{GPa}$, determine the vertical and horizontal displacements at the free end.

7.50 Find the true deflection and the vertical deflection for a 1.85 m length of equal-angle section mounted as shown in Fig. 7.47 when it supports a vertical load of 22.25 kN at its free end. Take $E = 207 \, \text{GPa}$.

Answer: 134 mm.

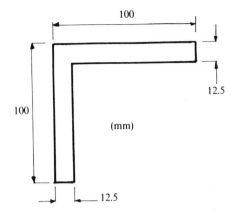

Figure 7.47

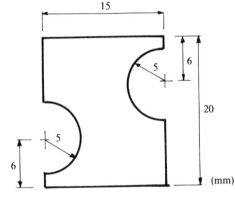

(mm)   **Figure 7.48**

7.51 A 1 m long beam with section in Fig. 7.48 carries a uniformly distributed weight of 75 N/m. It is simply supported over a length of 0.6 m with equal overhang. Find the magnitude and direction of the maximum deflection. Take $E = 80$ GPa.

Answer: 0.0875 mm at 27.2° from the vertical.

7.52 Two horizontal Z-section beams 0.8 m long are fixed at one of their ends to a rigid structure and connected together by several diaphragms that are rigid in their own planes and sufficient to prevent buckling of the flanges. A vertical force of 1 kN is applied mid-way between the sections at the free end. Taking Young's modulus as 210 GN/m², calculate the resultant displacement, and its direction, of the loaded point when the sections are arranged as in Fig. 7.49(a) and (b). (C.E.I. Pt. 2, 1976)

7.53 If one horizontal side of an equal-angle is used to support the loading given in Exercise 7.11 find the magnitude and direction of the maximum deflection given that $I = 36.6 \times 10^6$ mm⁴ about a principal axis at $+45°$ to x passing through the centroid and $I = 9.42 \times 10^6$ mm⁴ about a perpendicular axis.

7.54 A cantilever 0.8 m long with the triangular, tubular section in Fig. 7.50 is used to support a total distributed load of 64 N along its shorter horizontal side. Calculate the magnitude and direction of the maximum free-end deflection. Take $E = 80$ GPa.

Answer: 31.2 mm.

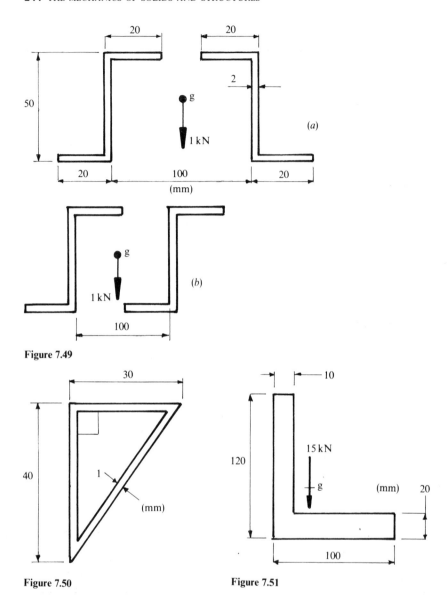

Figure 7.49

Figure 7.50   **Figure 7.51**

7.55 A carriage supporting a fixed load is to run along a pair of rails, each having the cross-section shown in Fig. 7.51. Each rail is simply supported at its ends and is 3 m long with $E = 208 \, \text{GN/m}^2$. Given that the carriage applies a force of 15 kN to each rail through the centroid g, as shown, find the vertical and horizontal deflections at mid-span. (C.E.I. Pt. 2, 1985)

7.56 A horizontal cantilever beam with the cross-section shown in Fig. 7.52 is rigidly built-in at one end and mounted with its web vertical. It is subjected to a vertical force of 1 kN at the free end of the beam, acting through the centroid of the section. If the beam is 1 m long and

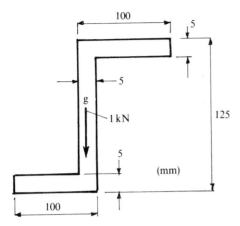

Figure 7.52

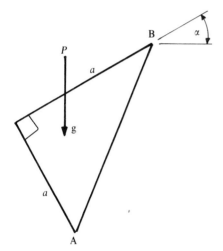

Figure 7.53

the modulus of the beam material is $80\,\text{GN/m}^2$, determine the vertical and horizontal displacements of the centroid at the free end when the force is applied. (C.E.I. Pt. 2, 1982)

7.57 A uniform cantilever has length l and the cross-section shown in Fig. 7.53. It has to carry a vertical downwards concentrated force P at its free end acting through the centroid g of the section. It is required that any resulting deflection will be vertically downwards with no horizontal movement and also that the deflection will be as small as possible. Find the necessary position of the beam in order to achieve these requirements (i.e. find the necessary angle α). Find an expression for the bending stress at points A and B on the section at the wall. (C.E.I. Pt. 2, 1984)

Principle of Superposition and Reciprocal Laws

7.58 In a laboratory test, an aircraft wing is loaded at eight different points on its top surface by concentrated forces. Measurements of load-point vertical displacements are made to enable the

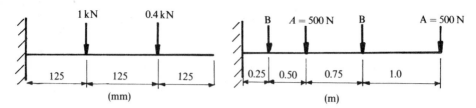

Figure 7.54   **Figure 7.55**

flexibility coefficients to be evaluated. Deduce the number of independent coefficients from the displacement–load expressions corresponding to each load point.

Answer: 36.

7.59 A uniform beam of length $3a$ is simply-supported in bearings at its ends. Use the reciprocal law to find (i) the vertical deflection at $z = a$ when a concentrated clockwise moment M is applied at $z = 2a$ and (ii) the slope at $z = 2a$ when a vertical force P is applied at $z = a$. (Use Macaulay's method to find the coefficient.)

Answer: $5a^2(M \text{ or } P)/18EI$.

7.60 Using an end load only, find the end deflection of the cantilever loaded as shown in Fig. 7.54. Take $E = 70\,\text{GPa}$ and $I = 3.5 \times 10^{-8}\,\text{m}^4$.

Answer: 2.55 mm.

7.61 Determine the magnitude of two equal loads in system B that meet the conditions given for the worked Example 7.23 when two equal loads $A = 500\,\text{N}$ are applied to the cantilever in Fig. 7.55. A and B act at the positions indicated. Take $EI = 25 \times 10^{-4}\,\text{MN}\,\text{m}^2$.

7.62 Table (i) below gives the deflections Δ(mm) and slopes θ(rad) for four positions in a structure that is loaded either by a concentrated moment $M(\text{N m})$ at points 1 and 2 or by a force $P(\text{N})$ at points 3 and 4 in four separate tests.

Table (i)

Test		Position 1	2	3	4
(1)	M	100	100	0	0
	θ	2×10^{-2}	1×10^{-2}	—	—
	Δ	—	—	3	5
(2)	M	100	200	0	0
	θ	?	2×10^{-2}	—	—
	Δ	—	—	5	-7
(3)	P	0	0	2000	0
	θ	?	?	—	—
	Δ	—	—	3	-6
(4)	P	0	0	2000	2000
	θ	?	?	—	—
	Δ	—	—	-6	9

Determine the slopes and deflections for the positions indicated by a question mark and also when $M_1 = 100\,\text{N m}$, $M_2 = 200\,\text{N m}$, $P_3 = 1000\,\text{N}$ and $P_4 = 500\,\text{N}$ act together. (K.P. 1977)

Answer: (2) $\theta_1 = 0.02\,\text{rad}$, (3) $\theta_1 = 0.02\,\text{rad}$, $\theta_2 = 0.04\,\text{rad}$, (4) $\theta_1 = 0.36\,\text{rad}$, $\theta_2 = -0.2\,\text{rad}$.

EIGHT

BUCKLING OF STRUTS AND PLATES

8.1 INTRODUCTION

In the consideration of the buckling behaviour of long thin members, i.e. struts under compression, both analytical and empirical approaches are presented for a variety of different end fixings. The additional examination of the buckling behaviour of long thin plates under direct compression is restricted here to relatively simple edge fixings. The common objective is that of finding the critical buckling load. Under purely elastic buckling conditions, where theoretical solutions are available, this load appears as a function of the length, elastic constants and some property of the resisting area, e.g. the second moment of area of a strut and an aspect ratio for a plate. Semi-empirical type solutions to the buckling load become necessary to account for the influence of plasticity when, with shorter members, the net section stress exceeds the yield stress.

8.2 BUCKLING THEORY FOR PERFECT STRUTS

Ideally, a perfectly straight strut would compress but not buckle under a purely axial compressive load. Observed buckling is therefore the result of imperfections arising, for example, from eccentricity of loading, lack of initial straightness and the presence of non-uniform residual stress. Their combined effects on the buckling load become predictable for long struts operating under elastic conditions because an equation governing instability behaviour applies. This approach is attributed to Leonhard Euler (1707–1783) and covers the following cases.

Pinned or Hinged Ends

The flexure equation (7.4) is applied to a point (z, v) on the deflected strut in Fig. 8.1 with v positive in the direction shown and hogging moments positive. Taking moments to the left of that point,

$$EI\, d^2v/dz^2 = -Pv \quad \text{(sagging)},$$

which is written as

$$d^2v/dz^2 + \alpha^2 v = 0, \tag{8.1}$$

where $\alpha^2 = P/EI$. The solution to Eq. (8.1) is

$$v = A \sin \alpha z + B \cos \alpha z \tag{8.2}$$

which must satisfy the following boundary conditions:

(i) $v = 0$ at $z = 0$, $\Rightarrow B = 0$
(ii) $v = 0$ at $z = L$, $\Rightarrow A \sin \alpha L = 0$.

In order to satisfy condition (ii) either $A = 0$ when, unrealistically, $v = 0$ for all z in Eq. (8.2), or, $\sin \alpha L = 0$ when $\alpha L = \pi, 2\pi, 3\pi, \ldots, n\pi$. Buckling of the strut will commence when the least of these values for αL is achieved. Thus,

$$\alpha L = \pi, \quad \Rightarrow \sqrt{(P/EI)}\, L = \pi,$$

from which the critical buckling load is

$$P_E = \pi^2 EI/L^2, \tag{8.3}$$

where I is the second moment of area about the axis in the cross-section that offers the least resistance to buckling, i.e. the minimum principal axis u. In a rectangular section, for example, I in Eq. (8.3) applies to an axis passing through the centroid parallel to the longer side. The following alternative form of Eq. (8.3) employs the definition that $I = Ak^2$, where k is the radius of gyration of the given section area A:

$$P_E = \pi^2 EA/(L/k)^2, \tag{8.4}$$

where the slenderness ratio L/k, becomes the determining factor in checking

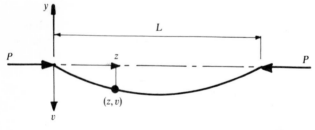

Figure 8.1 Deflection of a pinned-end strut.

the validity of elastic buckling analysis. Thus, from Eq. (8.4), the net section stress for an Euler strut is

$$\sigma_E = P_E/A = \pi^2 E/(L/k)^2. \tag{8.5}$$

It follows that the elastic Euler theory remains valid provided σ_E does not exceed the compressive yield stress σ_c. That is,

$$\pi^2 E/(L/k)^2 \leq \sigma_c \tag{8.6a}$$

and therefore the slenderness ratio must exceed some minimum value defined from Eq. (8.6a) as

$$L/k \geq \sqrt{(\pi^2 E/\sigma_c)} \tag{8.6b}$$

which is a constant for a given material. For example, for mild steel with $\sigma_c = 325\,\text{MPa}$ and $E = 207\,\text{GPa}$,

$$L/k \geq \sqrt{(\pi^2 \times 207\,000/325)} \simeq 80.$$

The semi-empirical forms discussed in Section 8.4 become more appropriate when L/k is less than the minimum Euler value supplied by Eq. (8.6b).

Other End Fixings with Axial Compressive Loads

It is possible to deduce the effect that different end fixings have on the buckling load without detailed derivation. Thus, in respect of Eq. (8.3), L is simply replaced by an effective length L_e, over which the previous pinned-end analysis would apply. Table 8.1 gives the effective buckling length and load for each of the five struts shown.

With imperfect fixed ends, an effective length of 0.6–0.8 is used to account for the degree of rotational restraint. When L is replaced by the appropriate L_e', Eq. (8.6b) again applies to the smallest slenderness ratio for which the Euler theory remains valid in each of these struts. This ensures that the resistance offered by direct compression is small in comparison with the flexural strength.

Example 8.1 A strut 2 m long has a tubular cross-section of 50 mm outside diameter and 44 mm inside diameter. Determine the Euler critical load when the ends are pinned. If the tube material has a compressive yield stress of 310 MPa, find the shortest length of this tube for which the theory applies. Repeat the calculation when the same strut has fixed ends. Take $E = 207\,\text{GPa}$.

$$I = \pi(d_o^4 - d_i^4)/64 = \pi[(50)^4 - (44)^4]/64 = 122.7 \times 10^3\,\text{mm}^4.$$

Table 8.1 Buckling predictions from Euler for other end fixings

End condition	Fixed–fixed	Pinned–fixed	Fixed–free	Fixed–fixed with lateral displacement	Fixed–pinned with lateral displacement
S					
T					
R					
U					
T					
L_e	$L/2$	$L/\sqrt{2}$	$2L$	L	$2L$
P_E	$4\pi^2 EI/L^2$	$2\pi^2 EI/L^2$	$\pi^2 EI/4L^2$	$\pi^2 EI/L^2$	$\pi^2 EI/4L^2$

Then, from Eq. (8.3), the pinned-end buckling load is

$$P_E = \pi^2 EI/L^2 = \pi^2 \times 207 \times 10^3 \times 122.7 \times 10^3/(2000)^2$$
$$= 62.7 \times 10^3\,\text{N} = 62.7\,\text{kN}$$

and from Eq. (8.6b) the shortest permissible length is

$$L = \sqrt{(\pi^2 Ek^2/\sigma_c)} = \sqrt{(\pi^2 EI/A\sigma_c)}$$
$$= \sqrt{[\pi^2 \times 207 \times 10^3 \times 122.7 \times 10^3/(\pi/4)(50^2 - 44^2) \times 310]}$$
$$= 1352\,\text{mm} = 1.352\,\text{m}.$$

It follows that the Euler theory is valid for a 2 m length strut. Table 8.1 gives the fixed-end buckling load as

$$P_E = 4\pi^2 EI/L^2 = 4 \times 62.7 = 250.8\,\text{kN},$$

and the shortest permissible length as

$$L = \sqrt{(4\pi^2 EI/\sigma_c)} = 2\sqrt{(\pi^2 EI/A\sigma_c)}$$
$$= 2 \times 1.352 = 2.704\,\text{m},$$

and therefore the Euler theory is invalid in this case.

Example 8.2 If the cross-section given in Fig. 1.16 is to be used as a strut with one end pinned and the other end fixed, calculate the buckling load for the minimum permissible length of an Euler strut given that $E = 207\,\text{GPa}$ and $\sigma_c = 310\,\text{MPa}$ for the strut material. •

The problem is that of finding the value of the least second moment of area for the section, i.e. I_u or I_v, by any one of the methods outlined in Chapter 1. Either the analytical or the graphical approach (page 16) showed that for this section the lesser value (which is the lowest value for the section) was $I_u = 4.47 \times 10^4\,\text{mm}^4$. Then, from Table 8.1 and Eq. (8.6), the minimum permissible length is

$$L = \sqrt{2L_e} = \sqrt{(2\pi^2 EI/A\sigma_c)}$$
$$= \sqrt{(2\pi^2 \times 207 \times 10^3 \times 4.47 \times 10^4/310 \times 900)} = 809.1\,\text{mm},$$

and the corresponding buckling load is

$$P_E = 2\pi^2 EI/L^2 = 2\pi^2 \times 207 \times 10^3 \times 4.47 \times 10^4/(809.1)^2$$
$$= 279\,000\,\text{N} = 279\,\text{kN}.$$

8.3 IMPERFECT EULER STRUTS

Practical strut configurations may differ from perfect Euler struts in the ways outlined in the following cases. The onset of buckling is normally associated with the attainment of the compressive yield stress at the section subjected to the greatest bending moment.

Pinned-end Strut with Eccentric Loading

When the line of action of the compressive force P is applied with a measurable amount of eccentricity e, to the strut axis, as shown in Fig. 8.2, then the bending moment at any point (z, v) will be modified. That is, the

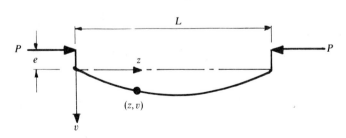

Figure 8.2 Deflection of an eccentrically loaded pinned-end strut.

governing flexure equation (7.4) becomes

$$EI \, d^2v/dz^2 = -P(e + v).$$

This now takes the form

$$d^2v/dz^2 + \alpha^2 v = -\alpha^2 e,$$

with the corresponding solution

$$v = A \cos \alpha z + B \sin \alpha z - e. \tag{8.7}$$

This solution may be employed to find the safe axial load for a given allowable stress σ. In general, for any point in the strut, the stress is composed of direct compressive stress and a bending stress that changes sense across the section. It follows from Eq. (6.18) that

$$\sigma = -P/A \pm My/I. \tag{8.8}$$

This stress is greatest on the compressive side of the section where the sagging bending moment is a maximum. That is, from Eq. (8.8),

$$\sigma = -P/A + M_{max}y/I, \tag{8.9}$$

where y is the perpendicular distance from the buckling axis to the furthest compressive edge. Now, from Fig. 8.2, at the central section

$$M_{max} = -P(e + v_{max}). \tag{8.10}$$

The maximum deflection at $z = L/2$ follows from Eq. (8.7), where the following boundary conditions apply:

(i) $v = 0$ at $z = 0$, $\Rightarrow A = e$,
(ii) $dv/dz = 0$ at $z = L/2$. This gives

$$-A\alpha \sin \alpha z + B\alpha \cos \alpha z = 0, \quad \Rightarrow B = e \tan (\alpha L/2),$$
$$\therefore \quad v_{max} = e \cos (\alpha L/2) + e \tan (\alpha L/2) \sin (\alpha L/2) - e,$$
$$v_{max} = e[\sec (\alpha L/2) - 1]. \tag{8.11}$$

Then, from Eqs (8.9) and (8.10), the maximum compressive stress is

$$M_{max} = -Pe \sec (\alpha L/2) \quad \text{(sagging)},$$
$$\sigma = -P/A - (Pey/I) \sec (\alpha L/2). \tag{8.12}$$

Since $\alpha = \sqrt{(P/EI)}$, this form is not convenient when P is required. However, we may employ Webb's approximation to the secant function,

$$\sec \theta \simeq \frac{(\pi/2)^2 + (\pi^2/8 - 1)(\theta)^2}{(\pi/2)^2 - (\theta)^2}. \tag{8.13}$$

Putting $\theta = \alpha L/2$ and introducing, from Eq. (8.3), the Euler buckling load

$P_E = \pi^2 EI/L^2$, Eq. (8.13) gives

$$\sec(\alpha L/2) \simeq [P_E + P(\pi^2/8 - 1)]/(P_E - P). \tag{8.14}$$

Substituting Eq. (8.14) into Eq. (8.12) supplies the quadratic

$$aP^2 + bP + c = 0,$$

in which P is the positive root with σ negative in the coefficients

$$a = eyA(\pi^2/8 - 1) - I, \qquad b = eyAP_E + IP_E - I\sigma A, \qquad c = I\sigma AP_E.$$

Example 8.3 A tubular section strut 75 mm o.d. and 65 mm i.d. is 3 m long. Calculate the maximum stress and deflection when the strut carries a compressive force of 50 kN offset by 3 mm from its longitudinal axis. Take $E = 200$ GPa.

$$A = (\pi/4)[(75)^2 - (65)^2] = 1099.56 \, \text{mm}^2,$$
$$I = (\pi/64)[(75)^4 - (65)^4] = 67.69 \times 10^4 \, \text{mm}^4,$$
$$\alpha = \sqrt{(P/EI)} = \sqrt{[(50 \times 10^3)/(200 \times 10^3 \times 67.69 \times 10^4)]}$$
$$= 6.0772 \times 10^{-4} \, \text{mm}^{-1}.$$

Since P is given here, the direct substitution into Eq. (8.12) gives the maximum compressive stress as

$$\sigma = -50 \times 10^3/1099.56 - (50 \times 10^3 \times 3 \times 37.5/67.69 \times 10^4)$$
$$\times \sec(3000 \times 6.0772 \times 10^{-4} \times 180/2\pi)$$
$$= -45.473 - 13.567 = -59.04 \, \text{MPa}.$$

From Eq. (8.11),

$$v_{max} = 3[\sec(6.0772 \times 10^{-4} \times 3000 \times 180/2\pi) - 1]$$
$$= 3[1.6327 - 1] = 1.898 \, \text{mm}.$$

Strut with One End Fixed, the Other End Free, Carrying an Eccentric Compressive Load

Let v_L be the free-end deflection with the origin of z,v coordinates lying at the fixed end as shown in Fig. 8.3. For any point (z, v) on the deflected strut axis, the flexure equation (7.4) now becomes

$$EI \, d^2v/dz^2 = P(e + v_L - v) \qquad \text{(hogging)},$$
$$d^2v/dz^2 + \alpha^2 v = \alpha^2(e + v_L),$$

and the solution is

$$v = A \cos \alpha z + B \sin \alpha z + (v_L + e).$$

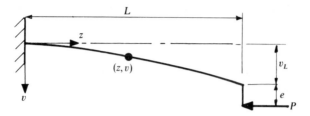

Figure 8.3 Deflection of a strut eccentrically loaded at its free end.

Applying the boundary conditions:

(i) $dv/dz = 0$ at $z = 0$, $\Rightarrow B = 0$,

(ii) $v = 0$ at $z = 0$, $\Rightarrow A = -(v_L + e)$

$$\therefore \quad v = (v_L + e)(1 - \cos \alpha z)$$

but $v = v_L$ at $z = L$:

$$\therefore \quad v_L = e(\sec \alpha L - 1). \tag{8.15}$$

The hogging bending moment is a maximum at the fixed end when

$$M_{\max} = P(e + v_L) = Pe \sec (\alpha L), \tag{8.16}$$

where, from Webb's approximation (Eq. (8.13)),

$$\sec \alpha L \simeq \frac{P_E + 4P(\pi^2/8 - 1)}{P_E - 4P}. \tag{8.17}$$

Substituting Eqs (8.16) and (8.17) into Eq. (8.9), P is the positive root of the quadratic for a given maximum compressive (negative) stress:

$$aP^2 + bP + c = 0, \tag{8.18}$$

where

$$a = 4[eyA(\pi^2/8 - 1) - I], \quad b = eyAP_E + IP_E - 4I\sigma A, \quad c = I\sigma AP_E.$$

Example 8.4 A 3 m long strut of tubular section 50 mm o.d. and 25 mm i.d. is loaded eccentrically at its free end with a compressive force at a radial distance of 75 mm from the axis. Find the maximum deflection and the safe axial load when (i) the maximum compressive stress is limited to 35 MPa and (ii) the net stress on the tensile side is to be zero at the fixed end. Take $E = 210$ GPa.

Part (i) involves the direct application of Eqs (8.15) and (8.18), with

$$A = (\pi/4)(50^2 - 25^2) = 1472.62 \text{ mm}^2,$$
$$I = (\pi/64)(50^4 - 25^4) = 28.76 \times 10^4 \text{ mm}^4,$$
$$P_E = \pi^2 EI/L^2 = \pi^2 \times 210\,000 \times 28.76 \times 10^4/(3000)^2 = 66\,236.23 \text{ N}.$$

Then, from Eq. (8.18) with $\sigma = -35\,\text{MPa}$ the quadratic coefficients are

$$a = 4[(75 \times 25 \times 1472.62 \times 0.2337) - (28.762 \times 10^4)]$$
$$= 1.431 \times 10^6\,\text{mm}^4,$$
$$b = 2.0787 \times 10^{11}\,\text{N}\,\text{mm}^4, \qquad c = -9.819 \times 10^{14}\,\text{N}^2\,\text{mm}^4.$$
$$\therefore \quad P^2 + (145.262 \times 10^3)P - (686.17 \times 10^6) = 0.$$

The positive root supplies the compressive load, $P = 4.58\,\text{kN}$. Then,

$$\alpha = \sqrt{(P/EI)} = \sqrt{[(4.58 \times 10^3)/(210\,000 \times 28.76 \times 10^4)]}$$
$$= 2.75 \times 10^{-4}\,\text{mm}^{-1}$$

and Eq. (8.15) provides the maximum displacement

$$v_L = 75[\sec(2.75 \times 10^{-4} \times 3000 \times 180/\pi) - 1] = 35.66\,\text{mm}.$$

In part (ii), zero stress on the tensile side of the fixed end, where $M = M_{\text{max}}$, is expressed from Eq. (8.8) and (8.16) as

$$\sigma = -P/A + [Pe\sec(\alpha L)]y/I = 0.$$

Substituting $\alpha^2 = P/EI$ leads to the compressive load:

$$P = (EI/L^2)[\sec^{-1}(I/Aye)]^2$$
$$= (207\,000 \times 28.76 \times 10^4/3000^2)$$
$$\quad \times [\sec^{-1}(28.76 \times 10^4/1472.62 \times 25 \times 75)]^2$$
$$= 6614.8[\sec^{-1}(0.1042)]^2 = 3.076\,\text{kN};$$
$$\alpha = \sqrt{(P/EI)} = \sqrt{[(3.076 \times 10^3)/(207\,000 \times 28.76 \times 10^4)]}$$
$$= 2.273 \times 10^{-4}\,\text{mm}^{-1};$$

and from Eq. (8.15) the maximum deflection is

$$v_L = 75[\sec(2.273 \times 10^{-4} \times 3000 \times 180/\pi) - 1] = 21.6\,\text{mm}.$$

Pinned-end Strut, Carrying a Lateral Load at Mid-span in Addition to an Axial Compressive Load

For a point (z, v) on the deflected strut in Fig. 8.4, the left-hand moment is composed of two sagging components (Pv) due to P and $(Wz/2)$ due to the support reaction. The flexure equation (7.4) becomes

$$EI\,\text{d}^2v/\text{d}z^2 = -Pv - Wz/2 \qquad \text{(sagging)},$$
$$\text{d}^2v/\text{d}z^2 + \alpha^2 v = -Wz/2EI.$$

The corresponding solution is

$$v = A\cos\alpha L + B\sin\alpha L - Wz/2P.$$

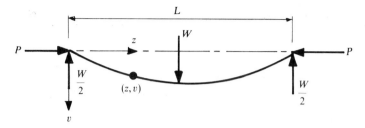

Figure 8.4 Pinned-end strut with axial and lateral loading.

Applying the boundary conditions:

(i) $v = 0$ at $z = 0$, $\Rightarrow A = 0$

(ii) $dv/dz = 0$ at $z = L/2$, $\Rightarrow B = W/2P\alpha \cos(\alpha L/2)$.

This leads to a maximum displacement at $z = L/2$ of

$$v_{max} = (W/2P)[(1/\alpha)\tan(\alpha L/2) - L/2], \tag{8.19}$$

and a maximum sagging moment at $z = L/2$ of

$$M_{max} = -Pv_{max} - WL/4 = -(W/2\alpha)\tan(\alpha L/2). \tag{8.20}$$

Equation (8.9) gives the maximum compressive stress in the central section:

$$\sigma = -P/A - (Wy/2I\alpha)\tan(\alpha L/2). \tag{8.21}$$

This equation may be more conveniently solved for P using the tangent approximation due to Webb:

$$\tan\theta = \frac{(\pi/2)^2\theta - (\pi^2/12 - 1)\theta^3}{(\pi/2)^2 - \theta^2}.$$

With $\theta = \alpha L/2$ and $P_E = \pi^2 EI/L^2$, this provides

$$\tan(\alpha L/2) = \frac{(\alpha L/2)[P_E - P(\pi^2/12 - 1)]}{(P_E - P)},$$

when, from Eq. (8.21), the following quadratic in positive P is found:

$$aP^2 + bP + c = 0,$$

where

$$a = 1, \quad b = (WLyA/4I)(\pi^2/12 - 1) + \sigma A - P_E, \quad c = -P_E A(\sigma + WLy/4I)$$

in which σ is the maximum compressive (negative) stress.

Example 8.5 A pinned-end strut 2 m long and 30 mm diameter supports an axial compressive force of 20 kN. If the maximum compressive stress is 200 MPa, calculate the additional load that may be applied laterally at mid-span. What is the maximum deflection and bending moment? Take $E = 207\,\text{GPa}$.

$$I = \pi d^4/64 = \pi(30)^4/64 = 39\,760.8\,\text{mm}^4,$$
$$A = \pi d^2/4 = \pi(30)^2/4 = 706.86\,\text{mm}^2,$$
$$\alpha = \sqrt{(P/EI)} = \sqrt{(20 \times 10^3/207 \times 10^3 \times 39\,760.8)} = 1.559 \times 10^{-3}\,\text{mm}^{-1}.$$

Substituting into Eq. (8.21),

$$-200 = -20 \times 10^3/706.86$$
$$- [(15W \times 10^3)/(2 \times 39\,760.8 \times 1.559)]\tan(1.559 \times 2 \times 180/2\pi)$$
$$= -28.494 - 10.256W.$$

From which $W = 16.72\,\text{N}$. This is low because, by the Euler theory, the strut is almost at the point of buckling under the axial load. That is,

$$P_E = \pi^2 EI/L^2 = \pi^2 \times 207\,000 \times 39\,760.8/2000^2 = 20.31\,\text{kN}.$$

From Eq. (8.19), the maximum deflection is

$$v_{max} = (16.72/2 \times 20 \times 10^3)[(10^3/1.559)\tan(1.559 \times 2 \times 180/2\pi) - 2000/2]$$
$$= 0.418 \times 10^{-3}(641.44 \tan 89.32° - 1000) = 22.17\,\text{mm},$$

and, from Eq. (8.20), the maximum central moment is

$$M_{max} = -[16.72 \times 10^3/(2 \times 1.559)]\tan(1.559 \times 2 \times 180/2\pi)$$
$$= -454.56 \times 10^3\,\text{N mm}, \quad = 454.56\,\text{Nm} \quad \text{(sagging).}$$

Pinned-end Strut Carrying a Uniformly Distributed Lateral Load in Addition to an Axial Compressive Load

In this case, Fig. 8.5 shows that the bending moment to the left of a point (z, v) is composed of two sagging components (Pv) and $(wLz/2)$ and a hogging component $(wz^2/2)$. The flexure equation (7.4) then becomes

$$EI\,d^2v/dz^2 = -Pv - wLz/2 + wz^2/2,$$
$$d^2v/dz^2 + \alpha^2 v = -(w/2EI)(Lz - z^2),$$

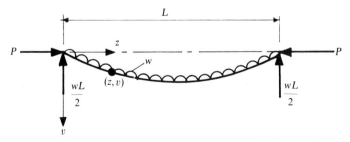

Figure 8.5 Pinned-end strut with laterally distributed and axial loading.

for which the solution is

$$v = A \cos \alpha z + B \sin \alpha z + (w/2P)(z^2 - Lz - 2EI/P).$$

Applying the boundary conditions:

(i) $v = 0$ at $z = 0,$ $\Rightarrow A = EIw/P^2$
(ii) $v = 0$ at $z = L,$ $\Rightarrow B = (EIw/P^2) \tan(\alpha L/2).$

The maximum displacement at $z = L/2$ is then

$$\begin{aligned}
v_{max} &= (EIw/P^2)\cos(\alpha L/2) + (EIw/P^2)\tan(\alpha L/2)\sin(\alpha L/2) \\
&\quad - (w/2P)(L^2/4 + 2EI/P) \\
&= (EIw/P^2)[\sec(\alpha L/2) - 1] - wL^2/8P.
\end{aligned}$$

The maximum, central sagging bending moment is, correspondingly,

$$\begin{aligned}
M_{max} &= -Pv_{max} - wL^2/4 + wL^2/8 \\
&= -(EIw/P)[\sec(\alpha L/2) - 1] + wL^2/8 - wL^2/4 + wL^2/8, \\
M_{max} &= -(EIw/P)[\sec(\alpha L/2) - 1]. \tag{8.22}
\end{aligned}$$

Equation (8.9) gives the maximum compressive stress at the central section:

$$\sigma = -P/A - (Ewy/P)[\sec(\alpha L/2) - 1] \tag{8.23}$$

If it is required to solve for P, then Webb's approximation to $\sec(\alpha L/2)$ in Eq. (8.14) results in the quadratic:

$$aP^2 + bP + c = 0,$$

where

$$a = 1, \qquad b = -(P_E - A\sigma), \qquad c = -A(\sigma P_E - Ewy\pi^2/8).$$

Note that if P in Fig. 8.5 is tensile the flexure equation becomes

$$EI\, d^2v/dz^2 = M = Pv - wLz/2 + wz^2/2,$$

and the solution gives the maximum central moment as

$$M_{max} = -(EIw/P)[1 - \text{sech}(\alpha L/2)].$$

Example 8.6 Find the maximum bending moment and the maximum compressive and tensile stresses for the strut in Fig. 8.5 when $P = 125\,\text{kN}$ (10 000 lbf), $w = 5\,\text{kN/m}$ (100 lbf/ft) and $L = 2.5\,\text{m}$ (20 ft). The I-section has an overall depth of 100 mm (6 in), area 2900 mm^2 (3.53 in^2) with $I = 3.5367 \times 10^6\,\text{mm}^4$ (21 in^4). Take $E = 210\,\text{GPa}$ (30 $\times 10^6$ lbf/in^2).

SOLUTION TO SI UNITS

$$\begin{aligned}
\alpha &= \sqrt{(P/EI)} = \sqrt{(125 \times 10^3/210 \times 10^3 \times 3.5367 \times 10^6)} \\
&= 4.103 \times 10^{-4}\,\text{mm}^{-1}.
\end{aligned}$$

Then, from Eq. (8.22),

$$M_{max} = -(210\,000 \times 3.5367 \times 10^6 \times 5/125\,000)$$
$$\times [\sec(4.103 \times 2.5 \times 10^{-1} \times 180/2\pi) - 1]$$
$$= -4.3856\,kN\,m,$$

and from Eq. (8.23), on the compressive side of the central section,

$$\sigma = -125\,000/2900 - (4.3856 \times 10^6 \times 50/3.5367 \times 10^6)$$
$$= -43.1 - 62.01 = -105.11\,MPa.$$

On the tensile side of the central section, Eq. (8.9) gives

$$\sigma = -43.1 + 62.01 = +18.91\,MPa.$$

SOLUTION TO IMPERIAL UNITS Note that the given quantities are not equivalent between units.

$$\alpha = \sqrt{(P/EI)} = \sqrt{[10\,000/(21 \times 30 \times 10^6)]} = 3.984 \times 10^{-3}\,in^{-1}$$

and, from Eq. (8.22),

$$M_{max} = -[(30 \times 10^6 \times 21 \times 100)/(12 \times 10\,000)]$$
$$\times [\sec(3.984 \times 10^{-3} \times 20 \times 12 \times 180/2\pi) - 1]$$
$$= -66\,300\,lbf\,in.$$

Then on the compressive side of the central section, Eq. (8.23) gives

$$\sigma = -10\,000/3.53 - 66\,300 \times 3/21 = -2832.86 - 9471.43$$
$$= -12\,304\,lbf/in^2 = -5.49\,tonf/in^2,$$

and on the tensile side

$$\sigma = -2832.86 + 9471.43$$
$$= 6638.57\,lbf/in^2 = 2.96\,tonf/in^2.$$

Laterally Loaded Struts with Fixed Ends

Figure 8.6(a) and (b) shows the preceding two cases, with the difference that each end is fixed. This results in a fixed-end moment M_o that modifies the corresponding moment expression used with Eq. (7.4). Thus, in Fig. 8.6(a) at a point (z, v) on the deflected strut,

$$EI\,d^2v/dz^2 = M = -Pv - Wz/2 + M_o.$$

Integration and the determination of the constants from the boundary conditions leads to central sagging and hogging end-moments of

$$M_c = -M_o = -(W/2\alpha)[\operatorname{cosec}(\alpha L/2) - \cot(\alpha L/2)].$$

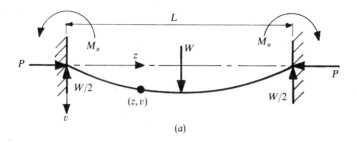

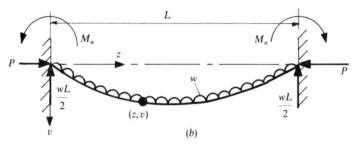

Figure 8.6 Encastre struts with lateral loading.

Similarly, in Fig. 8.6(b),

$$EI\,d^2v/dz^2 = M = -Pv - wLz/2 + wz^2/2 + M_o,$$

from which the corresponding central and end moments are

$$M_c = -(wL/2\alpha)[\operatorname{cosec}(\alpha L/2) - 2/\alpha L] \quad \text{(sagging)},$$
$$M_o = (wL/2\alpha)[\cot(\alpha L/2) - 2/\alpha L] \quad \text{(hogging)},$$

and the maximum stresses are again found from Eq. (8.9).

Pinned-end Strut with Initial Curvature

It is assumed that the initial curvature for the strut in Fig. 8.7 is described by the sinusoidal function $v_o = h\sin(\pi z/L)$, where h is the value of the maximum central deviation. Then with further deflection $v - v_o$ from the application of a compressive axial force P, the flexure Eq. (7.4) becomes

$$EI\,d^2(v - v_o)/dz^2 = -Pv,$$
$$EI\,d^2v/dz^2 + Pv = EI\,d^2v_o/dz^2 = (EIh)d^2(\sin \pi z/L)/dz^2,$$
$$d^2v/dz^2 + \alpha^2v = -(h\pi^2/L^2)\sin(\pi z/L).$$

The solution is

$$v = A\cos \alpha z + B\sin \alpha z + h\sin(\pi z/L)/[1 - (\alpha L/\pi)^2].$$

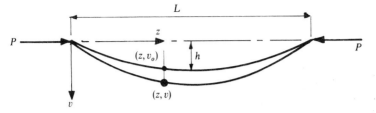

Figure 8.7 Deflection of an initially curved pinned-end strut.

Applying the boundary conditions $v = 0$ at $z = 0$ and $z = L$ leads to

$$v = h \sin(\pi z/L)/[1 - (\alpha L/\pi)^2],$$

which has its maximum at $z = L/2$:

$$v_{max} = h/[1 - (\alpha L/\pi)^2]. \tag{8.24}$$

Correspondingly, the maximum bending moment at this position is

$$M_{max} = -Pv_{max} = -Ph/[1 - (\alpha L/\pi)^2].$$

Since $(\alpha L/\pi)^2 = P/P_E$, the maximum compressive stress is, from Eq. (8.9),

$$\sigma = -P/A - Phy/I[1 - (\alpha L/\pi)^2] = -P/A - Phy/I(1 - P/P_E).$$

This equation may be directly solved for P in the resulting quadratic:

$$aP^2 + bP + c = 0, \tag{8.25}$$

where

$$a = 1, \qquad b = -(hyAP_E/I + P_E - \sigma A), \qquad c = -\sigma AP_E$$

in which σ is negative.

Example 8.7 Find P for a 2 m long strut in Fig. 8.7 that would just produce yielding in the mid-span 25 mm diameter circular section given that the maximum amplitude of initial curvature is 5 mm and the compressive yield stress is 300 MPa. What is the additional central deflection under this load? Take $E = 210$ GPa.

$$A = \pi d^2/4 = (\pi/4)25^2 = 490.87 \text{ mm}^2,$$
$$I = \pi d^4/64 = (\pi/64)25^4 = 19.17 \times 10^3 \text{ mm}^4,$$
$$P_E = \pi^2 EI/L^2 = \pi^2 \times 210\,000 \times 19.17 \times 10^3/(2000^2) = 9935.1 \text{ N}.$$

Substituting into Eq. (8.25) with $\sigma = -300$ MPa, $h = 5$ mm and $y = 12.5$ mm:

$$b = -[(5 \times 12.5 \times 490.87 \times 9935.1/19.17 \times 10^3) + 9935.1$$
$$- (-300 \times 490.87)] = -173.2 \times 10^3 \text{ N},$$
$$c = -\sigma AP_E = 300 \times 490.87 \times 9935.1 = 1.465 \times 10^9 \text{ N}^2.$$

Then in units of newtons,

$$P^2 - (173.2 \times 10^3)P + (1.465 \times 10^9) = 0,$$

from which $P = 8950$ N. Now, from Eq. (8.24), the maximum deflection is

$$v_{\max} = h/[1 - (\alpha L/\pi)^2] = h/(1 - P/P_E)$$
$$= 5/(1 - 8.95/9.935) = 51.1 \text{ mm}.$$

8.4 SEMI-EMPIRICAL APPROACHES

As the slenderness ratio decreases, so the Euler theory diverges from the observed behaviour. Figure 8.8 shows the region below and beyond the limiting L/k value of Eq. (8.6b) for this theory. The following empirical approaches, represented graphically in Fig. 8.8, attempt to account for the increasing effect that compressive yielding has on the buckling behaviour with decreasing L/k in the invalid Euler region (shown dotted).

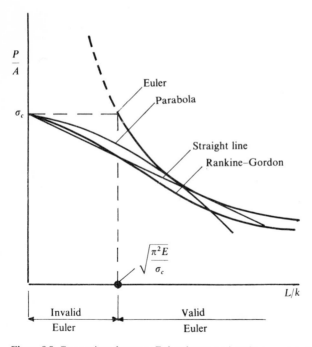

Figure 8.8 Comparison between Euler theory and various empirical approaches.

Rankine–Gordon

In the case of a pinned-end strut, the Euler buckling load P_E and the compressive yield load P_c are compined in reciprocal form:

$$\frac{1}{P_R} = \frac{1}{P_E} + \frac{1}{P_c},$$

from which the Rankine buckling load is

$$P_R = P_c/(1 + P_c/P_E).$$

Substituting $P_c = A\sigma_c$ and P_E from Eq. (8.3) gives

$$P_R = A\sigma_c/[1 + a(L/k)^2] \qquad (8.26)$$

where a is an empirical constant that replaces $\sigma_c/\pi^2 E$ in the derivation. This constant is found by matching Eq. (8.26) to test results. Typical values for a are given, along with compressive yield stress σ_c, in Table 8.2 for the pinned-end condition. For other end fixings the effective length concept may again be employed. Thus, L in Eq. (8.26) is replaced with the L_e expressions given in Table 8.1. Note, from the graphical representation of Eq. (8.26) in Fig. 8.8, that as the Rankine stress P_R/A does not exceed σ_c, it applies over the whole L/k range and lies on the 'safe' side of the Euler curve.

Example 8.8 Compare predicted values of the Euler and Rankine–Gordon buckling loads for a fixed 5 m (15 ft) length mild steel strut for the cross-section in Fig. 8.9. Show that the Euler value is invalid. Take $E = 210\,\text{GPa}$ $(30 \times 10^6\,\text{lbf/in}^2)$ and the Rankine constants from Table 8.2.

SOLUTION TO SI UNITS

$$A = 2[(300 \times 12) + 2(125 \times 20) + (160 \times 12)] = 21.04 \times 10^3\,\text{mm}^2.$$

Buckling occurs about the axis with the lesser I value, where

$$I_x = [(300 \times 224^3) - (300 \times 200^3) + 2(125 \times 200^3 - 113 \times 160^3)]/12$$
$$= 170.5 \times 10^6\,\text{mm}^4,$$

TABLE 8.2 Rankine–Gordon constants in Eq. (8.26)

Material	σ_c(MPa)	a^{-1}
Mild steel	325	7500
Wrought iron	247	9000
Cast iron	557	1600
Timber	35	750

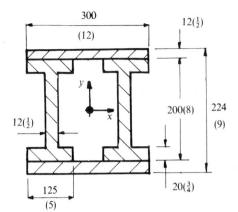

Figure 8.9 Strut section.

$$I_y = 2 \times 12 \times 300^3/12 + 2\{(2 \times 20 \times 125^3)/12 + (160 \times 12^3)/12$$
$$+ [(2 \times 125 \times 20) + (160 \times 12)] \times (150 - 62.5)^2\}$$
$$= 173.03 \times 10^6 \, \text{mm}^4,$$
$$L/k = L/\sqrt{(I_x/A)} = 2500/\sqrt{[(170.5 \times 10^6)/(21.04 \times 10^3)]}$$
$$= 2500/90.02 = 27.77.$$

Then, from Eq. (8.4),

$$P_E = \pi^2 EA/(L/k)^2 = \pi^2 \times 207\,000 \times 21.04 \times 10^3/(27.77)^2 = 56.21 \, \text{MN}.$$

and from Eq. (8.26),

$$P_R = 21.04 \times 10^3 \times 325/[1 + (27.77)^2/7500] = 6.2 \, \text{MN}.$$

Only the Rankine prediction is valid, since, from Eq. (8.6), Euler ceases to apply for lengths below

$$L = \sqrt{(\pi^2 Ek^2/\sigma_c)} = \sqrt{[\pi^2 \times 207\,000 \times (90.02)^2/325]} = 7.137 \, \text{m}.$$

SOLUTION TO IMPERIAL UNITS:

$$A = 2[(12 \times 0.5) + 2(5 \times 0.75) - (6.5 \times 0.5)] = 33.5 \, \text{in}^2.$$

The lesser I applies to the x-axis:

$$I_x = 2\{[5 \times 8^3/12 - 4.5 \times (6.5)^3/12]$$
$$+ [12 \times (0.5)^3/12 + (12 \times 0.5)(4.25)^2]\} = 437.85 \, \text{in}^4.$$
$$\therefore \quad L/k = 7.5 \times 12/\sqrt{(437.85/33.5)} = 249.$$

The invalid Euler prediction is, from Eq. (8.4),

$$P_E = \pi^2 EA/(L/k)^2 = \pi^2 \times 30 \times 10^6 \times 33.5/24.9^2$$
$$= 16 \times 10^6 \, \text{lbf} = 7142.9 \, \text{tonf}.$$

The valid Rankine prediction is found from Eq. (8.26), with $\sigma_c = 21\,\text{tonf/in}^2$,

$$P_R = 33.5 \times 21/[1 + (24.9)^2/7500] = 650\,\text{tonf}.$$

Example 8.9 The equal-angle section in Fig. 7.20 is used in a mild steel strut, 2 m in length with one end fixed and the other end pinned. Calculate the safe load from the Rankine–Gordon theory using a safety factor of 4 and the constants in Table 8.2.

It has previously been found in Example 7.17 that the least I value for this section is $I_n = 31.1 \times 10^4\,\text{mm}^4$:

$$A = (72 \times 12) + (60 \times 12) = 1584\,\text{mm}^2.$$

Now from Table 8.1, $L_e = L/\sqrt{2}$, when from Eq. (8.26),

$$P_R = A\sigma_w/[1 + (a/2)(L/k)^2],$$

where $\sigma_w = 325/4 = 81.25\,\text{MPa}$ and

$$k = \sqrt{(I/A)} = \sqrt{(31.1 \times 10^4/1584)} = 14.012\,\text{mm},$$
$$\therefore \quad P_R = 1584.81.25/[1 + (2000/14.012)^2/15\,000] = 54.58\,\text{kN}.$$

Straight-line Formula

Commonly used in the United States for a particular L/k range, the linear prediction shown in Fig. 8.8 has the form

$$P = A\sigma_c[1 - n(L/k)] \tag{8.27}$$

where n and σ_c are constants with typical values for a pin-ended strut.

Mild steel	$n = 0.004$,	$\sigma_c = 258\,\text{MPa}$	for $L/k \leq 150$
Cast iron	$n = 0.004$,	$\sigma_c = 234\,\text{MPa}$	for $L/k \leq 100$
Duralumin	$n = 0.009$,	$\sigma_c = 302\,\text{MPa}$	for $L/k \leq 85$
Oak	$n = 0.005$,	$\sigma_c = 37\,\text{MPa}$	for $L/k \leq 65$

These constants are determined from the two points through which the line passes, i.e. the compressive stress σ_c and a pre-determined point at intersection or tangency to the Euler curve at the upper limit of L/k.

Johnson's Parabolic Formula

Within the invalid Euler region the following parabolic form is employed with struts of smaller slenderness ratios:

$$P = A\sigma_c[1 - b(L/k)^2], \tag{8.28}$$

where b and σ_c are constants, with typical values; $b = (23 \text{ to } 30) \times 10^{-6}$ and $\sigma_c = 275\,\text{MPa}$ for $L/k \leq 150$ in mild steel. This form often takes σ_c to be less than the actual compressive yield stress of the strut material to offset the need for a safety factor. The particular form of Eq. (8.28) employed by the American Institute of Steel Construction is

$$P = A\sigma_c[1 - (1/2C_c^2)(L/k)^2] \qquad (8.29)$$

where C_c is determined from the intersection between Eq. (8.29) and the Euler curve at $\sigma_E = P/A = \sigma_c/2$. Then, from either Eq. (8.5) or (8.29), this condition gives $C_c = L/k = \sqrt{(2\pi^2 E/\sigma_c)}$. Alternatively, b in Eq. (8.28) may be found from the condition that a point of tangency with the Euler curve exists for a given L/k (typically 120).

Example 8.10 Determine the constant in Eqs (8.27) and (8.28) for a pinned-end strut given that each must pass through the Euler curve at

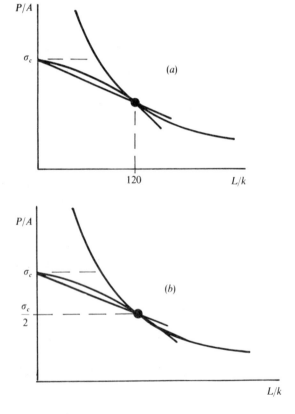

Figure 8.10 Euler, straight-line and Johnson curves.

(i) $L/k = 120$ and (ii) $\sigma_c/2$ as shown in Fig. 8.10. Take $E = 210\,\text{GPa}$, $\sigma_c = 320\,\text{MPa}$.

Condition (i) (Fig. 8.10a): Equating (8.5) and the straight-line formula (8.27) gives

$$P/A = \sigma_c[1 - n(L/k)] = \pi^2 E/(L/k)^2,$$
$$n = (L/k)^{-1}[1 - \pi^2 E/\sigma_c(L/k)^2]$$
$$= (1/120)[1 - \pi^2 \times 210 \times 10^3/320 = (120)^2] = 0.004\,59.$$

Equating (8.5) and Johnson's parabola (Eq. (8.28)):

$$P/A = \sigma_c[1 - b(L/k)^2] = \pi^2 E/(L/k)^2,$$
$$b = (L/k)^{-2}[1 - \pi^2 E/\sigma_c(L/k)^2],$$
$$= (1/120)^2[1 - \pi^2 \times 210 \times 10^3/320 \times (120)^2] = 38.21 \times 10^{-6}.$$

Condition (ii) (Fig. 8.10b): From Eq. (8.5) at the common point of intersection:

$$P/A = \sigma_c/2 = \pi^2 E/(L/k)^2,$$

from which

$$L/k = \sqrt{(2\pi^2 E/\sigma_c)} = \sqrt{(2 \times \pi^2 \times 210 \times 10^3/320)} = 113.82.$$

Substituting $P/A = 160\,\text{MPa}$, $\sigma_c = 320\,\text{MPa}$ and $L/k = 113.82$ in Eq. (8.27),

$$160 = 320(1 - 113.82n), \qquad \Rightarrow n = 0.0044.$$

Similar substitutions in Eq. (8.28) leads to

$$160 = 320[1 - (113.82)^2 b], \qquad \Rightarrow b = 38.6 \times 10^{-6}.$$

Example 8.11 Determine the corresponding constants when the straight line and parabolic strut formulae are to make tangents with the Euler curve at $L/k = 120$. Take $E = 207\,\text{GPa}$.

Here each representation must have the same gradient of their tangents at $L/k = 120$ (Fig. 8.11). The common stress ordinate is found from Eq. (8.5):

$$\sigma = \pi^2 E/(L/k)^2 = \pi^2 \times 207\,000/120^2 = 141.88\,\text{MPa}.$$

The gradients are found from Eqs (8.5), (8.27) and (8.28) to be

$$\text{Euler} \qquad d\sigma/d(L/k) = -2\pi^2 E/(L/k)^3, \qquad \text{(i)}$$
$$\text{Straight line} \qquad d\sigma/d(L/k) = -n\sigma_c, \qquad \text{(ii)}$$
$$\text{Johnson} \qquad d\sigma/d(L/k) = -2\sigma_c b(L/k). \qquad \text{(iii)}$$

Equating (i) and (ii):

$$-n\sigma_c = -2 \times \pi^2 \times 207 \times 10^3/120^3 = -2.365.$$

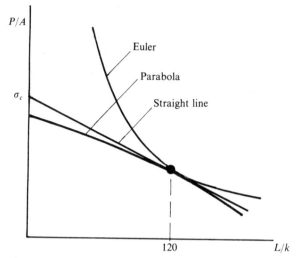

Figure 8.11 Strut bucking curves.

Substituting into Eq. (8.27):

$$141.88 = \sigma_c(1 - 120n) = 2.365(1/n - 120)$$
$$\therefore \quad n = 5.556 \times 10^{-3}, \qquad \sigma_c = 425.68 \text{ MPa}.$$

Equating (i) and (iii):

$$-2\sigma_c b \times 120 = -2.365,$$
$$\therefore \quad \sigma_c = (9.854 \times 10^{-3})/b.$$

Substituting into Eq. (8.28):

$$141.88 = \sigma_c[1 - (120)^2 b]$$
$$= 9.854 \times 10^{-3}[1/b - 14\,400],$$
$$\therefore \quad b = 3.473 \times 10^{-5}, \qquad \sigma_c = 283.78 \text{ MPa}.$$

Note that for this condition Fig. 8.11 shows that the intercept σ_c is dependent upon the chosen point of tangency and will not equal the compressive yield stress in general.

Perry–Robertson

This approach is recommended in BS 449: 1969, Pt. 2, for the determination of allowable compressive loads in structural steel columns. In common with the Rankine–Gordon method, the following expression represents a transition curve between pinned-end strut buckling at high L/k and short-column yielding for low L/k in the range $80 \le L/k \le 350$:

$$P = (A/2K)([\sigma_c + (\mu + 1)\sigma_E] - \sqrt{\{[\sigma_c + (\mu + 1)\sigma_E]^2 - 4\sigma_E\sigma_c\}}), \quad (8.30)$$

where σ_c is the minimum yield strength, and the load (safety) factor K is typically taken as 1.7–2.0. An account of initial sinusoidal curvature and eccentricity of loading appear indirectly in a deformation factor adopted by the British Standard: $\mu = a(L/k)$, where $a = 0.001$–0.003 based upon Robertson's experiments. Alternatively, the Dutheil factor

$$\mu = (0.3/\pi^2)(\sigma_c/E)(L/k)^2 \tag{8.31}$$

accorded with the lowest failure loads observed in his buckling experiments.

Fidler

A particular simplified form of Eq. (8.30) was developed by Fidler for the design of columns used in bridge construction:

$$P = (A/q)\{(\sigma_c + \sigma_E) - \sqrt{[(\sigma_c + \sigma_E)^2 - 2q\sigma_E\sigma_c]}\} \tag{8.32}$$

where q is a load factor with an average value of 1.2.

Engesser

This particular modification to the Euler theory attempts to account for inelastic buckling simply by replacing the elastic modulus with the tangent modulus. Then Eq. (8.5) becomes

$$\sigma = P/A = \pi^2 E_t/(cL/k)^2, \tag{8.33}$$

where c accounts for the effect of end fixing. In the range $50 \le L/k \le 100$ it is known that effect of end constraint on the plastic buckling load P is much

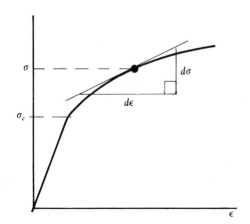

Figure 8.12 Tangent modulus.

less than the equivalent elastic constraint given in Table 8.1. The tangent modulus $E_t = d\sigma/d\epsilon$ is the gradient of the tangent to the uniaxial tensile or compressive stress–strain curve (Fig. 8.12). It follows from this definition of E_t that the buckling stress in Eq. (8.33) must satisfy the following condition:

$$\sigma = (\pi/c)^2 (L/k)^{-2} (d\sigma/d\epsilon), \tag{8.34}$$

which may be established either graphically or from a suitable empirical representation of the σ–ϵ curve, the simplest being the Hollomon law,

$$\sigma = B\epsilon^n, \tag{8.35}$$

where B and n are the respective hardening coefficient and exponent, which are constants for a given material. Now, from Eq. (8.35),

$$d\sigma/d\epsilon = (nB)\epsilon^{n-1} = (nB)(\sigma/B)^{(n-1)/n}. \tag{8.36}$$

Substituting Eq. (8.36) into Eq. (8.34) results in an equation that is soluble in σ.

Example 8.12 A pinned-end strut 1.5 m long has the cross-section shown in Fig. 8.13. Find the inelastic Engesser buckling load given that the law $\sigma/\sigma_c = 5\epsilon^{1/3}$ describes the compressive stress–strain curve in the region beyond the yield stress $\sigma_c = 300\,\text{MPa}$. Compare this with the corresponding Perry–Robertson and Fidler predictions with $E = 207\,\text{GPa}$.

ENGESSER

$$A = (100 \times 50) - (75 \times 25) = 3125\,\text{mm}^2,$$
$$I = (100 \times 50^3/12) - (75 \times 25^3/12) = 11.328 \times 10^6\,\text{mm}^4$$
$$k = \sqrt{(I/A)} = \sqrt{(11.328 \times 10^6/3125)} = 60.21\,\text{mm},$$
$$L/k = 1500/60.21 = 24.91.$$

Then from Eqs (8.34) and (8.36), with $c = 1$ and $B = 5\sigma_c$:

$$\sigma = 5\sigma_c[(1/3)(\pi/24.91)^2]^{1/3} = 261.56\,\text{MPa},$$

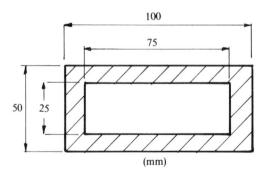

(mm)

Figure 8.13 Strut section.

from which the buckling load is

$$P = \sigma A = (261.56 \times 3125) = 817.2 \, \text{kN}.$$

PERRY–ROBERTSON

$$\sigma_E = \pi^2 E/(L/k)^2 = \pi^2 \times 207\,000/(24.91)^2 = 3292.48 \, \text{MPa}$$

and with the Dutheil deformation factor in Eq. (8.31),

$$\mu = (0.3/\pi^2)(\sigma_c/E)(L/k)^2$$
$$= (0.3/\pi^2)(300/207\,000)(24.91)^2 = 0.0273.$$

Then, from Eq. (8.30) with $K = 1$, the buckling load is

$$P = (3125/2)\big([300 + (1.0273 \times 3292.48)] - \sqrt{\{[300 + (1.0273 \times 3292.48)]^2}$$
$$- (4 \times 300 \times 3292.48)\}\big)$$
$$= 1562.5(3682.36 - 3099.81) = 910.23 \, \text{kN}.$$

FIDLER From Eq. (8.32) with $q = 1$:

$$P = (3125/1.0\{(3292.48 + 300) - \sqrt{[(3292.48 + 300)^2}$$
$$- (2 \times 1.0 \times 300 \times 3292.48)]\}$$
$$= 3125[3592.48 - 3306.12] = 894.88 \, \text{kN}.$$

It appears from these solutions that, without the introduction of load factors K and c in Eqs (8.30) and (8.32), the Engesser prediction is the most conservative.

8.5 BUCKLING THEORY OF PLATES

This introduction omits the full theoretical analysis that may be found in more specialist texts (see J. R. Vinson, *Structural Mechanics*, Wiley, 1974). Moreover the consideration of edge fixings is limited here to simple and clamped supports.

Plate as a Wide Strut with Long Edges Unsupported

When in a thin plate the width dimension b is large, i.e. of the same order as the length a, then biaxial in-plane stresses σ_x and σ_y will exist in the body of the plate (Fig. 8.14) ensuring that the cross-section remains rectangular during bending.

From Eq. (4.7), with no dimensional change in the y-direction,

$$\epsilon_y = 0 = (1/E)(\sigma_y - v\sigma_x),$$
$$\epsilon_x = (1/E)(\sigma_x - v\sigma_y) = (1 - v^2)\sigma_x/E. \tag{8.37}$$

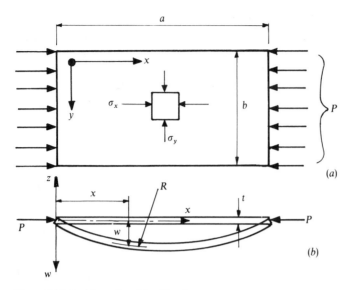

Figure 8.14 Buckling of a wide, thin plate.

The absence of σ_y in thin rectangular section struts of smaller width results in *anticlastic curvature*, when the section does not remain rectangular under bending but distorts owing to the opposite sense of the induced lateral strains between the tensile and compressive surfaces (Poisson effect). Here the strains in the length and width directions are simply $\epsilon_x = \sigma_x/E$ and $\epsilon_y = -\nu\sigma_x/E$. The difference in ϵ_x between small and large width struts shows that it is necessary to modify the flexure equation (7.4) by the factor $(1 - \nu^2)$ in the case of the wide strut. From Eq. (7.3), with x replacing z, w replacing v and z replacing y in Fig. 8.14:

$$\mathrm{d}^2w/\mathrm{d}x^2 = 1/R = \epsilon_x/z.$$

Substituting from Eq. (8.37) with $M/I = \sigma_x/z$, it follows that

$$\mathrm{d}^2w/\mathrm{d}x^2 = (1 - \nu^2)\sigma_x/Ez = (1 - \nu^2)M/EI.$$

Then, with $M = -Pw$, the critical buckling load for a plate with pinned ends is found by comparison with the solution to Eq. (8.1). That is,

$$P_{cr} = \pi^2 EI/(1 - \nu^2)a^2.$$

Substituting $P_{cr} = \sigma_{cr}bt$ and $I = bt^3/12$ leads to the more common form of expression in terms of aspect ratios t/b and $r = a/b$:

$$\sigma_{cr} = (1/12)(\pi/r)^2[E/(1 - \nu^2)](t/b)^2. \tag{8.38}$$

Other edge fixings may be accounted for by replacing the 1/12 factor for pinned ends in Eq. (8.38) with a buckling coefficient K.

Bi-directional Compressive Loading with Edges Supported

Consider a thin rectangular plate $a \times b$ with thickness t subjected to uniform compressive stresses σ_x acting normal to $b \times t$ and σ_y acting normal to $a \times t$ as shown in Fig. 8.15. It can be shown that when the edges are simply supported the equation governing the dependence of the lateral deflection w upon x and y is

$$D(\partial^2 w/\partial x^2 + \partial^2 w/\partial y^2)^2 = (\sigma_x t)\partial^2 w/\partial x^2 + (\sigma_y t)\partial^2 w/\partial y^2,$$

where the flexural stiffness $D = Et^3/12(1 - v^2)$. The solution is

$$w(x, y) = \sum_{m=1}^{\infty} \sum_{n=1}^{\infty} A_{mn} \sin(m\pi x/a) \sin(n\pi y/b),$$

in which the respective number of half-waves m and n for buckling in the x and y plate directions determine the number of terms and associated coefficients A_{mn}. In Fig. 8.15, for example, $m = 3$ and $n = 1$ is shown. When the stresses increase proportionately in the ratio $\alpha = \sigma_y/\sigma_x$, the actual number of half-waves of buckling is that which minimizes σ_x for given edge fixings. In general, this stress is expressed as

$$(\sigma_x)_{cr} = \frac{D\pi^2[(m/a)^2 + (n/b)^2]^2}{t[(m/a)^2 + \alpha(n/b)^2]}. \tag{8.39}$$

Example 8.13 Find the critical buckling stress for a square plate of side length a with simply supported edges subjected to equal in-plane loading.

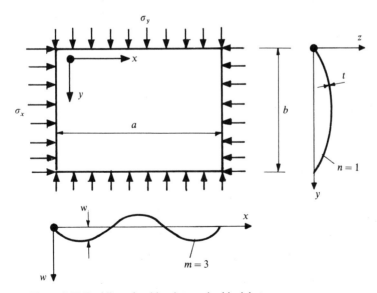

Figure 8.15 Buckling of a thin plate under biaxial stress.

With simply supported edges, i.e. no rotational restraint, a square plate buckles with one half-wave $m = n = 1$ in each direction. Substituting $r = a/b = 1$ and $\alpha = \sigma_y/\sigma_x = 1$ in Eq. (8.39) leads to the buckling stress:

$$(\sigma_x)_{cr} = (D\pi^2/ta^2)(1 + r^2)^2/(1 + r^2\alpha),$$
$$(\sigma_x)_{cr} = 2D\pi^2/ta^2$$
$$= (\pi^2/6)[E/(1 - v^2)](t/a)^2.$$

Example 8.14 If the plate in Fig. 8.15 has simply supported edges with $b \ll a$, find $(\sigma_x)_{cr}$ and the half-wavelength of the buckled shape in the direction of x, when (i) $\alpha = \sigma_y/\sigma_x < 1/2$ and (ii) as $\alpha \to 1/2$.

Taking $n = 1$ for the much smaller b dimension, Eq. (8.39) becomes

$$(\sigma_x)_{cr} = \frac{D\pi^2(m^2 + r^2)^2}{a^2t(m^2 + \alpha r^2)} = \frac{D\pi^2(m/r + r/m)^2}{b^2t[1 + \alpha(r/m)^2]}. \qquad (i)$$

Then m is found from the condition that $(\sigma_x)_{cr}$ is a minimum. That is,

$$d(\sigma_x)_{cr}/dm = [1 + \alpha(r/m)^2]2(m/r + r/m)(1/r - r/m^2)$$
$$- (m/r + r/m)^2 2\alpha(r/m)(-r/m^2) = 0,$$
$$[1 - (r/m)^2][1 + \alpha(r/m)^2] + \alpha(r/m)^2[1 + (r/m)^2] = 0,$$
$$\therefore \quad (r/m)^2(1 - 2\alpha) = 1.$$

This gives

$$m = r(1 - 2\alpha)^{1/2} \qquad (ii)$$

with the corresponding half-wavelength $a/m \simeq r/m = (1 - 2\alpha)^{-1/2}$. Substituting Eq. (ii) in (i)

$$(\sigma_x)_{cr} = \frac{D\pi^2[(1 - 2\alpha) + 2 + 1/(1 - 2\alpha)]}{bt^2[1 + \alpha/(1 - 2\alpha)]} = \frac{D\pi^2[(1 - 2\alpha)(3 - 2\alpha) + 1]}{b^2t(1 - \alpha)}.$$

It is apparent from this solution that, as $\alpha \to 1/2$, the half-wavelength in the x-direction approaches infinity as the buckling stress becomes

$$(\sigma_x)_{cr} \to 2D\pi^2/b^2t.$$

Unidirectional Compressive Loading

Simple supports With only a uniform compressive stress σ_x acting, the critical buckling stress is found by putting $\sigma_y = 0$ ($\alpha = 0$) in Eq. (8.39). This gives

$$(\sigma_x)_{cr} = (D\pi^2/t)[(m/a)^2 + (n/b)^2]^2(a/m)^2,$$
$$(\sigma_x)_{cr} = (D\pi^2/tb^2)(mb/a + n^2a/mb)^2. \qquad (8.40)$$

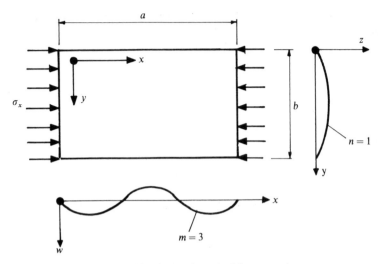

Figure 8.16 Buckling of a thin plate under uniaxial compression.

When the plate is simply supported along its unstressed sides, it buckles with one half-wave ($n = 1$) in the y-direction as shown in Fig. 8.16. Then, from Eq. (8.40) the buckling stress becomes

$$(\sigma_x)_{cr} = (D\pi^2/tb^2)(m/r + r/m)^2. \qquad (8.41)$$

The question then arises of what values of m minimize Eq. (8.41) for integral values of $r = a/b$? This condition is expressed in

$$d(\sigma_x)_{cr}/dm = 2(m/r + r/m)(1/r - r/m^2) = 0,$$
$$m/r^2 + 1/m - 1/m - r^2/m^3 = 0,$$
$$m^4 - r^4 = 0,$$
$$(m - r)(m + r)(m^2 + r^2) = 0.$$

Thus, $m = r$ implies that the plate will buckle into an integral number of square cells $a \times a$ under the same stress. That is, from Eq. (8.41),

$$(\sigma_x)_{cr} = 4D\pi^2/tb^2 = [\pi^2 E/3(1 - v^2)](t/b)^2. \qquad (8.42)$$

Furthermore when, for non-integral values of r, Eq. (8.41) is applied to particular values of m, the graph in Fig. 8.17 represents the buckling stress.

The trough in each curve corresponds to Eq. (8.42) and the points of intersection determine the r value for which m is increased by 1. This occurs when, from Eq. (8.41),

$$(m/r + r/m) = (m + 1)/r + r/(m + 1),$$
$$\therefore \quad r = \sqrt{[m(m + 1)]},$$

from which $r = \sqrt{2}$ for $m = 1$, $r = \sqrt{6}$ for $m = 2$, $r = \sqrt{12}$ for $m = 3$, etc.

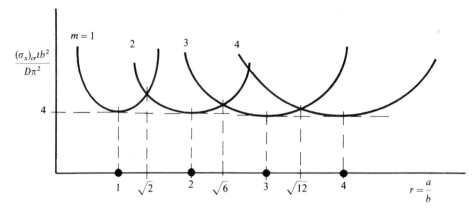

Figure 8.17 Effect of r and m on the uniaxial buckling stress.

Equation (8.42) should not be confused with buckling stress of a thin plate acting as a strut with its longer, parallel sides unsupported (Eq. (8.38)).

Example 8.15 Find the thickness of plate, with dimensions $a = 1.5$ m, $b = 1$ m simply supported along its edges, that will buckle elastically under a compressive stress of 210 MPa if $E = 70$ MPa, $v = 0.3$.

Figure 8.17 shows that with $r = a/b = 1.5$ the plate will buckle into $m = 2$ half-waves of length 0.75 m in the x-direction. From Eq. (8.41) with $D = Et^2/12(1 - v^2)$:

$$t = \sqrt{\left(\frac{12(1 - v^2)b^2(\sigma_x)_{cr}}{\pi^2 E(m/r + r/m)^2}\right)} = \sqrt{\left(\frac{12 \times 0.91 \times 1^2 \times 210}{\pi^2 \times 70 \times 10^3(2/1.5 + 1.5/2)^2}\right)}$$

$$= 0.0277 \text{ m} = 27.7 \text{ mm}.$$

Other edge fixings A number of approaches have been proposed. The simplest of these employs the analytical expression

$$(\sigma_x)_{cr} = K_r\pi^2D/tb^2,$$

where, for all possible edge fixings, the restraint coefficient is

$$K_r = (m/r)^2 + p + q(r/m)^2,$$

which contains Eq. (8.41) as a special case when the restraint factors are $p = 2$ and $q = 1$ for simple supports. The dependence of K_r upon the rotational edge restraints (p and q), the plate dimensions ($r = a/b$) and the buckling mode (m) has been established experimentally in certain cases.

Table 8.3 applies to the case of fixed sides, for example. As r increases, the effect of edge restraint lessens and K_r approaches a minimum value of

TABLE 8.3 Restraint coefficients for a plate with fixed sides

$r = a/b$	0.75	1.0	1.5	2.0	2.5	3.0
K_r	11.69	10.07	8.33	7.88	7.57	7.37

4 found in Eq. (8.42) for a plate with simply supported edges. An alternative graphical approach, favoured in design practice, employs the ESDU data sheet number 72019. This supplies the restraint coefficient for mixed, clamped, simply supported and free-edge conditions for elastic and plastic plate buckling. ESDU 71005 further deals with the buckling of flat plates under shear loading, from which it can be found that the critical elastic shear stress is

$$\tau_{cr} = K_s E(t/b)^2,$$

where the shear buckling coefficient K_s depends upon the edge fixing and b is the lesser side length.

EXERCISES

Euler Theory

8.1 Find the shortest length of tube, 50 mm o.d. with 3 mm wall thickness, for which the Euler theory would be applicable when it is made from (i) mild steel of $\sigma_c = 310$ MPa with pinned ends and (ii) high-tensile steel of $\sigma_c = 618$ MPa with fixed ends. Take $E = 207$ GPa.
Answer: 1.37 m; 1.94 m.

8.2 A 1 m long connecting rod of rectangular cross-section may be considered pinned at its end w.r.t. the stronger section axis but rigidly fixed about the weaker axis. If the rod is to be equally resistant to buckling when the breadth of section is limited to 20 mm determine the depth using a factor of safety of 4 for an axial compressive working load of 8 t. Take $E = 208$ GPa.

8.3 The ends of a thin-walled, elliptical, light alloy tube are fixed in the plane of its major 60 mm diameter and pinned in the plane of its minor 20 mm diameter. Given that the slenderness ratio is 100, determine the wall thickness that would support an axial compressive load of 50 kN. Take $E = 70$ GPa.

8.4 A strut is formed from two mild steel T-sections as shown in Fig. 8.18. If $A = 2430$ mm^2, $I_x = 1.05 \times 10^6$ mm^4, $I_y = 2.09 \times 10^6$ mm^4 and $\bar{y} = 18.82$ mm for one section, determine the maximum length of this strut for an end-load of 135 kN using a safety factor of 6. Take $E = 208$ GPa.
Answer: 3.1 m.

8.5 Find the buckling axis and the Euler load for the section in Fig. 8.19. The strut, 12 m long with pinned ends, is loaded through the centroid. Take $E = 207$ GPa.
Answer: 378 kN.

8.6 An I-section steel strut is attatched at its ends to a square shackle that fits closely to the web thickness with a pin passing through the centre of the web as shown in the two views of Fig. 8.20 (a) and (b). If the centre distance of 2.45 m defines the strut length, find the Euler buckling load considering buckling about axes x and y. Take $E = 207$ GPa.
Answer: 768 kN.

8.7 Figure 8.21 shows a horizontal beam of distributed mass 15 kg/m supported by two vertical members, A and B. Member A is pin-jointed to the beam and built in at its lower end. Member B is

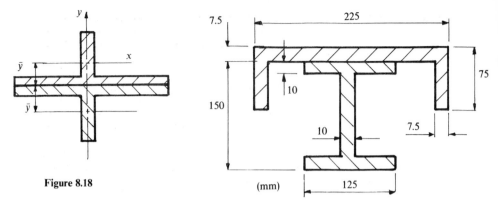

Figure 8.18

Figure 8.19

(mm)

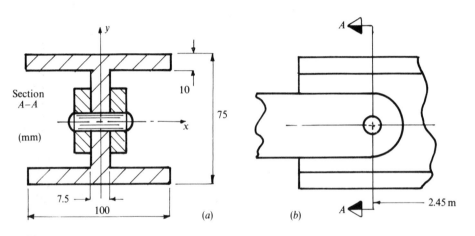

Section
A–A

(mm)

Figure 8.20

(a)

(b)

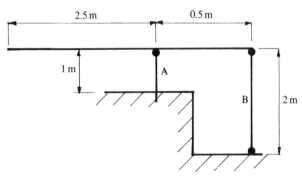

Figure 8.21

pin-jointed to both the beam and the lower anchor point. The least second moment of area of A is 420×10^3 mm^4 and that of B is 200×10^3 mm^4. Determine the maximum value of a movable vertical point force carried by the beam if collapse of the structure is to be avoided. Take $E = 200$ GPa. (C.E.I. Pt. 2, 1979)

8.8 The plane pin-jointed framework in Fig. 8.22 is designed to lift heavy loads at B and C. It takes the form of an isosceles triangle, with the angle at the support A being 90°. The strut member at the base is 3 m long and is made from a uniform solid circular steel rod. Given that each of the joints B and C supports a vertical load of 10 kN, find the smallest diameter of the strut BC to avoid buckling, and hence find the compressive stress therein. Take $E = 208$ GPa. (C.E.I. Pt. 2, 1985)

8.9 The vertical column of length L in Fig. 8.23 is pinned at the top and bottom. It consists of a lower part of length nL that may be regarded as rigid and this is fixed to an upper part which is slender and of stiffness EI. Show that, at the instability point, $\tan[(1-n)kL] = -nkL$, and determine the buckling load for $n = 1/2$.

8.10 Figure 8.24 shows two vertical struts, each of length $2L$, joined at their mid-point by a rigid horizontal connection. The struts have flexural rigidity of EI_1 and EI_2 about axes

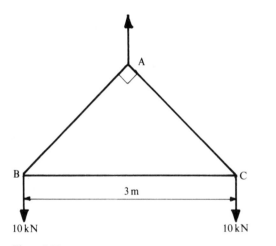

Figure 8.22

Figure 8.23

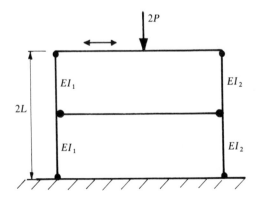

Figure 8.24

normal to the plane of the figure and are constructed from the same material. The struts are pin-jointed at one end to a rigid horizontal beam that slides in a frictionless guide, while the opposite ends are pin-jointed to the ground. Show that the condition for simultaneous buckling of both struts in the plane of the figure, when a central compressive force $2P$ is applied at mid-span of the beam, is given by $\alpha_1 \tan(\alpha_2 L) + \alpha_2 \tan(\alpha_1 L) = 2\alpha_1 \alpha_2 L$, where $\alpha_1^2 = P/EI_1$ and $\alpha_2^2 = P/EI_2$. (C.E.I. Pt. 2, 1980)

8.11 A straight strut of length L is fixed at one end and pinned at the other. If, in calculating the buckling load, it is assumed that the effective length is $2L/3$, what is the percentage error incurred?

Answer: 12.5 per cent.

8.12 A straight strut of length L, rigidly built-in at one end and free at the other, is subjected to an axial compressive force P and a side force S perpendicular to P, both applied at the free end. Show that the free-end deflection is given by $v_o = (S/P)[(1/\alpha)\tan(\alpha L) - L]$ where $\alpha^2 = P/EI$. Hence show that buckling failure occurs when the load $P \to \pi^2 EI/4L^2$. (C.E.I. Pt. 2, 1981)

8.13 A strut of length L has each end fixed in an elastic material which can exert a restraining moment μ/rad. Prove that the critical buckling load P is given by the following expression: $P + \mu\sqrt{(P/EI)}\tan[(L/2)\sqrt{(P/EI)}] = 0$. The designed buckling load of a 1 m long strut assuming the ends to be rigidly fixed was 2.5 kN. If during service, the ends were found to rotate, with each mounting exerting a restraining moment of 10 kN m/rad, show that the buckling load decreases by 20 per cent. (C.E.I. Pt. 2, 1974)

Imperfect Euler Struts

8.14 A tubular steel strut of 63.5 mm o.d. and 51 mm i.d. is 2.44 m long with pinned ends. The compressive end-load is applied parallel to the axis of the tube but eccentric to it. Find the maximum allowable eccentricity when the buckling load is 75 per cent of the ideal Euler load. Take the compressive yield stress as 310 MPa and $E = 207$ GPa.

Answer: 5.2 mm.

8.15 The 75×75 mm equal-angle section in Fig. 8.25 has a cross-sectional area of 1775 mm^2 and a least radius of gyration of 14.5 mm about a 45° axis passing through the centroid G and the apex. When used as a 1.5 m length strut the angle supports a compressive load through a point (0), 6.5 mm from G as shown. If at the mid-length cross-section the net compressive stress is not to exceed 123.5 MPa, find the maximum compressive load.

Answer: 90 kN.

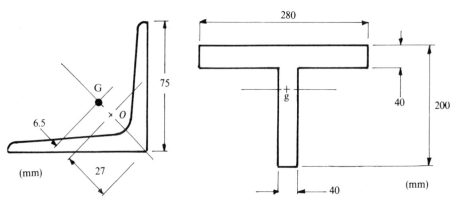

Figure 8.25   Figure 8.26

8.16 A 3 m length mild steel strut of the section given in Fig. 8.26 is built-in at one end and free at the other. A maximum eccentricity of 30 mm from the centroidal axis is permissible when supporting an axial compressive load of 200 kN. Calculate the additional allowable eccentricity at the free end due to the strut not being initially straight, if the maximum stress produced by the combined effects of compression and bending is limited to 80 MPa.

8.17 A tubular steel strut of 65 mm o.d. and 50 mm i.d. is 1 m long and fixed at one end. Find the maximum permissible eccentricity when a 130 kN compressive force is applied at the free end if the allowable stress for the strut material is 325 MPa. If the eccentricity is to be eliminated find (i) the length of strut that could support this force and (ii) the force that could be applied to a length of 1 m.

Answer: 15.56 mm; 440 kN; 1.507 m.

8.18 The axis of a 3 m upright column that is built-in at one end and free at the other deviates slightly from the vertical. The cross-section is an equal angle of 100 mm side length and 10 mm thickness. When the strut supports a vertical compressive load of 20 kN at its free end, a maximum eccentricity of 20 mm from the centroid is allowable. What is the greatest angular deviation of the strut w.r.t. the vertical axis if the maximum stress may not exceed 80 MPa? Take $E = 207$ GPa.

8.19 A vertical, 3 m long steel tube whose external and internal diameters are 76.2 mm and 63.5 mm respectively is fixed at the lower end and completely unrestrained at the top. The tube is subjected to a vertical compressive load at the upper end and eccentric by 6.35 mm from the centroidal axis. Determine the value of the instability load that gives a point of zero resultant stress at the base of the tube. Take $E = 207$ GPa.

Answer: 25.75 kN. (C.E.I. Pt. 2, 1967)

8.20 Determine the maximum compressive stress in a strut of solid circular section, diameter 150 mm, length 5 m, when a force of 15 kN is applied at a distance of 50 mm from the centroid of the section. Take $E = 207$ GPa.

8.21 A long vertical slender strut of uniform solid circular section is of length L and diameter d. The strut is rigidly built-in at the base and unrestrained at the top. Derive an expression for the free-end deflection of the strut when a vertical compressive force P is applied there at a point on the circumference. Hence determine the vertical force required to cause a horizontal end deflection of magnitude d in terms of the Euler buckling load P_E for the same strut. (C.E.I. Pt. 2, 1982)

8.22 A steel strut 25 mm in diameter and 1.5 m long supports an axial compressive load of 16 kN through pinned ends. What concentrated lateral force, when applied normally at mid-span, would cause the strut to buckle? Take $\sigma_c = 278$ MPa, $E = 207$ GPa.

Answer: 133 N.

8.23 A steel tie bar with pin-jointed ends is 2 m long and 25 mm in diameter. It carries an axial tensile load of 1.5 kN together with a concentrated lateral load of 0.3 kN, applied at mid-length. Working from first principles, use the flexure equation to calculate the maximum bending moment and the maximum axial stress in the bar. Take $E = 200$ GPa. (K.P. 1975)

8.24 A 100 × 31.75 mm rectangular section coupling rod supports an axial compressive thrust of 120 kN through pinned ends together with a uniformly distributed lateral load of 4.2 kN/m over its 2.5 m length. Find, for the mid-span position, the value of the bending moment and the tensile and compressive stresses. Estimate the effect that fixed ends would have on the maximum stress. Take $E = 208$ GPa.

Answer: 3.94 kN m; 109.46 MPa.

8.25 A horizontal beam of I-section with depth 150 mm, area 2280 mm² and $I = 8.75 \times 10^6$ mm⁴ is freely supported over a span of 6.1 m. The strut supports an axial compressive force of 44.5 kN and a total uniformly distributed load of 1.46 kN/m over its length. Find the greatest bending moment and stress in the beam. Take $E = 207$ GPa.

Answer: 86 MPa.

8.26 A long thin uniform bar of length L is lifted by $45°$ wires attatched to its ends and leading to a crane hook. If the mass of the bar is w/unit length and the ends are assumed to be pin-jointed show that the mid-span suspended deflection is $(4EI/wL^2)[\sec(\alpha L/2) - 1] - L/4$ where $\alpha = \sqrt{(wL/2EI)}$. Hence derive the mid-span bending moment. (C.E.I. Pt. 2, 1984)

8.27 A tie bar supports a total load of 8 kN uniformly distributed along its 4 m length together with an axial tensile load of 40 kN. The cross-section is symmetrical with $I = 10 \times 10^6$ mm^4, $A = 10 \times 10^3$ mm^2 and an overall depth of 200 mm. Determine the maximum bending moment and the stress in the cross-section. Take $E = 100$ GPa.
Answer: 3750 N m; 41.5 MPa.

8.28 A 3 m long strut of section depth of 100 mm with $I = 10 \times 10^6$ mm^4 and $A = 3000$ mm^2 carries a distributed loading varying linearly from zero at one end to a maximum of 1.5 kN at the other end. What magnitude of additional axial compressive thrust would cause the strut to buckle between pinned ends under a compressive yield stress of 280 MPa? Take $E = 210$ GPa.

8.29 A uniform horizontal strut with fixed ends carries a load that varies linearly from zero at one end to a maximum at the other. Show that the governing equation is $d^2M/dz^2 + (P/EI)M = z/2$ and hence solve it to find the fixing moments, knowing that $F = dM/dz$ at the fixed ends where the shear forces F are known.

8.30 A vertical cantilever strut of length 2.5 m is made from tubular section of 150 mm outside diameter and 100 mm inside diameter. The initial deformed shape is given by $v_o = C[1 - \cos(\pi z/2L)]$ where z is measured from the fixed end and C is a constant. When the strut carries a vertical force P at the centroid of the free end show that v_o increases by $CP/(P_E - P)$ where P_E is the Euler load for a straight vertical cantilever strut. Hence determine the allowable, initial tip displacement C under a vertical force of 500 kN if the maximum compressive stress is limited to 100 MPa. Take $E = 210$ GPa. (C.E.I. Pt. 2, 1976)

Semi-empirical Methods

8.31 A tubular steel strut of 50 mm o.d. and 3.2 mm thick is pinned at its ends. Find the length below which the Euler theory ceases to apply given a yield stress of 324 MPa. Compare the Euler and Rankine–Gordon buckling loads for a 1.83 m length of this strut. Take $E = 207$ GPa, $a = 1/7500$.
Answer: 1.34 m; 82.7 kN; 60.1 kN.

8.32 Compare graphically the Euler and Rankine–Gordon predictions to the buckling load for a mild steel pipe 4 m long with pinned ends, of o.d. $= 175$ mm and wall thickness 12 mm. Use a safety factor of 3 and take $a = 1/7500$, $\sigma_c = 320$ MPa, $E = 207$ GPa.
Answer: $P_E = 0.87$ MN; $P_R = 1.2$ MN.

8.33 A steel column of solid circular section is 1.83 m long. If this supports a compressive load of 50 kN with pinned ends, find the necessary strut diameter from the Euler theory using a safety factor of 6. What then is the Rankine buckling load with $\sigma_c = 324$ MPa and $a = 1/7500$? Take $E = 210$ GPa.
Answer: 56.6 mm; $P_R = 253$ kN.

8.34 An axially loaded steel strut has both ends built-in. If the length is 160 times the least radius of gyration calculate the Euler buckling stress. Is the Rankine prediction more realistic in this case? Take $E = 207$ GPa, $\sigma_c = 278$ MPa and $a = 1/7500$ for pinned ends.
Answer: 150 MPa.

8.35 A 5 m length of box section with outer dimensions 250 mm × 200 mm and respective thicknesses of 30 and 15 mm is to act as a pinned-end strut. Compare the Euler and Rankine buckling loads and show why one value is invalid. Take $E = 200$ GPa, $\sigma_c = 330$ MPa, $a = 1/7500$.
Answer: $P_E = 9.2$ MN, $P_R = 4.09$ MN.

8.36 A steel tube of 38 mm i.d. and 6.35 mm thick is rigidly bolted between two casings 1 m apart. The tube is then subjected to a temperature change of 10 °C without change in length. Find the factor of safety against buckling from the Rankine–Gordon theory. Take $E = 207$ GPa, $\alpha = 11.5 \times 10^{-6}/°C$, $\sigma_c = 324$ MPa and $a = 1/7500$.

Answer: 4.35.

8.37 A steel pipeline connects with two flanged plates each with a lateral stiffness of 883.3×10^6 N/m in the direction of the strut axis. If, when fluid flows through the pipe under room temperature conditions, the pipe material is unstressed, calculate the maximum allowable temperature rise of the fluid if buckling is to be avoided. Take $E = 200$ GPa, $\alpha = 10 \times 10^{-6}/°C$, $\sigma_c = 300$ MPa and $a = 1/7500$ (pinned ends). Show that the Euler theory is unrealistic in this case.

Answer: 306 °C.

8.38 A 6 m length of cast iron tube of 50 mm o.d. and 38 mm i.d. buckled between fixed ends under 136 kN. A much shorter length of tube material failed in compression under a load of 420 kN. Determine the Rankine constants and hence find the buckling load for a 1.5 m length of tube when used as a strut with pinned-ends. Would the Euler theory be valid in this case? Take $E = 92.6$ GPa.

Answer: $\sigma_c = 472$ MPa, $a = 1/1590$, 61.5 kN.

8.39 A 7.6 m length steel strut has a cross-sectional area of 9675 mm². It is required to carry an axial load of 900 kN with fixed ends. Find the least possible value of the radius of gyration according to the Rankine–Gordon theory with constants $\sigma_c = 324$ MPa, $a = 1/7500$.

Answer: 27.8 mm.

8.40 Compare the permissible Euler and Rankine–Gordon compressive buckling loads that can be applied to a tubular steel strut of 50 mm o.d. and 3.2 mm wall thickness when it is pinned at its ends over a length of 3 m. For what length of this strut does the Euler theory cease to apply. Take $\sigma_c = 324$ MPa, $a = 1/7500$ and $E = 207$ GPa.

Answer: 28.74 kN; 29.4 kN; 1.34 m.

8.41 A short piece of steel tube of 38 mm o.d. and 32 mm i.d. failed in compression under an axial load of 115 kN. When a 1.15 m length of the same tubing was tested as a fixed-end strut it buckled under a load of 90 kN. Determine the constants employed with the Rankine–Gordon theory and hence determine the buckling load for a 1.85 m length of this tubing when used as a strut with pinned ends. Compare with the Euler prediction where $E = 208$ GPa.

Answer: 1/7600; 329 MPa; 29.6 kN; 33 kN; 0.98 m.

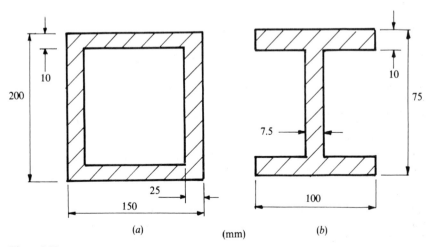

(a)   *(mm)*   *(b)*

Figure 8.27

8.42 Compare values of the Euler and Rankine–Gordon buckling loads for 5 m long fixed-end struts with the cross-sections given in Fig. 8.27(a) and (b). Take $E = 210$ GPa and the constants $\sigma_c = 325$ MPa, $a = 1/7500$.

Answer: (a) $P_E = 13.68$ MN; $P_R = 3.14$ MN. (b) $P_E = 553.5$ kN; $P_R = 355.6$ kN.

8.43 Find the axis about which buckling takes place for the section in Fig. 8.28. What is the safe axial load that may be applied to an 11 m length of this strut when the ends are fixed? Decide between the Euler and Rankine predictions given that the area properties for a single channel are $I_x = 34.5 \times 10^6$ mm^4, $I_y = 1.66 \times 10^4$ mm^4, $A = 3660$ mm^2, $\bar{x} = 18.8$ mm. Take $E = 207$ GPa, $a = 1/7500$, $\sigma_c = 324$ MPa.

Answer: 9880 kN; 3654 kN.

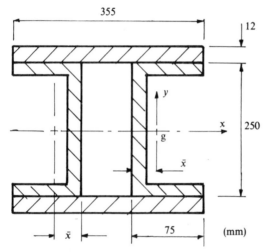

Figure 8.28

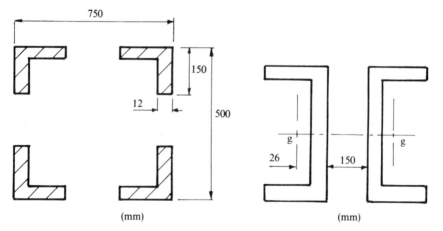

Figure 8.29　　　　　　　　　　　　　　　**Figure 8.30**

8.44 A steel strut 10 m long consists of four equal angles of 150 mm side length and 12 mm thick bolted together as shown in Fig. 8.29. Find the safe axial load that may be applied for a load factor of 6 when the ends are pinned and when they are fixed. Take $\sigma_c = 320$ MPa and $a = 1/7500$.

8.45 Given that rectangular sections of steel $b = 2d$ are available with $\sigma_c = 400$ MPa, $a = 1/9000$ to replace fixed ended tubular struts of 50 mm o.d. and 40 mm i.d. with $\sigma_c = 300$ MPa, $a = 1/7500$ and $L/k = 96$, determine the size of an appropriate rectangular section to the nearest millimetre.

8.46 A 150×200 mm tubular box-section with 10 mm wall thickness is to act as a steel strut of length 6 m with fixed ends. Find the safe axial compressive load using a load factor of 4 based upon the Rankine–Gordon, straight-line, parabolic and Perry–Robertson formulae. Where appropriate take $E = 207$ GPa, $\sigma_c = 325$ MPa, $a = 1/7500$, $\mu = 0.003L/k$. Let the straight line and parabola pass through σ_c and $L/k = 120$.
   Answer: $P_R = 1.63$ MN.

8.47 A 7.5 m length steel column is constructed by joining two channel sections together as shown in Fig. 8.30. If, for one section, the area is 6950 mm^2 and the greatest and least I values are 91.1×10^6 and 5.74×10^6 mm^4, compare the axial load the column may carry according to the straight-line, parabolic and Perry–Robertson predictions. Assume that these pass through a common yield point of 310 MPa and that the former two make tangents with the Euler curve at $L/k = 120$.

8.48 It is required to find the maximum compressive load that a brass strut with elliptical thin-walled section can support without plastic buckling. If the strut length is 825 mm, the major and minor axes are 80 mm and 30 mm, respectively, and the wall thickness is 3 mm, estimate the buckling load for the minor axis where the ends may be assumed pinned. The stress–strain behaviour of brass is $\sigma/\sigma_c = 7.83\epsilon^{1/3}$, where the yield stress $\sigma_c = 144$ MPa. (K.P. 1973)

Buckling of Plates

8.49 A steel plate $400 \times 250 \times 10$ mm is subjected to a compressive stress along its 200×10 mm simply supported edges. If the longer edges are unsupported, determine the critical buckling stress. Take $E = 207$ GPa, $v = 0.27$.

8.50 What thickness of aluminium plate can withstand a compressive force of 2 kN applied normally to its 100 mm simply supported sides when its longer 300 mm sides remain unsupported? Take $E = 71$ GPa, $v = 0.32$.

8.51 A vertical standing plate 1.25 m high carries a total 1.5 kN compressive force uniformly distributed along its 0.5 m breadth. If the plate is simply supported along all four sides, what thickness that would result in plate buckling? Take $E = 74$ GPa, $v = 0.32$.

8.52 Determine the buckling stress for a plate $300 \times 200 \times 5$ mm with all sides fixed when it is loaded in compression normal to its shorter sides. Take $E = 70$ GPa, $v = 0.33$.
   Answer: 200 MPa.

8.53 An aluminium alloy plate 1.5×0.5 m with clamped edges supports a compressive load of 650 kN on its shorter sides. Find the necessary thickness that will prevent buckling from occurring. If the allowable compressive stress is 210 MPa, does the calculated thickness ensure an elastic stress state? Take $E = 70$ GPa, and $v = 0.3$.

8.54 Determine the stress for which a $500 \times 250 \times 5$ mm steel plate buckles under the action of uniform compression along all four sides. Take $E = 207$ GPa, $v = 0.27$. How would you employ ETB to determine the critical buckling stress for this plate?
   Answer: 91.82 MPa.

8.55 Determine the elastic compressive stress that, when applied to the shorter sides of a thin plate $390 \times 200 \times 7.5$ mm will cause it to buckle in the presence of a constant stress of 80 MPa acting (i) in compression and (ii) in tension along its longer sides. Take $E = 72.5$ GPa and $v = 0.33$ with all sides simply supported.

Answer: 265 MPa; 455 MPa.

8.56 A plate $a \times b \times t$ is subjected to a normal tensile stress $\sigma_y = \sigma$ on its $a \times t$ sides and a compressive stress $\sigma_x = -\sigma$ on its $b \times t$ sides. If all edges are simply supported, show that the minimum value of critical buckling stress is given by $(\sigma_x)_{cr} = 8\pi^2 D/tb^2$ for one half-wave in the y-direction. What values of $r = a/b$ does this minimum apply?

Answer: $1/\sqrt{3}$; $2/\sqrt{3}$; etc.

THEORIES OF TORSION

9.1 TORSION OF CIRCULAR SECTIONS

The engineer's theory of torsion (ETT) relates the shear stress τ, and angular twist θ (radians), that are produced in a circular shaft or tube under an externally applied axial torque T, to the shaft length L, the material's rigidity modulus G, the section's radius r, and polar second moment of area J. Consider, for a solid shaft of outer radius R with one end fixed and the other end free, the cylindrical core element in Fig. 9.1(a) of radius r. It is apparent that when point A moves to A' at the free end, the shear strain γ is related to the angular distortion ϕ by the definitions in Eqs (3.7) and (3.8):

$$\gamma = \tau/G = \tan \phi \qquad (9.1)$$

where $\tan \phi = \mathrm{AA'}/\mathrm{OA}$. Provided deformation is elastic then ϕ is small and

$$\tan \phi \simeq \phi \simeq \mathrm{AA'}/L \text{ (rad)}. \qquad (9.2)$$

Since $\mathrm{AA'} = r\theta$, it follows from Eqs (9.1) and (9.2) that

$$\gamma = \tau/G = r\theta/L. \qquad (9.3)$$

The resisting torque exerted by the cross-section must equilibrate the applied torque. For a shear stress τ, acting on an elemental annular area of radius r and thickness δr, in a section of outer radius R (Fig. 9.1(b)), the equilibrium condition is given by

$$\delta T = 2\pi r\, \delta r \times \tau \times r,$$

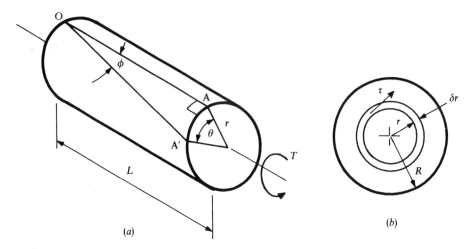

Figure 9.1 Core element and section of a shaft under torsion.

where from Eq. (9.3),

$$\delta T = 2\pi(G\theta/L)r^3 \, \delta r$$

and, therefore, for the whole cross-section,

$$T = 2\pi(G\theta/L) \int r^3 \, dr,$$

which is written as

$$T/J = G\theta/L, \tag{9.4}$$

where $J = 2\pi \int r^3 \, dr$ is the polar second moment of area. For the solid, uniform circular section in Fig. 9.1(b),

$$J = 2\pi \int_0^R r^3 \, dr = 2\pi |r^4/4|_0^R = \pi R^4/2 = \pi D^4/32.$$

If the section were tubular of inner and outer radii R_i and R_o respectively, then

$$J = 2\pi \int_{R_i}^{R_o} r^3 \, dr = 2\pi |r^4/4|_{R_i}^{R_o} = (\pi/2)(R_o^4 - R_i^4).$$

Combining Eqs (9.3) and (9.4) leads to the usual three part ETT:

$$T/J = G\theta/L = \tau/r, \tag{9.5}$$

in which the similarity with ETB (Eq. (6.4)) is apparent. The following examples show how Eq. (9.5) may be adapted to tapered, stepped and composite shafts transmitting power under non-uniform torque.

Shaft Transmitting Power

The power P transmitted by a shaft under torque T (N m) when rotating at N rev/min is

$$P(\text{watts}) = \text{torque}(\text{N m}) \times \text{angular speed}(\text{rad/s})$$

$$P = 2\pi N T/60 \quad (\text{watts}) \tag{9.6}$$

where T is supplied from Eq. (9.5) for a given τ.

Example 9.1 A hollow cylinder of 200 mm (8 in) o.d. and 100 mm (4 in) i.d. is used to transmit power while rotating at 90 rev/min. If the maximum shear stress is limited to 50 MPa (3.25 tonf/in²), determine the power and the angular twist on a length of 10 m (33 ft). If a solid shaft were used to transmit the same power with the same length and limiting stress what would be the percentage increase in weight of the material used? Take $G = 80\,000$ MPa (11.6×10^6 lbf/in²).

SOLUTION TO SI UNITS From Eq. (9.5) the torque is

$$T = J\tau/r = \pi(200^4 - 100^4) \times 50/(32 \times 100) = 73.63 \times 10^6 \text{ N mm}.$$

Then, from Eq. (9.6),

$$P = 2\pi \times 90 \times 73.63 \times 10^6/(60 \times 10^6) = 694 \text{ kW}.$$

The angular twist is found from Eq. (9.5). Since τ is proportional to r, the maximum shear stress applies to the outer radius R_o. Then,

$$\theta = \tau_{max} L/GR_o$$
$$= 50 \times 10 \times 10^3/(80 \times 10^3 \times 100) = 0.0625 \text{ rad} = 0.0625 \times 180/\pi = 3.58°.$$

If the solid shaft of outer radius R is used, then from Eqs (9.5) and (9.6),

$$P = (2\pi N/60)(J\tau/R) = (2\pi N/60)(\pi R^4 \tau_{max}/2R).$$

Then,

$$D = 2R = (480P/\pi^2 N\tau)^{1/3}$$
$$= [480 \times 694 \times 10^6/(\pi^2 \times 90 \times 50)]^{1/3} = 195.75 \text{ mm}.$$

Since weight is proportional to the cross-sectional area, the percentage weight increase is

$$= \frac{[\pi D^2 - \pi(D_o^2 - D_i^2)]100}{\pi(D_o^2 - D_i^2)} = \frac{[195.75^2 - (200^2 - 100^2)]}{(200^2 - 100^2)} = 27.7 \text{ per cent}.$$

SOLUTION TO IMPERIAL UNITS From Eqs (9.5) and (9.6):

$$T = J\tau/r = \pi(8^4 - 4^4)3.25/(32 \times 4) = 306.31 \text{ tonf in},$$
$$P = 2\pi \times 90 \times 306.31 \times 2240/(60 \times 12) = 53.89 \times 10^4 \text{ ft lbf/s};$$

and since $1\,\text{hp} \equiv 550\,\text{ft lbf/s}$ then $P = 979.8\,\text{hp}$.

$$\theta = 3.25 \times 2240 \times 33 \times 12/(11.6 \times 10^6 \times 4) = 0.062\,\text{rad}.$$

The solid shaft diameter is found from

$$P = [2\pi N/(12 \times 550 \times 60)](J\tau/R) = [2\pi N/(12 \times 33\,000)](\pi R^3 \tau/2)$$

with τ in pounds force per inch-squared and R in inches. Hence,

$$D = 2R = 2(12 \times 33\,000 P/\pi^2 N\tau)^{1/3}$$

$$\left[\text{units:} \left(\frac{\text{in lbf}}{\text{hp min}} \times \frac{\text{hp}}{\text{rad}} \times \frac{\text{min}}{\text{lbf}} \times \text{in}^2 \right)^{1/3} = \text{in} \right]$$

$$= 2[12 \times 33\,000 \times 979.8/(\pi^2 \times 90 \times 3.25 \times 2240)]^{1/3} = 7.83\,\text{in},$$

which gives a percentage increase of

$$\frac{7.83^2 - (8^2 - 4^2)}{(8^2 - 4^2)} = 27.7 \text{ per cent (as before).}$$

Tapered Shaft

In order to derive an expression for the angular twist in a solid circular shaft that tapers from radius a to b over length L in Fig. 9.2, the theory is applied to an element of length δl for which the angular twist is $\delta\theta$. Then, for an intermediate cross-section of elemental length δl, radius r at position l in the length as shown, a similar derivation to that which led to Eq. (9.5) now gives

$$\delta\theta = (T/JG)\,\delta l = (2T/\pi r^4 G)\,\delta l. \qquad (9.7)$$

This equation may be expressed in terms of the single variable r by substituting the relationship for a tapered shaft $\delta r/\delta l = (b-a)/L$. The twist θ, over length L, is then found from the integral

$$\theta = [2TL/G\pi(b-a)] \int_a^b r^{-4}\,dr = -[2TL/3G\pi(b-a)]|r^{-3}|_a^b,$$

$$\theta = [2TL/3G\pi(b-a)](1/a^3 - 1/b^3) = (2TL/3\pi G)(a^2 + ab + b^2)/a^3 b^3.$$

The maximum shear stress occurs in the outer fibres of the smaller diameter. That is, from Eq. (9.5),

$$\tau_{\text{max}} = Ta/J = 2Ta/\pi(a)^4 = 2T/\pi a^3.$$

Stepped Shaft

Since each section experiences the same torque T, the total twist θ_t is the sum of the twists for each bar. From Eq. (9.5), with subscripts 1 and 2 for a

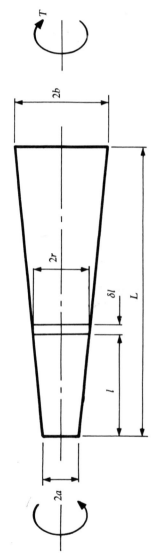

Figure 9.2 Tapered shaft under torsion.

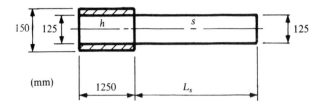

Figure 9.3 Welded shaft.

two-stepped shaft of outer radii r_1 and r_2,

$$\theta_t = (TL/JG)_1 + (TL/JG)_2,$$
$$\theta_t = (T/G)[(L/J)_1 + (L/J)_2](180/\pi) \quad \text{(rad)}. \tag{9.8}$$

The maximum shear stresses (at the outer radii) are

$$\tau_1 = Tr_1/J_1 \quad \text{and} \quad \tau_2 = Tr_2/J_2.$$

Example 9.2 The solid and hollow mild steel shafts in Fig. 9.3 are to be welded together at their ends. If the total angular twist is to be limited to 3° under an axial torque of 25 kN m, determine the required length of solid bar. Take the elastic rigidity modulus $G = 69$ GPa.

Transposing Eq. (9.8) the solid length is given by

$$L_s = [\pi\theta_t G/180T - (L/J)_h]J_s$$
$$= \frac{\pi \times 3 \times 69\,000 \times \pi(125)^4}{180 \times 25 \times 10^6 \times 32} - \frac{1250 \times 125^4}{(150^4 - 125^4)} = 3463.76 - 1164.3$$
$$= 2299 \text{ mm}.$$

Composite Shaft

Let subscripts 1 and 2 generally refer to the two different materials comprising the composite shaft section in Fig. 9.4. Provided that there is no slipping at the common diameter, the total torque carried is given by

$$T = T_1 + T_2 = T_1(1 + T_2/T_1), \tag{9.9}$$

where, from Eq. (9.4), at the outer diameter of each bar,

$$T_1 = \tau_1 J_1/r_1, \quad T_2 = \tau_2 J_2/r_2 \quad \text{and} \quad T_1/T_2 = G_1 J_1/G_2 J_2. \tag{9.10}$$

Substituting Eqs (9.10) in (9.9):

$$T = (2\tau_1/d_1)[J_1 + (G_2/G_1)J_2] = (2\tau_2/d_2)[J_2 + (G_1/G_2)J_1].$$

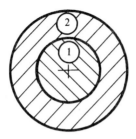

Figure 9.4 Composite shaft section.

Example 9.3 A 4 m length of brass tube is shrunk on a 75 mm steel core. If each material is to contribute equally to the total torque carried, determine the outer diameter of the brass bar. Find the maximum shear stress in each material and the angular twist under a total applied torque of 20 kN m. Take $G = 82$ GPa for steel and $G = 41$ GPa for brass.

Since the torques are the same, Eqs (9.9) and (9.10) become, on identifying 1 with steel and 2 with brass,

$$G_1 J_1/G_2 J_2 = (G_1/G_2)[(d_1/d_2)^4 - 1] = 1,$$
$$\therefore \quad d_1 = d_2(G_2/G_1 + 1)^{1/4} = 75(2 + 1)^{1/4} = 98.7 \text{ mm},$$

and, with the given total torque $T_1 = T_2 = 10$ kN m,

$$\tau_1 = T_1 r_1/J_1 = 10 \times 10^6 \times 37.5 \times 32/\pi(75^4) = 120.72 \text{ MPa},$$
$$\tau_2 = T_2 r_2/J_2 = 10 \times 10^6 \times 49.35 \times 32/[\pi(98.7^4 - 75^4)] = 79.46 \text{ MPa},$$
$$\theta_1 = (TL/JG)_1 = (10 \times 10^6 \times 4 \times 10^3 \times 32)/(\pi \times 75^4 \times 82 \times 10^3)$$
$$= 0.157 \text{ rad} = 9°;$$

Check:

$$\theta_2 = (TL/JG)_2 = (10 \times 10^6 \times 4 \times 10^3 \times 32)/[\pi(98.7^4 - 75^4) \times 41 \times 10^3]$$
$$= 0.157 \text{ rad}.$$

Varying Torque

Take the case of a multistepped shaft with fixed ends and concentrated torques applied at the stepped diameter positions as shown in Fig. 9.5. As there is no twist between the ends, it follows that

$$\theta_1 + \theta_2 + \theta_3 + \theta_4 + \cdots = 0. \tag{9.11a}$$

Now as the shaft is non-uniform, the twist in each parallel portion is found from Eq. (9.7). Integrate to find the relative twist between the right and left ends of each portion with T_1 assumed clockwise at the left end:

$$\theta_1 = T_1 l_1/G_1 J_1, \qquad\qquad \theta_2 = (T_1 + T_2)l_2/G_2 J_2,$$
$$\theta_3 = (T_1 + T_2 + T_3)l_3/G_3 J_3, \quad \theta_4 = (T_1 + T_2 + T_3 + T_4)l_4/G_4 J_4. \tag{9.11b}$$

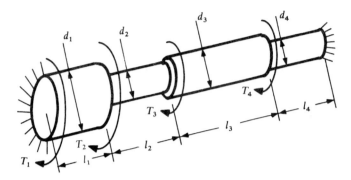

Figure 9.5 Stepped shaft under concentrated torques.

With a constant shaft diameter in the same material, Eq. (9.11) becomes

$$T_1 l_1 + (T_1 + T_2)l_2 + (T_1 + T_2 + T_3)l_3 + (T_1 + T_2 + T_3 + T_4)l_4 + \cdots = 0. \quad (9.12)$$

Example 9.4 Draw the torque and twist diagrams and find the maximum shear stress and angular twist for the shaft in Fig. 9.6(a). Take $G = 80\,\text{GPa}$.

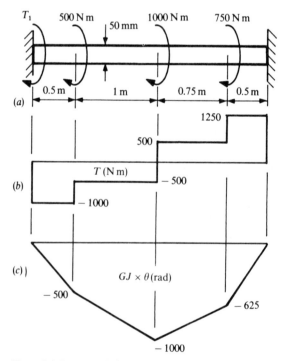

Figure 9.6 Encastre shaft.

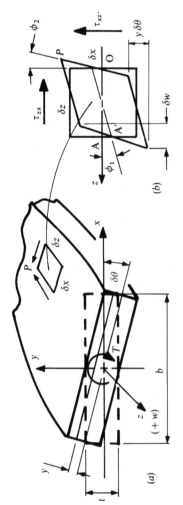

Figure 9.7 Torsion of a thin strip.

Substituting into Eq. (9.12),

$$0.5T_1 + (T_1 + 500) + (T_1 + 500 + 1000)0.75$$
$$+ (T_1 + 500 + 1000 + 750)0.5 = 0,$$
$$2.75T_1 + 2750 = 0, \qquad \Rightarrow T_1 = -1000 \,\mathrm{N\,m},$$

indicating that the direction of T_1 is reversed. Figure 9.6(b) shows the variation in T with shaft length, from which it is evident that the maximum torque is $1250 \,\mathrm{N\,m}$. Then,

$$\tau_{max} = T_{max} R/J = 1250 \times 10^3 \times 25 \times 32/\pi(50)^4 = 50.93 \,\mathrm{MPa}.$$

Equation (9.11b) supplies relative twists:

$$\theta_1 = -1000 \times 0.5/GJ = -500/GJ,$$
$$\theta_2 = -500 \times 1.0/GJ = -500/GJ,$$
$$\theta_3 = 500 \times 0.75/GJ = +375/GJ,$$
$$\theta_4 = 1250 \times 0.5/GJ = +625/GJ.$$

The maximum ordinate in the twist diagram (Fig. 9.6(c)) is

$$\theta_{max} = -1000/GJ$$
$$= -1000 \times 10^3 \times 10^3/[80 \times 10^3 \times \pi(50)^4/32] = 0.0203 \,\mathrm{rad} = 1.164°.$$

Alternatively, from Eq. (9.7),

$$\theta = (1/GJ) \int T \, dl$$

$$\therefore \quad \theta_{max} = (1/GJ)\left[\int_0^{0.5} (-1000) \, dl + \int_{0.5}^{1.5} (-500) \, dl \right]$$
$$= -(1/GJ)[(1000 \times 0.5) + 500(1.5 - 0.5)] = -1000/GJ.$$

9.2 TORSION OF THIN STRIPS AND OPEN SECTIONS

Consider the angular twist $\delta\theta$, produced in a thin strip of section $b \times t$, under the action of a torque T, applied about a centroidal axis z, as shown in Fig. 9.7(a). Unlike the circular section, plane cross-sections do not now remain plane. The distortion produced is due to warping in which the adjacent corner points of a section displace axially in the opposite sense with the edges remaining straight. The warping distortion δw and the chordwise distortion $AA' = y\,\delta\theta$ are shown in the plan view of an element $\delta z \times \delta x$, originally lying at a distance $+y$ from the x-axis in Fig. 9.7(b).

Over the length δz, the shear displacement is $y\delta\theta$. The total change in the angle between orginally perpendicular sides in Fig. 9.7(b) defines the shear strain γ, as

$$\gamma = \tan \phi_1 + \tan \phi_2, \tag{9.13}$$

where ϕ_1 is the angular rotation in the span and ϕ_2 is an equal rotation in the chord due to the complementary nature of the shear stress acting on the sides. Since distortion is elastic and therefore small, $\tan \phi_1 \simeq \phi_1 = \phi_2$ (rads). It follows from Fig. 9.7(b) that Eq. (9.13) becomes

$$\gamma = y\,\delta\theta/\delta z + \delta w/\delta x = 2y\,\delta\theta/\delta z, \tag{9.14}$$

where $\delta\theta/\delta z$ is the rate of angular twist w.r.t. length. Note that the centroidal surface retains its original rectangular shape, since $\gamma = 0$ for $y = 0$, and that the distortion in surfaces below this plane, i.e. with y negative, is the mirror image of Fig. 9.7(b). The horizontal shear stress acting on plane y is, from Eq. (9.14),

$$\tau = G\gamma = 2Gy\,\delta\theta/\delta z, \tag{9.15a}$$

which varies linearly across the thickness with maximum values at the edges (Fig. 9.8). The figure also shows the linear distribution in vertical shear stress τ' in which the maximum value τ'_{max} is assumed to be related to τ_{max} by $\tau'_{max} = t\tau_{max}/b$.

Then for any point (x, y) τ and τ' are related as

$$\tau' = 2x\tau'_{max}/b, \qquad \tau = 2y\tau_{max}/t.$$

Dividing these and substituting from Eq. (9.15a) gives an expression for the vertical shear stress distribution:

$$\tau' = 2Gx(t/b)^2(\delta\theta/\delta z). \tag{9.15b}$$

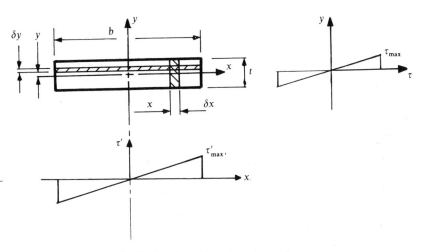

Figure 9.8 Shear stress distributions in a thin strip under torsion.

Since $\delta w/\delta x = y\,\delta\theta/\delta z$, the warping displacements w, are found from

$$w = \int y(\delta\theta/\delta z)\,dx. \tag{9.16}$$

An integration constant is unnecessary because $w = 0$ for all x at $y = 0$. The applied torque is equilibrated on all cross-sections by a resisting torque that is the resultant of the shear stress distribution. This is composed of two parts: the first is T_1 due to τ in Eq. (9.15a), where for the elemental strip of thickness δy, distance y from x in Fig. 9.8,

$$T_1 = \int (\tau b\,dy)y = 2Gb(\delta\theta/\delta z)\int_{-t/2}^{t/2} y^2\,dy = 2Gb(\delta\theta/\delta z)(t^3/12).$$

The remaining torque T_2 is that due to vertical shear stress τ' in Eq. (9.15b). For an elemental strip of thickness δx distance x from y in Fig. 9.8,

$$T_2 = \int (\tau' t\,dx)x = 2Gt(t/b)^2(\delta\theta/\delta z)\int_{-b/2}^{b/2} x^2\,\delta x$$

$$= 2Gb(\delta\theta/\delta z)(t^3/12) = T_1,$$

$$\therefore \quad T = T_1 + T_2 = 4Gb(\delta\theta/\delta z)(t^3/12). \tag{9.17}$$

Combining Eqs (9.15a) and (9.17),

$$G\,\delta\theta/\delta z = \tau/2y = T/(bt^3/3). \tag{9.18}$$

Writing the St Venant torsion constant in this equation as $J = bt^3/3$, the similarity with Eq. (9.5), for circular shafts, becomes clear. With a uniform thin section, the angular twist and maximum stress become

$$\theta = TL/JG \quad \text{and} \quad \tau_{\max} = 3T/bt^2. \tag{9.19a, b}$$

Equation (9.18) will also apply to the thin-walled open tubes shown in Fig. 9.9(a) and (b) provided the thickness t is small compared with the perimeter dimension b, in each case.

In the case of Fig. 9.9(c), where the thicknesses of the rectangular components vary, the torsion constant in Eq. (9.19a) is written as

$$J = (1/3)\sum(bt^3) = b_1 t_1^3/3 + b_2 t_2^3/3 + b_3 t_3^3/3 + \cdots.$$

When the thickness tapers gradually with the perimeter dimension s, then $t = t(s)$ and the torsion constant is found from the integral

$$J = (1/3)\int t^3\,ds. \tag{9.20}$$

For a uniform rectangular section, in which t is not small compared with b, Eqs (9.19a and b) are written as

$$\theta = TL/\beta bt^3 G \quad \text{and} \quad \tau_{\max} = T/\alpha bt^2, \tag{9.21a, b}$$

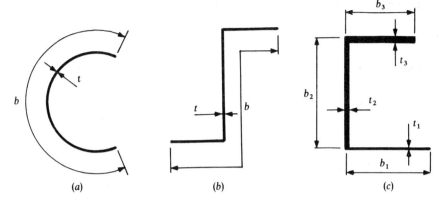

Figure 9.9 Thin-walled open sections under torsion.

where Table 9.1 shows that the coefficients α and β become functions of b/t that each approach $1/3$ as the ratio increases.

For standard, extruded I and channel sections and those fabricated from plates in which t is not usually small compared to b, Eqs (9.21) become

$$\theta = TL/\sum(\beta bt^3)G \quad \text{and} \quad \tau_{max} = T/\sum(\alpha bt^2), \quad \text{(9.22a, b)}$$

where

$$\sum(\beta bt^3) = (\beta bt^3)_1 + (\beta bt^3)_2 + (\beta bt^3)_3 + \cdots,$$
$$\sum(\alpha bt^2) = (\alpha bt^2)_1 + (\alpha bt^2)_2 + (\alpha bt^2)_3 + \cdots.$$

Since the same twist is experienced by each rectangle comprising the section

$$\theta = T_1 L/(\beta bt^3)_1 G = T_2 L/(\beta bt^3)_2 G = T_3 L/(\beta bt^3)_3 G, \quad \text{(9.23a, b, c)}$$

where $T = T_1 + T_2 + T_3$ and τ_{max} is the greatest of

$$\tau_1 = T_1/(\alpha bt^2)_1, \quad \tau_2 = T_2/(\alpha bt^2)_2, \quad \tau_3 = T_3/(\alpha bt^2)_3. \quad \text{(9.24a, b, c)}$$

Example 9.5 Figure 9.10 gives the dimensions of an extruded cross-section in light alloy that is to withstand a torque of $50\,\text{N}\,\text{m}$ applied about its centroidal axis. If $G = 30\,\text{GPa}$, calculate the position and magnitude of the maximum shear stress in each component rectangle and the angular twist for a length of $1\,\text{m}$.

Table 9.1 Coefficients α and β in Eqs (9.21)

b/t	1	2	4	6	8	10	∞
α	0.208	0.246	0.282	0.299	0.307	0.313	0.333
β	0.141	0.299	0.281	0.299	0.307	0.313	0.333

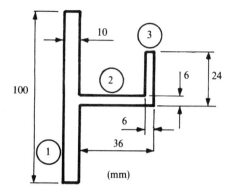

(mm)

Figure 9.10 Extruded section.

From Table 9.1 and Eq. (9.22a),

$$\sum(\beta bt^3) = (0.313 \times 100 \times 10^3) + (0.299 \times 36 \times 6^3) + (0.281 \times 24 \times 6^3)$$
$$= 35\,082\,\text{mm}^4,$$
$$\theta = 50 \times 10^3 \times 10^3/(35\,082 \times 30 \times 10^3) = 0.0475\,\text{rad} = 2.72°.$$

The torque contributions are, from Eq. (9.23a, b, c),

$$T_1 = \theta(\beta bt^3)_1 G/L = 0.0475 \times (0.313 \times 100 \times 10^3) \times 30 \times 10^3/(1 \times 10^3)$$
$$= 44\,602.5\,\text{N mm},$$
$$T_2 = \theta(\beta bt^3)_2 G/L = 0.0475 \times (0.299 \times 36 \times 6^3) \times 30 \times 10^3/(1 \times 10^3)$$
$$= 3313.16\,\text{N mm},$$
$$T_3 = \theta(\beta bt^3)_3 G/L = 0.0475 \times (0.281 \times 24 \times 6^3) \times 30 \times 10^3/(1 \times 10^3)$$
$$= 2075.8\,\text{N mm},$$

Then from Eq. (9.24a, b, c) and Table 9.1,

$$\tau_1 = 44\,602.5/(0.313 \times 100 \times 10^2) = 14.25\,\text{MPa},$$
$$\tau_2 = 3313.16/(0.299 \times 36 \times 6^2) = 8.55\,\text{MPa},$$
$$\tau_3 = 2075.8/(0.282 \times 24 \times 6^2) = 8.52\,\text{MPa},$$

which act along the longer sides of each rectangle. Compare θ and τ_3 with the approximate solutions from Eqs (9.19a and b), where

$$\theta = \frac{50 \times 10^3 \times 10^3}{(1/3)[(100 \times 10^3) + (36 \times 6^3) + (24 \times 6^3)] \times 30 \times 10^3} = 0.0443\,\text{rad}$$

and

$$\tau_{\text{max}} = 3T/bt^2 = 3 \times 50 \times 10^3/(100 \times 10^2) = 15\,\text{MPa}.$$

Example 9.6 The rectangular steel strip in Fig. 9.11 tapers along its 2 m length. If the maximum shear stress in the material is restricted to

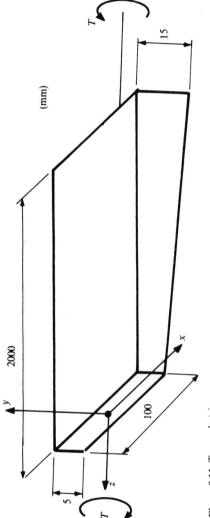

Figure 9.11 Tapered strip.

100 MPa, calculate the maximum torque T that can be carried and the angle through which the strip twists. Compare with the twist found for an average thickness over the 2 m length. Take $G = 83$ GPa.

The maximum torque is determined from Eq. (9.18) with $y = t/2$ at the outer edges:

$$T = J\tau_{max}/2y = (bt^3/3)\tau_{max}/2(t/2) = bt^2\tau_{max}/3.$$

It follows that if the allowable τ_{max} is not to be exceeded, T must be determined from the smaller end where $t_1 = 5$ mm. Then,

$$T = 100 \times 5^2 \times 100/3 = 83.33 \text{ N m.}$$

Now, from Eq. (9.18),

$$\theta = \int T \, dz/JG \tag{i}$$

where, in general terms, with breadth b and end-thicknesses t_1 and t_2 as shown,

$$J = bt^3/3 = b(2y)^3/3 = 8by^3/3, \tag{ii}$$

$$y = (t_2/2 - t_1/2)(z/L) + t_1/2 = [t_1 + (t_2 - t_1)z/L]/2. \tag{iii}$$

Substituting Eqs (ii) and (iii) into (i):

$$\theta = (3T/bG)\int_0^L [t_1 + (t_2 - t_1)z/L]^{-3} \, dz$$

$$= [3TL/bG(t_2 - t_1)]\int_0^L [t_1 + (t_2 - t_1)z/L]^{-3}(t_2 - t_1)\,dz/L$$

$$= -[3TL/2BG(t_2 - t_1)]|[t_1 + (t_2 - t_1)z/L]^{-2}|_0^L$$

$$= -[3TL/2BG(t_2 - t_1)](1/t_2^2 - 1/t_1^2) = 3TL(t_2 + t_1)/2bGt_1^2 t_2^2$$

$$= 3 \times 83.33 \times 10^3 \times 2000 \times 20/(2 \times 100 \times 83\,000 \times 25 \times 225)$$

$$= 0.107 \text{ rad} = 6.14°.$$

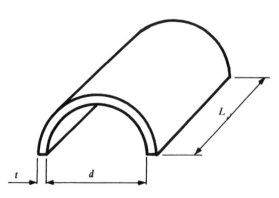

Figure 9.12 Semi-circular arc.

If the average thickness t_{av} is used as a constant in Eq. (i),

$$\theta = 3TL/(bt_{av}^3)G = 3 \times 83.33 \times 10^3 \times 2 \times 10^3/(100 \times 10^3 \times 83 \times 10^3)$$
$$= 0.0602 \, \text{rad} = 3.45°.$$

Example 9.7 Sheet steel of 3 mm thickness is bent into the semicircular arc in Fig. 9.12. What are the sheet dimensions for allowable values $\tau_{max} = 50 \, \text{MPa}$ and $\theta = 5°$ under a torque of 40 N m? Take $G = 80 \, \text{GPa}$.

Here $J = bt^3/3 = (\pi d/2)(t^3/3)$. Then from Eqs (9.18),

$$T = J\tau_{max}/2y = J\tau_{max}/t = \pi dt^2 \tau_{max}/6,$$
$$\therefore \quad d = 6T/\pi t^2 \tau_{max}$$
$$= 6 \times 40 \times 10^3/(\pi \times 3^2 \times 50) = 169.77 \, \text{mm},$$
$$\theta = TL/JG = 6TL/\pi dt^3 G,$$
$$\therefore \quad L = \pi dt^3 G\theta/6T = \pi \times 169.77 \times 3^3 \times 80 \times 10^3$$
$$\times (5 \times \pi/180)/(6 \times 40 \times 10^3) = 418.88 \, \text{mm}.$$

The dimensions are width $b = \pi d/2 = \pi \times 169.77/2 = 266.67 \, \text{mm}$ and length $L = 418.88 \, \text{mm}$.

Example 9.8 A light alloy strip gently tapers in the cross-section as shown in Fig. 9.13(a). Determine the torsion constant and find the angular twist on a length of 1 m when a torque of 10 N m is applied about the centroidal axis. What is the magnitude and position of the maximum shear stress in this section and the warping displacements at the four corners? Take $G = 30 \, \text{GPa}$.

With origin O' at the apex point of the section, the centroidal position $\bar{X}$ is found by taking moments about the longer vertical edge 187.5 mm from O'. That is,

$$(5 \times 187.5/2)(187.5/3) - (1 \times 37.5/2)(150 + 37.5/3)$$
$$= [(5 \times 187.5/2) - (1 \times 37.5/2)]\bar{X}$$
$$\bar{X} = 26\,250/450 = 58.33 \, \text{mm}.$$

In this case, as x is the perimeter dimension s, Eq. (9.20) is written as

$$J = (1/3) \int [t(x)]^3 \, dx,$$

where

$$y' = t/2 = (2.5/187.5)x, \tag{i}$$
$$\therefore \quad J = (1/3) \int_{37.5}^{187.5} (0.026\,67x)^3 \, dx$$
$$= (1/3)(0.026\,67)^3 |x^4/4|_{37.5}^{187.5} = 1950.7 \, \text{mm}^4.$$

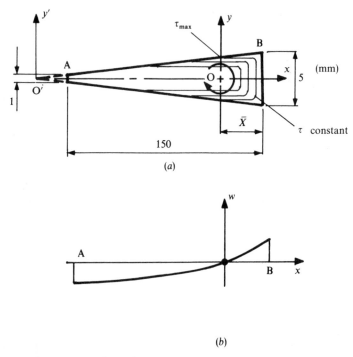

Figure 9.13 Tapered alloy strip.

Equation (9.19a) supplies the angular twist:

$$\theta = TL/GJ = 10 \times 10^3 \times 1 \times 10^3/(1950.7 \times 30 \times 10^3) = 0.1709\,\text{rad} = 9.79°.$$

When, $b/t > 10$, for a gently tapered thin strip, the maximum shear stress is sensibly constant along the longer edges as with a rectangular section. When b/t significantly decreases with the taper, the contours of constant shear stress break at the tapered edges (Fig. 9.13(a)), indicating that the shear stress is not constant. The rule of thumb is that the maximum shear stress occurs on the longer edges at a point closest to the centroid. This is consistent with τ_{max} occurring at the centre of each longer edge in a rectangular section for smaller b/t ratios. Then, from Fig. 9.13(a), putting $x = 187.5 - 58.33 = 129.17\,\text{mm}$ in Eq. (i) gives $y = 1.7223\,\text{mm}$. Equation (9.18) becomes

$$\tau_{\text{max}} = 2Ty/J = 2 \times 10 \times 10^3 \times 1.722/1950.7 = 17.66\,\text{MPa}.$$

The warping displacements are found from Eq. (9.16) w.r.t. the centre of twist (page 308). Assuming that this lies at the centroid O, the equation describing the top surface AB is

$$y = (2.5/187.5)x + 1.7223.$$

$$\delta\theta/\delta z = \theta/L = 0.1709 \times 10^{-3}\,\text{rad/mm},$$

when Eq. (9.16) becomes

$$\delta w/\delta x = y(\delta\theta/\delta z) = (0.1709 \times 10^{-3})(0.0133x + 1.7223),$$
$$w = (0.1709 \times 10^{-3})(0.006\,65x^2 + 1.7223x).$$

The parabolic dependence of w on x is the result of the tapered section:

At A: $x = -91.67\,\text{mm}$, $w_A = -0.0174\,\text{mm}$.

At B: $x = 58.33\,\text{mm}$, $w_B = +0.021\,04\,\text{mm}$.

And by deduction for the bottom surface $w_C = -0.021\,04\,\text{mm}$, $w_D = + 0.0174\,\text{mm}$. The top edge AB warps as shown in Fig. 9.13(b). The signs depend upon the direction of T w.r.t. the positive y and z (Fig. 9.7). Example 9.18 shows a linear w vs. x relationship for a rectangular section.

9.3 TORSION OF THIN-WALLED CLOSED TUBES OF ARBITRARY CROSS-SECTION

The Bredt–Batho torsion theory shows that closed, thin-walled, single and multicell tubular sections are able to withstand torsion by means of a constant shear flow in the wall.

Single-cell Tube

Let the closed tube of enclosed area A and variable thickness t in Fig. 9.14(a) be subjected to a torque T that produces shear stress τ in the wall. When axial stress σ_z is absent, the variation in stress for a tube wall element is as shown in Fig. 9.14(b).

In coordinates of circumferential distance s and axial length z, the following equation expresses vertical equilibrium for the element:

$$\tau(t\,\delta z) = \left(\tau + \frac{d\tau}{ds}\delta s\right)(t + \delta t)\delta z. \tag{9.25}$$

Since the thickness varies, a shear flow parameter $q = \tau t$ is introduced into Eq. (9.25) to give, when the product of infinitesimals is neglected,

$$\tau\frac{dt}{ds} + t\frac{d\tau}{ds} = 0$$

$$\therefore \quad \frac{d}{ds}(\tau t) = \frac{dq}{ds} = 0.$$

It is seen from this equation that q is constant around the cross-section regardless of its shape. Note that $\tau = q/t$ is constant only when t is constant.

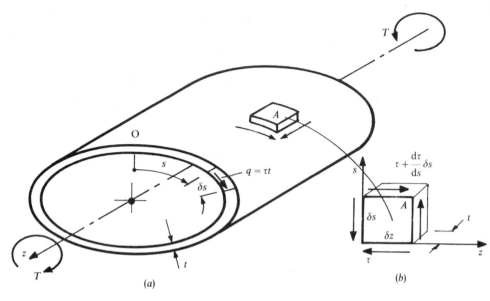

Figure 9.14 Torsion of a closed, single-cell tube.

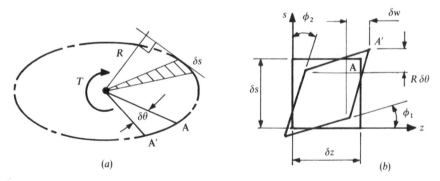

Figure 9.15 Shear flow and distortion in a single cell.

Figure 9.15(a) shows that the resisting or equilibrating torque exerted by the section with a constant shear flow around its mean perimeter is

$$T = \oint (q\,ds)R = q\oint R\,ds.$$

Because $\oint R\,ds$ is twice the shaded area shown, it follows that this path integral describes twice the area enclosed by the section. The torque is then simply

$$T = 2Aq = 2A\tau t. \tag{9.26}$$

Since closed arbitrary sections will, in general, warp under torsion, the rate of twist $\delta\theta/\delta z$ is found from a similar analysis to that employed previously in the St Venant theory. It is a consequence of the absence of axial strain that axial generators will not change their lengths and therefore every cross-section will warp in an identical manner.

For the distorted shape of the surface element in Fig. 9.15(b), the total shear strain is

$$\gamma = \tan\phi_1 + \tan\phi_2 \simeq \phi_1 + \phi_2 \quad \text{(radians)}. \tag{9.27}$$

Now the tangential component of displacement of a point from A on the circumference to A′ is $R\,\delta\theta$. Figure 9.15(b) shows that there is also a warping component of displacement δw associated with A′. Hence,

$$\phi_1 = R\,\delta\theta/\delta z, \qquad \phi_2 = \delta w/\delta s \quad \text{(radians)}. \tag{9.28a, b}$$

In the limit of Eq. (9.28), as δs and δz approach zero, Eq. (9.27) becomes

$$\gamma = R\,d\theta/dz + dw/ds,$$
$$\gamma\,ds = (R\,ds)\,d\theta/dz + (dw/ds)\,ds, \tag{9.29}$$

when from Eq. (9.26),

$$\gamma\,ds = (\tau/G)\,ds = q\,ds/Gt = (T/2A)(ds/Gt). \tag{9.30}$$

Substituting into Eq. (9.29) and integrating around the section:

$$(T/2A)\oint ds/Gt = (d\theta/dz)\oint R\,ds + \oint dw.$$

The final path integral will be zero, since integration starts and finishes at the same point. Then the rate of twist becomes

$$d\theta/dz = (T/4A^2)\oint ds/Gt = (q/2A)\oint ds/Gt. \tag{9.31a}$$

For a uniform section in the same material, Eq. (9.31a) may be integrated directly to give

$$\theta = (TL/4A^2G)\oint ds/t = (qL/2AG)\oint ds/t. \tag{9.31b}$$

An alternative derivation of this expression employs the unit load method in the principle of virtual work (see Example 12.11). Combining Eqs (9.26) and (9.31a) gives the complete theory

$$T = 4A^2(d\theta/dz)\left/ \oint (ds/Gt) = 2Aq.\right.$$

If G is constant, this becomes

$$T = GJ(d\theta/dz) = 2Aq, \tag{9.32}$$

where the torsion constant J is defined as

$$J = 4A^2 \bigg/ \oint ds/t.$$

In a tapered tube, J becomes a function of z because both A and s depend upon position in the length. When the cross-section is an n-sided polygon with each side of different thickness,

$$\oint ds/t \equiv \sum_i (s/t)_i. \tag{9.33}$$

The warping displacements w are found on substituting Eqs (9.30) and (9.31a) into Eq. (9.29) and then integrating from an initial value w_o that will be known from symmetry. That is,

$$dw = \gamma\,ds - (R\,ds)(d\theta/dz)$$

$$= (T/2A)(ds/Gt) - (R\,ds)(T/4A^2)\oint ds/Gt,$$

$$w - w_o = (T/2A)\int (ds/Gt) - (T/4A^2)\oint (ds/Gt)\int R\,ds,$$

$$w = (T/2A)\left[\int_0^s (ds/Gt) - (1/2A)\oint (ds/Gt)\int_0^s R\,ds\right] + w_o. \tag{9.34a}$$

If G is constant, and a point O on the periphery is chosen to give $w_o = 0$, Eq. (9.34a) is conveniently expressed as

$$w = (Ti/2AG)(i_{os}/i - A_{os}/A) \tag{9.34b}$$

where $i = \oint ds/t$, $i_{os} = \int_0^s ds/t$, A is the total area enclosed by the section, and A_{os} is the segment of area enclosed between the starting point O, the point on the periphery for which w is required and the centre of twist. The latter is that point in the enclosed section that does not suffer rotation or warping displacement. This can normally be found by inspection for symmetrical sections. Otherwise, Eq. (9.34b) is modified with the addition of w_o in respect of some arbitrary origin for the axis z. It follows from Eq. (9.34b) that a section will not warp when

$$i_{os}/i = A_{os}/A.$$

That is,

$$(1/i)\int_0^s ds/t = (1/2A)\int_0^s R\,ds,$$

from which, on differentiation,

$$Rt = 2A/i = \text{constant}.$$

Clearly a circular tube with uniform thickness satisfies this condition where $2A/i = 2\pi R^2/(2\pi R/t) = Rt$. It is also satisfied by a uniform-thickness triangular tube and a rectangular section in which the ratio of the side lengths equals the ratio of their thicknesses. These are all known as Neuber tubes.

Example 9.9 An alloy tube with the section given in Fig. 9.16(a) is 2 m long. Find the torque that can be applied, the angular twist when the maximum shear stress is limited to 27.5 MPa, and the warping displacements at the ends of the straight portions. Take $G = 27$ GPa.

$$A = \pi(40)^2 + (80)^2 = 11\,426.55\,\text{mm}^2,$$

$$i = \oint ds/t = (2 \times 80/1.5) + (\pi \times 80/1.5) = 274.22.$$

The torque is found from Eq. (9.26):

$$T = 2At\tau = 2 \times 11\,426.55 \times 1.5 \times 27.5 = 942.7\,\text{Nm}$$

and the angular twist from Eq. (9.31b):

$$\theta = qLi/2AG = (\tau t)Li/2AG$$
$$= (1.5 \times 27.5) \times 2 \times 10^3 \times 274.22/(2 \times 11\,426.55 \times 27 \times 10^3)$$
$$= 0.0367\,\text{rad} = 2.1°.$$

For the warping displacements at points A, B, C and D, it is obvious for this symmetrical section that the centroid g coincides with the centre of twist and that point O is a suitable starting position for which $w_o = 0$. Then Eq. (9.34b) gives the following.

At A: $A_{os} = (40 \times 40/2) + \pi(40)^2/4 = 2056.64\,\text{mm}^2,$

$i_{os} = (\Delta s/t)_{OA} = 2\pi \times 40/4 \times 1.5 = 13.33\pi,$

$$w_A = \left(\frac{942.7 \times 10^3 \times 274.22}{2 \times 11\,426.55 \times 27 \times 10^3} \right)\left(\frac{13.33\pi}{274.22} - \frac{2056.64}{11\,426.55} \right)$$
$$= (0.418\,95)(-0.027\,27) = -0.011\,41\,\text{mm}.$$

At B: $A_{os} = 2056.64 + (80 \times 40/2) = 3656.64\,\text{mm}^2,$

$i_{os} = 13.33\pi + 80/15 = 95.22,$

$w_B = 0.418\,95(95.22/274.22 - 3656.64/11\,426.55)$
$$= 0.01141\,\text{mm}.$$

At C: $A_{os} = 3656.64 + (2 \times 2056.64) = 7769.92\,\text{mm}^2,$

$i_{os} = 95.22 + (80\pi/2 \times 1.5) = 178.996,$

$w_C = 0.418\,95(178.996/274.22 - 7769.92/11\,426.55)$
$$= -0.011\,41\,\text{mm}.$$

At D: $w_D = +0.011\,41\,\text{mm}$ by inspection.

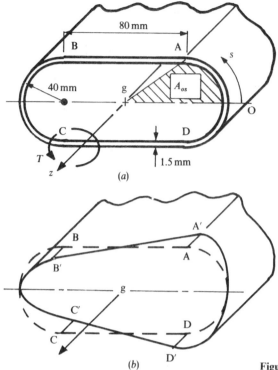

(a)

(b)

Figure 9.16 Closed alloy tube.

These displacements reveal that every cross-section warps in the manner shown in Fig. 9.16(b).

Example 9.10 A cantilever tube with the cross-section given in Fig. 9.17 is subjected to a uniformly distributed anticlockwise torque of 20 kN m per metre (2 tonf ft/ft) of its 2.5 m (8.25 ft) length. If $G = 26$ GPa (1690 tonf/in^2) and 18 GPa (1170 tonf/in^2) for the vertical and horizontal sides respectively, calculate the maximum shear stress and the manner in which the angular twist varies with length.

SOLUTION TO SI UNITS Since $T_{max} = 20 \times 2.5 = 50$ kN m at the fixed end, it follows from Eq. (9.26) that the maximum shear stress is given by

$$\tau_{max} = T_{max}/2At_{min}$$
$$= 50 \times 10^6/[2(1000 \times 250)1.2] = 83.33 \text{ MPa}.$$

When $T = T(z)$ and G is not constant, the angular twist is found from integrating Eq. (9.31a)

$$\theta = (1/4A^2) \oint ds/Gt \int T(z) \, dz, \qquad \text{(i)}$$

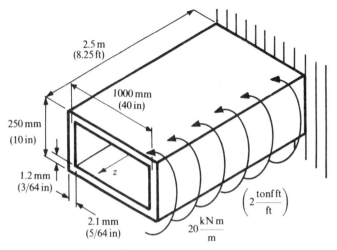

Figure 9.17 Cantilever tube.

where

$$\oint ds/Gt = [(2 \times 1000)/(18 \times 10^3 \times 1.2)] + [(2 \times 250)/(26 \times 10^3 \times 2.1)]$$

$$= (92.59 + 9.1575)10^{-3} = 101.75 \times 10^{-3}\,\text{mm}^2/\text{N}.$$

Taking the origin for z at the fixed end,

$$T(z) = (2.5 - z)20 \times 10^6\,\text{N mm}$$
$$= (2500 - z)20 \times 10^3\,\text{N mm} \qquad \text{(with } z \text{ in millimetres)}.$$

The net torque therefore varies linearly from zero at the free end to its maximum value at the fixed end. Substituting into Eq. (i),

$$\theta = \frac{101.75 \times 10^{-3} \times 20 \times 10^3}{4(1000 \times 250)^2} \int_0^z (2500 - z)\,dz$$

$$= 0.813 \times 10^{-8}(2500z - z^2/2)\,\text{rad}.$$

At the free end, where $z = 2500$ mm, $\theta = 0.0254$ rad $(1.45°)$.

SOLUTION TO IMPERIAL UNITS

$$\tau_{max} = T_{max}/2At_{min}$$
$$= 2 \times 12 \times 8.25/[2(10 \times 40)(3/64)] = 5.28\,\text{tonf/in}^2.$$

$$\oint ds/Gt = \{2 \times 40/[1170 \times (3/64)]\} + \{2 \times 10/[1690 \times (5/64)]\}$$

$$= 1.4587 + 0.1515 = 1.61\,\text{in}^2/\text{tonf}.$$

$$T(z) = (12 \times 8.25 - z)2 = (99 - z)2\,\text{tonf in} \qquad \text{(with } z \text{ in inches)}.$$

Substituting with Eq. (i) gives

$$\theta = \frac{1.61 \times 2}{4 \times (10 \times 40)^2} \int_0^z (99 - z)\,dz = 5.03 \times 10^{-6}(99z - z^2/2)\,\text{rad}.$$

At the free end where $z = 99$ in, $\theta = 0.0247\,\text{rad}\ (1.415°)$.

Example 9.11 An aircraft wing structure consists of the cross-section given in Fig. 9.18. Assuming that one end is fully built-in, calculate the maximum shear stress, the torsion constant and the angular twist at the free end when the wing is subjected to a uniformly distributed torque of 25 kN m per metre length. Take $G = 28.5\,\text{GPa} = \text{constant}$.

The enclosed area is

$$A = (1/2)(250 + 375)\sqrt{(400^2 - 62.5^2)} + (1/2)(375 + 125)\sqrt{(500^2 - 125^2)}$$
$$= (12.346 \times 10^4) + (12.1 \times 10^4) = 24.446 \times 10^4\,\text{mm}^2.$$

For the top right sloping side: $dt/ds = 1.5/500,\ \Rightarrow ds = 333.33\,dt$.
For the top left sloping side: $dt/ds = 1.5/400,\ \Rightarrow ds = 266.67\,dt$.
Then from Eq. (9.33),

$$ds/t = \sum(ds/t)_i = 125/3.5 + 2\int_3^{4.5} 333.33\,dt/t + 2\int_{4.5}^6 266.67\,dt/t + 250/3.5$$

$$= 107.14 + 666.67\ln(4.5/3) + 533.34\ln(6/4.5) = 530.88,$$

and from Eq. (9.26),

$$\tau_{max} = T_{max}/2At_{min}$$
$$= 6 \times 25 \times 10^6/(2 \times 24.446 \times 10^4 \times 3) = 102.3\,\text{MPa}.$$

Equation (9.32) gives the torsion constant as

$$J = \frac{4A^2}{i} = \frac{4A^2}{\oint ds/t} = \frac{4 \times (24.446 \times 10^4)^2}{530.88} = 450.28 \times 10^6\,\text{mm}^4,$$

and with the origin for z at the fixed end, $T(z) = 25 \times 10^3(6000 - z)$. The

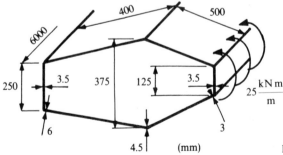

Figure 9.18 Aircraft wing.

twist at the free end is found from the integral

$$\theta = (1/GJ) \int T(z) \, dz$$

$$= (28.5 \times 10^3 \times 450.28 \times 10^6)^{-1} \times 25\,000 \int_0^{6000} (6000 - z) \, dz$$

$$= (28.5 \times 10^3 \times 450.28 \times 10^6)^{-1} \times 25\,000 \, |6000z - z^2/2|_0^{6000}$$

$$= 3.507 \times 10^{-2} \, \text{rad} = 2°.$$

Two-cell Tube

Let q_1 and q_2 be the respective shear flows around cells 1 and 2 in Fig. 9.19. In general, the thicknesses t_1, t_2 and t_{12} and the shear moduli G_1, G_2 and G_{12} will differ. (Subscripts 12 refer to the web.)

The applied torque T is resisted by the sum of the equilibrating torques in each cell. It then follows from Eq. (9.26) that

$$T = 2A_1 q_1 + 2A_2 q_2. \tag{9.35}$$

The two unknowns q_1 and q_2 in Eq. (9.35) render this structure statically indeterminate. Since both q_1 and q_2 act in opposing directions at the common web, their contributions to the rate of twist in Eq. (9.31a) become for each cell

$$(d\theta/dz)_1 = (q_1/2A_1) \oint_1 ds/Gt - (q_2/2A_1) \int_{12} ds/Gt,$$

$$(d\theta/dz)_2 = (q_2/2A_2) \oint_2 ds/Gt - (q_1/2A_2) \int_{12} ds/Gt.$$

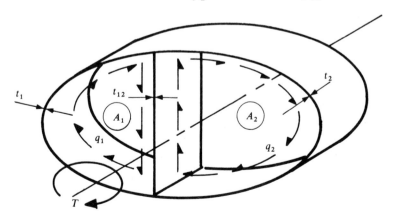

Figure 9.19 Shear flow in a two-cell tube.

The additional compatibility requirement, which provides a solution, is that $(d\theta/dz)_1 = (d\theta/dz)_2$. Hence,

$$2A_1(d\theta/dz) = q_1 \oint_1 ds/Gt - q_2 \int_{12} ds/Gt, \qquad (9.36a)$$

$$2A_2(d\theta/dz) = q_2 \oint_2 ds/Gt - q_1 \int_{12} ds/Gt. \qquad (9.36b)$$

Once q_1 and q_2 are solved from Eqs (9.35) and (9.36), the shear stresses in the walls are given by

$$\tau_2 = q_2/t_2, \qquad \tau_1 = q_1/t_1 \qquad \text{and} \qquad \tau_{12} = (q_1 - q_2)/t_{12}. \qquad (9.37a, b, c)$$

This problem has also been treated—as an example of the application of the principle of virtual work to a redundant structure (see page 488).

Example 9.12 The uniform two-cell tube in Fig. 9.20 is subjected to an axial torque of $T = 1\,\text{N m}$. If $G = 30\,\text{GPa} = \text{constant}$, find the shear stresses in the walls, the angular twist between the ends of the tube's 2 m length and the torsional stiffness.

Since $A_1 = A_2$ it follows from Eq. (9.36a and b) that

$$q_1 \oint_1 ds/Gt - q_2 \int_{12} ds/Gt = q_2 \oint_2 ds/Gt - q_1 \int_{12} ds/Gt,$$

$$q_1 \left(\oint_1 ds/Gt + \int_{12} ds/Gt \right) = q_2 \left(\oint_2 ds/Gt + \int_{12} ds/Gt \right).$$

Now $\oint_1 ds/Gt = \oint_2 ds/Gt$ and therefore $q_1 = q_2$, which will always apply to identical cells. For the cells in question,

$$A_1 = A_2 = 100 \times 120/2 = 6000\,\text{mm}^2,$$

$$\oint_1 ds/t = \oint_2 ds/t = (2 \times 130/2) + 100/4 = 155, \qquad \int_{12} ds/t = 100/4 = 25.$$

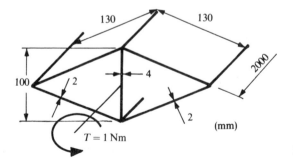

130   130

100   /2   —►◄— 4

2000

2

(mm)

$T = 1\,\text{Nm}$

Figure 9.20 Two-cell tube.

The equilibrium condition, Eq. (9.35), becomes $T = 4Aq_1$. Hence,

$$q_1 = T/4A = 1 \times 10^3/(4 \times 6000) = 1/24\,\text{N/mm} = q_2.$$

The shear stresses in the walls are (Eqs (9.37))

$$\tau_1 = 1/(24 \times 2) = 0.020\,83\,\text{MPa} = 20\,830\,\text{Pa} = \tau_2,$$

and since q_1 and q_2 are equal and opposite in the web, then $\tau_{12} = 0$ from Eq. (9.37c). Equation (9.36a) supplies the rate of twist:

$$d\theta/dz = \left(q_1 \oint_1 ds/t - q_2 \int_{12} ds/t \right) \Big/ 2A_1 G$$

$$= q_1 \left(\oint_1 ds/t - \int_{12} ds/t \right) \Big/ 2A_1 G$$

$$= (1/24)(155 - 25)/(2 \times 6000 \times 30 \times 10^3) = 1.505 \times 10^{-8}\,\text{rad/mm}.$$

The twist on a 2 m length is then

$$\theta = 1.505 \times 10^{-8} \times 2 \times 10^3 = 3.01 \times 10^{-5}\,\text{rad},$$

and the torsional stiffness is

$$T/(d\theta/dz) = 1 \times 10^3/(1.505 \times 10^{-8}) = 6.645 \times 10^{10}\,(\text{N mm})/(\text{rad/mm})$$

$$= 6.645 \times 10^4\,(\text{Nm})/(\text{rad/m}).$$

Example 9.13 Find the maximum shear stresses and the rate of angular twist under an axial torque of 10 N m for the uniform two-cell tube section given in Fig. 9.21. $G = 30\,\text{GPa}$.

$$A_1 = 50 \times 120 = 6000\,\text{mm}^2,$$
$$A_2 = 50\sqrt{(100^2 - 50^2)} = 4330\,\text{mm}^2.$$

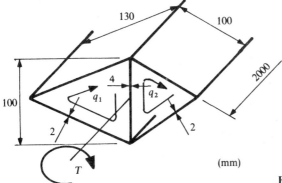

Figure 9.21 Two-cell tube.

From Eq. (9.35),

$$10 \times 10^3 = (12 \times 10^3 q_1) + (8.66 \times 10^3 q_2),$$
$$1 = 1.2q_1 + 0.866q_2, \qquad \text{(i)}$$

and, from Eqs (9.36a and b),

$$12G \times 10^3 (d\theta/dz) = q_1[(2 \times 130/2) + 100/4] - q_2(100/4)$$
$$= 155q_1 - 25q_2, \qquad \text{(ii)}$$

$$8.66G \times 10^3 (d\theta/dz) = q_2[(2 \times 100/2) + 100/4] - q_1(100/4)$$
$$= 125q_2 - 25q_1. \qquad \text{(iii)}$$

From Eqs (i), (ii) and (iii) (see also page 491),

$$111.858q_1 - 18.042q_2 = 125q_2 - 25q_1, \quad \Rightarrow q_1 = 1.045q_2,$$
$$1 = 1.2542q_2 + 0.866q_2, \quad \Rightarrow q_2 = 0.4717\,\text{N/mm}, \qquad q_1 = 0.4929\,\text{N/mm}.$$

Equations (9.37) give the maximum shear stresses in each cell as

$$(\tau_1)_{max} = q_1/t_{min} = 0.4929/2 = 0.247\,\text{MPa},$$
$$(\tau_2)_{max} = q_2/t_{min} = 0.4717/2 = 0.236\,\text{MPa},$$
$$\tau_{12} = (q_1 - q_2)/t = (0.4929 - 0.4717)/4 = 0.0053\,\text{MPa}.$$

Either Eq. (ii) or Eq. (iii) supplies the rate of twist. From Eq. (ii),

$$d\theta/dz = (155q_1 - 25q_2)/(12G \times 10^3)$$
$$= 64.607/(12 \times 30 \times 10^6) = 0.1795 \times 10^{-6}\,\text{rad/mm}.$$

Example 9.14 The uniform two-cell tube in Fig. 9.22 is subjected to a pure axial torque of 12 kN m. Find the rate of twist and the multiplying factor t in the dimensions of cell 2 in order that there is no shear stress in the vertical web. Take $G = 25$ GPa.

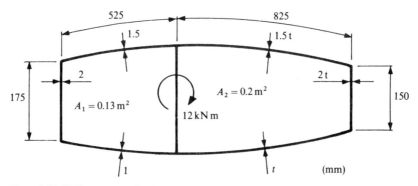

Figure 9.22 Uniform, two-cell tube.

Since the shear stress in the web is to be zero, it follows from Eq. (9.37c) that $q_1 - q_2 = 0$. That is $q_1 = q_2$, when from Eq. (9.35)

$$q_1 = T/2(A_1 + A_2) = 12 \times 10^6/[2(0.13 + 0.20) \times 10^6] = 18.182 \, \text{N mm.}$$

Then, from Eq. (9.36a and b),

$$d\theta/dz = (q_1/2A_1)\left(\oint_1 ds/Gt - \int_{12} ds/Gt\right)$$

$$= (q_2/2A_2)\left(\oint_2 ds/Gt - \int_{12} ds/Gt\right),$$

$$\therefore \quad (A_2/A_1)\left(\oint_1 ds/t - \int_{12} ds/t\right) = \left(\oint_2 ds/t - \int_{12} ds/t\right),$$

$$(0.20/0.13)(525/1.5 + 525/1 + 175/2) = (825/1.5t + 150/2t + 825/t),$$

$$1480.77 = 1450/t, \quad \Rightarrow t = 0.98.$$

The rate of twist is (from cell 1)

$$d\theta/dz = (q_1/2A_1 G)\left(\oint_1 ds/t - \int_{12} ds/t\right)$$

$$= [18.182/(2 \times 0.13 \times 10^6 \times 25 \times 10^3)](525/1.5 + 525/1 + 175/2)$$

$$= 2.69 \times 10^{-6} \, \text{rad/mm} = 0.002\,69 \, \text{rad/m} = 0.154°/\text{m.}$$

Example 9.15 The two-cell, semi-circular tube in Fig. 9.23 tapers linearly between end diameters of 50 and 100 mm over a length of 1 m. If the thickness is 1 mm, calculate the position and magnitude of the maximum shear stress, the torsional stiffness and the angular twist under a torque of 100 N m. Take $G = 27 \, \text{GPa.}$

For identical cells, $q_1 = q_2$ on any given section. For the smaller end, where τ is a maximum, Eq. (9.35) gives

$$q_1 = T/4A_1 = 100 \times 10^3/[4 \times (\pi/8)(50)^2] = 25.47 \, \text{N/mm.}$$

Then, from Eq. (9.37),

$$\tau_{\text{max}} = q_1/t = 25.47/1 = 25.47 \, \text{MPa.}$$

The rate of twist is found from Eq. (9.36a):

$$d\theta/dz = (q_1/2A_1 G)\left(\oint_1 ds/t - \int_{12} ds/t\right). \quad \text{(i)}$$

The bracket quantity indicates that the line integral need only be evaluated for the semi-circular part of the cell. With the origin of z at the smaller

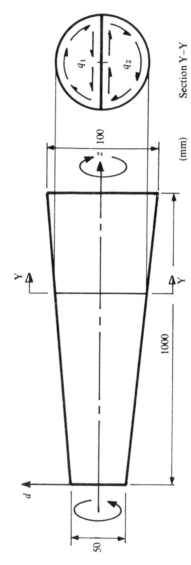

Section Y–Y

(mm)

Figure 9.23 Tapered two-cell tube.

end in Fig. 9.23, then d (mm) varies with z (mm) as

$$d/2 = 25 + (25/1000)z, \quad \Rightarrow d = 50 + 0.05z$$
$$\therefore \quad A_1(z) = (\pi/8)(50 + 0.05z)^2,$$
$$\left(\oint_1 ds/t - \int_{12} ds/t \right) = \pi d/2t = (\pi/2t)(50 + 0.05z),$$
$$q_1 = T/4A_1(z) = 100 \times 10^3/[4 \times (\pi/8)(50 + 0.05z)^2]$$
$$= 2 \times 10^5/\pi(50 + 0.05z)^2.$$

Substituting into Eq. (i), the twist rate is

$$d\theta/dz = 4 \times 10^5/\pi Gt(50 + 0.05z)^3 = 4.716/(50 + 0.05z)^3 \text{ rad/mm}$$

and the torsional stiffness becomes

$$T/(d\theta/dz) = (100 \times 10^3)(50 + 0.05z)^3/4.716$$
$$= (21.2 \times 10^3)(50 + 0.05z)^3 \text{ (N mm)/(rad/mm)}$$

Substituting $T = 100 \times 10^3$ N mm the angular twist is then

$$\theta = 4.716 \int_0^{1000} (50 + 0.05z)^{-3} \, dz$$
$$= -(4.716/2 \times 0.05)|(50 + 0.05z)^{-2}|_0^{1000} = -47.16[1/(100)^2 - 1/(50)^2]$$
$$= 0.014\,15 \text{ rad} = 0.811°.$$

That is, for a 1 m length of tube, $T/\theta = 7068$ N m/rad.

9.4 TORSION OF PRISMATIC BARS

Consider the solid shaft of any cross-section in Fig. 9.24(a), carrying a torque T, applied about an axis z, passing throught the centre of twist, O. Any point $P(x, y, z)$ in the cross-section (Fig. 9.24b), that is not coincident with O, will displace to P' by the respective amounts u and v in the x- and y-directions where

$$u = -r\theta \sin \alpha = -r\theta(y/r) = -y\theta, \tag{9.38a}$$

$$v = r\theta \cos \alpha = r\theta(x/r) = x\theta, \tag{9.38b}$$

in which θ is the angle of twist defining the movement of P to P' in an anticlockwise sense. Point P will also displace in the z-direction as the cross-section warps. The warping displacement w is assumed to be proportional to the rate of twist $\delta\theta/\delta z$, as with a tube (see Eqs (9.28) and (9.34)), and a function $\psi(x, y)$ that describes the variation in w over the cross-section. That is,

$$w = \psi(x, y) \, \delta\theta/\delta z, \tag{9.38c}$$

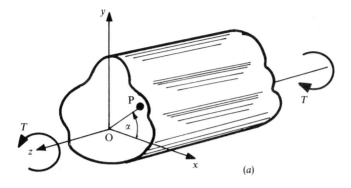

(a)

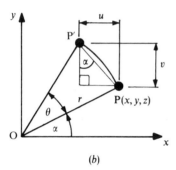

(b)

Figure 9.24 Prismatic bar under torsion.

where ψ is to be determined. In consideration of the strains for displacement PP', the distortion of the cross-section out of its own plane is defined in the following two shear strain components (page 567):

$$\gamma_{xz} = \partial u/\partial z + \partial w/\partial x$$
$$= -\partial(y\theta)/\partial z + (\partial\psi/\partial x)(\delta\theta/\delta z) = (\partial\psi/\partial x - y)\,\delta\theta/\delta z, \qquad (9.39a)$$

$$\gamma_{yz} = \partial v/\partial z + \partial w/\partial y$$
$$= \partial(x\theta)/\partial z + (\partial\psi/\partial y)(\delta\theta/\delta z) = (x + \partial\psi/\partial y)\,\delta\theta/\delta z. \qquad (9.39b)$$

It follows from Eqs (9.38) and Eqs (14.4) that direct strains $\epsilon_x = \partial u/\partial x$, $\epsilon_y = \partial v/\partial y$ and $\epsilon_z = \partial w/\partial z$ are zero and that since shear strain $\gamma_{xy} = \partial u/\partial y + \partial v/\partial x = -\theta + \theta = 0$ the cross-section does not distort in its own x, y plane. The only non-zero stresses are the components of shear stress associated with the non-zero shear strains. These are, from Eqs (9.39a and b),

$$\tau_{xz} = G\gamma_{xz} = G(\partial\psi/\partial x - y)\,\delta\theta/\delta z, \qquad (9.40a)$$

$$\tau_{yz} = G\gamma_{yz} = G(\partial\psi/\partial y + x)\,\delta\theta/\delta z, \qquad (9.40b)$$

which act together with their complements $(\tau_{xz} = \tau_{zx}, \tau_{yz} = \tau_{zy})$ in the directions shown in Fig. 9.25(a). From Eqs (9.40),

$$\partial\tau_{xz}/\partial y = G(\partial^2\psi/\partial x\partial y - 1)\delta\theta/\delta z, \tag{9.41a}$$

$$\partial\tau_{yz}/\partial x = G(\partial^2\psi/\partial x\partial y + 1)\delta\theta/\delta z. \tag{9.41b}$$

Subtracting Eq. (9.41a) from Eq. (9.41b) gives

$$\partial\tau_{yz}/\partial x - \partial\tau_{xz}/\partial y = 2G\,\delta\theta/\delta z. \tag{9.42}$$

Introducing the Prandtl stress function ϕ, where

$$\tau_{xz} = \partial\phi/\partial y, \qquad \tau_{yz} = -\partial\phi/\partial x, \tag{9.43a, b}$$

Eq. (9.42) becomes

$$\partial^2\phi/\partial x^2 + \partial^2\phi/\partial y^2 = -2G\,\delta\theta/\delta z. \tag{9.44}$$

Thus, the determination of the stress distribution over the cross-section consists of finding a function ϕ that satisfies Eq. (9.44). Once ϕ has been found, then the warping function ψ follows. Lines of constant ϕ in the section (Fig. 9.25(b)) imply trajectories of constant resultant shear stress $\tau = \sqrt{(\tau_{xz}^2 + \tau_{yz}^2)}$. The boundary is also a trajectory, usually associated with $\phi = 0$, since τ_{zy} will be zero for the exterior surface.

With reference to Fig. 9.25(a), the resisting torque exerted by the section is found from the double integral

$$T = \int_x \int_y (x\tau_{yz} - y\tau_{xz})\,\mathrm{d}x\,\mathrm{d}y,$$

which becomes, in terms of the stress function ϕ (from Eqs (9.43)),

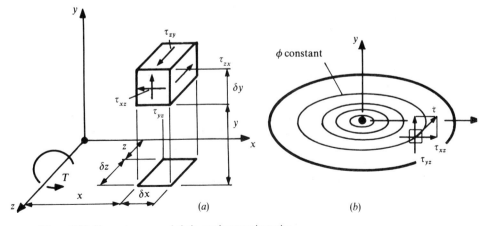

Figure 9.25 Shear stresses and their resultant trajectories.

$$T = - \iint (x \partial \phi / \partial x + y \partial \phi / \partial y) \, dx \, dy$$

$$= - \int dy \int (\partial \phi / \partial x) x \, dx - \int dx \int (\partial \phi / \partial y) y \, dy$$

$$= - \int dy \left(x \phi - \int \phi \, dx \right) - \int dx \left(y \phi - \int \phi \, dy \right)$$

$$= - \int \phi \, d(xy) + 2 \iint \phi \, dx \, dy$$

and, with $\phi = 0$ around the boundary this reduces to,

$$T = 2 \iint \phi \, dx \, dy = GJ(\delta \theta / \delta z), \tag{9.45}$$

where the final term is implied from the previous analyses given in Section 9.2 (page 298) for a particular prismatic bar—a thin strip. The key equations (9.43), (9.44) and (9.45) apply to any shape of section but may only be solved in closed form for relatively few sections. Numerical solutions are now readily available. In the past a membrane analogy has been made between ϕ in Eq. (9.44) and the deflection of a bubble pressurized from one side when covering a hole of the same shape as the bar cross-section. Experimental measurements of the gradient of tangents to the bubble along lines of constant deflection enabled the components of shear stress to be obtained by analogy with Eqs (9.43). From Eq. (9.45), the torque was identified with twice the volume beneath the bubble (see S. Timoshenko and J. N. Goodier, *Theory of Elasticity*, McGraw-Hill, 1970).

Example 9.16 Show that the function $\phi = C(1 - x^2/A^2 - y^2/B^2)$ provides a solution to the torsion of an elliptical shaft whose section, given in Fig. 9.26, is $x^2/A^2 + y^2/B^2 = 1$. Determine the stress distribution and the manner of warping and show that the equations reduce to those of a circular shaft.

The constant C is found from Eq. (9.44) when

$$- 2C/A^2 - 2C/B^2 = - 2G(\delta \theta / \delta z),$$
$$\therefore \quad C = A^2 B^2 G(\delta \theta / \delta z)/(A^2 + B^2). \tag{i}$$

The check on the function is made through Eq. (9.45) where

$$T = 2 \iint \phi \, dx \, dy = 2C \iint [(1 - x^2/A^2) - (y^2/B^2)] \, dx \, dy.$$

Substituting $dy = - (x/y)(B^2/A^2) \, dx$, where $y = B \sqrt{(1 - x^2/A^2)}$ gives

$$T = (8CB/3) \int_{-A}^{A} (1 - x^2/A^2)^{3/2} \, dx.$$

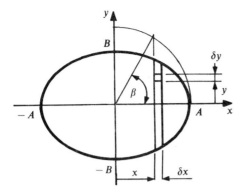

Figure 9.26 Elliptical section.

Then, with a further trigonometric substitution $x = A \cos \beta$, the integral is evaluated using the result in Eq. (i) to give

$$T = \pi ABC = [\pi A^3 B^3/(A^2 + B^2)]G(\delta\theta/\delta z), \qquad \text{(ii)}$$

The fact that the term in square brackets is the polar second moment of area (J) for an elliptical section confirms the validity of the function ϕ. The stresses are then found from Eq. (9.43):

$$\tau_{yz} = -\partial\phi/\partial x = -C\partial(1 - x^2/A^2 - y^2/B^2)/\partial x$$
$$= 2xB^2 G(\delta\theta/\delta z)/(A^2 + B^2), \qquad \text{(iii)}$$
$$\tau_{xz} = \partial\phi/\partial y = C\partial(1 - x^2/A^2 - y^2/B^2)/\partial y$$
$$= -2yA^2 G(\delta\theta/\delta z)/(A^2 + B^2). \qquad \text{(iv)}$$

The warping function is then found from either of Eqs (9.40):

$$\tau_{xz} = G(\partial\psi/\partial x - y)(\delta\theta/\delta z) = -2yA^2 G(\delta\theta/\delta z)/(A^2 + B^2),$$
$$\therefore \quad \partial\psi/\partial x - y = -2yA^2/(A^2 + B^2);$$
$$\partial\psi/\partial x = y[1 - 2A^2/(A^2 + B^2)] = y(B^2 - A^2)/(B^2 + A^2),$$
$$\therefore \quad \psi(x, y) = yx(B^2 - A^2)/(B^2 + A^2);$$

from which the warping displacements are given by

$$w = \psi(x, y)\,\delta x/\delta z = xy(\delta\theta/\delta z)(B^2 - A^2)/(B^2 + A^2). \qquad \text{(v)}$$

This may be checked from Eq. (9.40b). Thus, for a given rate of twist $\delta\theta/\delta z$, Eq. (v) describes the manner in which every cross-section warps. There is no warping at points along the x- and y-axes (for x and y zero). Putting $A = B = R$ in Eq. (ii) facilitates a circular section of radius R:

$$T = (\pi/2)R^4 G\,\delta\theta/\delta z = JG(\delta\theta/\delta z). \qquad \text{(vi)}$$

From Eqs (iii) and (iv) the resultant shear stress is

$$\tau = \sqrt{(\tau_{yz}^2 + \tau_{xz}^2)} = G(\delta\theta/\delta z)\sqrt{(x^2 + y^2)} = GR(\delta\theta/\delta z). \qquad \text{(vii)}$$

Combining Eqs (vi) and (vii) leads to the ETT (see Eq. (9.5)),

$$T/J = G(\delta\theta/\delta z) = \tau/R,$$

where it also becomes apparent from Eq. (v) that, as $w = 0$ for circular sections, then plane cross sections remain plane.

Example 9.17 Show that when a prismatic bar has a rectangular section for which breadth $b \gg$ thickness t (Fig. 9.27), the corresponding stress function confirms the thin-strip theory given in Section 9.2.

A considerable simplification arises when, for thin-strip curvature, $\partial^2\phi/\partial x^2 = 0$, for the x-direction. Equation (9.44) then becomes

$$\partial^2\phi/\partial y^2 = -2G\,\delta\theta/\delta z,$$

which may be integrated directly to give

$$\phi = -G(\delta\theta/\delta z)y^2 + Ay + B.$$

Taking the boundary $y = \pm t/2$ to be the contour $\phi = 0$ gives

$$0 = -G(\delta\theta/\delta z)t^2/4 + At/2 + B,$$
$$0 = -G(\delta\theta/\delta z)t^2/4 - At/2 + B.$$

Adding and subtracting gives $A = 0$ and $B = G(\delta\theta/\delta z)(t^2/4)$,

$$\therefore \quad \phi = G(\delta\theta/\delta z)(t^2/4 - y^2).$$

Then, from Eqs (9.43), the vertical shear stress component is

$$\tau_{yz} = -\partial\phi/\partial x = 0.$$

That is, the vertical shear stress is disregarded when $b \gg t$ in Fig. 9.27.

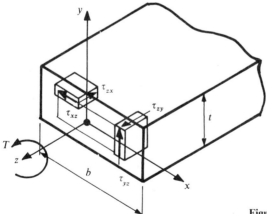

Figure 9.27 Rectangular section.

The non-zero horizontal shear stress is

$$\tau_{xz} = \partial\phi/\partial y = -2Gy(\delta\theta/\delta z), \qquad (i)$$

which displays a linear variation across the thickness having a maximum value along the longer sides for $y = \pm t/2$. The torque–twist relationship is supplied from Eq. (9.45):

$$T = 2\iint \phi \, dx \, dy = 2G(\delta\theta/\delta z)\int_{-t/2}^{t/2} (t^2/4 - y^2)\,dy \int_{-b/2}^{b/2} dx$$

$$= 2bG(\delta\theta/\delta z)\int_{-t/2}^{t/2} (t^2/4 - y^2)\,dy = 2bG(\delta\theta/\delta z)|t^2 y/4 - y^3/3|_{-t/2}^{t/2},$$

$$T = G(bt^3/3)(\delta\theta/\delta z) = GJ(\delta\theta/\delta z). \qquad (ii)$$

Combining Eqs (i) and (ii) confirms the previous derivation in Eq. (9.18).

Example 9.18 Find the warping function for the strip in Example 9.17. If the dimensions are $b = 100\,\text{mm}$, $t = 5\,\text{mm}$ (Fig. 9.28), find the twist rate and the manner in which the section warps under a torque of 10 N m. Take $G = 80\,\text{GPa}$.

Equations (i) in Example 9.17 and Eq. (9.40a), supply the warping function ψ:

$$\tau_{xz} = G(\delta\theta/\delta z)(\partial\psi/\partial x - y) = -2Gy(\delta\theta/\delta z),$$
$$\therefore \quad \partial\psi/\partial x = -y, \quad \Rightarrow \psi = -yx. \qquad (i)$$

This is readily confirmed from Eq. (9.40b). The integration applies to an origin O at the centre of twist where $w = 0$ for the x-, y-axes of symmetry.

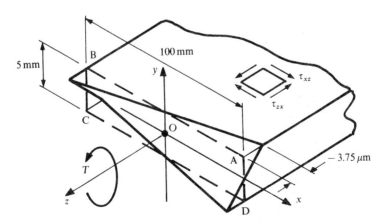

Figure 9.28 Warping of a rectangular section.

Note, from the comparison with Eq. (9.16), that the minus sign arises here because this derivation employs a positive anticlockwise torque. Now,

$$\delta\theta/\delta z = T/GJ = 10 \times 10^3/[(100 \times 5^3/3)80 \times 10^3] = 0.03 \times 10^{-3} \text{ rad/mm},$$

and the warping displacements are then found from Eqs (i) and (9.38c):

$$w = 0.03 \times 10^{-3}(-xy),$$

At A $(+50, +2.5)$, $w_A = -3.75 \times 10^{-3}$ mm,

At B $(+50, -2.5)$, $w_B = +3.75 \times 10^{-3}$ mm,

At C $(-50, -2.5)$, $w_C = -3.75 \times 10^{-3}$ mm,

At D $(-50, +2.5)$, $w_D = +3.75 \times 10^{-3}$ mm.

Along each axis of symmetry $x = y = 0$ and $w = 0$ (see Fig. 9.28).

9.5 PLASTIC TORSION OF CIRCULAR SECTIONS

Solid Bar

Consider a solid circular bar of radius b, subjected to an increasing torque T. While the cross-section is elastic the ETT applies to the ensuing deformation. When the outer fibres reach the shear yield stress k, the corresponding torque T_E for the limiting fully elastic section is supplied from Eq. (9.5) as

$$T_E = Jk/b = \pi b^4 k/2b = \pi b^3 k/2. \tag{9.46}$$

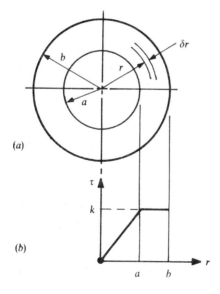

(a)

(b)

Figure 9.29 Elastic-plastic bar.

Increasing the torque beyond T_E results in the penetration of a plastic zone inwards from the outer radius. This results in the elastic-plastic section shown in Fig. 9.29(a), in which an elastic core is surrounded by a plastic annulus with a common interface radius a. Assuming that the material is elastic-perfectly plastic, i.e. that it does not work-harden (Fig. 9.29(b)), then k is constant for $a \le r \le b$. It follows that the linear variation in elastic shear stress τ within the core $0 \le r \le a$ is given by

$$\tau_e = k(r/a). \tag{9.47}$$

The corresponding elastic-plastic torque T_{ep}, is found from the following equilibrium condition:

$$T_{ep} = 2\pi \int_a^b kr^2 \, dr + 2\pi \int_0^a \tau_e r^2 \, dr \tag{9.48a}$$

$$= 2\pi k \int_a^b r^2 \, dr + 2\pi(k/a) \int_0^a r^3 \, dr$$

$$= 2\pi(k/3)|r^3|_a^b + 2\pi(k/4a)|r^4|_0^a$$

$$= (\pi k b^3/2)\{(4/3)[1 - (a/b)^3] + (a/b)^3\}. \tag{9.48b}$$

Substituting from Eq. (9.46)

$$T_{ep}/T_E = (4/3)[1 - (1/4)(a/b)^3]. \tag{9.49}$$

The angle of twist is found from applying the ETT to the elastic core where $\tau = k$ for $r = a$. That is, from Eq. (9.5)

$$\theta = kL/Ga = T_e L/GJ_e, \tag{9.50}$$

where $J_e = \pi a^4/2$ and T_e is the torque carried by the elastic core, i.e. the second integral in Eq. (9.48a), not Eq. (9.46). The section becomes fully plastic under a torque T_p that causes full plastic penetration to the bar centre. Putting $a = 0$ in Eq. (9.49) and substituting from Eq. (9.46) gives

$$T_P/T_E = 4/3 \quad \therefore \quad T_P = (2/3)(\pi b^3 k).$$

When the elastic-plastic bar in Fig. 9.30(a) is fully unloaded, elastic stresses recover to leave a residual stress distribution τ_R, which is found from

$$\tau_R = \tau - \tau_e, \tag{9.51}$$

where τ is the shear stress in either zone under the applied torque T_{ep} and τ_e is the wholly elastic stress that recovers on removal of T_{ep}. That is,

$$\tau_e = T_{ep} r/J. \tag{9.52}$$

It follows from Eqs (9.49), (9.51) and (9.52) that with $J = \pi b^4/2$ the

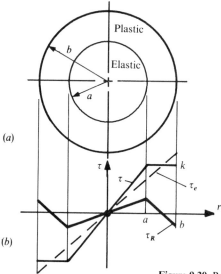

(a)

(b)

Figure 9.30 Residual stress in a partially plastic solid bar.

residual stress distribution in the plastic zone is given by

$$\tau_R = k - T_{ep}r/J = k[1 - (T_{ep}/T_E)(r/b)]$$
$$= k - (4/3)T_E[1 - (1/4)(a/b)^3](r/J)$$
$$= k\{1 - (4/3)(r/b)[1 - (1/4)(a/b)^3]\} \qquad (9.53\text{a})$$

and from Eqs (9.47), (9.49), (9.51) and (9.52) the residual stress in the elastic zone is

$$\tau_R = k(r/a) - T_{ep}r/J = k(r/a)[1 - (T_{ep}/T_E)(a/b)]$$
$$= k(r/a) - (4/3)k(r/b)[1 - (1/4)(a/b)^3]$$
$$= k(r/a)\{1 - (4/3)(a/b)[1 - (1/4)(a/b)^3]\}. \qquad (9.53\text{b})$$

The distributions for Eqs (9.53) are represented graphically in Fig. 9.30(b). The residual stresses are simply the difference in the ordinates between τ and τ_e within each zone, where the signs are determined from Eq. (9.51). It is seen that τ_R has the largest values at the outer and interface radii but the greater value will depend upon the depth of penetration in Fig. 9.30(a).

Hollow Cylinder

Figure 9.31(a) shows a partially yielded hollow cylinder with inner and outer radii c and b in which the plastic zone has penetrated to radius a, under torque T_{ep}. The corresponding distribution of shear stress is given in Fig. 9.31(b). Again, an elastically-perfectly plastic material is assumed in which the plastic zone spreads under a constant yield stress value k.

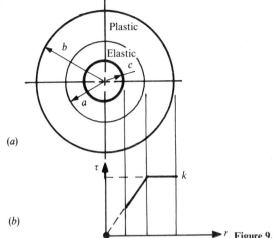

(a)

(b)

Figure 9.31 Elastic-plastic hollow cylinder.

Torque equilibrium is expressed by

$$T_{ep} = 2\pi k \int_a^b r^2 \, dr + 2\pi(k/a) \int_c^a r^3 \, dr \qquad (9.54a)$$

$$= (2\pi k/3)(b^3 - a^3) + (\pi k/2a)(a^4 - c^4). \qquad (9.54b)$$

The following three torques may be identified from Eq. (9.54b).
1. The fully elastic torque T_E, for $a = b$ in Eq. (9.54b):

$$T_E = (\pi k/2b)(b^4 - c^4). \qquad (9.55)$$

2. The dimensionless elastic-plastic torque T_{ep}, for $c < a < b$, by combining Eqs (9.54b) and (9.55):

$$T_{ep}/T_E = (4b/3)(b^3 - a^3)/(b^4 - c^4) + (b/a)(a^4 - c^4)/(b^4 - c^4)$$

$$= \frac{(4/3)[1 - (1/4)(a/b)^3] - (c/a)(c/b)^3}{1 - (c/b)^4}, \qquad (9.56)$$

which reduces to Eq. (9.49) for $c = 0$ in a solid cylinder.
3. The dimensionless fully plastic torque, $a = c$ in Eq. (9.56):

$$T_P/T_E = (4/3)[1 - (c/b)^3]/[1 - (c/b)^4]. \qquad (9.57)$$

Substituting Eqs (9.52) and (9.54b) into Eq. (9.51) with $J = \pi(b^4 - c^4)/2$ gives the residual shear stress in the plastic zone, $a \le r \le b$:

$$\tau_R = k[1 - (T_{ep}/T_E)(r/b)]$$

$$= k - (4kr/3)[(b^3 - a^3)/(b^4 - c^4) + 3(a^4 - c^4)/4a(b^4 - c^4)]. \qquad (9.58a)$$

The residual stress in the elastic zone $c \leq r \leq a$ is, from Eqs (9.47), (9.51), (9.52) and (9.54b),

$$\tau_R = k(r/a)[1 - (P_{ep}/T_E)(a/b)]$$
$$= k(r/a) - (4kr/3)[(b^3 - a^3)/(b^4 - c^4) + 3(a^4 - c^4)/4a(b^4 - c^4)], \quad (9.58b)$$

from which it is seen that when $c = 0$ in Eqs (9.58) they reduce to the solid-bar residuals in Eqs (9.53). Finally, Eq. (9.50) will again supply the twist in an elastic-plastic cylinder when based upon the inner elastic annulus. That is, $J_e = (\pi/2)(a^4 - c^4)$ for T_{ep} identified with the second integral in Eq. (9.54a). The reader should recognize the fact that when metallic materials work- (or strain-) harden this will increase the torque-carrying capacity above these non-hardening theoretical values.

Example 9.19 The 100 mm diameter bar in Fig. 9.32 is bored to 50 mm diameter over half its length. If, under an axially applied torque, the outer fibres of the solid section reach their shear yield stress, find (i) the depth of penetration in the hollow section and (ii) the ratio between the solid and hollow angular twists per unit length of the solid and hollow sections.

For the solid section, Eq. (9.46) gives the fully elastic torque:

$$T_E = \pi b^3 k/2 = (\pi/2)50^3 k. \quad (i)$$

Equation (9.54) applies to the elastic-plastic hollow section, giving

$$T_{ep} = (2\pi k/3)(b^3 - a^3) + (\pi k/2a)(a^4 - c^4)$$
$$= (\pi k/2)(4b^3/3 - a^3/3 - c^4/a). \quad (ii)$$

Since $T_E = T_{ep}$, then equating (i) and (ii) with $b = 50$ mm, $c = 25$ mm gives the following equation for the elastic-plastic interface radius a:

$$50^3 = (4/3)(50)^3 - a^3/3 - 25^4/a,$$
$$a^4 - (50)^3 a + 3(25)^4 = 0.$$

This quartic is satisfied by $a = 46.38$ mm. The radial penetration depth for the hollow cylinder is therefore 3.62 mm in from the outer diameter. In the solid cylinder, Eq. (9.50) supplies the rate of twist:

$$(\theta/L)_s = k/Gb = k/50G,$$

while, for the hollow cylinder, applying Eq. (9.50) to the elastic annulus gives its rate of twist as

$$(\theta/L)_h = k/Ga = k/46.38G,$$

and therefore the unit twist ratio is

$$(\theta/L)_h/(\theta/L)_s = 50/46.38 = 1.08.$$

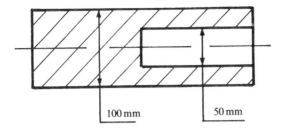

100 mm   50 mm   **Figure 9.32** Torsion bar.

Example 9.20 Find the torque required to cause yielding to the mean radius in a hollow steel cylinder of 200 mm (8 in) o.d. and 100 mm (4 in) i.d. (Fig. 9.33a) for a shear yield stress of 180 MPa (12 tonf/in²). Determine the distribution of shear stress remaining in the section after the torque has been removed.

SOLUTION TO SI UNITS With $a = 75$ mm, $b = 100$ mm and $c = 50$ mm, the direct substitution into Eq. (9.54) gives the corresponding elastic-plastic torque:

$$T_{ep} = (2\pi \times 180)(100^3 - 75^3)/3 + (\pi \times 180)(75^4 - 50^4)/(2 \times 75)$$
$$= 313.67 \times 10^6 \text{ N mm} = 313.67 \text{ kN m}.$$

Equation (9.58a) supplies the residual stresses in the plastic annulus:

$$\tau_R/k = 1 - (4r/3)[(b^3 - a^3)/(b^4 - c^4) + 3(a^4 - c^4)/4a(b^4 - c^4)]$$
$$= 1 - (4r/3)[0.006\,17 + 0.002\,71] = 1 - (11.837 \times 10^{-3})r, \quad \text{(i)}$$

and in the inner elastic annulus Eq. (9.58b) gives
$$\tau_R/k = r/a - (4r/3)[(b^3 - a^3)/(b^4 - c^4) + 3(a^4 - c^4)/4a(b^4 - c^4)]$$
$$= r/75 - (11.837 \times 10^{-3})r. \qquad \text{(ii)}$$

Clearly the distributions represented by Eqs (i) and (ii) are both linear in the radius r. Figure 9.33(b) shows that the greatest value, $\tau_R/k = -0.184$, occurs at the outer radius, $r = 100$ mm.

SOLUTION TO IMPERIAL UNITS Substituting $a = 3$ in, $b = 4$ in and $c = 2$ in into Eq. (9.54b):

$$T_{ep} = (2\pi \times 12/3)(4^3 - 3^3) + (\pi \times 12/2 \times 3)(3^4 - 2^4) = 1338.3 \text{ tonf in.}$$

Equation (9.58) supply the normalized residual stress expressions:

$$\tau_R/k = 1 - (4r/3)(0.154 + 0.0677)$$
$$= 1 - 0.2956r \qquad \text{for } 3 \le r \le 4 \text{ in,}$$

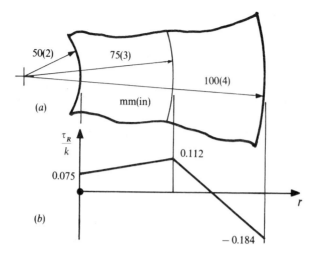

Figure 9.33 Hollow steel cylinder.

$$\tau_R/k = r/a - (4r/3)(0.154 + 0.0677)$$
$$= 0.0377r \qquad \text{for } 2 \leq r \leq 3 \text{ in.}$$

These residuals agree at $r = 3$ in when $\tau_R/k = 0.113$.

Composite Shaft

The fully elastic torque capacity T_E for a composite shaft is determined from the material that is the first to reach the shear yield stress at its outer fibres. Thus, in Fig. 9.4, with shear yield stresses k_1 and k_2 for the inner (1) and outer (2) components respectively, it follows from Eqs (9.9 and 9.10) that T_E is the lesser of

$$T_E = (k_1/r_1)[J_1 + (G_2/G_1)J_2], \qquad (9.59a)$$

$$T_E = (k_2/r_2)[J_2 + (G_1/G_2)J_1]. \qquad (9.59b)$$

Assuming that the lesser T_E is determined from Eq. (9.59b) with k_2 first being attained in the outer fibres of the annulus, then the core remains elastic. The maximum shear stress $\tau_1 < k_1$, at the outer diameter of the core, follows from the compatibility condition that both materials suffer the same shear strain at the interface. That is,

$$(\tau/G)_1 = (\tau/G)_2,$$
$$\therefore \quad \tau_1 = k_2(r_1/r_2)(G_1/G_2) = (T_2 r_1/J_2)(G_1/G_2). \qquad (9.60)$$

Alternatively, if the lesser T_E is found from k_1 first being reached in the core (Eq. (9.59a)) then the annulus remains elastic. The maximum stress $\tau_2 < k_2$, at the outer diameter of the annulus, is again found from equality of shear strain at the interface. That is,

$$(\tau_2/G_2)(r_1/r_2) = k_1/G_1, \quad \Rightarrow \tau_2 = k_1(r_2/r_1)(G_2/G_1),$$
$$\therefore \quad \tau_2 = (T_1 r_1/J_1)(r_2/r_1)(G_2/G_1) = (T_1 r_2/J_1)(G_2/G_1).$$

With increasing torque beyond T_E, the plastic zone inwardly penetrates either the shaft or the annulus independently until the remaining component first becomes plastic. As plastic penetration then occurs in both components simultaneously, the foregoing elastic-plastic analyses again apply in which the total torque is the sum of the component torques.

Example 9.21 The composite bar in Fig. 9.4 is composed of a central brass core 75 mm in diameter, firmly surrounded by a steel annulus of 100 mm outer diameter. Determine the maximum elastic torque the section can sustain. Which material is understressed and by how much? Find also the torque to produce a fully plastic steel annulus. Take for brass (1) $G_1 = 41\,\text{GPa}$, $k_1 = 60\,\text{MPa}$, and for steel (2) $G_2 = 82\,\text{GPa}$, $k_2 = 120\,\text{MPa}$.

$$J_1 = \pi(75^4)/32 = 3.106 \times 10^6\,\text{mm}^4,$$
$$J_2 = \pi(100^4 - 75^4)/32 = 6.711 \times 10^6\,\text{mm}^4.$$

Then from Eqs (9.59),

$$T_E = (60/37.5)[3.106 + (2 \times 6.711)] \times 10^6$$
$$= 26.445 \times 10^6\,\text{N mm} = 26.445\,\text{kN m}$$
$$T_E = (120/50)[6.711 + (0.5 \times 3.106)] \times 10^6$$
$$= 19.834 \times 10^6\,\text{N mm} = 19.834\,\text{kN m}.$$

The lesser T_E determines the maximum elastic torque. Thus the steel is at the point of yield at its outer diameter while the brass is understressed. The maximum stress in the brass is, from Eq. (9.60),

$$\tau_1 = (G_1/G_2)(r_1/r_2)k_2 = (41/82)(37.5/50)120 = 45\,\text{MPa}.$$

When the steel becomes fully plastic, Eq. (9.57) is applied to give

$$T_P/T_E = (4/3)[1 - (37.5/50)^3]/[1 - (37.5/50)^4] = 1.1276$$

in which Eq. (9.55) supplies T_E for a hollow cylinder:

$$T_E = (\pi \times 120/2 \times 50)(50^4 - 37.5^4) = 16.107\,\text{kN m},$$
$$\therefore \quad (T_P)_2 = 1.1276 \times 16.107 = 18.162\,\text{kN m}.$$

The shear strain in the steel interface is then $k_2/G_2 = 1.463 \times 1^{-3}$. Since this is also the shear strain k_1/G_1 at which the brass interface yields, it follows that the brass is at its fully elastic condition. From Eq. (9.46), for a solid bar

$$(T_E)_1 = \pi \times 37.5^3 \times 60/2 = 4.97 \times 10^6\,\text{N mm} = 4.97\,\text{kN m}.$$

The net torque is then

$$(T_P)_1 + (T_E)_2 = 18.162 + 4.97 = 23.132 \, \text{kN m}.$$

Note that composite bar problems are complicated by the mismatch in shear strain that arises at the interface when $k_1/G_1 \neq k_2/G_2$ for non-hardening component materials. The amount of simultaneous penetration in each material must then be determined from these limiting shear strain values.

EXERCISES

Torsion of Circular Sections

9.1 A solid steel shaft, 50 mm in diameter, can transmit a torque of 2.17 kN m. What diameter steel shaft can transmit 45 kW when rotating at 240 rev/min?

Answer: 47.5 mm.

9.2 A tubular extension of 38 mm o.d., 32 mm i.d. and 2.5 m long is fitted over the end of a 32 mm diameter solid spindle. If, when transmitting torque, the maximum shear stress in the spindle is 31 MPa, calculate the maximum shear stress and the angle of twist for the tube. Take $G = 82.7$ GPa.

Answer: 34 MPa; 3.04°.

9.3 Find the maximum power that can be transmitted by a 150 mm diameter shaft running at 240 rev/min, given that the allowable shear stress is 55 MPa. The shaft is coupled to a motor with a flanged joint containing six bolts on a 265 mm pitch circle diameter. Find the required bolt diameter when the maximum shear stress for the bolt material is restricted to 100 MPa.

Answer: 962 kW; 24.5 mm.

9.4 A hollow drive shaft with diameter ratio 3:5 is to transmit 448 kW at 120 rev/min. If the shear stress must not exceed 62 MPa and the angle of twist is restricted to 1° over a 2.5 m length, calculate the necessary external diameter of the shaft. Take $G = 82.7$ GPa.

Answer: 163 mm.

9.5 Compare the ratio of the weights for solid and hollow shafts in the same material when transmitting the same torque if, for the hollow shaft, the outer diameter is twice the inner. What power can a hollow shaft of 305 mm o.d. and 205 mm i.d. transmit at a speed of 210 rev/min when the maximum shear stress is restricted to 62 MPa?

Answer: 1.29 per cent, 6065 kW.

9.6 The constraints on a shaft transmitting 932.5 kW at 240 rev/min are (i) that it should not twist more than 1° on 15 diameters and (ii) that the maximum shear stress should not exceed 54 MPa. Calculate the necessary shaft diameter and the working shear stress. Take $G = 82.7$ GPa.

Answer: 158 mm; 48 MPa.

9.7 A hollow drive shaft of 203 mm o.d. with a 25 mm wall thickness is to transmit 933 kW at 125 rev/min. Calculate the working shear stress for the shaft material based upon a safety factor of 3.

Answer: 190 MPa.

9.8 A solid drive shaft of diameter d is replaced by a hollow tube of mean diameter D that can sustain the same torque without exceeding the maximum shear stress. Show that these conditions are achieved when $D = \sqrt{(d^3/8t)}$ and that the ratio between the solid and hollow twist rates is equal to d/D under a given torque.

9.9 A solid steel shaft of diameter 60 mm and length 1.5 m is subjected to a torque of 2 kN m. Find the angle of twist on a length of 1.5 m and the greatest shear stress. Take $G = 80$ GPa.

Answer: 1.69°; 47.17 MPa.

9.10 A hollow steel shaft is to transmit 30 kN m without the twist exceeding more than 3° on a length of 4 m. If the maximum shear stress is restricted to 80 MPa, find the shaft diameters that will minimize the weight. Take $G = 80$ GPa.

Answer: 159 mm; 135 mm.

9.11 A hollow steel shaft of 90 mm o.d. and 6 mm wall thickness is used to transmit 240 kW at 300 rev/min. Calculate the shear stresses at the inner and outer diameters and the angular twist on a length of 2 m. Take $G = 80$ GPa.

9.12 One degree of twist is measured for a propeller shaft, 305 mm in diameter and 4.5 m in length. If the shaft rotates at 300 rev/min, determine the power transmitted by the shaft and the maximum shear stress induced in it. If a hollow shaft with diameter ratio 1:4 is used to transmit the same torque at the same maximum shear stress, determine the percentage saving in weight. Take $G = 80.2$ GPa.

Answer: 8.13 MW; 46.6 MPa; 58 per cent.

9.13 The transmission shaft of a motor car is required to transmit 18.65 kW at 250 rev/min. If the shaft is hollow of 76 mm o.d. and 63.5 mm i.d., calculate the maximum shear stress in the shaft when the maximum torque exceeds the mean torque by 20 per cent. Find the greatest angle of twist for a 1.8 m length of shaft. Note that T in Eq. (9.6) is the mean torque. Take $G = 82.7$ GPa.

Answer: 19 MPa; 0.63°.

9.14 What power can be transmitted by a solid steel shaft 255 mm in diameter at 180 rev/min if the maximum torque exceeds the mean torque by 10 per cent when the maximum shear stress is restricted to 62 MPa? Find the length of this shaft over which the angle of twist is 1.5°. Take $G = 82.7$ GPa.

Answer: 34 kW; 4.45 m.

9.15 Determine the torque that can be transmitted by a solid shaft of 100 mm diameter, when the maximum shear stress is restricted to 48 MPa. Through what angle would this shaft twist in a length of 3.65 m if $G = 82.7$ GPa? What is the percentage saving in weight for a hollow shaft of equal strength with a diameter ratio of 1.5?

Answer: 10 kN m; 2.4°.

9.16 A 100 mm diameter solid steel shaft transmits 180 kW at 800 rev/min with the maximum torque exceeding the mean torque by 50 per cent. What is the maximum shear stress in the shaft? If this shaft contains a coupling with six bolts equispaced on a pitch circle diameter of 180 mm, calculate the required minimum diameter of the bolts when their allowable shear stress value is 27.5 MPa.

Answer: 15.5 MPa; 165 mm.

9.17 A wrought iron shaft has a diameter of 63.5 mm and is 7.5 m long. What will be the twist in the bar when the maximum shear stress reaches 77 MPa? Take $G = 72.4$ GPa.

Answer: 14°.

9.18 A solid circular shaft is connected to the drive shaft of an electric motor with a flanged coupling. The drive is taken by eight bolts each 12.5 mm in diameter on a pitch circle diameter of 230 mm. Calculate the shaft diameter if the maximum shear stress in the shaft is to equal that in the bolts.

Answer: 84 mm.

9.19 A hollow steel shaft of 205 mm i.d. and 305 mm o.d. is to be replaced by a solid alloy shaft. If their polar second moments of area (J) are equal, calculate the solid shaft diameter and the ratio between their stiffnesses. Take G(steel) $= 2.4G$(alloy). What would be the ratio between their J values when the stiffnesses remain the same?

Answer: 283 mm; 2.58, 0.49.

9.20 A hollow steel shaft of 150 mm o.d. and 12.5 mm thick is coupled to a solid 125 mm diameter steel shaft. If the length of the hollow shaft is 1.25 m, find the length of solid shaft that limits the total twist to 3° under a torque of 25 kN m. Take $G = 69$ GPa.

9.21 A solid shaft 200 mm in diameter is subjected to a torque of 37.5 kN m. This shaft is to be replaced in the same material by a hollow shaft with diameter ratio 2 for the same torque and maximum shear stress. Compare the ratio of their weights and angles of twist over a length of 3 m. Take $G = 80$ GPa for both shafts.
Answer: solid:tubular—1.283:1; 0.981:1.

9.22 A hollow steel shaft is required to transmit 6 MW at 110 rev/min. If the allowable shear stress is 60 MPa and the i.d. = (3/5)(o.d.), calculate the dimensions of the shaft and the angle of twist on a 3 m length. Take $G = 86$ GPa.
Answer: o.d. = 372 mm; i.d. = 222 mm.

9.23 A tubular turbine shaft is to transmit power at 240 rev/min. If the shaft is of 1 m o.d. and 25 mm thick find the power transmitted at a maximum shear stress of 69 MPa. What is the diameter of the equivalent solid shaft and the percentage saving in weight of the hollow shaft?
Answer: 66 MW; 580 mm; 70 per cent.

9.24 One end of a 1 m long aluminium alloy rod is joined to the end of a 0.75 m long nimonic alloy rod. The respective allowable shear stresses are 35 MPa and 75 MPa and the rigidity moduli are 30 GPa and 80 GPa. If the ends are rigidly fixed, find the magnitude and position in the length where an externally applied torque must be applied in order to fully stress both materials.

9.25 If the materials and allowable shear stresses given in Exercise 9.24 form a composite shaft consisting of two concentric cylinders with dimensions: nimonic, 10 mm i.d., 20 mm o.d.; Al alloy, 20 mm i.d., 25 mm o.d., what is the maximum torque that can be carried?

9.26 A 20 mm steel bar is firmly surrounded by an aluminium alloy tube. If their respective allowable shear stresses are not to exceed 100 and 50 MPa under an axial torque of 1 kN m, determine the composite shaft diameters and the angle of twist over a 1.5 m length. Take $G(\text{Al}) = 28$ GPa and $G(\text{steel}) = 80$ GPa.

9.27 The diameter of a 0.5 m long steel shaft increases uniformly from 50 mm at one end to 75 mm at the other end. If the maximum shear stress in the shaft is limited to 95 MPa, find the torque that may be transmitted and the angle of twist between the ends. Take $G = 80$ GPa.

9.28 A solid steel shaft tapers for a length of 1.5 m between end diameters, of which one is twice the other. If the maximum shear stress and the angle of twist are limited to 60 MPa and 4° respectively under a torque of 10 kN m, determine the diameters. Take $G = 80$ GPa.

9.29 A steel shaft of 100 mm outer diameter is solid for 1 m and hollow at 60 mm inner diameter for its remaining 1 m length. Find the greatest torque the shaft can withstand when the maximum shear stress is limited to 100 MPa. What is the angle of twist between the ends? Take $G = 80$ GPa.
Answer: 17.1 kN m; 2.68°.

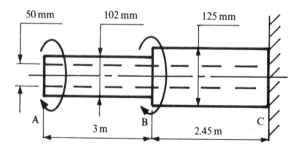

Figure 9.34

9.30 The hollow stepped shaft in Fig. 9.34 is subjected to clockwise torques of 12.65 kN m at sections A and B and is built-in at section C. Determine the maximum shear stress and the angle of twist between A and C, given $G = 75$ GPa.

Answer: 65 MPa; 4.8°.

9.31 The two steel shafts in Fig. 9.35 are connected by gears. If the end of shaft 1 is fixed, determine the angle of twist at the free end of shaft 2 when a torque of 400 N m is applied at this position. Take $G = 80$ GPa.

Answer: 10.1°.

9.32 Determine the torque diagram and the maximum shear stress in the shaft of Fig. 9.36 when opposing torques of 200 N m are applied through pulleys at A and B.

Answer: 65.2 MPa.

9.33 The stepped shaft in Fig. 9.37 is fixed at its ends and is subjected to concentrated torques at the positions shown. Determine the maximum shear stress and twist. Take $G = 80$ GPa.

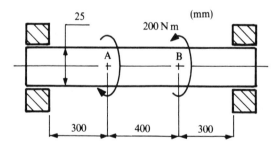

Figure 9.35

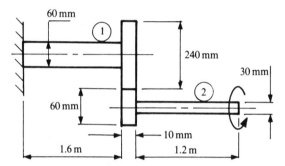

Figure 9.36

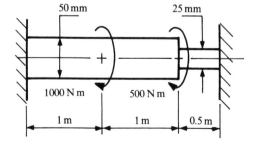

Figure 9.37

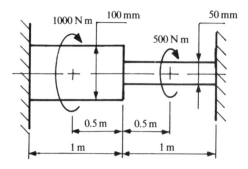

Figure 9.38

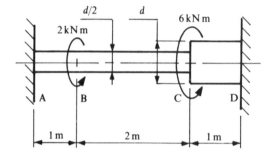

Figure 9.39

9.34 Draw the torque and twist diagrams for the stepped shaft in Fig. 9.38 showing the maximum values.

9.35 Figure 9.39 shows a stepped shaft ABCD, which is rigidly fixed at both ends. The shaft is subjected to applied torques of 2 kN m and 6 kN m at B and C respectively. The diameter of the portion AC is one-half that of portion CD. Calculate the torque distribution in the shaft. Determine the diameters of the shaft if it is required that neither the maximum shear stress nor the angle of twist should exceed 50 N/mm² and 1° respectively. Assume $G = 80 \times 10^3$ N/mm². (C.E.I. Pt. 2, 1973)

Torsion of Thin Strips and Open Sections

9.36 A thin rectangular strip 50 mm wide × 2 mm thick is subjected to a pure axial torque of 10 N m. Determine the maximum shear stress and the angle of twist over a 0.5 m length.

9.37 A steel sheet 5 mm thick is formed into a 270° circular arc that is to support a torque of 50 N m without the maximum shear stress exceeding 55 MPa while restricting the angular twist to 3°. Find the necessary width and length of the sheet.

9.38 A right-angled section 5 mm thick with outer dimensions of 100 mm × 75 mm carries a torque of 15 N m. Calculate the rate of twist and the position and magnitude of the maximum shear stress in the section. Take $G = 82$ GPa.

9.39 A strip of metal of rectangular section, 10 mm × 60 mm, length 0.3 m and shear modulus 80 GPa, is clamped in a torsion testing machine and subjected to a steadily increasing torque. If the material yields when the shear stress exceeds 100 MPa, determine the torque required just to cause yielding in the material, and also determine the corresponding relative angular rotation of the two ends of the strip. (C.E.I. Pt. 2, 1977)

9.40 A long, straight, uniform bar of rectangular section is to act as a door by twisting along its length. If the maximum shear stress is to be 100 MPa and the modulus of rigidity is 80 GPa, find the torque and the length if the angle of twist of the bar is to be $\pi/2$ rad. The width of the bar is 20 mm and the thickness is 1.5 mm (take $\alpha = \beta = 0.32$ in Table 9.1). If this bar is replaced by a circular bar of equal length and cross-sectional area, find the maximum shear stress and torque to twist it through $\pi/2$ rad. (C.E.I. Pt. 2, 1986)

9.41 A beam of thin channel section is made from an aluminium alloy with $G = 25$ GPa. The flanges are both 100 mm wide and of thickness t and the web is 200 mm long and of thickness $2t$. If the section is to transmit an axial torque of 75 N m without the twist in 1 m length and the shear stress exceeding $25°$ and 40 MPa respectively, find the least value of thickness t to satisfy both of these requirements. (C.E.I. Pt. 2, 1985)

9.42 A torsion member used for stirring a chemical process is made of a circular tube to which are welded four rectangular strips as shown in Fig. 9.40. The tube has inner and outer diameters of 94 and 100 mm respectively, each strip is 50 mm by 18 mm, and the stirrer is 3 m in length. If the maximum shearing stress in any part of the cross-section is limited to 56 MN/m², neglecting

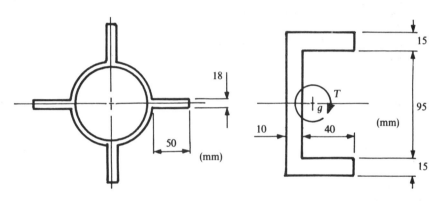

Figure 9.40   **Figure 9.41**

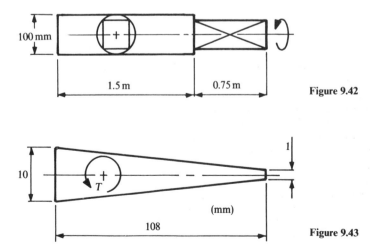

Figure 9.42

Figure 9.43

any stress concentration, calculate the maximum torque that can be carried by the stirrer and the resulting twist over the full length. (Take the coefficients in Eqs (9.21) as $\alpha = 0.264$, $\beta = 0.258$, and $G = 83$ GPa.) (C.E.I. Pt. 2, 1971)

9.43 Find the torque T that can be withstood by the steel channel section in Fig. 9.41 when the greatest shear stress is limited to 120 MPa. What angle would this section twist through in a length of 3 m? Take $G = 83$ GPa and use the coefficients given in Table 9.1.

9.44 Determine the torsional stiffness for the steel drive shaft in Fig. 9.42 when 0.75 m of the length is a square section whose diagonal is equal to the remaining 100 mm circular shaft diameter. Assume that the square section is free to warp for the coefficients in Table 9.1. Take $G = 80$ GPa.

Answer: 481.6 kN m/rad.

9.45 Determine the St Venant torsion constant for the thin tapered section in Fig. 9.43. Hence derive an expression for the shear stress acting along the longer edges in terms of the torque T when the thickness (mm) $10 \leq t \leq 1$. If $T = 15$ N m estimate the position and magnitude of the maximum shear stress from a consideration of contours of constant shear stress. (K.P. 1974)

9.46 The equal-angle, light-alloy section in Fig. 9.44 tapers along its inner edges. Calculate the rate of twist and the maximum shear stress under an applied torque of 75 N m. Take $G = 29$ GPa.

9.47 Figure 9.45 shows the free-end of an equal angle cantilever bar section. The dimensions increase uniformly over its 500 mm length such that at the fixed end all dimensions become 50 per cent greater. If the maximum shear stress is limited to 100 MPa, determine the allowable free-end torque and the angle of twist between the ends.

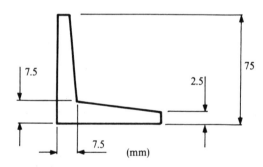

Figure 9.44

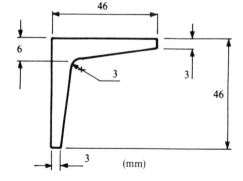

Figure 9.45

Torsion of Thin-walled Closed Sections

9.48 Find the maximum torque that can be carried by a 5 mm thick tube with equilateral triangular section of 0.2 m mean side length if the allowable shear stress is 50 MPa.

9.49 Calculate the angle of twist and thickness t for a thin-walled circular steel tube of mean radius $R = 0.2$ m and length $L = 1.5$ m that is required to withstand a torque of 75 kN m when the shear stress is limited to 35 MPa. Examine the possibility of torsional buckling given the critical stress, $\tau_{cr} = 0.272E(t/R)^{3/2}/(1 - v^2)^{3/4}$. Take $E = 208$ GPa, $v = 0.27$.

9.50 A steel tube of 30 mm mean diameter and 1 mm wall thickness is required to sustain a torque of 80 N m. Calculate the maximum shear stress and the safety factor when the working stress is 124 MPa.

Answer: 61 MPa; 2.

9.51 A rectangular box-section cantilever has walls 5 mm thick. The 200 mm depth is constant but the breadth varies over its 3 m length from 100 mm at the free end to 200 mm at the fixed end. Find the maximum shear stress and derive an expression for the twist as a function of length when a torque of 100 N m is applied at the free end. Take $G = 28$ GPa.

9.52 A thin-walled equilateral triangular section is to withstand an axial torque of 500 N m without the shear stress exceeding 40 MPa. If the material cost in pence per square millimetre of tube section, per metre length, is given by $C = 15 + s/10t$, determine the tube side length s and thickness t that will minimize cost whilst sustaining this torque.

9.53 The material cost of a thin-walled square tube is $(0.4 + s/100t)$ pence per square millimetre of cross-sectional area, per metre length. Determine the side length s and the thickness t to minimize cost when the applied torque is 660 N m for a maximum shear stress of 75 MPa.

Answer: 43.1 mm; 2.16 mm.

9.54 Steel plates 100 mm × 10 mm in section are welded together to form a closed tube of triangular section and an I-section. Calculate the ratio between the maximum shear stresses and the ratio between the torsional stiffnesses when each section is subjected to the same torque. (T.C.D. 1978)

9.55 Compare the torsional stiffnesses according to the Batho and ETT for a circular tube whose wall thickness is 1/20 of the mean wall diameter.

9.56 A 120 mm wide × 300 deep I-section, with web and flange thicknesses of 10 and 4 mm respectively, sustains a torque of 300 N m over a length of 3 m. Find the angle of twist in degrees and the maximum shear stress. Compare with the corresponding values for the box section formed by splitting the web into two 5 mm plates. Take $G = 80$ GPa. (T.C.D. 1981)

9.57 Develop a relationship between torque and angle of twist for a closed uniform tube of thin-walled, non-circular section and use this to derive the twist per unit length for a strip of thin rectangular cross-section. Use the above relationship to show that for the same torque the ratio of angular twist per unit length for a closed square-section tube to that for the same section but opened by a longitudinal slit and free to warp is approximately $4t^2/3b^2$, where t, the material thickness is much less than the mean width b, of the cross-section. (C.E.I. Pt. 2, 1969.) (*Hint*: In the derivation, apply Batho to an elementary tube of thickness δy to find δT and $\delta\theta/\delta z$ noting that $y \ll b$.)

9.58 Calculate the twist in a 2 m length of the section shown in Fig. 9.46, when it is subjected to a torque of 100 N m. Find the position and magnitude of τ_{max}. Take $G = 30$ GPa. (K.P. 1975)

9.59 Determine the torsion constant for the thin-walled section shown in Fig. 9.47. If the maximum shear stress is limited to 30 MPa, determine the angular twist over a length of 1.5 m. Take $G = 30$ GPa.

Answer: 78×10^3 mm^4; 3.7°.

9.60 Determine the warping displacements at the four corners of the box section in Fig. 9.48 when it is subjected to a pure torque of 100 N m. Take $G = 27$ GPa.

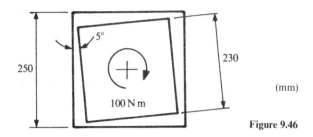

(mm)

Figure 9.46

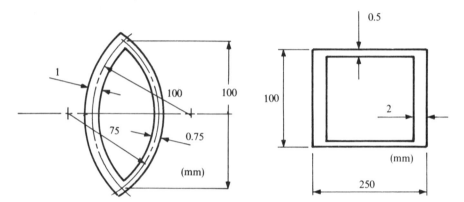

Figure 9.47

Figure 9.48

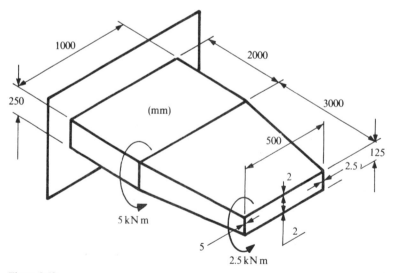

Figure 9.49

9.61 The two-bay alloy wing structure in Fig. 9.49 is subjected to concentrated torques at the two positions shown. Calculate the position and magnitude of the maximum shear stress in the skin when the given free-end thicknesses are uniform throughout. Take $G = 28$ GPa. (K.P. 1975)

9.62 Figure 9.50 shows the cross-section of an aircraft fuselage. Calculate the shear stress for each thickness of section and the angular twist over a 25 m length for an applied torque of 2 MN m. How would the addition of a luggage hold (dotted), to make a two-cell tube, affect the shear stresses? Take $G = 30$ GPa.

 Answer: 197.8 MPa; 98.9 MPa; 49.8 MPa; 3.34°.

9.63 The aluminium alloy wing section in Fig. 9.51 is subjected to a pure torque of 100 kN m over its 10 m length. Calculate the angle of twist and the shear stress in each thickness of section assuming that it is free to warp. Examine the effect on the stress and angular twist when a 1 mm thick web of the same material is placed in the dotted position shown. Take $G = 30$ GPa.

 Answer: 10°; 188.5 MPa, 94.3 MPa single cell; 0.31°, 86.07 MPa, 34.8 MPa two-cell. (K.P. 1976)

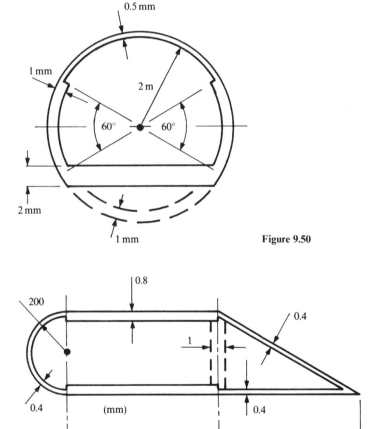

Figure 9.50

Figure 9.51

9.64 Figure 9.52 shows the uniform cross-section of a 5 m long tailplane. If the tube is subjected to a torque of 50 kN m, determine the shear stress in each of the webs and the angle of twist, assuming that the centre of twist lies at the centroid. Take $G = 30$ GPa.

Answer: 55.56 MPa; 111.11 MPa; 37.04 MPa; 4.39°.

9.65 The closed, symmetric double-wedge section shown in Fig. 9.53 is made from different thicknesses of steel (t_s) and aluminium (t_a). The section is that of a 1.25 m long cantilever that

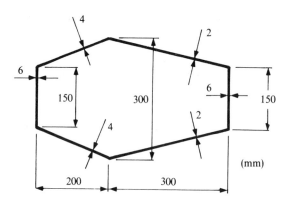

Figure 9.52

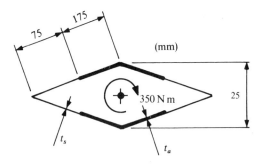

Figure 9.53

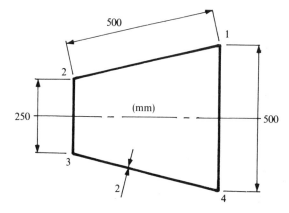

Figure 9.54

is to support a torque of 350 N m applied at the free end without the twist exceeding 0.035 rad. If the maximum shear stress in the steel is also restricted to 18 MPa, determine t_a and t_s. Take $G_s = 3G_a = 81$ GPa.

Answer: $t_a = 2.56$ mm; $t_s = 1.55$ mm.

9.66 Figure 9.54 shows the basically similar cross-sections of two thin-walled uniformly thick tubes that differ in that one has a closed section while the other is open, being slit axially at the centre of web 2–3. Each tube is subjected to a constant torque and is constrained to twist about an axis passing through the middle of web 1–4. Calculate the warping of each cross-section in terms of the rate of twist, illustrating the distribution with a sketch on which principal values are indicated. Assume that the shear stress in the closed tube is given by Batho and in the open tube by St Venant. (C.E.I. Pt. 2, 1973)

9.67 The long tube in Fig. 9.55 is subjected to a torque of 200 N m. The tube has a double-cell, thin-walled effective cross-section. Assuming that no buckling occurs, and that the twist per unit length of the tube is constant, determine the maximum shear stress in each wall of the tube. (C.E.I. Pt. 2, 1972)

9.68 Determine the position and magnitude of the maximum shear stress and the rate of twist when the two-cell tube in Fig. 9.56 is subjected to a pure torque of 900 N m. Take for the vertical web $G = 81$ GPa, and for the remaining material $G = 27$ GPa.

Answer: 12.025 MPa; 0.26°/m. (K.P. 1977)

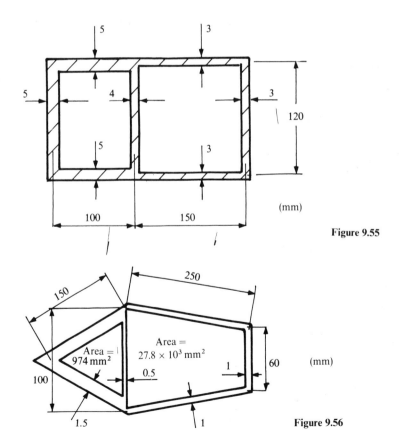

Figure 9.55

Figure 9.56

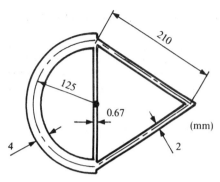

Figure 9.57

9.69 The two-cell tube in Fig. 9.57 supports a torque of 2 kN m. Calculate the position and magnitude of the maximum shear stress and rate of twist. What web thickness would ensure zero shear stress? Compare torsional stiffnesses per metre length for sections with and without the web. Take $G = 79$ GPa.

Answer: 10.14 MPa; 0.0522°/m; 38.31 kN m/degree; 37.07 kN m/degree.

Torsion of Prismatic Bars

9.70 Show that the Prandtl stress function $\phi = K(x^2/R^2 + y^2/R^2)$ defines the shear stress in a solid circular section of radius R. Confirm ETT and the absence of warping.

9.71 The function $\phi = G(\delta\theta/\delta z)(t^2/4 - y^2)$ represents torsion of a thin rectangular strip where $b \gg t$. Determine the maximum shear stress under a torque of 2 N m for $b = 50$ mm and $t = 2$ mm. How is the theory altered when t is increased to 10 mm? Take $G = 80$ GPa.

Answer: 30 MPa.

9.72 Find the rate of twist under a torque of 175 N m for a solid shaft of elliptical section with major and minor diameters $2a = 50$ mm and $2b = 35$ mm respectively. Assume the Prandtl stress function $\phi = D(x^2/a^2 - y^2/b^2 - 1)$, where D is a constant and take $G = 80$ GPa.

9.73 Show that the Prandtl stress function $\phi = A[(x^2 + y^2)/2 - (x^3 - 3xy^2)/2a - 2a^2/27]$ provides a solution to the torsion of a solid equilateral triangular section of height a. Derive from the constant A the torsion constant J and plot the distribution of shear stress τ_{yz} along this height, locating the maximum value. If the side length is 50 mm, determine the angle of twist on a 2 m length under a pure axial torque of 100 N m. Take $G = 80$ GPa.

9.74 Derive the warping function for the solid triangular section in Exercise 9.73 and hence establish the manner in which this cross-section warps.

Answer: $\psi = (\delta\theta/\delta z)(y^3 - 3x^2 y)/2a$.

Plastic Torsion

9.75 Determine the ratio T_{ep}/T_E for a solid circular shaft when the plastic zone has penetrated through the section to one-half the radial depth. If $k = 150$ MPa, determine the shear strain and the rate of twist on a 25 mm diameter.

Answer: 1.29.

9.76 Examine graphically the variation in T_{ep}/T_E with θ_{ep}/θ_E in the range $4 \le \theta_{ep}/\theta_E \le 0$ for elastic-plastic torsion of a solid circular bar. Note that here θ is the twist/unit length (unit twist) and θ_E is the external diameter unit twist value at the start of yielding.

9.77 Find the torque that will just initiate yielding in a 25 mm diameter solid bar given that the shear yield stress is 175 MPa.

9.78 A solid, circular steel shaft 100 mm in diameter is subjected to a torque of 30 kN m. If the shear yield stress is 125 MPa, find the fully plastic torque and show that the shaft is in an elastic-plastic condition. Determine the angular twist over a 3 m length and find the radial depth of plastic penetration. What is the maximum residual stress in the shaft corresponding to the fully unloaded condition? Take $G = 82$ GPa.

Answer: 6.5°; 10.5 mm.

9.79 Determine the torque required to produce an elastic-plastic interface at the mean radius of a hollow tube of 25 mm i.d. and 100 mm o.d. Find the corresponding residual angle of twist/unit length and the stress distribution that remains upon removal of the torque. Take $k = 225$ MPa and $G = 78$ GPa.

9.80 A solid circular shaft 90 mm in diameter is drilled to 40 mm diameter for one half of its length. If, under an applied torque, the outer fibres of the solid shaft have reached the yield point, find the depth of yielding in the tubular section and the ratio between the angular twist per unit length of each section.

9.81 Give the principal reason for pre-straining of torsion bars and state why care has to be taken when this process is applied to the torsion bar suspension systems of motor vehicles. A cylindrical torsion bar measuring 20 mm in diameter and 0.85 m long is pre-strained by being twisted through an angle of 1.50 radians before being released. Assuming that strain-hardening effects may be neglected, (i) sketch, but do not calculate, the residual stress distribution across the cross-section of the bar, and (ii) calculate the depth to which yielding takes place and the maximum torque required during pre-straining, given that the shear yield stress is 625 N/mm^2 and the modulus of rigidity is 80 kN/mm^2. (C.E.I. Pt. 2, 1975)

9.82 A solid shaft of non-hardening material is 0.25 m long and tapers uniformly between end diameters of 45 mm and 48 mm. Find the torque required to produce yielding at the outer surface of the smaller diameter. If this torque is then increased by 15 per cent, determine the length over which yielding has occurred at the outer surface and the radial depth of yielding at the smaller end.

9.83 The diameter of a tapered circular steel bar increases uniformly from 50 mm at one end to 55 mm at the other end over a length of 500 mm. Determine the torque that initiates yielding in the bar and that which produces full plasticity at the smaller end. What is the angular twist in each case and the length of surface yielding?

9.84 The dimensions of a bonded-composite tubular section are brass, 50 mm i.d., 75 mm o.d. and steel, 75 mm i.d., 100 mm o.d. Find the maximum torque that may be applied without causing yielding. What then are the maximum shear stresses? What magnitude of applied torque would cause the steel annulus to become fully plastic? For the brass and steel take shear yield stresses of 60 and 120 MPa, and rigidity moduli of 41 and 82 GPa respectively.

Answer: 19.097 kN m; 120 MPa; 45 MPa; 22.15 kN m.

9.85 A solid, cylindrical composite shaft 1 m long consists of a solid copper core of 50 mm diameter surrounded by a well-fitting steel sleeve of 65 mm external diameter. If the steel is elastic-perfectly plastic, determine the torque that will cause yielding to penetrate the shaft through to the common diameter. Calculate the residual shear stress at the outer steel diameter after removal of the torque. Neglect interference stresses and assume no slipping at the interface. For steel, take $G = 82.7$ GPa and $k = 124$ MPa; for copper, take $G = 44.8$ GPa and $k = 75.8$ MPa. (C.E.I. Pt. 2, 1968)

FLEXURAL SHEAR FLOW

10.1 INTRODUCTION

Up to now, the manner in which a material becomes stressed under the action of direct loading (i.e. tension, compression and shear), bending and torsional effects have been considered. A further action which induces shear stress is that of flexural shear. It was seen in Chapter 5 that a vertical shear force accompanied a bending moment in a loaded beam. The present chapter examines the manner in which the beam cross-section resists that shear force. This becomes an important additional design consideration for all beam sections that are employed in practice, e.g., I- and channel sections used in construction. Flexural shearing in a thin-walled section becomes the major issue where reduction in weight is of prime consideration as with an aircraft structure. Here the position at which the shear force is applied, the symmetry of the section, and whether it is open or closed, are the controlling factors. As with the concept of shear flow used in torsion of thin-walled sections (Chapter 9), the use of shear flow again simplifies the analysis of the shear stress distribution resulting from flexure of thin-walled structures.

10.2 SHEAR STRESS DUE TO SHEAR FORCE IN BEAMS

In Fig. 10.1(a) an element δz in the length of a beam is shown over which the shear force and bending moment vary by δF and δM respectively. These produce corresponding variations in the shear and bending stresses of $\delta \tau$

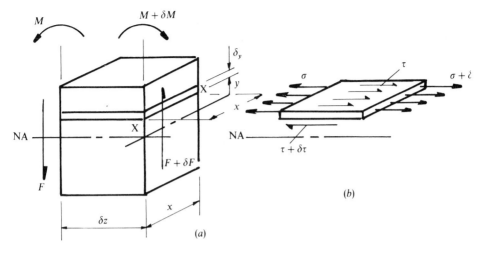

Figure 10.1 Flexural shear in a beam element.

and $\delta\sigma$, for any given slice, $\delta z \times \delta y \times x$, lying at a distance y, from the neutral axis as shown in Fig. 10.1(b).

In order to find an expression for the shear stress τ, acting along XX, it is only necessary to consider the horizontal equilibrium between its perpendicular complementary actions shown, and the bending stress σ. That is,

$$(\sigma + \delta\sigma)x\,\delta y + \tau x\,\delta z = \sigma x\,\delta y + (\tau + \delta\tau)x\,\delta z,$$
$$(\delta\sigma)x\,\delta y = (\delta\tau)x\,\delta z,$$
$$\therefore \quad \delta\tau = (\delta\sigma)x\,\delta y/x\,\delta z. \tag{10.1}$$

Now from ETB in Eq. (6.4), with the hogging action shown, $\delta\sigma = \delta M\,y/I$, where from Eq. (5.2), $\delta F = \delta M/\delta z$. Substituting into Eq. (10.1) gives

$$\delta\tau = (F/Ix)yx\,\delta y, \tag{10.2}$$

and therefore the shear stress acting along any fibre in the cross-section, parallel to the centroidal axis and distance y from it, is found from integrating Eq. (10.2). This gives

$$\tau = (F/Ix) \int_{y}^{y_1} yx(y)\,\mathrm{d}y. \tag{10.3}$$

Since the shear stress is zero along the top free surface the limits of y apply to the shaded area above the section. Thus, the lower limit y is the height in the section at which the shear stress is required, while upper limit y_1 defines the height of free surface. The width x outside the integral applies to the chosen section, while that within $x(y)$ accounts for variable breadth

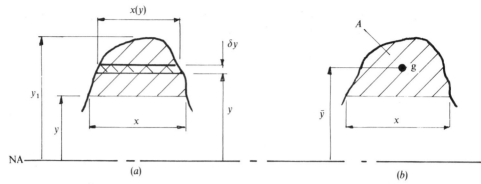

Figure 10.2 Effective section area in shear.

within the shaded area (Fig. 10.2(a)). For a section with constant breadth, the x's in Eq. (10.3) will cancel.

Recognizing that the integral in Eq. (10.3) is, by definition, the first moment of the shaded area A, about the centroidal axes in Fig. 10.2(b), leads to the equivalent alternative expression for τ:

$$\tau = F(A\bar{y})/Ix, \tag{10.4}$$

where $\bar{y}$ is the distance between the centroid of area A and the centroid of the whole cross-section as shown in Fig. 10.2(b). Note that I in Eq. (10.4) is the second moment of area for the whole cross-section for which positive F is a vertically downwards acting shear force.

Example 10.1 Show that the maximum shear stress in a rectangular section $b \times d$ (Fig. 10.3(a)) is given by $3F/2bd$ acting at the neutral axis. Plot the variation in τ across the depth. A simply supported timber beam of length $l = 3\,\mathrm{m}$ is to be designed to carry a uniformly distributed load $w = 18\,\mathrm{N/mm}$. Allowable stresses are $\sigma_{max} = 10\,\mathrm{MPa}$ for bending and

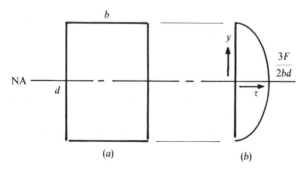

Figure 10.3 Rectangular section.

$\tau_{max} = 0.75\,\text{MPa}$, parallel to the grain, for shear. If the rectangular beam section has a nominal depth of $d = 300\,\text{mm}$, what width is required?

Substituting $I = bd^3/12$, $x = b$, $y_1 = d/2$ and leaving the lower limit as the variable y, Eq. (10.3) becomes

$$\tau = (12F/bd^3) \int_y^{d/2} y\,dy$$

$$= (6F/bd^3)|y^2|_y^{d/2} = (6F/bd^3)(d^2/4 - y^2).$$

This parabolic variation of τ with y is shown in Fig. 10.3(b). Clearly, the maximum value, $\tau_{max} = 3F/2bd$, occurs at the NA where $y = 0$. This result is readily confirmed from Eq. (10.4), where

$$\tau_{max} = \frac{F(bd/2)(d/4)}{(bd^3/12)b} = \frac{3F}{2bd}.$$

Now the maximum shear force and bending moment for the beam are $wl/2$ and $wl^2/8$ respectively (see Fig. 5.3(b)). Thus, a design based upon the allowable shear stress gives

$\tau_{max} = 3F_{max}/2bd = 3(wl/2)/2bd$

$\therefore\quad b = 3wl/4\tau_{max}d = 3 \times 18 \times 3 \times 10^3/(4 \times 0.75 \times 300) = 180\,\text{mm}.$

For a design based upon the allowable bending stress, the ETB gives

$$\sigma_{max} = M_{max}y/I = (wl^2/8)(d/2)/(bd^3/12),$$

$$b = 3wl^2/4\sigma_{max}d^2,$$

$$\therefore\quad b = 3 \times 18 \times (3 \times 10^3)^2/(4 \times 10 \times 300^2) = 135\,\text{mm}.$$

Hence shear governs the design.

Example 10.2 Determine the distribution of shear stress for the trapezoidal section in Fig. 10.4(a) when it is subjected to a vertical shear force of 20 kN.

The position of the neutral axis is found from taking first moments of area about the base. Then, if g is $\bar{Y}$ above the base,

$$(50 \times 25 \times 25) + 2(12.5 \times 50/2)(50/3) = [(25 \times 50) + (12.5 \times 50)]\bar{Y}$$

from which $\bar{Y} = 22.22\,\text{mm}$. Noting, for a right-angled triangle base b and height h, that $I = bh^3/12$ about the base, the total I about the base of the trapezoidal section becomes

$$I = (25 \times 50^3/3) + (2 \times 12.5 \times 50^3/12) = 1.3021 \times 10^6\,\text{mm}^4.$$

Transferring this to the centroid using the parallel axis theorem:

$$I_x = I - A\bar{Y}^2 = (1.3021 \times 10^6) - 1875(22.22)^2 = 37.636 \times 10^4\,\text{mm}^4.$$

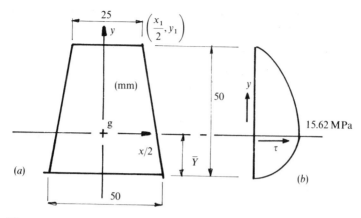

Figure 10.4 Trapezoidal section.

In applying Eq. (10.3) it is necessary to find the function $x(y)$ in this case. For the $x/2, y$-coordinates shown, the equation of the right sloping side passing through the top corner point $(x_1/2, y_1)$ is

$$y - y_1 = m(x/2 - x_1/2),$$
$$y - 27.78 = -(50/12.5)(x/2 - 12.5),$$
$$\therefore \quad x = 38.89 - y/2.$$

Substituting into Eq. (10.3) with $y_1 = 27.78$ mm,

$$\tau = (F/I_x x) \int_y^{27.78} (38.89 - y/2) y \, dy,$$

$$= (F/I_x x) | 38.89 y^2/2 - y^3/6 |_y^{27.78} \tag{i}$$
$$= (F/I_x x)[11\,433.16 - (38.89 y^2/2 - y^3/6)],$$

where $F/I_x = 20 \times 10^3/(37.636 \times 10^4) = 53.14 \times 10^{-3}$ N/mm^4. Then

$\tau = 0$ at the top free surface where $y = 27.78$ mm,

$\tau = 15.14$ MPa is an intermediate value when $y = 10$ mm, $x = 33.89$ mm,

$\tau = 15.62$ MPa is the maximum value at the centroidal axis where
 $y = 0$, $x = 38.89$ mm.

In applying the integral to the area beneath the centroidal axis, with $y_1 = -22.22$ mm;

$$\tau = (F/I_x x) \int_y^{-22.2} (38.89 - y/2) y \, dy$$

$$= (F/I_x x) | 38.89 y^2/2 - y^3/6 |_y^{-22.2}$$

This shows that Eq. (i) also applies to negative y. The variation across the depth is illustrated in Fig. 10.4(b), where

$\tau = 0$ at the bottom free surface when $y = -22.22$ mm,

$\tau = 11.29$ MPa is an intermediate value when $y = -10$ mm, $x = 43.89$ mm,

$\tau = 15.62$ MPa at the centroidal axis when $y = 0$, $x = 38.89$ mm.

Example 10.3 The section in Fig. 10.5 is subjected to a shear force of 30 kN. Find the maximum shear stress and the shear stress for the upper web–flange join.

Let the position of the NA be $\bar{Y}$ from the base. Then

$$(105 \times 10 \times 5) + (80 \times 5 \times 50) + (55 \times 10 \times 95)$$
$$= [(105 \times 10) + (80 \times 5) + (55 \times 10)]\bar{Y}, \quad \Rightarrow \bar{Y} = 38.75 \text{ mm}.$$

Using the $I = bd^3/3$ form for rectangles with their bases lying on the NA,

$$I = (105 \times 38.75^3/3) - (2 \times 50 \times 28.75^3/3) + (55 \times 61.25^3/3)$$
$$- (2 \times 25 \times 51.25^3/3) = 3.2135 \times 10^6 \text{ mm}^4.$$

Here the application of Eq. (10.4) is illustrated. The maximum shear stress is associated with the whole area above the NA. That is,

$$A\bar{y} = (55 \times 10 \times 56.25) + (5 \times 51.25 \times 25.625) = 37.504 \times 10^3 \text{ mm}^3,$$
$$\tau_{\text{max}} = F(A\bar{y})/Ix = 30 \times 10^3 \times 37.504 \times 10^3/(3.2135 \times 10^6 \times 5)$$
$$= 70.025 \text{ MPa}.$$

For the area above the web–flange join,

$$A\bar{y} = 55 \times 10 \times 56.25 = 30.938 \times 10^3 \text{ mm}^3.$$

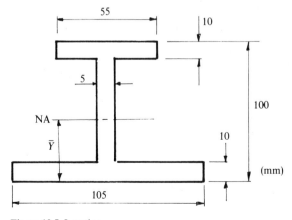

Figure 10.5 I-section.

The fact that the breadth x at the join is either 5 or 55 mm reveals how the stress discontinuity arises at this position. Thus, at the web top where $x = 5$ mm,

$$\tau = F(A\bar{y})/Ix = 30 \times 10^3 \times 30.938 \times 10^3/(3.2135 \times 10^6 \times 5) = 57.765 \text{ MPa}$$

and at the flange bottom where $x = 55$ mm, $\tau = 55.765 \times 5/55 = 5.25$ MPa. This is an average value, since free surfaces are unstressed.

Example 10.4 An I-section beam 300 mm (12 in) deep, 200 mm (8 in) wide with 25 mm (1 n) thick flanges and a 10 mm ($\frac{1}{2}$ in) thick web, sustains at a certain section a shear force F(N or tonf) and a bending moment M(Nm or tonf in). Derive expressions for the distribution of shear stress in the web and flange and show them diagrammatically when $F = 30$ kN (3 tonf). What percentage of F is carried by the web and what percentage of M is carried by the flanges?

SOLUTION TO SI UNITS

$$I = 200 \times 300^3/12 - 2 \times 95 \times 250^3/12 = 202.5 \times 10^6 \text{ mm}^4.$$

Applying Eq. (10.3) to a fibre in the top flange where $x(y) = 200$ mm, $y_1 = 150$ mm and $x = 200$ mm gives

$$\tau_f = (F/I \times 200) \int_y^{150} 200y \, dy = (F/I) \int_y^{150} y \, dy$$

$$= (F/I)|y^2/2|_y^{150} = (F/2I)[150^2 - y^2]. \tag{i}$$

The distribution is parabolic in y. When $F = 30 \times 10^3$ N in Eq. (i),

$$\tau_f = (0.7407 \times 10^{-4})[(2.25 \times 10^4) - y^2],$$
$$\tau_f = 0 \text{ at the flange top where } y = 150 \text{ mm},$$
$$\tau_f = 0.51 \text{ MPa at the flange bottom, where } y = 125 \text{ mm}.$$

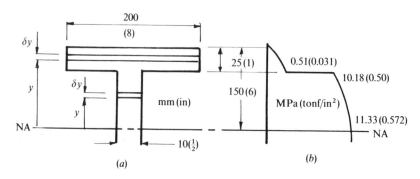

Figure 10.6 I-section under shear and bending.

In applying Eq. (10.3) to a fibre in the web above the NA, both the web and flange areas above this section contribute to τ. That is, with $x = 10$ mm:

$$\tau_w = (F/10I)\left[\int_{125}^{150} 200\, y\, dy + \int_{y}^{125} 10y\, dy\right]$$

$$= (F/10I)[200(150^2 - 125^2)/2 + 10(125^2 - y^2)/2]$$

$$= (F/I)[(7.6563 \times 10^4) - y^2/2]. \qquad (ii)$$

This distribution is again parabolic in y. When $F = 30 \times 10^3$ N in Eq. (ii),

$$\tau_w = (1.48 \times 10^{-4})[(7.6563 \times 10^4) - y^2/2],$$

$$\tau_w = 10.18\,\text{MPa with } y = 125\,\text{mm},$$

$$\tau_w = 11.33\,\text{MPa with } y = 0.$$

With the symmetrical top half-section in Fig. 10.6(a), these values are mirrored in the bottom half. Figure 10.6(b) shows the discontinuity in shear stress that arises at the junction between the flange and web. For the elemental web strip in Fig. 10.6(a), the shear force δF_w is

$$\delta F_w = 10\tau_w\, \delta y.$$

Substituting from Eq. (ii) gives

$$dF_w = 10(F/I)\int_{-125}^{125} [(7.6563 \times 10^4) - y^2/2]\, dy.$$

Integrating for the whole length of web gives

$$F_w = (10F/I)|(7.6563 \times 10^4)y - y^3/6|_{-125}^{125},$$

$$= (184.9 \times 10^6)(F/I)$$

$$F_w/F = 184.9/202.5 = 0.913.$$

That is, the web carries 91.3 per cent of the applied shear force. For the elemental flange strip shown in Fig. 10.6(a), the bending moment δM_f is

$$\delta M_f = (200\delta y)\sigma y = (200\, \delta y)(My/I)y.$$

Integrating this across each flange depth gives

$$M_f = 2 \times 200(M/I)\int_{125}^{150} y^2\, dy$$

$$= 2 \times 200(M/3I)[150^3 - 125^3] = 2(94.792 \times 10^6)M/I,$$

$$M_f/M = 2 \times 94.792/202.5 = 0.936.$$

That is, the flanges carry 93.6 per cent of the applied bending moment.

SOLUTION TO IMPERIAL UNITS

$$I = 8 \times 12^3/12 - 7.5 \times 10^3/12 = 527 \, \text{in}^4.$$

The shear stress in the flange is, from Eq. (10.3):

$$\tau_f = (F/I \times 8) \int_y^6 8y \, dy.$$

$$= (F/2I)(36 - y^2).$$

This gives $\tau_f = 0$ for $y = 6 \, \text{in}$ and $\tau_f = 0.031 \, \text{tonf/in}^2$ for $y = 5 \, \text{in}$ in Fig. 10.6(b). Now in the web the shear stress is, from Eq. (10.3):

$$\tau_w = (F/I \times 0.5)\left(\int_y^5 0.5y \, dy + \int_5^6 8y \, dy \right)$$

$$= (F/2I)(201 - y^2).$$

This gives $\tau_w = 0.5 \, \text{tonf/in}^2$ for $y = 5 \, \text{in}$ and $\tau_w = 0.572 \, \text{tonf/in}^2$ for $y = 0$ in Fig. 10.6(b). The shear force carried by the web is found from:

$$F_w = \int_{-5}^5 (\tfrac{1}{2})\tau_w \, dy = (F/4I)\int_{-5}^5 (201 - y^2) \, dy = 481.67 F/I.$$

$$\therefore \quad \% F_w/F = (481.67/527) \times 100 = 91.5\%.$$

The moment carried by the flanges is found from:

$$M_f = 2 \int_5^6 (My/I)8y \, dy = 0.92 M$$

$$\therefore \quad \% M_f/M = 92\%.$$

This example illustrates that I-sections exhibit the desirable combination of high strength and low weight; hence their common use as beams in structural work. It is shown in Section 10.6, from the fact that the web carries most of the shear force and the flange most of the bending moment, that a structural idealization can simplify the analysis of shear stress distributions.

10.3 THE SHEAR CENTRE AND FLEXURAL AXIS

It can be seen from the foregoing examples that when the line of action of the vertical shear force passes through the centroid there will be no twisting effect. If longitudinal twisting is to be avoided, for a section that is not symmetrical about a vertical axis, then the shear force must be displaced to pass through a point called the shear centre. This ensures that the shear stress distribution, which is statically equivalent to the applied shear force,

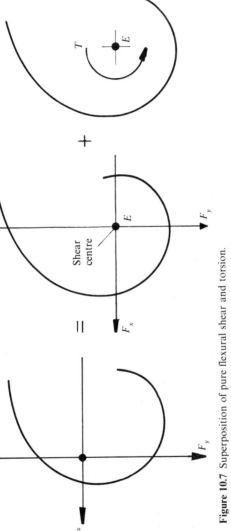

Figure 10.7 Superposition of pure flexural shear and torsion.

has zero moment about any point in the line of that force. Since the shear force usually varies with length in a loaded beam, twisting is avoided when the forces lie on a flexural axis that is the locus of the shear centres for all cross-sections. The shear centre is a property of the section not generally coincident with the centroid. It is seen from the above examples that the shear centre of a doubly or multiply symmetric section does lie at the centroid. The flexural axis is then the centroidal axis. The shear centre of a singly symmetric section lies on its axis of symmetry and the flexural axis will lie in the plane of symmetry. The shear centre will lie at the intersection of the limbs in T, angle and crucifix sections since this is the point of intersection between the force resultants of the shear stress distributions in those limbs.

Where, in practice, the applied shear forces are not concurrent with the shear centre, then shear stresses due to torsion as well as flexural shear are induced. The principle of superposition enables the net shear stress to be found from adding the separate effects of pure flexural shear and pure torsion as illustrated in Fig. 10.7. Note that shear stress due to torsion in the non-circular open section illustrated, has been dealt with previously in Chapter 9.

The following discussion illustrates how the shear flow distribution and the position of the shear centre may be found in the case of open and closed thin-walled sections.

10.4 SHEAR FLOW IN THIN-WALLED OPEN SECTIONS

Symmetrical Sections (x and y are principal axes)

Consider any thin-walled cantilever beam section, typically with curved sides and varying thickness (Fig. 10.8(a)), in which x and y are principal axes. Let transverse shear forces F_x and F_y both act through the shear centre E in the negative x- and y-directions as shown. At any point in the plane of the cross-section there will be a component of shear stress τ_{sz} parallel to the mid-section centre line, where the material is assumed to be concentrated, and one normal to this mid-line, τ_{yz}. Because the latter component is zero at the two free surfaces, τ_{yz} is normally ignored, while τ_{sz} is assumed to be uniform in sections where the thickness variation is small. Consequently, the concept of shear flow, $q = t\tau_{sz}$, will conveniently account for variations in τ_{sz} with t around the section in the positive s direction shown. Contrast this with the linear variation in torsional shear stress through the thickness of an open section (see Fig. 9.8).

Now, as mid-line complementary shear flow, $q = t\tau_{zs}$, also exists together with a bending stress σ for the z-direction, the equilibrium condition for an element $\delta s \times \delta z$ of wall in Fig. 10.8(b) becomes

$$[\sigma + (\partial\sigma/\partial z)\,\delta z](\delta s \times t) - \sigma(\delta s \times t) = [q + (\partial q/\partial s)\,\delta s]\,\delta z - q\,\delta z.$$

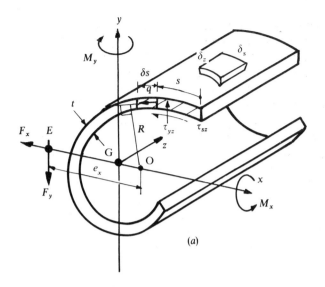

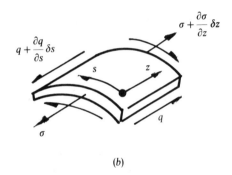

Figure 10.8 Open section in shear.

This leads to the following equilibrium equation, where q and σ increase with positive s and z respectively:

$$\partial q/\partial s = t\, \partial\sigma/\partial z. \tag{10.5}$$

With positive hogging moments M_x and M_y, in respect of the first x, y quadrant in Fig. 10.8(a), the ETB gives

$$\sigma = M_x y/I_x + M_y x/I_y. \tag{10.6}$$

Substituting Eq. (10.6) into Eq. (10.5) and integrating gives

$$q = (1/I_x) \int (dM_x/dz)(yt\, ds) + (1/I_y) \int (dM_y/dz)(xt\, ds). \tag{10.7}$$

Now, with the given origin for z in Fig. 10.8(a), the hogging moments for any section are $M_x = F_y z$, $M_y = F_x z$, therefore

$$dM_x/dz = F_y \quad \text{and} \quad dM_y/dz = F_x. \tag{10.8a, b}$$

That is, F_x and F_y are the positive shear forces associated with these hogging moments. Because Eqs (10.8) are general relationships (see page 95) Eq. (10.7) may further be applied to any beam loading where F is known at the section z. Substituting Eqs (10.8a and b) into Eq. (10.7) gives

$$q = (F_y/I_x)D_x + (F_x/I_y)D_y, \tag{10.9}$$

where $D_x = \int yt\,ds$ and $D_y = \int xt\,ds$ are the first moments of the shaded area about the x- and y-axes in Fig. 10.8(a) respectively. Positive q will result, for the given (positive) F_x and F_y directions, when the peripheral distance s is measured anticlockwise from the free surface. Equation (10.9) is not restricted to these positive force directions because the sign of q will indicate its true direction relative to the chosen direction for s. The appropriate term is omitted from Eq. (10.9) in the case of a symmetrical section under a single shear force F_x or F_y. The principle employed to find the shear centre E is that the moments due to q about any point in the plane of the section are statically equivalent to the moments due to the applied forces about that point. For example, with a moment centre O, on the horizontal axis of symmetry in Fig. 10.8(a):

$$F_y e_x = \int qR\,ds, \tag{10.10}$$

where R is the perpendicular distance of q from O. Clearly, the effect of F_x is eliminated when E and O here lie along the axis of F_x.

Asymmetric Sections (x and y are not principal axes)

When owing to the asymmetric nature of the cross-section, a single vertical shear force F does not coincide with a principal axis of area, Eq. (10.9) remains valid provided the shear force is applied through the shear centre. Equivalent shear force components $\hat{F}_x$ and $\hat{F}_y$ are then defined as for asymmetric deflection of beams. That is, from Eqs (7.28c) and (7.29c):

$$\hat{F}_x = -F/(I_x/I_{xy} - I_{xy}/I_y), \qquad \hat{F}_y = F/(1 - I_{xy}^2/I_x I_y).$$

In the more general case where both vertical (S_y) and horizontal (S_x) shear forces are applied to an asymmetric section in the negative x- and y-directions, the principle of superposition gives equivalent shear force components $\hat{F}_x$ and $\hat{F}_y$ in Eq. (10.9) as

$$\hat{F}_x = \frac{-S_y}{(I_x/I_{xy} - I_{xy}/I_y)} + \frac{S_x}{(1 - I_{xy}^2/I_x I_y)}$$

$$= \frac{S_x - (I_{xy}/I_x)S_y}{(1 - I_{xy}^2/I_x I_y)}, \tag{10.11a}$$

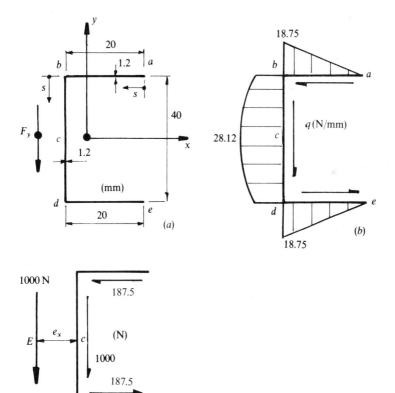

Figure 10.9 Shear flow in a channel section.

$$\hat{F}_y = \frac{S_y}{(1 - I_{xy}^2/I_xI_y)} - \frac{S_x}{(I_y/I_{xy} - I_{xy}/I_x)}$$

$$= \frac{S_x - (I_{xy}/I_y)S_x}{(1 - I_{xy}^2/I_xI_y)}. \tag{10.11b}$$

Example 10.5 Determine the shear flow and the maximum shear stress for the section in Fig. 10.9(a) when a downward shear force of 1 kN is applied parallel to the vertical side. Where must this force be applied if the section is not to be subjected to simultaneous torque?

This is an example of a singly symmetric section with a single force $F_y = +100\,\text{N}$ applied parallel to a principal axis (S_x and S_y are used for non-principal axes). Equation (10.9) then becomes, simply

$$q = (F_y/I_x)D_x,$$

where

$$I_x = (1.2 \times 40^3/12) + 2(20 \times 1.2 \times 20^2) = 25\,600\,\text{mm}^4$$
$$\therefore \quad F_y/I_x = 3.906 \times 10^{-2}\,\text{N/mm}^4.$$

For any point, distance s from a, in the flange ab

$$D_x = \int yt\,\mathrm{d}s = 20 \times 1.2 \int_0^s \mathrm{d}s = 24s,$$

$$\therefore \quad q_{ab} = 3.906 \times 10^{-2}(24s).$$

It follows that the variation in q_{ab} is linear, being zero at a and reaching a maximum value at b of

$$q_b = 3.906 \times 10^{-2} \times 24 \times 20 = 18.75\,\text{N/mm}.$$

The maximum shear stress in the top flange is then

$$\tau_b = q_b/t = 18.75/1.2 = 15.63\,\text{MPa}.$$

Working from b to point c ensures that y remains positive upwards. Now, for any point, distance s from b in the web, $y = 20 - s$ and the shear flow is

$$q_{bc} = (F_y/I_x)D_x + q_b$$

$$= (3.906 \times 10^{-2})1.2 \int_0^s (20 - s)\,\mathrm{d}s + 18.75$$

$$= 0.046\,87(20s - s^2/2) + 18.75.$$

The variation in q_{bc} is now parabolic with extreme values of

$$q_b = 18.75\,\text{N/mm} \qquad\qquad \text{for } s = 0 \text{ at } b,$$
$$q_c = 9.374 + 18.75 = 28.12\,\text{N/mm} \qquad \text{for } s = 20\,\text{mm at } c.$$

Beneath c, the shear flow mirrors that found above. This may be confirmed, either by continuing from c in an anticlockwise direction, or by working clockwise from the free surface at e. The resulting shear flow is that shown in Fig. 10.9(b), from which it follows that the greatest shear stress in the section occurs at c with magnitude $\tau_c = q_c t = 28.12 \times 1.2 = 23.43\,\text{MPa}$. The resultant applied shear forces along each side are then

$$F_{ab} = \int q_{ab}\,\mathrm{d}s = (F_y/I_x) \int_0^{20} 24s\,\mathrm{d}s$$

$$= 3.906 \times 10^{-2} \times 24(20^2/2) = 187.5\,\text{N} = F_{de},$$

$$F_{bc} = \int q_{bc}\,\mathrm{d}s = \int_0^{20} [0.046\,87(20s - s^2/2) + 18.75]\,\mathrm{d}s = 500\,\text{N},$$

$$\therefore \quad F_y = 2F_{bc} = 1000\,\text{N}.$$

The resultant forces are exerted by the cross-section in the same direction as q, shown in Fig. 10.9(c). If the section is not to be subjected to torque, then F_y must act at a distance e_x from the shear centre E such that the applied torque at any point in the plane of the section is balanced by the moment effects of the resultant forces due to q at that point. For convenience, take moments about point c so that Eq. (10.10) again applies:

$$1000e_x = (187.5 \times 20) + (187.5 \times 20),$$
$$\therefore \quad e_x = 7.5 \text{ mm}.$$

Example 10.6 Determine the shear flow distribution and the position of the shear centre for the thin-walled semi-circular section in Fig. 10.10(a) when it is loaded through the shear centre by shear forces F_x and F_y lying parallel to the principal axes x and y as shown.

Here both F_x and F_y are conventionally positive when acting in the negative x- and y-directions. It has previously been shown (page 14) that $I_x = I_y = \pi a^3 t/2$. Since the position of the centroid g (found from taking first moments of area about the Y-axis) is $\bar{X} = 2a/\pi$ from Y, it follows from parallel axes that

$$I_y = I_Y - A\bar{X}^2$$
$$= \pi a^3 t/2 - (\pi at)(2a/\pi)^2 = a^3 t(\pi/2 - 4/\pi).$$

Now let s and δs subtend angles θ and $\delta\theta$ at O as shown in Fig. 10.10(a).

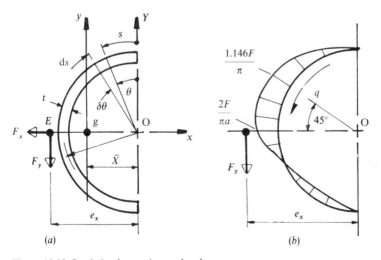

Figure 10.10 Semi-circular section under shear.

The first moments are then

$$D_x = \int_0^s ty\, ds = \int_0^\theta t(a\cos\theta)(a\, d\theta) = a^2 t \int_0^\theta \cos\theta\, d\theta = a^2 t \sin\theta$$

$$D_y = \int_0^s tx\, ds = \int_0^\theta t(2a/\pi - a\sin\theta)(a\, d\theta)$$

$$= a^2 t |2\theta/\pi + \cos\theta|_0^\theta = a^2 t(2\theta/\pi + \cos\theta - 1).$$

Substituting into Eq. (10.9) results in the following expression for q in terms of θ,

$$q = (2F_y/\pi a)\sin\theta + 2\pi F_x(2\theta/\pi + \cos\theta - 1)/a(\pi^2 - 8),$$

which is positive in the direction of s, with a typical polar variation around the periphery shown in Fig. 10.10(b) for $F_x = F_y = F$. The shear flow distribution is statically equivalent to the two applied shear forces provided that they act through the shear centre E, which in this case will lie along the x-axis, say distance e_x from O. Thus, by taking moments about O it is only necessary to consider moment equivalence between F_y and its associated shear flow q. That is $q = (2F_y/\pi a)\sin\theta$ acting in that sense given in Fig. 10.10(b). Then, from Eq. (10.10),

$$F_y e_x = \int qa\, ds,$$

where the integral is applied around the periphery. Substituting $\delta s = a\delta\theta$, the integral is conveniently applied between limits of 0 and π rad to give

$$F_y e_x = \int_0^s qa\, ds = a^2 \int_0^\theta q\, d\theta$$

$$= (2aF_y/\pi) \int_0^\pi \sin\theta\, d\theta = -(2aF_y/\pi)|\cos\theta|_0^\pi = 4aF_y/\pi,$$

$$\therefore \quad e_x = 4a/\pi.$$

Example 10.7 Determine the shear flow distribution and the position of the shear centre for the 1.5 mm thick extrusion in Fig. 10.11(a) when it is subjected to a downward vertical force of 1 kN.

Since x is a principal axis of symmetry and F_x is absent, the problem becomes that of applying $q = (F_y/I_x)D_x$ to each leg of the section identified as 1–5 in Fig. 10.11(a). Now:

$F_y = +1000$ N (acting in the negative y-direction),

$I_x = 2[(20 \times 1.5)35^2 + (15 \times 1.5)25^2] + (1.5 \times 70^3/12) = 144.5 \times 10^3$ mm^4.

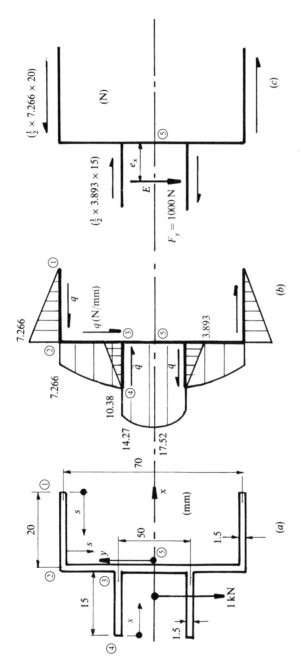

Figure 10.11 Shear flow distribution in an extruded section.

First establish the $D_x = \int yt \, ds$ distribution, which is later converted to q with the multiplying factor $F_y/I_x = 6.92 \times 10^{-3} \, \text{N/mm}^4$.

For flange $1 \rightarrow 2$, $y = 35 \, \text{mm}$, $t = 1.5 \, \text{mm}$:

$$D_x = (35 \times 1.5) \int_0^s ds = 52.5s,$$

At 1, $\quad s = 0, \quad \therefore \quad D_{x1} = 0,$

At 2, $\quad s = 20 \, \text{mm}, \quad \therefore \quad D_{x2} = (52.5 \times 20) = 1050 \, \text{mm}^3.$

For web $2 \rightarrow 3$, $y = 35 - s$, $t = 1.5 \, \text{mm}$:

$$D_x = 1.5 \int_0^s (35 - s) \, ds + D_{x2} = 1.5(35s - s^2/2) + 1050,$$

At 3, $\quad s = 10 \, \text{mm}, \quad \therefore \quad D_{x3} = 1500 \, \text{mm}^3.$

For flange $4 \rightarrow 3$, $y = 25 \, \text{mm}$, $t = 1.5 \, \text{mm}$:

$$D_x = (25 \times 1.5) \int_0^s ds = 37.5s,$$

At 4, $\quad s = 0, \quad \therefore \quad D_{x4} = 0,$

At 3, $\quad s = 15 \, \text{mm}, \quad \therefore \quad D_{x3} = (37.5 \times 15) = 562.5 \, \text{mm}^3,$

$$\sum D_{x3} = 1500 + 562.5 = 2062.5 \, \text{mm}^3.$$

For web $3 \rightarrow 5$, $y = 25 - s$, $t = 1.5 \, \text{mm}$:

$$D_x = 1.5 \int_0^s (25 - s) \, ds + \sum D_{x3} = 1.5(25s - s^2/2) + 2062.5,$$

At 5, $\quad s = 25 \, \text{mm}, \quad \therefore \quad D_{x5} = 2531.25 \, \text{mm}^3.$

The corresponding q distribution is that given in Fig. 10.11(b). To find the shear centre let F_y act at distance e_x from the web and take moments about point 5. This will eliminate the torque produced by q in the web. The resultant coupling forces are those given in Fig. 10.11(c), from which Eq. (10.10) becomes

$$1000e_x = (0.5 \times 7.266 \times 20 \times 70) - (0.5 \times 3.893 \times 15 \times 50),$$

$$\therefore \quad e_x = 3.626 \, \text{mm}.$$

Example 10.8. Derive an expression for the shear flow distribution along the mid-line of the equal angle section in Fig. 10.12(a).

Since x and y are not principal axes, the equivalent forces $\hat{F}_x$ and $\hat{F}_y$ are used with Eq. (10.9).

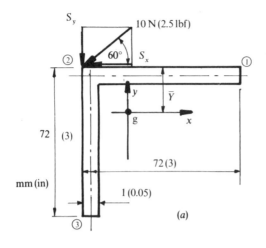

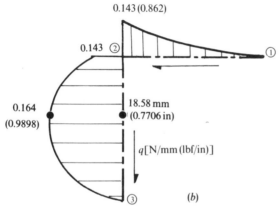

Figure 10.12 Shear flow in an equal-angle section.

The centroid position g is found from

$$(71 + 72)\bar{Y} = (72 \times 1 \times 1/2) + (71 \times 1 \times 36.5), \quad \Rightarrow \bar{Y} = \bar{X} = 18.37 \, \text{mm}.$$

Then

$$I_x = I_y = (72 \times 17.87^2) + (71 \times 18.13^2) + (1 \times 71^3/12) = 76.156 \times 10^3 \, \text{mm}^4,$$
$$I_{xy} = (72 \times 17.63 \times 17.87) + [71(-18.13)(-17.87)] = 45.94 \times 10^3 \, \text{mm}^4.$$

The 10 N force is applied at the shear centre, giving

$$S_x = 10 \cos 60° = +5 \, \text{N (in the } -x\text{-direction)},$$
$$S_y = 10 \sin 60° = +8.66 \, \text{N (in the } -y\text{-direction)}$$

when, from Eqs (10.11),

$$\hat{F}_x = \frac{5 - 8.66(45.94/76.16)}{1 - (45.94/76.16)^2} = -0.352 \, \text{N},$$

$$\hat{F}_y = \frac{8.66 - 5(45.94/76.16)}{1 - (45.94/76.16)^2} = +8.872 \, \text{N}$$

$$\hat{F}_x/I_y = -4.625 \times 10^{-6} \, \text{N/mm}^4, \qquad \hat{F}_y/I_x = 116.5 \times 10^{-6} \, \text{N/mm}^4.$$

For the flange $1 \to 2$, $y = 17.87 \, \text{mm}$, $x = 53.63 - s$, $t = 1 \, \text{mm}$. Equation (10.9) becomes

$$q = (\hat{F}_y/I_x) \int_0^s 17.87 \, ds + (\hat{F}_x/I_y) \int_0^s (53.63 - s) \, ds$$

$$= (116.5 \times 10^{-6} \times 17.87)s - (4.625 \times 10^{-6})(53.63s - s^2/2),$$

$$q = (1833.82s + 2.3128s^2) \times 10^{-6} \, \text{N/mm}.$$

The distribution is therefore parabolic, varying from zero to a maximum value of $0.143 \, \text{N/mm}$ at point 2 ($s = 71.5 \, \text{mm}$).

For the web $2 \to 3$, $y = -(s - 17.87)$, $x = -17.87 \, \text{mm}$, $t = 1 \, \text{mm}$. Equation (10.9) becomes

$$q = (\hat{F}_y/I_x) \int_0^s -(s - 17.87) \, ds + (\hat{F}_x/I_y) \int_0^s -17.87 \, ds + q_2,$$

$$= (116.5 \times 10^{-6})(17.87s - s^2/2) + (4.625 \times 10^{-6} \times 17.87)s + 0.143,$$

$$q = 0.143 + (2164.51s - 58.25s^2) \times 10^{-6} \, \text{N/mm}.$$

The web q distribution is again parabolic, varying from $+0.143 \, \text{N/mm}$ at 2 to a maximum of $0.164 \, \text{N/mm}$ at $s = 18.58 \, \text{mm}$. At the free surface 3, where $s = 71.5 \, \text{mm}$, $q_3 \simeq 0$. Figure 10.12(b) shows these shear flow variations graphically. Note that point 2 is the shear centre for an equal-angle section.

SOLUTION TO IMPERIAL UNITS For the centroid $g(\bar{X}, \bar{Y})$:

$$(3 \times 0.05 \times 0.025) + (2.95 \times 0.05 \times 1.525) = 0.2975\bar{X},$$

$$\therefore \quad \bar{X} = \bar{Y} = 0.769 \, \text{in}.$$

$$I_x = I_y = 3 \times 0.05 \times (0.744)^2 + (0.05 \times 2.95^3/12)$$

$$+ (2.95 \times 0.05)(2.231 - 1.475)^2 = 0.274 \, \text{in}^4,$$

$$I_{xy} = (3 \times 0.05 \times 0.731 \times 0.744) + (2.95 \times 0.05)(-0.756)(-0.744)$$

$$= 0.1645 \, \text{in}^4,$$

$$S_x = 2.5 \cos 60° = 1.25 \, \text{lbf}, \qquad S_y = 2.5 \sin 60° = 2.165 \, \text{lbf}.$$

The equivalent forces are, from Eq. (10.11):

$$\hat{F}_x = \frac{1.25 - (0.1645/0.274)2.165}{1 - (0.1645)^2/(0.274)^2} = -0.0779\,\text{lbf},$$

$$\hat{F}_y = \frac{2.165 - (0.1645/0.274)1.25}{1 - (0.1645)^2/(0.274)^2} = 2.212\,\text{lbf},$$

$$\therefore \quad \hat{F}_x/I_y = -0.2843\,\text{lbf/in}^4, \qquad \hat{F}_y/I_x = 8.072\,\text{lbf/in}^4.$$

The shear flow in the flange 1–2 is, from Eq. (10.9),

$$q = 8.072 \int_0^s (0.744 \times 0.05)\,\text{d}s - 0.2843 \int_0^s (2.231 - s)0.05\,\text{d}s$$

$$= 0.3003s - 0.0142(2.231s - s^2/2)$$

$$= 0.2686s + 0.0071s^2 \quad (\text{lbf/in}).$$

At point 2, where $s = 2.975\,\text{in}$, $q_2 = 0.862\,\text{lbf/in}$.

For the web 2–3, the shear flow is

$$q = 8.072 \int_0^s -(s - 0.744)0.05\,\text{d}s - 0.2843 \int_0^s -(0.744 \times 0.05)\,\text{d}s + 0.87$$

$$= 0.4036(0.744s - s^2/2) + 0.010\,58s + 0.862$$
$$= 0.862 + 0.311s - 0.2018s^2.$$

The corresponding flange and web shear flows are again shown in Fig. 10.12(b). As a check, when $s = 2.975\,\text{in}$, $q \approx 0$ at point 3.

10.5 SHEAR FLOW IN THIN-WALLED CLOSED SECTIONS

Here again, pure flexural shear flow arises when only transverse shear forces act through shear centre. In general, when these forces do not act through this centre, the resulting shear flow will be composed of flexural and torsional components. This is usually dealt with by a more convenient method than that employed for an open section in Fig. 10.7. Since a closed section has no free surface for which $q = 0$, a modification to Eq. (10.9) is necessary for direct calculation of the shear flow q resulting from any shear loading. Thus, in Fig. 10.13(a), with the two components of shear force S_x and S_y applied to an asymmetric section through an arbitrary point P, Eq. (10.9) becomes

$$q = q_b + q_o, \tag{10.12a}$$

where

$$q_b = (\hat{F}_x/I_y)D_{yb} + (\hat{F}_y/I_x)D_{xb}, \tag{10.12b}$$

in which $\hat{F}_x$ and $\hat{F}_y$ are defined in Eqs (10.11) and q_o is equal to the constant value of q at the origin for s. Both the position of this origin and the position of point P will govern the magnitude of q_o. With a fixed origin for s, the torsional effects are accounted for solely in the effect that the position of P has on q_o. Now q_b is actually the pure flexural shear flow for the equivalent tube when opened at $s = 0$. This must correspond to a shift in the position of applied forces S_x and S_y to become concurrent with the shear centre for the opened section. In order to restore the original loading, a torque must also act together with the translated forces at the shear centre of the open section. It follows that q_o is effectively the constant shear flow produced by this torque. The reasoning becomes analogous to that for Fig. 10.7, recognizing that different opening positions are associated with particular q_o values, because of the resulting change in position of the shear centre. The method amounts to converting a structure with a single degree of redundancy into a statically determinate one. A similar appoach may be employed to determine the shear flow distribution in a closed, multicell tube (see page 491). Note that, in order to ensure that the following elastic shear flow analysis is valid, it becomes necessary to employ a yield criterion (see page 711).

Initially, it is convenient to assume that $q_o = 0$ for $s = 0$, in the determination of the q_b distribution. The value of q_o is later determined from the equilibrium condition that the moments produced by the net q and the applied shear forces are statically equivalent for any point O in the plane of the cross-section. That is, for the given directions in Fig. 10.13(b):

$$S_y p_x - S_x p_y = \oint qR \, ds.$$

Substituting from Eq. (10.12a),

$$S_y p_x - S_x p_y = \oint q_b R \, ds + q_o \oint R \, ds, \tag{10.13}$$

where $\oint R \, ds$ is twice the area enclosed by the mid-line of the wall section. When point O coincides with P, the L.H.S. of Eq. (10.13) is zero.

A problem arises in the determination of q_o by this method when it is known that the forces S_x and S_y are applied at the shear centre E in Fig. 10.13(b), since this position will initially be unknown. It then becomes necessary to employ the additional compatibility condition that the rate of twist will be zero. A similar derivation to that which led to Eq. (9.31b) again applies, but we must note that q now varies with s. This gives

$$d\theta/dz = (1/2AG) \oint q \, ds/t. \tag{10.14}$$

For pure flexural shear q_e becomes the particular q_o value in Eq. (10.12). Substituting into Eq. (10.14) and equating to zero gives

$$\oint q_b \, ds/t + q_e \oint ds/t = 0. \tag{10.15}$$

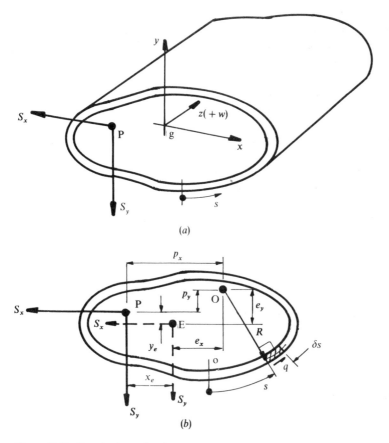

Figure 10.13 Closed tube under shear.

Once q_e is found from Eq. (10.15), the position of the shear centre E, is found by taking moments about any convenient point O. That is, with the corresponding coordinates e_x and e_y in Fig. 10.13(b),

$$S_y e_x - S_x e_y = \oint q_b R \, ds + q_e \oint R \, ds. \tag{10.16}$$

When E lies on a horizontal axis of symmetry, and O is chosen to lie along this axis, then $e_y = 0$ and e_x may be determined solely from S_y (or F_y for a symmetric section).

The unrestrained warping that accompanies twisting of the section, when shear forces are applied away from the shear centre, may readily be deduced from the prevous derivation leading to Eq. (9.34). Replacing the Bredt–Batho shear flow $q = T/2A$ in this equation by the net shear flow distribution $q = q(s)$

gives the relative warping displacement:

$$\Delta w = \int_0^s q \, ds/Gt - (A_{os}/A) \oint q \, ds/Gt, \qquad (10.17)$$

where A_{os} is the area enclosed between the centre of twist (normally coincident with the shear centre) a peripheral origin o, that may itself warp, and the chosen point on the periphery, distance s from o. If, however, there is no warping for $s = 0$, usually at a point of symmetry, then Eq. (10.17) will supply absolute w-values. Note that when warping is prevented at an encastre support, a system of self-equilibrating induced forces further modifies the shear flow distribution (see T.H.G. Megson, *Aircraft Structures For Engineering Students*, Edward Arnold, 1972).

Example 10.9 The thin-walled tubular section $r = 50$ mm, $t = 1$ mm in Fig. 10.14(a) is subjected to a vertical shear force of $F_y = 500$ N as shown. Determine the shear flow distribution, the maximum shear stress, the rate of twist and the free warping distribution. Take $G = 80$ GPa.

The section is symmetric with F_y positive and $F_x = 0$. Hence, from Eq. (10.12), the net shear flow is

$$q = (F_y/I_x)D_{xb} + q_o.$$

Now, as $y = r \sin \theta$ and $ds = r \, d\theta$ in Fig. 10.14(a), then

$$I_x = 2 \int_0^{\pi r} y^2 t \, ds = 2r^3 t \int_0^{\pi} \sin^2 \theta \, d\theta = \pi r^3 t,$$

$$D_{xb} = \int_0^s yt \, ds = r^2 t \int_0^{\theta} \sin \theta \, d\theta = r^2 t(1 - \cos \theta),$$

$$\therefore \quad q_b = (F_y/I_x)D_{xb} = (F_y/\pi r)(1 - \cos \theta).$$

From Eq. (10.13), with the moment centre O at the centre of the circle,

$$F_y r = (F_y r/\pi) \oint (1 - \cos \theta) \, d\theta + 2(\pi r^2)q_o = 2F_y r + 2\pi r^2 q_o,$$

$$\therefore \quad q_o = -F_y/2\pi r.$$

Hence the net shear flow, $q = q_o + q_b$, becomes,

$$q = (F_y/\pi r)(1 - \cos \theta) - F_y/2\pi r = (F_y/2\pi r)(1 - 2\cos \theta).$$

This has the distribution in Fig. 10.14(b). For the given r, t and F_y values, the greatest q occurs at $\theta = \pi$ rad giving $\tau_{max} = q/t = 4.775$ MPa. Clearly, here, the centroid O is the shear centre. The section will therefore twist

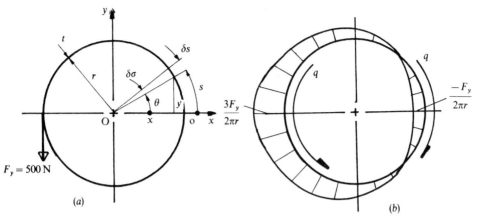

Figure 10.14 Thin-walled tubular section.

under F_y at the rate supplied by Eq. (10.14):

$$d\theta/dz = (1/2\pi r^2 G)(F_y/2\pi r)\oint (1 - 2\cos\theta)\,d\theta = F_y/2\pi r^2 Gt$$

$$= 500/(2\pi \times 50^2 \times 80\,000 \times 1) = 0.3978 \times 10^{-6}\,\text{rad/mm}.$$

Applying Eq. (10.17) to find the warping distribution with the given origin for s in Fig. 10.14(a):

$$\int_0^s q\,ds = (F_y/2\pi r)\int_0^\theta (1 - 2\cos\theta)r\,d\theta = (F_y/2\pi)(\theta - 2\sin\theta),$$

$$\oint q\,ds = (F_y/2\pi)|\theta - 2\sin\theta|_0^{2\pi} = F_y,$$

$$A_{os} = (1/2)\int_0^s r\,ds = (r^2/2)\int_0^\theta d\theta = r^2\theta/2,$$

$$\therefore \quad Gt\,\Delta w = (F_y/2\pi)(\theta - 2\sin\theta) - (r^2\theta/2\pi r^2)F_y,$$

$$\Delta w = -(F_y/\pi Gt)\sin\theta,$$

which shows that all points on the periphery displace axially relative to $+z$ in Fig. 10.13(a) excepting those on the horizontal diameter.

Example 10.10 Find the pure flexural shear flow distribution and the shear centre for the thin-walled, semi-circular, closed tube of uniform thickness t and radius a in Fig. 10.15(a)

With F_y applied through the shear centre, Eq. (10.12) is written as

$$q = q_b + q_e. \tag{i}$$

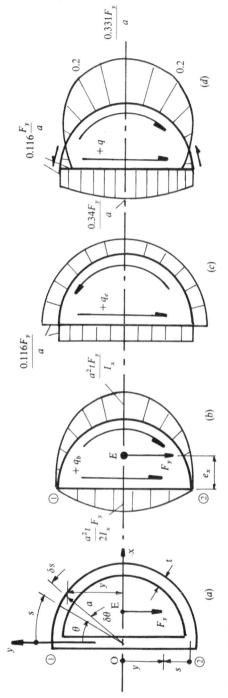

Figure 10.15 Thin-walled closed tube.

Since, for this singly symmetric section, $F_x = 0$ and x is a principal axis,

$$q_b = (F_y/I_x)D_{xb}, \qquad \text{(ii)}$$

$$I_x = \pi a^3 t/2 + t(2a)^3/12 = a^3 t(\pi/2 + 2/3) \qquad \text{(see page 14)}.$$

With the origin at 1 and s measured clockwise from $1 \to 2$ in the semi-circular portion,

$$D_{xb} = \int_0^s yt\,ds = \int_0^\theta (a\cos\theta)t(a\,d\theta) = a^2 t \int_0^\theta \cos\theta\,d\theta = a^2 t \sin\theta.$$

At 1: $\theta = 0$, $D_{xb1} = 0$,

At 2: $\theta = \pi$ rad, $D_{xb2} = 0$,

At $\theta = \pi/2$ rad, $D_{xb} = a^2 t$, a maximum.

and for $2 \to 1$ in the vertical web,

$$D_{xb} = \int_0^s yt\,ds + D_{xb2} = \int_0^s (s-a)t\,ds + 0 = t|s^2/2 - as|_0^s = ts(s/2 - a).$$

At 2: $s = 0$, $D_{xb2} = 0$,

At 1: $s = 2a$, $D_{xb1} = 0$,

At $s = a$, $D_{xb} = -a^2 t/2$.

Correspondingly, the q_b shear flow distribution in Fig. 10.15(b) is found from Eq. (ii):

$$q_b = \frac{F_y \sin\theta}{a(\pi/2 + 2/3)} \qquad \text{for } 1 \to 2,$$

$$q_b = \frac{F_y s(s/2 - a)}{a^3(\pi/2 + 2/3)} \qquad \text{for } 2 \to 1.$$

Because the latter gives negative q_b for $2a < s < 0$, this q_b direction is reversed as shown in Fig. 10.15(b). Substituting for q_b in Eq. (10.15) to find q_e:

$$\oint q_b\,ds/t + q_e \oint ds/t = 0,$$

$$[F_y/at(\pi/2 + 2/3)]\int_0^\pi \sin\theta(a\,d\theta) + [F_y/a^3 t(\pi/2 + 2/3)]\int_0^{2a} (s^2/2 - as)\,ds$$

$$+ q_e\left[\int_0^\pi (a\,d\theta)/t + \int_0^{2a} ds/t\right] = 0,$$

$$\frac{2F_y}{t(\pi/2 + 2/3)} - \frac{2F_y}{3t(\pi/2 + 2/3)} + \frac{q_e a(\pi + 2)}{t} = 0,$$

$$\therefore \quad q_e = -\frac{F_y}{a(1 + 3\pi/4)(1 + \pi/2)} = -\frac{0.116 F_y}{a},$$

which represents a constant shear flow around the section shown in Fig. 10.15(c). For the semi-circular portion, the net shear flow in Fig. 10.15(d) is found from Eq. (i):

$$q = \frac{F_y \sin \theta}{a(\pi/2 + 2/3)} - \frac{F_y}{a(1 + 3\pi/4)(1 + \pi/2)}$$

$$= (F_y/a)(0.447 \sin \theta - 0.116).$$

For the web:

$$q = \frac{F_y s(s/2 - a)}{a^3(\pi/2 + 2/3)} - \frac{F_y}{a(1 + 3\pi/4)(1 + \pi/2)}$$

$$= (F_y/a)[0.447(s/a^2)(s/2 - a) - 0.116].$$

It is seen that, provided the direction chosen for s remains the same in the determination of q_b and q_e, then q follows from their direct addition. Now apply Eq. (10.16) to find the shear centre with F_x and F_y replacing S_x and S_y. Taking moments about the web centre O, in Fig. 10.15(d) eliminates F_x (when present) and q in the web to give

$$F_y e_x = \int_0^s qa \, ds$$

$$= [F_y/a(\pi/2 + 2/3)] \int_0^\pi \sin \theta (a^2 \, d\theta)$$

$$- [F_y/a(1 + 3\pi/4)(1 + \pi/2)] \int_0^\pi a^2 \, d\theta,$$

$$F_y e_x = 2aF_y/(\pi/2 + 2/3) - \pi a F_y/(1 + \pi/2)(1 + 3\pi/4),$$

$$\therefore \quad e_x = a(3 + \pi/2)/(1 + \pi/2)(1 + 3\pi/4) = 0.53a.$$

Note that it not necessary to calculate the x-position of the centroid. Moreover, if only the shear centre is asked for, the magnitude of the applied shear forces are not required and the shear flow in the web need be considered only as far as it is necessary to establish q_e.

Example 10.11 Calculate the distribution of shear flow, the rate of twist and the shear centre (w.r.t. the longer vertical web) when the wing box structure in Fig. 10.16(a) is shear loaded as shown. Take $G = 27\,\text{GPa}$.

Here x is a principal axis of symmetry, so that

$$q_b = (F_y/I_x)D_{xb},$$

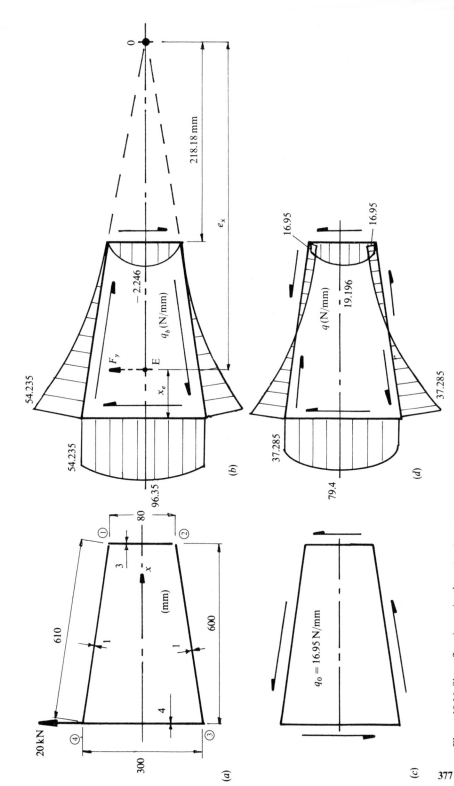

Figure 10.16 Shear flow in a wing box structure.

377

where $F_y = -20\,000$ N and, from Example 1.10, with $\sin \alpha = 110/610$,

$$I_x = 4(300)^3/12 + 3(80)^3/12 \quad \text{(for web)}$$
$$+ (1 \times 610)[80^2/2 + (80 \times 610)(110/610) + (2 \times 610^2/3)(110/610)^2]$$
$$\text{(for sides)}$$

$$= (9 + 0.128 + 12.24)10^6 = 21.37 \times 10^6 \text{ mm}^4.$$

$1 \rightarrow 2$, $y = 40 - s$, $t = 3$ mm:

$$D_{xb} = \int yt\,ds = 3\int_0^s (40 - s)\,ds = 3(40s - s^2/2).$$

At 1: $s = 0$, $D_{xb1} = 0$,

At 2: $s = 80$ mm, $D_{xb2} = 0$,

At $s = 40$ mm, $D_{xb} = 2400$ mm^3 (a maximum).

$2 \rightarrow 3$, $y = -(40 + 110s/610)$, $t = 1$ mm:

$$D_{xb} = -\int_0^s (40 + 110s/610)\,ds + D_{xb1} = -(40s + 11s^2/122) + 0.$$

At 2: $s = 0$, $D_{xb2} = 0$

At 3: $s = 610$ mm, $D_{xb3} = -57\,950$ mm^3 (a maximum).

$3 \rightarrow 4$, $y = s - 150$, $t = 4$ mm:

$$D_{xb} = 4\int_0^s (s - 150)\,ds + D_{xb3} = 2s^2 - 600s - 57\,950.$$

At 3: $s = 0$, $D_{xb3} = -57\,950$ mm^3,

At 4: $s = 300$ mm, $D_{xb4} = -57\,950$ mm^3,

At $s = 150$ mm, $D_{xb} = -102\,950$ mm^3 (a maximum).

$4 \rightarrow 1$, $y = 150 - 110s/610$, $t = 1$ mm.

$$D_{xb} = \int_0^s (150 - 110s/610)\,ds + D_{xb4} = 150s - 11s^2/122 - 57\,950.$$

At 4: $s = 0$, $D_{xb4} = -57\,950$ (a maximum),

At 1: $s = 610$ mm, $D_{xb} = 0$.

The q_b distribution in Fig. 10.16(b) is found from multiplying D_{xb} by the constant $F_y/I_x = -9.359 \times 10^{-4}$ N/mm^4. The given directions correspond to positive q_b except for web 1–2. In the application of Eq. (10.13) to find q_o it is convenient to take the moment centre O, where the cover lines intersect so that no moments are exerted by the shear flow along them. For the webs, Eq. (10.13) becomes

$$F_y p_x = \oint q_b R \, ds + 2Aq_o, \qquad \text{(i)}$$

$$20 \times 10^3 (600 + 218.18) =$$

$$-9.359 \times 10^{-4} \int_0^{300} (2s^2 - 600s - 57\,950)(818.18) \, ds$$

$$\text{(clockwise for } q_b \; 3 \rightarrow 4)$$

$$-\left[-9.359 \times 10^{-4} \int_0^{80} 3(40s - s^2/2)(218.18) \, ds \right]$$

$$\text{(anticlockwise for } q_b \; 1 \rightarrow 2)$$

$$+ (300 + 80)600q_o \qquad \qquad (q_o \text{ assumed clockwise}).$$

$$1636.36 \times 10^4 = -0.7657 |2s^3/3 - 300s^2 - 57\,950s|_0^{300}$$

$$+ 0.6126|20s^2 - s^3/6|_0^{80} + (22.8 \times 10^4)q_o$$

$$= (2020.3 \times 10^4) + (2.6137 \times 10^4) + (22.8 \times 10^4)q_o$$

$$\therefore \quad q_o = -16.95 \text{ N/mm} \qquad \text{(anticlockwise in Fig. 10.16(c))}.$$

Note that particular care must be taken with signs. Remember that the q_b expressions were derived for a given s-direction. This direction controls the sense of the moments $\oint q_b R \, ds$ in Eq. (i) before numerical values are substituted. These moments may be equated to $F_y p_x$ in this way provided it is recognized that there is static equivalence with the moments due to the net $q = q_b + q_o$ distribution. Alternatively, with the given q_b directions in Fig. 10.16(b), the moments about O may be found by multiplying the net web forces by their perpendicular distances. Since

Force in 12 is $(2 \times 80 \times 2.246)/3 = 119.79$ N,

Force in 34 is $[54.235 + 2(96.35 - 54.235)/3]300 = 24\,693.93$ N,

then

$$(20 \times 10^3 \times 818.18) = (24\,693.93 \times 818.18) + (119.79 \times 218.18)$$

$$+ (22.8 \times 10^4)q_o.$$

Hence, $q_o = -16.95$ N/mm and the net positive shear flow $q = q_b + q_o$ is given in Fig. 10.16(d). Applying Eq. (10.14) for the rate of twist:

$$d\theta/dz = (1/2AG) \oint q \, ds/t$$

$$= (1/2AG) \left[\oint q_b \, ds/t + q_o \oint ds/t \right] \qquad \text{(ii)}$$

where

$$A = (300 + 80)600/2 = 114 \times 10^3 \, \text{mm}^2,$$

$$q_o \oint ds/t = -16.95[80/3 + 610/1 + 300/4 + 610/1]$$

$$= -16.95 \times 1321.67 = -22.403 \times 10^3 \, \text{N/mm},$$

$$\oint q_b \, ds/t = (F_y/I_x)\left[\int_0^{80} 3(40s - s^2/2)(ds/3) \right.$$

$$+ \int_0^{610} -(40s + 11s^2/122)(ds/1)$$

$$+ \int_0^{300} (2s^2 - 600s - 57\,950)(ds/4)$$

$$\left. + \int_0^{610} (150s - 11s^2/122 - 57\,950)(ds/1) \right]$$

$$= -(9.359 \times 10^{-4})[|20s^2 - s^3/6|_0^{80} - |20s^2 + 11s^3/366|_0^{610}$$

$$+ (1/4)|2s^3/3 - 300s^2 - 57\,950|_0^{300}$$

$$+ |150s^2/2 - 11s^3/366 - 57\,950s|_0^{610}]$$

$$= -(9.359 \times 10^{-4})(4.2666 - 1426.3833$$

$$- 659.6228 - 1426.3833) \times 10^4$$

$$= -(9.359 \times 10^{-4})(-3508.1228 \times 10^4)$$

$$= 32.833 \times 10^3 \, \text{N/mm}.$$

Substituting into Eq. (ii):

$$d\theta/dz = (-22.403 + 32.833)10^3/(2 \times 114 \times 10^3 \times 27 \times 10^3)$$

$$= 1.6943 \times 10^{-6} \, \text{rad/mm} = 0.0971 \, \text{degrees/m}.$$

In following the method used in the previous example to find the shear centre E, with F_y acting through this point, Eq. (10.15) gives

$$\oint q_b \, ds/t + q_e \oint ds/t = 0,$$

where q_e accounts for the transference of F_y to E with q_b and the path integrals remaining unchanged. Thus, from the previous calculations,

$$(32.833 \times 10^3) + 1321.67 q_e = 0,$$

$$\therefore \quad q_e = -24.842 \, \text{N/mm},$$

and taking moments about the same cover intersection point O in

Fig. 10.16(b) distance e_x from E, Eq. (10.16) gives

$$(20e_x)10^3 = (119.79 \times 218.18) + (24\,693.93 \times 818.18)$$
$$+ [2 \times 114(-24.842) \times 10^3],$$

from which $e_x = 728.09$ mm. That is, the shear centre E lies on the x-axis 90 mm to the right of the longer vertical web. Alternatively, by the reciprocal law (page 230), the position of the shear centre may be found from the pure torque T required to produce the same rate of twist as under F_y. From Eq. (9.31a),

$$(T/4A^2G) \oint ds/t = 1.6943 \times 10^{-6} \, \text{rad/mm},$$

$$T = 1.6943 \times 10^{-6} \times 4 \times (114 \times 10^3)^2 \times 27 \times 10^3/1321.67$$
$$= 1.80 \times 10^6 \, \text{N mm}.$$

This is counterbalanced to give zero twist rate when F_y is translated distance x_e to the shear centre. That is,

$$F_y x_e = 1.80 \times 10^6,$$
$$\therefore \quad x_e = 1.80 \times 10^6/(20 \times 10^3) = 90 \, \text{mm, as before.}$$

10.6 WEB–BOOM IDEALIZATION FOR THIN-WALLED STRUCTURES

Symmetrical Sections (x and y are principal axes)

The previous examples 10.3 and 10.4 showed how little the flanges contributed to the vertical shear force in I-section beams. Moreover, Fig. 10.6(b) shows that the shear stress does not vary greatly over the depth of the web. This leads to the idealization given in Fig. 10.17 for thin web sections.

The flange areas above and below the neutral axis in Fig. 10.17(a), which effectively resist the bending moment, may be considered to be concentrated as "booms" separated by the depth d of the shear web as shown in Fig. 10.17(b). This is a particularly useful approximation to make for the analysis of the effects of longitudinal stiffeners and stringers on the shear flow in thin-walled aircraft structures. Consider all the symmetrical, idealized, open and closed structures given in Figs 10.17–10.20. A single shear force F_y is applied parallel to the principal y-axis where x is an axis of symmetry. Equation (10.9) gives the web or skin shear flow as

$$q = (F_y/I_x)D_x + q_o \tag{10.18a}$$

where, with the contribution to D_x from both the web and boom areas A_r

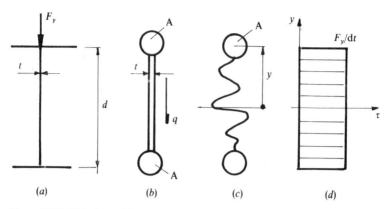

Figure 10.17 Web–boom idealization for an I-beam.

$(r = 1, 2, 3, \ldots n)$, the first moment of area becomes

$$D_x = \int_0^s yt \, ds + \sum_{r=1}^n A_r y_r. \tag{10.18b}$$

The total second moment of area is given by

$$I_x = I_x \text{ (for web or skin)} + \sum_{r=1}^n A_r y_r^2 \text{ (for booms)}.$$

The greater change in shear flow will be caused by the booms when

$$\sum A_r y_r > \int yt \, ds,$$

which is normally the case for thin-walled sections under any flexural shear loading. In the simplest of idealizations, $\int_0^s yt \, ds$ in Eq. (10.18b) is replaced by lumping the web or skin area into that of the booms. Equation (10.18a) then becomes

$$q = (F_y/I_x) \sum_{r=1}^n (A_r y_r) + q_o. \tag{10.19}$$

This gives a constant shear flow over any shape of web connecting two adjacent boom areas. In Fig. 10.17(c), for example, with $Ay = Ad/2$, $I = 2A(d/2)^2$ and $q_o = 0$, for an open section, the web shear flow becomes

$$q = F_y(Ad/2)/2A(d/2)^2 = F/d,$$

acting in a direction parallel to its mid-line. Here $\tau = q/t = F_y/dt$ is the average shear stress for the web as shown in Fig. 10.17(d). In the application of Eq. (10.19) to Fig. 10.18(a), $\sum_{r=1}^n (A_r y_r)$ is the sum of the first moments of those boom areas about the NA lying above the web being considered.

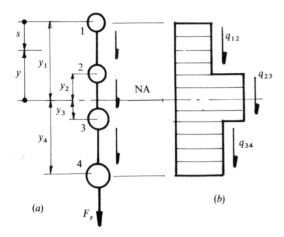

Figure 10.18 Web–boom idealization showing constant q between booms.

Thus, with $I = A_1y_1^2 + A_2y_2^2 + A_3y_3^2 + A_4y_4^2$ in Fig. 10.18(b),

$$q_{12} = (F_y/I)A_1y_1,$$
$$q_{23} = (F_y/I)(A_1y_1 + A_2y_2) = (F_y/I)(-A_4y_4 - A_3y_3),$$
$$q_{34} = (F_y/I)(-A_4y_4) = (F_y/I)(A_1y_1 + A_2y_2 - A_3y_3).$$

If the section supports a second shear force F_x at right angles to F_y, where x and y are principal axes, the effect on the web shear flow is additive. Thus, with F_x and F_y acting in the negative x- and y-directions, the superposition of similar derivations that lead to Eq. (10.19) gives

$$q = (F_y/I_x) \sum_{r=1}^{n} (A_r y_r) + (F_x/I_y) \sum_{r=1}^{n} (A_r x_r) + q_o, \qquad (10.20)$$

where $\sum(Ay)$ and $\sum(Ax)$ are sums of the first moments of those boom areas, lying 'above' and 'below' the web being considered, about the principal centroidal axes. In Fig. 10.19, for example, with the boom areas balanced w.r.t. x, Eq. (10.20) will supply the mean line web shear flow provided there is no twisting of the section. That is, from Eq. (10.20), with $q_o = 0$, for an open section,

$$q_{12} = (F_y/I_x)(A_1y_1) + (F_x/I_y)(-A_1x_1)$$
$$= (F_y/I_x)(-A_1y_1) + (F_x/I_y)[2(A_2x_2) + 2(A_3x_3) - (A_1x_1)],$$
$$q_{23} = (F_y/I_x)[(A_1y_1) + (A_2y_2)] + (F_x/I_y)[-(A_1x_1) + (A_2x_2)]$$
$$= (F_y/I_x)[-(A_1y_1) - (A_2y_2)] + (F_x/i_y)[2(A_3x_3) + (A_2x_2) - (A_1x_1)],$$
$$q_{33} = (F_y/I_x)[(A_1y_1) + (A_2y_2) + (A_3y_3)]$$
$$\quad + (F_x/I_y)[-(A_1x_1) + (A_2x_2) + (A_3x_3)],$$

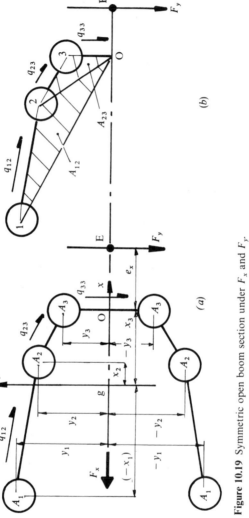

Figure 10.19 Symmetric open boom section under F_x and F_y.

where

$$I_x = 2[A_1 y_1^2 + A_2 y_2^2 + A_3 y_3^2], \qquad I_y = 2[A_1 x_1^2 + A_2 x_2^2 + A_3 x_3^2].$$

In this case, the shear centre E lies along the x-axis distance e_x from the vertical web as shown in Fig. 10.19(a). The moment centre O, is chosen to lie at the centre of the vertical web, since neither F_x nor the shear flow for that web will then exert a moment. Equation (10.10) becomes

$$F_y e_x = \int qR \, ds = 2\sum (qA). \tag{10.21}$$

The summation follows from q being constant along each web when $\int R \, ds$ is twice the area enclosed by the web and O as shown for q_{12} and q_{23} in Fig. 10.19(b). Thus, for the relevant areas above and below the neutral axis in Fig. 10.19(b),

$$\sum (qA) = 2[q_{12}A_{12} + q_{23}A_{23}].$$

In the application of Eq. (10.19) to the closed tube in Fig. 10.20(a), $q_o \neq 0$ and the summation term accounts for the boom areas passed by the peripheral coordinate s, when taken clockwise (say) from its origin at boom 1 (Fig. 10.20b). For example, in the determination of q_{34},

$$\sum A_r y_r = A_1 y_1 - A_2 y_2 - A_3 y_3.$$

The position of the shear centre for an idealized symmetrical closed section is supplied by Eq. (10.16), where with a summation similar to Eq. (10.21) for the path integral,

$$F_y e_x - F_x e_y = 2\sum (q_b A) + 2A q_e.$$

A particular example of shear flow in a symmetrical multicell tube appears on page 493 within the general consideration of the application of the principle of virtual work to redundant structures.

Asymmetric Sections (x and y are not principal axes)

Equations (10.20) also applies when shear forces S_x and S_y are applied to a section in the negative x- and y-directions where x and y are not principal axes. It again becomes necessary to determine the equivalent shear forces $\hat{F}_x$ and $\hat{F}_y$ from Eqs (10.11) to replace F_x and F_y in Eq. (10.20).

Example 10.12 Determine the shear centre and the shear flow for the flanges and the web in the unsymmetrical channel section of Fig. 10.21(a) when shear forces act at as shown.

The centroidal position $\bar{X}$ of the y-axis is found from

$$(500 \times 100) + (500 \times 200) = (4 \times 500)\bar{X}, \quad \Rightarrow \bar{X} = 75 \, \text{mm},$$

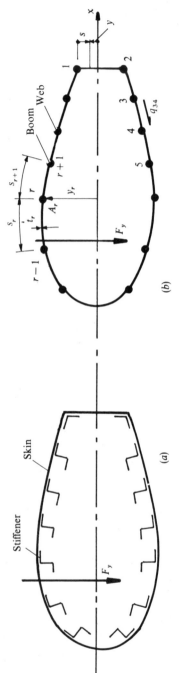

Figure 10.20 Idealization of a single-cell closed tube.

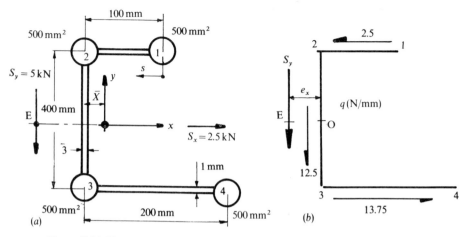

Figure 10.21 Unsymmetrical channel section.

and since the x-axis passes through the centre of the depth,

$$I_x = 4 \times 500 \times 200 = 80 \times 10^6 \text{ mm}^4,$$
$$I_y = (2 \times 75^2 \times 500) + (25^2 \times 500) + (125^2 \times 500) = 13.75 \times 10^6 \text{ mm}^4,$$
$$I_{xy} = (500 \times 200 \times 25) - (500 \times 75 \times 200) + (500 \times 200 \times 75)$$
$$- (500 \times 125 \times 200) = -10 \times 10^6 \text{ mm}^4.$$

Now, from Eqs (10.11), with $S_y = +5\,\text{kN}$ and $S_x = -2.5\,\text{kN}$,

$$\hat{F}_x = \frac{-2.5 - 5(-10/80)}{1 - 10^2/(80 \times 13.75)} = -2.0625\,\text{kN},$$

$$\hat{F}_y = \frac{5 + 2.5(-10/13.75)}{1 - 10^2/(80 \times 13.75)} = 3.5\,\text{kN},$$

$$\therefore \quad \hat{F}_x/I_y = -2.0625 \times 10^3/(13.75 \times 10^6) = -150 \times 10^{-6}\,\text{N/mm}^4,$$
$$\hat{F}_y/I_x = 3.5 \times 10^3/80 \times 10^6 = 43.75 \times 10^{-6}\,\text{N/mm}^4.$$

In the application of Eq. (10.20) with $q_o = 0$, the shear flow between booms 1 and 2, i.e. from 1 to 2, is found from applying the summation terms to boom 1 only. That is

$$q_{12} = (\hat{F}_y/I_x)(A_1 y_1) + (\hat{F}_x/I_y)(A_1 x_1)$$
$$= (F_y/I_x)(500 \times 200) + (\hat{F}/I_y)(500 \times 25)$$
$$= 2.50\,\text{N/mm, acting in the sense shown in Fig. 10.21(b)}.$$

The shear flow between booms 2 and 3 is found from applying the summations to booms 1 and 2. That is, working anticlockwise from 1,

$$q_{23} = (\hat{F}_y/I_x)[(500 \times 200) + (500 \times 200)]$$
$$+ (\hat{F}_x/I_y)[(500 \times 25) - (500 \times 75)]$$
$$= 12.5 \, \text{N/mm, positive in the sense shown.}$$

Alternatively, working clockwise from 4;

$$q_{23} = (\hat{F}_y/I_x)[-(500 \times 200) - (500 \times 200)]$$
$$+ (\hat{F}_x/I_y)][(500 \times 125) - (500 \times 75)]$$
$$= -12.5 \, \text{N/mm,}$$

confirming the positive direction found previously.

Working anticlockwise from 1, the shear flow between 3 and 4 is given by

$$q_{34} = (\hat{F}_y/I_x)[-(500 \times 200) + (500 \times 200) + (500 \times 200)]$$
$$+ (\hat{F}_x/I_y)[-(500 \times 75) - (500 \times 75) + (500 \times 25)] = 13.75 \, \text{N/mm.}$$

Alternatively, working clockwise from 4,

$$q_{34} = (\hat{F}_y/I_x)[-(500 \times 200)] + (\hat{F}_x/I_y)(500 \times 125) = -13.75 \, \text{N/mm.}$$

The magnitude of q and the directions shown in Fig. 10.21(b) must be statically equivalent to the applied forces. It is only necessary to multiply q by the respective web lengths in order to check this, because $q = \tau t$. Then,

$$S_x = (2.5 \times 100) - (13.75 \times 200) = -2500 \, \text{N}, \quad S_y = (12.5 \times 400) = 5000 \, \text{N}.$$

Taking moments about the web centre O, in Fig. 10.21(b) eliminates the effects of S_x and the web shear flow. That is, from Eq. (10.10),

$$(5 \times 10^3)e_x = (2.5 \times 100 \times 200) + (13.75 \times 200 \times 200),$$
$$\therefore \quad e_x = 120 \, \text{mm.}$$

Calculation of Boom Areas

A structure under flexural shear will also carry direct stress due to bending. In the idealized structure, the boom areas carry this direct stress. Their areas reflect the combined effects that the skin or web, the flanges and longitudinal stiffners (Figs 10.20 and 10.21) have on resisting bending. The following simplifications are made in idealizing a structure.

1. The boom centroids lie in the plane of the web or skin mid-line.
2. The skin or web carries only shear stress while the booms carry only bending stress. Note, however, that boom areas abruptly alter the first moment of area, thereby interrupting the shear flow.

3. The shear stress is uniform across the web or skin thickness and the direct stress is constant across the boom area.

In addition, the net value of I about the neutral axis of the section must remain unchanged. Thus, in Fig. 10.20, for example, when the booms are to represent the finite stiffener areas at their given positions in the original structure, then at the rth boom,

$$(A_r y_r^2)_{\text{boom}} = (I_{xr})_{\text{stiffener}},$$
$$\therefore \quad A_r = I_{xr}/y_r^2 \qquad\qquad (10.22)$$

from which it follows that A_r will approximate to the actual stiffener area only when y_r is large. The web or skin makes an additional contribution to I_x by increasing the net area of the boom. This is assessed from the following consideration of bending of the section about its neutral axis.

The panel element in Fig. 10.22(a) is idealized into two booms connected by a shear web of zero bending stiffness in Fig. 10.22(b). The following conditions apply.

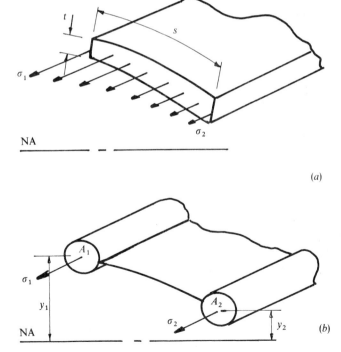

Figure 10.22 Web–boom idealization of a single channel.

1. The net axial force must be the same. With a linear variation in bending stress assumed between the two booms,

$$st(\sigma_1 + \sigma_2)/2 = A_1\sigma_1 + A_2\sigma_2, \tag{10.23}$$

where from the ETB the stress is proportional to the distance from the neutral axis:

$$\sigma_1/\sigma_2 = y_1/y_2. \tag{10.24}$$

2. The moments about the neutral axis must be the same.

$$\sigma_2 st(y_1 + y_2)/2 + [(\sigma_1 - \sigma_2)st/2][y_2 + 2(y_1 - y_2)/3] = A_1\sigma_1 y_1 + A_2\sigma_2 y_2. \tag{10.25}$$

Combining Eqs (10.23)–(10.25) leads to the boom area contributions:

$$A_1 = (st/6)(2 + \sigma_2/\sigma_1), \tag{10.26a}$$

$$A_2 = (st/6)(2 + \sigma_1/\sigma_2). \tag{10.26b}$$

The reader should verify that it is implicit in these conditions that I_x for the panel will remain unchanged. In extending Eqs (10.26) to the multiboom idealization in Fig. 10.20(b), the area of the rth boom is influenced by both the adjacent webs r and $r+1$. Then,

$$A_r = (s_r t_r/6)(2 + \sigma_{r-1}/\sigma_r) + (s_{r+1}t_{r+1}/6)(2 + \sigma_{r+1}/\sigma_r), \tag{10.27}$$

where

$$\sigma_{r-1}/\sigma_r = y_{r-1}/y_r \quad \text{and} \quad \sigma_{r+1}/\sigma_r = y_{r+1}/y_r.$$

The net boom areas are then found from adding Eqs (10.22) and (10.27). In practice, the skin or web thickness construction may not be fully effective in resisting direct stress—where there is a risk of buckling, for example. This leads to the use of an effective skin thickness in Eqs (10.26) and (10.27) that differs from the actual thickness, for the assessment of its contribution to the boom area. However, the following examples illustrate that the actual thickness is normally retained in the path integral $\oint ds/t$ when determining warping displacements and the position of the shear centre.

Example 10.13 Determine the web shear flow for Fig. 10.11 using a four-boom idealization.

First, the flanges are assumed to be concentrations of area at the web centre line as shown in Fig. 10.23(a). The contributions to these areas from the web are found from Eqs (10.26) and (10.27). The former equation deals with the end boom,

$$A_1 = (s_1 t_1/6)(2 + y_2/y_1) = (10 \times 1.5/6)(2 + 25/35) = 786 \text{ mm}^2 = A_4,$$

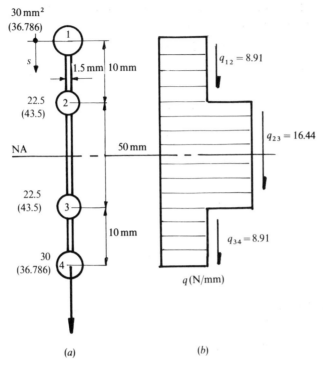

Figure 10.23 Shear flow for a four-boom idealization.

and the latter deals with the intermediate booms areas in the table:

Position	1	2	3	4
s (mm)	—	10	50	10
t (mm)	—	1.5	1.5	1.5
y (mm)	35	25	-25	-35

Then

$$A_2 = (s_2 t_2/6)(2 + y_1/y_2) + (s_3 t_3/6)(2 + y_3/y_2)$$
$$= (10 \times 1.5/6)(2 + 35/25) + (50 \times 1.5/6)(2 - 25/25) = 21 \text{ mm}^2.$$
$$A_3 = (s_3 t_3/6)(2 + y_2/y_3) + (s_4 t_4/6)(2 + y_4/y_3)$$
$$= (50 \times 1.5/6)[2 + 25/(-25)] + (10 \times 1.5/6)(2 + 35/25) = 21 \text{ mm}^2.$$

The net boom areas are those bracketed in Fig. 10.23(a), from which

$$I_x = 2[(36.786 \times 35^2) + (43.5 \times 25^2)]$$
$$= 144.5 \times 10^3 \text{ mm}^4 \text{ (as before, page 364)}$$

and

$$F_y/I_x = (144.5)^{-1} \, \text{N/mm}^4.$$

Now, from Eq. (10.19), with $q_0 = 0$ for an open section,

$$q = (F_y/I_x)\sum A_r y_r,$$

where $\sum A_r y_r$ and q are

$$
\begin{aligned}
1 \to 2 \quad & A_1 y_1 = 36.786 \times 35 = 1287.5 \, \text{mm}^3, \\
& q_{12} = 8.91 \, \text{N/mm}, \\
2 \to 3 \quad & A_1 y_1 + A_2 y_2 = 1287.5 + (43.5 \times 25) = 2375 \, \text{mm}^2, \\
& q_{23} = 16.4 \, \text{N/m}, \\
3 \to 4 \quad & A_1 y_1 + A_2 y_2 + A_3 y_3 = 2375 + 43.5(-25) = 1287.5 \, \text{mm}^3, \\
& q_{34} = 8.91 \, \text{N/mm}.
\end{aligned}
$$

The web shear flow in Fig. 10.23(b) is thus a reasonable approximation to that shown in Fig. 10.11(b).

Example 10.14 Using a four-boom idealization (Fig. 10.24(a)) for the tube in Fig. 10.16, determine the shear flow distribution, the position of the shear centre and the warping displacements at the four corners. Take $G = 30 \, \text{GPa}$.

With reference to the chosen boom positions in Fig. 10.24(a), the following table applies:

Boom	1	2	3	4
y(mm)	+40	−40	−150	+150
s(mm)	610	80	610	300
t(mm)	1	3	1	4

Since there are no stiffeners to deal with in this case, the boom areas are directly supplied by Eq. (10.27):

$$
\begin{aligned}
A_1 &= (s_1 t_1/6)(2 + y_4/y_1) + (s_2 t_2/6)(2 + y_2/y_1) \\
&= (610 \times 1/6)(2 + 150/40) + (80 \times 3/6)[2 + (-40)/40] \\
&= 625 \, \text{mm}^2 = A_2, \\
A_3 &= (s_3 t_3/6)(2 + y_2/y_3) + (s_4 t_4/6)(2 + y_4/y_3) \\
&= (610 \times 1/6)[2 + (-40)/(-150)] + (300 \times 4/6)[2 + 150/(-150)] \\
&= 430 \, \text{mm}^2 = A_4, \\
I_x &= 2[(625 \times 40^2) + (430 \times 150^2)] = 21.36 \times 10^6 \, \text{mm}^4,
\end{aligned}
$$

agreeing with the previous value for Fig. 10.16. From Eq. (10.19),

$$q_b = (F_y/I_x)D_{xb} = (F_y/I_x)(\sum A_r y_r)_b,$$

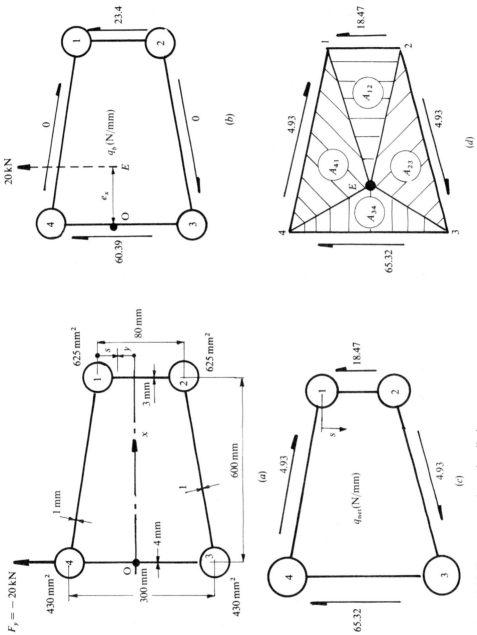

Figure 10.24 Shear centre and warping displacements.

where

$$F_y/I_x = -20\,000/(21.36 \times 10^6) = -9.363 \times 10^{-4}\,\text{N/mm}^4.$$

D_{xb} and q_b becomes

$$1 \to 2 \qquad D_{xb} = 625 \times 40 = 2.5 \times 10^4\,\text{mm}^3,$$
$$\therefore \quad q_b = -9.363 \times 2.5 = -23.4\,\text{N/mm},$$
$$2 \to 3 \qquad D_{xb} = (2.5 \times 10^3) - (625 \times 40) = 0,$$
$$\therefore \quad q_b = 0,$$
$$3 \to 4 \qquad D_{xb} = 0 - (430 \times 150) = -6.45 \times 10^4\,\text{mm}^3,$$
$$\therefore \quad q_b = (-9.363)(-6.45) = 60.39\,\text{N/mm}.$$

This establishes the q_b distribution shown in Fig. 10.24(b). For the determination of q_o it is convenient to take moments about the centre O, of the left-hand web. Equation (10.13) becomes, for the (+ve) s-direction:

$$0 = \oint q_b R\,\mathrm{d}s + q_o \oint R\,\mathrm{d}s = \oint q_b R\,\mathrm{d}s + 2Aq_o,$$

$$= -(23.4 \times 80 \times 600) + (300 + 80)600q_o, \quad \Rightarrow q_o = +4.93\,\text{N/mm}.$$

Thus, the net shear flow, $q = q_b + q_o$, is that shown in Fig. 10.24(c). For the shear centre E, with F_y translated a distance e_x to E in Fig. 10.24(b), Eq. (10.15) becomes

$$\oint q_b\,\mathrm{d}s/t + q_e \oint \mathrm{d}s/t = 0,$$

$$[-(23.4 \times 80/3) + 0 + (60.39 \times 300/4) + 0]$$
$$+ q_e(80/3 + 610/1 + 300/4 + 610/1) = 0.$$

Hence, $q_e = -2.955\,\text{N/mm}$. Taking moments about O, Eq. (10.16) becomes

$$F_y e_x = \oint q_b R\,\mathrm{d}s + q_e \oint R\,\mathrm{d}s = \oint q_b R\,\mathrm{d}s + 2Aq_e,$$

$$-(20 \times 10^3)e_x = -(23.4 \times 80 \times 600) - 600(300 + 80)2.955.$$

Again $e_x = 90\,\text{mm}$ (page 381). The warping displacements are found from applying Eq. (10.17) to the net q shear flow distribution in Fig. 10.24(c). Starting at position 1, where there will be a warping displacement (w_1), Eq. (10.17) is applied by measuring A_{os} w.r.t. the shear centre E. Now,

$$\oint q\,\mathrm{d}s/t = (-18.47 \times 80/3) + (4.93 \times 610/1) + (65.32 \times 300/4)$$

$$+ (4.93 \times 610/1) = 10\,421.1.$$

$$A = (300 + 80)600/2 = 114 \times 10^3 \, \text{mm}^2,$$

$$\therefore \quad G\Delta w = \int_0^s q \, ds/t - (91.4 \times 10^{-3})A_{os}.$$

$1 \to 2$
$$A_{os} = A_{12} = 80 \times 510/2 = 20.4 \times 10^3 \, \text{mm}^2,$$

$$\int_0^s q \, ds/t = (-18.47/3) \int_0^s ds = -6.158s,$$

$$\text{At 2,} \quad s = 80 \, \text{mm}, \quad \int_o^s = -492.5,$$

$$\therefore \quad G(w_2 - w_1) = -492.5 - (91.4 \times 20.4)$$
$$= -2.357 \times 10^3 \, \text{N/mm}.$$

$2 \to 3$
$$A_{23} = [114 - 20.4 - (3 \times 9/2)]10^3/2 = 40.05 \times 10^3 \, \text{mm}^2,$$

$$A_{os} = A_{12} + A_{23} = (20.4 + 40.05)10^3 = 60.45 \times 10^3 \, \text{mm}^2,$$

$$\int_0^s q \, ds/t = -492.5 + (4.93/1) \int_0^s ds = -492.5 + 4.93s,$$

$$\text{At 3,} \quad s = 610 \, \text{mm}, \quad \int_o^s = 2514.8,$$

$$\therefore \quad G(w_3 - w_1) = 2514.8 - (91.4 \times 60.45) = -3.01 \times 10^3 \, \text{N/mm}.$$

$3 \to 4$
$$A_{34} = (300 \times 90)/2 = 13.5 \times 10^3 \, \text{mm}^2,$$
$$A_{os} = A_{12} + A_{23} + A_{34} = (20.4 + 40.05 + 13.5)10^3$$
$$= 73.95 \times 10^3 \, \text{mm}^2,$$

$$\int_0^s q \, ds/t = 2514.8 + (65.32/4) \int_0^s ds = 2514.8 + 16.33s,$$

$$\text{At 4,} \quad s = 300 \, \text{mm}, \quad \int_o^s = 7413.8,$$

$$\therefore \quad G(w_4 - w_1) = 7413.8 - (91.4 \times 73.95)$$
$$= +0.655 \times 10^3 \, \text{N/mm}.$$

$4 \to 1$
$$A_{os} = 114 \times 10^3 \, \text{mm}^2,$$

$$\int_0^s q \, ds/t = 7413.8 + (610/1) \int_0^s ds = 7413.8 + 610s,$$

$$\text{At 1,} \quad s = 610 \, \text{mm}, \quad \int_o^s = 10421.1 = \oint q \, ds/t,$$

$$\therefore \quad G(w_1 - w_1) = 10421.1 - (91.4 \times 114) \simeq 0 \quad \text{(check)}.$$

Assuming that the x-axis of symmetry does not warp and noting that

Δw is linear in s throughout,

$$Gw_1 = +2.357 \times 10^3/2 = +1.179 \times 10^3 \, \text{N/mm},$$
$$Gw_2 = (-2.357 + 1.179)10^3 = -1.79 \times 10^3 \, \text{N/mm},$$
$$Gw_3 = (-3.01 + 1.179)10^3 = -1.832 \times 10^3 \, \text{N/mm},$$
$$Gw_4 = (+0.655 + 1.179)10^3 = +1.833 \times 10^3 \, \text{N/mm}.$$

Finally, with $G = 30 \times 10^3 \, \text{MPa}$, the warping displacements are: $w_1 = +0.039$, $w_2 = -0.039$, $w_3 = -0.061$ and $w_4 = +0.061$ mm. Points above the x-axis displace axially in the positive z-direction (see Fig. 10.13a), while those points below the x-axis displace axially in the opposite sense.

EXERCISES

Shear Stress in Beams

10.1 Show that the maximum flexural shear stress in a rectangular beam is 50 per cent greater than the mean value for the section.

10.2 An I-beam is simply supported over a span of 3.65 m and carries a total uniformly distributed load of 8 t over its length. Find the maximum shear stress due to shear force and state where in the length and cross-section this occurs. What is the percentage shear force carried by the web?

Answer: 26,4 MPa; 95 per cent.

10.3 A cantilever 1.5 m long is formed from bolting two 150 × 100 mm timbers together to give a cross-section 150 mm wide × 200 mm deep. When clamping bolts of 12.5 mm diameter are spaced 150 mm apart, calculate the shear stress in each bolt caused by a concentrated vertical load of 6.5 kN at the free end

Answer: 64.2 MPa.

10.4 A rectangular channel section 505 mm wide, 305 mm deep and 25 mm thick is used as a beam in the form of a trough. Derive equations in which the shear stress at any layer in the vertical sides may be found under a vertical shear force of 8.9 kN. Plot the stress distribution showing major values. Find the fraction of the shear force that is carried by the vertical sides.

Answer: 0.8 MPa; 0.72 MPa; 94.5 per cent.

10.5 A bar section with 25 mm hexagonal sides is used as a cantilever with one diagonal lying horizontally. Plot the distribution of shear stress through the depth when a concentrated vertical force of 20 kN is applied at the free end. Show where the shear stress is a maximum and what is its value at the neutral axis?

Answer: 14.3 MPa.

10.6 The section of an I-beam is subjected to a vertical shear force of 40 kN. Derive an equation for which the shear stress at any layer in the web may be determined. Calculate the shear stress acting at the top of the web and the maximum shear stress value. What percentage of the shear force is carried by the web? Section details are, flanges 175 mm wide × 25 mm deep, total depth 225 mm, web thickness 10 mm.

Answer: 19 MPa; 20.6 MPa; 88 per cent.

10.7 Derive expressions for the shear stress due to shear force F in each of the sections given in Fig. 10.25(a–c). Find also the ratio between the maximum shear stress and the mean shear stress in each case.

Answer: $3F/2bd$; $(4F/3\pi R^2)\cos^2 \theta$; F/a^2; 3/2; 4/3; 1.

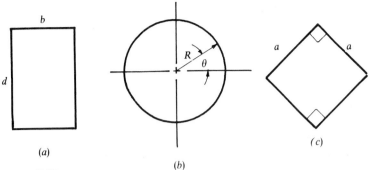

(a) (b) (c)

Figure 10.25

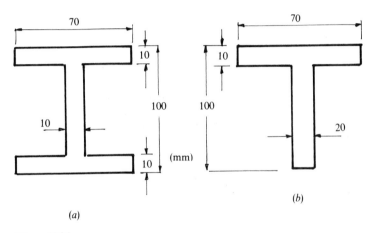

(a) (b)

Figure 10.26

10.8 Draw the distribution of shear stress due to a shear force of 30 kN for the two sections in Fig. 10.26(a) and (b).

10.9 Plot the distribution of vertical and horizontal shear stress for the channel sections in Fig. 10.27(a) and (b) when they are each subjected to a vertical shear force of 44.5 kN. Determine the shear centres for each channel.

Answer: (a) 28.12 MPa; 39.15 MPa; 61.8 mm. (b) 18 MPa; 150 MPa; 23.31 MPa; 60.3 mm.

10.10 Calculate for the cross-sections shown in Fig. 10.28(a), the maximum shear stress, the shear flow carried by the rivets and the load per rivet when they are spaced at a 40 mm pitch in the length. If the section is modified as shown in Fig. 10.28(b) describe how the shear flow is then calculated.

Answer: (a) 88.5 MPa; 125.8 N/mm; 5.03 kN.

10.11 Find an expression for the shear stress at any point in the flange and the web for the I-section in Fig. 10.29. Plot the distribution of shear stress through the depth, inserting major values corresponding to a vertical shear force of 400 kN.

Answer: 3.5; 44.5; 59.3 MPa.

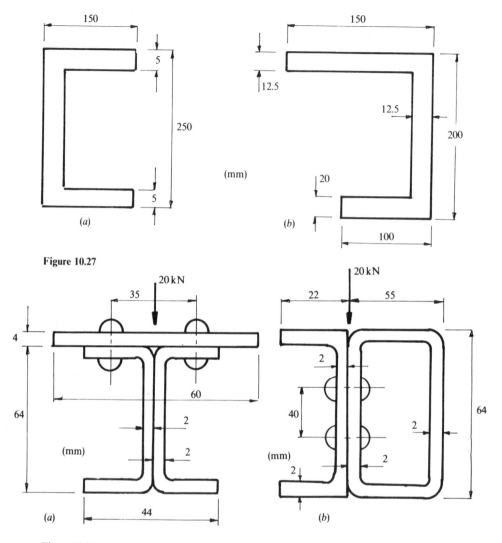

Figure 10.27

Figure 10.28

10.12 Draw the distribution of vertical shear stress for the cross-section shown in Fig. 10.30, corresponding to a vertical shear force of 300 kN.

Answer: 52.65; 579.15; 702.02 (max); 663.4; 31.6 MPa.

10.13 Find the maximum shear stress when the I-section in Fig. 10.31 is subjected to a vertical shear force of 8 t. What is the percentage error when this stress is taken to be the mean value calculated from force/area?

Answer: 81.1 MPa; 1.5 per cent.

10.14 The welded composite beam in Fig. 10.32 is subjected to a vertical force of 250 kN. Find the magnitude of the shear flow at the two welded flange joints if, for the channel section, $A = 3000\,\text{mm}^2$, $\bar{y} = 12.5\,\text{mm}$ and $I = 150 \times 10^6\,\text{mm}^4$ as shown.

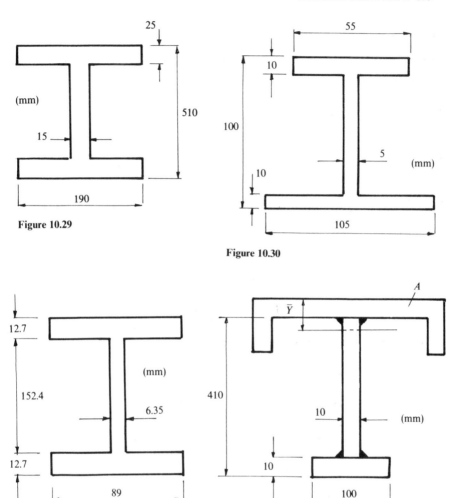

Figure 10.29

Figure 10.30

Figure 10.31

Figure 10.32

Shear Flow in Thin-walled Open Sections

10.15 A 0.5 m long cantilever with the channel section in Fig. 10.33 carries an inclined force of 25 N through the shear centre E as shown. Determine the position of E and the magnitude and the position of the maximum shear stress.

Answer: 13 mm from vertical web; 4.2 MPa.

10.16 The section in Fig. 10.34 is symmetrical about the x-axis and has a constant thickness of 1 mm. Calculate and plot the shear flow distribution when a vertical force of 1 kN is applied downward, through the shear centre. Where does the shear centre lie?

Answer: 1.071 N/mm; 1.429 N/mm; 6.26 mm from vertical web.

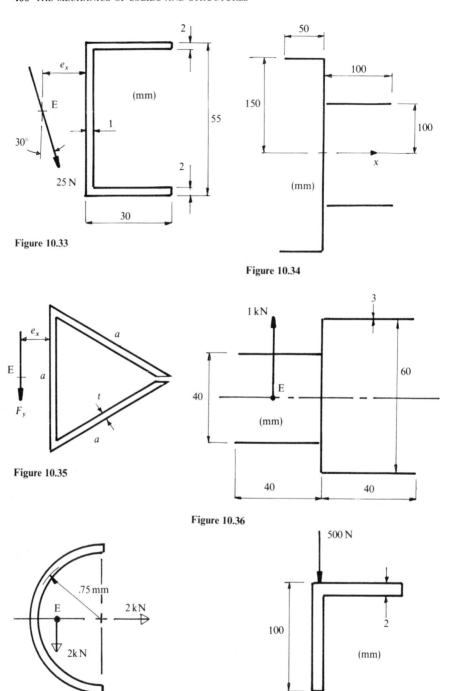

Figure 10.33

Figure 10.34

Figure 10.35

Figure 10.36

Figure 10.37

Figure 10.38

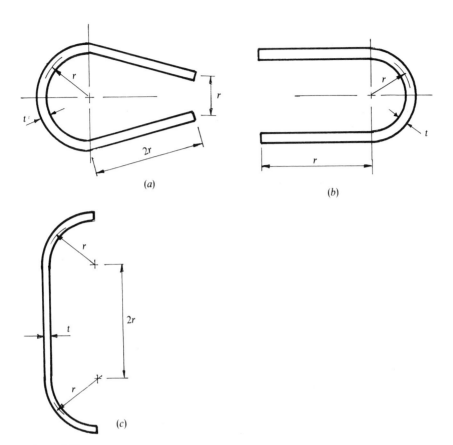

(a)

(b)

(c)

Figure 10.39

10.17 A thin tube of equilateral triangular section with side length a and thickness t is opened by a longitudinal slit as shown in Fig. 10.35. Determine the shear flow distribution along the sloping sides and the position of the shear centre when a vertical shear force F_y is applied at this point.

Answer: $q = F_y s^2 / a^3$; $0.289a$.

10.18 The beam section in Fig. 10.36 is subjected to an upward vertical shear force of $1\,\mathrm{kN}$ through the shear centre. Determine the maximum shear flow and the distance of the shear centre from the vertical web.

Answer: $2.69\,\mathrm{N/mm}$, $6.56\,\mathrm{mm}$.

10.19 Draw the shear flow distribution and find the position of the shear centre E when the open semi-circular tube in Fig. 10.37 is loaded through this point as shown.

10.20 Establish the shear flow distribution for the equal-angle section in Fig. 10.38 when a vertical shear force of $500\,\mathrm{N}$ is applied downwards along the vertical web.

10.21 Find the position of the shear centre for each of the open tubes in Fig. 10.39(a–c).

10.22 A uniform cantilever has a thin tubular cross-section of mean radius r and thickness t with a narrow slit cut through the wall at the right-hand side of the horizontal diameter as shown in Fig. 10.40. Show that the twisting moment set up under a vertical shear force Q is

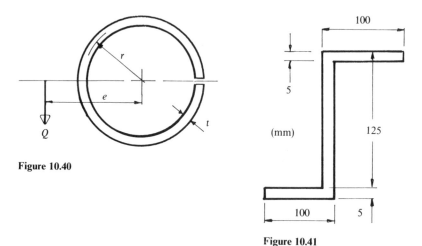

Figure 10.40

Figure 10.41

$2Qr^4\pi t/I$ and hence that, in order to prevent twisting taking place, the force should be placed at $e = 2r$ from the centre as shown. (C.E.I. Pt. 2, 1986)

10.23 Determine the distribution of shear stress in the flanges and web and their maxima for the Z-section in Fig. 10.41, when a downward, shear force of 1 kN is applied along the web.

Shear Flow in Thin-walled Closed Sections

10.24 The thin-walled tubes in Fig. 10.42(a) and (b) each have a downward vertical shear force of 500 N applied along their left sides. Determine the shear flow distribution around the section, the rate of twist and the position of the shear centre.

10.25 Find the position and magnitude of the maximum shear stress in the tubular parallelogram section in Fig. 10.43 when it is loaded vertically through the shear centre with a 1 kN downwards shear force.

10.26 Plot the distribution of shear flow, inserting principal values, for the tubular rectangular section in Fig. 10.44 when a vertical shear force of 6 kN acts downwards along the left web.
Answer: in N/mm; 5.4, 7.9, 19.6, 24.6, 19.6.

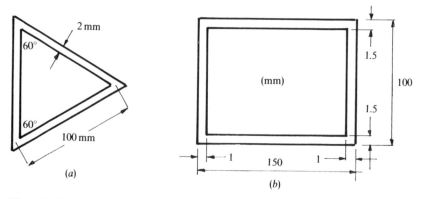

(a)

(b)

Figure 10.42

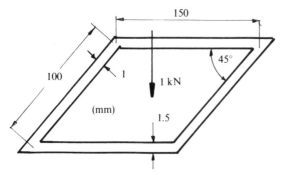

Figure 10.43

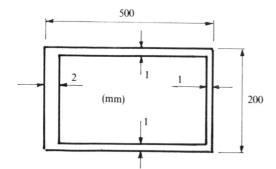

Figure 10.44

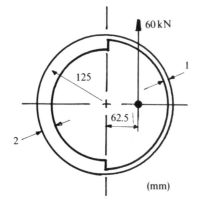

Figure 10.45

10.27 A 2.5 m long uniform cantilever with the cross-section in Fig. 10.45 supports a 60 kN vertical upwards force at its free end, offset to the right by 62.5 mm. Determine the distribution of shear flow over the section, the maximum shear stress and the angular twist at the free end. Take $G = 30\,\text{GPa}$.

10.28 Determine the shear flow distribution for the tube section in Fig. 10.46 when two vertical shear forces are applied as shown. Determine the position of the shear centre.

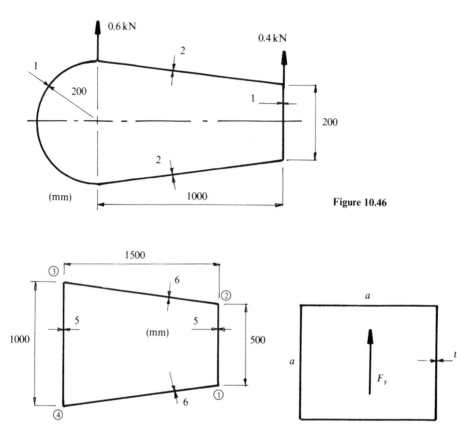

Figure 10.46

Figure 10.47

Figure 10.48

10.29 Determine the position of the shear centre for the trapezoidal tubular section given in Fig. 10.47.

10.30 The square-section tube in Fig. 10.48 is loaded through the centroid by an upwards vertical shear force F_y. Determine the warping displacements at the four corners.

Web–Boom Idealization for Thin-walled Structures

10.31 Determine the web shear flow for the open section in Fig. 10.34 using a simple four-boom idealization.

10.32 Confirm the position of the shear centre using a suitable idealization of the section in Fig. 10.36.

10.33 A 50 kN vertical shear force acts downwards through the shear centre of the section in Fig. 10.49. Determine the flxural shear flow in each web and the position e_x of the shear centre.
 Answer: 49.65 N/mm, 69.5 N/mm, 86.06 N/mm and 94.3 mm.

10.34 The cross-section of a beam is simplified into the web–boom idealization given in Fig. 10.50. Calculate the flxural shear flow distribution corresponding to a vertical downward

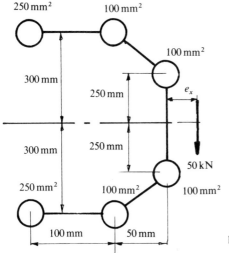

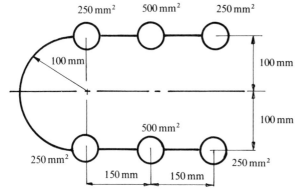

Figure 10.49

Figure 10.50

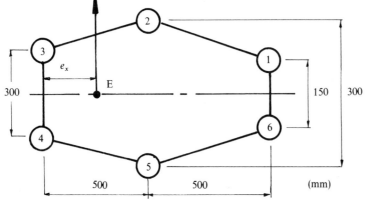

Figure 10.51

shear force of 200 kN applied through the shear centre. What is the position of the centre w.r.t. the open end?

Answer: 607 mm. (K.P. 1976)

10.35 Determine the position of the shear centre for the section in Fig. 10.51 assuming that the booms carry direct stress and the walls carry shear stress. Take $A_1 = A_6 = 500$ mm²; $A_2 = A_5 = 200$ mm²; $A_3 = A_4 = 800$ mm²; $t_{12} = t_{23} = t_{45} = t_{56} = 1.5$ mm; $t_{16} = 2$ mm; $t_{34} = 4$ mm.

Answer: 327 mm from left.

10.36 Determine the shear flow distribution, the rate of twist, the position of the shear centre and the warping displacements at the booms when the idealized section in Fig. 10.52 is subjected to an upward shear force of 2.5 kN. The enclosed areas are $A_1 = 3 \times 10^3$ mm², $A_2 = 5 \times 10^3$ mm².

Answer: 10.09 N/mm, 1.02 N/mm, 3.8 N/mm; and 0.0056°/m, 104 mm to the right of the applied force.

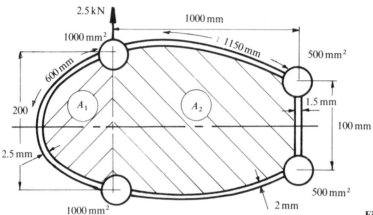

Figure 10.52

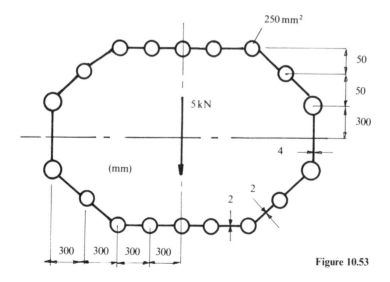

Figure 10.53

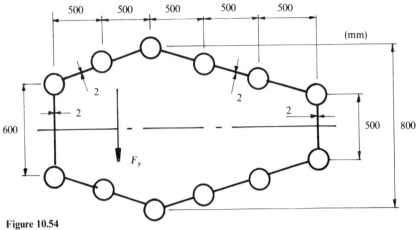

Figure 10.54

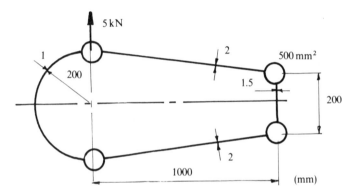

Figure 10.55

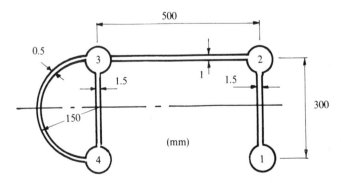

Figure 10.56

10.37 Determine the shear flow distribution and the warping displacements when the box section in Fig. 10.53 supports a 5 kN downward force along its centre line.

10.38 Determine the position of the shear centre for the idealized wing-section in Fig. 10.54, when the 200 mm² booms carry direct stress and the 2 mm thick webs carry only shear stress.

10.39 The idealized, singly symmetric, closed tube in Fig. 10.55 carries a 5 kN upwards vertical force at the position shown. Determine the shear flow distribution and the unrestrained warping displacements resulting from this force.

10.40 The cross-section of a wing at the position of the undercarriage is shown in Fig. 10.56. The front cell is closed and the rear cell is open, the section being symmetrical apart from the missing skin 1–4. Walls 1–2, 2–3, 3–4 are straight, wall 3–4 is curved. All walls have the same shear modulus and are assumed to be effective in carrying only shear stress. Direct stress is carried by four equal booms 1–4, each of area 1000 mm². The enclosed area of the front cell is 120×10^3 mm². Calculate the position of the shear centre of the tube cross-section. (C.E.I. Pt. 2, 1974)

10.41 Idealize the box section in Fig. 10.47 when the given thicknesses are all effective in shear. Webs 1–2 and 3–4 carry no direct stress but covers 1–4 and 2–3 carry direct stress equivalent to a thickness of 5 mm. Hence determine the position of the shear centre.

10.42 The given thicknesses for the tube section in Fig. 10.55 are those effective in shear. The boom areas, which carry direct stress, are further increased by a contribution from the upper and lower panels, whose effective thicknesses are both 4 mm. Establish the net boom areas and the position of the shear centre.

ELEVEN

STRAIN ENERGY AND APPLICATIONS

11.1 ELASTIC STRAIN ENERGY STORED

The first law of thermodynamics states that energy is conserved within a closed system in the following manner:

$$\delta Q - \delta W = \delta U \tag{11.1}$$

where δQ and δW are respectively heat and work transfers to and from the system that result in a change δU in its internal energy. Consider the system as a body being loaded under forces in equilibrium. With an increase in these forces, external work is done on the body $(-\delta W)$ that will result in a positive change in its internal energy δU. When this process is adiabatic $(\delta Q = 0)$ it follows from Eq. (11.1) that

$$\int dW = \int dU \tag{11.2}$$

i.e. the total work done by these forces is equal to the increase in internal energy. When the forces are slowly applied, such that kinetic energy due to the rate of deformation is negligible, δU becomes the increase in the elastic strain energy stored within the body. It is then possible to derive an expression for δU. In the general case, consider a volume element $\delta V = \delta x \times \delta y \times \delta z$ for any solid or structure as in Fig. 11.1, that is subjected to a generalized stress system σ_{ij}. The equivalence between the stress components in the tensor notation $(i, j = 1, 2, 3)$ and the engineering notation $(i, j = x, y, z)$ is expressed in the three independent normal components: $\sigma_{11} = \sigma_x$, $\sigma_{22} = \sigma_y$ and $\sigma_{33} = \sigma_z$, and in the three independent shear stresses: $\sigma_{12} = \tau_{xy}$, $\sigma_{23} = \tau_{yz}$ and $\sigma_{31} = \tau_{zx}$.

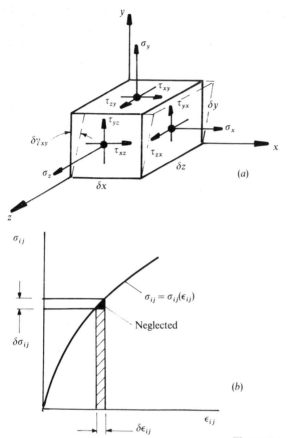

Figure 11.1 Volume element under stress.

Dependent complements of shear stress $\sigma_{21} = \tau_{yx}$, $\sigma_{32} = \tau_{zy}$ and $\sigma_{31} = \tau_{zx}$ also exist to give a 3×3 symmetrical stress matrix $(\sigma_{ij} = \sigma_{ji})$ in either notation:

$$\sigma_{ij} = \begin{bmatrix} \sigma_{11} & \sigma_{12} & \sigma_{13} \\ \sigma_{21} & \sigma_{22} & \sigma_{23} \\ \sigma_{31} & \sigma_{32} & \sigma_{33} \end{bmatrix} \equiv \begin{bmatrix} \sigma_x & \tau_{xy} & \tau_{xz} \\ \tau_{yx} & \sigma_y & \tau_{yz} \\ \tau_{zx} & \tau_{zy} & \delta_z \end{bmatrix}.$$

Let the corresponding strain increments $\delta\epsilon_{ij}$ $(i, j = 1, 2, 3)$ be produced by an elastic stress change $\delta\sigma_{ij}$, where

$$\delta\epsilon_{ij} = \begin{bmatrix} \delta\epsilon_{11} & \delta\epsilon_{12} & \delta\epsilon_{13} \\ \delta\epsilon_{21} & \delta\epsilon_{22} & \delta\epsilon_{23} \\ \delta\epsilon_{31} & \delta\epsilon_{32} & \delta\epsilon_{33} \end{bmatrix} \equiv \begin{bmatrix} \delta\epsilon_x & \delta\gamma_{xy}/2 & \delta\gamma_{xz}/2 \\ \delta\gamma_{yx}/2 & \delta\epsilon_y & \delta\gamma_{yz}/2 \\ \delta\gamma_{zx}/2 & \delta\gamma_{zy}/2 & \delta\epsilon_z \end{bmatrix},$$

from which it is seen that each component of engineering shear strain (γ_{xy}, γ_{yz}, γ_{xz}) is twice the corresponding tensor component (ϵ_{12}, ϵ_{23}, ϵ_{13}). Work is done by the normal stresses when forces $(\sigma_x + \delta\sigma_x)(\delta y \times \delta z)$, $(\sigma_y + \delta\sigma_y)(\delta x \times \delta z)$ and $(\sigma_z + \delta\sigma_z)(\delta x \times \delta y)$ move their points of application through their respective extensions $(\delta\epsilon_x \times \delta x)$, $(\delta\epsilon_y \times \delta y)$ and $(\delta\epsilon_z \times \delta z)$. Work is done by the shear forces: $(\tau_{xy} + \delta\tau_{xy})(\delta x \times \delta z)$, $(\tau_{yz} + \delta\tau_{yz})(\delta y \times \delta x)$ and $(\tau_{zx} + \delta\tau_{zx}) \times (\delta z \times \delta y)$ as they move their points of application through the respective shear displacements, $(\delta\gamma_{xy}/2)\delta y$, $(\delta\gamma_{yz}/2)\delta z$ and $(\delta\gamma_{zx}/2)\delta x$. In addition, the complementary shear forces $(\tau_{yx} + \delta\tau_{yx})(\delta y \times \delta z)$, $(\tau_{zy} + \delta\tau_{zy})(\delta x \times \delta z)$ and $(\tau_{xz} + \delta\tau_{xz})(\delta y \times \delta x)$ do work against their respective displacements; $(\delta\gamma_{yx}/2)\delta x$, $(\delta\gamma_{zy}/2)\delta y$ and $(\delta\gamma_{xz}/2)\delta z$. The net increase in strain energy is, by superposition, the sum of the products of these forces and the displacements along their lines of action. That is,

$$
\begin{aligned}
\delta U = {} & (\sigma_x + \delta\sigma_x)(\delta y \times \delta z)(\delta\epsilon_x \times \delta x) + (\sigma_y + \delta\sigma_y)(\delta x \times \delta z)(\delta\epsilon_y \times \delta y) \\
& + (\sigma_z + \delta\sigma_z)(\delta x \times \delta y)(\delta\epsilon_z \times \delta z) + (\tau_{xy} + \delta\tau_{xy})(\delta x \times \delta z)(\delta\gamma_{xy}/2)\delta y \\
& + (\tau_{yz} + \delta\tau_{yz})(\delta y \times \delta z)(\delta\gamma_{yz}/2)\delta z + (\tau_{zx} + \delta\tau_{zx})(\delta z \times \delta y)(\delta\gamma_{zx}/2)\delta x \\
& + (\tau_{yx} + \delta\tau_{yx})(\delta y \times \delta z)(\delta\gamma_{yx}/2)\delta x + (\tau_{zy} + \delta\tau_{zy})(\delta x \times \delta z)(\delta\gamma_{zy}/2)\delta y \\
& + (\tau_{xz} + \delta\tau_{xz})(\delta y \times \delta x)(\delta\gamma_{xz}/2)\delta z.
\end{aligned}
$$

Since elastic strains are small, the infinitesimal products $\delta\sigma \times \delta\epsilon$ may be neglected to give a simpler form of δU, which further embodies the properties of complementary shear, $\tau_{xy} = \tau_{yx}$, $\delta\gamma_{xy} = \delta\gamma_{xy}/2 + \delta\gamma_{yx}/2$, etc. That is,

$$
\delta U = (\sigma_x\delta\epsilon_x + \sigma_y\delta\epsilon_y + \sigma_z\delta\epsilon_z + \tau_{xy}\delta\gamma_{xy} + \tau_{yz}\delta\gamma_{yz} + \tau_{zx}\delta\gamma_{zx})\,\delta V. \quad (11.3a)
$$

Note that, in the tensor notation, Eq. (11.3a) is conveniently written as,

$$
\delta U = (\sigma_{ij}\delta\epsilon_{ij})\,\delta V \quad (11.3b)
$$

because, by the summation convention,

$$
\begin{aligned}
\sigma_{ij}\delta\epsilon_{ij} = {} & \sigma_{1j}\delta\epsilon_{1j} + \sigma_{2j}\delta\epsilon_{2j} + \sigma_{3j}\delta\epsilon_{3j} \\
= {} & [\sigma_{11}\delta\epsilon_{11} + \sigma_{12}\delta\epsilon_{12} + \sigma_{13}\delta\epsilon_{13}] + [\sigma_{21}\delta\epsilon_{21} + \sigma_{22}\delta\epsilon_{22} + \sigma_{23}\delta\epsilon_{23}] \\
& + [\sigma_{31}\delta\epsilon_{31} + \sigma_{32}\delta\epsilon_{32} + \sigma_{33}\delta\epsilon_{33}].
\end{aligned}
$$

Figure 11.1(b) shows the change in all the stress and strain components when an elastic relationship exists between them. Clearly, the shaded area also represents the change in strain energy per unit volume, i.e. the strain energy density $\delta U/\delta V = \sigma_{ij}\delta\epsilon_{ij}$. Elasticity implies that strains and the strain energy are wholly recoverable. When the same stress–strain path is followed between loading and unloading, the formulation $\sigma_{ij} = \sigma_{ij}(\epsilon_{ij})$, admits materials that are elastic and non-linear. Only in the particular case where Hooke's law is obeyed will the stress be linearly related to strain through the usual elastic constants. The total strain energy stored in a body may be found from integrating Eq. (11.3b) along the strain path and over the whole volume:

$$
U = \int_V \int_{\epsilon_{ij}} (\sigma_{ij}\mathrm{d}\epsilon_{ij})\,\mathrm{d}V. \quad (11.3c)
$$

Particular stress systems are now considered for Hookean materials in which each component of $\delta\epsilon_{ij}$ in Eq. (11.3c) depends upon all acting stresses, by the constitutive laws given in Eqs (3.3), (3.8) and (4.14).

Direct Stress

The strain energy stored in a bar of uniform cross-section (Fig. 11.2), when it is elastically loaded in tension or compression, is found by substituting the stress σ and strain increment $d\epsilon = d\sigma/E$ into Eq. (11.3c) with $\delta V = A\,\delta z$. Then,

$$U = (1/2E) \int_0^L \sigma^2 A\,dz = (W^2/2E) \int_0^L dz/A \qquad (11.4a)$$

$$= \sigma^2 AL/2E = W^2 L/2AE. \qquad (11.4b)$$

In non-uniform sections, the strain energy can also be found from Eq. (11.4a) provided the function $A = A(z)$ is known. This is illustrated with the following two examples.

Example 11.1 The steel (s) and brass (b) bars comprising the stepped shaft in Fig. 11.3 are to suffer the same displacement under an axial tensile force of 40 kN. Determine the diameter of the brass bar, d_b, the stress in each material and the total strain energy stored. Take $E_s = 207\,\text{GPa}$, $E_b = 82.7\,\text{GPa}$.

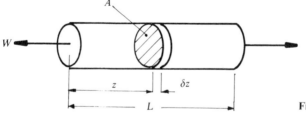

Figure 11.2 Bar in tension.

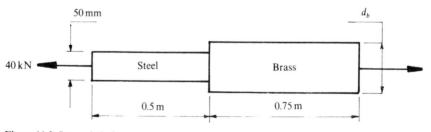

Figure 11.3 Stepped shaft.

Since the displacements in the two materials are the same,

$$(WL/AE)_s = (WL/AE)_b, \quad \Rightarrow (WL/d^2E)_s = (WL/d^2E)_b,$$

$$\therefore \quad d_b^2 = (d^2E/WL)_s(WL/E)_b$$

$$= (50^2 \times 207/40 \times 0.5)(40 \times 0.75/82.7)$$

$$= 93.86 \times 10^2 \,\text{mm}^2,$$

$$d_b = 96.88 \,\text{mm},$$

$$\therefore \quad \sigma_s = (W/A)_s = 40 \times 10^3/(\pi/4)(50)^2 = 20.37 \,\text{MPa},$$

$$\sigma_b = (W/A)_b = 40 \times 10^3/(\pi/4)(96.88)^2 = 2.33 \,\text{MPa}.$$

From Eq. (11.4),

$$U = (1/2E_s) \int_0^{L_s} \sigma^2 A \, \mathrm{d}z + (1/2E_b) \int_0^{L_b} \sigma^2 A \, \mathrm{d}z$$

$$= (W^2/2)[(L/AE)_s + (L/AE)_b] = (2W^2/\pi)[(L/d^2E)_s + (L/d^2E)_b]$$

$$= [2 \times (40 \times 10^3)^2/\pi]\{[(0.5 \times 10^3)/(50^2 \times 207 \times 10^3)]$$

$$+ [(0.75 \times 10^3)/(96.88^2 \times 82.7 \times 10^3)]\} = 1968.4 \,\text{N mm} = 1.97 \,\text{J}.$$

Example 11.2 Show that for the tapered shaft under the axial compressive force W (Fig. 11.4), the strain energy stored is $U = 2W^2L/\pi E d_1 d_2$ and deflection is $\delta = 4WL/\pi E d_1 d_2$.

In the z, y coordinates shown, the function $A = A(z)$ in Eq. (11.4) becomes

$$A(z) = (\pi/4)(2y)^2 = \pi[d_1/2 + (d_2 - d_1)z/2L]^2$$

$$= (\pi d_1^2/4)\{1 + [(d_2/d_1) - 1](z/L)\}^2.$$

Substituting into Eq. (11.4),

$$U = (2W^2/\pi E d_1^2) \int_0^L \mathrm{d}z/\{1 + [(d_2/d_1) - 1](z/L)\}^2$$

$$= \frac{-2W^2L}{\pi E d_1^2[(d_2/d_1) - 1]} \left| \frac{1}{1 + [(d_2/d_1) - 1](z/L)} \right|_0^L$$

$$= 2W^2L/\pi E d_1 d_2.$$

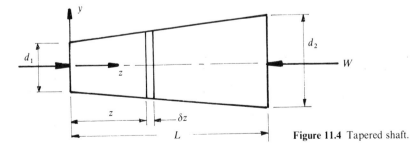

Figure 11.4 Tapered shaft.

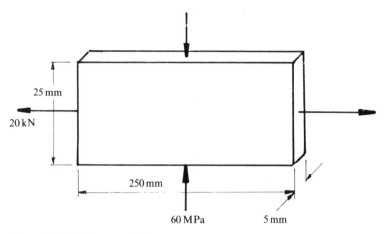

Figure 11.5 Biaxially stressed plate.

Equating U to the external work done:

$$W\delta/2 = 2W^2 L/\pi E d_1 d_2,$$
$$\therefore \quad \delta = 4WL/\pi E d_1 d_2,$$

which agrees with the direct integration method (see Example 3.3).

Example 11.3 Find the strain energy stored in the steel plate in Fig. 11.5 when it is subjected to an in-plane, longitudinal tensile force of 20 kN together with a compressive stress of 60 MPa on its remaining edges. Take $E = 207$ GPa, $v = 0.28$.

For the principal 1, 2 directions shown, Eq. (4.14) is written as

$$\delta\epsilon_1 = (\delta\sigma_1 - v\,\delta\sigma_2)/E \quad \text{and} \quad \delta\epsilon_2 = (\delta\sigma_2 - v\,\delta\sigma_1)/E.$$

Substituting into Eq. (11.3b),

$$\delta U = (\sigma_1 \delta\epsilon_1 + \sigma_2 \delta\epsilon_2)\delta V,$$

$$\therefore \quad U = \int_V \int_\sigma (\sigma_1\,\delta\sigma_1 - v\sigma_1\,\delta\sigma_2 + \sigma_2\,\delta\sigma_2 - v\sigma_2\,\delta\sigma_1)\delta V/E$$

$$= \int_V \int_\sigma [\sigma_1\,\delta\sigma_1 - v\,\delta(\sigma_1\sigma_2) + \sigma_2\,\delta\sigma_2]\,\delta V/E$$

$$= \int_V (\sigma_1^2/2 - v\sigma_1\sigma_2 + \sigma_2^2/2)\,\mathrm{d}V/E$$

$$= (\sigma_1^2 - 2v\sigma_1\sigma_2 + \sigma_2^2)V/2E.$$

Substituting $\sigma_1 = (20 \times 10^3/25 \times 5) = 160\,\text{MPa}$, $\sigma_2 = 60\,\text{MPa}$, $V = (25 \times 5 \times 250) = 31.25 \times 10^3\,\text{mm}^3$:

$$U = \frac{[160^2 - (2 \times 0.28 \times 160 \times 60) + 60^2](31.25 \times 10^3)}{2 \times 207 \times 10^3}$$

$$= 1798.3\,\text{N mm} = 1.79\,\text{J}.$$

Torsion

For a uniform shaft under torsion the only non-zero stress is τ. Hence, $\sigma_{ij}\,d\epsilon_{ij} = \gamma\,d\gamma$ in Eq. (11.3c), where from Eq. (9.5); $\tau = Tr/J$ and $d\gamma = d\tau/G$. For a constant torque this gives

$$U = (1/G) \int_V \left[\int_\tau (\tau\,d\tau) \right] dV = (1/2G) \int_V \tau^2\,dV$$

$$= (T^2/2J^2G) \int_V r^2\,dV, \tag{11.5a}$$

but

$$dV = dA \times dz = (2\pi r\,dr)\,dz \qquad \text{(see Fig. 11.6a)},$$

$$\therefore \quad \int_V r^2\,dV = 2\pi \int_0^r r^3\,dr \int_0^L dz = J \int_0^L dz$$

and Eq. (11.5a) becomes

$$U = (T^2/2JG) \int_0^L dz. \tag{11.5b}$$

When the torque $T = T(z)$ varies with z in a non-uniform shaft $J = J(z)$, comprising more than one material, i.e. when G varies, a more general form of Eq. (11.5b) applies:

$$U = (1/2) \int_0^L (T^2/GJ)\,dz. \tag{11.5c}$$

Example 11.4 The solid shaft, with diameter $d = 180\,\text{mm}$, in Fig. 11.6(a) is to be replaced by a hollow shaft of the same material, length and weight in Fig. 11.6(b). Find the diameters of the hollow shaft in order to make its store of strain energy 20 per cent greater than that of the solid rod when transmitting torque at the same maximum shear stress.

For the same weight and length the cross-sectional areas of the two bars are equal:

$$\pi(D_1^2 - D_2^2)/4 = \pi d^2/4,$$
$$D_1^2 - D_2^2 = d^2. \tag{i}$$

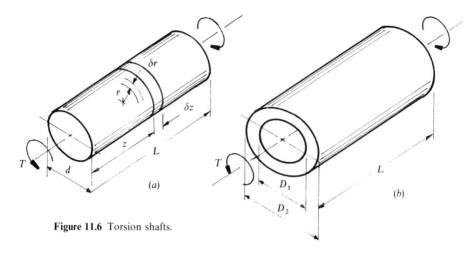

Figure 11.6 Torsion shafts.

Now, from Eq. (11.5b) with $\tau_{max} = T(d/2)/J$ and $J = \pi d^4/32$, for the solid shaft,

$$U_s = T^2 L/2JG = \tau_{max}^2 LJ/2G(d/2)^2,$$
$$U_s = (\tau_{max}^2/4G)V. \tag{ii}$$

For the hollow shaft, with $\tau_{max} = T(D_1/2)/J$ and $J = \pi(D_1^4 - D_2^4)/32$, Eq. (11.5b) becomes

$$U_h = T^2 L/2JG = \tau_{max}^2 LJ/2G(D_1/2)^2,$$
$$U_h = \tau_{max}^2 \pi L(D_1^4 - D_2^4)/16D_1^2 G,$$
$$U_h = (\tau_{max}^2/4G)(D_1^2 + D_2^2)V/D_1^2. \tag{iii}$$

$U_h = 1.2U_s$ when, from Eqs. (ii) and (iii),

$$D_1^2 + D_2^2 = 1.2D_1^2. \tag{iv}$$

Substituting from Eq. (i), with $d = 180\,\text{mm}$, into Eq. (iv),

$$D_1^2 + (D_1^2 - 180^2) = 1.2D_1^2,$$
$$0.8D_1^2 = 180^2,$$
$$\therefore \quad D_1 = 201.25\,\text{mm and } D_2 = \sqrt{(D_1^2 - 180^2)} = 90\,\text{mm}.$$

Shear

In the presence of a shear stress τ, together with its complement, then $\sigma_{ij}\,d\epsilon_{ij} = \tau\,d\gamma$, where $d\gamma = d\tau/G$ as with simple torsion. Equation (11.3c) becomes, on integration,

$$U = (1/2G)\int_V \tau^2 \, dV. \tag{11.6a}$$

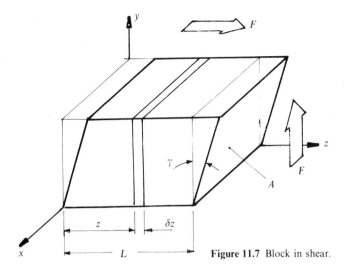

Figure 11.7 Block in shear.

Particular expressions may be substituted for τ. In the case of direct shearing forces F, slowly applied to side areas A, of a block with volume $V = AL$ in Fig. 11.7, then $\tau = F/A$. The net stored energy, from Eq. (11.6), is

$$U = (F^2/2A^2G)V = F^2L/2AG. \qquad (11.6b)$$

In the case of flexural shearing in a beam, where $F = F(z)$ and $\tau = \tau(y)$, substitution from Eq. (10.3) into (11.6) leads to the expression for the total strain energy stored that can admit a non-uniform section $A(z)$ and a change in material (G) in both length and section. However, unless the beam is very short, U in shear is normally small and can often be neglected in comparison

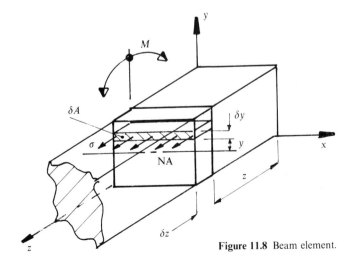

Figure 11.8 Beam element.

with the simultaneous energy that is stored owing to beam bending (see Example 11.5). For this reason it is convenient to consider U in bending separately.

Bending

Consider an element $\delta A (= x\,\delta y) \times \delta z$ of a uniform section beam, distance y from the NA in Fig. 11.8, carrying a bending moment $M = M(z)$. With shearing absent, $\sigma_{ij} d\epsilon_{ij} = \sigma\, d\epsilon$ where, from Eq. (6.4), $\sigma = My/I$ and $d\epsilon = d\sigma/E$. Integrating Eq. (11.3c) with $\sigma = E\epsilon$ leads to:

$$U = (1/2) \int_V \sigma\epsilon\, dV$$

$$= (1/2EI^2) \int_V (M^2 y^2)\, dV$$

$$= (1/2EI^2) \int_0^L M^2 \int_A (y^2\, dA)\, dz$$

$$= (1/2EI) \int_0^L M^2\, dz, \tag{11.7a}$$

where, by definition, $I = \int_A y^2\, dA$. To account for a non-uniform section $I = I(z)$, and a change in material (E), the general expression is written as

$$U = (1/2) \int_0^L M^2\, dz/EI. \tag{11.7b}$$

Example 11.5 Find the total strain energy stored for the stepped cantilever in Fig. 11.9(a) owing to bending and shear effects. Is it possible to find the free-end deflection in this case? Take the moduli as $E = 100\,\mathrm{GPa}$ and $G = 38\,\mathrm{GPa}$.

The shear force $F(z)$ and bending moment $M(z)$ diagrams are shown in Fig. 11.9(b) and (c). The shear strain energy is found from applying Eq. (11.6) to each part of the beam. That is, for parts 1 and 2, identified as shown,

$$U_s = (1/2GA_1) \int F_1^2\, dz + (1/2GA_2) \int F_2^2\, dz$$

$$= (2 \times 38 \times 10^3 \times 20^2)^{-1} \int_{0.5}^{1.0} 15^2\, dz$$

$$+ (2 \times 38 \times 10^3 \times 20 \times 25)^{-1} \int_0^{0.5} (25)^2\, dz$$

$$= (0.0037 + 0.0039)10^{-3}\,\mathrm{N\,m} = 0.0076 \times 10^{-3}\,\mathrm{J}.$$

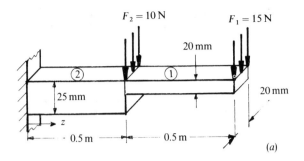

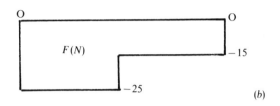

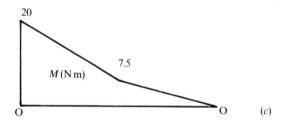

Figure 11.9 Stepped cantilever.

Applying Eq. (11.7), the strain energy in bending is,

$$U_b = (1/2EI_1) \int M_1^2 \, dz + (1/2EI_2) \int M_2^2 \, dz,$$

where

$$EI_1 = 100 \times 10^3 \times 20^4/12 = 13.33 \times 10^8 \, \text{N mm}^2 = 1333 \, \text{N m}^2,$$
$$EI_2 = 100 \times 10^3 \times 20 \times 25^3/12 = 52.08 \times 10^8 \, \text{N mm}^2 = 5208 \, \text{N m}^2,$$

and, with the origin of z at the fixed end:

$$M_1 = 15(1 - z) \quad (\text{N m}) \qquad \text{for} \quad 0.5 \le z \le 1 \, \text{m},$$
$$M_2 = (1 - z)15 + (0.5 - z)10 = 5(4 - 5z)(\text{N m}) \qquad \text{for} \qquad 0 \le z \le 0.5 \, \text{m},$$

$$\therefore \quad U_b = (2 \times 1333)^{-1} 225 \int_{0.5}^{1.0} (1 - z)^2 \, dz$$

$$+ (2 \times 5208)^{-1} 25 \int_{0}^{0.5} (4 - 5z^2) \, dz$$

$$= 0.0844 | z - z^2 + z^3/3 |_{0.5}^{1.0} + 0.0024 | 16z - 20z^2 + 25z^3/3 |_{0}^{0.5}$$

$$= (3.516 + 9.7) 10^{-3} \, \text{N m} = 13.22 \times 10^{-3} \, \text{J}.$$

The total strain energy $U_t = U_s + U_b = 13.2276 \times 10^{-3}$ J, though clearly the contribution from shear is negligible.

The application of Eq. (11.2) would enable the free-end deflection to be found if this were the only point under load. In the presence of the two forces the external work done for an increment of displacement is $\delta W = F_1 \delta v_1 + F_2 \delta v_2$. The equation contains two unknown displacements that are each dependent upon F_1 and F_2. It is shown in Section 11.5 and Example 11.16, how Eq. (11.2) may be developed further by the method of Castigliano to separate these displacements. Alternatively, the virtual work principle, as outlined in Chapter 12, may be employed.

Example 11.6 The U-shaped, 20 mm diameter steel cantilever in Fig 11.10 carries a vertical end force of $F = 50$ N. Determine the displacement v, under F, from a consideration of the strain energy stored in each limb under bending, shear and torsional effects. Take $E = 200$ GPa and $G = 80$ GPa.

Here

$$A = (\pi/4)(0.02)^2 = 0.3142 \times 10^{-3} \, \text{m}^2,$$
$$I = (\pi/64)(0.02)^4 = 0.7854 \times 10^{-8} \, \text{m}^4,$$
$$J = 2I = 1.571 \times 10^{-8} \, \text{m}^4.$$

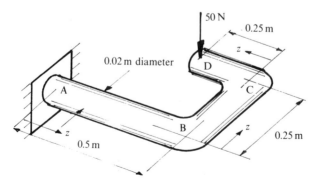

Figure 11.10 U-cantilever.

With the origin for z at A, limb AB is subjected to the three effects:

$$M = 50(0.25 - z) \, \text{N m} \qquad \text{for} \quad 0 \le z \le 0.25 \, \text{m},$$
$$T = 50 \times 0.25 = 12.5 \, \text{N m} \qquad \text{for} \quad 0 \le z \le 0.5 \, \text{m},$$
$$F = 50 \, \text{N} \qquad \text{for} \quad 0 \le z \le 0.25 \, \text{m}.$$

Substituting these into Eqs. (11.5)–(11.7) and adding, the net strain energy stored in AB is

$$U_{AB} = \int_0^{0.25} (50)^2 (0.25 - z)^2 \, \text{d}z/2EI + \int_0^{0.5} (12.5)^2 \, \text{d}z/2GJ$$
$$+ \int_0^{0.25} (50)^2 \, \text{d}z/2GA. \tag{i}$$

Limb BC is also subjected to all three effects. With the origin for z at B, they are

$$M = 50(0.25 - z) \, \text{N m} \qquad \text{for} \quad 0 \le z \le 0.25,$$
$$T = 50 \times 0.25 = 12.5 \, \text{N m} \qquad \text{for} \quad 0 \le z \le 0.25 \, \text{m},$$
$$F = 50 \, \text{N} \qquad \text{for} \quad 0 \le z \le 0.25 \, \text{m}.$$

$$\therefore \quad U_{BC} = \int_0^{0.25} (50)^2 (0.25 - z)^2 \, \text{d}z/2EI + \int_0^{0.25} (12.5)^2 \, \text{d}z/2GJ$$
$$+ \int_0^{0.25} (50)^2 \, \text{d}z/2GA. \tag{ii}$$

Limb CD is subjected to bending and shear. With the origin for z at C,

$$M = 50(0.25 - z) \, \text{N m} \qquad \text{for} \quad 0 \le z \le 0.25,$$
$$F = 50 \, \text{N} \qquad \text{for} \quad 0 \le z \le 0.25 \, \text{m},$$

$$\therefore \quad U_{CD} = \int_0^{0.25} (50)^2 (0.25 - z)^2 \, \text{d}z/2EI + \int_0^{0.25} (50)^2 \, \text{d}z/2GA. \tag{iii}$$

Adding Eqs (i), (ii) and (iii) to give the total strain energy stored,

$$U_T = 3\left[\int_0^{0.25} (50)^2 (0.25 - z)^2 \, \text{d}z/2EI + \int_0^{0.25} (12.5)^2 \, \text{d}z/2GJ \right.$$
$$\left. + \int_0^{0.25} (50)^2 \, \text{d}z/2GA \right] = (3/2)(50)^2 \left[\int_0^{0.25} (0.25 - z)^2 \, \text{d}z/EI \right.$$
$$\left. + \int_0^{0.25} \text{d}z/16GJ + \int_0^{0.25} \text{d}z/GA \right]$$

$$= (12.434 + 46.627 + 0.0373) \times 10^{-3} \, \text{N m} = 59.1 \times 10^{-3} \, \text{J},$$

in which the contribution from shear is negligible. The external work done is $W = \int F \, dv$, and when a linear law $F = Kv$ (K is a constant stiffness) is followed, the integration leads to $W = Fv/2$, i.e. the triangular area beneath the F vs. v diagram for loading. Then, from Eq. (11.2), $Fv/2 = U_T$ and the deflection is

$$\therefore \quad v = 2U_T/F = 2 \times 59.1/50 = 2.36 \, \text{mm}.$$

11.2 DEFLECTION DUE TO SHEAR

There is associated with the shear stress distribution in a beam, a shear strain and hence a deflection due to this distortion. These are usually negligibly small except for short, deep beams, where the magnitude of the shear deflection is comparable with the bending deflection. For this situation, the shear strain energy enables the contribution to the total deflection from shear to be acceptably estimated for beams with single concentrated and uniformly distributed loading. The shear deflection δ_s follows from Eq. (11.2) with $W = F \delta_s/2$ and U defined in Eq. (11.6a). That is, for a volume element $\delta V = dx \, \delta y \, \delta z$,

$$F \delta_s/2 = (1/2G) \int_{x,y,z} \tau^2 (dx \, dy \, dz), \tag{11.8}$$

where τ is the expression for the shear stress due to shear force in Eq. (10.3). In the general case, τ is a function of x, y and z because the shear force F varies with the length z, and the first moment integral varies with x and y, in a section with variable breadth.

Example 11.7 Find an expression for the shear deflection of a cantilever of length L with rectangular section $b \times d$ carrying a vertical concentrated force F at its free end.

It has previously been shown (page 351) that at every position in the length the shear stress due to shear force varies parabolically with the depth y of section from the neutral axis according to

$$\tau = (6F/bd^3)(d^2/4 - y^2),$$

where here the shear force ($= F$) remains constant for the length of beam (see Fig. 5.3(c)). Since τ varies only with y, substitution into Eq. (11.8) gives

$$F \delta_s/2 = (1/2G)(6F/bd^3)^2 bL \int_{-d/2}^{d/2} (d^2/4 - y^2)^2 \, dy$$

$$= (18F^2L/Gbd^6)|d^4y/16 - d^2y^3/6 + y^5/5|_{-d/2}^{d/2}$$

$$= 3F^2L/5bdG,$$

$$\therefore \quad \delta_s = 6FL/5bdG.$$

The free-end bending deflection δ_b derived in Chapter 7 (page 198) is also readily confirmed from Eq. (11.2). Substituting from Eq. (11.7a),

$$F\,\delta_b/2 = (1/2EI)\int_0^L M^2\,dz,$$

where, with the origin for z at the fixed end (see Fig. 7.3.), $M = F(L - z)$. Hence,

$$F\,\delta_b/2 = (F^2/2EI)\int_0^L (L^2 - 2Lz + z^2)\,dz$$

$$= F^2L^3/6EI,$$

$$\therefore \quad \delta_b = FL^3/3EI.$$

The total free-end deflection is then

$$\delta_t = \delta_s + \delta_b = 6FL/5bdG + FL^3/3EI.$$

Example 11.8 Determine the maximum deflection due to shear in a simply supported beam of length L, carrying a uniformly distributed load w.

In this case, as the shear force F varies with z, then τ is a function of both z and y. Using the result of the previous example, the strain energy stored in a rectangular section $b \times d \times \delta z$ is

$$F\,\delta_s/2 = \int_0^L 3F^2\,dz/5bdG.$$

With the origin at the beam centre, $F = wz$ for $0 \le z \le L/2$ in Fig. 5.3(b). Applying Eq. (11.8) to whole beam:

$$2(F\delta_s/2) = 2\int_0^{L/2} (3F^2\,dz/5bdG),$$

$$\therefore \quad \delta_s = (6/5bdG)\int_0^{L/2} F\,dz$$

$$= (6w/5bdG)|z^2/2|_0^{L/2},$$

$$\delta_s = 3wL^2/20bdG.$$

In a similar way, the shear deflection for a simply supported rectangular section beam, with central concentrated load F is $\delta_s = 3FL/10bdG$, and that for uniformly distributed loading (w/unit length) of a rectangular-section cantilever is $\delta_s = 3wL^2/5bdG$. The approach may be extended to more complex sections as in the following example.

Example 11.9 Find the total shear resilience for the I-section beam in Fig. 10.6 when it acts as a 1.25 m long cantilever carrying an end-load of 30 kN. Hence determine the deflection due to shear. Take $G = 82$ GPa.

The shear stress in the flange was previously derived as

$$\tau_f = (0.7407 \times 10^{-4})[(2.25 \times 10^4) - y^2] = 1.667 - (0.7407 \times 10^{-4})y^2.$$

Applying Eq. (11.6a) to a volume element $\delta V = (b_f \times \delta y \times L)$ in the top flange and noting that τ_f varies only with y:

$$U_f = (b_f L/2G) \int_{125}^{150} \tau_f^2 \, dy$$

$$= (200 \times 1250/2 \times 82 \times 10^3) \int_{125}^{150} [2.7789 - (2.47 \times 10^{-4})y^2$$

$$+ (0.5486 \times 10^{-8})y^4] \, dy$$

$$= 1.5244 |2.7789y - (0.8233 \times 10^{-4})y^3 + (0.1097 \times 10^{-8})y^5|_{125}^{150}$$

$$= 1.5244[222.274 - 220.04] = 3.4 \text{ N mm}.$$

For a volume element $\delta V = b_w \times \delta y \times L$ in the upper half of the web,

$$U_w = (b_w L/2G) \int_0^{125} \tau_w^2 \, dy$$

$$= (10 \times 1250/2 \times 82 \times 10^3) \int_0^{125} [128.37 - (16.768 \times 10^{-4})y^2$$

$$+ (0.5476 \times 10^{-8})y^4] \, dy$$

$$= 0.0762 |128.37y - (5.5893 \times 10^{-4})y^3 + (0.1095 \times 10^{-8})y^5|_0^{125}$$

$$= 0.0762[16\,046.25 - 1091.66 + 33.423] = 1142.09 \text{ N mm}.$$

The total strain energy stored in the beam is

$$\sum U = 2(U_f + U_w) = 2(3.4 + 1142.09) = 2290.98 \text{ N mm}.$$

Applying Eq. (11.8) to the load point gives

$$F \delta_s / 2 = \sum U$$

$$\therefore \quad \delta_s = 2 \sum U / F = 2 \times 2290.98 / (30 \times 10^3) = 0.153 \text{ mm}.$$

11.3 APPLICATION TO SPRINGS

Close-coiled Helical Spring

A particularly useful application of Eq. (11.2) is to determine the elastic deformation behaviour of closely wound springs subjected to either tensile or torsional loading (Figs 11.11(a) and (b)).

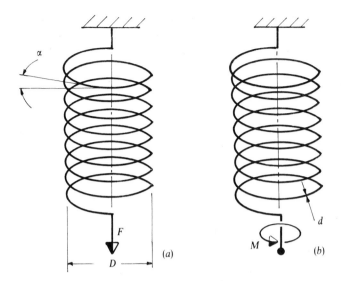

Figure 11.11 Tensile and torsional loading of a close-coiled helical spring.

It is assumed that when the helix angle α is less than $10°$ the deformation in the spring wire is the result of one or other mode, i.e. there is no interaction between tension and torsion.

Spring under axial force F Let the wire diameter in Fig. 11.11(a) be d, the mean coil diameter D, the number of turns N, and the coil length $L = \pi DN$. The basis of the solution lies in recognizing that F will twist the spring wire with a constant torque $T = FD/2$ all along the spring length. If δ is the deflection beneath F, then, as with Example 11.6, the external work done is $W = F\delta/2$ and the strain energy in torsion (Eq. (11.5b)) is that stored in the wire, $U = T^2L/2GJ$. It follows from Eq. (11.2) that

$$F\delta/2 = T^2L/2GJ.$$

Substituting for T, L and $J = \pi d^4/32$ leads to the following relationships:

$$G = 8FD^2L/\pi\delta d^4 = 8FD^3N/\delta d^4 \qquad (11.9a)$$

and the spring stiffness K is

$$K = F/\delta = \pi G d^4/8D^2L = Gd^4/8D^3N. \qquad (11.9b)$$

The angular twist in radians between the ends of the wire is found from ETT (Eq. (9.5)):

$$\theta = TL/GJ = 16NFD^2/Gd^4. \qquad (11.10)$$

Moreover, Eq. (9.5) also supplies the maximum torsional shear stress in the wire:

$$\tau = Tr/J = 8FD/\pi d^3. \tag{11.11}$$

This shear stress is enhanced, since every wire cross-section (of area A) is further subjected to a shear force F. The shear stress produced is, nominally,

$$\tau = F/A = 4F/\pi d^2. \tag{11.12}$$

The net shear stress is then the sum of Eqs (11.11) and (11.12):

$$\tau = (8FD/\pi d^3)(1 + d/2D). \tag{11.13}$$

A Wahl correction factor $m = D/d$ accounts for the increasing contribution from shear force and a greater amount of shear strain at the inner coil diameter than is provided by torsion theory, as m decreases. Correspondingly, Eq. (11.13) has been modified empirically to become

$$\tau = (8FD/\pi d^3)[(4m - 1)/(4m - 4) + 0.615/m].$$

In theory, the principles outlined in Chapter 9 could account for non-circular sections, particularly where square-section wire is wound into springs.

Example 11.10 A close-coiled helical spring is made from steel wire of 12 mm (1/2 in) diameter. The mean diameter of the coils is 150 mm (3 in) and the number of effective coils is 21. Find the elongation of the spring when it carries an axial load of 25 N (60 lbf) and the maximum torsional shear stress in the wire. Take $G = 85$ GPa (12×10^6 lbf/in^2).

SOLUTION TO SI UNITS From Eq. (11.9) and (11.11),

$$\delta = 8FD^3N/Gd^4$$
$$= 8 \times 25 \times 150^3 \times 21/(85 \times 10^3 \times 12^4) = 8.04 \text{ mm},$$
$$\tau = 8FD/\pi d^3 = 8 \times 25 \times 150/(\pi \times 12^3) = 5.53 \text{ MPa}.$$

SOLUTION TO IMPERIAL UNITS

$$\delta = 8FD^3N/Gd^4$$
$$= 8 \times 60 \times 21 \times 6^3/[12 \times 10^6 \times (0.5)^4]$$
$$= 2.905 \text{ in},$$
$$\tau = 8FD/\pi d^3$$
$$= 8 \times 60 \times 6/[\pi \times (0.5)^3]$$
$$= 7333 \text{ lbf/in}^2.$$

Example 11.11 A close-coiled helical spring is required to have a stiffness of 1 N/mm when carrying a maximum axial force of 40 N without the shear stress exceeding 125 MPa. If the solid length of the spring is 45 mm, when the spring coils are touching, find the diameter of the spring wire, the mean coil diameter and the number of coils required. Take $G = 41.5\,\text{GPa}$.

Noting that the solid length $l_s = Nd$, the stiffness (Eq. (11.9b)) becomes

$$K = F/\delta = Gd^4/8D^3N = Gd^5/8D^3l_s,$$
$$1 = 41.5 \times 10^3 \times d^5/(8 \times 45 \times D^3),$$
$$D^3 = 115.28d^5. \tag{i}$$

Then, from Eq. (11.11),

$$\tau = 8FD/\pi d^3,$$
$$125 = 8 \times 40 \times D/\pi d^3,$$
$$D = 1.227d^3. \tag{ii}$$

Finally, from Eqs (i) and (ii),

$$(1.227d^3)^3 = 115.28d^5,$$
$$d = 2.81\,\text{mm},$$
$$\therefore \quad D = 1.227(2.81)^3 = 27.24\,\text{mm},$$
$$N = l_s/d = 45/2.81 \approx 16.$$

Spring under axial moment or torque M The same symbols again apply when the spring in Fig. 11.11(b) is subjected to an axial twisting moment or torque M that is reacted by every wire cross-section. The strain energy stored in bending (Eq. (11.7)) is the result of the external work $W = \int M\,\mathrm{d}\phi$, where ϕ is the angular twist in radians accompanying M. With a linear relationship between M and ϕ, Eq. (11.2) gives

$$M\phi/2 = M^2L/2EI,$$
$$\therefore \quad \phi = ML/EI. \tag{11.14}$$

Equation (11.14) and ETB (Eq. (6.4)) may now be applied according to the shape of the wire cross-section.

(i) Circular section of diameter, d

$$I = \pi d^4/32 \quad \text{and} \quad L = \pi DN,$$
$$\phi = 64M(\pi DN)/E\pi d^4,$$
$$E = 64MDN/\phi d^4, \tag{11.15}$$
$$\sigma = My/I = M(d/2)/(\pi d^4/64) = 32M/\pi d^3 \text{ (max.)}. \tag{11.16}$$

(ii) Square section side length a

$$I = a^4/12 \quad \text{and} \quad L = \pi DN,$$
$$\phi = 12M(\pi DN)/Ea^4,$$

$$E = 12M\pi DN/\phi a^4, \tag{11.17}$$

$$\sigma = M(a/2)/(a^4/12) = 6M/a^3 \quad \text{(max.)}. \tag{11.18}$$

Example 11.12 A close-coiled helical spring is subjected to an axial torque of 8 N m, which produces an axial twist of 60°. The mean coil diameter is 30 mm and the bending stress in the wire is restricted to 415 MPa. Calculate the wire size and the number of effective coils if the section is (i) circular and (ii) square. Take $E = 207$ GPa.

(i) Circular section

The wire diameter is found from Eq. (11.16):

$$d^3 = (32M/\pi\sigma),$$
$$d = [32 \times 8 \times 10^3/(\pi \times 415)]^{1/3} = 5.8 \text{ mm},$$

and the number of coils is, from Eq. (11.15),

$$N = E\phi d^4/64MD$$
$$= \frac{(207 \times 10^3)(60 \times \pi/180)(5.8)^4}{64 \times 8 \times 10^3 \times 30} \approx 16 \quad \text{(rounded up)}.$$

(ii) Square section

Equation (11.18) supplies the square size:

$$a^3 = 6M/\sigma,$$
$$a = (6 \times 8 \times 10^3/415)^{1/3} = 4.87 \text{ mm},$$

and Eq. (11.17) gives the number of coils:

$$N = E\phi a^4/12M\pi D$$
$$= \frac{(207 \times 10^3)(60 \times \pi/180)(4.87)^4}{12 \times 8 \times 10^3 \times \pi \times 30} = 14 \quad \text{(rounded up)}$$

Open-coiled Helical Springs

When the angle between the wire axis and a plane perpendicular to the spring axis (i.e. the helix angle α in Fig. 11.12) increases beyond 10° then both bending and torsion at each section of the wire occur under either an applied axial load F or a twisting moment M.

Axial load Consider a portion AB, of the helix formed by the axis of the wire in Fig. 11.12. When A lies in a vertical plane, the moment $M = FD/2$, due to the axial load, is represented by the horizontal vector at A with the

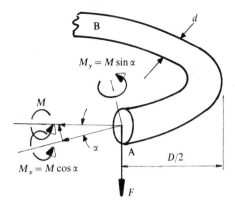

Figure 11.12 Spring under tension.

two components:

$M_x = M \cos \alpha$, parallel to the helix axis at A, which twists the wire; and
$M_y = M \sin \alpha$, perpendicular to the helix axis at A, which bends the wire.

The net strain energy stored in the spring is the sum of Eq. (11.5) and Eq. (11.7). Equating to the external work $F\delta/2$ at the load point gives

$$F\delta/2 = \int_0^L M_x^2 \, dL/2GJ + \int_0^L M_y^2 \, dL/2EI.$$

Since M_x and M_y remain constant along the helical length L of the spring, the integration of this equation leads to

$$\delta = (FD^2L/4)[(1/GJ)\cos^2 \alpha + (1/EI)\sin^2 \alpha].$$

Now $L = \pi DN/\cos \alpha$. Substituting, with $J = \pi d^4/32$ and $I = \pi d^4/64$ gives

$$\delta = (8FD^3N/d^4 \cos \alpha)[(1/G)\cos^2 \alpha + (2/E)\sin^2 \alpha]. \qquad (11.19)$$

Axial twisting moment In this case, the applied moment M acts at the vertical section A in the direction shown in Fig. 11.13. Resolution gives the components

$$M_x = M \sin \alpha, \quad \text{twisting the wire; and}$$
$$M_y = M \cos \alpha, \quad \text{bending the wire.}$$

The external work is $M\phi/2$, where ϕ is the rotation (rad) produced by M. Thus, Eq. (11.2) becomes

$$M\phi/2 = \int_0^L M_x^2 \, dL/2GJ + \int_0^L M_y^2 \, dL/2EI.$$

Substituting for M_x and M_y gives, after integration,

$$\phi = (32MDN/d^4 \cos \alpha)[(1/G)\sin^2 \alpha + (2/E)\cos^2 \alpha]. \qquad (11.20)$$

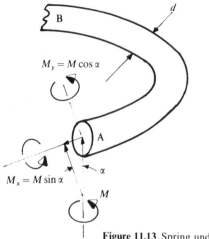

Figure 11.13 Spring under torsion.

Note that under either loading condition the shear stress τ and the twist θ (rad) in the wire are found from ETT:

$$\tau = M_x(d/2)/J, \qquad \theta = M_x L/JG. \qquad (11.21a, b)$$

The bending stress and radius of curvature R (m) of the wire are found from ETB:

$$\sigma = M_y(d/2)/I, \qquad R = EI/M_y. \qquad (11.22a, b)$$

Example 11.13 An open-coiled helical spring with wire diameter 6.5 mm and helix angle of 30° has 10 coils each of 50 mm mean diameter. Determine the axial force necessary to extend the spring by 12.5 mm. What magnitude of shear and bending stresses are induced? What magnitude of axial torque would cause the bending stress in the wire to reach 55 MPa? Also find the corresponding axial rotation under this torque. Take $E = 207$ GPa, $G = 82.7$ GPa.

Transposing Eq. (11.19),

$$F = \frac{\delta d^4 \cos \alpha}{8 D^3 N [(1/G)\cos^2 \alpha + (2/E)\sin^2 \alpha]}$$

$$= \frac{12.5 \times 6.5^4 \cos 30°}{8 \times 50^3 \times 10[(1/82\,700)\cos^2 30° + (2/207\,000)\sin^2 30°]} = 168.26 \text{ N}.$$

Now, from Eq. (11.21a),

$$\tau = (FD/2)(\cos \alpha)(d/2)/J$$
$$= (168.26 \times 50/2)(\cos 30°) \times 32 \times 6.5/[2\pi(6.5)^4]$$
$$= 67.56 \text{ MPa},$$

and from Eq. (11.22a),

$$\sigma = (FD/2)(\sin \alpha)(d/2)/I$$
$$= (168.26 \times 50/2)(\sin 30°) \times 64 \times 6.5/[2\pi(6.5)^4]$$
$$= 78 \, \text{MPa}.$$

With the spring under an axial moment M, that component, M_y, responsible for the bending stress of $\sigma = 55 \, \text{MPa}$, is found from Eq. (11.22a):

$$M_y = 2\sigma I/d = M \cos \alpha,$$
$$\therefore \quad M = 2\sigma I/d \cos \alpha = \sigma \pi d^3/(32 \cos \alpha)$$
$$= 55 \times \pi(6.5)^3/(32 \cos 30°)$$
$$= 1712.3 \, \text{N mm} = 1.71 \, \text{N m}.$$

The accompanying rotation is found from Eq. (11.20):

$$\phi = [32 \times 1712.3 \times 50 \times 10/(6.5^4 \cos 30°)][(1/82 \, 700) \sin^2 30°$$
$$+ (2/207 \, 000) \cos^2 30°]$$
$$= 0.178 \, \text{rad} = 10.18°.$$

The Leaf Spring

The fabricated arrangement of steel plates shown in Fig. 11.14(a) approximates to a beam in which the maximum bending stress is constant over the length. It is constructed from a diamond-shaped steel plate, bent to the arc of a circle, then split into a number n of smaller leaves with parallel portions of equal width b and thickness t. Contact between the leaves is made only at their tapered ends, so that they are free to slide over one another without excessive friction under load. In practice, some further variations in geometry are incorporated to improve performance in vehicles, where considerable deflection is required under moderate loading.

When the top plate is subjected to a central vertical force F, the stress at the outer fibres for a section of breadth $2x$, distance z from one end, is

$$\sigma = My/I = (Fz/2)(t/2)/(2xt^3/12) = 3Fz/2xt^2. \tag{11.23a}$$

Clearly, as x is proportional to z, then σ is maintained constant. It is convenient to refer this stress to the central section where $z = l/2$ and $x = b/2$. Then,

$$\sigma = 3Fl/2bt^2. \tag{11.23b}$$

All other leaves are subjected to four-point loading (Fig. 11.14(b)). The parallel portion of width b within the region of contact from the upper leaf, is therefore subjected to a constant bending moment $M = Fl/4$. Hence, the bending stress, $\sigma = My/I$, is constant in this region and equal to that in the

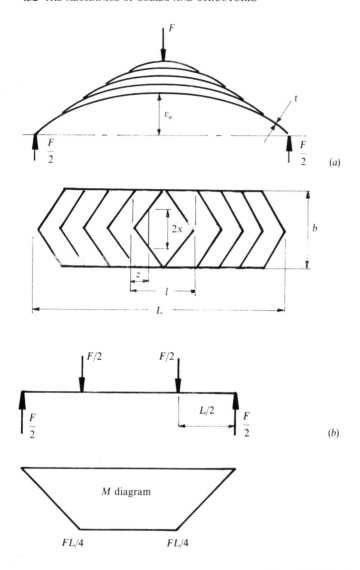

Figure 11.14 Loading on a leaf spring.

tapered ends, where Eq. (11.23a) again applies. In this way the outer fibre stress is maintained constant along the length, everywhere equal to Eq. (11.23b). If there are n leaves, and the length of the longest leaf is L, then $l = L/n$, and from Eq. (11.23b),

$$\sigma = 3FL/2nbt^2. \tag{11.23c}$$

It follows that the linear bending stress gradient through the thickness of a

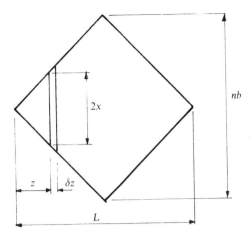

Figure 11.15 Equivalent plate for leaf.

single plate will be reduced for a large number of thin leaves. The mid-span deflection is found from Eq. (11.2) where the external work is $W = F\delta/2$ and the net strain energy stored U, is the sum of the energies of bending within each leaf. For the purpose of finding U, the leaves may be represented by a single equivalent leaf of length L and diagonal breadth nb (Fig. 11.15), which sustains the same maximum stress in Eq. (11.23c).

With the given origin for z, x in the plane of the equivalent plate:

$$I = (2x)t^3/12, \qquad M = Fz/2 \qquad \text{for } 0 \le z \le L/2.$$

Then, from Eq. (11.7b), for the whole plate,

$$U = 2(3F^2/4Et^3) \int_0^{L/2} (z^2/x)\,\mathrm{d}z.$$

Substituting $x = z(bn/L)$ and integrating gives

$$U = 3F^2L^3/16Et^3bn = F\delta/2,$$
$$\therefore \quad \delta = 3FL^3/8Et^3bn. \tag{11.24a}$$

Equation (11.24a) may also be found from the geometry of the deflected arc or from integration of the flexure equation (7.4). This deflection identifies the force (of proof load) that is required to flatten the spring. That is, when $\delta = v_o$ (see Fig. 11.14a), Eq. (11.14) gives

$$F = 8Et^3bnv_o/3L^3. \tag{11.24b}$$

Example 11.14 A leaf spring is made from seven steel plates each 6.5 mm ($\frac{1}{4}$ in) thick. It is required to deflect 38 mm ($1\frac{1}{2}$ in) when subjected to a central concentrated force of 5 kN ($\frac{1}{2}$ tonf). If the maximum stress in the plates is not to exceed 230 MPa (15 tonf/in^2), calculate the span of the main plate and a suitable common width for the plates. Take $E = 207\,\text{GPa}\ (13\,400\,\text{tonf/in}^2)$.

SOLUTION TO SI UNITS Transposing Eq. (11.23c) for b,

$$b = 3FL/2\sigma nt^2,$$

and from Eq. (11.24a),

$$L^3 = (8Et^3 n\delta/3F)b = (8Et^3 n\delta/3F)(3FL/2\sigma nt^2),$$
$$L^2 = 4Et\delta/\sigma$$
$$= 4 \times 207 \times 10^3 \times 6.5 \times 38/230 = 88.92 \times 10^4\,\text{mm}^2,$$
$$\therefore\quad L = 943\,\text{mm},$$
$$b = 3 \times 5 \times 10^3 \times 943/(2 \times 230 \times 7 \times 6.5^2)$$
$$= 104\,\text{mm}.$$

SOLUTION TO IMPERIAL UNITS Again:

$$L^2 = 4Et\delta/\sigma$$
$$= 4 \times 13\,400 \times \tfrac{1}{4} \times 1\tfrac{1}{2}/15 = 1340$$
$$L = 36.6\,\text{in}.$$
$$b = 3FL/2\sigma nt^2$$
$$= 3 \times \tfrac{1}{2} \times 36.6/2 \times 15 \times 7 \times (\tfrac{1}{4})^2$$
$$= 4.18\,\text{in}.$$

11.4 IMPACT LOADING

In the previous consideration of stored strain energy U, the assumption was made that the application of the loading was slow enough to be quasistatic. A greater amount of energy will be imparted to an elastic body under rapid loading and under impact conditions. For the latter, where a mass is initially accelerating or moving with constant velocity, the effect of an elastic impact with a body may be simply estimated from an energy balance for the equivalent static force. Consider a mass of known weight P falling freely through height h to cause an instantaneous deflection δ on impact. The loss in potential energy for P is equated to the work done by an equivalent static force P_E, that would produce the same displacement. That is,

$$P(h + \delta) = P_E\delta/2 = U, \tag{11.25}$$

where U is the strain energy stored under the application of P_E. For a given body under elastic conditions, δ is a known linear function in P_E. Hence Eq. (11.25) results in a quadratic in either P_E or δ. The instantaneous stress is then found from the static force P_E in the normal way, as the following example illustrates.

Example 11.15 Find the instantaneous deflection, the equivalent static load and the maximum stress produced, when a weight $P = 135\,\text{N}$ falls freely through a height $h = 50\,\text{mm}$, to make elastic impact in each of the following ways. Take $E = 207\,\text{GPa}$ and $G = 80\,\text{GPa}$ where appropriate.

(a) Axial loading of a circular section bar $d = 63.5\,\text{mm}$, $L = 3\,\text{m}$

Here P_E is first found from Eqs (11.2) and (11.4):

$$P_E \delta/2 = (\sigma^2/2E)V,$$
$$\therefore \quad \delta = \sigma^2 V/EP_E$$
$$= (P_E/A)^2(AL)/EP_E,$$
$$P_E = AE\delta/L,$$

which could have been found directly from $\delta = \epsilon L = \sigma L/E$. Substituting into Eq. (11.25),

$$P(h + \delta) = AE\delta^2/2L,$$

which gives the quadratic in δ:

$$(AE)\delta^2 - (2LP)\delta - (2LPh) = 0. \tag{i}$$

Substituting for L, P, E, h and $A = 3167\,\text{mm}^2$ in Eq. (i) gives $\delta = 0.249\,\text{mm}$,

$$\therefore \quad P_E = AE\delta/L = 3167 \times 207 \times 10^3 \times 0.249/3000 = 54\,412\,\text{N},$$
$$\sigma = P_E/A = 54\,412/3167 = 17.18\,\text{MPa}.$$

Compare with the static stress under P of $\sigma = 135/3167 = 0.0426\,\text{MPa}$. The alternative solution from Eq. (11.25) employs the corresponding quadratic in P_E:

$$P_E^2 - 2PP_E - 2AEPh/L, \tag{ii}$$

from which $P_E = 54.4\,\text{kN}$ directly. Note that the static pinned-end buckling load is 720 N.

(b) Transverse loading of the central 1.25 m span of a 12.5 mm square-section, simply-supported, steel beam

The central deflection is found from Eqs (11.2) and (11.7a):

$$P_E \delta/2 = 2(1/2EI)\int_0^{L/2} (P_E z/2)^2 \, dz,$$
$$\delta = P_E L^3/48EI.$$

Then Eq. (11.25) supplies the following quadratic in P_E:

$$(L^3/96EI)P_E^2 - (PL^3/48EI)P_E - Ph = 0,$$
$$P_E^2 - (2P)P_E - 96PIEh/L^3 = 0. \tag{iii}$$

Substituting for P, h, E, $L = 1.25$ m and $I = 2034.5$ mm^4 in Eq. (iii) gives $P_E = 532.43$ N,

$$\therefore \ \delta = P_E L^3/48EI = 532.43 \times 1250^3/(48 \times 207\,000 \times 2034.5) = 51.44\,\text{mm},$$
$$\sigma = My/I = P_E Ly/4I$$
$$= 532.43 \times 1250 \times 6.25/(4 \times 2034.5) = 511.13\,\text{MPa}.$$

Note that the collision is elastic provided this stress does not exceed the static yield stress. For low-carbon steels, with yield stress in the range 250–350 MPa, this collision would involve plastic deformation.

(c) Transverse loading of the central 1 m span of a seven-leaf spring with plates 65 mm wide × 6.5 mm thick

Putting $F = P_E$ in Eq. (11.24) and substituting into Eq. (11.25):

$$P_E^2 - 2PP_E - 16Et^3bnPh/3L^3 = 0. \tag{iv}$$

Substituting for P, h, E, $L = 1$ m, $n = 7$, $b = 65$ mm and $t = 6.5$ mm in Eq. (iv) gives $P_E = 1109.4$ N. Then, from Eqs (11.23c) and (11.24):

$$\delta = 3P_E L^3/8Et^3 bn = 3 \times 1109.4 \times 10^9/(8 \times 207\,000 \times 6.5^3 \times 65 \times 7)$$
$$= 16.08\,\text{mm},$$
$$\sigma = 3P_E L/2nbt^2 = 3 \times 1109.4 \times 1000/(2 \times 7 \times 65 \times 6.5^2) = 86.56\,\text{MPa}.$$

(d) Transverse loading of the end of a 200 mm long arm, attached at right angles to a 20 mm diameter × 500 mm long torsion bar as shown in Fig. 11.16. Neglect bending effects.

From Eqs (11.2) and (11.5c):

$$T\theta/2 = (1/2GJ) \int_0^L T^2\, dz.$$

Putting $T = P_E R$ for the equivalent static load gives

$$\theta = \delta/R = P_E RL/GJ,$$
$$\delta = P_E R^2 L/GJ.$$

Then from Eq. (11.25) the following quadratic in P_E results:

$$P_E^2 - 2PP_E - 2GJPh/R^2 L = 0. \tag{v}$$

Substituting for P, h, G, $R = 200$ mm, $L = 500$ mm and $J = 15\,708$ mm^4 in Eq. (v) gives $P_E = 1065.8$ N,

$$\therefore \quad \delta = P_E R^2 L/GJ = 1065.8 \times 200^2 \times 500/(80 \times 10^3 \times 15\,708)$$
$$= 16.96\,\text{mm},$$

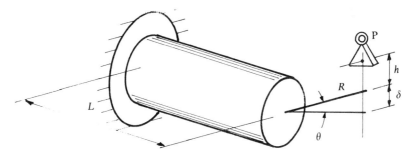

Figure 11.16 Torsion bar.

$$\tau = Tr/J = (P_E R)r/J$$
$$= 1065.8 \times 200 \times 10/15\,708 = 135.7\,\text{MPa}.$$

(e) Axial loading of a close-coiled helical spring, 3 m long with wire and coil diameters of 6.5 and 60 mm respectively

Putting $F = P_E$ and $L = \pi DN$ in Eq. (11.9a) and substituting into Eq. (11.25) gives

$$P_E^2 - 2PP_E - \pi d^4 GPh/4D^2 L = 0. \tag{vi}$$

Substituting for P, h, G, $d = 6.5\,\text{mm}$, $D = 60\,\text{mm}$, and $L = 3\,\text{m}$ gives $P_E = 432.2\,\text{N}$,

$$\therefore \quad \delta = 8P_E D^2 L/\pi d^4 G = 8 \times 432.2 \times 60^2 \times 3000/(\pi \times 6.5^4 \times 80 \times 10^3)$$
$$= 83.2\,\text{mm},$$

and from Eq. (11.11) the maximum torsional stress is

$$\tau = 8P_E D/\pi d^3 = 8 \times 432.2 \times 60/(\pi \times 6.5^3) = 240.46\,\text{MPa}.$$

(f) Transverse loading of the end of a 75 mm long arm, attached at right angles to the axis of a closely coiled helical spring of length 3 m, with wire and coil diameters of 12.5 and 60 mm respectively

With a similar arrangement as in Fig. 11.16, Eq. (11.14) gives

$$\phi = \delta/R = ML/EI,$$
$$\delta = P_E LR^2/EI,$$

and from Eq. (11.25) the corresponding quadratic in P_E is

$$P_E^2 - 2PP_E - 2EIPh/LR^2 = 0. \tag{vii}$$

Substituting for P, h, E, $D = 60\,\text{mm}$, $L = 3\,\text{m}$, $R = 75\,\text{mm}$ and $I = 1198.42\,\text{mm}^4$ in Eq. (vii) gives $P_E = 600.5\,\text{N}$,

$$\therefore \quad \delta = P_E LR^2/EI = 600.5 \times 3000 \times 75^2/(207\,000 \times 1198.42)$$
$$= 40.85\,\text{mm},$$

and from Eq. (11.16):

$$\sigma = 32(P_E R)/\pi d^3 = 32 \times 600.5 \times 75/(\pi \times 12.5^3) = 234.88 \text{ MPa.}$$

The reader should note that Eqs (ii)–(vii) all have the common form:

$$P_E^2 - 2PP_E - 2PhK = 0,$$

where $K = P_E/\delta$ is the elastic stiffness of the particular body under the equivalent static load.

11.5 CASTIGLIANO'S THEOREMS

The young Italian railway engineer A. Castigliano (1847–1884) proposed a two-part theorem, based upon strain energy, that has been widely used since its publication in 1879. The first part follows the application of Eq. (11.2) to an equilibrium structure carrying n external forces F_i $(i = 1, 2, 3, \ldots, n)$, for which the associated in-line displacements are Δ_i. Let $\delta\Delta_i$ correspond to a change δF_i in these forces. Then, from Eq. (11.2),

$$\int dU = \int dW,$$

$$U = \int_0^{\Delta_i} F_i \, d\Delta_i,$$

in which the following summation is implied:

$$U = \int_0^{\Delta_1} F_1 \, d\Delta_1 + \int_0^{\Delta_2} F_2 \, d\Delta_2 + \int_0^{\Delta_3} F_3 \, d\Delta_3 + \cdots + \int_0^{\Delta_n} F_n \, d\Delta_n.$$

It can be seen that if all but one displacement be held fixed, then

$$\partial U/\partial \Delta_i = \partial \left(\int_0^{\Delta_i} F_i \, d\Delta_i \right) \Big/ \partial \Delta_i,$$

or

$$F_i = \partial U/\partial \Delta_i. \tag{11.26}$$

That is, the force F_i at point i may be obtained by differentiating U partially w.r.t. its corresponding in-line displacement Δ_i.

The second part of the theorem, which leads to a more useful result than Eq. (11.26), employs the concepts of complementary energy and work. The external work integral $W = \int_\Delta F \, \delta\Delta$ is readily identified with the area beneath the F vs. Δ diagram in Fig. 11.17. The complementary work increment $\delta W^* = \Delta \, \delta F$ is that area W^* in Fig. 11.17 defined as

$$W^* = \int_0^F \Delta \, dF. \tag{11.27a}$$

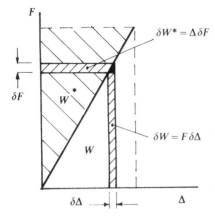

Figure 11.17 External W and complementary W^* work.

In physical terms W^* may be identified with the work done when the point of application of a variable force F, undergoes a displacement Δ. This is stored as complementary energy U^* such that Eq. (11.27a) is

$$U^* = W^* = \int_0^F \Delta \, dF. \tag{11.27b}$$

In the case of a number of externally applied forces F_i with corresponding displacements Δ_i, Eq. (11.27b) becomes

$$U^* = \int_0^{F_i} \Delta_i \, dF_i$$

and, if all but one force remains fixed, then

$$\partial U^*/\partial F_i = \partial \left(\int_0^{F_i} \Delta_i \, dF_i \right) \Big/ \partial F_i,$$

or

$$\Delta_i = \partial U^*/\partial F_i. \tag{11.28}$$

That is, the deflection Δ_i, at a given load point i, may be obtained by differentiating U^* w.r.t. the force F_i. Equation (11.28) applies to any non-linear elastic body. In the particular case of a linear Hookean material, where $F = K \Delta$, the strain and complementary energies become

$$U = \int_0^{\Delta} F \, d\Delta = K \int_0^{\Delta} \Delta \, d\Delta = K\Delta^2/2 = F\Delta/2,$$

$$U^* = \int_0^F \Delta \, dF = (1/K) \int_0^F F \, dF = F^2/2K = F \Delta/2.$$

Thus, $U = U^*$ and Eq. (11.28) appears in its more usual form:

$$\Delta_i = \partial U / \partial F_i. \tag{11.29}$$

The theorem may be extended to include the rotations ϕ_i under externally applied moments M_i. Of interest here is the equivalent expression to Eq. (11.29), where, following a similar proof for a Hookean material,

$$\phi_i = \partial U / \partial M_i. \tag{11.30}$$

Application to Beams, Rings and Arches

The following examples illustrate the wide range of application of Castigliano's theorem. It is, however, subject to the limitation that displacements cannot readily be found at points where no loads act. In this situation, the virtual work principle, as outlined in Chapter 12, is often preferred.

Example 11.16 Determine the load point displacements for the beam in Fig. 11.9, given that the shear strain energy is negligible compared to the strain energy of bending.

Putting $F_1 = 15$ N and $F_2 = 10$ N, the moments are

$$M_1 = (1 - z)F_1 \quad \text{(N m)} \qquad \text{for } 0.5 \le z \le 1, \tag{i}$$

$$M_2 = (1 - z)F_1 + (0.5 - z)F_2 \quad \text{(N m)} \qquad \text{for } 0 \ge z \ge 0.5. \tag{ii}$$

The net strain energy in bending is, from Eq. (11.7a),

$$U = (1/2EI_1) \int M_1^2 \, dz + (1/2EI_2) \int M_2^2 \, dz.$$

Applying Eq. (11.29) for Δ_1,

$$\Delta_1 = \partial U / \partial F_1 = (1/EI_1) \int M_1 (\partial M_1 / \partial F_1) \, dz + (1/EI_2) \int M_2 (\partial M_2 / \partial F_1) \, dz.$$

Substituting from Eqs (i) and (ii):

$$\Delta_1 = (1/EI_1) \int_{0.5}^{1.0} (1 - z)^2 F_1 \, dz + (1/EI_2) \int_0^{0.5} [(1 - z)^2 F_1$$
$$+ (1 - z)(0.5 - z)F_2] \, dz$$
$$= (15/EI_1)|z - z^2 + z^3/3|_{0.5}^{1.0}$$
$$+ (1/EI_2)(15|z - z^2 + z^3/3|_0^{0.5} + 10|0.5z - 1.5z^2/2 + z^3/3|_0^{0.5})$$
$$= 15 \times 0.0417/1333 + [(15 \times 0.2917) + (10 \times 0.1042)]/5208$$
$$= 1.51 \times 10^{-3} \, \text{m} = 1.51 \, \text{mm}.$$

Applying Eq. (11.29) for Δ_2:

$$\Delta_2 = \partial U/\partial F_2 = (1/EI_1)\int M_1(\partial M_1/\partial F_2)\,dz + (1/EI_2)\int M_2(\partial M_2/\partial F_2)\,dz.$$

Substituting from Eqs (i) and (ii):

$$\Delta_2 = (1/EI_2)\int_0^{0.5} [(1-z)(0.5-z)F_1 + (0.5-z)^2 F_2]\,dz$$

$$= (1/EI_2)(15|0.5z - 1.5z^2/2 + z^3/3|_0^{0.5}$$
$$+ 10|0.25z - z^2/2 + z^3/3|_0^{0.5})$$
$$= [(15 \times 0.1042) + (10 \times 0.041\,67)]/5208 = 3.8 \times 10^{-4}\,\text{m} = 0.38\,\text{mm}.$$

Example 11.17 The thin strip in Fig. 11.18 is fixed at C and free to slide without friction at A. Calculate the horizontal reaction F_A at A, when the strip supports a vertical force F_B at point B, as shown. EI is constant.

Assume that all the strain energy stored is due to bending. With θ measured clockwise from A, and taking clockwise moments as positive, the moment expressions become

$$M_{AB} = F_A R(1 - \cos\theta) \qquad \text{for } 0 \le \theta \le \pi/2, \tag{i}$$

$$M_{BC} = F_A R[1 + \sin(\theta - \pi/2)] - F_B R[1 - \cos(\theta - \pi/2)] \tag{ii}$$
$$= F_A R(1 - \cos\theta) - F_B R(1 - \sin\theta) \qquad \text{for } \pi/2 \le \theta \le 5\pi/4.$$

The total strain energy of bending is, from Eq. (11.7a),

$$U = U_{AB} + U_{BC}$$

$$= (1/2EI)\int M_{AB}^2\,ds + (1/2EI)\int M_{BC}^2\,ds.$$

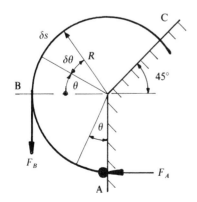

Figure 11.18 Thin strip.

Deflection in line with F_A is prevented. Thus, from Eq. (11.29),

$$\Delta_A = 0 = \partial U / \partial F_A$$

$$= (1/EI) \int M_{AB}(\partial M_{AB} / \partial F_A) \, ds + (1/EI) \int M_{BC}(\partial M_{BC} / \partial F_A) \, ds.$$

Substituting from Eqs (i) and (ii) with $ds = R \, d\theta$,

$$(R^3/EI) \int_0^{\pi/2} F_A(1 - \cos\theta)^2 \, d\theta + (R^3/EI) \int_{\pi/2}^{5\pi/4} [F_A(1 - \cos\theta)^2$$

$$- F_B(1 - \sin\theta)(1 - \cos\theta)] \, d\theta = 0,$$

$$F_A|3\theta/2 - 2\sin\theta + (1/4)\sin 2\theta|_0^{\pi/2} + F_A|3\theta/2 - 2\sin\theta + (1/4)\sin 2\theta|_{\pi/2}^{5\pi/4}$$

$$- F_B|\theta + \cos\theta - \sin\theta - (1/4)\cos 2\theta|_{\pi/2}^{5\pi/4} = 0,$$

$$F_A(3\pi/4 - 2) + F_A[(15\pi/8 + \sqrt{2} + 1/4) - (3\pi/4 - 2)]$$

$$- F_B[(5\pi/4 - 1/\sqrt{2} + 1/\sqrt{2}) - (\pi/2 - 1 + 1/4)] = 0,$$

$$0.3562F_A + 7.1985F_A - 3.1062F_B = 0,$$

$$F_A = 3.1062F_B / 7.5547 = 0.411F_B.$$

Example 11.18 The thin semi-circular arch in Fig. 11.19 supports a central vertical load P. Derive expressions for the vertical deflection beneath P when the arch is (a) fixed at A and B and (b) fixed at A and supported on rollers at B. What is the horizontal deflection at B and the horizontal reaction in each case?

In the more general case of (a), the positive clockwise moments acting about any section in the region defined by $0 \le \theta \le \pi/2$ are

$$M = HR\sin\theta - (PR/2)(1 - \cos\theta). \qquad (i)$$

The strain energy stored in bending for the whole beam becomes

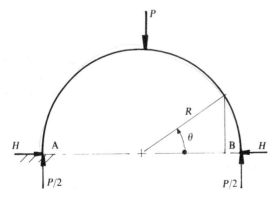

Figure 11.19 Semi-circular arch.

$$U = 2(1/2EI) \int M^2 \, ds, \qquad \text{(ii)}$$

when, from Eq. (11.29), the deflection beneath P is

$$\Delta_P = \partial U / \partial P = (2/EI) \int M(\partial M / \partial P) \, ds.$$

Substituting from Eq. (i), with $ds = R \, d\theta$:

$$\Delta_P = (R^3/EI) \int_0^{\pi/2} [(P/2)(1 - \cos \theta)^2 - H \sin \theta (1 - \cos \theta)] \, d\theta$$

$$= (R^3/EI)[(P/2)|3\theta/2 - 2 \sin \theta + (1/4) \sin 2\theta|_0^{\pi/2}$$
$$- H|-\cos \theta + (1/4) \cos 2\theta|_0^{\pi/2}]$$
$$= (R^3/EI)[(P/2)(3\pi/4 - 2) - H/2],$$
$$\Delta_P = (R^3/2EI)[P(3\pi/4 - 2) - H]. \qquad \text{(iii)}$$

The horizontal deflection is found from Eq. (ii), with $ds = R \, d\theta$:

$$\Delta_H = \partial U / \partial H = (2/EI) \int M(\partial M / \partial H) \, ds,$$

$$\Delta_H = (2R^3/EI) \int_0^{\pi/2} [H \sin^2 \theta - (P/2) \sin \theta (1 - \cos \theta)] \, d\theta$$

$$= (2R^3/EI)[H]\theta/2 - (1/4) \sin 2\theta|_0^{\pi/2}$$
$$- (P/2)|-\cos \theta + (1/4) \cos 2\theta|_0^{\pi/2}]$$
$$= (2R^3/EI)(H\pi/4 - P/4)$$
$$\Delta_H = (R^3/2EI)(H\pi - P). \qquad \text{(iv)}$$

For (a), with B fixed $\Delta_B = \Delta_H = 0$, when Eqs (iii) and (iv) give

$$H = P/\pi, \qquad \Delta_P = (PR^3/EI)(3\pi/8 - 1 - 1/2\pi);$$

and for (b), with B on rollers, $H = 0$, and Eqs (iii) and (iv) give

$$\Delta_P = (PR^3/EI)(3\pi/8 - 1), \qquad \Delta_H = -PR^3/2EI.$$

Example 11.19 Derive an expression for the vertical displacement at the free end C, of the davit in Fig. 11.20, when vertical and horizontal forces, F_V and F_H, are applied at this point. Do not neglect the strain energy stored in compression. Take area A, E, and I to be constants.

With the origins for θ and z shown, the bending moments are

$$M_{BC} = F_V R \sin \theta - F_H R (1 - \cos \theta) \qquad \text{for } 0 \le \theta \le \pi/2, \quad \text{(i)}$$
$$M_{AB} = F_V R - F_H (R + h - z) \qquad \text{for } 0 \le z \le h. \quad \text{(ii)}$$

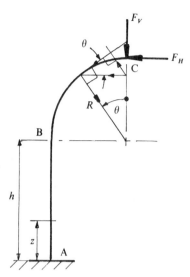

Figure 11.20 Davit displacements.

Each section of the davit is subjected to F_V and F_H. These may be resolved in tangential directions to give the compressive forces

$$F_{BC} = -F_V \sin\theta - F_H \cos\theta \qquad \text{for} \quad 0 \le \theta \le \pi/2, \quad \text{(iii)}$$

$$F_{AB} = -F_V \qquad\qquad\qquad \text{for} \quad 0 \le z \le h. \quad \text{(iv)}$$

From Eqs (11.4a) and (11.7a) the net strain energy is

$$U = (1/2EI)\int M^2\,ds + (1/2AE)\int F^2\,ds,$$

where from Eq. (11.29), the vertical deflection is

$$\Delta_V = (1/EI)\int M_{BC}(\partial M_{BC}/\partial F_V)\,ds + (1/EI)\int M_{AB}(\partial M_{AB}/\partial F_V)\,ds$$

$$+ (1/AE)\int F_{BC}(\partial F_{BC}/\partial F_V)\,ds + (1/AE)\int F_{AB}(\partial F_{AB}/\partial F_V)\,ds.$$

Substituting from Eqs (i)–(iv):

$$\Delta_V = (R^3/EI)\int_0^{\pi/2} [F_V\sin^2\theta - F_H\sin\theta(1-\cos\theta)]\,d\theta$$

$$+ (R/EI)\int_0^h [F_V R - F_H(R+h-z)]\,dz$$

$$+ (R/AE)\int_0^{\pi/2} (F_V\sin^2\theta + F_H\sin\theta\cos\theta)\,d\theta$$

$$+ (1/AE)\int_0^h (-F_V)(-1)\,dz,$$

$$\Delta_V = (R^3/EI)[(F_V/2)|\theta - (1/2)\sin 2\theta|_0^{\pi/2} - F_H| - \cos\theta + (1/4)\cos 2\theta|_0^{\pi/2}]$$
$$+ (R/EI)|(F_V R - F_H R - F_H h)z + F_H z^2/2|_0^h$$
$$+ (R/AE)[(F_V/2)|\theta - (1/2)\sin 2\theta|_0^{\pi/2} - (F_H/4)|\cos 2\theta|_0^{\pi/2}]$$
$$+ (F_V/AE)|z|_0^h,$$
$$\Delta_V = (R^3/EI)(\pi F_V/4 - F_H/2) + (R/EI)[(F_V - F_H)Rh + 3F_H h^2/2]$$
$$+ (R/2AE)(\pi F_V/2 + F_H) + F_V h/AE.$$

The first two terms are the contributions from bending and the final two terms ae the contributions from compression.

Example 11.20 During the elastic calibration of the proving ring in Fig. 11.21(a), the in-line deflection is measured under a vertical diametral force F. Establish the theoretical stiffness factor.

To find the strain energy it is only necessary to consider one half of the ring in Fig. 11.21(b). However, as the induced moment M_o renders the problem statically indeterminate, an additional compatibility condition requires that there must be no rotation ϕ under M_o. The strain energy for the whole ring is given by Eq. (11.7a):

$$U = 2(1/2EI)\int M^2 \, ds.$$

Under F the deflection and slope are, from Eqs (11.29) and (11.30)

$$\Delta = (2/EI)\int M(\partial M/\partial F)\,ds, \tag{i}$$

$$\phi = (2/EI)\int M(\partial M/\partial M_o)\,ds = 0, \tag{ii}$$

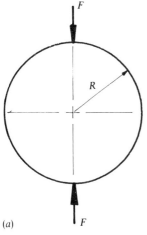

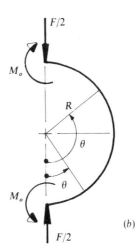

(a)

(b)

Figure 11.21 Proving ring calibration.

where from Fig. 11.21(b), with the given origin for θ,

$$M = (FR/2)\sin\theta - M_o \qquad \text{for} \quad 0 \le \theta \le \pi/2,$$
$$M = (FR/2)\cos(\theta - \pi/2) - M_o = (FR/2)\sin\theta - M_o \qquad \text{for} \quad \pi/2 \le \theta \le \pi,$$

i.e. a single expression for M applies. Substituting this into Eq. (ii):

$$\int_0^\pi [(FR/2)\sin\theta - M_o](-1)R\,d\theta = 0,$$

$$|M_o\theta + (FR/2)\cos\theta|_0^\pi = M_o\pi - FR = 0,$$

$$\therefore \quad M_o = FR/\pi.$$

Substituting for M and M_o in Eq. (i):

$$\Delta = (2/EI)\int_0^\pi [(FR/2)\sin\theta - FR/\pi](R/2)\sin\theta\,(R\,d\theta)$$

$$= (FR^3/2EI)\int_0^\pi [\sin^2\theta - (2/\pi)\sin\theta]\,d\theta$$

$$= (FR^3/2EI)|\theta/2 - (1/4)\sin 2\theta + (2/\pi)\cos\theta|_0^\pi$$

$$= (FR^3/2EI)[(\pi/2 - 0 - 2/\pi) - (2/\pi)] = (FR^3/EI)(\pi/4 - 2/\pi),$$

$$\therefore \quad K = F/\Delta = EI/R^3(\pi/4 - 2/\pi).$$

Application to Pin-jointed Structures

Statically determinate structures The total strain energy stored is the sum of the strain energies stored in each bar under tension or compression. That is, from Eq. (11.4b), with a constant force P induced in each bar of length L and cross-sectional area A,

$$U = \sum(P^2L/2AE).$$

In the case where the forces P are induced by a single external force F, the deflection Δ beneath F is found from equating the external work to U. That is, from Eq. (11.2),

$$F\Delta/2 = \sum(P^2L/2AE). \tag{11.31}$$

If a unit force replaces F to produce bar forces k, it follows that $P = kF$ and Eq. (11.31) becomes

$$F\Delta/2 = (F^2/2)\sum(k^2L/AE),$$
$$\Delta = F\sum(k^2L/AE). \tag{11.32}$$

The problem becomes more complicated when several external forces F are applied to the frame. Say the deflection Δ_A is required at a node point A where the force is F_A. It is convenient to take the induced forces P as the sum of the two components; $P = P' + P''$, where P' are the bar forces in the absence of F_A but in the presence of all other external forces, and P'' are the bar forces when F_A acts in isolation. If a unit force, when acting in isolation at A, induces bar forces k, then $P'' = kF_A$. The total strain energy is then

$$U = \sum [(P' + P'')^2 L / 2AE] = \sum [(P' + kF_A)^2 L / 2AE].$$

Then, from Eq. (11.29), the deflection in the direction of F_A is

$$\Delta_A = \partial U / \partial F_A = \sum 2(P' + kF_A)(kL/2AE),$$
$$\Delta_A = \sum (P'kL/AE) + F_A \sum (k^2 L/AE). \tag{11.33}$$

Note from the units of the R.H.S. that k is a dimensionless force coefficient. Moreover, it is seen, from Eq. (11.33), that if $F_A = 0$ it is still possible to obtain the deflection at A from the first term. This is illustrated in the following example.

Example 11.21 Find the deflection at point B for the structure in Fig. 2.19 given that all bar areas are $500\,\text{mm}^2$ and $E = 207\,\text{GPa}$.

Since there is no force applied at point B, the bar forces previously calculated on page 36 correspond to P' in Eq. (11.33). To find k, let a unit force act vertically downwards at point B in Fig. 11.22. From joint equilibrium:

At A: $+\uparrow$,

$$-1/2 + k_{AE} \cos 30° = 0, \quad \Rightarrow k_{AE} = 1/\sqrt{3} = k_{CD};$$

$+\rightarrow$,

$$k_{AB} + k_{AE} \sin 30° = 0, \quad \Rightarrow k_{AB} = -k_{AE}/2 = -1/(2\sqrt{3}) = k_{CB}.$$

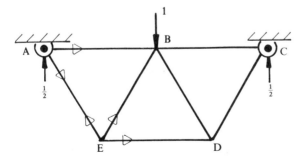

Figure 11.22

At E: $+\uparrow$,

$$k_{EA}\cos 30° + k_{EB}\cos 30° = 0, \quad \Rightarrow k_{EB} = -k_{EA} = -1/\sqrt{3} = k_{BD};$$

$+\rightarrow$,

$$-k_{EA}/2 + k_{BE}/2 + k_{DE} = 0, \quad \Rightarrow k_{DE} = k_{AE}/2 - k_{BE}/2 = 1/\sqrt{3}.$$

The P' and k forces are:

Bar	P' (t)	k (no units)	$P'k$
AB	-1.4434	-0.2887	$+0.4167$
BC	-2.0207	-0.2887	$+0.5834$
CD	$+4.0415$	$+0.5774$	$+2.3336$
DE	$+1.7321$	$+0.5774$	$+1.0000$
BE	-0.5774	-0.5774	$+0.3334$
BD	$+0.5774$	-0.5774	-0.3334
AE	$+2.8868$	$+0.5774$	$+1.6668$
			$\sum(P'k) = +6.0006$

Then, from Eq. (11.33), since $F_B = 0$,

$$\Delta_B = \sum(P'kL/AE) = (L/AE)\sum(P'k)$$
$$= 10 \times 10^3 \times 6.0006 \times 10^3 \times 9.81/(500 \times 207\,000) = 5.69 \text{ mm}.$$

Statically indeterminate (redundant or hyperstatic) structures When the bar forces cannot be found from equilibrium considerations alone, a further compatibility condition is necessary. For example, the structure in Fig. 11.23(a) is redundant because either of the bars AC or BD could be removed without lessening the ability of the structure to support external forces. Castigliano postulated, in a second theorem, that the forces in the members of a redundant structure adjust themselves to minimize the strain energy. To apply this theorem, first remove the redundant bar AC (or BD) to render the structure statically determinate, as shown in Fig. 11.23(b). Determine the bar forces P' under the externally applied loading. Then remove all external forces and apply a unit tensile force k_o at either joint A or joint C in the direction of the missing bar. Establish the bar forces k under the unit force. If now the redundant force R replaces k_o, it follows that the bar forces become $P'' = kR$.

Since the net bar forces are $P = P' + P''$ the total strain energy of the frame in Fig. 11.23(b) becomes

$$U = \sum_{1}^{n-1} (P' + kR)^2 L/2AE, \tag{11.34}$$

where n is the number of bars in Fig. 11.23(a). From Eqs (11.29) and (11.34),

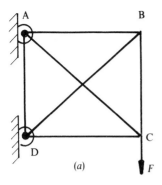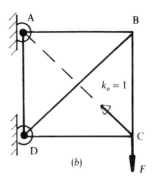

Figure 11.23 Redundant structure.

the deflection of C in the line of R is

$$\Delta_C = \partial U/\partial R = \sum_{1}^{n-1} (P' + kR)kL/AE,$$

$$\Delta_C = (1/E)\left[\sum_{1}^{n-1} (LkP'/A) + R\sum_{1}^{n-1} (k^2L/A)\right]. \tag{11.35}$$

The free extension of the redundant bar under R is

$$x = RL/AE = Rk_o^2 L/AE \tag{11.36}$$

since $k_o = 1$. When the redundant bar is replaced, it must fit exactly between A and C. This condition is ensured from Eqs (11.35) and (11.36) when

$$(1/E)\left[\sum_{1}^{n-1} (LkP'/A) + R\sum_{1}^{n-1} (k^2L/A)\right] + Rk_o^2 L/AE = 0,$$

$$\sum_{1}^{n-1} (LkP'/A) + R\left[\sum_{1}^{n-1} (Lk^2/A) + Lk_o^2/A\right] = 0,$$

$$R = \frac{-\sum_{1}^{n-1} (LkP'/A)}{\sum_{1}^{n-1} (Lk^2/A) + Lk_o^2/A} = \frac{-\sum_{1}^{n} (LkP'/A)}{\sum_{1}^{n} (Lk^2/A)}, \tag{11.37}$$

where the second expression follows from $P' = 0$ for the redundant bar.

Example 11.22 If for the structure in Fig. 11.23(a) $F = 2\,\text{t}$, AB = DC = 10 m, AD = BC = 8 m and for all bars $A = 500\,\text{mm}^2$, $E = 210\,\text{GPa}$, determine the bar forces and the vertical deflection at C.

The reader should first check the P' and k values given in the following table by any one of the methods outlined in Chapter 2.

Member	L (m)	P' (t)	k	LkP'	Lk^2	$P'' = kR$	$P = P' + P''$
AB	10	2.5	−0.78	−19.500	6.084	−1.428	1.072
BC	8	2	−0.624	−9.984	3.115	−1.142	0.858
CD	10	0	−0.780	0	6.084	−1.428	−1.428
BD	12.82	−3.21	1.0	−41.152	12.820	1.830	−1.380
AD	8	2	−0.624	−9.984	3.115	−1.142	0.858
AC	12.82	0	$1.0(k_o)$	0	12.820	1.830	1.830

$$\Sigma = -80.62 \quad \Sigma = 44.038$$

Following the procedure outlined above, the fifth and sixth columns in the table are completed first. Equation (11.37) is then applied to their sums, given at the foot of each column. Since the bar areas are constant,

$$R = - \sum_{1}^{n} (LkP') \Big/ \sum_{1}^{n} (Lk^2) = 80.62/44.038 = 1.83 \, \text{t}.$$

This enables the determination of the bar forces P, given in the final column. With a single external force F, it follows that Eq. (11.32) will supply the vertical deflection at point C. Here k, the bar forces when a unit vertical load is applied at C, are found by dividing P in the above table by $F = 2 \, \text{t}$. That is,

Member	AB	BC	CD	BD	AD	AC
k	0.536	0.429	−0.714	−0.690	0.429	0.915
L (m)	10	8	10	12.82	8	12.82
k^2L	2.873	1.472	5.098	6.104	1.4723	10.780

$\therefore$ $\Sigma k^2 L = 27.80 \, (\text{m})$ and, for constant bar areas in the same material, Eq. (11.32) becomes

$$\Delta_C = (F/AE) \sum (k^2 L)$$
$$= 2 \times 10^3 \times 9.81 \times 27.8 \times 10^3 / (500 \times 210 \times 10^3) = 5.2 \, \text{mm}.$$

Equation (11.33) also enables the deflections to be found for a redundant structure under a number of externally applied forces. If the deflection at some force point A is required, it follows that the force in the redundant bar must be calculated twice from Eq. (11.37), under each of the following loading conditions:

(i) for all external forces present except F_A to give P' in Eq. (11.33);
(ii) with only F_A present to give k in Eq. (11.33).

However, it is seen from Eq. (11.37) that associated with each redundant force calculation there will be separate P' and k bar force calculations.

In the following chapter it is shown how the alternative unit load method may be employed to simplify the finding of deflections in structures with one redundancy or more.

EXERCISES

Strain Energy

11.1 Calculate the stress and strain energy stored in a bronze buffer 250 mm long and 125 mm in diameter when the compressive axial displacement is limited to 0.25 mm. Take $E = 207$ GPa.

Answer: 83.4 MPa; 121 N m.

11.2 A 50 mm diameter rod, 5 m long is reduced to 25 mm diameter for a 2 m length. Calculate the net strain energy stored in the bar when it carries an axial tensile load of 30 kN. Take $E = 200$ GPa.

Answer: 1 N m.

11.3 A 25 mm diameter bar, 150 mm long is subjected to a gradually applied load of 40 kN. A bar of the same material and initial dimensions is turned down to 22.5 mm for 50 mm of its length and a load is gradually applied. If the maximum stress in both bars is the same, find the strain energy in each bar. Take $E = 200$ GPa.

Answer: $\simeq 1.14$; $\simeq 0.74$ N m.

11.4 An axial force of 40 kN is gradually applied to a brass bar 50 mm in diameter and 510 mm in length. Find the strain energy stored in this bar when the central 125 mm length is reduced to 25 mm diameter. Take $E = 84$ GPa.

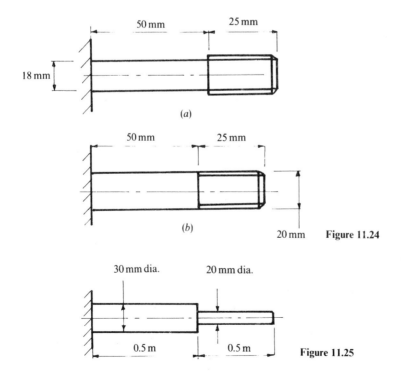

(a)

(b)

20 mm   **Figure 11.24**

Figure 11.25

11.5 Compare the strain energy stored for the bolt shanks in Fig. 11.24(a) and (b) when they are to resist an axial tensile force of 10 kN. Comment on the differences found in respect of tensile shock loading. Take $E = 207$ GPa.

11.6 A solid steel shaft 50 mm in diameter is to be replaced by a solid phosphor bronze shaft of the same length that will transmit the same torque and store the same strain energy per unit volume. Find the diameter of the new shaft and the ratio between the maximum shear stress in each shaft. Take $G = 80$ GPa for steel and $G = 47$ GPa for phosphor bronze.

11.7 The stepped circular shaft in Fig. 11.25 is built in at the L.H. end and is to carry separately, a torque of 100 N m and a moment of 100 N m at its free end. Determine the respective free-end rotations from the energy stored. Under which mode of loading is the shaft the stiffer? Take $E = 2.6G = 100$ GPa.

Answer: $\theta = 5.68°$; $\phi = 4.36°$; bending.

11.8 The 25 mm square × 1 mm thick box section cantilever in Fig. 11.26 supports separately a vertical force and an axial tensile force of 10 N at its free end. Determine the in-line deflection in each case from the energy stored. Under which mode of loading is the cantilever more flexible? Take $E = 100$ GPa.

Answer: 3.604 mm; 1×10^{-3} mm; bending.

11.9 Use Eq. (11.2) to find the maximum deflection in each of the following beams:

    (a) simply supported, carrying a central concentrated force W (*Answer:* $WL^3/48EI$);

    (b) simply supported, carrying a uniformly distributed load w/unit length (*Answer:* $5wL^4/384EI$);

    (c) cantilever carrying a concentrated force W at its free end (*Answer:* $WL^3/3EI$);

    (d) cantilever carrying a uniformly distributed load w/unit length (*Answer:* $wL^4/8EI$);

    (e) simply supported, length L, carrying a concentrated load W distance a from L.H. end and b from R.H. end (*Answer:* $Wa^2b^2/3EIL$ beneath W, not a maximum);

    (f) encastre with central concentrated force W (*Answer:* $WL^2/192EI$).

11.10 A solid steel shaft of diameter 63.5 mm, 0.75 m long rotates at 200 rev/min. Calculate the power transmitted if the energy stored is 67.5 J. Take $G = 80$ GPa.

Answer: $\simeq 100$ kW.

11.11 A steel beam 5 m long, simply supported at its ends, carries a concentrated force of 30 kN at 3.5 m from one end. Given that $I = 78 \times 10^4$ mm^4 and $E = 208$ GPa, employ an energy method to find the deflection beneath the force.

Answer: 3.3 mm.

11.12 A circular section cantilever tapers from diameter d_1 at the fixed-end to diameter d_2 at the free end ($d_1 > d_2$). Derive an expression for the free-end deflection when a vertical force F is applied at that end.

11.13 If, for the cantilver in Exercise 11.2, $d_1 = 10$ mm, $d_2 = 30$ mm, $F = 50$ N and the length is 200 mm, determine the free-end vertical deflection and the maximum stress induced in bending. Take $E = 60$ GPa.

Answer: 0.19 mm, 7.5 MPa.

11.14 Derive the maximum deflection due to shear in each of the following beams, when the cross-section is rectangular of area A.

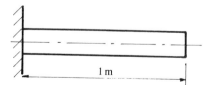

1 m

Figure 11.26

(a) simply supported, carrying a central concentrated force F (*Answer*: $3FL/10AG$);

(b) cantilever carrying a uniformly distributed load w (*Answer*: $3wL^2/5AG$).

11.15 A rectangular section cantilever of length L and depth d carries a concentrated force F at its free end. Show that the ratio between the maximum shear and bending deflections is $(3/10)(E/G)d^2/L^2$. Hence determine the least value of d/L if the shear deflection is not to exceed 1 per cent of the total deflection. Take $E = 207\,\text{GPa}$, $G = 80\,\text{GPa}$.

Answer: 0.114.

11.16 A steel cantilever 100 mm long with 25 mm square section is arranged with one section diagonal lying horizontally. Calculate the total end deflection due to bending and shear effects. Take $E = 207\,\text{GPa}$, $v = 0.3$.

Answer: 0.054 mm; 0.002 27 mm.

11.17 The shear deflection for an I-beam is commonly calculated from assuming that the shear stress is uniformly distributed over a rectangle formed from the web thickness and the overall depth of the cross-section. Determine the percentage error in this assumption for a 304 mm × 152.5 mm I-section with web and flange thicknesses of 12.5 and 25 mm respectively.

Answer: 6 per cent.

11.18 An I-section cantilever 150 mm deep × 125 mm wide with 12.5 mm thick flanges and web is 1.22 m long and carries an end load of 3 t. Find the total shear resilience of the cantilever and the deflection due to shear. Take $G = 82\,\text{GPa}$.

Answer: 3.5 N m; 0.24 mm.

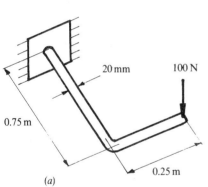

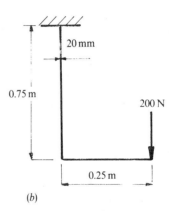

Figure 11.27

Figure 11.28

11.19 Calculate the deflection beneath the force for each of the cantilevered brackets in Fig. 11.27(a) and (b). Take $E = 200\,\text{GPa}$, $G = 80\,\text{GPa}$.

Answer: 13 mm; 6.64 mm.

11.20 The tubular structure in Fig. 11.28 is fixed at one end and is to be loaded separately in the x-, y- and z-directions at the free end. Decide on the induced effects that make the major contributions to the store of energy U in each case and the mode of loading that results in maximum stiffness (i.e. least U).

Answer: z-loading.

Springs

11.21 A close-coiled helical spring has a mean diameter of 100 mm with a 6.5 mm wire diameter. If the spring is to extend 25 mm under an axial force of 90 N, how many coils should the spring have? Take $G = 76\,\text{GPa}$.

Answer: 4.5 coils.

11.22 Determine the safe maximum load and the deflection of close-coiled helical spring with 10 coils, mean diameter 50 mm and wire diameter 6.5 mm, when the maximum torsional shear stress is limited to 345 MPa. Take $G = 83\,\text{GPa}$.

Answer: 680 N; 53 mm.

11.23 The mean diameter of a close-coiled helical spring is six times that of the wire diameter. If the spring is to deflect 25 mm under an axial load of 220 N and the maximum torsional shear stress is not to exceed 413 MPa, find the wire diameter and the nearest number of complete coils. Take $G = 83\,\text{GPa}$.

Answer: 2.9 mm; 116.

11.24 A close-coiled helical spring supports a load of 160 N. The mean coil diameter is 50 mm, the wire diameter is 6.5 mm, there are 15 coils and the spring stiffness is 8.4 N/mm. Find the rigidity modulus for the spring material and the maximum load the spring can support when (i) the maximum torsional shear stress is limited to 275 MPa and (ii) the total shear stress is limited to 275 MPa.

Answer: 81 GPa; 545 N.

11.25 A close-coiled helical spring absorbs 70 N m of energy when compressed 50 mm. The coil diameter is eight times the wire diameter. If there are 10 coils, calculate the diameters of the coil and the wire, the maximum torsional shear stress and the total shear stress achieved based upon the Wahl correction factor. Take $G = 86.5\,\text{GPa}$.

Answer: 25 mm, 203 mm; 84 MPa.

11.26 A close-coiled helical spring with an axial stiffness of 0.88 N/mm, is to support a maximum axial load of 40 N without the maximum torsional shear stress exceeding 124 MPa. If the solid height of the spring is 46 mm, find the wire and mean coil diameters and the number of coils. Take $G = 42\,\text{GPa}$.

Answer: 3 mm; 30 mm; 16.

11.27 A close-coiled helical spring has a mean coil radius of 90 mm with 25 effective coils. If the spring extends by 80 mm when carrying an axial load of 300 N, determine the wire diameter and the maximum shear stress in the wire. Take $G = 85\,\text{GPa}$.

Answer: 15 mm, 20 MPa.

11.28 A close-coiled helical spring is to be subjected to an axial couple of 0.45 N m. If, for six effective coils the angular rotation is to be 25° without the maximum bending stress in the wire exceeding 300 MPa, find the wire and coil diameters. Take $E = 207\,\text{GPa}$.

Answer: 2.47 mm; 19.7 mm.

11.29 A close-coiled helical spring with 12 coils of 54 mm mean diameter is made from square-section steel wire of 4.75 mm side length. The spring is subjected to an axial torque of

2.75 N m. Calculate the maximum bending stress and the angular rotation between the ends. Take $E = 207$ GPa.

11.30 A close-coiled helical torsion spring is to twist by $45°$ under an axial couple of 550 N m without the maximum bending stress exceeding 350 MPa. If there are eight coils, determine the wire and coil diameters when the section is (i) circular and (ii) square. Take $E = 210$ GPa.

11.31 A close-coiled helical spring is subjected to an axial torque of 12 N m that produces $80°$ of twist. If the mean coil diameter 45 mm, calculate the wire size and the number of effective coils for (i) a circular section with a maximum bending stress of 420 MPa and (ii) a square section with a maximum bending stress of 500 MPa. Take $E = 207$ GPa.

Answer: 6.5 mm; 15.

11.32 A close-coiled helical spring twists 1 radian under a torque of 3 N m and extends 130 mm under an axial load of 220 N. If $v = 0.26$ for the spring material, find the wire and coil diameters when the bending stress is restricted to 200 MPa.

Answer: 75 mm; 5.3 mm.

11.33 A close-coiled helical spring, wound from 6.5 mm diameter steel wire, has a mean coil diameter of 108 mm and nine effective coils. It extends 90 mm under an axial load of 125 N and twists through an angle of $95°$ under an axial moment of 9 N m. Determine the moduli of elasticity and rigidity for the spring material.

Answer: 208.5 GPa; 78 GPa.

11.34 An open-coiled helical spring (o.c.h.s.) is made from steel wire of 7.5 mm diameter with a mean diameter of 50 mm and a $30°$ helix angle. If the displacement under a maximum axial load is 15 mm whilst the stress due to bending is restricted to 103 MPa, find the load and the number of coils required. Take $E = 207$ GPa, $G = 83$ GPa.

Answer: 354 N; 11.

11.35 The bending and shear stresses in a 10 coil open-coiled helical spring must not exceed 69 and 96.5 MPa respectively, when it is subjected to an axial load of 225 N. If the mean coil diameter is 10 times the wire diameter, find these dimensions together with the axial displacement and helix angle. Take $E = 207$ GPa, $G = 83$ GPa.

Answer: 8 mm; 23.8°.

11.36 Calculate the load required to extend an open-coiled helical spring by 46 mm given that the wire and coil diameters are 9.5 and 76 mm respectively. The helix angle is $30°$ and there are 12 coils. What are the stresses due to bending and twisting of the wire?

11.37 An open-coiled helical spring has 12 coils each with a $35°$ helix angle, a mean diameter of 75 mm and a wire diameter of 10 mm. Calculate the axial load that will deflect the spring by 20 mm and find the corresponding stresses due to bending and twisting.

11.38 An open-coiled helical spring has 12 coils wound with a helix angle of $30°$. If the coil and wire diameters are 102 mm and 12.5 mm respectively, calculate the axial couple necessary to produce an angular twist of 2 radians. Take $E = 207$ GPa, $G = 83$ GPa.

Answer: 112.5 N m.

11.39 A leaf spring is to span 0.75 m and carry a central concentrated load of 10 kN for a plate width of 75 mm. If the permissible stress due to bending is 465 MPa and the central deflection is not to exceed 50 mm, calculate the thickness of the plates and the number required. What mass could be dropped from a height of 510 mm and not exceed the same central deflection? Take $E = 207$ GPa.

Answer: 6.5 mm; 8; 453 N.

11.40 A steel leaf spring spans 915 mm and lies flat under a central concentrated load of 3.3 kN. If the width of the plates is 57 mm and they are each 4.75 mm thick, find the number of plates required and the central deflection for a maximum stress of 525 MPa.

11.41 What are the advantages in a leaf spring construction? The design requirements for a leaf spring are that plates, 6.5 mm thick and 75 mm wide, should not deflect more than 51 mm under a central concentrated load of 10 kN. If the maximum bending stress is 463 MPa, calculate the number of plates and the support span. Take $E = 207$ GPa.

Answer: 8; 0.75 m.

11.42 Show that, by clamping the normal leaf spring at its centre, the end deflection and maximum stress, for a cantilever leaf spring construction, are given by $\delta = 6FL^3/nbt^3E$ and $\sigma = 6FL/nbt^2$ respectively.

Impact Loading

11.43 A bar of 25 mm square section, 3 m long, is fixed at the top and held vertically. A load of 450 N is dropped a distance of 150 mm on to a collar attached to the bottom of the bar (Fig. 11.29). Determine (i) the maximum tensile stress induced in the bar and (ii) the extension of the bar. Take $E = 207$ GPa.

Answer: 121 MPa; 1.8 mm.

11.44 A load of 2.2 kN falls freely through a height of 19 mm on to a stop at the lower end of a vertical bar 4.6 m long and 645 mm² in cross-sectional area (Fig. 11.29). Determine the stress in the bar at impact and the instantaneous tensile deflection. Take $E = 200$ GPa.

Answer: 1.8 mm; 79.3 MPa.

11.45 A 5 N force falls through 0.5 mm on to a shoulder at the end of a vertical rod 0.8 m long and 10 mm diameter (Fig. 11.29). Find the maximum direct stress induced in the rod for this and the following cases: (i) when the rod diameter is doubled; (ii) when the rod length is doubled; and (iii) when $E = 200$ GPa is doubled.

Answer: 4.1 MPa.

11.46 A weight of 133.5 N falls through a height of 127 mm to strike a collar on the end of a bar 20 mm in diameter and 1.25 m in length (Fig. 11.29). Assuming that the elongation produced is small compared to the height through which the weight falls, calculate (i) the maximum stress induced in the bar and (ii) the maximum tensile elongation of the bar. Take $E = 205$ GPa.

Answer: 142 MPa; 0.84 mm.

11.47 A 90 N weight falls through a height of 265 mm to strike the spring-loaded cylinder shown in Fig. 11.30. The impact causes the end of the cylinder to make contact with the base plate. If the spring stiffness is 0.35 N/mm, determine the maximum compressive stress induced in the cylinder due to impact. Take $E = 207$ GPa.

Answer: 36.5 MPa.

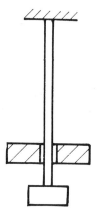

Figure 11.29

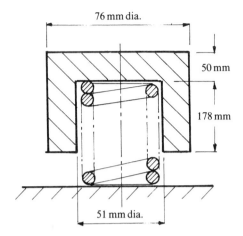

76 mm dia.

50 mm

178 mm

51 mm dia.

Figure 11.30

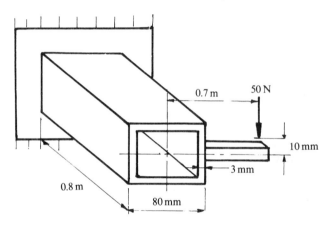

0.7 m

50 N

10 mm

3 mm

0.8 m

80 mm

Figure 11.31

11.48 The square, tubular-section, torsion bar in Fig. 11.31 is mounted as a cantilever with a 0.7 m rigid torsion arm as shown. Calculate the maximum shear stress in the tube when a force of 50 N falls 10 mm to impact with the arm end. Take $E = 2.5G = 200$ GPa.

Answer: 13.2 MPa.

11.49 A close-coiled helical spring made from square section twists 60° under an axial torque of 9 N m. If the mean coil diameter is 32 mm and the maximum bending stress in the wire is 414 MPa, find the necessary wire size and the number of coils required. For the same spring a weight of 6.5 N falls freely through 50 mm on to the end of a torsion arm 102 mm long. Find the initial deflection at the point of impact. Take $E = 207$ GPa.

Answer: 5 mm; 14 mm; 37.5 mm.

11.50 A close-coiled helical spring with 12 coils of 102 mm mean diameter stretches by 38 mm under an axial load of 220 N. Find the wire diameter. If this load is dropped onto the spring from a height of 50 mm, calculate the maximum deflection and the shear stress in the wire. Neglect loss of energy on impact. Take $G = 80$ GPa.

Answer: 9.2 mm; 111 mm; 212 GPa.

Castigliano's Theorems

11.51 Derive an expression for the horizontal displacement at the free end of the davit in Fig. 11.20.

11.52 The circular steel ring in Fig. 11.32, with mean diameter 100 mm and 6 mm section diameter, has a narrow gap in the position shown. Determine the force required to open the gap by a further 0.5 mm. Take $E = 200$ GPa.

11.53 The split steel piston ring in Fig. 11.33 has a mean diameter of 200 mm with a 10 mm square cross-section. Determine the equal opposing forces necessary to prise open the cut by 5 mm when they are applied tangentially at this point. Take $E = 210$ GPa.
Answer: 92.8 N.

11.54 The tubular body of the bow saw in Fig. 11.34 is 30 mm in outer diameter and 2 mm thick. The blade section is 25 mm × 1.5 mm. How much shorter than the nominal length of 1 m should the blade be, so that when mounted in the frame the tensile stress does not exceed 20 MPa? Take $E = 200$ GPa for both tube and blade materials.

11.55 The light alloy frame in Fig. 11.35 has rigid joints at B and C and is fixed at D. Determine the horizontal deflection at point A under the 10 N force. Take $E = 70$ GPa.

11.56 The lantern of the lamp-post in Fig. 11.36 weighs 1.75 kN. Find the vertical deflection due to this weight if $I = 210 \times 10^6$ mm^4 and $E = 13.8$ GPa.
Answer: 12.33 mm.

11.57 Derive expressions for the bending moment M_o and the deflection in the direction of the

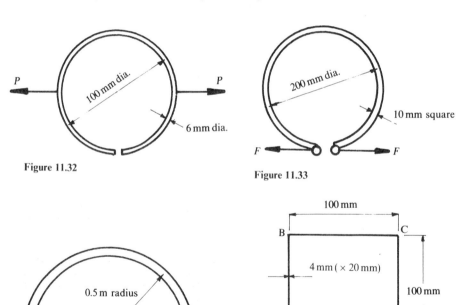

Figure 11.32

Figure 11.33

Figure 11.34

Figure 11.35

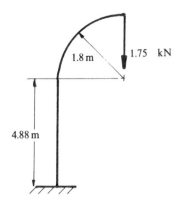

Figure 11.36

Figure 11.37

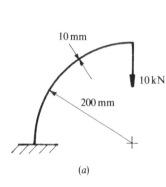

(a)

(b)   **Figure 11.38**

vertical load F_V, when the free end of the cantilevered quadrant in Fig. 11.37 is restrained from horizontal displacement.

Answer: $M_o = 2F_V R/\pi$; $\delta_V = (F_V R^3/EI)(\pi/4 - 2/\pi)$.

11.58 The cantilever in Fig. 11.38(a) is made from a solid 10 mm diameter rod by bending into the quadrant of a circle of mean radius 200 mm. Find the slope θ and deflection in the plane of an oblique 10 kN force applied at its free end as shown in Fig. 11.38(b). Take $E = 210$ GPa and $G = 84$ GPa.

Answer: 1.54 mm; 0.17°.

11.59 Find the force H that is necessary to eliminate horizontal deflection at A and B for each of the parabolic steel arches shown in Fig. 11.39(a) and (b). Draw the bending moment diagram in M, θ coordinates, inserting major values in each case. What is the increase in H if the additional horizontal displacement due to a temperature rise of 50 °C is to be prevented. Take $I = 300 \times 10^6$ mm^4, $\alpha = 11 \times 10^{-6}/$°C.

Answer: 65 kN; 83.3 kN; 83.8 kN m; 178 kN m.

11.60 Determine the displacement in the direction of the load for each of the frames in Figs 11.40(a), (b), (c) and (d) given that all bar areas are 1280 mm^2 and $E = 207$ GPa.

Answer: (b) 1.7 mm; (c) 2.96 mm; (d) 8.9 mm.

11.61 Determine the vertical displacement beneath each load for the frames in Figs 11.41(a) and (b) when all bar areas are 1280 mm^2 and $E = 207$ GPa.

460

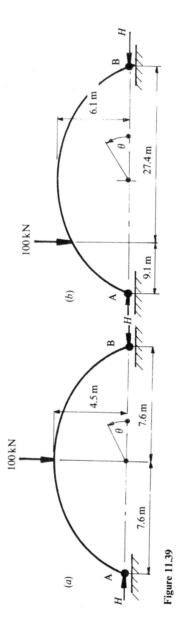

Figure 11.39

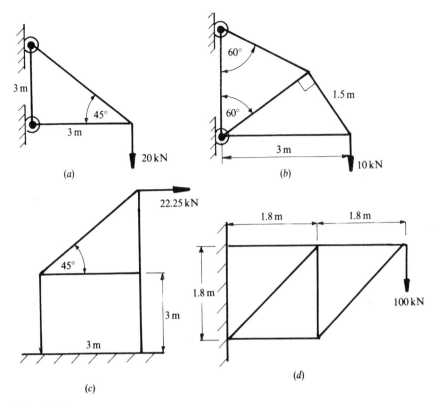

Figure 11.40

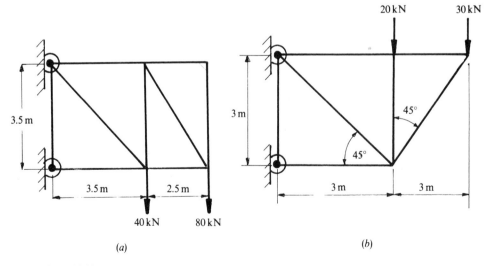

Figure 11.41

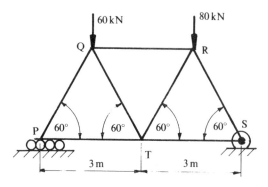

Figure 11.42

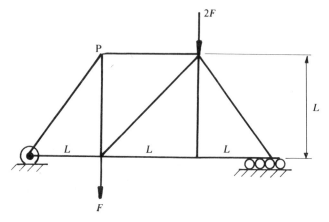

Figure 11.43

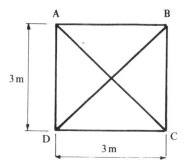

Figure 11.44

11.62 Find the vertical deflection at points Q and T for the 60° Warren bay girder in Fig. 11.42, when it is hinged at S and supported on rollers at P. The cross-sectional areas for all the bars are 1280 mm². Take $E = 208$ GPa.

Answer: 1.36 mm; 1.51 mm.

11.63 Show, for the frame in Fig. 11.43, that the vertical deflection at point P is given by $\delta = 6.22FL/EA$ when EA is constant throughout the frame.

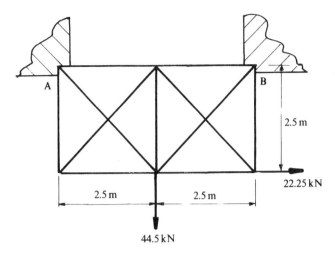

Figure 11.45

11.64 Calculate the forces for the frame in Fig. 11.44 given that bar **BD**, which was initially 3 mm short, was elastically strained into position. Take the area $A = 970\,\text{mm}^2$ and Young's modulus $E = 204\,\text{GPa}$ to be constants throughout.

Answer: $AB = BC = CD = AD \simeq 30\,\text{kN}$; $AC = BD = +42.7\,\text{kN}$.

11.65 Find the forces in each bar of the redundant structure in Fig. 11.45, when the walls react both vertical and horizontal forces at points A and B. Take $A = 500\,\text{mm}^2$ and $E = 200\,\text{GPa}$ to be constants.

TWELVE

ENERGY METHODS

12.1 THE PRINCIPLE OF VIRTUAL WORK (PVW)

It was shown in Chapter 2 that when a system of coplanar forces is in equilibrium (Fig. 2.1) then the resultant force is zero. Assume that a similar system of forces F_k ($k = 1, 2, 3, \ldots, n$) in Fig. 12.1 acts on the body, which does not itself deform. Let both the body and the forces be displaced by the virtual in-line displacements Δ_k^v, under the action of some external agency that does not change the magnitude of F_k.

Clearly, since no external work is done by the forces, the principle of virtual work states that the virtual work is zero under a virtual displacement as a consequence of the equilibrium state. That is,

$$F_1\Delta_1^v + F_2\Delta_2^v + F_3\Delta_3^v + F_4\Delta_4^v + \cdots + F_n\Delta_n^v = 0, \qquad (12.1a)$$

$$F_k\Delta_k^v = 0. \qquad (12.1b)$$

Equations (12.1a, b) express the virtual work principle in its simplest mathematical form for a rigid body. Now consider the deformable (non-rigid) volume element in Fig. 11.1(a). The above principle will further apply where each element in a body is subjected to an equilibrium system of internal stresses and external forces that each contribute to the net work done. The expression (11.3c) previously derived for the strain energy stored U, represents the work done *on* the element. It follows that the internal work done W_I, *by* the stresses is identical to U but with a change in sign. That is,

$$W_I = -\int_V \int_{\epsilon_{ij}} (\sigma_{ij}\, d\epsilon_{ij})\, dV. \qquad (12.2)$$

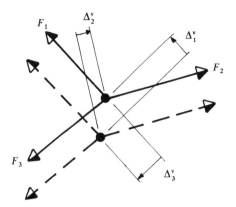

Figure 12.1 Virtual work principle.

Let the stress system σ_{ij} be produced by a system of external forces F_k and the strain increment system $\delta\epsilon_{ij}$ be produced by the virtual incremental displacements $\delta\Delta_k$. When the agency responsible for the latter is temperature, for example, it follows in the general case that σ_{ij} and $\delta\epsilon_{ij}$ are independent. Now, since the external work done is

$$W_E = \int_0^{\Delta_k} F_k \, d\Delta_k \tag{12.3}$$

and the virtual work is zero, as a consequence of equilibrium, it follows from Eqs (12.2) and (12.3) that

$$W_E + W_I = 0,$$

$$\int_0^{\Delta_k} F_k \, d\Delta_k - \int_V \int_{\epsilon_{ij}} (\sigma_{ij} \, d\epsilon_{ij}) \, dV = 0. \tag{12.4a}$$

Because of the non-dependence between F_k and $\delta\Delta_k$, σ_{ij} and $\delta\epsilon_{ij}$, Eq. (12.4a) integrates directly to give

$$F_k\Delta_k - \int_V (\sigma_{ij}\epsilon_{ij}) \, dV = 0, \tag{12.4b}$$

in which the summation convention is implied over like subscripts. Alternatively, the corresponding matrix form of Eq. (12.4b) is

$$[F]\{\Delta\} - \int_V [\sigma]\{\epsilon\} \, dV = [\Delta]\{F\} - \int_V [\epsilon]\{\sigma\} \, dV = 0, \tag{12.4c}$$

where [ ] and { } are respectively row and column matrices of the components involved. The virtual work principle is stated thus:

A system of forces in equilibrium does zero virtual work when moved through a virtual displacement due to deformation of the body on which those forces act.

Note that the principle was derived from the application of a virtual displacement (strain) in the presence of a real or actual force (stress) system. Applications of this approach (i.e. the principle of virtual displacements) are considered in Section 12.5. Equation (12.4b) could also have been derived in consideration of the following.

1. A virtual, equilibrium, force–stress system with a real but independent, compatible, displacement–strain system. This is the principle of virtual forces, which is particularly useful for the determination of displacements and forces in redundant structures.
2. Both systems real as with Castigliano's theorems and SPE (section 12.6).
3. Both systems virtual—which is of limited use.

Applications of the particular principles of virtual forces and displacements are now considered.

12.2 THE PRINCIPLE OF VIRTUAL FORCES (PVF)

The real strain (ϵ_{ij})–displacement (Δ_k) system is produced by the actual loading. Since the virtual force–stress system is independent, it applies when the actual loading is removed and a single virtual or dummy force F^v is applied at the point where the actual displacement Δ is required. Equations (12.4b and c) then become

$$F^v \Delta - \int_V (\sigma_{ij}^v \epsilon_{ij}) \, dV = 0, \qquad (i, j = 1, 2, 3), \qquad (12.5a)$$

or

$$F^v \Delta - \int_V [\sigma^v]\{\epsilon\} \, dV = 0, \qquad (12.5b)$$

where the second term in Eqs (12.5) has been re-expressed as $-W_I$ (PVW) in Table 12.1 for the more common states of virtual stress and real strain. The derivations are similar to those previously given in Chapter 11 (pages 412–422) for the corresponding strain energy expressions.

Table 12.1 Common forms of $W_I = -\int_V (\sigma_{ij}^v \epsilon_{ij}) \, dV = -\int_V [\sigma^v]\{\epsilon\} \, dV$

Loading	σ_{ij}^v	ϵ_{ij}	dV	$-W_I$(PVW)	$-W_I$(ULM)
Direct stress	$\sigma^v = P^v/A$	$\epsilon = P/AE$	$dA\,dz$	$P^v PL/AE$	pPL/AE
Torsion	$\tau^v = T^v r/J$	$\gamma = Tr/J$	$2\pi r\,dr\,dz$	$\int T^v T\,dz/GJ$	$\int tT\,dz/GJ$
Shear	$\tau^v = F^v/A$	$\gamma = F/GA$	$dA\,dz$	$\int F^v F\,dz/GA$	$\int fF\,dz/GA$
Bending	$\sigma^v = M^v y/I$	$\epsilon = My/IE$	$dA\,dz$	$\int M^v M\,dz/EI$	$\int mM\,dz/EI$
Shear flow	$\tau^v = q^v/t$	$\gamma = q/tG$	$t\,ds \times 1$	$\oint q^v q\,ds/tG$	$\oint q_1 q\,ds/Gt$

Example 12.1 Find the deflection at the free end of a cantilever in Fig. 12.2(a) when the temperature varies linearly in both the y- and z-directions as shown.

A virtual force F^v applied at the free end (Fig. 12.2b) produces the virtual uniaxial bending stress

$$\sigma_{ij}^v \equiv \sigma = M^v y/I = F^v(L-z)y/I.$$

The actual or real uniaxial strain produced by a temperature variation in both y and z is given by

$$\epsilon_{ij} \equiv \alpha \, \Delta T(y,z) = \alpha[\Delta T_0 y(L-z)/hL].$$

The volume element of beam is here $dV = dA \times dz$. Substituting into Eq. (12.5a):

$$F^v \Delta - \int\int [F(L-z)y/I]\alpha[\Delta T_0 y(L-z)/hL](dA \times dz) = 0,$$

$$F^v \Delta - (F^v \alpha \, \Delta T_0/IhL)\int_0^L (L-z)^2 \, dz \int_A y^2 \, dA = 0,$$

but $I = \int_A y^2 \, dA$,

$$\therefore \quad F^v \Delta - (F^v \alpha \, \Delta T_0/hL)\int_0^L (L-z)^2 \, dz = 0,$$

$$F^v \Delta - F^v \alpha \, \Delta T_0 L^2/3h = 0,$$

$$\Delta = \alpha \, \Delta T_0 L^2/3h.$$

Example 12.2 Determine the slope and deflections at the free end of the cantilever structure in Fig. 12.3(a) from a consideration of bending only. What is the largest source of error from neglect of other effects?

Table 12.1 supplies the appropriate form of Eq. (12.5) to give both the vertical (Δ_y) and horizontal (Δ_x) deflection at A. That is,

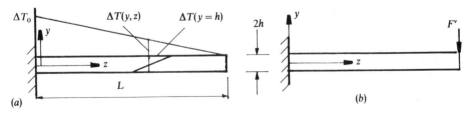

(a)

(b)

Figure 12.2 Cantilever beam.

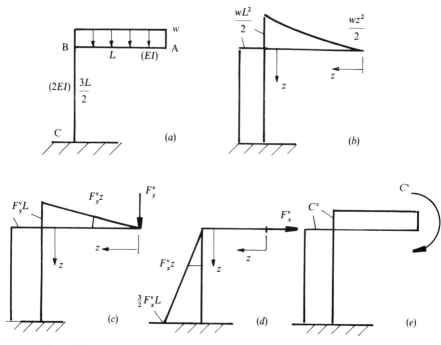

Figure 12.3 Cantilever structure.

$$F^v\Delta - \int M^v M \, dz/EI = 0. \tag{i}$$

For Δ_y, use is made of the actual M diagram under the distributed loading in Fig. 12.3(b) and the M_y^v diagram (Fig. 12.3c) under an isolated virtual vertical force F_y^v at A. Applying the integral in Eq. (i) to limbs AB and BC separately,

$$F_y^v\Delta_y - \left[\int_0^L M_y^v M \, dz/EI + \int_0^{3L/2} M_y^v M \, dz/EI \right] = 0,$$

$$F_y^v\Delta_y - \left[\int_0^L (F_y^v z)(wz^2/2) dz/EI + \int_0^{3L/2} (F_y^v L)(wL^2/2) \, dz/2EI \right] = 0,$$

$$F_y^v\Delta_y - [(wF_y^v/2EI)|z^4/4|_0^L + (wF_y^v L^3/4EI)|z|_0^{3L/2}] = 0,$$

$$\Delta_y = wL^4/2EI.$$

For Δ_x, Fig. 12.3(b) is used in conjunction with the M_x^v diagram (Fig. 12.3d) under a virtual horizontal force F_x^v at A. Equation (i) becomes

$$F_x^v\Delta_x - \left[\int_0^L M_x^v M \, dz/EI + \int_0^{3L/2} M_x^v M \, dz/EI \right] = 0,$$

$$F^v_x \Delta_x - \left[\int_0^L (0)(wz^2/2)\,dz/EI + \int_0^{3L/2} (F^v_x z)(wL^2/2)\,dz/2EI \right] = 0,$$

$$F^v_x \Delta_x - (wF^v_x L^2/4EI)|z^2/2|_0^{3L/2} = 0,$$

$$\Delta_x = 9wL^4/32EI.$$

The following form of Eq. (12.5) determines the free-end slope ϕ when a virtual couple C^v is applied at that point. This gives

$$C^v \phi - \int M^v M \, dz/EI = 0. \tag{ii}$$

Applying the integral in Eq. (ii) separately to each limb:

$$C^v \phi - \left[\int_0^L M^v M \, dz/EI + \int_0^{3L/2} M^v M \, dz/EI \right] = 0,$$

in which M^v are the virtual moments produced in each limb under C^v (Fig. 12.3e) and M are the actual moments in Fig. 12.3(b). Then,

$$C^v \phi - \left[\int_0^L C^v (wz^2/2)\,dz/EI + \int_0^{3L/2} C^v (wL^2/2)\,dz/2EI \right] = 0,$$

$$C^v \phi - [(wC^v/2EI)|z^3/3|_0^L + (wC^v L^2/4EI)|z|_0^{3L/2}] = 0,$$

$$C^v \phi - [(wC^v L^3/6EI) + (3wC^v L^3/8EI)] = 0,$$

$$\phi = 13wL^3/24EI.$$

Previous examples have shown that contributions from shear force are usually negligible. The largest source of error in Δ_y therefore results from ignoring compression in BC.

Example 12.3 Derive an expression for the angle of twist and the vertical deflection under the end force P for the cantilever arc in Fig. 12.4(a).

Ignoring shear force effects, the deformation becomes a combination of both bending and torsion. The particular form of Eq. (12.5), required for the angular twist, follows from Table 12.1:

$$C^v \theta - \left[\int M^v M \, ds/EI + \int T^v T \, ds/GJ \right] = 0, \tag{i}$$

where s, the peripheral distance, replaces z and C^v is a virtual couple applied at the free end. The virtual moment M^v and torque T^v produced by C^v are found from resolution in Fig. 12.4(b):

$$M^v = C^v \cos \alpha, \qquad T^v = C^v \sin \alpha. \tag{ii}$$

The actual moment M and the torque T produced by the applied force

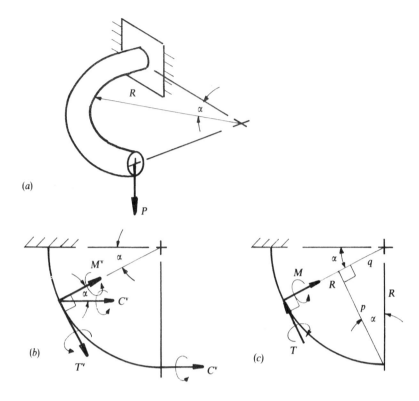

(a)

(b)

(c)

Figure 12.4 Cantilever arc.

P at any section defined by R and α are, from Fig. 12.4(c),

$$M = Pp = PR \cos \alpha, \qquad T = -P(R - q) = -PR(1 - \sin \alpha). \qquad \text{(iii)}$$

The moment vectors show that M and M^v act in the same sense but the torque vectors T and T^v oppose. Hence the need for the minus sign in the T expression. Substituting Eqs (ii) and (iii) into (i) with $ds = R\,d\alpha$:

$$C^v\theta - \left[\int (C^v \cos \alpha)(PR \cos \alpha)(R\,d\alpha)/EI \right.$$

$$\left. - \int (C^v \sin \alpha)PR(1 - \sin \alpha)(R\,d\alpha)/EI \right] = 0,$$

$$C^v\theta - \left[(C^v PR^2/EI) \int_0^{\pi/2} \cos^2 \alpha\,d\alpha \right.$$

$$\left. - (C^v PR^2/GJ) \int_0^{\pi/2} (\sin \alpha - \sin^2 \alpha)\,d\alpha \right] = 0,$$

$$C^v\theta - [(C^vPR^2/2EI)|\alpha + (1/2)\sin 2\alpha|_0^{\pi/2}$$
$$- (C^vPR^2/GJ)|(1/4)\sin 2\alpha - \cos\alpha - \alpha/2 + |_0^{\pi/2}] = 0,$$
$$\theta - [(PR^2/2EI)(\pi/2) - (PR^2/GJ)(1 - \pi/4)] = 0,$$
$$\therefore \quad \theta = PR^2[(\pi/4EI) - (1 - \pi/4)/GJ].$$

The vertical deflection Δ is supplied from Eq. (12.5) and Table 12.1:

$$F^v\Delta - \left[\int M^vM \, ds/EI + \int T^vTds/GJ\right] = 0, \qquad \text{(iv)}$$

where Eq. (iii) again gives M and T under the actual loading. The virtual force F^v applied at the free end produces the virtual moment and torque

$$M^v = F^vR\cos\alpha, \qquad T^v = -F^vR(1 - \sin\alpha). \qquad \text{(v)}$$

Substituting Eqs (iii) and (v) into (iv):

$$F^v\Delta - \left[(PF^vR^3/EI)\int_0^{\pi/2}\cos^2\alpha \, d\alpha + (PF^vR^3/GJ)\int_0^{\pi/2}(1 - \sin\alpha)^2 \, d\alpha\right] = 0,$$

$$\Delta - [(PR^3/2EI)|\alpha + (1/2)\sin 2\alpha|_0^{\pi/2}$$
$$+ (PR^3/GJ)|3\alpha/2 + 2\cos\alpha - (1/4)\sin 2\alpha|_0^{\pi/2}] = 0,$$

$$\Delta - [(\pi PR^3/4EI) + (3\pi/4 - 2)PR^3/GJ] = 0,$$

$$\therefore \quad \Delta = PR^3[(\pi/4EI) + (3\pi/4 - 2)/GJ].$$

Example 12.4 Determine the vertical deflection at point Q for the Warren bay girder in Fig. 12.5 when it is hinged at S and supported on rollers at P. Take $E = 208$ GPa (13 500 tonf/in^2) with each bar length and area to be constant values of 10 m (10 ft) and 1280 mm^2 (2 in^2) respectively.

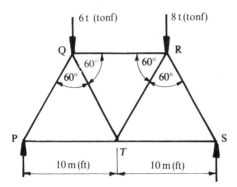

Figure 12.5 Warren bay girder.

For n bars in the frame under direct stress, the particular combination of Eq. (12.5) and Table 12.1 becomes

$$F^{\rm v}\Delta - \sum_{j=1}^{n} (P^{\rm v} PL/AE)_j = 0, \qquad\qquad (\text{i})$$

where $P^{\rm v}$ are the bar forces under a virtual vertical force $F^{\rm v}$ applied at Q, and P are the bar forces under the actual loading. Note that for the latter the force at Q is not removed as in Castigliano's approach. The reader should check the P and $P^{\rm w}$ bar force values given (in either units) in Table 12.2 by any suitable method

Table 12.2

Bar	$P^{\rm v}$	P	$P^{\rm v}P$
PQ	$-0.864F^{\rm v}$	-7.505	$6.4840F^{\rm v}$
QT	$-0.291F^{\rm v}$	0.5774	$-0.1680F^{\rm v}$
QR	$-0.288F^{\rm v}$	-4.042	$1.1641F^{\rm v}$
PT	$0.432F^{\rm v}$	3.753	$1.6213F^{\rm v}$
RT	$0.288F^{\rm v}$	-0.5774	$-0.1663F^{\rm v}$
RS	$-0.288F^{\rm v}$	-8.66	$2.4941F^{\rm v}$
TS	$0.144F^{\rm v}$	4.33	$0.6235F^{\rm v}$
			$12.0527F^{\rm v}$

SOLUTION TO SI UNITS $\sum P^{\rm v}P = 12.0527F^{\rm v}$ (t)^2 and Eq. (i) becomes

$$F^{\rm v}\Delta_Q = (L/AE)\sum P^{\rm v}P$$
$$= 10 \times 10^3 \times 12.0527F^{\rm v} \times 10^3 \times 9.81/(1280 \times 10^3)$$
$$\therefore \quad \Delta_Q = 4.44\,\text{mm.}$$

SOLUTION TO IMPERIAL UNITS $\sum P^{\rm v}P = 12.0527F^{\rm v}$ (tonf)^2 and Eq. (i) becomes

$$F^{\rm v}\Delta_Q = (L/AE)\sum P^{\rm v}P$$
$$= 10 \times 12 \times 12.0527F^{\rm v}/2 \times 13\,500$$
$$\therefore \quad \Delta_Q = 0.0536\,\text{in.}$$

12.3 THE UNIT LOAD METHOD (ULM)

It will be seen that in each of the above examples the virtual force $F^{\rm v}$ cancels within the virtual work expression. It follows that $F^{\rm v}$ could be replaced by a unit virtual load (i.e. a force, moment or torque) applied at the point where the displacement Δ, rotation ϕ or twist θ is required. Equation (12.5) then becomes

$$\Delta(\text{or } \phi, \theta) - \int_V (\sigma^{\text{v}}_{ij}\epsilon_{ij})\,\mathrm{d}V = 0. \tag{12.6}$$

In order to distinguish between the virtual forces, moments and torques, etc., that are produced by the unit load, the corresponding lower-case symbols appear with $- W_l(\text{ULM})$ in Table 12.1. The following examples illustrate this widely used simplification to the PVF.

Example 12.5 Determine the vertical deflection at the free end of the cantilever in Fig. 12.6(a) when, for the cross-section, $I = 10^9\,\text{mm}^4$ and $E = 207\,\text{GPa}$.

Assuming that deformation is solely due to bending, Eq. (12.6) becomes

$$\Delta - \int mM\,\mathrm{d}z/EI = 0, \tag{i}$$

where from Fig. 12.6(b) under a unit force at the free end,

$$m = z \quad \text{for} \quad 0 \le z \le 10,$$

and from Fig. 12.6(c) under the applied loading,

$$
\begin{aligned}
M &= 0 & &\text{for} \quad 0 \le z \le 4\,\text{m} \\
M &= 10(z - 4)\,(\text{N m}) & &\text{for} \quad 4 \le z \le 6 \\
M &= 10(z - 4) + 5(z - 6)\,(\text{N m}) & &\text{for} \quad 6 \le z \le 10.
\end{aligned}
$$

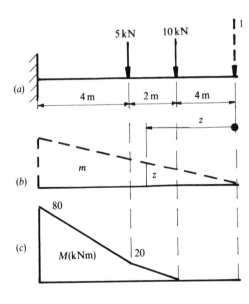

(a)

(b)

(c)

Figure 12.6 Cantilever deflection.

$$\frac{2}{EA} = 1 \times 10^{-3}.$$

Substituting into Eq. (i)

$$EI\Delta = \int_4^6 10(z-4)z\,dz + \int_6^{10} [10(z-4) + 5(z-6)]z\,dz$$

$$= 10|z^3/3 - 2z^2|_4^6 + 10|z^3/3 - 2z^2|_6^{10} + 5|z^3/3 - 3z^2|_6^{10}$$

$$= 10(10.667) + 10(133.33) + 5(33.33 + 36)$$

$$= 1786.62\,\text{kN m}^3,$$

$$\therefore \quad \Delta = 1786.62 \times 10^{12}/(207 \times 10^3 \times 10^3)$$

$$\Delta = 8.66\,\text{mm}.$$

Example 12.6 Determine the horizontal displacement at the free end of the cantilevered arc in Fig. 12.7. Take $EI = 2.1 \times 10^{12}\,\text{N mm}^2$ (2250 tonf ft^2).

Equation (12.6) is written as

$$\Delta - \int mM\,ds/EI = 0, \tag{i}$$

where for a unit horizontal force at the free end,

$$m = R(1 - \cos\theta)(\text{clockwise} + \text{ve}) \quad \text{for} \quad 0 \le \theta \le 3\pi/2.$$

Under the applied loading in Fig. 12.7,

$$M = PR(1 - \cos\theta) \qquad\qquad\qquad \text{for} \quad 0 \le \theta \le \pi/2$$

$$M = PR(1 - \cos\theta) - WR[1 - \cos(\theta - \pi/2)]$$

$$= PR(1 - \cos\theta) - WR(1 - \sin\theta) \qquad \text{for} \quad \pi/2 \le \theta \le 3\pi/2.$$

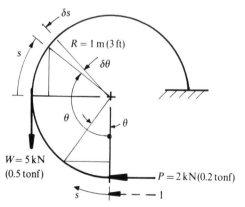

Figure 12.7 Cantilever arc.

Substituting into Eq. (i) with $ds = R\,d\theta$,

$$\Delta = (PR^3/EI) \int_0^{\pi/2} (1 - \cos\theta)^2\,d\theta + (R^3/EI)$$

$$\times \int_{\pi/2}^{3\pi/2} P(1 - \cos\theta)^2 - W(1 - \cos\theta)(1 - \sin\theta)\,d\theta,$$

$$\Delta = (PR^3/EI)|3\theta/2 - 2\sin\theta + \cos 2\theta|_0^{\pi/2}$$
$$+ (R^3/EI)[P|3\theta/2 - 2\sin\theta + \cos 2\theta|_{\pi/2}^{3\pi/2}$$
$$- W|\theta - \sin\theta + \cos\theta - (1/4)\cos 2\theta|_{\pi/2}^{3\pi/2}]$$
$$= (PR^3/EI)(3\pi/4 - 2) + (R^3/EI)[P(3\pi/2 + 4) - W(\pi + 2)]$$
$$= [(9\pi/4 + 2)P - (\pi + 2)W]R^3/EI.$$

IN SI UNITS Substituting $P = 2\,\text{kN}$, $W = 5\,\text{kN}$ and $R = 1\,\text{m}$:

$$\Delta = [(9.0685 \times 2) - (5.1417 \times 5)]10^9 \times 10^3/(2.1 \times 10^{12})$$
$$= -3.605\,\text{mm} \qquad \text{(opposing the direction of } P\text{)}.$$

IN IMPERIAL UNITS Substituting $P = 0.2\,\text{tonf}$, $W = 0.5\,\text{tonf}$ and $R = 3\,\text{ft}$.

$$\Delta = [(9.0685 \times 0.2) - (5.1417 \times 0.5)]3^3/2250$$
$$= -0.0909\,\text{ft} = -0.109\,\text{in}.$$

Example 12.7 Derive expressions for both the vertical and horizontal displacements at the load point C for the davit in Fig. 12.8.

Ignoring the effects of shear and compression, the vertical deflection is found from

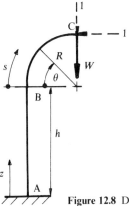

Figure 12.8 Davit displacement.

$$\Delta_y - \int m_y M \, ds/EI = 0, \tag{i}$$

where, under a unit vertical load $\downarrow$ at point C,

$m_y = 1 \times R$ clockwise for $0 \le z \le h$ in the straight portion AB,

$m_y = 1 \times R \cos \theta$ clockwise for $0 \le \theta \le \pi/2$ in the curved portion BC,

and, correspondingly, under the actual load W, the clockwise moments are

$$M = WR \text{ in AB} \qquad \text{and} \qquad M = WR \cos \theta \text{ in BC.}$$

Substituting into Eq. (i) with $ds = dz$ for AB and $ds = R \, d\theta$ for BC:

$$\Delta_y = \int_0^h WR^2 \, dz/EI + \int_0^{\pi/2} WR^3 \cos^2 \theta \, d\theta/EI$$

$$= WR^2 h/EI + (WR^3/2EI)|\theta + (1/2)\sin 2\theta|_0^{\pi/2}$$

$$= WR^2 h/EI + \pi WR^3/4EI$$

$$= (WR^2/EI)(h + \pi R/4) \quad \text{downwards} \downarrow.$$

The horizontal deflection is found from

$$\Delta_x - \int m_x M \, ds/EI = 0, \tag{ii}$$

where, under a unit horizontal load $\leftarrow$ at C,

$m_x = -(R + h - z)$ anticlockwise for $0 \le z \le h$ in AB,

$m_x = -R(1 - \sin \theta)$ anticlockwise for $0 \le \theta \le \pi/2$ in BC.

Substituting, together with M in Eq. (ii):

$$\Delta_x = -\int_0^h (R + h - z)(WR) \, dz/EI - \int_0^{\pi/2} WR^3(1 - \sin \theta)\cos \theta \, d\theta/EI$$

$$= -(WR/EI)|Rz + hz - z^2/2|_0^h - (WR^3/EI)|\sin \theta + (1/4)\cos 2\theta|_0^{\pi/2}$$

$$= -(WR/EI)(Rh + h^2 - h^2/2) - (WR^3/2EI)$$

$$= -WR(R + h)^2/2EI \quad \text{forwards} \rightarrow.$$

Example 12.8 Determine the vertical deflection at B for the simply supported beam in Fig. 12.9, given $EI = 40 \times 10^3 \, \text{kN m}^2$.

The ULM provides an alternative to the Macaulay method for the determination of beam deflections where

$$\Delta = \int mM \, dz/EI. \tag{i}$$

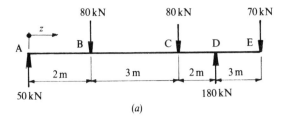

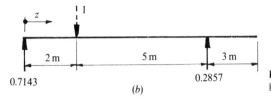

Figure 12.9 Simply supported beam.

With the reactions under the actual and virtual loading in Figs 12.9(a) and (b) and the origin for z at the L.H. ends, the moments are

AB $(0 \leq z \leq 2)$ $\quad M = -50z,$
$\quad\quad\quad\quad\quad\quad\quad\quad m = -0.7143z;$

BC $(2 \leq z \leq 5)$ $\quad M = -50z + 80(z-2),$
$\quad\quad\quad\quad\quad\quad\quad\quad m = -0.7143z + 1(z-2);$

CD $(5 \leq z \leq 7)$ $\quad M = -50z + 80(z-2) + 80(z-5),$
$\quad\quad\quad\quad\quad\quad\quad\quad m = -0.7143z + 1(z-2);$

DE $(7 \leq z \leq 10)$ $\quad M = -50z + 80(z-2) + 80(z-5) - 180(z-7),$
$\quad\quad\quad\quad\quad\quad\quad\quad m = -0.7143z + 1(z-2) - 0.2857(z-7) = 0.$

Substituting into Eq. (i):

$$EI\Delta = 35.715 \int_0^2 z^2 \, dz + \int_2^5 (0.2857z - 2)(30z - 160) \, dz$$

$$+ \int_5^7 (0.2857z - 2)(110z - 560) \, dz + 0$$

$$= 35.715|z^3/3|_0^2 + 8.571|z^3/3|_2^5 - 105.712|z^2/2|_2^5 + 320|z|_2^5$$
$$+ 31.43|z^3/3|_5^7 - 78|z^2/2|_5^7 + 1120|z|_5^7,$$

$$EI\Delta = 243.34 \, \text{kN m}^3$$

$$\therefore \quad \Delta = 243.34 \times 10^{12}/(40 \times 10^{12}) = 6.08 \, \text{mm}.$$

Example 12.9 The cranked cantilever in Fig. 12.10 carries an inclined force F at its tip. Taking bending and direct stress into account, determine

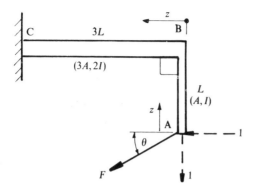

Figure 12.10 Cranked cantilever.

the horizontal and vertical components of the free-end deflection. Ignoring the effect of direct stress, what is the inclination θ when the tip deflection is aligned with F?

The horizontal deflection is found from

$$\Delta_x = \int M m_x \, dz/EI + P p_x L/AE, \qquad (i)$$

where M and P are found from the actual loading, m_x and p_x for a unit force $\leftarrow$ at A. That is,

for AB, $M = Fz \cos \theta,$ $m_x = 1 \times z$ (clockwise)

 $P = F \sin \theta,$ $p_x = 0$ (tensile)

for BC, $M = FL \cos \theta + Fz \sin \theta,$ $m_x = 1 \times L$ (clockwise)

 $P = -F \cos \theta,$ $p_x = -1$ (both compressive).

Substituting into Eq. (i):

$$\Delta_x = \int_0^L Fz^2 \cos \theta \, dz/EI + \int_0^{3L} (FL \cos \theta + Fz \sin \theta)L \, dz/2EI$$

$$+ F \cos \theta (3L)/3AE$$

$$= FL^3 \cos \theta/3EI + 3FL^3 \cos \theta/2EI + 9FL^3 \sin \theta/4EI + FL \cos \theta/AE. \qquad (ii)$$

The vertical deflection is found from

$$\Delta_y = \int M m_y \, dz/EI + P p_y L/AE, \qquad (iii)$$

where, with a unit force $\downarrow$ at A,

 $m_y = 0,$ $p_y = +1$ for AB   and $m_y = z,$ $p_y = 0$ for BC.

Substituting with the previous M and P into Eq. (iii):

$$\Delta_y = \int_0^{3L} (FL\cos\theta + Fz\sin\theta)z\, dz/2EI + FL\sin\theta/AE$$

$$= 9FL^3\cos\theta/4EI + 9FL^3\sin\theta/2EI + FL\sin\theta/AE. \quad \text{(iv)}$$

Dividing Eq. (iv) by (ii) and omitting the final direct stress term:

$$\tan\theta = \frac{\Delta_y}{\Delta_x} = \frac{(9/4) + (9/2\tan\theta)}{(11/16) + (9/4)\tan\theta},$$

$$27\tan^2\theta - 32\tan\theta - 27 = 0,$$

$$\tan\theta = 1.755 \quad \text{or} \quad -0.569,$$

$$\theta = 60.33° \quad \text{or} \quad -29.68°.$$

Only the first value corresponds with the $\Delta_x(\leftarrow)$ and $\Delta_y(\downarrow)$ directions.

Example 12.10 Find the vertical and horizontal displacements at the force point for the cantilever frame in Fig. 12.11. EI is constant.

Since bar deformation is caused by direct stress, the ULM gives the displacements:

$$\Delta_x = \sum p_x PL/AE \quad \text{and} \quad \Delta_y = \sum p_y PL/AE \quad \text{(i)}$$

where, in Table 12.3, p_x and p_y are the bar forces under unit loads applied at C in the horizontal ($\leftarrow$) and vertical ($\downarrow$) directions respectively and P are the bar forces under the actual loading.

Then, from Eqs (i),

$$\Delta_x = -3WL/AE, \quad \Delta_y = 12.657WL/AE.$$

The minus indicates that the direction of Δ_x is $\rightarrow$.

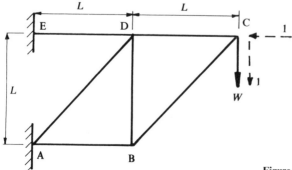

Figure 12.11 Cantilever frame.

Table 12.3

Bar	p_x	p_y	P	L	$p_x PL$	$p_y PL$
AB	0	-1	$-W$	L	0	WL
BC	0	$-\sqrt{2}$	$-\sqrt{2}W$	$\sqrt{2}L$	0	$2\sqrt{2}WL$
CD	-1	1	W	L	$-WL$	WL
BD	0	1	W	L	0	WL
AD	0	$-\sqrt{2}$	$-\sqrt{2}W$	$\sqrt{2}L$	0	$2\sqrt{2}WL$
DE	-1	2	$2W$	L	$-2WL$	$4WL$
					$\sum = -3WL$	$\sum = 12.657WL$

Example 12.11 Use the ULM to determine the rate of twist in a thin-walled tube of any cross-section (see Fig. 9.15) subjected to an axial torque.

This alternative derivation of Eq. (9.31) employs the ULM corresponding to deformation under torsion. That is, from Eq. (12.5a),

$$T^v \theta - \int \tau^v \gamma \, dV = 0, \tag{i}$$

where, from the Bredt–Batho theory (Eq. 9.26);

$$\tau^v = T^v/2At \qquad (T^v = 1 \text{ in the ULM}),$$
$$\gamma = \tau/G = T/2AGt,$$
$$dV = t \times ds \times dz.$$

Putting $\theta = \int (d\theta/dz) \, dz$ and substituting into Eq. (i) gives

$$\int_0^z (d\theta/dz) \, dz = \int_0^z \oint_s (1/2At)(T/2AGt)(t \, ds \, dz).$$

Differentiating w.r.t. z:

$$d\theta/dz = \oint_s (T/4A^2Gt) \, ds = (T/4A^2) \oint_s ds/Gt.$$

12.4 REDUNDANT STRUCTURES

The ULM in PVF provides an alternative to Castigliano's theorem for the solution to redundant pin-jointed structures. Moreover, the unit load method may be adapted to many other forms of redundant structure, i.e. where the stresses cannot be determined from equilibrium alone. Here, the use of additional compatibility condition(s) between internal strain and imposed displacement is necessary. The ULM is then applied as follows.

1. Render the structure statically determinate by removing the redundancies, i.e. by cutting the structure.
2. Determine the virtual stresses σ_{ij}^v corresponding to the separate isolated application of unit loads to each cut in turn. These stresses produce relative displacements and rotations at each cut.
3. With the introduction of each unknown redundant load (force, torque or moment) determine the strain ϵ_{ij} expression under the applied loading.
4. Apply the compatibility conditions necessary to eliminate the relative displacements. These correspond to zero virtual work at each cut. That is, from Eq. (12.6),

$$\int_V \sigma_{ij}^v \epsilon_{ij} \, dV = 0, \tag{12.7}$$

which enables the redundant loading to be found. The particular form of Eq. (12.7) depends upon the applied loading (see Table 12.1).

The following examples illustrate this common approach for a variety of different redundant structures.

Propped Cantilever

A propped cantilever (Fig. 12.12) has three unknowns M_A, R_A and R_B that cannot be determined from two equations expressing force and moment equilibrium. Recall from Chapter 7 that in the moment–area method this problem was solved from prior knowledge of the prop displacement, which is now used as the necessary compatibility condition in the ULM.

Example 12.12 Determine the deflection at the centre of the propped cantilever in Fig. 12.12(a) when the end displacement is zero and EI is constant.

Remove the redundant prop (reaction R_B). Then, for the actual distributed loading at distance z from B, the moment is

$$M_0 = wz^2/2,$$

and, with a unit upward force applied at B in the absence of w

$$m = -1 \times z \quad \text{for} \quad 0 \leq z \leq L \text{ (see Fig. 12.12b).}$$

With the introduction of R_B, the net moment at any section becomes

$$M = mR_B + M_0$$
$$= -R_B z + wz^2/2 \quad \text{for} \quad 0 \leq z \leq L.$$

Now, from Eq. (12.7) and Table 12.1, zero displacement at B is ensured when

$$\int mM \, dz/EI = 0. \tag{i}$$

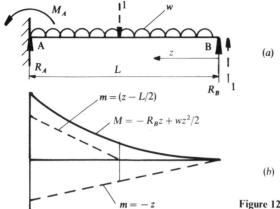

Figure 12.12 Propped cantilever.

Substituting for m and M into Eq. (i):

$$-\int_0^L (-R_B z + wz^2/2)z\,dz/EI = 0,$$

$$|wz^4/8 - R_B z^3/3|_0^L = 0,$$

$$R_B = 3wL/8.$$

Displacements are found from

$$\Delta = \int mM\,dz/EI. \tag{ii}$$

Now, with a unit downward force applied at the centre (see Fig. 12.12b),

$$m = 0 \qquad\qquad\text{for}\quad 0 \le z \le L/2,$$
$$m = 1 \times (z - L/2) \qquad\text{for}\quad L/2 \le z \le L,$$
$$M = -3wLz/8 + wz^2/2 \qquad\text{for}\quad 0 \le z \le L.$$

Substituting into Eq. (ii):

$$\Delta = \int_{L/2}^L (z - L/2)(-3wLz/8 + wz^2/2)\,dz/EI$$

$$= (1/2EI)\int_{L/2}^L (wz^2 - 3wLz/4)(z - L/2)\,dz,$$

$$= (1/2EI)\int_{L/2}^L (wz^3 - 5wLz^2/4 + 3wL^2z/8)\,dz$$

$$= (1/2EI)|(wz^4/4 - 5wLz^3/12) + (3wL^2z^2/16)|_{L/2}^L$$

$$= (1/2EI)(wL^4/64 - 5wL^4/96 + 3wL^4/64)$$

$$= wL^4/192EI.$$

Note that if the origin for z was at the L.H. end, the compatibility condition is that of zero slope ϕ at A. Equation (i) must then be coupled with $\phi = \int mM \, dz/EI = 0$ where m is the moment under a unit couple applied at A and $M = M_A - R_A z + wz^2/2$.

Pin-jointed Structures

With a single degree of redundancy, let p be the bar forces under a unit load replacing the redundant bar. Now, if the bar forces under the applied loading in the absence of the redundant bar are P_o and the force in the redundant bar is R, the net bar forces are $P = P_o + Rp$. It follows that

$$\sigma_{ij}^v = p/A, \qquad \epsilon_{ij} = P/AE = (P_o + Rp)/AE.$$

Substituting into Eq. (12.7);

$$\int \sigma_{ij}^v \epsilon_{ij} dV = \sum [(p/A)(P_o + Rp)/AE](AL) = 0,$$

$$\sum p(P_o + Rp)L/AE = 0,$$

$$\sum (pP_o L/AE) + R \sum p^2 L/AE = 0,$$

$$R = -\frac{\sum (pP_o L/AE)}{\sum (p^2 L/AE)}. \tag{12.8}$$

Clearly, with E constant throughout the structure, Eq. (12.8) is identical to Eq. (11.37) from Castigliano's theorem.

With two or more redundancies, let P_o be the statically determinate bar forces, let $p_1, p_2, p_3, \ldots$, be the bar forces under separately applied unit loads replacing the removed bars and let $R_1, R_2, R_3, \ldots$, be the redundant bar forces. The net bar forces and the strains are then

$$P = P_o + R_1 p_1 + R_2 p_2 + R_3 p_3 + \cdots, \quad \Rightarrow \epsilon_{ij} = P/AE$$

and the virtual stresses are

$$(\sigma_{ij}^v)_1 - p_1/A, \qquad (\sigma_{ij}^v)_2 = p_2/A, \qquad (\sigma_{ij}^v)_3 = p_3/A, \qquad \cdots.$$

Substituting into Eq. (12.7) leads to the system of virtual work expressions:

$$\sum(P_o p_1 L/AE) + R_1 \sum(p_1^2 L/AE) + R_2 \sum(p_1 p_2 L/AE) + R_3 \sum(p_1 p_3 L/AE) + \cdots = 0,$$
$$\sum(P_o p_2 L/AE) + R_1 \sum(p_1 p_2 L/AE) + R_2 \sum(p_2^2 L/AE) + R_3 \sum(p_2 p_3 L/AE) + \cdots = 0$$
$$\sum(P_o p_3 L/AE) + R_1 \sum(p_1 p_3 L/AE) + R_2 \sum(p_2 p_3 L/AE) + R_3 \sum(p_3^2 L/AE) + \cdots = 0$$

$$\cdots \cdots \cdots \cdots \cdots \cdots \cdots \cdots \cdots \cdots$$

$$\tag{12.9}$$

which are solved simultaneously for the redundant bar forces $R_1, R_2, R_3, \ldots$.

Example 12.13 Determine the forces in the bars of the frame in Fig. 12.13 when the lengths and areas are those given in Table 12.4. Find also the in-line vertical deflection beneath the 2.4 kN force. Take $E = 70$ GPa.

Cut BD and let this redundant bar force be R. The forces under the resulting statically determinate structure are those in the P_o column of Table 12.4. With unit loads acting in tension at B and D in the line of the missing bar BD, the remaining bar forces are those in the p column.

Table 12.4

Bar	L (mm)	A (mm²)	P_o (kN)	p	$P_o pL/A$	$p^2 L/A$	$P = P_o + Rp$(kN)
AB	500	62.5	+3.2	0	0	0	3.2
BC	500	125	+3.2	−0.8	−10.24	2.56	5.48
AE	625	62.5	−4.0	0	0	0	−4.0
BE	375	62.5	0	−0.6	0	2.16	1.71
CE	625	62.5	+4.0	1	40	10	1.15
DE	500	62.5	−6.4	−0.8	40.96	5.12	−4.12
BD	625	125	0	1	0	5	−2.85
					$\sum = 70.72$	$\sum = 24.84$	

Then, from Eq. (12.8), $R = -70.72/24.84 = -2.85$ kN, which enables the final column to be completed.

The vertical deflection Δ_y follows from

$$F^v \Delta_y - \sum pPL/AE = 0. \qquad (i)$$

The bar forces P follows from the above table and p are now the bar forces under a unit force ($F^v = 1$) at A. Since these are $p = P/2.4$, Eq. (i) gives

$$\Delta_y = (1/2.4E)\sum P^2 L/A. \qquad (ii)$$

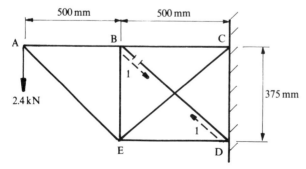

Figure 12.13 Frame forces.

Table 12.5

Bar	AB	BC	AE	BE	CE	DE	BD
P (kN)	3.2	5.48	−4	1.71	1.15	−4.12	−2.85
P^2	10.24	30.03	16	2.924	1.323	16.974	8.123
P^2L/A	81.92	120.12	160	17.54	13.23	135.79	40.62

From Table 12.5 $\sum P^2 L/A = 569.22\,\text{kN}^2/\text{mm}$. Then from Eq. (ii),

$$\Delta_y = 569.22 \times 10^3/(2.4 \times 70 \times 10^3) = 3.34\,\text{mm},$$

which is less than would be found in the absence of bar BD.

Example 12.14 Determine the forces in each bar of the space frame in Fig. 12.14 and the vertical deflection at A. Bars AC, AD and AE are each $200\,\text{mm}^2$ $(0.25\,\text{in}^2)$ in area while AB is $1200\,\text{mm}^2$ $(1.50\,\text{in}^2)$. Take $E = 200\,\text{GPa}$ $(13\,000\,\text{tonf/in}^2)$. (K.P. 1976)

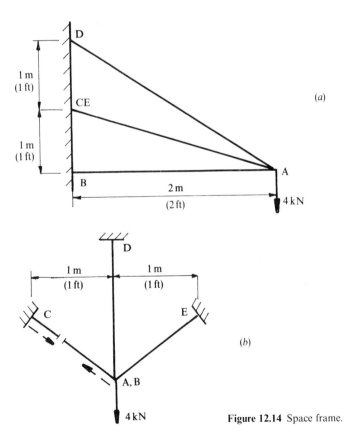

Figure 12.14 Space frame.

SOLUTION TO SI UNITS   Let AC be the redundant bar. The P_o and p bar forces given in Table 12.6 are conveniently found from tension coefficients:

Table 12.6

Bar	L (mm)	A (mm²)	P_o(kN)	p	$P_o p L / A$	$p^2 L / A$	$P = P_o + Rp$	$P^2 L / A$
AB	2000	1200	−4	−0.8165	5.443	1.111	−5.596	52.192
AC	2449.5	200	0	1	0	12.248	1.955	46.810
AD	2828.4	200	5.657	−1.155	−92.38	18.866	3.400	163.48
AE	2449.5	200	0	1	0	12.248	1.955	46.81
					$\sum = -86.937$	$\sum = 44.473$		$\sum 309.292$

From Eq. (12.8), $R = 86.937/44.473 = 1.955 \, \text{kN}$, which enables the column of net bar forces $P = P_o + Rp$ to be completed. As with the previous example, the vertical deflection at A is given by

$$\Delta = (1/4E)\sum P^2 L / A$$
$$= 309.29 \times 10^3/(4 \times 200 \times 10^3) = 0.386 \, \text{mm.}$$

Alternatively, Δ may be calculated from the above P forces in the simplest statically determinate frame, say AB and AD. With a downward unit force at A, $p_{AB} = -1$, $p_{AD} = \sqrt{2}$ and,

$$\Delta = (pPL/AE)_{AB} + (pPL/AE)_{AD}$$
$$= [(-1)(-5.596)2000/(1200E)] + [\sqrt{2} \times 3.4 \times 2828.4/(200E)]$$
$$= 0.386 \, \text{mm.}$$

SOLUTION TO IMPERIAL UNITS   In imperial units, Table 12.7 applies:

Table 12.7

Bar	L(in)	A (in²)	P_o(tonf)	p	$P_o p L / A$	$p^2 L / A$	$P = P_o + Rp$	$P^2 L / A$
AB	24	1.5	−4	−0.8165	52.256	10.667	−5.5966	501.151
AC	29.39	0.25	0	1	0	117.56	1.955	449.317
AD	33.94	0.25	5.657	−1.155	−887.033	181.107	3.400	1569.386
AE	29.39	0.25	0	1	0	117.56	1.955	449.317
					$\sum = -834.78$	$\sum = 426.894$		$\sum 2969.17$

Again, $R = 834.78/426.89 = 1.955 \, \text{tonf}$. The bar forces P then give

$$\Delta = (1/4E)\sum P^2 L / A$$
$$= 2969.17/(4 \times 13\,000) = 0.057 \, \text{in.}$$

Example 12.15 Determine the bar forces for the pin-jointed structure in Fig. 12.15(a). EI is constant.

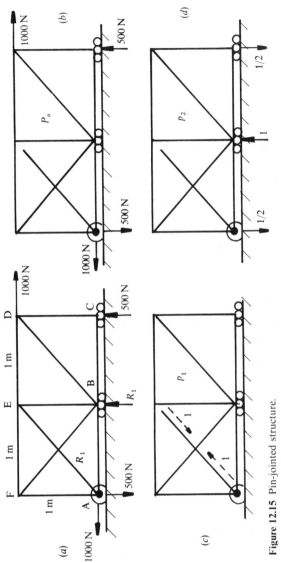

Figure 12.15 Pin-jointed structure.

The two redundancies present in this structure may be deduced from the application of moment and force equilibrium equations. Taking these to be the force in bar $AE(R_1)$ and the reaction at $B(R_2)$, Eq. (12.9) becomes

$$\sum(P_o p_1 L/AE) + R_1 \sum(p_1^2 L/AE) + R_2 \sum(p_1 p_2 L/AE) = 0, \quad \text{(i)}$$

$$\sum(P_o p_2 L/AE) + R_1 \sum(p_1 p_2 L/AE) + R_2 \sum(p_2^2 L/AE) = 0 \quad \text{(ii)}$$

where the bar forces apply to Figs 12.15(b), (c) and (d) as tabulated in Table 12.8.

Table 12.8

Bar	L	P_o	p_1	p_2	$P_o p_1 L$	$P_o p_2 L$	$p_1 p_2 L$	$p_1^2 L$	$p_2^2 L$	$P(N)$
AB	1	1000	−0.707	0	−707	0	0	0.5	0	597.5
AE	$\sqrt{2}$	0	1	0	0	0	0	1.414	0	569.2
AF	1	500	−0.707	0.5	−353.6	250	−0.354	0.5	0.25	195.3
BC	1	0	0	0	0	0	0	0	0	0
BD	$\sqrt{2}$	707	0	−0.707	0	−707	0	0	0.707	568.7
BE	1	0	−0.707	0	0	0	0	0.5	0	−402.5
BF	$\sqrt{2}$	−707	1	−0.707	−1000	707	−1	1.414	0.707	−276
CD	1	−500	0	0.5	0	−250	0	0	0.25	−402.2
DE	1	500	0	0.5	0	250	0	0	0.25	597.8
EF	1	500	−0.707	0.5	−353.6	250	−0.354	0.5	0.25	195.3
					$\sum = -2414$	500	−1.708	4.828	2.4142	

Substituting into Eqs (i) and (ii):

$$-2414 + 4.828R_1 - 1.7080R_2 = 0,$$
$$500 - 1.708R_1 + 2.4142R_2 = 0$$

from which $R_1 = 569.2$ N and $R_2 = 195.6$ N. Hence the final column of net bar forces is found from $P = P_o + 569.2p_1 + 195.6p_2$.

In order to find the deflection at any joint, it is only necessary to take a statically determinate part of the frame and apply

$$\Delta = \sum pPL/AE,$$

where P are the actual bar forces in the table and p are the bar forces under a unit load applied at the joint in the required direction. Note that, where necessary, the reactions that equilibrate the unit load must be included in the calculation of p.

Torsion of a Thin-walled Multicell Tube

Though the two-cell problem has been solved previously in Chapter 9 (page 313), by the separate application of equilibrium and compatibility, it is instructive to confirm this from the general approach to redundant structures. Cut cell 2 as shown in Fig. 12.16(a) to give the statically determinate Batho shear flow in cell 1 as

$$q_o = T/2A_1. \tag{12.10}$$

This will cause a relative displacement between the cut faces. By applying isolated, opposing unit shear forces at the cut, a unit shear flow will be induced in cell 2 (Fig. 12.16b).

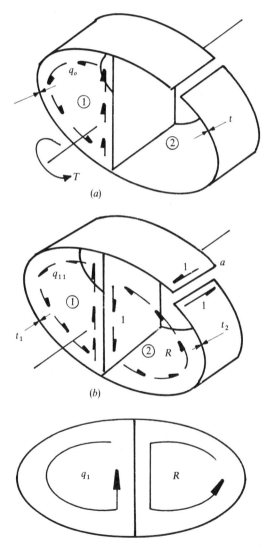

Figure 12.16 Torsional shear flow in a two-cell tube of unit length.

The unit shear flow will be reacted by a shear flow q_{11} in cell 1 according to the equilibrium condition $T = 0$ in Eq. (9.35). That is,

$$0 = 2A_1 q_{11} + 2A_2, \quad \Rightarrow q_{11} = -A_2/A_1. \tag{12.11}$$

If q_1 is the net shear flow in cell 1 and $R(= q_2)$ is the shear flow in the redundant cell 2 (Fig. 12.16c), then

$$q_1 = q_o + q_{11} R. \tag{12.12}$$

The actual and virtual shear flows for each cell are shown in Fig. 12.17(a) and (b).

Compatibility is restored, with zero relative displacement at the cut, through the application of Eq. (12.7). That is, for zero virtual work,

$$\int_V \sigma^v_{ij} \epsilon_{ij} \, dV = \int_V \tau^v \gamma \, dV = 0. \tag{12.13a}$$

Now, from Eq. (9.26), $\tau^v = q^v/t$, $\gamma = q/tG$ and, for a unit length, $dV = t \, ds$ (Fig. 9.14). Substituting into Eq. (12.13a) for each tube:

$$q^v q \oint_1 ds/Gt + q^v q \oint_2 ds/Gt = 0. \tag{12.13b}$$

Define $\Delta_1 = \int_1 ds/Gt$ for the common path over which q_1 and q_{11} are constant, $\Delta_2 = \int_2 ds/Gt$ the common path for R and 1 and $\Delta_{12} = \int_{1,2} ds/Gt$ for the path (web) over which $\pm(q_1 - R)$ and $\pm(q_{11} - 1)$ are constant (Fig. 12.17(a) and (b)). Equation (12.13b) becomes

$$q_1 q_{11} \Delta_1 + (q_1 - R)(q_{11} - 1)\Delta_{12} + R(1)\Delta_2 + (R - q_1)(1 - q_{11})\Delta_{12} = 0. \tag{12.13c}$$

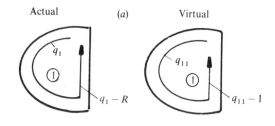

Actual   (a)   Virtual

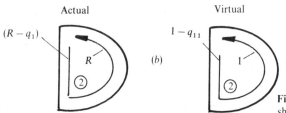

Actual   Virtual

(b)

Figure 12.17 Actual and virtual shear flow in each cell.

Substituting from Eq. (12.12) and expanding leads to

$$q_o[q_{11}\Delta_1 + 2(q_{11} - 1)\Delta_{12}] + R[q_{11}^2\Delta_1 + 2(q_{11} - 1)^2\Delta_{12} + (1)^2\Delta_2] = 0.$$
$$(12.13d)$$

Writing, for brevity, q_{1a} as the components of the virtual shear flow in each cell, Eq. (12.13d) becomes

$$\sum q_o q_{1a}\Delta + R\sum q_{1a}^2\Delta = 0, \qquad (12.13e)$$

from which $R(= q_2)$ can be found and the similarity in form with Eq. (12.8) is evident. Though this approach may be extended to multicell tubes, it is more convenient to solve these from the fact that each cell will suffer the same rate of twist (see Examples 9.13 and 9.14). The simultaneous solution for the shear flows more readily follows from the appropriate extensions to Eqs (9.35) and (9.36) than from virtual work.

Example 12.16 Confirm the shear flows found for Fig. 9.21 using the unit load method.

From Eqs (12.10) and (12.11),

$$q_o = T/2A_1 = 10 \times 10^3/(2 \times 6000) = 0.8333 \text{ N/mm},$$
$$q_{11} = -A_2/A_1 = -4330/6000 = -0.7217 \text{ N/mm},$$
$$\Delta_1 = 2 \times 130/2 = 130, \qquad \Delta_{12} = 100/4 = 25 \qquad \text{and}$$
$$\Delta_2 = 2 \times 100/2 = 100.$$

Substituting into Eq. (12.13d):

$$0.8333[(-0.7217)130 + 2(-0.7217 - 1)25]$$
$$+ R[(-0.7217)2130 + 2(-0.7217 - 1)225 + 100] = 0$$
$$\therefore \quad -149.92 + 315.92R = 0,$$
$$R = 0.474 \text{ N/mm} = q_2,$$
$$\therefore \quad q_1 = q_o + Rq_{11} = 0.8333 - (0.474 \times 0.7217) = 0.491 \text{ N/mm}.$$

Flexural Shear Flow for Idealized Multicell Tubular Structures

The unit load method is particularly useful for determining shear flow in multicell, idealized tubes when a transverse shear force F acts through the shear centre. The single-cell tube that has one degree of shear flow redundancy R is again solved by making a single cut in one web. If q_o is the resulting shear flow under F and $q^v = 1$ is the virtual shear flow due to opposing unit forces, acting in isolation at the cut, then the net shear flow is

$$q = q_o + Rq^v$$

where the redundant force R is found from eliminating the relative displacement at the cut with zero virtual work. That is, from Eq. (12.7),

$$\int_V \sigma_{ij}^v \epsilon_{ij} \, dV = \sum \oint q^v q \, ds/Gt = 0,$$

$$\sum q^v(q_o + Rq^v) \oint ds/Gt = 0,$$

$$\sum q^v q_o \oint ds/Gt + R \sum (q^v)^2 \oint ds/Gt = 0,$$

which is clearly similar to Eq. (12.13e) for a singly redundant, two-cell tube under torsion. The summations account for variations in q_o and the path integral $\oint ds/Gt$ within each web. In the case of the two-cell tube under flexural shear in Fig. 12.18(a), two cuts a and b are necessary for static determinancy.

The two degrees of shear flow redundancy are now denoted by R_a and R_b respectively, giving the net shear flow in any cell as

$$q = q_o + R_a q_a + R_b q_b, \tag{12.14}$$

where $q_a = 1$ and $q_b = 1$ are the unit virtual shear flows arising from the application of unit forces to each cut in isolation (Figs 12.18(b) and (c)). Compatibility is ensured when the cut displacements are eliminated through Eq. (12.7):

$$\sum q_o q_a \oint ds/Gt + R_a \sum q_a^2 \oint ds/Gt + R_b \sum q_a q_b \oint ds/Gt = 0, \tag{12.15a}$$

$$\sum q_o q_b \oint ds/Gt + R_a \sum q_a q_b \oint ds/Gt + R_b \sum q_b^2 \oint ds/Gt = 0, \tag{12.15b}$$

which are solved simultaneously for R_a and R_b.

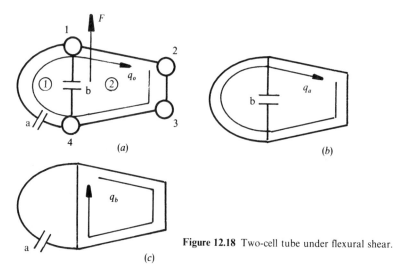

Figure 12.18 Two-cell tube under flexural shear.

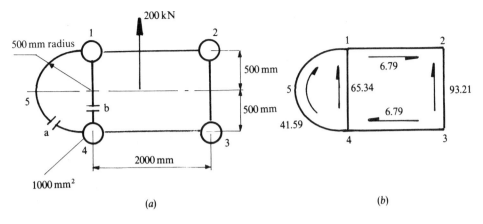

Figure 12.19 Wing box structure.

Example 12.17 Determine the shear flows for the idealized wing-box structure in Fig. 12.19(a) when a shear force of 200 kN acts vertically upwards through the shear centre. Take G and t as constants throughout.

The cuts correspond to points a and b in Fig. 12.18(a). Working clockwise from the 'free surface' at 1, the following q_o values are found from $q_o = (F_y/I_x)\sum_r A_r y_r$, where $F_y/I_x = -200 \times 10^3/(4 \times 1000 \times 500^2) = -200 \times 10^{-6}\,\text{N/mm}^4$:

1–2, $q_o = -200 \times 10^{-6}(1000 \times 500) = -100\,\text{N/mm}$,

2–3, $q_o = -200 \times 10^{-6}[(1000 \times 500) + (1000 \times 500)] = -200\,\text{N mm}$,

3–4, $q_o = -200 \times 10^{-6}[(1000 \times 500) + (1000 \times 500) - (1000 \times 500)]$
$= -100\,\text{N/mm}$.

1–4 and 1–5–4, $q_o = 0$.

The minus signs indicate that the q_o direction is anticlockwise. Taking q_a and q_b to be anticlockwise, Table 12.9 is constructed for the components in Eqs (12.15).

Table 12.9

Web	q_o	q_a	q_b	s (mm)	$q_o q_a s$	$q_o q_b s$	$q_a^2 s$	$q_b^2 s$	$q_a q_b s$
1–2	100	1	1	2000	2×10^5	2×10^5	2000	2000	2000
2–3	200	1	1	1000	2×10^5	2×10^5	1000	1000	1000
3–4	100	1	1	2000	2×10^5	2×10^5	2000	2000	2000
1–4	0	0	1	1000	0	0	0	1000	0
1–5–4	0	1	0	1571	0	0	1571	0	0
					$\sum = 6 \times 10^5$	6×10^5	6571	6000	5000

Substituting into Eqs (12.15a and b):

$$(6 \times 10^5) + 6571R_a + 5000R_b = 0,$$
$$(6 \times 10^5) + 5000R_a + 6000R_b = 0,$$

from which $R_a = -41.59$, $R_b = -65.34$ N/mm. Equation (12.14) supplies the final shear flows, illustrated with positive anticlockwise directions in Fig. 12.19(b):

1–2,	$q = 100 - 41.45 - 65.34 = -6.79$ N/mm,
2–3,	$q = 200 - 41.45 - 65.34 = 93.21$ N/mm,
3–4,	$q = 100 - 41.45 - 65.34 = -6.79$ N/mm,
1–4,	$q = 0 - 0 - 65.34 = -65.34$ N/mm,
1–5–4,	$q = 0 - 41.59 - 0 = -41.59$ N/mm.

12.5 THE PRINCIPLE OF VIRTUAL DISPLACEMENTS (PVD)

Recall that the virtual work principle in Eq. (12.4) was derived using a virtual displacement system with real loads. That is, for externally applied forces F_k and virtual displacements Δ_k^v:

$$F_k \Delta_k^v - \int_V (\sigma_{ij} \epsilon_{ij}^v) \, dV = 0. \tag{12.16a}$$

Alternatively, when external moments M_E comprise the loading, the principle employs a virtual rotation θ^v:

$$M_k \theta_k^v - \int_V (\sigma_{ij} \epsilon_{ij}^v) \, dV = 0. \tag{12.16b}$$

The PVD or a unit displacement method could be employed for finding forces analytically in structures, including those that are redundant, but the PVF or the unit load method are more often used for this purpose. The PVD is however particularly useful for providing the internal forces and displacements numerically from an assumed displacement mode, this being the basis for the 'displacement' or 'stiffness' methods of finite element analysis (see page 615). The following examples illustrate these applications.

Example 12.18 Determine the force in the bar CD for the frame in Fig. 12.20(a), given that EI is constant.

Impose a virtual displacement Δ^v at C on CD such that the lengths of all other bars remain unchanged. It follows that C is displaced to C′ along an arc of radius AC. The geometry is such that B is displaced by Δ^v to B′ as

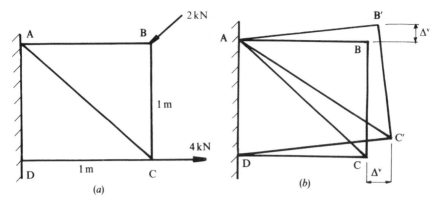

Figure 12.20 Displacements of a plane frame.

shown in Fig. 12.20(b). Equation (12.16a) becomes

$$F_k\Delta_k^{\text{v}} - \sum(P/A)(\delta^{\text{v}}/L)(AL) = 0,$$
$$\therefore \quad F_k\Delta_k^{\text{v}} = \sum P\delta^{\text{v}}, \tag{i}$$

where in general, δ^{v} are the displacements of the bars necessary to accommodate Δ^{v}. The external work (LHS of Eq. (i)) is found from the product of the applied forces F and their corresponding in-line Δ^{v}. Then, with a minus sign to denote opposing in-line directions,

$$F_k\Delta_k^{\text{v}} = (4 \times \Delta^{\text{v}}) + (-2\sin 45° \times \Delta^{\text{v}}) = 2.586\Delta^{\text{v}}.$$

The internal work (RHS of Eq. (i)) is simply $P_{CD}\delta^{\text{v}} = P_{CD}\Delta^{\text{v}}$, since no other bar is strained.

$$\therefore \quad P_{CD} = 2.586\,\text{kN}.$$

Note that since Δ^{v} cancels, a unit displacement could have replaced it.

Example 12.19 Find the central deflection v_c of a simply supported beam of length L when a concentrated moment M_o is applied at one end (Fig. 12.21a). Assume a deflected shape of the form $v = v_c \sin(\pi z/L)$. Compare with the true value from the PVF. Take EI as constant. (K.P. 1976)

The form assumed for the actual displacement v clearly satisfies the boundary conditions that $v = 0$ for $z = 0$ and $z = L$. The virtual displacement function v^{v} may be any shape that matches the boundary conditions. Here it is taken to have the same form as v:

$$v^{\text{v}} = v_c^{\text{v}} \sin(\pi z/L).$$

Now from ETB and the flexure theory (Eq. (7.4)), the PVD

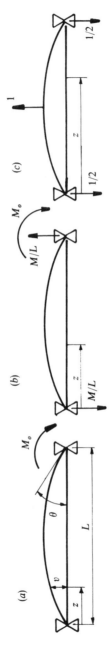

Figure 12.21 Deflection of a simply supported beam.

equation (12.16b) becomes

$$M\theta^v - \int_0^L M^v M \, dz/EI = 0,$$

$$M\theta^v - EI \int_0^L (d^2v^v/dz^2)(d^2v/dz^2) \, dz = 0, \qquad (i)$$

where here

$$d^2v^v/dz^2 = -v_c^v(\pi/L)^2 \sin(\pi z/L), \qquad (ii)$$

$$d^2v/dz^2 = -v_c(\pi/L)^2 \sin(\pi z/L), \qquad (iii)$$

and since

$$\theta^v = dv^v/dz = v_c^v(\pi/L) \cos(\pi z/L),$$

the virtual rotation to be applied at the end point $z = L$, where the external moment is M_o, is

$$\theta^v = -v_c^v \pi/L. \qquad (iv)$$

Substituting Eqs (ii), (iii) and (iv) in (i):

$$M_o(-v_c^v \pi/L) = EIv_c v_c^v(\pi/L)^4 \int_0^L \sin^2(\pi z/L) \, dz,$$

$$M_o = -v_c \pi^3 EI/2L^3,$$

$$\therefore \quad v_c = -M_o L^2/(\pi^3/2)EI \text{ (upwards)}. \qquad (v)$$

The true value may be found from the unit load method:

$$\Delta = \int_0^L mM \, dz/EI, \qquad (vi)$$

where, from Figs 12.21(b) and (c),

$$M = M_o z/L,$$
$$m = z/2 \qquad \text{for} \quad 0 \le z \le L/2,$$
$$m = z/2 - (z - L/2) \times 1 \qquad \text{for} \quad L/2 \le z \le L.$$

Substituting into Eq. (vi):

$$\Delta = \int_0^{L/2} (z/2)(M_o z/L) \, dz/EI + \int_{L/2}^L [z/2 - (z - L/2)](M_o z/L) \, dz/EI$$

$$= (M_o/2EIL) \int_0^{L/2} z^2 \, dz + (M_o/EIL) \int_{L/2}^L z[z/2 - (z - L/2)] \, dz$$

$$= (M_o/2EIL)|z^3/3|_0^{L/2} + (M_o/EIL)(|z^3/6|_{L/2}^L - |z^3/3 - z^2 L/4|_{L/2}^L)$$

$$= M_o L^2/48EI + (M_o/EIL)\{(L^3/6 - L^3/48)$$
$$- [(L^3/3 - L^3/4) - (L^3/24 - L^3/16)]\}$$
$$= M_o L^2/16EI.$$

Since $\pi^3/2 = 15.5$ in the denominator of the approximation expression (v), it is seen that the PVD approach yields an acceptable solution.

Example 12.20 Derive an expression for the force in bar BD for the redundant pin-jointed frame in Fig. 12.22(a) when a vertical load F is applied at point D and the lengths are in the ratio BD/BC/CD = 3/4/5, given that the bar material obeys (i) the non-linear elastic law $\sigma = a\epsilon(1 - b\epsilon^2)$ and (ii) Hooke's law. All bar areas A are constant.

Employing the PVD equation (12.16a) for part (i):

$$F\Delta^v = \int_V (\sigma_{ij}\epsilon_{ij}^v)\,dV = \sum P\delta^v. \tag{i}$$

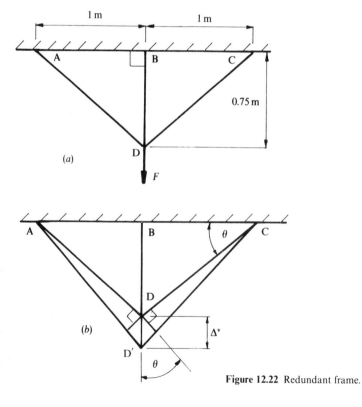

(a)

(b)

Figure 12.22 Redundant frame.

Applying Δ^v at D (Fig. 12.22b) produces the virtual bar displacements δ^v:

$$\delta^v_{CD} = \delta^v_{AD} = \Delta^v \sin \theta = (3/5)\Delta^v,$$
$$\delta^v_{BD} = \Delta^v.$$

Substituting into Eq. (i):

$$F\Delta^v = (P\delta^v)_{AD} + (P\delta^v)_{BD} + (P\delta^v)_{CD}$$
$$= (3/5)P_{AD}\Delta^v + P_{BD}\Delta^v + (3/5)P_{CD}\Delta^v,$$
$$\therefore \quad F = (3/5)P_{AD} + P_{BD} + (3/5)P_{CD}, \tag{ii}$$

where the bar forces P may be expressed in terms of the actual bar displacements $\delta = \epsilon L$ from the particular σ, ϵ law. Here,

$$P = \sigma A = a\epsilon A(1 - b\epsilon^2) = aA(\delta/L)[1 - b(\delta/L)^2],$$
$$\therefore \quad P_{AD} = aA(\delta/L)_{AD}[1 - b(\delta/L)^2_{AD}] = P_{CD},$$
$$P_{BD} = aA(\delta/L)_{BD}[1 - b(\delta/L)^2_{BD}].$$

Now if Δ is the actual displacement under F:

$$\delta_{CD} = \delta_{AD} = (3/5)\Delta, \qquad \delta_{BD} = \Delta,$$
$$\therefore \quad P_{AD} = aA(3/5)(\Delta/L_{AD})[1 - b(9/25)(\Delta/L_{AD})^2] = P_{CD},$$
$$P_{BD} = aA(\Delta/L_{BD})[1 - b(\Delta/L_{BD})^2].$$

Substituting into Eq. (ii):

$$F = (18/25)aA(\Delta/L_{AD})[1 - b(9/25)(\Delta/L_{AD})^2]$$
$$+ aA(\Delta/L_{BD})[1 - b(\Delta/L_{BD})^2], \tag{iii}$$
$$\therefore \quad P_{BD}/F = \{1 + (54/125)[1 - b(9/25)(\Delta/L_{AD})^2]/[1 - b(\Delta L_{BD})^2]\}^{-1}. \tag{iv}$$

Thus, for given values of A, F, a, b and L, Eqs (iii) and (iv) may be solved to provide a numerical value for the ratio P_{BD}/F. This is unecessary for a Hookean material, when for $b = 0$, Eq. (iv) becomes

$$P_{BD}/F = [1 + (54/125)]^{-1} = 0.698.$$

A similar PVD approach may be employed for any redundant structure. In general, the number of simultaneous equations to be solved with this method equates to the total number of degrees of freedom for the frame. This is the sum of the degrees of freedom for each joint. In Fig. 12.22 the number is only one since joint D moves in the y-direction. The choice between PVD and PVF for the determination of bar forces, depends upon whether the frame has a smaller number of degrees of freedom in PVD than the number of redundancies in PVF. In Fig. 12.13, for example, since the two degrees of freedom at each of joints A, B and E total 6 in this singly redundant frame, PVF is the better choice.

Example 12.21 A thin pipe is simply supported at length intervals L when carrying fluid of density ρ at pressure p and velocity V (Fig. 12.23a). Assuming a deflected shape $v = a[1 - \cos(2\pi z/L)]$ between any pair of supports, derive an expression that will enable the prediction of instability. The weight of the pipe is negligible compared to the fluid.

When external vertical loading w/unit length is distributed over a given length L, Eq. (12.16a) becomes

$$\int_0^L wv^v\,dz = EI \int_0^L (d^2v^v/dz^2)(d^2v/dz^2)\,dz, \qquad \text{(i)}$$

in which the derivation of the RHS is given in Example 12.19.

Now, for a unit length of the deflected pipe shown, w is composed of inertia and pressure effects. That is, from Fig. 12.23(b),

$$w = m(V^2/R) + pA\,\delta\theta = (\rho A)(V^2/R) + pA\,\delta\theta.$$

Recall, from Eqs (7.2) and (7.3) that

$$1/R = d^2v/dz^2 \quad \text{and} \quad \delta\theta = 1/R \quad \text{when} \quad \delta z = 1,$$
$$\therefore \quad w = (\rho A V^2 + pA)\,d^2v/dz^2.$$

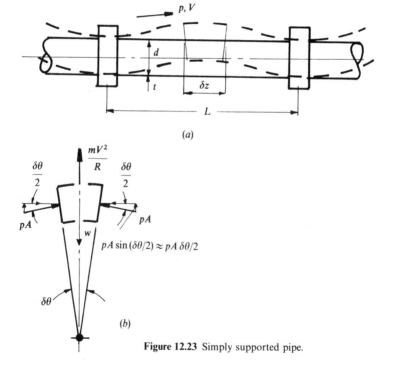

(a)

(b)

Figure 12.23 Simply supported pipe.

Substituting into Eq. (i) gives

$$-\int_0^L (\rho A V^2 + pA)(\mathrm{d}^2 v/\mathrm{d}z^2) v^{\mathrm{v}}\,\mathrm{d}z = EI \int_0^L (\mathrm{d}^2 v^{\mathrm{v}}/\mathrm{d}z^2)(\mathrm{d}^2 v/\mathrm{d}z^2)\,\mathrm{d}z,$$

(ii)

in which the directions of v^{v} and w oppose each other. Assuming that the virtual deflection has the same form as v:

$$v = a[1 - \cos(2\pi z/L)], \qquad \mathrm{d}^2 v/\mathrm{d}z^2 = (4\pi^2/L^2)a\cos(2\pi z/L),$$
$$v^{\mathrm{v}} = a^{\mathrm{v}}[1 - \cos(2\pi z/L)], \qquad \mathrm{d}^2 v^{\mathrm{v}}/\mathrm{d}z^2 = (4\pi^2/L^2)a^{\mathrm{v}}\cos(2\pi z/L).$$

Substituting into Eq. (ii) and integrating leads to the condition

$$V^2\rho + p = 4\pi^2 EI/AL^2,$$

and thus fluid properties V, ρ and p are related to the properties E, A and I for the beam section at the lowest point of instability.

12.6 STATIONARY POTENTIAL ENERGY (SPE)

Since $\delta W_I = -\delta U$ from Eq. (12.2), the principle of virtual work (Eq. (12.4a)) may be rewritten in the following differential form:

$$\delta U - \delta W_E = 0. \tag{12.17}$$

Now, as potential energy (V) is the energy stored in the body for its displaced position under the external forces, it follows that

$$\delta V = -\delta W_E. \tag{12.18}$$

Combining Eqs (12.17) and (12.18):

$$\delta U + \delta V = 0.$$

That is, for a conservative force system where a complete cycle of loading–unloading results in zero net work, the principle of SPE becomes

$$\delta(U + V) = 0. \tag{12.19}$$

Taking U to mean potential energy (PE) of strain, Eq. (12.19) indicates that a stationary value of the total potential energy ($U + V$) exists when the system is in equilibrium. The principle states that

> Of all the possible compatible displacement systems that satisfy the boundary conditions, that which also satisfies the equilibrium conditions gives a stationary value to the PE of the system.

Equation (12.19) employs real force and displacement systems. It is normally applied together with Eq. (12.18) in the following form:

$$\delta[U - \sum(F_k\Delta_k)] = 0. \tag{12.20}$$

It is particularly useful for finding good approximations to displacement functions, where exact solutions may not be available.

Example 12.22 Find the deflection v at any point in the length z of a simply supported beam of length L that carries a vertical concentrated force P at a distance c from the end (Fig. 12.24). Assume the deflected shape function $v = \sum a_n \sin(n\pi z/L)$ for $n = 1, 2, \ldots, \infty$. Compare with the accepted expression when $z = c = L/2$.

The boundary conditions $v = 0$ for $z = 0$ and $z = L$ are satisfied. Now,

$$\sum F_k\Delta_k = Pv_c$$

and from Eqs (7.4) and (11.7b),

$$U = \int M^2\, dz/2EI = (EI/2)\int_0^L (d^2v/dz^2)^2\, dz,$$

$$\therefore \quad U + V = -Pv_c + (EI/2)\int_0^L (d^2v/dz^2)^2\, dz$$

$$= -P\sum a_n \sin(n\pi c/L)$$

$$\quad + (EI/2)\int_0^L [\sum a_n(n\pi/L)^2 \sin(n\pi z/L)]^2\, dz$$

$$= -P\sum a_n \sin(n\pi c/L) + (EI/2)\sum a_n^2(n\pi/L)^4(L/2).$$

As a_n controls the magnitude of $U + V$, for equilibrium, Eq. (12.20) becomes

$$\delta(U + V)/\delta a_n = 0 = -P\sum \sin(n\pi c/L) + (EI)\sum a_n(n\pi/L)^4(L/2),$$

$$\therefore \quad a_n = P\sum \sin(n\pi c/L)/(EI)\sum(n\pi/L)^4(L/2)$$

$$= (2PL^3/EI\pi^4)\sum n^{-4} \sin(n\pi c/L),$$

$$\therefore \quad v = (2PL^3/EI\pi^4)\sum n^{-4} \sin(n\pi c/L)\sin(n\pi z/L). \tag{i}$$

Equation (i) supplies the deflection at any point. Putting $z = c = L/2$ gives

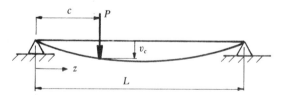

Figure 12.24 Simply supported beam.

the deflection beneath a central concentrated load. That is,

$$v = (2PL^3/EI\pi^4)(1 + 3^{-4} + 5^{-4} + \cdots)$$
$$= PL^3/48.035EI,$$

which compares with the exact value of 48 in the denominator.

Example 12.23 Given that the radial and hoop strains are $\epsilon_r = \partial u/\partial r$ and $\epsilon_\theta = u/r$ in the wall of a thick cylinder under internal pressure p, with zero outer pressure, determine the constants A and B for the assumed radial displacement function $u = A(r + B/r)$ (see Figs 14.6 and 15.6).

$$u = A(r + B/r), \tag{i}$$

$$\therefore \quad \epsilon_r = A(1 - B/r^2), \qquad \epsilon_\theta = A(1 + B/r^2). \tag{ii}$$

The constitutive relations are found from Eq. (4.15):

$$\sigma_\theta = E(\epsilon_\theta + v\epsilon_r)/(1 - v^2), \tag{iii}$$

$$\sigma_r = E(\epsilon_r + v\epsilon_\theta)/(1 - v^2). \tag{iv}$$

Equations (ii) and (iv) must satisfy the boundary condition $\sigma_r = 0$ for $r = r_2$:

$$\partial u/\partial r + vu/r = A(1 - B/r_2^2) + Av(1 + B/r_2^2) = 0,$$
$$\therefore \quad B = r_2^2(1 + v)/(1 - v). \tag{v}$$

Though A could be found from the second boundary condition, $\sigma_r = -p$ for $r = r_1$, the SPE method will now be used for this purpose. U in biaxial stress $(\sigma_r, \sigma_\theta)$ was previously derived in Example 11.3. With V' for volume,

$$U = \int_{V'} (\sigma_r \epsilon_r/2 + \sigma_\theta \epsilon_\theta/2)\,dV'.$$

and from Eqs (iii) and (iv),

$$U = [E/2(1 - v^2)] \int_{V'} [\epsilon_r^2 + 2v\epsilon_r\epsilon_\theta + \epsilon_\theta^2]\,dV'.$$

Now, since $\sum F_k \Delta_k = 2\pi r_1 p u_1$ and $\delta V' = r\,\delta\theta\,\delta r$ (per unit length), Eq. (12.20) becomes

$$U + V = -2\pi r_1 p u_1 + [E/2(1 - v^2)] \int_0^{2\pi} \int_{r_1}^{r_2} [\epsilon_r^2 + 2v\epsilon_r\epsilon_\theta + \epsilon_\theta^2] r\,d\theta\,dr.$$

Substituting from Eqs (i) and (ii):

$$U + V = - 2\pi r_1 pA(r_1 + B/r_1) + [2\pi E A^2/(1 - v^2)]$$
$$\times \int_{r_1}^{r_2} [(1 + B^2/r^4) + v(1 - B^2/r^4)]r\, dr$$
$$= - 2\pi r_1 pA(r_1 + B/r_1) + [\pi E A^2/(1 - v^2)]$$
$$\times |r^2 - B/r^2 + v(r^2 + B/r^2)|_{r_1}^{r_2}. \tag{vi}$$

Now, from Eq. (12.20), since B is a constant, the stationary value will be found from Eq. (vi) corresponding to

$$d(U + V)/dA = - 2\pi r_1 p(r_1 + B/r_1) + [2\pi E A/(1 - v^2)]$$
$$\times |r^2 - B/r^2 + v(r^2 + B/r^2)|_{r_1}^{r_2} = 0,$$
$$\therefore \quad A = \frac{pr_1(r_1 + B/r_1)(1 - v^2)}{E|r^2 - B/r^2 + v(r^2 + B/r^2)|_{r_1}^{r_2}}$$
$$= \frac{pr_1(r_1 + B/r_1)(1 - v^2)}{E[(1 + v)(r_2^2 - r_1^2) - B(1 - v)(1/r_2^2 - 1/r_1^2)]}.$$

Substituting for B from Eq. (v) finally leads to

$$A = pr_1^2(1 - v)/E(r_2^2 - r_1^2).$$

12.7 STATIONARY COMPLEMENTARY ENERGY (SCE)

It is also possible to formulate a further stationary principle based upon the corresponding complements of δU and δV in Eq. (12.19). Thus, if δU^* is the complementary strain energy and $\delta V^* = - \delta W^*$, where δW^* is the complementary work of the external forces, then the total complementary energy for the system is $U^* + V^*$. The stationary value in $U^* + V^*$ becomes equivalent to the principle of virtual complementary work:

$$\delta(U^* + V^*) = 0. \tag{12.21}$$

The SCE principle is then stated as:

Of all the equilibrium force (stress) systems that satisfy the boundary conditions, that which also satisfies the compatibility condition gives a stationary value to the complementary energy of the system.

For Hookean materials, where $U = U^*$ and $\delta W^* = \Delta\, \delta F$, Eq. (12.21) reduces to Eq. (12.20):

$$\delta(U - \sum \Delta_k F_k) = 0. \tag{12.20}$$

SCE can be used as an alternative to the unit load method for solving redundant structures. A further simplification is often made that in the presence of small boundary displacements the summation term in Eq. (12.20) is negligibly small, though this will only strictly apply to a rigid boundary. Equation (12.22) then becomes

$$\delta U = \delta U^* = 0, \tag{12.22}$$

which is particularly useful for obtaining approximate stress distributions where exact solutions are not available. This reduced form of SCE (Eq. (12.22)) is often referred to as the principle of least work.

Example 12.24 The cantilever in Fig. 12.25 is propped at mid-position when carrying an end load P. Determine from SCE the prop reaction R_A when the prop height is (i) Δ_A upwards and (ii) zero. EI is constant.

Taking case (i) generally,

$$U^* = U = \int M^2 \, dz/2EI$$

$$= \int_0^{l/2} (Pz)^2 \, dz/2EI + \int_{l/2}^{l} [Pz - R_A(z - l/2)]^2 \, dz/2EI,$$

$$V^* = -\sum \Delta_k F_k = -R_A \Delta_A - Pv.$$

The total complementary energy is then

$$U^* + V^* = \int_0^{l/2} (Pz)^2 \, dz/2EI$$

$$+ \int_{l/2}^{l} [Pz - R_A(z - l/2)]^2 \, dz/2EI - R_A \Delta_A - Pv.$$

This has a stationary value in respect of the single redundancy R_A. That

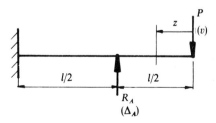

Figure 12.25 Propped cantilever.

is, from Eq. (12.21),

$$d(U^* + V^*)/dR_A = \int_{l/2}^{l} [Pz - R_A(z - l/2)](z - l/2)\,dz/EI - \Delta_A = 0,$$

$$\int_{l/2}^{l} [P(z^2 - lz/2) - R_A(z^2 - lz + l^2/4)]\,dz + EI\Delta_A = 0,$$

$$P|z^3/3 - lz^2/4|_{l/2}^{l} - R_A|z^3/3 - lz^2/2 + l^2z/4|_{l/2}^{l} + EI\Delta_A = 0,$$

$$P[(l^3/3 - l^3/4) - (l^3/24 - l^3/16)] - R_A[(l^3/3 - l^3/2 + l^3/4)$$
$$- (l^3/24 - l^3/8 + l^3/8)] + EI\Delta_A = 0,$$

$$5Pl^3/48 - R_A l^3/24 + EI\Delta_A = 0,$$

$$R_A = 5P/2 + 24EI\Delta_A/l^3.$$

When in part (ii) $\Delta_A = 0$, $R_A = 5P/2$. For a structure with m redundancies Eq. (12.20) may be applied to each redundancy in turn to yield m simultaneous equations.

Example 12.25 A linear increase in hoop stress $\sigma_\theta = A + Br$ is assumed from the inner (r_1) to outer (r_2) radii in a thick-walled annular disc under internal pressure p (Fig. 12.26). Determine from SCE the constants A and B when the effect of radial stress is ignored.

The boundary condition is satisfied by equating forces along a horizontal diameter for the upper half of the disc shown in Fig. 12.26 to give

$$p(2r_1)t = 2\int_{r_1}^{r_2} \sigma_\theta\,dr\,t = 2\int_{r_1}^{r_2} (A + Br)t\,dr,$$

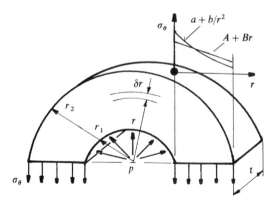

Figure 12.26 Thick-walled disc.

where p acts on the projected area $2r_1 t$. This gives

$$A = p r_1/(r_2 - r_1) - B(r_2 + r_1)/2, \qquad \text{(i)}$$

$$\therefore \quad \sigma_\theta = p r_1/(r_2 - r_1) + B[r - (r_1 + r_2)/2]. \qquad \text{(ii)}$$

Assuming a rigid boundary the total complementary energy is simply $U^* = U$ from Eq. (12.22). Since the stress system is assumed to be uniaxial, Eq. (11.4) gives

$$U^* = \int_V (1/2E)\sigma_\theta^2 \, dV.$$

The stationary value applies in respect of B:

$$dU^*/dB = (1/E) \int_V \sigma_\theta (d\sigma_\theta/dB) \, dV = 0.$$

Substituting from Eq. (ii) with $dV = (2\pi r \, dr)t$:

$$\int_{r_1}^{r_2} \{ p r_1/(r_2 - r_1) + B[r - (r_1 + r_2)/2] \} [r - (r_1 + r_2)/2] r \, dr = 0.$$

Integrating and introducing the radius ratio $K = r_2/r_1$ leads to

$$B = \frac{6p(K + 1)^2 - 8p(K^2 + K + 1)}{r_1 [6(K^4 - 1) - 8(K^3 - 1)(K + 1) + 3(K^2 - 1)(K + 1)^2]}$$

and, from Eq. (i), A is written as

$$A = p/(K - 1) - B r_1 (K + 1)/2.$$

Satisfying equilibrium and compatibility in this way provides the most acceptable linear hoop stress distribution. It is compared in Fig. 12.26 with Lamé's exact solution; $\sigma_\theta = a + b/r^2$ (see Eqs (14.28) and (15.9)).

EXERCISES

The Principle of Virtual Forces and the Unit Load Method

12.1 Apply the PVF or the ULM to confirm Castigliano's answers to the loaded arcs given in Figs 11.32–11.37.

12.2 Find the vertical deflection at points Q and T for the frame in Fig. 11.42 and compare with the method of Castigliano.

12.3 A semi-circular arch of 3.5 m radius supports a central vertical load of 10 kN. If the arch is pinned at one end and rests on rollers at the other end, find the horizontal deflection at the rollers. If both ends are pinned what are the horizontal reactions? Sketch the polar shear force and bending moment diagrams, inserting the major values. Take $EI = 4725$ kNm^2.

Answer: 45.4 mm, 3.18 kN.

12.4 Examine the effect of replacing the semi-circular arch in Exercise 12.3 with a parabolic arch of 2 m rise over the same base length and under the same loading.

12.5 If the cantilever in Fig. 12.4 carries, instead of P, a uniformly distributed load w throughout its length, determine the vertical deflection at the free end due to bending and torsional effects.
  Answer: $(wR^4/2)[(1/EI) + (1 - \pi - \pi^2/4)/GJ]$.

12.6 Find the vertical deflection at the free end of the cantilevered bracket in Fig. 12.27. Take $I = 42 \times 10^3$ mm^4, $E = 207$ GPa.
  Answer: 1.85 mm.

12.7 Determine the vertical deflection at C for the frame in Fig. 12.28.
  Answer: 0.202 mm.

12.8 Find the vertical deflection at point B for the stepped section beam in Fig. 12.29. Take $E = 207$ GPa, $I = 50 \times 10^6$ mm^4.
  Answer: 5.87 mm.

12.9 The cantilever quadrant in Fig. 12.30 is loaded at its free end through two cables under equal tensions T. The tip is not to deflect vertically such that the horizontal tension T remains in position. At what angle β must the second tension be applied to ensure this condition? EI is constant. (K.P. 1976)

12.10 Derive expressions for the vertical and horizontal components of deflection at the load point in each of the cantilever arcs in Fig. 12.31(a) and (b).
  Answer: (a) $\delta_v = \pi W R^3/4EI$; $\delta_H = W R^3/2EI$. (b) $\delta_v = W R^3(\pi - 2)/EI$.

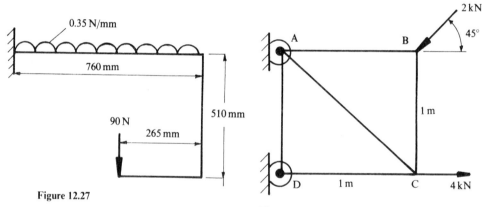

Figure 12.27

Figure 12.28

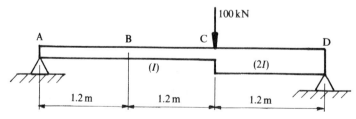

Figure 12.29

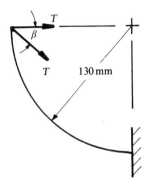

Figure 12.30

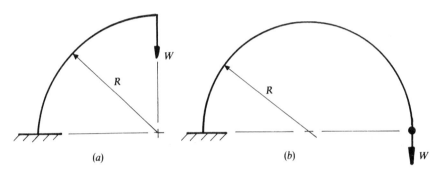

(a)   (b)

Figure 12.31

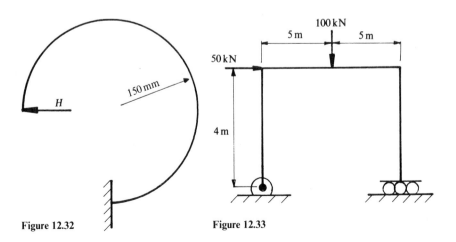

Figure 12.32   Figure 12.33

12.11 The 3/4 circular arc in Fig. 12.32 with mean radius 150 mm, deflects 10 mm horizontally at its tip under the action of a horizontal force H applied as shown. If the maximum bending stress in the material is limited to 90 MPa, find (i) the cross-section dimensions when the breadth is five times the thickness, (ii) the force H and, (iii) the vertical deflection at the tip. Take $E = 210$ GPa.

Answer: 4.54 mm; 46.8 N; 2.12 mm.

12.12 Calculate the horizontal displacement at the rollers for the frame in Fig. 12.33 if, for the vertical cross-section, $A = 9000$ mm^2, $I = 1 \times 10^8$ mm^4 and, for the horizontal section, $A = 8500$ mm^2, $I = 3 \times 10^8$ mm^4. Take $E = 200$ GPa.

Answer: 203.3 mm.

12.13 Find the horizontal deflection at A for the pin-jointed frame in Fig. 12.34 when for all bars, $A = 200$ mm^2 and $E = 200$ GPa.

Answer: 1.21 mm.

12.14 The bar lengths for the pin-jointed frame in Fig. 12.35 are DC = CE = AB = BE = $\sqrt{2}$ m and DE = CB = EA = 2 m. If all bar areas are 100 mm^2, with $E = 100$ GPa, determine the vertical and horizontal deflections at point E when a vertical force of 2 kN is applied at B.

Answer: 0.84; 0.24 mm.

12.15 The arch in Fig. 12.36 is loaded by two vertical forces at B and C and pinned at A and D as shown. If the section is a tube of 40 mm o.d. with $I = 3.37 \times 10^4$ mm^4 and $E = 207$ GPa, determine, from the polar bending moment diagram, the position and magnitude of the maximum bending stress when A and D are free to move apart. What horizontal force would be required at A and D to prevent this movement?

Answer: 692.6 MPa; 1.83 kN. (K.P. 1977)

12.16 The structure in Fig. 12.37 consists of a vertical portal frame resting on two parallel, horizontal simply supported beams. A uniformly distributed load w acts outward throughout as shown. Determine the horizontal reaction at the feet of the frame based upon bending effects only. EI is constant.

Answer: $p(8a^4 - 5L^4)/(20a^3 + 4L^3)$.

12.17 The Warren girder in Fig. 12.38 is built in to a wall at C and D. The structure is used to lift a weight of 40 kN with a wire passing over pulleys at A, B and C. Determine the vertical

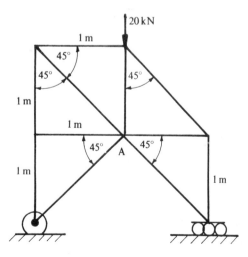

Figure 12.34

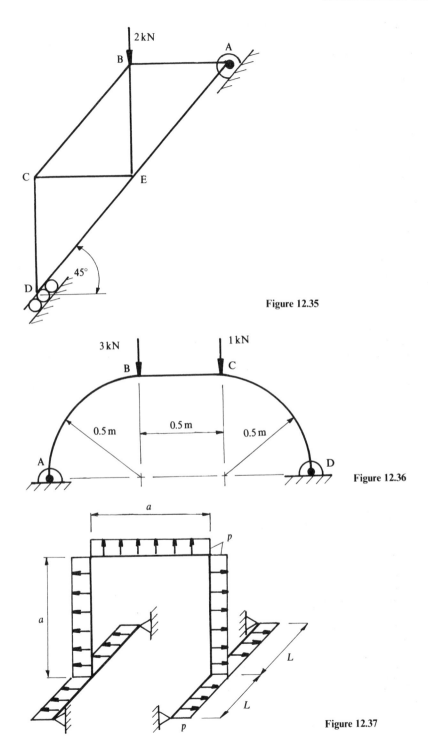

Figure 12.35

Figure 12.36

Figure 12.37

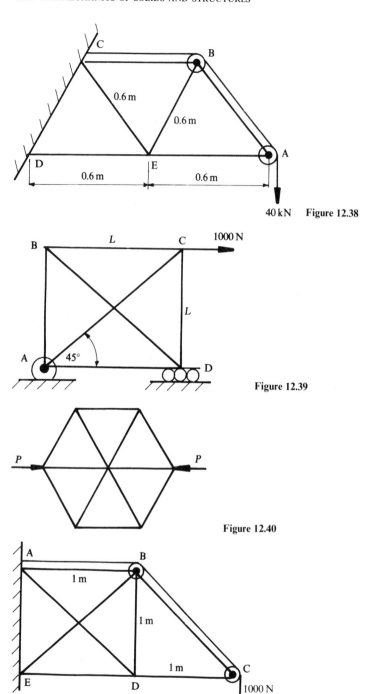

40 kN   **Figure 12.38**

Figure 12.39

Figure 12.40

Figure 12.41

deflection at A and the force acting in the direction AB that will eliminate this deflection. Bar areas are 180 mm² and $E = 70$ GPa.

Answer: 4.44 mm. (K.P. 1977)

Redundant Structures

12.18 Find the forces in all the bars of the pin-jointed structure in Fig. 12.39 when EA is constant throughout.

Answer: AB = BC = AD = 397 N; CD = 603; BD = 561 N; AC = 853 N.

12.19 Find the forces in all the members of the plane, pin-jointed, hexagonal frame in Fig. 12.40 when A, E and L for each member are constant.

Answer: outers $-P/6$; inner horizontals $-5P/6$; others $+P/6$.

12.20 The plane frame in Fig. 12.41 is loaded through a cable passing over frictionless pulleys at B and C. If E and A are constant determine the force in each bar.

Answer: AB, 673 N; AD, 462 N; BC, 414 N; BD, -372 N; BE, -952 N; CD, -1414 N; DE, -1327 N.

12.21 Steel tubes of length 2 m and 100 mm² in area comprise the outer bars of the frame in Fig. 12.42. Determine the minimum area for steel wires to remain taut when connecting points A to C and B to D as shown. E is constant.

Answer: 58 mm².

12.22 Find the forces in the bars of the pin-jointed frame in Fig. 12.43 when the expansion of AD is prevented by built-in hinges and the area of bar AC is twice that of the remaining bars.

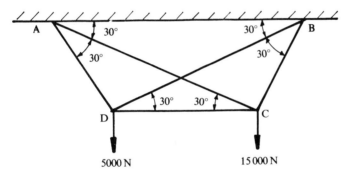

Figure 12.42

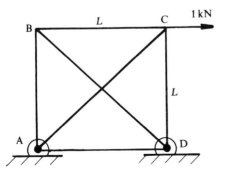

Figure 12.43

(*Hint*: there are two redundancies, say the force in BD and a horizontal reaction.)

    Answer: AB, BC 333 N; CD − 667 N; AC 943 N; BD − 471 N; AD 0.

12.23 Calculate the force in each member of the redundant structure in Fig. 12.44 and examine the possibility of unstable Euler strut failure when each bar is made from 20 mm square section with a constant $E = 200$ GPa. What then is the vertical deflection at A?

    Answer: AB − 26.8 kN; BC − 26.8 kN; BD − 13.4 kN; AD 11.8 kN; AE 13.2 kN.

12.24 Calculate the horizontal deflection at the load point for the protal frame in Fig. 12.45 when A is built-in and D is (a) hinged and (b) built-in. Take *EI* as constant.

    Answer: (a) $1.836 PL^3/EI$. (b) $1.315 PL^3/EI$.

12.25 Find the horizontal base reactions for the frame in Fig. 12.46 when, for the verticals, $A = 9000$ mm^2, $I = 10^8$ mm^4 and, for the horizontal bar, $A = 8500$ mm^2, $I = 3 \times 10^8$ mm^4.

    Answer: 42.4 kN.

12.26 The aircraft fin in Fig. 12.47 is built-in along its base with all other connections free to transmit horizontal shear forces only. Determine the bending moments at the built-in connections

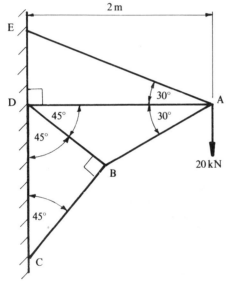

Figure 12.44

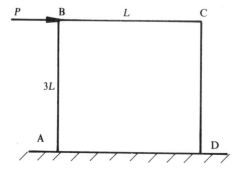

Figure 12.45

when a horizontal force F is applied at the central connection normal to the plane of the fin. EI is constant throughout.

Answer: 0.32*FL* outer; 0.364*FL* inner.

12.27 Find the shear stress in the walls and the torsional stiffness when the two-cell fuselage in Fig. 12.48 is subjected to a torque of 2 MN m over a 25 m length. Take $G = 30$ GPa.

12.28 Find the shear stress in the walls and the torsional stiffness for the two-cell tube in Fig. 12.49, when it is subjected to a torque of 100 N mm. Take $G = 27$ GPa as constant.

Answer: 2085 Pa; 0 centre spar; 1043 (Nm)/(degree/m).

12.29 Show, by extending Eq. (9.36), by the PVF to a three-cell tube under torsion:

(a) $(d\theta/dz)_1 = (q_1/2A_1)\oint_1 ds/Gt - (q_2/2A_1)\int_{12} ds/Gt;$
(b) $(d\theta/dz)_2 = (q_2/2A_2)\oint_2 ds/Gt - (q_1/2A_2)\int_{12} ds/Gt - (q_3/2A_2)\int_{23} ds/Gt;$
(c) $(d\theta/dz)_3 = (q_3/2A_3)\oint_3 ds/Gt - (q_2/2A_3)\int_{23} ds/Gt.$

Then, with the appropriate extension to Eq. (9.35), determine the shear flows for the tube in Fig. 12.50 when it carries a pure torque of 10 kN m. What is the relative angle of twist in a 10 m length? Take $G = 30$ GPa.

Answer: $q_1 = 6.59$ N/mm; $q_2 = 8.99$ N/mm; $q_3 = 4.95$ N/mm.

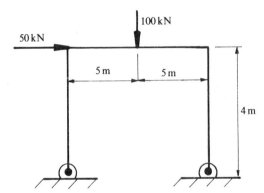

50 kN

100 kN

5 m   5 m

4 m

Figure 12.46

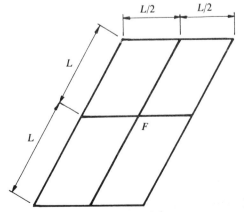

L/2   L/2

L

L

F

Figure 12.47

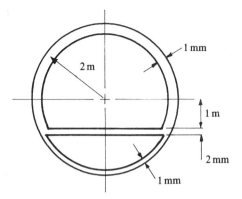

Figure 12.48

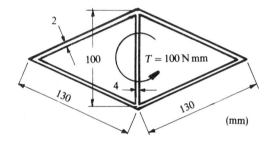

(mm)

Figure 12.49

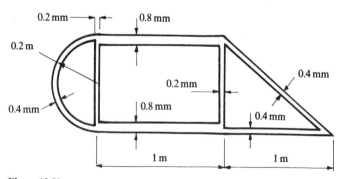

Figure 12.50

12.30 Using the ULM determine the shear flow in each web of the idealized structure in Fig. 12.51 when a vertically upward shear force of 20 kN is applied through the shear centre.
Answer: left vertical 57.44 N/mm; right vertical − 26.36 N/mm; others − 2.955 N/mm.

Principle of Virtual Displacements

12.31 Find the free-end deflection v_f for a cantilever of length L when it supports a concentrated end-load P. Assume a deflected shape of the form $v = v_f[1 - \cos(\pi z/L)]$ and take EI to be

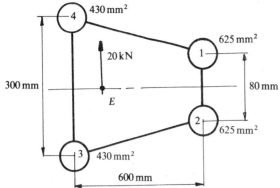

Figure 12.51

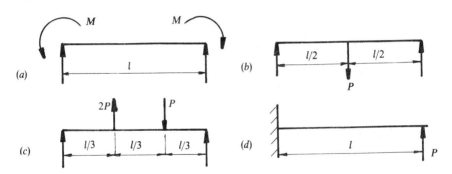

Figure 12.52

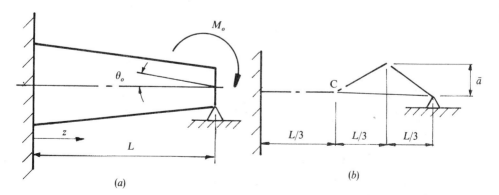

Figure 12.53

constant. Compare with the exact value and that value found from assuming a virtual displacement, where the cantilever deflects rigidly about a stiff hinge at its centre.

Answer: $v = 32PL^3/\pi^4 EI$; $v = 2\sqrt{2}PL^3/\pi^2 EI$.

12.32 Decide on suitable displacement functions in each of Figs 12.52(a)–(d) and hence derive the maximum displacement in each case. (EI in constant.)

12.33 Determine the deflection at quarter and half length positions for a simply supported beam of length L with a central concentrated force W, assuming a displacement function of the form $v = a \sin(\pi z/L)$ and a virtual function of the same form. Compare with the exact values.

Answer: $WL^3/68.9EI$; $WL^3/48.7EI$.

12.34 An encastre beam of length L rests on an elastic support of stiffness K (N/m) at its centre, where it carries a central concentrated load P. Assuming a deflected shape of the form $v = \sum a_n[\cos(2\pi nz/L) - 1]$, where $n = 1, 2, \ldots, N$, show that the Rayleigh–Ritz method supplies the coefficients a_n in the form of the linear simultaneous equation $(n^4 a_n)4\pi^4 EI/L^3 + 2K\sum a_n + p$, where n is odd. What value for N would be acceptable?

12.35 Use the PVD to find the deflection at $z = L/2$ and $z = 3L/2$ positions for a beam of length $2L$ when a clockwise moment M is applied at its centre. Assume that a sine wave describes the deflected shape. Compare with the corresponding exact values. EI is constant.

Answer: $ML^2/32EI$; $-ML^2/32EI$.

12.36 A strut of length L is encastre at one end and carries an axial compressive force P at its free end. If the section varies from I for $0 \leq z \leq L/2$ to $2I$ for $L/2 \leq z \leq L$ where z is measured from the free end, determine the critical buckling load. Use the PVD assuming a deflected shape of the form $v = \Delta[1 - \cos(\pi z/2L)]$, where Δ is the lateral deflection at the free end.

Answer: $P_{cr} = 2\pi^2 EI/(3\pi - 2)L^2$.

12.37 A beam is simply supported at its centre and ends. The beam is partially constrained against rotation θ at its centre by a torque spring with a torsional stiffness $K = T/\theta$. Determine the rotation at the centre when a external moment M acts against this spring. Take EI to be constant.

Answer: $\theta = M/(3EI/L + K)$.

12.38 A beam material obeys the non-linear law $\sigma = E(\epsilon - \beta\epsilon^3)$. Determine the deflection beneath a central concentrated load of 12.5 kN when the beam is simply supported over a length of 5 m given that $E = 70.3$ GPa, $\beta = 10^4$ and $I = 1.83 \times 10^6$ mm⁴. Assume a deflected shape of the form $v = a \sin(\pi z/L)$. If the linear law $\sigma = E\epsilon$ applies during gradual unloading, determine the deflection remaining in the beam for its unloaded state.

Answer: 261.5 mm; 11.5 mm.

12.39 The tapered beam in Fig. 12.53(a) is propped and loaded by a concentrated moment M_o at its free end. Assuming a deflected shape of the form $v = a[(z/L)^3 - (z/L)^2]$ determine the rotation θ_o under M_o when the stiffness varies as $EI = K(2 - z/L)^3$. Take the virtual deflection (i) to have the same form as v and (ii) as shown in Fig. 12.53(b), where C is a point of contraflexure.

Answer: $M_o L/9K$; $M_o L/9.48K$.

12.40 Determine, from a consideration of the number of redundancies and the total number of degrees of freedom for each frame in Figs 12.54(a) and (b), whether the bar forces would best be found from the PVF or the PVD. Outline the solutions in each case.

12.41 Find the forces in all the bars of the plane frame in Fig. 12.55 using the PVD. Take EA as a constant for all bars.

Answer: AD, 0.74P; BD, 0.16P; CD, $-0.58P$.

Stationary Potential and Complementary Energy

12.42 In SPE the series $v = \sum_n b_n \cos(n\pi z/L)$ with n odd is to represent the deflected shape of a simply supported beam carrying a uniformly distributed load (w) over its length L. Taking the

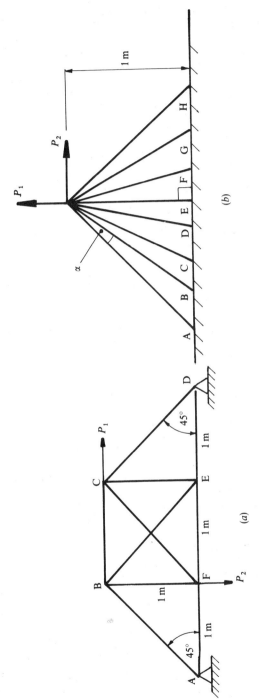

Figure 12.54

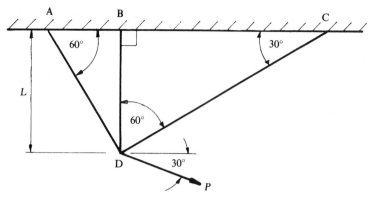

Figure 12.55

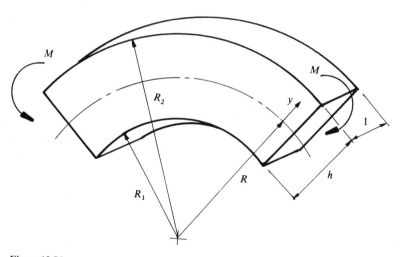

Figure 12.56

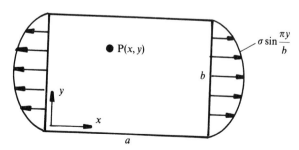

Figure 12.57

origin for z to be at mid-length derive the expression for the displacement at any point. Compare with the accepted value at the mid-length.

Answer: $(4wL^4/\pi^5 EI)(1 - 1/3^5 - 1/5^5 - 1/7^5 \ldots)$.

12.43 The expression $u = K(1 - r)$ is assumed to represent the radial displacement at any radius r when a thin annular disc is held rigidly around its outer radius r_2 and subjected to a pressure p around its inner radius r_1. Apply the principle of SPE to determine A in terms of E, v, r_1, r_2 and p. (I.C. 1966)

Answer: $A = pr_1(1 - v^2)/E(1 - r_2)\ln(r_2/r_1)$.

12.44 Use the principle of SCE to find the forces in the frame of Fig. 12.55.

12.45 Confirm that if the expression $\sigma = R(y + Dy^3)$ is used to represent the stress distribution for a beam under a bending moment M, then the application of the principle of SCE leads to the engineer's theory of bending.

12.46 Use the principle of SCE to determine the shear stress due to shear force F in the rectangular section ($b \times d$) of a beam, assuming that this is approximated by $\tau = A + By + Cy^2$. Compare graphically with the exact solution.

12.47 If the expression $\tau = Mr^2 + Nr^3$ approximates to the shear stress distribution for a solid cylindrical bar of radius R under a pure torque, find M and N from the principle of SCE. Compare with the exact solution from engineer's theory of torsion and show that a 6.6 per cent error arises in τ for $r = R$.

Answer: $M = 45T/8\pi R^5$; $N = -90T/24\pi R^6$.

12.48 The distribution of hoop stress $\sigma_\theta = C_o + C_1 y + C_2 y^2$ is assumed for a sharply curved beam of mean radius $R = (R_1 + R_2)/2$ (from which y is measured) with depth h and unit thickness, when it is subjected to a pure bending moment M, tending to decrease its curvature (Fig. 12.56). Ignoring the effect of radial stress, determine C_o, C_1 and C_2 from the principle of SCE and compare graphically with that found from Eq. (6.14), when $R = h$ and $R_2/R_1 = 3$. (I.C. 1968)

Answer: $C_o = M/hR$; $C_1 = 12M/h^3$; $C_2 = -12M/h^3 R$.

12.49 Determine the σ_x and τ_{xy} stress distributions for a point $P(x, y)$ within the plate subjected to sinusoidal external loading shown in Fig. 12.57. Use horizontal equilibrium to find B and the principle of SCE to find A when the σ_x distribution is assumed to be $\sigma_x = A\sin(\pi x/a) + B\sin(\pi y/a)$. Neglect σ_y and note that σ_x and τ_{xy} are related in the equilibrium condition $\partial\sigma_x/\partial x + \partial\tau_{xy}/\partial y = 0$. (I.C. 1970)

Answer: $A = 4\sigma(\pi - 4)/\pi(3\pi - 8)$.

THIRTEEN

STRESS AND STRAIN TRANSFORMATION

13.1 TWO-DIMENSIONAL STRESS TRANSFORMATION

Given the state of stress on two perpendicular reference planes, the problem is to determine the normal and shear stresses $(\sigma_\theta, \tau_\theta)$ for any plane inclined at θ to the reference planes. Analytical, graphical and matrix methods are available. Conventionally, reference planes aligned with the x- and y-directions are subjected to normal and shear stress components σ_x, σ_y and τ_{xy}, either acting together or in isolation.

Analytical Method

The following four stress systems may be readily identified in practice. They each represent a particular state of stress occurring at a point in a body under externally applied loading. A knowledge of the latter enables the stress components σ_x, σ_y and τ_{xy} to be found.

Simple tension and compression When a bar or plate is subjected to uniaxial stress σ_x (Fig. 13.1(a)), the normal plane AB suffers σ_x while AC suffers σ_θ and τ_θ as shown. Assuming that the bar is rectangular with unit thickness, the forces produced by these stresses are resolved parallel and perpendicular to AC (Fig. 13.1(b)). Noting that $AB = AC \cos \theta$, this gives

$$\sigma_\theta AC = \sigma_x AB \cos \theta, \qquad \tau_\theta AC = \sigma_x AB \sin \theta,$$
$$\sigma_\theta AC = \sigma_x AC \cos^2 \theta, \qquad \tau_\theta AC = \sigma_x AC \sin \theta \cos \theta,$$
$$\sigma_\theta = \sigma_x \cos^2 \theta, \qquad \tau_\theta = (\sigma_x/2) \sin 2\theta.$$

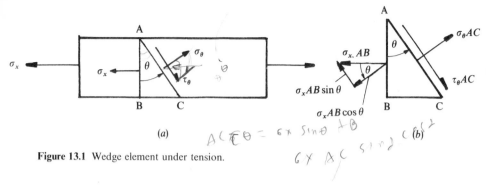

Figure 13.1 Wedge element under tension.

Thus, $\tau_\theta = \sigma_x/2$ is a maximum for the $\theta = 45°$ and $135°$ planes where $\sigma_\theta = \sigma_x/2$. The same result applies to compression provided a minus sign accompanies the numerical value for σ_x. That is, the plane of maximum shear stress lies at $\pm 45°$ to the applied stress axis. The normal and shear stress magnitudes for the maximum shear plane are each half the applied stress.

Pure shear A state of pure shear, where $\sigma_x = \sigma_y = 0$, is typical of a point on the surface of a shaft in torsion (Fig. 13.2a). Recall from Chapter 4, page 77 that when τ_{yx} acts alone in the y-direction then a complementary shear stress τ_{xy}, of equal magnitude, acts in the x-direction as shown in Fig. 13.2(b). Force resolution has been previously dealt with in Fig. 4.5. For the plane AC this leads to

$$\sigma_\theta = \tau_{xy} \sin 2\theta, \qquad \tau_\theta = \tau_{xy} \cos 2\theta.$$

Thus when $\theta = 45°$,

$$\sigma_{45°} = \tau_{xy}, \qquad \tau_{45°} = 0,$$

and when $\theta = 135°$,

$$\sigma_{135°} = -\tau_{xy}, \qquad \tau_{135°} = 0.$$

Figure 13.2 Element under shear.

Note that as the 45° and 135° planes are free of shear stress they are called principal planes. The respective tensile and compressive stresses acting normally to these planes, with magnitudes numerically equal to τ_{xy}, are the principal stresses. The numerically greater value is the major principal stress (σ_1) while the lesser value is the minor principal stress (σ_2). The normal planes on which they act are called the major and minor principal planes respectively. Here $\sigma_1 = \sigma_{45°} = \tau_{xy}$, $\sigma_2 = \sigma_{135°} = -\tau_{xy}$, from which the equivalent principal stress system in Fig. 4.4(b) was found. That is, pure shear produces tension and compression on planes inclined at $\pm 45°$ to the shear planes.

Two normal stresses When the applied stresses σ_x and σ_y ($\sigma_x > \sigma_y$) in Fig. 13.3 act without shear, they are, by definition, principal stresses. That is, σ_x is the major principal stress (σ_1) and σ_y is the minor principal stress (σ_2). AB and BC are the major and minor principal planes respectively. Converting the wedge stresses into the force system in Fig. 13.3(b), assuming unit thickness, the forces may be resolved parallel and perpendicular to AC as shown. Noting that $AB = AC \cos \theta$ and $BC = AC \sin \theta$, the corresponding equilibrium equations are

$$\tau_\theta AC = \sigma_x AB \sin \theta - \sigma_y BC \cos \theta$$
$$= \sigma_x AC \sin \theta \cos \theta - \sigma_y AC \sin \theta \cos \theta,$$
$$\tau_\theta = \tfrac{1}{2}(\sigma_x - \sigma_y) \sin 2\theta,$$
$$\sigma_\theta AC = \sigma_x AB \cos \theta + \sigma_y BC \sin \theta,$$
$$= \sigma_x AC \cos^2 \theta + \sigma_y AC \sin^2 \theta,$$
$$\sigma_\theta = \sigma_x \cos^2 \theta + \sigma_y \sin^2 \theta.$$

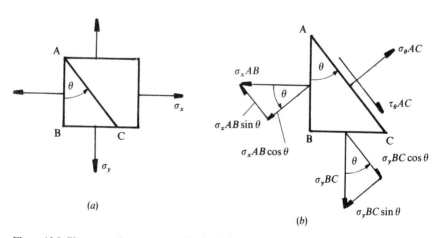

(a)

(b)

Figure 13.3 Element under two normal (principal) stresses.

The following observations apply.

1. $\tau_\theta = \frac{1}{2}(\sigma_x - \sigma_y)$ is a maximum for $\theta = 45°$ and $135°$, i.e. the maximum shear stress is half the difference between the principal stresses and acts on planes at 45 and $135°$ to the principal planes.
2. The normal stress acting on the $\theta = 45$ and $135°$ planes of maximum shear stress is $\sigma_\theta = \frac{1}{2}(\sigma_x + \sigma_y)$.
3. The values of θ, giving maximum and minimum values of σ_θ, correspond to $d\sigma_\theta/d\theta = 0$. That is,

$$- 2\sigma_x \cos\theta \sin\theta + 2\sigma_y \sin\theta \cos\theta = 0,$$
$$- (\sigma_x - \sigma_y)\sin 2\theta = 0.$$

Hence $\sin 2\theta = 0$ when $\theta = 0°$ or $90°$, giving $\sigma_{0°} = \sigma_x$ and $\sigma_{90°} = \sigma_y$. This shows that the major and minor principal stresses are respectively the maximum and minimum normal stress values for this and any other stress system.

General stress system Here σ_x and σ_y $(\sigma_x > \sigma_y)$ together with shear stresses τ_{xy} and τ_{yx} act on the perpendicular planes AB and BC as shown in Fig. 13.4(a). Complementary shear $(\tau_{xy} = \tau_{yx})$ will exist from the consideration of moment equilibrium about any point. Assuming unit thickness, ABC is converted to the force system in Fig. 13.4(b). The forces are resolved parallel and perpendicular to AC to give the respective equilibrium equations where $AB = AC \cos\theta$ and $BC = AC \sin\theta$:

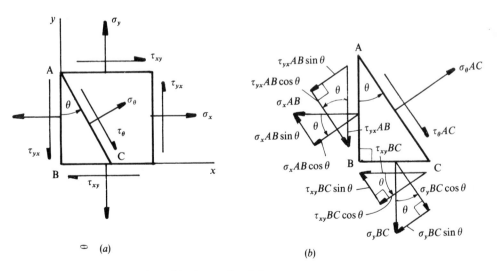

(a)

(b)

Figure 13.4 Element with combined normal and shear stresses.

Perpendicular

$$\sigma_\theta AC = \sigma_x AB \cos\theta + \tau_{yx} AB \sin\theta + \sigma_y BC \sin\theta + \tau_{xy} BC \cos\theta,$$
$$\sigma_\theta AC = AC\sigma_x \cos^2\theta + AC\tau_{xy} \sin\theta\cos\theta + AC\sigma_y \sin^2\theta + AC\tau_{xy} \sin\theta\cos\theta,$$
$$\sigma_\theta = \sigma_x \cos^2\theta + \sigma_y \sin^2\theta + \tau_{xy} \sin 2\theta. \tag{13.1a}$$

Parallel

$$\tau_\theta AC = \sigma_x AB \sin\theta - \tau_{yx} AB \cos\theta - \sigma_y BC \cos\theta + \tau_{xy} BC \sin\theta,$$
$$\tau_\theta AC = AC\sigma_x \sin\theta\cos\theta - AC\tau_{yx} \cos^2\theta - AC\sigma_y \sin\theta\cos\theta + AC\tau_{xy} \sin^2\theta,$$
$$\tau_\theta = (\sigma_x - \sigma_y)\sin\theta\cos\theta - \tau_{xy}(\cos^2\theta - \sin^2\theta)$$
$$\tau_\theta = \tfrac{1}{2}(\sigma_x - \sigma_y)\sin 2\theta - \tau_{xy}\cos 2\theta. \tag{13.1b}$$

Equations (13.1a and b) contain stress systems (a–c) as special cases. They could therefore have been obtained directly by superimposing systems of pure shear and two normal stresses in which opposing τ_θ components are subtracted to give Eq. (13.1b). Since $\tau_\theta = 0$ for the principal planes, it follows from Eq. (13.1b) that their inclinations are found from

$$\tan 2\theta = 2\tau_{xy}/(\sigma_x - \sigma_y). \tag{13.2}$$

The principal stresses σ_1 and σ_2 may then be found by eliminating θ between Eqs (13.1a) and (13.2). Alternatively if AP in Fig. 13.5(a) is a principal plane, the following vertical and horizontal force resolution leads to reduced equilibrium equations.

Horizontal Equilibrium	*Vertical Equilibrium*
$\sigma_x AB + \tau_{xy} BP = \sigma AP \cos\theta,$	$\sigma AP \sin\theta = \tau_{yx} AB + \sigma_y BP,$
$\sigma_x AP \cos\theta + \tau_{xy} AP \sin\theta$	$\sigma AP \sin\theta = \tau_{yx} AP \cos\theta$
$\quad = \sigma AP \cos\theta,$	$\quad\quad + \sigma_y AP \sin\theta,$
$\sigma - \sigma_x = \tau_{xy}\tan\theta.$	$\sigma - \sigma_y = \tau_{yx}\cot\theta.$

Multiplying these equations and solving for σ with $\tau_{xy} = \tau_{yx}$ leads to

$$(\sigma - \sigma_x)(\sigma - \sigma_y) = \tau_{yx}^2,$$
$$\sigma^2 - (\sigma_x + \sigma_y)\sigma + (\sigma_x\sigma_y - \tau_{yx}^2) = 0, \tag{13.3}$$
$$\sigma_{1,2} = \tfrac{1}{2}(\sigma_x + \sigma_y) \pm \tfrac{1}{2}\sqrt{[(\sigma_x - \sigma_y)^2 + 4\tau_{xy}^2]}.$$

The major stress value (σ_1), for the positive discriminant in Eq. (13.3), acts on a θ-plane defined by Eq. (13.2). The minor value, with the negative discriminant, acts on a $\theta + 90°$ plane. The inclination of the induced principal system is as shown in Fig. 13.5(b). Here the previous analysis for two normal stresses applies. In particular, it follows from Eq. (13.3) that the maximum shear stress is given by

$$\tau_{max} = \tfrac{1}{2}(\sigma_1 - \sigma_2) = \tfrac{1}{2}\sqrt{[(\sigma_x - \sigma_y)^2 + 4\tau_{xy}^2]}, \tag{13.4}$$

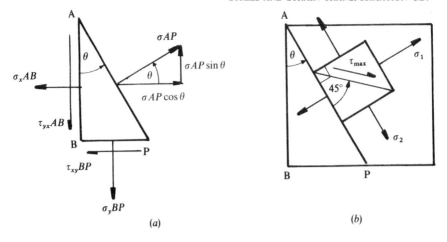

Figure 13.5 Principal and maximum shear planes under general stress.

which acts along the 45° plane shown. Further points to note are as follows.

1. From Eq. (13.3), $\sigma_1 + \sigma_2 = \sigma_x + \sigma_y$, i.e. the sum of the principal stresses equals the sum of the applied normal stresses.
2. If σ_1 is a predetermined design stress value, then vertical force equilibrium gives the inclination of the major principal plane from

$$\tan \theta = (\sigma_1 - \sigma_x)/\tau_{xy}.$$

3. Respective zero or negative substitutions for σ_x or σ_y in Eqs (13.1)–(13.3) should be made when one or other normal stress may be absent or acts in a compressive sense.
4. If the directions of τ_{xy} and τ_{yx} are reversed in Fig. 13.4(a), then these too become negative in the foregoing equations. However, this should not be necessary with the correct choice of plane AB.

Example 13.1 A beam of circular section is supported in bearings 2 m (6 ft) apart while rotating at $N = 500$ rev/min. If it also carries a central concentrated load of $W = 20$ kN (2 tonf), (Fig. 13.6), determine the shaft diameter when the maximum shear stress in the material is restricted to 75 MPa (5 tonf/in²) when transmitting power $P = 900$ W (1200 hp).

SOLUTION TO SI UNITS The applied stresses σ_x and τ_{xy} are due to bending and torsion respectively. These are greatest for the outer surface fibres. Taking a point on the bottom surface of the central section where σ_x is a maximum in tension, for a bar of diameter d,

$$\sigma_x = \frac{My}{I} = \frac{M(d/2)}{\pi d^4/64} = \frac{32M}{\pi d^3}, \tag{i}$$

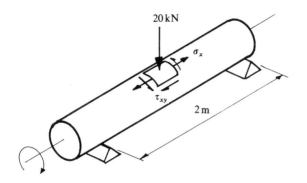

Figure 13.6 Rotating beam.

where $M = WL/4 = 20 \times 2/4 = 10\,\text{kN}\,\text{m}$.

$$\therefore \quad \sigma_x = 32 \times (10 \times 10^6)/\pi d^3 = 320 \times 10^6/\pi d^3 \quad (\text{MPa}), \text{(ii)}$$

$$\tau_{xy} = \frac{Tr}{J} = \frac{T(d/2)}{\pi d^4/32} = \frac{16T}{\pi d^3}. \tag{iii}$$

Now $P = 2\pi N T/60, \Rightarrow T = 60P/2\pi N = 60 \times 900/(2\pi \times 500) = 17.19\,\text{kN}\,\text{m}$,

$$\therefore \quad \tau_{xy} = 16 \times 17.19 \times 10^6/\pi d^3 = 275.02 \times 10^6/\pi d^3 \quad (\text{MPa}). \text{(iv)}$$

Substituting Eqs (ii) and (iv) into Eq. (13.4) with $\sigma_y = 0$ and $\tau_{\max} = 75\,\text{MPa}$

$$75 = \tfrac{1}{2}\sqrt{[(320 \times 10^6/\pi d^3)^2 + 4(275.02 \times 10^6/\pi d^3)^2]}.$$

Squaring both sides leads to

$$d^6 = (10^6/75 \times 2)^2[(320/\pi)^2 + 4(275.02/\pi)^2]$$

$$\therefore \quad d = 90.47\,\text{mm}.$$

The same result applies to the top surface, since compressive σ_x is squared in Eq. (13.4). However, the principal stresses in Eqs (13.3) will differ between these points.

SOLUTION TO IMPERIAL UNITS

$$\sigma_x = 32M/\pi d^3$$
$$= 32 \times 3 \times 12/\pi d^3 = 1152/\pi d^3 \quad (\text{tonf/in}^2).$$
$$P = 2\pi N T/33\,000,$$
$$\therefore \quad T = 33\,000 \times 1200/(2\pi \times 500 \times 2240)$$
$$= 5.627\,\text{tonf}\,\text{ft}; \quad \left(\frac{\text{lbf}\,\text{ft}}{\text{hp}\,\text{min}} \times \frac{\text{hp}\,\text{min}}{\text{rad}} \times \frac{\text{tonf}}{\text{lbf}} \equiv \text{tonf}\,\text{ft}\right).$$

$$\tau_{xy} = 16T/\pi d^3$$
$$= 16 \times 5.627 \times 12/\pi d^3 = 1080.44/\pi d^3 \quad (\text{tonf/in}^2).$$

Substituting into Eq. (13.4) with $\tau_{max} = 5\,\text{tonf/in}^2$

$$5 = \tfrac{1}{2}\sqrt{[(1152/\pi d^3)^2 + 4(1080.44/\pi d^3)^2]},$$

which gives $d = 4.27\,\text{in}$.

Note that τ_{max} and σ_1, under combined bending and torsion of a solid circular shaft, may be directly solved by the substitution of Eqs (i) and (iii) into Eqs (13.3 and 13.4). This gives

$$\tau_{max} = 16T_E/\pi d^3, \qquad \sigma_1 = 32M_E/\pi d^3,$$

where the equivalent twisting and bending moments are

$$T_E = \sqrt{(M^2 + T^2)}, \qquad M_E = \tfrac{1}{2}[M + \sqrt{(M^2 + T^2)}].$$

Example 13.2 Figure 13.7(a) shows the stress states for two inclined planes. Determine τ_θ, τ_{max}, σ_1 and σ_2 and the planes on which they act.

The complete stress state for wedge element ABC involves a further unknown normal stress σ_y acting on BC along which a complementary shear stress of magnitude $3\,\text{MPa}$ will also act (Fig. 13.7b). Thus w.r.t. ABC in Fig. 13.7(a) $\sigma^x = -5\,\text{MPa}$, $\tau_{xy} = \tau_{yx} = 3\,\text{MPa}$, $\sigma_{35^\circ} = 2\,\text{MPa}$, $\tau_{35^\circ} = ?$ and $\sigma_y = ?$ Substitution into Eqs (13.1) gives

$$2 = -5\cos^2 35^\circ + \sigma_y \sin^2 35^\circ + 3\sin 70^\circ = 0.329\sigma_y - 0.536,$$

$$\sigma_y = 7.71\,\text{MPa (tensile)},$$

$$\tau_{35^\circ} = \tfrac{1}{2}(-5 - \sigma_y)\sin 70^\circ - 3\cos 70^\circ = \tfrac{1}{2}(-12.71)\sin 70^\circ - 3\cos 70^\circ,$$

$$\tau_{35^\circ} = -7.0\,\text{MPa (minus indicates that the direction is reversed)}.$$

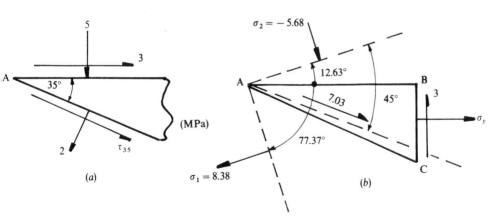

Figure 13.7 Stress states on inclined planes.

Then, from Eqs (13.2)–(13.4),

$$\tan 2\theta = 2 \times 3/(-5 - 7.71) = -0.472,$$
$$2\theta = -25.26° \quad \text{or} \quad 154.73°.$$

The positive value $\theta = 77.37°$ locates the major principal plane while the negative value, $\theta = -12.63°$, locates the minor principal plane. The principal stresses are

$$\sigma_{1,2} = \tfrac{1}{2}(-5 + 7.71) \pm \tfrac{1}{2}\sqrt{[(-5 - 7.71)^2 + 4(3)^2]},$$
$$\sigma_1 = 8.38\,\text{MPa} \quad \text{and} \quad \sigma_2 = -5.68\,\text{MPa},$$
$$\tau_{\text{max}} = \tfrac{1}{2}\sqrt{[(-5 - 7.71)^2 + 4(3)^2]}$$
$$= 7.03\,\text{MPa (at } 45° \text{ to the principal planes in Fig. 13.7(b).}$$

Example 13.3 A weight of 100 N falls freely through 20 mm to impact a point 0.70 m from the axis at the free end of a square-section cantilevered tube in Fig. 13.8. Neglecting deformation of the torque arm, determine the major principal stress induced in the tube. Take $E = 2.5\,G = 210\,\text{GPa}$.

The equivalent static load P_E is found from Eq. (11.25),

$$P(h + \delta) = P_E\delta/2 \tag{i}$$

where δ, the total static deflection under P_E (N), is composed of the following two parts:
   (a) Deflection due to bending:

$$\delta_B = P_E L^3/3EI,$$

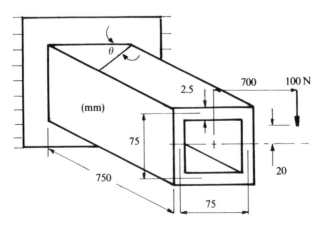

Figure 13.8 Impact loading.

where $I = (75^4 - 70^4)/12 = 635.9 \times 10^3$ mm^4,

$$\therefore \quad \delta_B = P_E \times 750^3/(3 \times 210 \times 10^3 \times 635.9 \times 10^3)$$
$$= 1.0531 \times 10^{-3} P_E \quad \text{(mm)}. \tag{ii}$$

(b) Deflection due to torsion (δ_T). For a non-circular section, the twist is, from Eq. (9.31b),

$$\theta = (TL/4A^2G) \oint ds/t,$$

where here $T = P_E R$ and $\theta = \delta_T/R$ (rad).

$$\frac{\delta_T}{700} = \frac{P_E \times 700 \times 750(4 \times 75/2.5)}{4(75^2)^2 \times (210 \times 10^3/2.5)},$$

$$\therefore \quad \delta_T = 4.148 \times 10^{-3} P_E \quad \text{(mm)}. \tag{iii}$$

Hence, from Eqs (ii) and (iii),

$$\delta = \delta_B + \delta_T = 5.2 \times 10^{-3} P_E \quad \text{(mm)}. \tag{iv}$$

Substituting Eq. (iv) into Eq. (i) with $P = 100$ N, $h = 20$ mm,

$$100[20 + (5.2 \times 10^{-3})P_E] = (2.6 \times 10^{-3})P_E^2,$$
$$P_E^2 - 200P_E - (76.92 \times 10^4) = 0,$$

from which $P_E = 982.74$ N. The state of stress for a point on the top surface of the fixed end is found from the ETB. That is,

$$\sigma_x = My/I = (P_E L)y/I$$
$$= 982.74 \times 750 \times (37.5 + 1.25)/(635.9 \times 10^3)$$
$$= 44.91 \text{ MPa}.$$

The Batho theory (Eq. (9.26)) supplies the shear stress in the tube:

$$\tau_{xy} = T/2At = (P_E R)/2At$$
$$= 982.74 \times 700/(2 \times 75^2 \times 2.5)$$
$$= 24.46 \text{ MPa}.$$

Then, from Eq. (13.3) with $\sigma_y = 0$, the major principal stress is

$$\sigma_1 = \sigma_x/2 + (1/2)\sqrt{(\sigma_x^2 + 4\tau_{xy}^2)}$$
$$= 44.91/2 + (1/2)\sqrt{[(44.91)^2 + 4(24.46)^2]} = 55.66 \text{ MPa},$$

with an inclination of the major principal plane (Eq. (13.2)) of

$$\theta = (1/2)\tan^{-1}(2\tau_{xy}/\sigma_x)$$
$$= (1/2)\tan^{-1}(2 \times 24.46/44.91)$$
$$= 23.72° \text{ in the direction indicated}.$$

Example 13.4 An I-section beam (Fig. 13.9(a)) with flanges 100 mm wide, 200 mm deep and 8 mm thick throughout is simply supported over a span of $l = 4$ m and carries a uniformly distributed load of $w = 15$ kN/m. Determine, for a cross-section of the beam $z = 1$ m from one support, the magnitude and direction of the principal stresses (i) at the neutral axis and (ii) at a point in the web 40 mm from the neutral axis on the compressive side.

$$I = (100 \times 200^3/12) - (92 \times 184^3/12) = 19 \times 10^6 \text{ mm}^4,$$
$$M = wz^2/2 - wlz/2 = (w/2)(z^2 - lz)$$
$$= (15/2)[1^2 - (4 \times 1)] = -22.5 \text{ kN m} \quad \text{(sagging)}.$$

(i) At the neutral axis (Fig. 13.9(a)) the bending stress is zero while the shear stress due to shear force F is a maximum. From Eq. (10.4),

$$\tau_{xy} = F(A\bar{y})/Ix$$

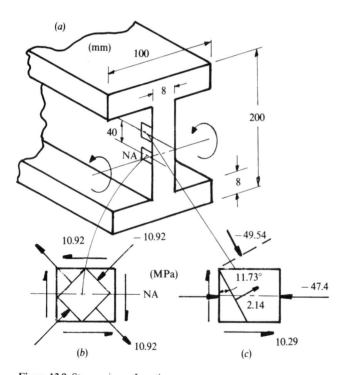

Figure 13.9 Stresses in an I-section.

where

$$F = dM/dz = (w/2)(2z - 1) = -15\,\text{kN},$$

$$x = 8\,\text{mm} \quad \text{and}$$

$$A\bar{y} = (100 \times 8 \times 96) + (92 \times 8 \times 46) = 110.656 \times 10^3\,\text{mm}^3,$$

$$\therefore \quad \tau_{xy} = -15 \times 10^3 \times 110.656 \times 10^3/(19 \times 10^6 \times 8) = -10.92\,\text{MPa}.$$

With its complement τ_{yx} the stress system is identified with pure shear. That is, tensile and compressive stresses, each of magnitude 10.92 MPa, act in 45° directions at all points along the neutral axis (Fig. 13.9(b)).

(ii) At the position $y = 40\,\text{mm}$ in the web (Fig. 13.9(a)):

$$\sigma_x = My/I = -22.5 \times 10^6 \times 40/(19 \times 10^6) = -47.4\,\text{MPa},$$

$$A\bar{y} = (52 \times 8 \times 66) + (100 \times 8 \times 96) = 104.256 \times 10^3\,\text{mm}^3,$$

$$\tau_{xy} = -15 \times 10^3 \times 104.256 \times 10^3/(19 \times 10^6 \times 8) = -10.29\,\text{MPa}.$$

The stress system acting at this point is shown in Fig. 13.9(c). The principal stresses are found from Eq. (13.3) with $\sigma_y = 0$:

$$\sigma_{1,2} = (\sigma_x/2) \pm (1/2)\sqrt{(\sigma_x^2 + 4\tau_{xy}^2)}$$

$$= (-47.4/2) \pm (1/2)\sqrt{[(-47.4)^2 + 4(-10.29)^2]} = -23.7 \pm 25.837.$$

$$\sigma_1 = 2.14\,\text{MPa} \quad \text{and} \quad \sigma_2 = -49.54\,\text{MPa}.$$

$$\theta = \tfrac{1}{2}\tan^{-1}(2\tau_{xy}/\sigma_x) = \tfrac{1}{2}\tan^{-1}[2 \times (-10.29)/(-47.4)].$$

Hence $\theta = 11.73°$, which locates the principal planes given in Fig. 13.9(c).

Graphical Method—Mohr's Circle and the Focus

Stress states $(\sigma_\theta, \tau_\theta)$ existing on an inclined plane AC for any stress system may also be found from a graphical construction due to Otto Mohr (1850). By squaring and adding Eqs (13.1a and b) in the general case it can be shown that

$$[\sigma_\theta - \tfrac{1}{2}(\sigma_x + \sigma_y)]^2 + \tau_\theta^2 = \tfrac{1}{4}(\sigma_x - \sigma_y)^2 + \tau_{xy}^2.$$

This is the equation of a circle, $(X - A)^2 + Y^2 = C^2$, with centre $(A, 0) \equiv [\tfrac{1}{2}(\sigma_x + \sigma_y), 0]$ and radius; $C \equiv \tau_{\text{max}} = [\tfrac{1}{4}(\sigma_x - \sigma_y)^2 + \tau_{xy}^2]^{1/2}$. The circle can be constructed from the stress state on perpendicular planes according to the following convention.

Normal stresses σ_x and σ_y are plotted to the right (positive X) when tensile and to the left when compressive (negative X). Shear stress τ_{xy} is plotted upwards (positive Y) when acting in a clockwise sense and downwards (negative Y) when acting in an anticlockwise sense.

Thus, for the element in Fig. 13.10(a), the coordinate points for AB and BC will lie on opposite ends of a diameter, enabling the circle to be drawn as shown in Fig. 13.10(b). The stress state on AC is found by projecting its

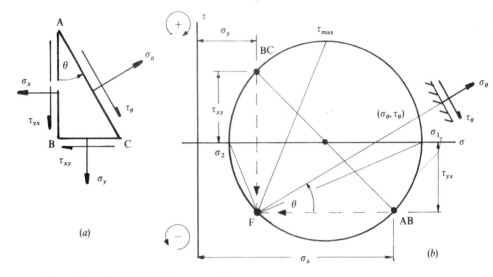

Figure 13.10 Mohr's circle for a general plane stress state.

normal direction through the focus F. The latter is first located by projecting the normal to plane AB through the corresponding stress point AB in the circle. Since the singular point F is the focus of all such normal projections, it may also be located from the normal to plane BC. From the circle, as both σ_θ and τ_θ are positive, their directions are as shown in Fig. 13.10(a).

The normal to other planes of interest may readily be projected from the focus. For example, the major and minor principal planes lie normal to $F\sigma_1$ and $F\sigma_2$ where σ_1 and σ_2 are the principal stress points on the circle (zero shear stress). The line obtained from joining F to the top point (τ_{max}) is the normal to the maximum shear plane. Note that for a given plane, the focus normal is parallel to the normal stress component for that plane.

Example 13.5 Two perpendicular tensile stresses, with magnitudes of 30 and 120 MPa, act together with a shear stress of 75 MPa as shown in Fig. 13.11(a). Determine, graphically, the principal stresses, the maximum shear stress and the planes on which they act. Confirm the answers from the respective formulae.

The circle in Fig. 13.11(b) has been drawn to a common stress scale for σ_x, σ_y and τ_{xy}. The magnitudes of σ_1, σ_2 and τ_{max} are indicated together with the planes on which they act. These enable the inset diagram (a) to be constructed showing the stresses and the inclination of their planes.

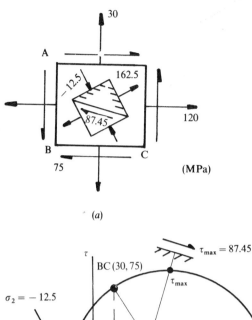

(a)

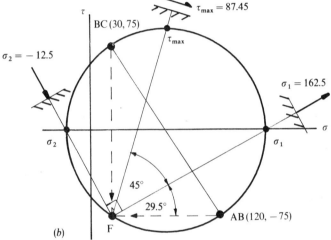

(b)

Figure 13.11 Mohr's circle construction.

The following check is made from Eqs (13.2)–(13.4):

$$\sigma_{1,2} = [(\sigma_x + \sigma_y)/2] \pm (1/2)\sqrt{[(\sigma_x - \sigma_y)^2 + 4\tau_{xy}^2]}$$
$$= [(120 + 30)/2] \pm (1/2)\sqrt{[(120 - 30)^2 + 4(75)^2]} = 75 \pm 87.45,$$
$$\sigma_1 = 162.45 \quad \text{and} \quad \sigma_2 = -12.45 \,\text{MPa};$$
$$\tan 2\theta = 2\tau_{xy}/(\sigma_x - \sigma_y) = 2 \times 75/(120 - 30) = 1.667,$$
$$\theta = 29.5° \text{ and } 119.5°;$$
$$\tau_{max} = (1/2)\sqrt{[(\sigma_x - \sigma_y)^2 + 4\tau_{xy}^2]}$$
$$= (\sigma_1 - \sigma_2)/2 = (1/2)[162.45 - (-12.45)]$$
$$= 87.45 \,\text{MPa at } 45° \text{ to the principal planes or } 74.5° \text{ to AB.}$$

Example 13.6 A duralumin bar 60 mm in diameter is simultaneously subjected to an axial compressive load $W = 200$ kN, a bending moment $M = 600$ N m and a torque $T = 400$ N m. For a point subjected to the greatest net compressive stress, determine graphically the state of stress on a plane inclined at 40° to the bar axis, the maximum shear stress, and the principal stresses and their planes.

Owing to torsion, the shear stress is

$$\tau_{xy} = \frac{Tr}{J} = \frac{400 \times 10^3 \times 30}{\pi(60)^4/32} = 28.3 \text{ MPa}.$$

On the compressive side, the bending stress is

$$\sigma_x = -W/A - My/I$$
$$= -\frac{200 \times 10^3}{\pi(60)^2/4} - \frac{600 \times 10^3 \times 30}{\pi(60)^4/64}$$
$$= -70.736 - 28.294 = -99.03 \text{ MPa}.$$

The stress state is that shown in Fig. 13.12(a), where BC is parallel to the shaft axis. The circle construction in Fig. 13.12(b) gives $\sigma_1 = 7.52$, $\sigma_2 = -106.54$ and $\tau_{\max} = 57.03$ MPa acting on the planes shown. The stress state found for the 50° plane AC given in (Fig. 13.12(a))

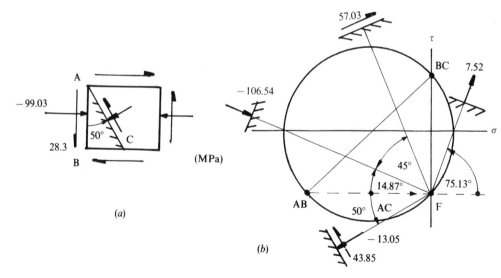

Figure 13.12 Mohr's circle construction for principal stresses.

may be checked from Eqs (13.1) with $\sigma_y = 0$. That is,

$$\sigma_\theta = \sigma_y \sin^2 \theta + \sigma_x \cos^2 \theta + \tau_{xy} \sin 2\theta$$
$$= -99.03 \cos^2 50° + 28.3 \sin 100° = -13.05 \, \text{MPa},$$
$$\tau_\theta = \tfrac{1}{2}(\sigma_x - \sigma_y) \sin 2\theta - \tau_{xy} \cos 2\theta$$
$$= -(99.03/2) \sin 100° - 28.3 \cos 100° = -43.85 \, \text{MPa}.$$

Matrix Method

The reader should note the similarity between the foregoing stress transformation, Eqs (13.1)–(13.3), and those of Eqs (1.2)–(1.4), expressing transformation of second moments of area. It follows from Eq. (1.5) that the matrix form of the plane stress transformation is

$$\mathbf{S}' = \mathbf{L}\mathbf{S}\mathbf{L}^\mathsf{T} \tag{13.5a}$$

where $\mathbf{S}$ is a 2×2 matrix containing σ_x, σ_y and $\tau_{xy} = \tau_{yx}$, and $\mathbf{S}'$ is the matrix of stress components referred to a pair of inclined perpendicular axes x', y' (Fig. 13.13). The components (l_{ij}) of the 2×2 matrix of direction cosines $\mathbf{L}$ are then $l_{11} = l_{x'x} = \cos\theta$, $l_{12} = l_{x'y} = \cos(90° - \theta) = \sin\theta$, $l_{21} = l_{y'x} = \cos(\theta + 90°) = -\sin\theta$ and $l_{22} = l_{y'y} = \cos\theta$. The first subscript refers to the primed axis and the second to the unprimed axis. Thus, from Eq. (13.5a),

$$\begin{bmatrix} \sigma_{x'} & \tau_{xy'} \\ \tau_{yx'} & \sigma_{y'} \end{bmatrix} = \begin{bmatrix} \cos\theta & \sin\theta \\ -\sin\theta & \cos\theta \end{bmatrix} \begin{bmatrix} \sigma_x & \tau_{xy} \\ \tau_{yx} & \sigma_y \end{bmatrix} \begin{bmatrix} \cos\theta & -\sin\theta \\ \sin\theta & \cos\theta \end{bmatrix}. \tag{13.5b}$$

The principal stresses (Eqs (13.3)) follow from expanding the determinant:

$$\det \begin{vmatrix} \sigma_x - \sigma & \tau_{xy} \\ \tau_{yx} & \sigma_y - \sigma \end{vmatrix} = 0. \tag{13.6}$$

Example 13.7 Determine the state of stress for an element inclined at 30° to the given element in Fig. 13.14(a). What are the principal stress values?

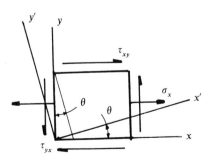

Figure 13.13 Rotation in stress axes.

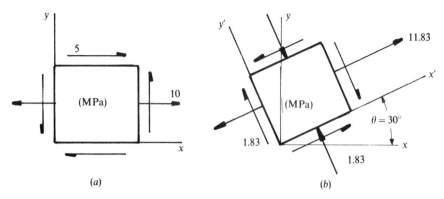

Figure 13.14 Plane stress transformation.

Substituting into Eq. (13.5b) for $\theta = 30°, \sigma_x = 10, \tau_{xy} = \tau_{yx} = 5, \sigma_y = 0$ gives

$$S' = \begin{bmatrix} \sqrt{3}/2 & 1/2 \\ -1/2 & \sqrt{3}/2 \end{bmatrix} \begin{bmatrix} 10 & 5 \\ 5 & 0 \end{bmatrix} \begin{bmatrix} \sqrt{3}/2 & -1/2 \\ 1/2 & \sqrt{3}/2 \end{bmatrix}$$

$$= \begin{bmatrix} \sqrt{3}/2 & 1/2 \\ -1/2 & \sqrt{3}/2 \end{bmatrix} \begin{bmatrix} 11.6 & -0.67 \\ 4.33 & -2.5 \end{bmatrix} = \begin{bmatrix} 11.83 & -1.83 \\ -1.83 & -1.83 \end{bmatrix},$$

which act as shown in Fig. 13.14(b). Note that the matrix supplies the complete stress state for the rotated element, i.e. for the planes θ and $\theta + 90°$ in Eqs (13.1). The principal stresses follow from Eq. (13.6), where

$$\det \begin{vmatrix} 10 - \sigma & 5 \\ 5 & 0 - \sigma \end{vmatrix} = 0,$$

$$-10\sigma + \sigma^2 - 25 = 0,$$

$$\sigma = 10/2 \pm (1/2)\sqrt{[(-10)^2 - 4(-25)]},$$

$$\sigma_1 = 12.05, \qquad \sigma_2 = -2.07 \, \text{MPa} \qquad \text{(check from Eq. (13.3))}.$$

13.2 TWO-DIMENSIONAL STRAIN TRANSFORMATION

Though particular strain systems may be analysed separately, as for shear on page 78, it is convenient to consider the strain produced by the general stress system σ_x, σ_y and τ_{xy} in Fig. 13.4. Thus, with the resulting distortion of the element ABDE shown in Figs. 13.15(a) and (b), the following 'applied' strains can be defined for the sides provided deformations are small.

Normal strains: $\epsilon_x = (B'D' - BD)/BD$, $\epsilon_y = (A'B' - AB)/AB$.
Shear strain: $\gamma_{xy} = \tan \phi \simeq \phi$ (ϕ is the change in the right angle in radians).

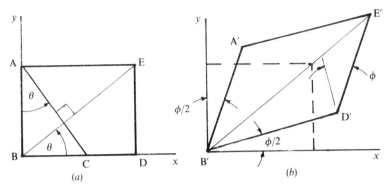

Figure 13.15 Normal and shear distortion of a rectangular element.

Now, associated with the plane AC, there will be normal (ϵ_θ) and shear (γ_θ) strain. These are respectively the strain in the normal (BE) direction and the rotation of AC in radians. For the normal strain apply the cosine rule to triangle B'D'E':

$$(B'E')^2 = (B'D')^2 + (D'E')^2 - 2(B'D')(D'E')\cos(\pi/2 + \phi).$$

Substituting $B'E' = BE(1 + \epsilon_\theta)$, $B'D' = BD(1 + \epsilon_x)$, $D'E' = DE(1 + \epsilon_y)$ together with $\cos(\pi/2 + \phi) = -\sin\phi \simeq -\phi(\text{rad}) = -\gamma_{xy}$ leads to

$$(BE)^2(1 + \epsilon_\theta)^2 = (BD)^2(1 + \epsilon_x)^2 + (DE)^2(1 + \epsilon_y)^2$$
$$- 2(BD)(DE)(1 + \epsilon_x)(1 + \epsilon_y)(-\gamma_{xy}).$$

Neglecting strain-squared terms:

$$(BE)^2(1 + 2\epsilon_\theta) = (BD)^2(1 + 2\epsilon_x) + (DE)^2(1 + 2\epsilon_y) + 2(BD)(DE)\gamma_{xy}.$$

Now $BD = BE\cos\theta$, $DE = BE\sin\theta$ and $(BE)^2 = (BD)^2 + (DE)^2$. This gives

$$\epsilon_\theta = \epsilon_x\cos^2\theta + \epsilon_y\sin^2\theta + \gamma_{xy}\sin\theta\cos\theta,$$
$$\epsilon_\theta = \epsilon_x\cos^2\theta + \epsilon_y\sin^2\theta + (\gamma_{xy}/2)\sin 2\theta. \tag{13.7a}$$

The change in the right angle formed by the intersection between AC and BE is the shear strain γ_θ. The rotation in BE is composed of contributions from ϵ_x, ϵ_y and γ_{xy}, shown separately in Fig. 13.16 (a), (b), (c) respectively. Then, net clockwise positive rotation in BE $= -\angle GBE + \angle EBF + \angle EBE'$ is

$$+ \overset{\curvearrowright}{\text{(BE}} \simeq -\frac{EG}{BE}\cos\theta + \frac{EF}{BE}\sin\theta + \frac{EE'}{BE}\sin\theta \quad \text{(radians)}$$

but $\epsilon_x = EF/BD$, $\epsilon_y = EG/ED$ and $\gamma_{xy} = EE'/ED$, $ED = BE\sin\theta$ and $BD = BE\cos\theta$.

$$\therefore \quad + \overset{\curvearrowright}{\text{(BE}} \simeq -\epsilon_y\sin\theta\cos\theta + \epsilon_x\sin\theta\cos\theta + \gamma_{xy}\sin^2\theta.$$

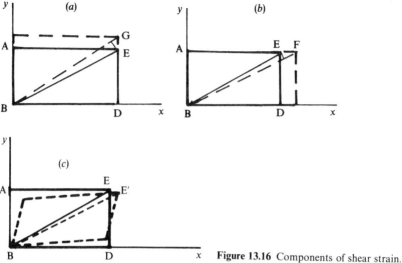

Figure 13.16 Components of shear strain.

The rotation in AC is found by replacing θ with $\theta + \pi/2$. This gives

$$+ \stackrel{\curvearrowright}{(AC)} \simeq + \epsilon_y \sin\theta\cos\theta - \epsilon_x \sin\theta\cos\theta + \gamma_{xy}\cos^2\theta.$$

Then,

$$\gamma_\theta = [+\stackrel{\curvearrowright}{(BE)}] - [+\stackrel{\curvearrowright}{(AC)}] = 2(\epsilon_x - \epsilon_y)\sin\theta\cos\theta - \gamma_{xy}(\cos^2\theta - \sin^2\theta)$$

$$= (\epsilon_x - \epsilon_y)\sin 2\theta - \gamma_{xy}\cos 2\theta,$$

$$\gamma_\theta/2 = (1/2)(\epsilon_x - \epsilon_y)\sin 2\theta - (\gamma_{xy}/2)\cos 2\theta. \tag{13.7b}$$

It is seen that Eqs (13.7a and b) are identical in form to Eqs (13.1a and b), indicating that stress and strain have the same transformation properties. It is, however, necessary to associate τ with $\gamma/2$. The multiplication factor of $1/2$ arises with the shear strain terms when, in the engineering definition of shear strain, $\gamma = \phi$ in Fig. 13.15(b) is the net change in the right-angle. Note that the mathematical tensor definition of shear strain $\epsilon_{xy} = \phi/2$ in Fig. 13.15(b) would avoid the $1/2$ factor. Equations (13.7a and b) may be reduced to simpler strain systems in Table 13.1(a)–(c) to complete the comparison between strain and stress transformation. The quoted principal strains ϵ_1 and ϵ_2 act on those perpendicular planes in each system where shear strain is absent.

The Mohr's circle can similarly be drawn for strain, and a focus located with the axes of Fig. 13.17. Strains are positive when they accompany positive stress. Thus, for the inset stressed element shown, ϵ_x is positive, γ_{xy} is negative for plane AB, while both ϵ_y and γ_{xy} are positive for the plane BC. These locate two diametrically opposite points on the circle that enable its construction. The focus F is then found from projecting the normals to the planes AB or

Table 13.1 Stress and strain transformation equations

System	Stress	Strain
(a)	$\sigma_\theta = \epsilon_x \cos^2\theta$ $\tau_\theta = \frac{1}{2}\sigma_x \sin 2\theta$ $\tau_{max} = \frac{1}{2}\sigma_x$ $\sigma_{45°} = \frac{1}{2}\sigma_x$	$\epsilon_\theta = \epsilon_x \cos^2\theta$ $\frac{1}{2}\gamma_\theta = \frac{1}{2}\epsilon_x \sin 2\theta$ $\gamma_{max} = \epsilon_x$ $\epsilon_{45°} = \frac{1}{2}\epsilon_x$
(b)	$\sigma_\theta = \tau_{xy} \sin 2\theta$ $\tau_\theta = \tau_{xy} \cos 2\theta$ $\sigma_1 = \tau_{xy}$ $\sigma_2 = -\tau_{xy}$	$\epsilon_\theta = \frac{1}{2}\gamma_{xy} \sin 2\theta$ $\frac{1}{2}\gamma_\theta = \frac{1}{2}\gamma_{xy} \cos 2\theta$ $\epsilon_1 = \frac{1}{2}\gamma_{xy}$ $\epsilon_2 = -\frac{1}{2}\gamma_{xy}$
(c)	$\sigma_\theta = \sigma_y \sin^2\theta + \sigma_x \cos^2\theta$ $\tau_\theta = \frac{1}{2}(\sigma_x - \sigma_y) \sin 2\theta$ $\tau_{max} = \frac{1}{2}(\sigma_x - \sigma_y)$ $\sigma_{45°} = \frac{1}{2}(\sigma_x + \sigma_y)$	$\epsilon_\theta = \epsilon_y \sin^2\theta + \epsilon_x \cos^2\theta$ $\frac{1}{2}\gamma_\theta = \frac{1}{2}(\epsilon_x - \epsilon_y) \sin 2\theta$ $\gamma_{max} = \epsilon_x - \epsilon_y$ $\epsilon_{45°} = \frac{1}{2}(\epsilon_x + \epsilon_y)$
(d)	$\sigma_\theta = \sigma_y \sin^2\theta + \sigma_x \cos^2\theta + \tau_{xy} \sin 2\theta$ $\quad = \frac{1}{2}(\sigma_x + \sigma_y) + \frac{1}{2}(\sigma_x - \sigma_y) \cos 2\theta + \tau_{xy} \sin 2\theta$ $\tau_\theta = \frac{1}{2}(\sigma_x - \sigma_y) \sin 2\theta - \tau_{xy} \cos 2\theta$ $\sigma_{1,2} = \frac{1}{2}(\sigma_x + \sigma_y) \pm \frac{1}{2}\sqrt{[(\sigma_x - \sigma_y)^2 + 4\tau_{xy}^2]}$ $\tan 2\theta = 2\tau_{xy}/(\sigma_x - \sigma_y)$ $\tau_{max} = \frac{1}{2}(\sigma_1 - \sigma_2) = \frac{1}{2}\sqrt{[(\sigma_x - \sigma_y)^2 + 4\tau_{xy}^2]}$	$\epsilon_\theta = \epsilon_y \sin^2\theta + \epsilon_x \cos^2\theta + \frac{1}{2}\gamma_{xy} \sin 2\theta$ $\quad = \frac{1}{2}(\epsilon_x + \epsilon_y) + \frac{1}{2}(\epsilon_x - \epsilon_y) \cos 2\theta + \frac{1}{2}\gamma_{xy} \sin 2\theta$ $\frac{1}{2}\gamma_\theta = \frac{1}{2}(\epsilon_x - \epsilon_y) \sin 2\theta - \frac{1}{2}\gamma_{xy} \cos 2\theta$ $\epsilon_{1,2} = \frac{1}{2}(\epsilon_x + \epsilon_y) \pm \frac{1}{2}\sqrt{[(\epsilon_x - \epsilon_y)^2 + \gamma_{xy}^2]}$ $\tan 2\theta = \gamma_{xy}/(\epsilon_x - \epsilon_y)$ $\gamma_{max} = \epsilon_1 - \epsilon_2 = \sqrt{[(\epsilon_x - \epsilon_y)^2 + \gamma_{xy}^2]}$

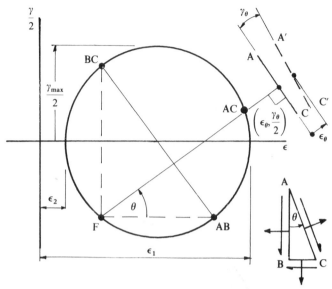

Figure 13.17 Mohr's strain circle with focus point F.

BC through the corresponding point on the strain circle. For any plane AC the state of strain $(\epsilon_\theta, \gamma_\theta/2)$ is found from projecting the normal to AC through F as shown. The intercept point shows tensile strain (ϵ_θ) in a direction normal to AC accompanied by a net clockwise angular change of γ_θ between AC and its normal. The circle also provides the principal strains ϵ_1, ϵ_2, the maximum shear strain γ_{max} and their planes w.r.t. F. Since the normals to planes AB, BC and AC are the respective directions in which ϵ_x, ϵ_y and ϵ_θ act, the use of strain directions can be employed directly for the construction.

> **Example 13.8** At a point in a strained body, the normal strains are $\epsilon_x = 350 \times 10^{-6}$ and $\epsilon_y = 50 \times 10^{-6}$ together with an unknown shear strain γ_{xy}. If the major principal strain is $\epsilon_1 = 420 \times 10^{-6}$, find, both graphically and analytically, γ_{xy} and ϵ_2 and their directions relative to x. What are the principal stresses? Take $E = 210\,\text{GPa}$, $v = 0.3$.

The circle in Fig. 13.18 is drawn with centre $C = (50 + 350)/2 = 200 \times 10^{-6}$ and radius $(420 - 200) \times 10^{-6}$ as shown. The ordinate associated with either $\epsilon_x = 50 \times 10^{-6}$ or $\epsilon_y = 350 \times 10^{-6}$ from the respective points P and Q on the circle defines $\gamma_{xy}/2$. This gives $\gamma_{xy} = 322 \times 10^{-6}$. The minor principal strain lies at the zero-shear strain point R of magnitude $\epsilon_2 = -20 \times 10^{-6}$. The focus F is found from projecting the direction of ϵ_y through P or the direction of ϵ_x through Q. The major and minor

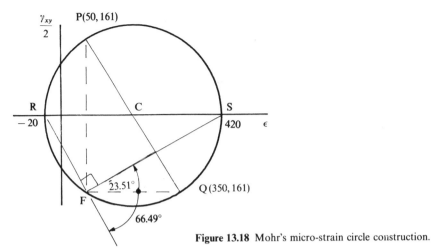

Figure 13.18 Mohr's micro-strain circle construction.

principal strain directions are then parallel to FS and FR respectively, with the given orientations relative to x (FQ). In the analytical solution, direct substitution into the major principal strain expression (Table 13.1) will yield both γ_{xy} and ϵ_2:

$$\epsilon_1 = [(\epsilon_x + \epsilon_y)/2] + (1/2)\sqrt{[(\epsilon_x - \epsilon_y)^2 + (\gamma_{xy})^2]},$$
$$420 = 200 + (1/2)\sqrt{[(300)^2 + \gamma_{xy}^2]},$$
$$\therefore \quad \gamma_{xy} = \pm 321.9 \times 10^{-6};$$
$$\epsilon_2 = [(\epsilon_x + \epsilon_y)/2] - (1/2)\sqrt{[(\epsilon_x - \epsilon_y)^2 + (\gamma_{xy})^2]}$$
$$= 200 - (1/2)\sqrt{[(300)^2 + (321.9)^2]} = -20 \times 10^{-6}.$$

The directions of the major and minor principal planes of strain are coincident with those of stress. From Table 13.1:

$$\tan 2\theta = 2(\gamma_{xy}/2)/(\epsilon_x - \epsilon_y),$$
$$\theta = \tfrac{1}{2}\tan^{-1}[\pm 321.9/(50 - 350)].$$

Then $\theta = +23.51°$ (minor principal plane) and $-66.49°$ (major principal plane) w.r.t. the y-direction. The principal strain thus acts in the normal directions, i.e. $23.51°$ (major) and $-66.49°$(minor) w.r.t. x, consistent with Fig. 13.18. In converting the principal strains to principal stresses, it is necessary to employ the inverted plane constitutive equations (4.15a and b) (page 82).

$$\sigma_1 = E(\epsilon_1 + v\epsilon_2)/(1 - v^2)$$

$$= \frac{210 \times 10^3[420 - (0.3 \times 20)] \times 10^{-6}}{(1 - 0.09)} = 95.54 \, \text{MPa};$$

$$\sigma_2 = E(\epsilon_2 + v\epsilon_1)/(1 - v^2)$$
$$= \frac{210 \times 10^3 [-20 + (0.3 \times 420)] \times 10^{-6}}{(1 - 0.09)} = 24.46 \, \text{MPa}.$$

Example 13.9 Normal strains of 600μ, -200μ and 250μ ($1\mu = 1 \times 10^{-6}$) are measured with a strain gauge rosette at $0°$, $60°$ and $120°$ anticlockwise from the x-axis. Find the magnitude and direction of the principal strains graphically and analytically.

For the construction in Fig. 13.19, the three gauges are rotated so that any one lies in the vertical position as shown inset in (a). The recorded strains are measured horizontally from the vertical semi-shear strain axis and parallel lines 1, 2 and 3 drawn as shown in (b). Any focus point F is chosen along the line representing the vertically aligned gauge (3). From F lines are drawn parallel to the repositioned directions 1 and 2 of the remaining two gauges to intersect their strain lines in P and Q respectively. The intersection between the perpendicular bisectors of chords FP and FQ locates the centre lying on the normal strain axis. The extremities of the circle along this axis are the principal strains; $\epsilon_1 = 680\mu$, $\epsilon_2 = -246\mu$. Their respective directions in the inset diagram lie parallel to $F\epsilon_1$ and $F\epsilon_2$. The true directions are found by rotating this diagram into the original configuration, giving ϵ_2 at $73°$ and ϵ_1 at $-17°$ from x. For the analytical solutions, substitute for θ and

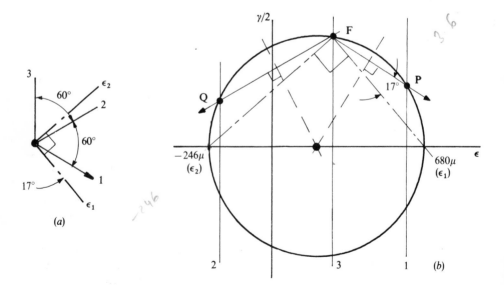

Figure 13.19 Mohr's circle for three direct rosette strains.

ε_θ in Eq. (13.7a) to give the three simultaneous equations:

$$600\mu = \epsilon_x, \tag{i}$$

$$-200\mu = \epsilon_y \sin 260° + \epsilon_x \cos 260° + (\gamma_{xy}/2)\sin 120°,$$
$$0.75\epsilon_y + 0.433\gamma_{xy} = -350\mu, \tag{ii}$$

$$250\mu = \epsilon_y \sin^2 120° + \epsilon_x \cos^2 120° + (\gamma_{xy}/2)\sin 240°,$$
$$0.75\epsilon_y - 0.433\gamma_{xy} = 100\mu. \tag{iii}$$

The solution to Eqs (i), (ii) and (iii) results in the x- and y-strains; $\epsilon_x = 600\mu$, $\epsilon_y = -166.67\mu$ and $\gamma_{xy} = -519.60\mu$. Substitution into the principal strain expression (Table 13.1) then gives

$$\epsilon_{1,2} = [(\epsilon_x + \epsilon_y)/2] \pm (1/2)\sqrt{[(\epsilon_x - \epsilon_y)^2 + (\gamma_{xy})^2]}$$
$$= [(600 - 166.67)/2] \pm (1/2)\sqrt{[(600 + 166.67)^2 + (519.63)^2]},$$

$$\epsilon_1 = 679.75\mu \quad \text{and} \quad \epsilon_2 = -246.42\mu;$$

$$\theta = \tfrac{1}{2}\tan^{-1}[\gamma_{xy}/(\epsilon_x - \epsilon_y)] = \tfrac{1}{2}\tan^{-1}(-519.6/766.67),$$

$$\theta = -17.07° \quad \text{and} \quad \underline{72.94°} \quad \text{(principal planes relative to } y)$$

$$\text{or } -17.07° \quad \text{and} \quad 72.94° \quad \text{(principal directions relative to } x).$$

The matrix representation of plane (or two-dimensional strain) transformation from matrix **E** to **E'** is identical in form to Eq. (13.5) provided the mathematical shear strains $\epsilon_{xy} = \epsilon_{yx} = \gamma_{xy}/2$ are employed. That is

$$\mathbf{E'} = \mathbf{LEL}^T$$

$$\begin{bmatrix} \epsilon'_x & \epsilon'_{xy} \\ \epsilon'_{yx} & \epsilon'_y \end{bmatrix} = \begin{bmatrix} l_{xx} & l_{xy} \\ l_{yx} & l_{yy} \end{bmatrix}\begin{bmatrix} \epsilon_x & \epsilon_{xy} \\ \epsilon_{yx} & \epsilon_y \end{bmatrix}\begin{bmatrix} l_{xx} & l_{yx} \\ l_{xy} & l_{yy} \end{bmatrix},$$

where direction cosines apply to x, y and x', y' as before (see Fig. 13.13).

13.3 THREE-DIMENSIONAL STRESS TRANSFORMATION

Direction Cosines

Consider a tetrahedron OABC in Fig. 13.20, set in a Cartesian coordinate frame x, y, z in which six known independent stress components σ_x, σ_y, σ_z, $\tau_{xy} = \tau_{yx}$, $\tau_{xz} = \tau_{zx}$ and $\tau_{yz} = \tau_{zy}$ act on the back three triangular faces OAB, OBC and OAC as shown. Note that ABC is an oblique plane cutting the volume element in Fig. 11.1(a). On the front three faces of the element in Fig. 11.1(a) it can be seen that the stress components act in the positive coordinate directions. Thus, the negative directions apply to the back three orthogonal faces in Fig. 13.20. The objective is to find, for Fig. 13.20(a), the stress state (σ, τ) on the front triangular face ABC in both magnitude and direction. For this purpose it is first necessary to find the areas of each back face. Thus, in

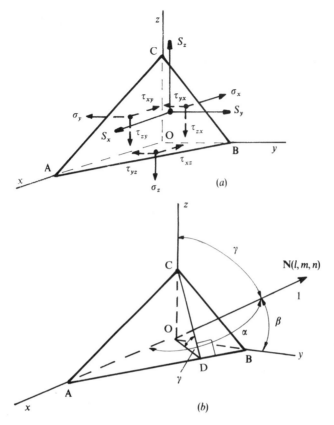

Figure 13.20 3-D stress state for an oblique plane ABC.

Fig. 13.20(b) let the area ABC be unity and construct CD perpendicular to AB and join OD. The normal vector **N**, to plane ABC is defined by the respective direction cosines w.r.t. x, y and z:

$$l = \cos \alpha, \qquad m = \cos \beta \qquad \text{and} \qquad n = \cos \gamma. \qquad (13.8\text{a}, \text{b}, \text{c})$$

Then as Area ABC $= \frac{1}{2}$AB $\times$ CD and Area OAB $= \frac{1}{2}$AB $\times$ OD,

$$\frac{\text{Area OAB}}{\text{Area ABC}} = \frac{\text{OD}}{\text{CD}} = \cos \gamma = n.$$

Hence, Area OAB $= n$ and, similarly, Area OBC $= l$ and Area OAC $= m$. The direction cosines are not independent. Their relationship follows from the vector **N** equation:

$$\mathbf{N} = N_x \mathbf{u}_x + N_y \mathbf{u}_y + N_z \mathbf{u}_z, \qquad (13.9\text{a})$$

where $\mathbf{u}_x$, $\mathbf{u}_y$ and $\mathbf{u}_z$ are unit vectors and N_x, N_y and N_z are scalar intercepts with the coordinates x, y and z as seen in Fig. 13.21(a).

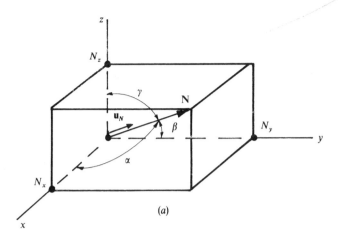

(a)

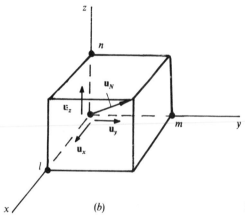

(b)

Figure 13.21 Normal vector intercepts.

The unit vector $\mathbf{u}_N$ for the normal direction is found by dividing Eq. (13.9a) by the magnitude $|\mathbf{N}|$ of $\mathbf{N}$.

$$\mathbf{u}_N = (N_x/|\mathbf{N}|)\mathbf{u}_x + (N_y/|\mathbf{N}|)\mathbf{u}_y + (N_z/|\mathbf{N}|)\mathbf{u}_z, \qquad (13.9b)$$

but from Eqs (13.8) $l = \cos\alpha = N_x/|\mathbf{N}|$, $m = \cos\beta = N_y/|\mathbf{N}|$, $n = \cos\gamma = N_z/|\mathbf{N}|$. Hence Eq. (13.9b) becomes

$$\mathbf{u}_N = l\mathbf{u}_x + m\mathbf{u}_y + n\mathbf{u}_z. \qquad (13.9c)$$

Hence l, m and n are the intercepts that the unit normal vector $\mathbf{u}_N$ makes with x, y and z (Fig. 13.21b). Furthermore, since

$$(N_x)^2 + (N_y)^2 + (N_z)^2 = |\mathbf{N}|^2,$$
$$(N_x/|\mathbf{N}|)^2 + (N_y/|\mathbf{N}|)^2 + (N_z/|\mathbf{N}|)^2 = 1$$
$$\therefore \quad l^2 + m^2 + n^2 = 1. \qquad (13.10)$$

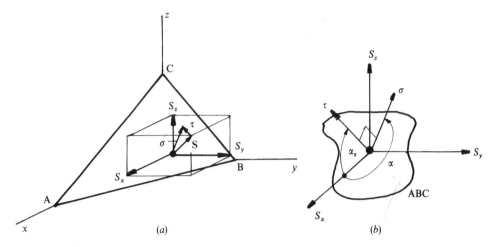

Figure 13.22 Stress state for the oblique plane ABC.

Normal (σ) and Shear (τ) Stress on Plane ABC

Magnitudes Let S_x, S_y and S_z (Fig. 13.22a) be the coordinate components for the resultant force **S** acting on plane ABC in Fig. 13.22(a). **S** equilibrates the forces due to the applied stress components on the back faces. The following three equilibrium equations apply from Fig. 13.20(a):

$$S_x = l\sigma_x + m\tau_{xy} + n\tau_{xz}, \tag{13.11a}$$

$$S_y = m\sigma_y + n\tau_{yz} + l\tau_{yx}, \tag{13.11b}$$

$$S_z = n\sigma_z + l\tau_{zx} + m\tau_{zy}. \tag{13.11c}$$

As the area of ABC is unity, σ (Fig. 13.22(a)) is the sum of the S_x, S_y and S_z force components resolved into the normal direction. That is,

$$\sigma = S_x \cos\alpha + S_y \cos\beta + S_z \cos\gamma = S_x l + S_y m + S_z n,$$

where, from Eqs (13.11), this becomes

$$\sigma = \sigma_x l^2 + \sigma_y m^2 + \sigma_z n^2 + 2(lm\tau_{xy} + mn\tau_{yz} + ln\tau_{zx}). \tag{13.12a}$$

The resultant force **S** on ABC in Fig. 13.22(a) is expressed in two ways:

$$S^2 = S_x^2 + S_y^2 + S_z^2 = \sigma^2 + \tau^2,$$

$$\therefore \quad \tau^2 = S^2 - \sigma^2 = S_x^2 + S_y^2 + S_z^2 - \sigma^2, \tag{13.12b}$$

where, substituting from Eqs (13.11 and 13.12(a)), τ can be found.

Directions The direction of σ is defined by l, m and n since it acts in a direction parallel to **N**. The direction of τ in the plane ABC is defined by the direction cosines $l_s = \cos \alpha_s$, $m_s = \cos \beta_s$ and $n_s = \cos \gamma_s$ (Fig. 13.22(b)). Now S_x, S_y and S_z are resultant forces for the x-, y- and z-components of σ and τ. That is,

$$S_x = \sigma \cos \alpha + \tau \cos \alpha_s = l\sigma + l_s \tau,$$
$$S_y = \sigma \cos \beta + \tau \cos \beta_s = m\sigma + m_s \tau,$$
$$S_z = \sigma \cos \gamma + \tau \cos \gamma_s = n\sigma + n_s \tau,$$

where, it follows that

$$l_s = (S_x - l\sigma)/\tau, \tag{13.13a}$$

$$m_s = (S_y - m\sigma)/\tau, \tag{13.13b}$$

$$n_s = (S_z - n\sigma)/\tau. \tag{13.13c}$$

Example 13.10 The state of stress at a point is given by $\sigma_x = 14$, $\sigma_y = 10$, $\sigma_z = 35$, $\tau_{xy} = \tau_{yx} = 7$, $\tau_{xz} = \tau_{zx} = -7$ and $\tau_{yz} = \tau_{zy} = 0$ MPa. Determine the normal and shear stresses for a plane whose normal is defined by the direction cosines $l = 2/\sqrt{14}$, $m = -1/\sqrt{14}$ and $n = 3/\sqrt{14}$. What is the direction of the shear stress acting in this plane?

Substituting the stress components into Eq. (13.12a):

$$\sigma = \sigma_x l^2 + \sigma_y m^2 + \sigma_z n^2 + 2(lm\tau_{xy} + mn\tau_{yz} + ln\tau_{zx})$$
$$= (14 \times 4/14) + (10 \times 1/14) + (35 \times 9/14)$$
$$+ 2[(-2/14)7 + (-3/14)0 + (6/14)(-7)]$$
$$= 4 + 5/7 + 45/2 + 2(-1 - 0 - 3) = 19.21 \text{ MPa}.$$

From Eqs (13.11), the x-, y- and z-resultants are

$$S_x = l\sigma_x + m\tau_{xy} + n\tau_{xz}$$
$$= (2/\sqrt{14})14 - (1/\sqrt{14})7 - (3/\sqrt{14})7 = 0,$$
$$S_y = m\sigma_y + n\tau_{yz} + l\tau_{yx}$$
$$= -(1/\sqrt{14})10 + (3/\sqrt{14})0 + (2/\sqrt{14})7 = 4/\sqrt{14},$$
$$S_z = n\sigma_z + l\tau_{zx} + m\tau_{zy}$$
$$= (3/\sqrt{14})35 - (2/\sqrt{14})7 - (1/\sqrt{14})0 = 91/\sqrt{14}.$$

Then, from Eq. (13.12b), the shear stress acting on the plane is

$$\tau^2 = S_x^2 + S_y^2 + S_z^2 - \sigma^2 = 0 + (16/14) + (91^2/14) - 19.21^2 = 223.62,$$
$$\therefore \quad \tau = 14.95 \text{ MPa}.$$

The direction cosines for this shear stress are found from Eqs (13.13):

$$l_s = (S_x - l\sigma)/\tau = [0 - (2/\sqrt{14}) \times 19.21]/14.95 = -0.687,$$
$$m_s = (S_y - m\sigma)/\tau = [4/\sqrt{14} - (-1/\sqrt{14}) \times 19.21]/14.95 = 0.415,$$
$$n_s = (S_z - n\sigma)/\tau = [91/\sqrt{14} - (3/\sqrt{14}) \times 19.21]/14.95 = 0.597.$$

The corresponding inclinations of τ to x, y and z are

$$\alpha_s = \cos^{-1}(-0.687) = 133.4°,$$
$$\beta_s = \cos^{-1}(0.415) = 65.5°,$$
$$\gamma_s = \cos^{-1}(0.597) = 53.37°.$$

Principal Stresses and Invariants

Magnitudes of principal stresses When for the plane ABC in Fig. 13.20(a) the shear stress τ is absent, then, by definition, σ becomes a principal stress. Force resolution in the x-, y- and z-directions then gives Eqs (13.11) as

$$S_x = l\sigma = l\sigma_x + m\tau_{xy} + n\tau_{xz},$$
$$S_y = m\sigma = m\sigma_y + n\tau_{yz} + l\tau_{yx},$$
$$S_z = n\sigma = n\sigma_z + l\tau_{zx} + m\tau_{zy}.$$

That is,

$$l(\sigma_x - \sigma) + m\tau_{xy} + n\tau_{xz} = 0,$$
$$l\tau_{yx} + m(\sigma_y - \sigma) + n\tau_{yz} = 0, \tag{13.14a}$$
$$l\tau_{zx} + m\tau_{zy} + n(\sigma_z - \sigma) = 0.$$

By Cramar's rule, the solution to σ is found from the determinant

$$\det \begin{vmatrix} (\sigma_x - \sigma) & \tau_{xy} & \tau_{xz} \\ \tau_{yx} & (\sigma_y - \sigma) & \tau_{yz} \\ \tau_{zx} & \tau_{zy} & (\sigma_z - \sigma) \end{vmatrix} = 0. \tag{13.14b}$$

Expanding Eq. (13.14b) leads to the cubic equation

$$(\sigma_x - \sigma)[(\sigma_y - \sigma)(\sigma_z - \sigma) - \tau_{yz}\tau_{zy}] - \tau_{xy}[\tau_{yx}(\sigma_z - \sigma) - \tau_{yz}\tau_{zx}]$$
$$+ \tau_{xz}[\tau_{yx}\tau_{zy} - \tau_{zx}(\sigma_y - \sigma)] = 0,$$
$$\sigma^3 - (\sigma_x + \sigma_y + \sigma_z)\sigma^2 + (\sigma_x\sigma_y + \sigma_y\sigma_z + \sigma_z\sigma_x - \tau_{xy}^2 - \tau_{yz}^2 - \tau_{zx}^2)\sigma$$
$$- (\sigma_x\sigma_y\sigma_z + 2\tau_{xy}\tau_{yz}\tau_{zx} - \sigma_x\tau_{yz}^2 - \sigma_y\tau_{zx}^2 - \sigma_z\tau_{xy}^2) = 0. \tag{13.15a}$$

The three roots (eigenvalues) σ_1, σ_2 and σ_3 to this characteristic equation give the principal stress magnitudes. Equation (13.15a) is usually written as

$$\sigma^3 - J_1\sigma^2 + J_2\sigma - J_3 = 0. \tag{13.15b}$$

Since there is a unique set of principal stresses for any given applied stress system, it follows that the coefficients J_1, J_2 and J_3 will be independent of the coordinate frame chosen to define the applied stresses. Thus, J_1, J_2 and J_3 are invariants of the general stress tensor σ_{ij}. The latter, which is shown in Fig. 11.1, is expressible in either of the 3×3 matrices on page 410. Since equation (13.15a) includes the case where the x, y, z frame coincides with the principal stress directions 1, 2 and 3, the invariants may be written in the general (subscripts x, y, z) or principal (subscripts 1, 2, 3) forms:

$$J_1 = \sigma_1 + \sigma_2 + \sigma_3 = \sigma_x + \sigma_y + \sigma_z = \sigma_{ii}, \tag{13.15c}$$

$$J_2 = \sigma_1\sigma_2 + \sigma_2\sigma_3 + \sigma_1\sigma_3 = \sigma_x\sigma_y + \sigma_y\sigma_z + \sigma_z\sigma_x - \tau_{xy}^2 - \tau_{yz}^2 - \tau_{zx}^2$$

$$= \tfrac{1}{2}(\sigma_{ii}\sigma_{jj} - \sigma_{ij}\sigma_{ji}), \tag{13.15d}$$

$$J_3 = \sigma_1\sigma_2\sigma_3 = \sigma_x\sigma_y\sigma_z + 2\tau_{xy}\tau_{yz}\tau_{zx} - \sigma_x\tau_{yz}^2 - \sigma_y\tau_{zx}^2 - \sigma_z\tau_{xy}^2 = \det[\sigma_{ij}], \tag{13.15e}$$

where, in the shorthand tensor notation, repeated subscripts on a single symbol, or within a term, denote summation for i and $j = 1, 2, 3$. Where there are exact roots, the principal stresses are more conveniently found from expanding the determinant in Eq. (13.14b) with numerical values substituted. Otherwise, the major (σ_1), intermediate (σ_2) and minor (σ_3) principal stresses ($\sigma_1 > \sigma_2 > \sigma_3$) must be found from the solution to the characteristic cubic equation (13.15).

Direction cosines of principal stresses Substituting for the applied stresses into Eq. (13.14a) with $\sigma = \sigma_1(l_1, m_1, n_1)$ leads to three simultaneous equations in l_1, m_1 and n_1, of which only two equations are independent, together with the relationship $l_1^2 + m_1^2 + n_1^2 = 1$ from Eq. (13.10). A similar deduction is made for the separate substitutions of $\sigma_2(l_2, m_2, n_2)$ and $\sigma_3(l_3, m_3, n_3)$. Now it follows from Eq. (13.9c) that the principal sets of direction cosines (l_1, m_1, n_1), (l_2, m_2, n_2) and (l_3, m_3, n_3) define unit vectors in the principal directions as

$$\mathbf{u}_1 = l_1\mathbf{u}_x + m_1\mathbf{u}_y + n_1\mathbf{u}_z, \tag{13.16a}$$

$$\mathbf{u}_2 = l_2\mathbf{u}_x + m_2\mathbf{u}_y + n_2\mathbf{u}_z, \tag{13.16b}$$

$$\mathbf{u}_3 = l_3\mathbf{u}_x + m_3\mathbf{u}_y + n_3\mathbf{u}_z. \tag{13.16c}$$

These are orthogonal when the dot product of any two vectors is zero. That is, for the 1 and 2 directions,

$$\mathbf{u}_1 \cdot \mathbf{u}_2 = (l_1\mathbf{u}_x + m_1\mathbf{u}_y + n_1\mathbf{u}_z) \cdot (l_2\mathbf{u}_x + m_2\mathbf{u}_y + n_2\mathbf{u}_z) = 0.$$

Now $\mathbf{u}_x \cdot \mathbf{u}_x = \mathbf{u}_y \cdot \mathbf{u}_y = \mathbf{u}_z \cdot \mathbf{u}_z = 1$ and $\mathbf{u}_x \cdot \mathbf{u}_y = \mathbf{u}_x \cdot \mathbf{u}_z = \mathbf{u}_y \cdot \mathbf{u}_z = 0$. Similar relationships for the 1, 3 and 2, 3 directions determines orthogonality:

$$l_1l_2 + m_1m_2 + n_1n_2 = 0,$$
$$l_2l_3 + m_2m_3 + n_2n_3 = 0,$$
$$l_1l_3 + m_1m_3 + n_1n_3 = 0.$$

Since these three relationships are the only conditions that satisfy the simultaneous equations (13.14a), they confirm that the principal stresses directions and their associated planes are orthogonal.

Example 13.11 At a point in a loaded material, a resultant stress of magnitude 216 MPa makes angles of $\alpha_r = 43°$, $\beta_r = 75°$ and $\gamma_r = 50.88°$

with the coordinates x, y and z respectively. Find the normal and shear stress on a plane whose direction cosines are $l = 0.387$, $m = 0.866$ and $n = 0.3167$. Given that the applied shear stresses are $\tau_{xy} = 23$, $\tau_{yz} = -3.1$ and $\tau_{xz} = 57$ MPa, determine σ_x, σ_y, σ_z and the principal stresses for the system.

Resolve the resultant force $\mathbf{S}$ (stress S acting on unit area ABC) in the x-, y- and z-directions (Fig. 13.22a) to give

$$S_x = Sl_r = S\cos\alpha_r, \qquad S_y = Sm_r = S\cos\beta_r \quad \text{and} \quad S_z = Sn_r = S\cos\gamma_r.$$

The normal stress in Eq. (13.12a) is then

$$\sigma = S_x l + S_y m + S_z n = S(l\cos\alpha_r + m\cos\beta_r + n\cos\gamma_r)$$
$$= 216(0.387\cos 43° + 0.866\cos 75° + 0.3167\cos 50.88°) = 152.71\,\text{MPa}.$$

Now, from Eq. (13.12b), the shear stress is

$$\tau^2 = S^2 - \sigma^2,$$
$$\tau = \sqrt{(216^2 - 152.71^2)} = 152.76\,\text{MPa}.$$

Substituting into Eqs (13.11):

$$S_x = l\sigma_x + m\tau_{xy} + n\tau_{xz},$$
$$216\cos 43° = 0.387\sigma_x + (0.866 \times 23) + (0.3167 \times 57),$$
$$\sigma_x = 310.1\,\text{MPa}.$$

$$S_y = m\sigma_y + n\tau_{yz} + l\tau_{yx},$$
$$216\cos 75° = 0.866\sigma_y - (0.3167 \times 3.1) + (0.387 \times 23),$$
$$\sigma_y = 55.41\,\text{MPa}.$$

$$S_z = n\sigma_z + l\tau_{zx} + m\tau_{zy},$$
$$216\cos 50.88° = 0.3167\sigma_z + (0.387 \times 57) - (0.866 \times 3.1),$$
$$\sigma_z = 369.15\,\text{MPa}.$$

Substituting into Eqs (13.15c–e), the invariants are

$$J_1 = \sigma_x + \sigma_y + \sigma_z = 310.1 + 55.41 + 369.15 = 734.66,$$
$$J_2 = \sigma_x\sigma_y + \sigma_y\sigma_z + \sigma_z\sigma_x - \tau_{xy}^2 - \tau_{yz}^2 - \tau_{zx}^2$$
$$= (310.1 \times 55.41) + (55.41 \times 369.15)$$
$$\quad + (369.15 \times 310.1) - (23)^2 - (-3.1)^2 - (57)^2$$
$$= 152\,110.66 - 3787.61 = 148\,323.05,$$

$$J_3 = \det \begin{vmatrix} 310.1 & 23 & 57 \\ 23 & 55.41 & -3.1 \\ 57 & -3.1 & 369.15 \end{vmatrix}$$

$$= 310.1[(55.41 \times 369.15) - (3.1)^2] - 23[(23 \times 369.15) - (-3.1 \times 57)]$$
$$\quad + 57[-(23 \times 3.1) - (57 \times 55.41)] = 5\,956\,556.22.$$

The principal stress cubic (Eq. (13.15b)) becomes

$$\sigma^3 - 734.66\sigma^2 + 148\,323.05\sigma - 5\,956\,556.22 = 0. \qquad \text{(i)}$$

Using Newton's approximation to find the roots of Eq. (i):

$$f(\sigma) = \sigma^3 - 734.66\sigma^2 + 148\,323.05\sigma - 5\,956\,556.22, \qquad \text{(ii)}$$

$$f'(\sigma) = 3\sigma^2 - 1469.32\sigma + 148\,323.05. \qquad \text{(iii)}$$

One root lies between 50 and 60. Take an approximation $\alpha = 55$ so that the numerical values of Eqs (ii) and (iii) are $f(\alpha) = 145\,240.03$ and $f'(\alpha) = 76\,585.45$. A closer approximation is then given by

$$\alpha - f(\alpha)/f'(\alpha) = 55 - 145\,240.03/76\,585.45 = 53.1.$$

Again, from Eqs (ii) and (iii): $f(\alpha) = -2335.66$, $f'(\alpha) = 78\,761$, giving an even closer approximation to one root of

$$\alpha = 53.1 - (-2335.66/78\,761) = 53.13\,\text{MPa}.$$

The remaining roots are found from the quadratic $a\sigma^2 + b\sigma + c = 0$, where

$$(\sigma - 53.13)(a\sigma^2 + b\sigma + c) = \sigma^3 - 734.66\sigma^2 + 148\,323.05\sigma - 5\,956\,556.22,$$
$$a\sigma^3 - (53.13a - b)\sigma^2 + (c - 53.13b)\sigma - 53.13c$$
$$= \sigma^3 - 734.66\sigma^2 + 148\,323.05\sigma - 5\,956\,556.22.$$

Equating coefficients of σ^3, σ^2, σ gives

$$a = 1,$$
$$53.13a - b = 734.66, \quad \Rightarrow b = -681.53,$$
$$c - 53.13b = 148\,323.05, \quad \Rightarrow c = 112\,113.36.$$

Thus

$$\sigma^2 - 681.53\sigma + 112\,113.36 = 0,$$

for which the roots are 404.07 and 277.46. Hence, the principal stresses are

$$\sigma_1 = 404.07, \qquad \sigma_2 = 277.46 \quad \text{and} \quad \sigma_3 = 53.13\,\text{MPa}.$$

Applied Principal Stresses

When the applied stresses are the principal stresses $\sigma_1 > \sigma_2 > \sigma_3$, the coordinate axes become aligned with the orthogonal principal directions 1, 2 and 3 as shown in Fig. 13.23. Because shear stress is then absent on faces ACO, ABO and BCO, the expressions for the normal and shear stress acting on the oblique plane ABC are simplified.

Replacing x, y and z in Eqs (13.11) and (13.12) by 1, 2 and 3 respectively, and setting $\tau_{xy} = \tau_{yz} = \tau_{xz} = 0$ gives the reduced forms

$$S_1 = l\sigma_1, \qquad S_2 = m\sigma_2, \qquad S_3 = n\sigma_3, \qquad \text{(13.17a, b, c)}$$

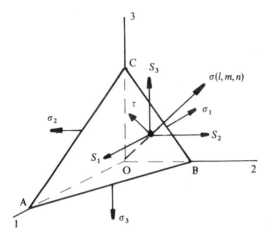

Figure 13.23 Principal stress axes.

which also follow from the equilibrium in Fig. 13.23. Now

$$\sigma = S_1 l + S_2 m + S_3 n$$
$$= \sigma_1 l^2 + \sigma_2 m^2 + \sigma_3 n^2, \tag{13.18a}$$

$$\tau^2 = S^2 - \sigma^2 = (S_1^2 + S_2^2 + S_3^2) - \sigma^2$$
$$= (l\sigma_1)^2 + (m\sigma_2)^2 + (n\sigma_3)^2 - (\sigma_1 l^2 + \sigma_2 m^2 + \sigma_3 n^2)^2,$$
$$\therefore \quad \tau = \sqrt{[(l\sigma_1)^2 + (m\sigma_2)^2 + (n\sigma_3)^2 - (\sigma_1 l^2 + \sigma_2 m^2 + \sigma_3 n^2)^2]} \tag{13.18b}$$

with associated direction cosines, from Eq. (13.13),

$$l_s = (S_1 - l\sigma)/\tau = l(\sigma_1 - \sigma)/\tau, \tag{13.19a}$$

$$m_s = (S_2 - m\sigma)/\tau = m(\sigma_2 - \sigma)/\tau, \tag{13.19b}$$

$$n_s = (S_3 - n\sigma)/\tau = n(\sigma_3 - \sigma)/\tau. \tag{13.19c}$$

Maximum shear stress It can be shown that the maximum shear stresses act on planes inclined at $45°$ to two principal planes and perpendicular to the remaining plane. In the 1–2 plane of Fig. 13.24, for example, the normal **N** to the $45°$ plane shown, has the direction cosines $l = m = \cos 45° = 1/\sqrt{2}$ and $n = \cos 90° = 0$.

Substituting l, m and n into Eq. (13.18b) the magnitude is

$$\tau_{12}^2 = \sigma_1^2/2 + \sigma_2^2/2 - (\sigma_1/2 + \sigma_2/2)^2 = \tfrac{1}{4}(\sigma_1^2 + \sigma_2^2 - 2\sigma_1\sigma_2) = \tfrac{1}{4}(\sigma_1 - \sigma_2)^2,$$
$$\therefore \quad \tau_{12} = \pm \tfrac{1}{2}(\sigma_1 - \sigma_2). \tag{13.20a}$$

Similarly, for the plane inclined at $45°$ to the 1 and 3 directions ($l = n = 1/\sqrt{2}$ and $m = 0$), the shear stress is

$$\tau_{13} = \pm \tfrac{1}{2}(\sigma_1 - \sigma_3) \tag{13.20b}$$

and, for the plane inclined at $45°$ to the 2 and 3 directions, where $n = m = 1/\sqrt{2}$

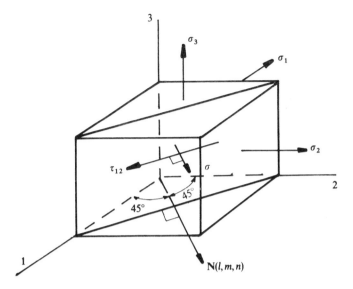

Figure 13.24 Maximum 45° shear plane.

and $1 = 0$,

$$\tau_{23} = \pm \tfrac{1}{2}(\sigma_2 - \sigma_3).$$ (13.20c)

The greatest shear stress for the system $\sigma_1 > \sigma_2 > \sigma_3$ is $\tau_{max} = \tau_{13}$. When the 45° shear planes are constructed in all four quadrants, they form a rhombic dodecahedron. The normal stress acting on the planes of maximum shear stress is found from Eq. (13.18a). For example, with $l = m = 1/\sqrt{2}$ and $n = 0$ for the 45° plane in Fig. 13.24, $\sigma = \tfrac{1}{2}(\sigma_1 + \sigma_2)$.

Octahedral plane It follows from Eq. (13.10) that the direction cosines for the normal to the plane equally inclined to the principal directions are $l = m = n = 1/\sqrt{3}$ ($\alpha = \beta = \gamma = 54.7°$). Substituting these into Eq. (13.18a) gives the octahedral normal stress (σ_o):

$$\sigma_o = \sigma_1 l^2 + \sigma_2 m^2 + \sigma_3 n^2 = \sigma_1(1/\sqrt{3})^2 + \sigma_2(1/\sqrt{3})^2 + \sigma_3(1/\sqrt{3})^2,$$

$$\sigma_o = (\sigma_1 + \sigma_2 + \sigma_3)/3.$$ (13.21a)

Since σ_o is the average of the principal stresses, it is also called the mean or hydrostatic stress σ_m (Eq. (4.5)). The octahedral shear stress τ_o, is found by substituting $l = m = n = 1/\sqrt{3}$ into Eq. (13.18b):

$$\tau_o^2 = (\sigma_1/\sqrt{3})^2 + (\sigma_2/\sqrt{3})^2 + (\sigma_3/\sqrt{3})^2 - [\sigma_1(1/\sqrt{3})^2 + \sigma_2(1/\sqrt{3})^2 + \sigma_3(1/\sqrt{3})^2]^2$$

$$= (\sigma_1 + \sigma_2 + \sigma_3)/3 - [(\sigma_1 + \sigma_2 + \sigma_3)/3]^2$$

$$= (2/9)(\sigma_1^2 + \sigma_2^2 + \sigma_3^2 - \sigma_1\sigma_2 - \sigma_1\sigma_3 - \sigma_2\sigma_3)$$

$$= (1/9)[(\sigma_1 - \sigma_2)^2 + (\sigma_2 - \sigma_3)^2 + (\sigma_1 - \sigma_3)^2],$$

$$\tau_o = (1/3)\sqrt{[(\sigma_1 - \sigma_2)^2 + (\sigma_2 - \sigma_3)^2 + (\sigma_1 - \sigma_3)^2]}$$ (13.21b)

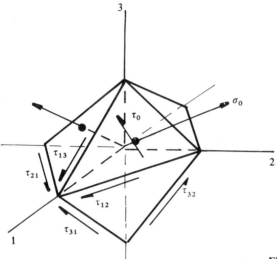

Figure 13.25 Octahedral planes.

or, from Eqs (13.20),

$$\tau_o = (2/3)\sqrt{(\tau_{12}^2 + \tau_{23}^2 + \tau_{13}^2)}.$$ (13.21c)

Equations (13.19) supplies the direction cosines for τ_o:

$$l_o = (\sigma_1 - \sigma_o)/\sqrt{3}\tau_o,$$ (13.22a)

$$m_o = (\sigma_2 - \sigma_o)/\sqrt{3}\tau_o,$$ (13.22b)

$$\tau_o = (2/3)\sqrt{(\tau_{12}^2 + \tau_{23}^2 + \tau_{13}^2)}.$$ (13.22c)

When the octahedral planes in all four quadrants are considered, they form the faces of the regular octahedron in Fig. 13.25. Here σ_o and τ_o act on each plane while τ_{12}, τ_{23} and τ_{13} act along the edges. The deformation under any stress state may be examined on an octahedral basis. Since σ_o acts with equal inclination and intensity, it causes an elastic volume change that is recoverable irrespective of the principal stress magnitudes. Superimposed on this is the distortion produced by τ_o. As the magnitude of τ_o, given in Eq. (13.21b), depends upon differences between the principal stresses, a critical value of τ_o will determine whether the deformation will be elastic or elastic-plastic. In Chapter 16 it is shown that a yield criterion may be formulated on this basis.

Example 13.12 The following matrix of stress components

$$\sigma_{ij} = \begin{bmatrix} 6 & 2 & 2 \\ 2 & 0 & 4 \\ 2 & 4 & 0 \end{bmatrix}$$

(MPa) describes the stress state at a point. Find in magnitude and direction, the normal and shear stress on a plane whose unit normal vector equation is $\mathbf{u}_N = 0.53\mathbf{u}_x + 0.35\mathbf{u}_y + 0.77\mathbf{u}_z$. Determine the principal stresses, the greatest shear stress and the stress state on the octahedral plane. (T.C.D. 1983)

Substituting $\sigma_x = 6$, $\sigma_y = 0$, $\sigma_z = 0$, $\tau_{xy} = 2$, $\tau_{xz} = 2$ and $\tau_{yz} = 4$ MPa in Eq. (13.12a) with $l = 0.53$, $m = 0.35$ and $n = 0.77$ gives

$$\sigma = \sigma_x l^2 + \sigma_y m^2 + \sigma_z n^2 + 2(lm\tau_{xy} + mn\tau_{yz} + ln\tau_{xz})$$
$$= 6(0.53)^2 + 2[(0.53 \times 0.35 \times 2) + (0.35 \times 0.77 \times 4) + (0.53 \times 0.77 \times 2)]$$
$$= 6.216 \text{ MPa}.$$

Now from Eqs (13.11) the x, y, z stress resultants are

$$S_x = l\sigma_x + m\tau_{xy} + n\tau_{xz}$$
$$= (0.53 \times 6) + (0.35 \times 2) + (0.77 \times 2) = 5.42 \text{ MPa},$$
$$S_y = m\sigma_y + l\tau_{xy} + n\tau_{yz}$$
$$= (0.35 \times 0) + (0.77 \times 4) + (0.53 \times 2) = 4.14 \text{ MPa},$$
$$S_z = n\sigma_z + l\tau_{xz} + m\tau_{yz}$$
$$= (0.77 \times 0) + (0.53 \times 2) + (0.35 \times 4) = 2.46 \text{ MPa}.$$

Then, from Eq. (13.12b), the shear stress is

$$\tau^2 = S_x^2 + S_y^2 + S_z^2 - \sigma^2$$
$$= 5.42^2 + 4.14^2 + 2.46^2 - 6.216^2 = 13.93,$$
$$\tau = 3.732 \text{ MPa}.$$

The direction of τ is defined with direction cosines supplied from Eqs (13.13):

$$l_s = (S_x - l\sigma)/\tau = [5.42 - (0.53 \times 6.216)]/3.732 = 0.57,$$
$$m_s = (S_y - m\sigma)/\tau = [4.14 - (0.35 \times 6.216)]/3.732 = 0.526,$$
$$n_s = (S_z - n\sigma)/\tau = [2.46 - (0.77 \times 6.216)]/3.732 = -0.623.$$

Check from Eq. (13.10):

$$l_s^2 + m_s^2 + n_s^2 = 1 \quad \checkmark$$

The unit vector in the direction of τ w.r.t. x, y and z is then

$$\mathbf{u}_s = l_s\mathbf{u}_x + m_s\mathbf{u}_y + n_s\mathbf{u}_z = 0.57\mathbf{u}_x + 0.526\mathbf{u}_y - 0.623\mathbf{u}_z.$$

The principal stresses are found from the determinant

$$\det \begin{vmatrix} 6 - \sigma & 2 & 2 \\ 2 & 0 - \sigma & 4 \\ 2 & 4 & 0 - \sigma \end{vmatrix} = 0.$$

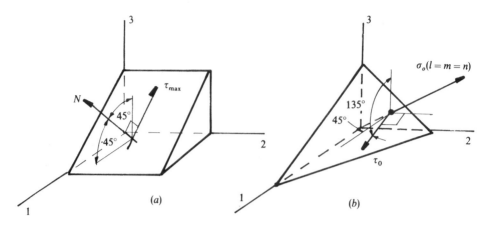

Figure 13.26 Principal, maximum shear and octahedral planes.

Expanding leads to:

$$(6 - \sigma)[(0 - \sigma)^2 - 16] - 2[2(0 - \sigma) - 8] + 2[8 - 2(0 - \sigma)] = 0,$$
$$(6 - \sigma)(\sigma - 4)(\sigma + 4) + 4(\sigma + 4) + 4(\sigma + 4) = 0,$$
$$(\sigma + 4)(10\sigma - \sigma^2 - 16) = (\sigma + 4)(2 - \sigma)(\sigma - 8) = 0,$$
$$\therefore \quad \sigma_1 = 8, \qquad \sigma_2 = 2 \quad \text{and} \quad \sigma_3 = -4 \, \text{MPa}.$$

The greatest shear stress is, from (Eq. (13.20b)),

$$\tau_{max} = \tfrac{1}{2}(\sigma_1 - \sigma_3) = \tfrac{1}{2}[8 - (-4)] = 6 \, \text{MPa},$$

which acts along the plane defined by the normal N ($l = 1/\sqrt{2}$, $m = 0$, $n = 1/\sqrt{2}$) w.r.t. the principal directions 1, 2 and 3 (Fig. 13.26a).
The normal stress acting on the octahedral plane is found from Eq. (13.21a)

$$\sigma_o = (\sigma_1 + \sigma_2 + \sigma_3)/3 = (8 + 2 - 4)/3 = 2 \, \text{MPa},$$

which acts in the direction of the normal $l = m = n = 1/\sqrt{3}$. The octahedral shear stress is found from Eq. (13.21b)

$$\tau_o = (1/3)\sqrt{[(\sigma_1 - \sigma_2)^2 + (\sigma_2 - \sigma_3)^2 + (\sigma_1 - \sigma_3)^2]}$$
$$= (1/3)\sqrt{[(8 - 2)^2 + (2 + 4)^2 + (8 + 4)^2]} = 4.9 \, \text{MPa},$$

with directions cosines, from Eqs (13.22),

$$l_o = (\sigma_1 - \sigma_o)/\sqrt{3}\tau_o = (8 - 2)/(\sqrt{3} \times 4.9) = 0.707, \qquad \text{(i.e. } 45° \text{ to 1)}$$
$$m_o = (\sigma_2 - \sigma_o)/\sqrt{3}\tau_o = (2 - 2)/(\sqrt{3} \times 4.9) = 0, \qquad \text{(i.e. } 90° \text{ to 2)}$$
$$n_o = (\sigma_3 - \sigma_o)/\sqrt{3}\tau_o = (-4 - 2)/(\sqrt{3} \times 4.9) = -0.707 \qquad \text{(i.e. } 135° \text{ to 3)}$$

These stresses and directions are shown in Fig. 13.26(b).

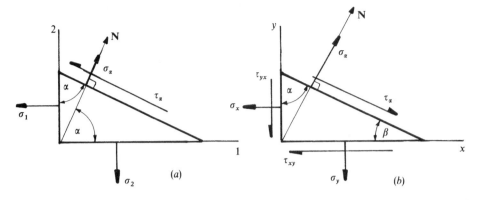

Figure 13.27 Reduction to plane stress states.

Reduction to plane stress The plane stress transformation equations given in Table 13.1 are particular cases of the foregoing 3D equations. For example, with applied principal stresses σ_1 and σ_2 the oblique plane in Fig. 13.27(a) has normal direction cosines; $l = \cos\alpha$, $m = \cos(90° - \alpha) = \sin\alpha$, $n = 0$, w.r.t. 1, 2 and 3. Substituting into Eq. (13.18a and b):

$$\sigma_\alpha = \sigma_1 \cos^2\alpha + \sigma_2 \sin^2\alpha = \sigma_1(1 + \cos 2\alpha)/2 + \sigma_2(1 - \cos 2\alpha)/2$$
$$= \tfrac{1}{2}(\sigma_1 + \sigma_2) + \tfrac{1}{2}(\sigma_1 - \sigma_2)\cos 2\alpha,$$
$$\tau_\alpha^2 = \sigma_1^2 \cos^2\alpha + \sigma_2^2 \sin^2\alpha - (\sigma_1 \cos^2\alpha + \sigma_2 \sin^2\alpha)^2$$
$$= (\sigma_1 - \sigma_2)^2 \sin^2\alpha \cos^2\alpha,$$
$$\tau_\alpha = \tfrac{1}{2}(\sigma_1 - \sigma_2)\sin 2\alpha.$$

In Fig. 13.27(b), with the applied stresses σ_x, σ_y and τ_{xy}, the direction cosines are again $l = \cos\alpha$, $m = \sin\alpha$ and $n = 0$, w.r.t. x, y and z. Substituting into Eq. (13.12a):

$$\sigma_\alpha = \sigma_x \cos^2\alpha + \sigma_y \sin^2\alpha + 2\tau_{xy}\cos\alpha\sin\alpha$$
$$= \sigma_x \cos^2\alpha + \sigma_y \sin^2\alpha + \tau_{xy}\sin 2\alpha,$$
$$\sigma_\alpha = \tfrac{1}{2}(\sigma_x + \sigma_y) + \tfrac{1}{2}(\sigma_x - \sigma_y)\cos 2\alpha + \tau_{xy}\sin 2\alpha,$$

and from Eqs (13.11) and (13.12b),

$$S_x = \sigma_x \cos\alpha + \tau_{xy}\sin\alpha, \qquad S_y = \sigma_y \sin\alpha + \tau_{yx}\cos\alpha \qquad \text{and } S_z = 0,$$
$$\therefore \quad \tau_\alpha^2 = S_x^2 + S_y^2 + S_z^2 - \sigma_\alpha^2$$
$$= (\sigma_x \cos\alpha + \tau_{xy}\sin\alpha)^2 + (\sigma_y \sin\alpha + \tau_{yx}\cos\alpha)^2$$
$$\quad - [\sigma_x \cos^2\alpha + \sigma_y \sin^2\alpha + \tau_{xy}\sin 2\alpha]^2$$
$$= (\sigma_x^2 - \sigma_x\sigma_y + \sigma_y^2)\sin^2\alpha\cos^2\alpha$$
$$\quad + \tau_{xy}\sin 2\alpha[\sigma_x(1 - 2\cos^2\alpha) + \sigma_y(1 - 2\sin^2\alpha)]$$
$$\quad + \tau_{xy}^2(\sin^2\alpha + \cos^2\alpha - \sin^2 2\alpha),$$

$$\tau_\alpha^2 = \tfrac{1}{4}(\sigma_x - \sigma_y)^2 \sin^2 2\alpha - \tau_{xy}(\sigma_x - \sigma_y)\sin 2\alpha \cos 2\alpha + \tau_{xy}^2 \cos^2 2\alpha$$
$$= [\tfrac{1}{2}(\sigma_x - \sigma_y)\sin 2\alpha - \tau_{xy}\cos 2\alpha]^2,$$
$$\tau_\alpha = \tfrac{1}{2}(\sigma_x - \sigma_y)\sin 2\alpha - \tau_{xy}\cos 2\alpha.$$

From Eqs (13.15c–e) the invariants become

$$J_1 = \sigma_x + \sigma_y, \qquad J_2 = \sigma_x \sigma_y - \tau_{xy}^2 \qquad \text{and} \qquad J_3 = 0,$$

and the principal stress cubic Eq. (13.15b) reduces to the quadratic

$$\sigma^2 - (\sigma_x + \sigma_y)\sigma + (\sigma_x \sigma_y - \tau_{xy}^2) = 0,$$

giving the principal stresses (Eq. (13.3)) as roots:

$$\sigma_{1,2} = \tfrac{1}{2}(\sigma_x + \sigma_y) \pm \tfrac{1}{2}\sqrt{[(\sigma_x - \sigma_y)^2 + 4\tau_{xy}^2]}.$$

Geometrical representation When the applied stresses are the principal stresses σ_1, σ_2 and σ_3, a further construction due to Mohr (1914) enables σ

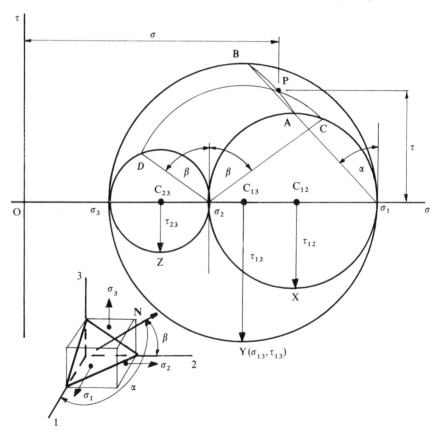

Figure 13.28 Mohr's circles for applied principal stresses.

and τ to be found for a plane whose normals are defined by $\alpha = \cos^{-1} l$, $\beta = \cos^{-1} m$ and $\gamma = \cos^{-1} n$ w.r.t. the 1, 2 and 3 directions. Assuming $\sigma_1 > \sigma_2 > \sigma_3$ are all tensile, Eqs (13.10 and 13.12) combine graphically as shown in Fig. 13.28. In constructing the circle the following steps apply.

1. Erect σ- and τ-axes.
2. Fix points σ_1, σ_2 and σ_3 to scale from the origin along the σ axis.
3. With centres C_{12}, C_{23} and C_{13} draw circles of radii τ_{12}, τ_{23} and τ_{13} respectively.
4. Draw the line $\sigma_1 AB$ inclined at α from the vertical through σ_1.
5. Draw the lines $\sigma_2 D$ and $\sigma_2 C$ with inclinations of β on either side of the vertical through σ_2.
6. With centres C_{13} and C_{23} draw the arcs DC and AB.
7. The intersection point P has coordinates σ and τ as shown.

Note that for all planes, point P will lie in either of the outer regions. Points X, Y and Z represent the state of stress for the maximum shear planes. Point Y gives the greatest shear stress:

$$\tau_{13} = \tfrac{1}{2}(\sigma_1 - \sigma_3) \qquad \text{with} \qquad \sigma_{13} = \tfrac{1}{2}(\sigma_1 + \sigma_3).$$

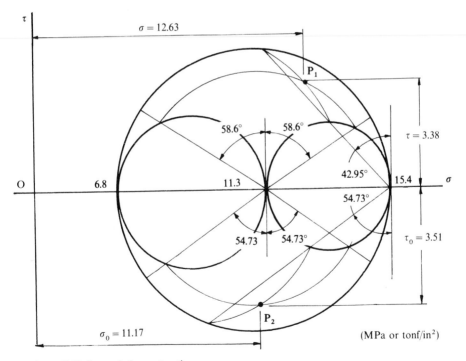

Figure 13.29 Stress circle construction.

Example 13.13 Given the principal applied stress system, $\sigma_1 = 15.4$, $\sigma_2 = 11.3$ and $\sigma_3 = 6.8$ (MPa or tonf/in²), determine graphically the normal and shear stresses for a plane whose direction cosines are $l = 0.732$ and $m = 0.521$. Find also the stress state for the octahedral plane. Check the answers numerically.

In SI Units With the required angles $\alpha = \cos^{-1} 0.732 = 42.95°$ and $\beta = \cos^{-1} 0.521 = 58.6°$, the scaled construction in Fig. 13.29 gives $\sigma = 12.63$, $\tau = 3.38$ MPa. The resultant stress $S = \sqrt{(\sigma^2 + \tau^2)}$ is given by the length of $OP_1 = 13.15$ MPa. The octahedral plane is defined by $\alpha = \beta = 54.73°$, which, when applied to Fig. 13.29, gives another intersection point P_2 with coordinates; $\sigma_o = 11.17$, $\tau_o = 3.51$ MPa. The greatest shear stress $\tau_{13} = 4.3$ MPa is the radius of the largest circle. Checking from Eqs (13.18) and (13.21):

$$n = \sqrt{[1 - (l^2 + m^2)]} = \sqrt{[1 - (0.732^2 + 0.521^2)]} = 0.439,$$

$$\sigma = \sigma_1 l^2 + \sigma_2 m^2 + \sigma_3 n^2$$
$$= 15.4(0.732)^2 + 11.3(0.521)^2 + 6.8(0.439)^2 = 12.63 \text{ MPa},$$

$$\tau^2 = (\sigma_1 l)^2 + (\sigma_2 m)^2 + (\sigma_3 n)^2 - \sigma^2$$
$$= (15.4 \times 0.732)^2 + (11.3 \times 0.521)^2 + (6.8 \times 0.439)^2 - 12.63^2$$
$$= 170.98 - 159.57,$$

$$\tau = 3.38 \text{ MPa},$$

$$\sigma_o = (\sigma_1 + \sigma_2 + \sigma_3)/3$$
$$= (15.4 + 11.3 + 6.8)/3 = 11.17 \text{ MPa},$$

$$\tau_o = (1/3)\sqrt{[(\sigma_1 - \sigma_2)^2 + (\sigma_2 - \sigma_3)^2 + (\sigma_1 - \sigma_3)^2]}$$
$$= (1/3)\sqrt{[(15.4 - 11.3)^2 + (11.3 - 6.8)^2 + (15.4 - 6.8)^2]} = 3.51 \text{ MPa},$$

$$\tau_{max} = \tau_{13} = \tfrac{1}{2}(\sigma_1 - \sigma_3) = \tfrac{1}{2}(15.4 - 6.8) = 4.3 \text{ MPa}.$$

In Imperial Units The same solutions apply to imperial units when the SI stress unit of MPa is here replaced by tonf/in².

General Matrix and Tensor Transformations for Stress

The matrix equation (13.5a), $S' = LSL^T$, will also apply to the transformation of a 3×3 stress matrix S in axes x_1, x_2 and x_3 (Fig. 13.30(a)) to a 3×3 matrix S' in axes x_1', x_2' and x_3' (Fig. 13.30(b)). Note that here x_1, x_2, and x_3 can be either the generalized coordinates x, y and z or the principal coordinates $1, 2$ and 3. The components l_{ij} $(i, j = 1, 2, 3)$ of the matrix L are the directions cosines defining the primed direction w.r.t. the unprimed direction. That is, $l_{ij} = \cos(x_i' x_j)$ where, for example, $l_{11} = \cos(x_1' x_1)$, $l_{12} = \cos(x_1' x_2)$, $l_{13} = \cos(x_1' x_3)$ define the direction cosines for x_1' w.r.t. x_1, x_2 and x_3 as shown.

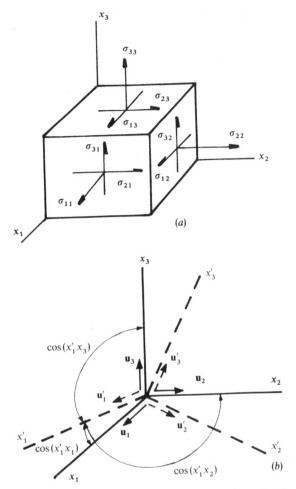

Figure 13.30 Generalized stress components with rotation in orthogonal axes.

Alternatively, from Eqs (13.9b and c), these direction cosines are contained within the unit vectors expressions for the x'_1, x'_2 and x'_3 directions:

$$\mathbf{u}'_1 = l_{11}\mathbf{u}_1 + l_{12}\mathbf{u}_2 + l_{13}\mathbf{u}_3,$$
$$\mathbf{u}'_2 = l_{21}\mathbf{u}_1 + l_{22}\mathbf{u}_2 + l_{23}\mathbf{u}_3,$$
$$\mathbf{u}'_3 = l_{31}\mathbf{u}_1 + l_{32}\mathbf{u}_2 + l_{33}\mathbf{u}_3.$$

The proof of $\mathbf{S}' = \mathbf{LSL}^T$ employs the general transformation law that defines any tensor of second order, of which stress is a member. That is, in double subscript tensor notation:

$$\sigma'_{ij} = l_{ip}l_{jq}\sigma_{pq}. \tag{13.23a}$$

Before the conversion of Eq. (13.23a) to matrix form can be made, similar subscripts must appear adjacent within each term to become consistent with matrix multiplication:

$$\sigma'_{ij} = l_{ip} l_{jq} \sigma_{pq} = l_{ip} \sigma_{pq} l_{jq} = l_{ip} \sigma_{pq} l^{T}_{qj}, \quad \Rightarrow \mathbf{S}' = \mathbf{LSL}^{T}.$$

In full, this transformation law is written;

$$\begin{bmatrix} \sigma'_{11} & \sigma'_{12} & \sigma'_{13} \\ \sigma'_{21} & \sigma'_{22} & \sigma'_{23} \\ \sigma'_{31} & \sigma'_{32} & \sigma'_{33} \end{bmatrix} = \begin{bmatrix} l_{11} & l_{12} & l_{13} \\ l_{21} & l_{22} & l_{23} \\ l_{31} & l_{32} & l_{33} \end{bmatrix} \begin{bmatrix} \sigma_{11} & \sigma_{12} & \sigma_{13} \\ \sigma_{21} & \sigma_{22} & \sigma_{23} \\ \sigma_{31} & \sigma_{32} & \sigma_{33} \end{bmatrix} \begin{bmatrix} l_{11} & l_{21} & l_{31} \\ l_{12} & l_{22} & l_{32} \\ l_{13} & l_{23} & l_{33} \end{bmatrix}.$$

$$(13.23b)$$

The result is that the normal and shear stresses are found for three orthogonal planes in the x'_i ($i = 1, 2, 3$) frame. In the analytical method, the stress state for a single oblique plane ABC (Fig. 13.20) was found. We can identify ABC with the plane lying normal to x'_1 (say) with direction cosines l_{11}, l_{12} and l_{13}. The stress components associated with this plane are $\sigma'_{11}, \sigma'_{12}$ and σ'_{13}. In the anlytical solution, the normal and shear stress referred to in Eq. (13.12) become $\sigma = \sigma'_{11}$ and $\tau = \sqrt{[(\sigma'_{12})^2 + (\sigma'_{13})^2]}$. Clearly, τ is the resultant shear stress acting on plane ABC for which $\sigma'_{12}, \sigma'_{13}$ are its components.

A *deviatoric* stress tensor σ^{D}_{ij} is the remaining part of the absolute tensor σ_{ij} after the mean or hydrostatic stress σ_m has been subtracted from its normal stress components (σ_{11}, σ_{22} and σ_{33}). In general:

$$\sigma^{D}_{ij} = \sigma_{ij} - \delta_{ij} \sigma_m \qquad (13.24)$$

where, in order to ensure that the shear stress components remain unaffected, the Kronecker delta δ_{ij} is unity for $i = j$ and zero for $i \neq j$. For example, Eq. (13.24) gives with $i = 1$ and $j = 1, 2$ and 3,

$$\sigma^{D}_{11} = \sigma_{11} - \sigma_m = \sigma_{11} - (\sigma_{11} + \sigma_{22} + \sigma_{33})/3,$$
$$\sigma^{D}_{12} = \sigma_{12}, \qquad \sigma^{D}_{13} = \sigma_{13}.$$

It follows that the resulting tensor σ^{D}_{ij} in Eq. (13.24) is composed of $\sigma^{D}_{11}, \sigma^{D}_{22}$ and σ^{D}_{33} and the original shear stresses σ_{12}, σ_{13} and σ_{23}. When σ^{D}_{ij} is expressed in a functional form $f(\sigma^{D}_{ij}) = $ constant, known as a yield criterion, it governs the inception of plasticity under any multiaxial stress state (see Chapter 16). One such function is the expression (13.21b or c) for τ_o.

Example 13.14 Given the following matrix **S** of stress components

$$\mathbf{S} = \begin{bmatrix} 1 & 5 & -5 \\ 5 & 0 & 0 \\ -5 & 0 & -1 \end{bmatrix},$$

determine **S'** when x'_1 and x'_2 are aligned with the vectors $\mathbf{A} = \mathbf{u}_1 + 2\mathbf{u}_2 + 3\mathbf{u}_3$ and $\mathbf{B} = \mathbf{u}_1 + \mathbf{u}_2 - \mathbf{u}_3$.

Divide each vector equation by its respective magnitude $|\mathbf{A}| = \sqrt{14}$, $|\mathbf{B}| = \sqrt{3}$ to give the unit vectors:

$$\mathbf{u}'_1 = (1/\sqrt{14})\mathbf{u}_1 + (2/\sqrt{14})\mathbf{u}_2 + (3/\sqrt{14})\mathbf{u}_3,$$
$$\mathbf{u}'_2 = (1/\sqrt{3})\mathbf{u}_1 + (1/\sqrt{3})\mathbf{u}_2 - (1/\sqrt{3})\mathbf{u}_3,$$

when from Eq. (13.9b),

$$l_{11} = 1/\sqrt{14}, \qquad l_{12} = 2/\sqrt{14}, \qquad l_{13} = 3/\sqrt{14},$$
$$l_{21} = 1/\sqrt{3}, \qquad l_{22} = 1/\sqrt{3}, \qquad l_{23} = -1/\sqrt{3}.$$

The direction cosines for the third orthogonal direction x'_3 are found from the cross-product. Thus, if a vector $\mathbf{C}$ lies in x'_3 then, by definition,

$$\mathbf{C} = \mathbf{A} \times \mathbf{B} = \begin{bmatrix} \mathbf{u}_1 & \mathbf{u}_2 & \mathbf{u}_3 \\ A_1 & A_2 & A_3 \\ B_1 & B_2 & B_3 \end{bmatrix} = \begin{bmatrix} \mathbf{u}_1 & \mathbf{u}_2 & \mathbf{u}_3 \\ 1 & 2 & 3 \\ 1 & 1 & -1 \end{bmatrix},$$

$$\mathbf{C} = (-2-3)\mathbf{u}_1 - (-1-3)\mathbf{u}_2 + (1-2)\mathbf{u}_3 = -5\mathbf{u}_1 + 4\mathbf{u}_2 - \mathbf{u}_3,$$

$$\therefore \quad \mathbf{u}'_3 = \mathbf{C}/|\mathbf{C}| = (-5/\sqrt{42})\mathbf{u}_1 + (4/\sqrt{42})\mathbf{u}_2 - (1/\sqrt{42})\mathbf{u}_3,$$

$$\therefore \quad l_{31} = -5/\sqrt{42}, \qquad l_{32} = 4/\sqrt{42}, \qquad l_{33} = -1/\sqrt{42}.$$

Substituting into Eq. (13.23b):

$$S' = \begin{bmatrix} 1/\sqrt{14} & 2/\sqrt{14} & 3/\sqrt{14} \\ 1/\sqrt{3} & 1/\sqrt{3} & -1/\sqrt{3} \\ -5/\sqrt{42} & 4/\sqrt{42} & -1/\sqrt{42} \end{bmatrix} \begin{bmatrix} 1 & 5 & -5 \\ 5 & 0 & 0 \\ -5 & 0 & -1 \end{bmatrix} \begin{bmatrix} 1/\sqrt{14} & 1/\sqrt{3} & -5/\sqrt{42} \\ 2/\sqrt{14} & 1/\sqrt{3} & 4/\sqrt{42} \\ 3/\sqrt{14} & -1/\sqrt{3} & -1/\sqrt{42} \end{bmatrix}$$

$$S' = \begin{bmatrix} 1/\sqrt{14} & 2/\sqrt{14} & 3/\sqrt{14} \\ 1/\sqrt{3} & 1/\sqrt{3} & -1/\sqrt{3} \\ -5/\sqrt{42} & 4/\sqrt{42} & -1/\sqrt{42} \end{bmatrix} \begin{bmatrix} -4/\sqrt{14} & 11/\sqrt{3} & 20/\sqrt{42} \\ 5/\sqrt{14} & 5/\sqrt{3} & -25/\sqrt{42} \\ -8/\sqrt{14} & -4/\sqrt{3} & 26/\sqrt{42} \end{bmatrix}$$

$$= \begin{bmatrix} -1.286 & 1.389 & 1.980 \\ 1.389 & 6.667 & -2.762 \\ 1.980 & -2.762 & -5.381 \end{bmatrix}$$

13.4 THREE-DIMENSIONAL STRAIN TRANSFORMATION

In the analysis of the distortion produced by the stress components σ_x, σ_y, σ_z, τ_{xy}, τ_{xz} and τ_{yz} in Fig. 13.20(a), there are two effects: (i) direct strain ϵ_x, ϵ_y and ϵ_z from σ_x, σ_y and σ_z, and (ii) angular distortions e_{xy}, e_{xz} and e_{yz}. These distortions are composed of shear strain due to τ_{xy}, τ_{xz} and τ_{yz} and rigid body rotations due to the differences in direct strains. It is necessary to subtract rotations from the angular distortions to establish the shear strain that is responsible for shape change, i.e. the angular change in the right angle.

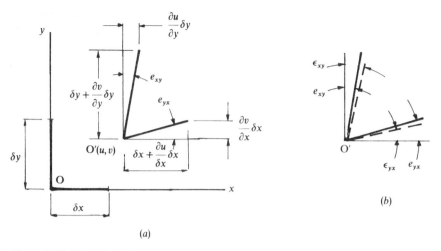

(a)

(b)

Figure 13.31 Distortion of a corner in the x–y plane.

In Fig. 13.31(a), for example, the angular distortions e_{xy} and e_{yx} for one corner of an element $\delta x \times \delta y \times \delta z$ in the x, y plane is shown. With the simultaneous presence of direct strains ϵ_x and ϵ_y, the corner point O is displaced to O' in the x- and y-directions by amounts u and v that are functions of the coordinates: $u = u(x, y, z)$ and $v = v(x, y, z)$. Now δx and δy change their lengths in proportion to the displacement gradients, i.e. the direct strains,

$$\epsilon_x = \partial u/\partial x \qquad \text{and} \qquad \epsilon_y = \partial v/\partial y. \qquad (13.25\text{a, b})$$

Because it has not been stipulated that $\delta x = \delta y$ and $\epsilon_x = \epsilon_y$ then, in general $e_{xy} \neq e_{yx}$ where $e_{xy} = \partial u/\partial y$ and $e_{yx} = \partial v/\partial x$. The engineering shear strain is defined as the total change in the right angle,

$$\gamma_{xy} = \partial u/\partial y + \partial v/\partial x. \qquad (13.25\text{c})$$

Note that, for plane deformation, Eqs (13.7) may be derived directly from Eqs (13.25) by chain-rule differentiation with displacement functions of the form given in Eqs (1.2). Rotating the distorted corner so that it becomes equally inclined to the x- and y-directions (Fig. 13.31b), the tensor shear strains become

$$\epsilon_{xy} = \epsilon_{yx} = \gamma_{xy}/2 = (1/2)(\partial u/\partial y + \partial v/\partial x)$$

and the corresponding rotations are

$$\omega_{xy} = e_{xy} - \epsilon_{xy} = \partial u/\partial y - (1/2)(\partial u/\partial y + \partial v/\partial x) = (1/2)(\partial u/\partial y - \partial v/\partial x)$$
$$\omega_{yx} = e_{yx} - \epsilon_{yx} = \partial v/\partial x - (1/2)(\partial u/\partial y + \partial v/\partial x) = -(1/2)(\partial u/\partial y - \partial v/\partial x).$$

With an additional displacement function $w = w(x, y, z)$ for the z-direction, further strain–displacement relationships, similar to Eqs (13.25), apply to normal and shear distortion in the xz and yz planes. Consequently, the complete distortion of an element $\delta x \times \delta y \times \delta z$ may be expressed as the sum of corresponding strain and rotation tensors in the following form:

$$e_{ij} = \epsilon_{ij} + \omega_{ij} \quad (\text{for } i, j = x, y, z), \tag{13.26a}$$

or, by the corresponding matrix summation:

$$\begin{bmatrix} e_x & e_{xy} & e_{xz} \\ e_{yx} & e_y & e_{yz} \\ e_{zx} & e_{zy} & e_z \end{bmatrix} = \begin{bmatrix} \epsilon_x & \epsilon_{xy} & \epsilon_{xz} \\ \epsilon_{yx} & \epsilon_y & \epsilon_{yz} \\ \epsilon_{zx} & \epsilon_{zy} & \epsilon_z \end{bmatrix} + \begin{bmatrix} \omega_x & \omega_{xy} & \omega_{xz} \\ \omega_{yx} & \omega_y & \omega_{yz} \\ \omega_{zx} & \omega_{zy} & \omega_z \end{bmatrix}, \tag{13.26b}$$

where $\omega_x = \omega_y = \omega_z = 0$. When the strain and rotation components for these matrices are expressed in terms of their displacement rates from Eqs (13.25), then Eq. (13.26b) becomes

$$\begin{bmatrix} \dfrac{\partial u}{\partial x} & \dfrac{\partial u}{\partial y} & \dfrac{\partial u}{\partial z} \\[2ex] \dfrac{\partial v}{\partial x} & \dfrac{\partial v}{\partial y} & \dfrac{\partial v}{\partial z} \\[2ex] \dfrac{\partial w}{\partial x} & \dfrac{\partial w}{\partial y} & \dfrac{\partial w}{\partial z} \end{bmatrix} = \begin{bmatrix} \dfrac{\partial u}{\partial x} & \dfrac{1}{2}\left(\dfrac{\partial u}{\partial y}+\dfrac{\partial v}{\partial x}\right) & \dfrac{1}{2}\left(\dfrac{\partial u}{\partial z}+\dfrac{\partial w}{\partial x}\right) \\[2ex] \dfrac{1}{2}\left(\dfrac{\partial u}{\partial y}+\dfrac{\partial v}{\partial x}\right) & \dfrac{\partial v}{\partial y} & \dfrac{1}{2}\left(\dfrac{\partial v}{\partial z}+\dfrac{\partial w}{\partial y}\right) \\[2ex] \dfrac{1}{2}\left(\dfrac{\partial u}{\partial z}+\dfrac{\partial w}{\partial x}\right) & \dfrac{1}{2}\left(\dfrac{\partial v}{\partial z}+\dfrac{\partial w}{\partial y}\right) & \dfrac{\partial w}{\partial z} \end{bmatrix}$$

$$+ \begin{bmatrix} 0 & \dfrac{1}{2}\left(\dfrac{\partial u}{\partial y}-\dfrac{\partial v}{\partial x}\right) & \dfrac{1}{2}\left(\dfrac{\partial u}{\partial z}-\dfrac{\partial w}{\partial x}\right) \\[2ex] -\dfrac{1}{2}\left(\dfrac{\partial u}{\partial y}-\dfrac{\partial v}{\partial x}\right) & 0 & \dfrac{1}{2}\left(\dfrac{\partial v}{\partial z}-\dfrac{\partial w}{\partial y}\right) \\[2ex] -\dfrac{1}{2}\left(\dfrac{\partial u}{\partial z}-\dfrac{\partial w}{\partial x}\right) & -\dfrac{1}{2}\left(\dfrac{\partial v}{\partial z}-\dfrac{\partial w}{\partial y}\right) & 0 \end{bmatrix}.$$

This shows that the strain matrix is symmetric, i.e. $\epsilon_{ij} = \epsilon_{ji}$ and the rotation matrix is skew symmetric, i.e. $\omega_{ij} = -\omega_{ji}$. Thus, it is deduced that the transformation properties of strain ϵ_{ij}, are identical to those of stress, since the latter was also a symmetric 3×3 matrix ($\sigma_{ij} = \sigma_{ji}$). The strain transformation equivalent to Eq. (13.23b) is then

$$\epsilon'_{ij} = l_{ip}l_{jq}\epsilon_{pq}, \quad \Rightarrow \mathbf{E}' = \mathbf{LEL}^{\mathsf{T}}. \tag{13.27}$$

The 3D strain expressions are identical in form to the corresponding stress expressions provided τ is identified with the tensor definition of shear strain ($\gamma/2$); e.g. the normal strain on an oblique plane in terms of coordinate strains ϵ_{ij} follows from converting Eq. (13.12a):

$$\epsilon = l^2\epsilon_x + m^2\epsilon_y + n^2\epsilon_z + 2(lm\epsilon_{xy} + mn\epsilon_{yz} + ln\epsilon_{xz}), \tag{13.28a}$$

$$= l^2\epsilon_x + m^2\epsilon_y + n^2\epsilon_z + lm\gamma_{xy} + mn\gamma_{yz} + ln\gamma_{zx}. \tag{13.28b}$$

The principal strain cubic may be deduced from Eq. (13.15):

$$\epsilon^3 - I_1\epsilon^2 + I_2\epsilon - I_3 = 0, \tag{13.29}$$

where the strain invariants are

$$I_1 = \epsilon_x + \epsilon_y + \epsilon_z = \epsilon_{ii},$$
$$I_2 = \epsilon_x\epsilon_y + \epsilon_y\epsilon_z + \epsilon_z\epsilon_x - \epsilon_{xy}^2 - \epsilon_{yz}^2 - \epsilon_{zx}^2 = \tfrac{1}{2}(\epsilon_{ii}\epsilon_{jj} - \epsilon_{ij}\epsilon_{ji}),$$
$$I_3 = \epsilon_x\epsilon_y\epsilon_z + 2\epsilon_{xy}\epsilon_{yz}\epsilon_{zx} - \epsilon_x\epsilon_{yz}^2 - \epsilon_y\epsilon_{zx}^2 - \epsilon_z\epsilon_{xy}^2$$
$$= \epsilon_x\epsilon_y\epsilon_z + \tfrac{1}{4}\gamma_{xy}\gamma_{yz}\gamma_{zx} - \tfrac{1}{4}\epsilon_x\gamma_{yz}^2 - \tfrac{1}{4}\epsilon_y\gamma_{zx}^2 - \tfrac{1}{4}\epsilon_z\gamma_{xy}^2 = \det[\epsilon_{ij}].$$

The exception to direct conversion from stress applies to generalized shear strain, where perpendicular directions must be specified. Any one tensor shear strain component in Eq. (13.27) (i.e. with $i \neq j$) provides the simplest method of finding the shear strain between given perpendicular lines. With $i = 1$ and $j = 2$, Eq. (13.27) becomes $\epsilon'_{12} = l_{1p}l_{2q}\epsilon_{pq}$. This is equivalent to the abbreviated matrix multiplication:

$$\gamma/2 = \begin{bmatrix} l_1 & m_1 & n_1 \end{bmatrix} \begin{bmatrix} \epsilon_x & \epsilon_{xy} & \epsilon_{xz} \\ \epsilon_{xy} & \epsilon_y & \epsilon_{yz} \\ \epsilon_{xz} & \epsilon_{yz} & \epsilon_z \end{bmatrix} \begin{Bmatrix} l_2 \\ m_2 \\ n_2 \end{Bmatrix}, \tag{13.30a}$$

$$\gamma/2 = [(l_1\epsilon_x + m_1\epsilon_{xy} + n_1\epsilon_{xz})(l_1\epsilon_{xy} + m_1\epsilon_y + n_1\epsilon_{yz})(l_1\epsilon_{xz} + m_1\epsilon_{yz} + n_1\epsilon_z)] \begin{Bmatrix} l_2 \\ m_2 \\ n_2 \end{Bmatrix}$$

$$= (l_1\epsilon_x + m_1\epsilon_{xy} + n_1\epsilon_{xz})l_2 + (l_1\epsilon_{xy} + m_1\epsilon_y + n_1\epsilon_{yz})m_2$$
$$+ (l_1\epsilon_{xz} + m_1\epsilon_{yz} + n_1\epsilon_z)n_2$$
$$= l_1l_2\epsilon_x + m_1m_2\epsilon_y + n_1n_2\epsilon_z + (l_1m_2 + l_2m_1)\epsilon_{xy}$$
$$+ (m_1n_2 + m_2n_1)\epsilon_{yz} + (l_1n_2 + l_2n_1)\epsilon_{xz}.$$

Alternatively, the generalized shear strain (Eq. (13.30a)) is given as

$$\gamma/2 = \mathbf{u}'_1 \cdot \mathbf{E} \cdot \mathbf{u}'_2 \tag{13.30b}$$

where $\mathbf{u}'_1$ and $\mathbf{u}'_2$ are the unit vectors for the given pair of orthogonal directions 1′ and 2′. These are expressed from Eq. (13.9c) as

$$\mathbf{u}'_1 = l_1\mathbf{u}_x + m_1\mathbf{u}_y + n_1\mathbf{u}_z \quad \text{and} \quad \mathbf{u}'_2 = l_2\mathbf{u}_x + m_2\mathbf{u}_y + n_2\mathbf{u}_z$$

where l_1, m_1, n_1 defines direction 1′ and l_2, m_2, n_2 defines direction 2′ w.r.t. x, y and z. The direct conversion from τ in the stress equations (13.12b) to $\gamma/2$ is equivalent to Eq. (13.30) when l_2, m_2 and n_2 are identified with the direction cosines l_s, m_s and n_s in Eqs (13.13) and l_1, m_1 and n_1 define the normal to the oblique plane. That is, Eqs (13.30) then give the tensor shear strain between perpendicular directions originally aligned with the normal and the resultant shear stress on an oblique plane.

Because of the identical nature of stress and strain transformations, the directions of the principal stress and strain will be coincident. Moreover, geometrical similarity exists between Mohr's circles for stress and strain. Thus, with known principal applied strains $\epsilon_1 > \epsilon_2 > \epsilon_3$, an identical construction to Fig. 13.28 will apply with an ordinate of $\gamma/2$ replacing τ and abscissa ϵ replacing σ.

Example 13.15 Find the principal strains and their directions for the following plane strain state.

$$\frac{\epsilon_{ij}}{10^{-4}} = \begin{bmatrix} 65 & 33 & 0 \\ 33 & -73 & 0 \\ 0 & 0 & 4 \end{bmatrix}$$

Show that these directions are orthogonal.

Substituting $\epsilon_x = 65 \times 10^{-4}$, $\epsilon_y = -73 \times 10^{-4}$, $\epsilon_z = 4 \times 10^{-4}$, $\epsilon_{xy} = \epsilon_{yx} = 33 \times 10^{-4}$ and $\epsilon_{xz} = \epsilon_{yz} = 0$ into Eq. (13.29) leads to the principal strain cubic:

$$\epsilon^3 + (4 \times 10^{-4})\epsilon^2 - (5049.3 \times 10^{-8})\epsilon + (20\,069.2 \times 10^{-12}) = 0. \quad \text{(i)}$$

For the plane strain condition, the absence of shear strain ϵ_{xz} and ϵ_{yz} is a consequence of the absence of associated shear stresses τ_{xz} and τ_{yz}. Thus, the normal strain $\epsilon_z = 4 \times 10^{-4}$ is a principal strain and hence a root of Eq. (i). It follows from Eq. (i) that

$$[\epsilon - (4 \times 10^{-4})](a\epsilon^2 + b\epsilon + c) = \epsilon^3 + (4 \times 10^{-4})\epsilon^2 - (5049.3 \times 10^{-8})\epsilon$$
$$+ (20\,069.2 \times 10^{-12}).$$

Equating coefficients:

$[\epsilon^3]$; $a = 1$,
$[\epsilon^2]$; $b - (4 \times 10^{-4}) = 4 \times 10^{-4}$, $\Rightarrow b = 8 \times 10^{-4}$,
$[\epsilon]$; $c - (4 \times 10^{-4})b = -5049.3 \times 10^{-8}$, $\Rightarrow c = -5017.3 \times 10^{-8}$.

Then the remaining two principal strains are roots of the quadratic

$$\epsilon^2 + (8 \times 10^{-4})\epsilon - (5017.3 \times 10^{-8}) = 0,$$

giving $\epsilon = -75 \times 10^{-4}$ and 67×10^{-4}. The principal strains $\epsilon_1 > \epsilon_2 > \epsilon_3$ are then

$$\epsilon_1 = 67 \times 10^{-4}, \qquad \epsilon_2 = 4 \times 10^{-4} \qquad \text{and} \qquad \epsilon_3 = -75 \times 10^{-4}.$$

Since the direction of ϵ_2 ($= \epsilon_z$) is parallel to the z-direction, it follows that the direction cosines are $l_2 = 0$, $m_2 = 0$ and $n_2 = 1$. The strain equivalent to Eq. (13.14a), together with Eq. (13.10), enables calculation of the direction cosines for the remaining two directions. That is,

$$2l(\epsilon_x - \epsilon) + m\epsilon_{xy} + n\epsilon_{xz} = 0,$$
$$l\epsilon_{xy} + 2m(\epsilon_y - \epsilon) + n\epsilon_{yz} = 0, \qquad \text{(ii)}$$
$$l\epsilon_{zx} + m\epsilon_{zy} + 2n(\epsilon_z - \epsilon) = 0,$$

of which only two equations are independent. Then for $\epsilon = 67 \times 10^{-4}$ and the given strain components, Eq. (ii) gives the simultaneous equations

$$-4l_1 + 33m_1 = 0, \qquad 33l_1 - 280m_1 = 0 \qquad \text{and} \qquad -126n_1 = 0.$$

Thus, $n_1 = 0$ and $l_1 = 8.25m_1$. Since $l_1^2 + m_1^2 + n_1^2 = 1$, this gives direction cosines for the major principal strain:

$$l_1 = 0.993, \qquad m_1 = 0.120, \qquad n_1 = 0. \qquad \text{(iii)}$$

For $\epsilon = -75 \times 10^{-4}$, Eq. (ii) gives

$$280l_3 + 33m_3 = 0, \qquad 33l_3 + 4m_3 = 0 \qquad \text{and} \qquad 158n_3 = 0.$$

Thus, $n_3 = 0$ and $m_3 = -(8.2513)l_3$. Substituting into $l_3^2 + m_3^2 + n_3^2 = 1$ gives direction cosines for the minor principal strain of

$$l_3 = 0.120, \qquad m_3 = -0.93, \qquad n_3 = 0, \qquad \text{(iv)}$$

and substituting Eqs (iii) and (iv) into Eqs (13.16) shows that the principal directions are orthogonal. Alternatively, the dot product of any two unit vectors in the 1, 2 and 3 directions will be zero. It follows from Eq. (13.9c) that the unit vector equations are

$$\mathbf{u}_1 = 0.993\mathbf{u}_x + 0.120\mathbf{u}_y, \qquad \mathbf{u}_2 = \mathbf{u}_z \qquad \text{and} \qquad \mathbf{u}_3 = 0.120\mathbf{u}_x - 0.993\mathbf{u}_y.$$

Clearly, $\mathbf{u}_1 \cdot \mathbf{u}_2 = \mathbf{u}_1 \cdot \mathbf{u}_3 = \mathbf{u}_2 \cdot \mathbf{u}_3 = 0$.

Example 13.16 Find for the given strain tensor:

$$\mu\epsilon_{ij} = \begin{bmatrix} 100 & 100 & -100 \\ 100 & 200 & 200 \\ -100 & 200 & 200 \end{bmatrix}$$

$$(1\mu = 1 \times 10^{-6}),$$

(a) the normal strain in a direction defined by $l = 2/3$, $m = 1/3$ and $n = 2/3$; (b) the shear strain between the normal direction and a perpendicular direction defined by $l = -0.25$, $m = 0.942$ and $n = -0.221$; (c) the principal strains; (d) the greatest shear strain; (e) the octahedral normal and shear strains; and (f) the deviatoric strains.

(a) Substituting $\mu\epsilon_x = 100$, $\mu\epsilon_y = 200$, $\mu\epsilon_z = 200$, $\mu\epsilon_{xy} = 100$, $\mu\epsilon_{xz} = -100$ and $\mu\epsilon_{yz} = 200$ into Eq. (13.28a) gives the normal strain:

$$\epsilon = l^2\epsilon_x + m^2\epsilon_y + n^2\epsilon_z + 2(lm\epsilon_{xy} + mn\epsilon_{yz} + ln\epsilon_{xz}),$$
$$\mu\epsilon = (2/3)^2 100 + (1/3)^2 200 + (2/3)^2 200 + 2[(2/3)(1/3)100 + (1/3)(2/3)200$$
$$+ (2/3)^2(-100)] = 196.7.$$

(b) Substituting into Eq. (13.30a) with $l_1 = 2/3$, $m_1 = 1/3$, $n_1 = 2/3$, $l_2 = -0.25$, $m_2 = 0.942$ and $n_2 = -0.221$ for the given strain components:

$$\mu\gamma/2 = (2/3)(-0.25)100 + (1/3)(0.942)200 + (2/3)(-0.221)200$$
$$+ [(2/3)0.942 + (1/3)(-0.25)]100 + [(1/3)(-0.2213) + (2/3)0.942]200$$
$$+ [(2/3)(-0.2213) + (2/3)(-0.25)](-100) = 213.05.$$

(c) The principal strains may be found from Eq. (13.29) or, more conveniently, for exact roots, from expanding the determinant. That is,

$$\begin{bmatrix} 100-\epsilon & 100 & -100 \\ 100 & 200-\epsilon & 200 \\ -100 & 200 & 200-\epsilon \end{bmatrix} = 0,$$

$$(100-\epsilon)[(200-\epsilon)^2 - 200^2] - 100^2[(200-\epsilon)+200]$$
$$- 100^2[200 + (200-\epsilon)] = 0,$$
$$(\epsilon - 400)(\epsilon - \epsilon^2 + 200) = 0,$$
$$(\epsilon - 400)(\epsilon + 100)(\epsilon - 200) = 0,$$
$$\epsilon_1 = 400\mu, \qquad \epsilon_2 = 200\mu \qquad \text{and} \qquad \epsilon_3 = -100\mu.$$

(d) The direct conversion from Eq. (13.20b) provides the greatest shear strain expression when $\epsilon_1 > \epsilon_2 > \epsilon_3$:

$$\gamma_{\max}/2 = (\epsilon_1 - \epsilon_3)/2,$$
$$\gamma_{\max} = \epsilon_1 - \epsilon_3 = 400 + 100 = 500\mu.$$

(e) Similarly, from the correspondence with the octahedral stress expressions in Eqs (13.21), the mean or hydrostatic strain for the normal direction is

$$\epsilon_o = (\epsilon_1 + \epsilon_2 + \epsilon_3)/3$$
$$= (400 + 200 - 100)/3 = 166.7\mu;$$

and the octahedral shear strain is

$$\gamma_o/2 = (1/3)\sqrt{[(\epsilon_1 - \epsilon_2)^2 + (\epsilon_2 - \epsilon_3)^2 + (\epsilon_1 - \epsilon_3)^2]}$$
$$= (1/3)\sqrt{[(400 - 200)^2 + (200 + 100)^2 + (400 + 100)^2]} = 205.5\mu,$$
$$\therefore \quad \gamma_o = 411\mu,$$

which applies to perpendicular directions aligned with the normal to the octahedral plane and the direction (Eq. 13.22) of τ_o for that plane.

(f) The deviatoric strains are what remains of the normal strain components when the mean or hydrostatic strain ϵ_m, which quantifies elastic dilatation, has been subtracted. The direct correspondence with Eq. (13.24) leads to,

$$\epsilon_{ij}^{\ D} = \epsilon_{ij} - \delta_{ij}\epsilon_m,$$

$$\epsilon_x^{\,D} = \epsilon_x - \epsilon_m = 100 - 166.7 = -66.7\mu,$$

$$\epsilon_y^{\,D} = \epsilon_y - \epsilon_m = 200 - 166.7 = 33.3\mu = \epsilon_z^{\,D}.$$

These, together with the given shear strain (since $\delta_{ij} = 0$ for $i \neq j$), constitute a strain tensor $\epsilon_{ij}^{\,D}$ of six independent deviatoric components that define asymmetrical distortion.

EXERCISES

Plane Stress Transformation

13.1 Working from first principles, find the magnitude and direction of the principal stresses and the maximum shear stress for each element in Fig. 13.32(a)–(d).
Answer: (a) 342.8 MPa, -3.1 MPa, 31.75°, 173 MPa, 76.75°; (b) 171.7 MPa, -48.17 MPa, 38.5°, 64.9 MPa, 83.5°; (c) -102.1 MPa, 86.6 MPa, 27.6°, 94.4 MPa, 72.6°; (d) 197.6 MPa, -135.9 MPa, 11.32°, 166.8 MPa, 56.32°.

13.2 A 30 mm diameter bar is subjected to an axial force of 50 kN. Calculate the normal and tangential stresses on planes that make angles of 8°, 38° and 68° with the bar axis.
Answer: [σ_θ, τ_θ(MPa)] 1.37, 9.8; 26.8, 34.7; 61, 24.5.

13.3 Plane stresses $\sigma_x = 62$ MPa, $\sigma_y = -46$ MPa and $\tau_{xy} = 39$ MPa act at a point. Determine, the magnitude and direction of the maximum shear stress and the major and minor principal stresses.
Answer: 74.2 MPa, -58.8 MPa, 18°; and 77 MPa, 63°.

13.4 Given the applied principal stresses $\sigma_1 = 40$ MPa and $\sigma_2 = -20$ MPa, find σ_θ and θ for a plane on which $\tau_\theta = 15$ MPa.

13.5 At a certain point a tensile stress of 123.5 MPa acts at 90° to a compressive stress of 154.4 MPa in the same plane. If the major principal stress is limited to 208.5 MPa, determine the maximum permissible shear stress that may act in these directions. What is the maximum shear stress and the orientation of the major principal plane?
Answer: 133.9 MPa; 193 MPa; 67°; 22°.

13.6 Given that the major principal tensile stress is 4 MPa and that a normal compressive stress of 1 MPa acts on a plane inclined at 60° to the major principal plane, determine both graphically and analytically, the shear stress on the inclined plane and the minor principal stress.
Answer: 2.9 MPa; -2.7 MPa.

13.7 Determine the principal stresses, the maximum shear stress and their orientations for each of the following general, plane stress systems (MPa):
(a) $\sigma_x = 50$, $\sigma_y = 10$, $\tau_{xy} = 20$;
(b) $\sigma_x = -20$, $\sigma_y = 30$, $\tau_{xy} = 15$;
(c) $\sigma_x = 45$, $\sigma_y = 30$, $\tau_{xy} = 20$;
(d) $\sigma_x = -30$, $\sigma_y = 50$, $\tau_{xy} = 40$.
What for (a) is the state of stress on a plane making 75° to the plane on which σ_x acts, and what in (b) is the angle between the σ_y plane and the plane on which the shear stress is 25 MPa?

13.8 The principal stresses at a point are 30 and 50 kPa in tension. Determine, for a plane inclined at 40° to the major principal plane (a) the normal and shear stresses, and (b) the magnitude and direction of the resultant stress.
Answer: 4.18 kPa; 0.985 kPa; 4.3 kPa; 53.5°.

13.9 Normal stresses, $\sigma_x = 80$ kPa in tension and $\sigma_y = 60$ kPa in compression, are applied to an elastic material. If the major principal stress is limited to 100 kPa what magnitude of shear stress may be applied to the given planes and what is the magnitude of the maximum shear stress?
Answer: 56.6 kPa; 90 kPa.

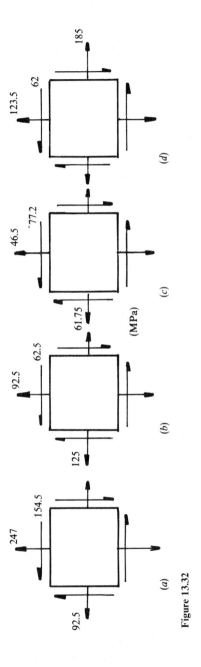

(MPa)

Figure 13.32

13.10 A glass-fibre sheet is loaded on a plane lying at 25° to the fibre axis with combined direct and shear stresses of 15 MPa tensile and 5 MPa respectively. If the allowable working stresses are 60 MPa tension for a plane normal to the fibre axis and 10 MPa in parallel shear, determine the safety factor used in the design.

Answer: 1.12.

13.11 A block of brittle material with rectangular section, 50 mm wide × 100 mm deep, is axially loaded under a compressive force of 71.2 kN. What are the normal and shear stresses on an oblique plane making 60° with the vertical depth? If the ultimate compressive and shear strengths are 50 and 20 MPa respectively, determine the compressive fracture force and the failure planes.

Answer: 10.68 MPa; 6.17 MPa; 200 kN; 45°, 135°.

13.12 A thin-walled steel tube of 150 mm i.d. and thickness 2 mm contains gas at a pressure of 11 bar. Find the tensile and shear stresses on a helical seam inclined at 25° to the cross-section. (1 MPa = 10 bar.)

Answer: 24.4 MPa; 7.92 MPa.

13.13 The prism in Fig. 13.33 has a resultant stress of 77 MPa on face AB and a principal stress of 123.5 MPa on face AC. Determine θ, the state of stress on plane BC and the other principal stress.

13.14 A plane of brittle material, 6.5 mm thick, is subjected to the forces in Fig. 13.34. Find the normal and shear stresses that act on the 30° plane AC. If the ultimate compressive and shear strengths are 30 MPa and 17 MPa respectively, find the planes along which the material would fail when the applied stresses are increased in proportion. State the condition for compressive failure.

Answer: 11 MPa; 9.65 MPa; 45°; 135°; USS > UCS/2.

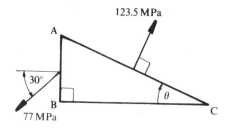

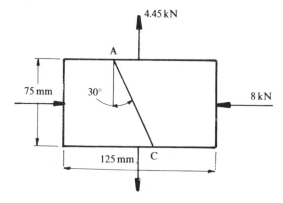

Figure 13.33

Figure 13.34

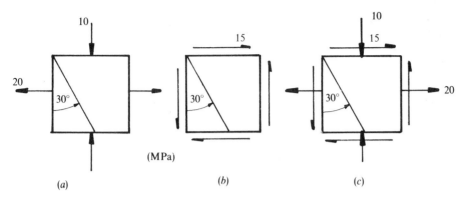

(MPa)

(a)   (b)   (c)

Figure 13.35

13.15 Draw the Mohr's circles for the stress systems (MPa) in Fig. 13.35(a)–(c). Locate the focus and find the normal and shear stress on the 30° planes shown. Determine the principal stresses in (b) and (c) and the planes on which they act. Check your answers numerically.

> *Answer*: (a) 12.5 MPa, 13 MPa; (b) 13 MPa, 7.5 MPa, ± 15°, ± 45°; (c) 25.5 MPa, 5.5 MPa, 26.2 MPa, 22.5°, − 16.2 MPa, 112.5°.

13.16 The states of stress (MPa) for two elements with a common plane are shown in Fig. 13.36. Determine the stress state for this plane and comment on the result.

> *Answer*: 1 and 5 MPa—uniform stress for all parallel planes.

13.17 If, for the rotating shaft in Fig. 13.37, the major principal stress is not to exceed 80 MPa, determine the shaft diameter based upon bending and torsional effects, assuming an encastre bearing support.

> *Answer*: 14.2 mm.

13.18 Calculate θ for the cantilevered arc in Fig. 13.38 in order that the maximum bending and torsional stresses are equal. If, for this condition $W = 10$ N and $d = 10$ mm, find R when the maximum shear stress is limited to 80 MPa.

> *Answer*: 53°, 880 mm.

13.19 A flat, brass plate, stressed in plane perpendicular directions, gave respective extensions of 0.036 mm and 0.0085 mm over a 50 mm length. Find the inclination θ to the length for a plane whose normal stress is 60.4 MPa. Take $E = 82.7$ GPa and $v = 0.3$.

> *Answer*: 30°.

13.20 A cylindrical vessel is formed by welding steel plate along a helical seam inclined at 30° to the cross-section. Determine the seam stress state when the mean diameter and thickness are 1.5 m and 20 mm at a test pressure of 27.5 bar.

> *Answer*: 68.9 MPa; 23.87 MPa.

13.21 At a point on the surface of a shaft the axial tensile stress due to bending is 77 MPa and the maximum shear stress due to torsion is 31 MPa. Determine, graphically, the magnitude and direction of the maximum shear stress and the principal stresses.

13.22 A shaft 100 mm in diameter transmits 375 kW at 300 rev/min while simultaneously supporting an axial thrust of 400 kN (compression). Find graphically the principal stresses induced.

> *Answer*: 38.1; − 87.7 MPa.

13.23 A steel shaft is to transmit 225 kW at 250 rev/min without the major principal stress exceeding 123.5 MPa. If the shaft also carries a bending moment of 3.04 kN m, find the shaft diameter. What is the maximum shear stress and its plane relative to the axis?

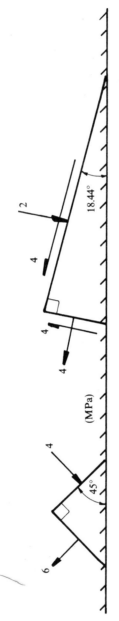

Figure 13.36

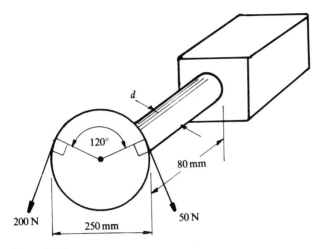

Figure 13.37

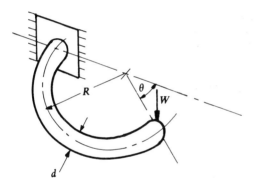

Figure 13.38

13.24 A 250 mm propeller shaft transmits a torque of 162.8 kN m. In addition, stresses of 9.65 MPa in compression and ± 5.5 MPa in bending arise owing to the respective actions of axial thrusts and self-weight. If the allowable direct and shear stresses are 165.4 and 103.35 MPa, determine the safety factor used in the shaft design.

    Answer: 10.9; 13.6.

13.25 A 20 mm diameter shaft is subjected to a bending moment of 25 N m and an axial torque of 15 N m. Determine, graphically for the o.d., the principal stresses and the angles between the maximum shear planes and the shaft axis. Check your answers analytically.

13.26 The bending stress in a tube with diameter ratio of 5:1 is not to exceed 90 MPa under a moment of 80 kN m. Determine the tube diameters. If the moment is replaced by a torque of 80 kN m, find the major principal stress.

    Answer: 130 mm; 650 mm; 45 MPa.

13.27 A steel bar withstands the simultaneous actions of a 30 kN compressive force, a 5 kN m bending moment and a torque of 1 kN m. Determine, for a point at the position of the greatest

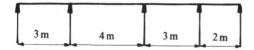

Figure 13.39

compressive bending stress, (a) the normal stress on a plane inclined at 30° to the shaft axis and (b) the shear stress on a plane inclined at 60° to the shaft axis.

13.28 A solid steel propeller shaft 152.5 mm in diameter absorbs 447.5 kW at 180 rev/min in developing a thrust of 1.5 MN. Determine the principal stresses, the maximum shear stress in the shaft and the planes on which they act.

Answer: 12.35 MPa; −94.2 MPa; 53.3 MPa; 19.92°; 64.92°.

13.29 A shaft is required to transmit a torque of 36.4 kN m and a bending moment of 15.2 kN m. Given that the major principal stress must not exceed 92.5 MPa, find the minimum external shaft diameter when the shaft section is (a) solid and (b) hollow, with diameter ratio 0.6.

Answer: 143.8 mm; 90.3 mm.

13.30 Calculate the diameter of a solid steel shaft that is required to transmit 89.5 kW at 200 rev/min if the angle of twist must not exceed 0.13°/m. If the maximum bending stress is 155 MPa, calculate the principal stresses and the orientation of their planes for the compressive side. What is the maximum shear stress induced in the shaft. Take $G = 77.2$ GPa.

Answer: 70 mm; 177.6 MPa; −23.2 MPa; 20°; 110°; 100.4 MPa.

13.31 A fibre-reinforced tube of 240 mm i.d. and 300 mm o.d. has a density of 1 kg/m³. The pipe is simply supported as shown in Fig. 13.39 when carrying fluid of density 12 kg/m³. If the allowable axial and bending stresses in the pipe are 150 and 8 MPa respectively, check that these are nowhere exceeded.

13.32 The bracket in Fig. 13.40 carries a load of 4 kN on a pin passing through the hole. Neglecting the effect of the hole, find that cross-section which is subjected to the greatest bending stress. What is the maximum induced shear stress at this section due to the combined effects of bending and shear?

Answer: 10 mm from wall; 7.3 MPa.

13.33 A 1.5 m long steel cantilever of I-section has flanges 150 mm wide and 25 mm thick. The overall depth is 305 mm and the web is 25 mm thick. If the cantilever carries a uniformly distributed load of 202.5 kN/m, determine, for the fixed end, the principal stresses at the top

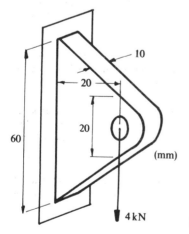

Figure 13.40

and bottom of the web and at the neutral axis of bending. Indicate with sketches the planes on which they act.

13.34 A simply-supported, I-section beam, 4 m long, carries a uniformly distributed load of 65 kN/m. Calculate, for a section of the beam 0.5 m from the LH support, the principal stresses at the neutral axis and at the top of the web on the compressive side. The flanges are 100 mm wide, overall depth 200 mm, with web and flange thicknesses of 7.5 mm.

13.35 The fixed end of an I-section cantilever is subjected to a bending moment of 190 kN m. What is the value of the vertical shear force at this position, given that the major principal tensile stress is 92.3 MPa at the web–flange interface. The flanges are each 152.5 mm wide × 50 mm deep and the web is 25 mm thick × 200 mm deep.

Answer: 312 kN.

13.36 An I-section cantilever is 150 mm long. If the flanges are 50 mm wide × 5 mm thick and the web is 65 mm deep × 5 mm thick, what is the maximum uniformly distributed load the beam can carry when the bending stress is not to exceed 95 MPa? Determine, for the section at which the shear force is a maximum, the principal stresses and planes for (a) the flange top, (b) the web top, and (c) at the neutral axis of bending.

Answer: 164.7 kN/m; 95 MPa; 114 MPa; -31 MPa; 27.5°; ± 77 MPa at $\pm 45°$.

13.37 An I-section simply supported, 3.05 m long beam with flanges 152.5 × 25 mm and web 255 × 12.5 mm, carries a central concentrated load of 300 kN. Find the principal stresses at the top of the tension side of the web at a section 1.22 m from a support.

Answer: 146.8 MPa; -9.8 MPa.

13.38 An I-section beam is 255 mm deep with $I = 50.8 \times 10^6$ mm^4. At a section where the bending moment is 36.4 kN m, the shear force produces shear stresses of 23.3 MPa at the neutral axis and 19.3 MPa at 100 mm from the neutral axis. Find the principal stresses and planes at each point on the tensile side.

Answer: 77.8 MPa; 4.83 MPa; 14.03°; 23.3 MPa; $\pm 45°$.

2D Strain Transformation

13.39 A point in a material is under the strains $\epsilon_x = 400\mu$, $\epsilon_y = 200\mu$ and $\gamma_{xy} = 350\mu$. Determine analytically and graphically the magnitude of the principal strains and the directions of their axes. Also find the state of strain for an axis inclined at 45° to the x-axis.

Answer: 500μ, 30° to x; 100μ, 120° to x; 470μ; 212μ.

13.40 If the strain state at a point is given by $\epsilon_x = -250\mu$, $\epsilon_y = 375\mu$ and $\gamma_{xy} = 400\mu$, calculate the magnitude and direction of the principal strains. Determine the state of strain for perpendicular directions aligned with a clockwise rotation of 50° in the x–y coordinates.

Answer: 433.5μ; -308.5μ; 73.7°; -80μ; 205μ; -685μ.

13.41 In plane strain line elements in the x- and y-directions increase by 17 per cent and 22 per cent respectively and change their right angle by 7°. What is the percentage increase in an element of line originally at 45° to x and y and by how much has this line rotated?

Answer: 27 per cent; 2.4°.

13.42 A strain gauge records 800×10^{-6} when it is bonded in the direction of the major principal strain for a 25 mm diameter shaft under torsion. What will be the change in this strain when a 150 mm length of this bar is simultaneously subjected to an axial force causing a 0.25 mm length change and a 0.0125 mm diameter reduction?

Answer: 1300×10^{-6}.

13.43 A 45° strain gauge rosette measures direct strains of 250μ, 200μ and 350μ for 0°, 45° and 90° directions passing through a point. What is the percentage change in the right angle in radians between the 0° and 90° directions?

Answer: 200μ.

13.44 Normal strains of 600×10^{-6}, -200×10^{-6} and 250×10^{-6} are measured in directions of $0°$, $60°$ and $120°$ anticlockwise w.r.t. the horizontal direction. Find the magnitude and direction of both the principal strains and principal stresses. Take $E = 207\,\text{GPa}$ and $v = 0.3$.

   Answer: 680μ at $-17°$; -246μ at $73°$; $137.7\,\text{MPa}$; $-9.73\,\text{MPa}$.

13.45 A $60°$ strain gauge rosette records direct strains of 500μ, 200μ and -300μ from its three arms at the surface of a loaded structure. Calculate the principal stresses given $E = 100\,\text{GPa}$ and $v = 0.3$ for the structure material.

   Answer: $57\,\text{MPa}$; $-16.6\,\text{MPa}$.

13.46 A $60°$ strain gauge rosette measures direct strains of -50μ, 250μ and 500μ in directions inclined at $0°$, $60°$ and $120°$ to the horizontal x-direction. Determine the complete state of strain for the x- and y-directions.

   Answer: $\epsilon_x = -500\mu$; $\epsilon_y = 515\mu$; $\gamma_{xy} = -289\mu$.

13.47 Three strain gauges A, B and C, with A and C at $30°$ on either side of B, form a rosette that is bonded to a structural member made from an aluminium alloy having a modulus of rigidity $G = 25\,\text{GN/m}^2$. When the structure is loaded, gauges A, B and C read 5×10^{-4}, 1×10^{-4} and -4×10^{-4} respectively. Find the maximum shear stress in the material at the point of application of the rosette. (C.E.I. Pt.2, 1985)

13.48 A strain gauge rosette is bonded to the surface of a 20 mm thick plate. For the three gauges A, B and C: gauge B is at $60°$ to gauge A and gauge C is at $120°$ to gauge A, both measured in the same direction. When the plate is loaded, the strain values from A, B and C are 0.5×10^{-4}, 7×10^{-4} and 3.75×10^{-4} respectively. Assuming that the stress and strain do not vary with the thickness of plate, find the principal stresses and the change in plate thickness at the rosette point. Take $E = 200\,\text{GPa}$ and $v = 0.25$. (C.E.I. Pt.2, 1986)

13.49 A rosette of three strain gauge elements spaced $120°$ apart is bonded to the web of a loaded beam. The elements record tensile strains of 3.22×10^{-4}, 2.18×10^{-4} and 0.21×10^{-4}. Determine the magnitude of the principal strains and stresses and the direction of the maximum principal stress relative to the direction of the maximum strain reading. Take $E = 207\,\text{GPa}$, $v = 0.28$. (C.E.I. Pt.2, 1981)

13.50 A strain gauge rosette is bonded to the cylindrical surface of a 40 mm drive shaft of a motor. There are three gauges A, B and C and the centre-line of gauge A lies at $45°$ to the centre line of the shaft. The angles between gauges A and B, B and C are both $60°$, all angles measured anticlockwise. When the motor runs and transmits constant torque T, the readings of gauges A, B and C are found to be 6.25×10^{-4}, -3.125×10^{-4} and -3.125×10^{-4} respectively. If $E = 208\,\text{GN/m}^2$ and $v = 0.3$, calculate T, the principal stresses and the maximum shear stress in the shaft. (C.E.I. Pt.2, 1984)

13.51 The $60°$ strain gauge rosette in Fig. 13.41 measures strains $\epsilon_a = 4 \times 10^{-4}$, $\epsilon_b = 3 \times 10^{-4}$ and $\epsilon_c = 2 \times 10^{-4}$. Determine the shear strain between the x- and y-directions and the normal strain in the y-direction.

   Answer: $\gamma_{xy} = -1.16 \times 10^{-4}$; $\epsilon_y = 2 \times 10^{-4}$.

13.52 A strain gauge rosette consisting of three gauges a, b and c is arranged in an equilateral triangle as shown in Fig. 13.41, and attached to an isotropic sheet. Find the principal strains

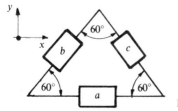

Figure 13.41

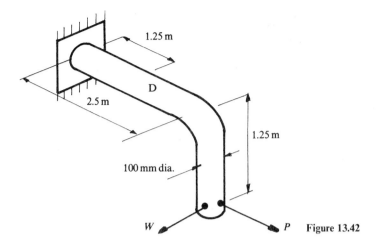

Figure 13.42

and their inclination to gauge *a* if, when the sheet is in a state of plane stress, the strains are $\epsilon_a = \epsilon$, $\epsilon_b = 2\epsilon$ and $\epsilon_c = 3\epsilon$. (C.E.I. Pt.2, 1972)

13.53 The right-angled cantilever steel pipe in Fig. 13.42 is of 100 mm o.d. with 6 mm wall thickness. When forces *P* and *W* are applied at the free end in the directions shown, a 60° strain gauge rosette (Fig. 13.41) fixed to the centre point D of the horizontal portion give the strains $\epsilon_a = -2.75 \times 10^{-4}$, $\epsilon_b = 1.98 \times 10^{-4}$ and $\epsilon_c = -2.08 \times 10^{-4}$ when gauge *a* is aligned with the pipe axis. Calculate *P* and *W* given $E = 200 \, \text{GN/m}^2$, $v = 0.3$. The relationship in Eq. (13.7a) may be assumed. (C.E.I. Pt.2, 1976)

13.54 A certain material is linear elastic, has a Young's modulus of $2 \, \text{GN/m}^2$ and a Poisson's ratio of 0.3. A thin sheet of the material is marked in the unstressed condition with three crosses as shown in Fig. 13.43, each at a different orientation relative to a particular *x*-axis. The legs of each cross are perpendicular and of unit length. Determine the lengths of the legs of each cross and the angle between them when a uniaxial stress of $200 \, \text{MN/m}^2$ is applied in the plane of the sheet in the *x*-direction. (C.E.I. Pt.2, 1983)

3D Stress Transformation

13.55 The state of stress at a point is given by $\sigma_x = 13.78$, $\sigma_y = 26.2$, $\sigma_z = 41.34$, $\tau_{xy} = 27.56$, $\tau_{yz} = 6.89$ and $\tau_{xz} = 10.34 \, \text{MPa}$. Determine the stress resultants and the normal and shear stresses for a plane whose normal makes angles of $50°70'$ and $60°$ with the directions of *x* and *y* respectively.

Answer: 23.45; 30.12; 29.67; 40.14; 48.35 (MPa).

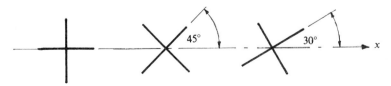

Figure 13.43

13.56 At a point in a material the state of stress is given by the components $\sigma_x = 12.8$, $\sigma_y = 27$, $\sigma_z = 51.3$, $\tau_{xy} = 23.4$, $\tau_{xz} = 11$ and $\tau_{yz} = -6.24$ (MPa). Determine the normal and shear stresses on a plane whose normal makes angles of $48°$ and $71°$ to the x- and y-axes respectively. Also find the direction of the shear stress w.r.t. those axes. (C.E.I. Pt. 2, 1967).

Answer: 48.8 MPa; 12.21 MPa; 138.25°; 68.8°; 55°.

13.57 A stress resultant of 140 MPa makes angles of $43°$, $75°$ and $50°53'$ with the x-, y- and z-axes. Determine the normal and shear stresses on an oblique plane whose normal makes respective angles of $67°13'$, $30°$ and $71°34'$ with these axes.

13.58 A resultant stress of 170 MPa is inclined at $23°$ and $72°$ to the x- and y-directions respectively. This resultant stress acts on a plane whose normal direction cosines are 0.656 and 0.259 w.r.t. x and y respectively. Determine the normal and shear stresses on this plane. Given component shear stresses $\tau_{xy} = 35.5$ MPa and $\tau_{xz} = -47.86$ MPa, find the components when the invariant $J_1 = 926.4$ MPa.

Answer: 144.7; 89.9; 276.4; 497.5; 152.6 (MPa).

11.59 Find the principal stresses and directions for the components of the stress tensor: $\sigma_x = 2$, $\sigma_y = 2$, $\sigma_z = 1$, $\tau_{xy} = 2$, $\tau_{xz} = \tau_{yz} = 0$.

Answer: 4, 1, 0; $1/\sqrt{2}$, $1/\sqrt{2}$, 0; 0, 0, 1; $-1/\sqrt{2}$, $1/\sqrt{2}$, 0.

13.60 Find the invariants and principal stresses for the following stress components $\sigma_x = 9.68$, $\sigma_y = 14.32$, $\sigma_z = 17.28$, $\tau_{xy} = 2.44$, $\tau_{xz} = 10.21$, $\tau_{yz} = 7.38$ (MPa). If $\tau_{xy} = \tau_{yz} = 0$, what then will be the principal stresses? Show that the principal stress directions are orthogonal.

Answer: 28.37, 10.82, 2.09, 24.31, 14.36, 2.61 (MPa).

13.61 Given that the partial state of stress at a point is $\tau_{xy} = 30$, $\tau_{xz} = 10$, $\tau_{yz} = 30$, $\sigma_x = 20$ MPa, $\sigma_y = 20$ and that $J_1 = 50$ MPa, determine the remaining normal stress, the principal stresses and the maximum shear stress.

Answer: 10; 64, 6, -20; 42 (MPa).

13.62 The stress components at a point are $\sigma_x = 5$, $\sigma_y = 7$, $\sigma_z = 6$, $\tau_{xy} = 10$, $\tau_{xz} = 8$ and $\tau_{yz} = 12$ MPa. Find the magnitudes of the principal stresses and the maximum shear stress. What are the direction cosines for the normal to the major principal plane?

Answer: 26.2; -2.7; -5.5; 15.9 (MPa).

13.63 Principal stresses of 77, 31 and -46 MPa act in directions 1, 2 and 3. Determine the normal and shear stresses for a plane whose normal lies at $30°$ to the 3 axis. The normal projection on the 1, 2 plane is inclined at $55°$ to the 1, 3 plane.

Answer: -23.2; 41.4 (MPa).

13.64 Given the principal stresses $\sigma_1 = 7.5$, $\sigma_2 = 3.1$ and $\sigma_3 = 1.4$ MPa, find graphically and analytically the maximum shear stresses in each principal plane and the normal and octahedral shear stresses.

Answer: 2.2; -3.05; 0.85; 4.0; 2.57 (MPa).

13.65 Given principal stresses $\sigma_1 = 130$, $\sigma_2 = 40$ and $\sigma_3 = 30$ MPa, determine the normal and shear stresses on a plane whose normal makes an angle of $60°$ with the 3 axis. The projection of the normal in the 1–2 plane is inclined at $45°$ to the 1 axis.

Answer: 71.25; 91.25 (MPa).

13.66 If $\sigma_1 = 60$ MPa, $\sigma_2 = 20$ MPa and $\tau_{max} = 25$ MPa, find the remaining principal stress and the resultant, normal and shear stresses on a plane whose normal is inclined at $45°$, $60°$ and $60°$ to the 1, 2 and 3 principal stress directions.

Answer: 10; 43.9; 37.5; 22.8 (MPa).

13.67 The major principal stress in a material is 6 MPa. If $J_1 = 4$ and $J_2 = -48$ find:

(a) the remaining principal stresses;

(b) the normal and shear stresses on a plane whose normal direction cosines are 0.6, 0.3 and 0.742;

(c) the state of stress existing on the maximum shear and octahedral planes; and

(d) the unit vectors which define the planes in (c) w.r.t. the principal directions.

Answer: 2, -4; 0.138, 4.53; 5, 1, 4.11, 1.33, $\mathbf{u}_N = (1/\sqrt{2})\mathbf{u}_1 + \mathbf{u}_2$, $\mathbf{u}_N = (1/\sqrt{3})(\mathbf{u}_1 + \mathbf{u}_2 + \mathbf{u}_3)$.
(T.C.D. 1981)

13.68 Find the principal stresses in magnitude and direction and the invariants for the given stress matrix (MPa):

$$\sigma_{ij} = \begin{bmatrix} 3 & 1 & 1 \\ 1 & 0 & 2 \\ 1 & 2 & 0 \end{bmatrix}.$$

Determine the direction cosines for the two greatest principal stresses. Show that their directions are orthogonal.

Answer: 4; 1; -2; 3; -6; -8; $\sqrt{(2/3)}$; $1/\sqrt{6}$; $1/\sqrt{6}$; $1/\sqrt{3}$; $-1/\sqrt{3}$; $-1/\sqrt{3}$.

13.69 At a point in a stressed material the state of stress (MPa) is defined for the given matrix components:

$$\sigma_{ij} = \begin{bmatrix} 1 & 1 & -1 \\ 1 & 2 & 2 \\ -1 & 2 & 2 \end{bmatrix}.$$

Find (a) the magnitude and direction of the normal and shear stresses acting on a plane whose direction cosines are 0.67 and 0.326 w.r.t. x and y, (b) the principal stresses, (c) the maximum shear stress, (d) the normal and shear stresses acting on the octahedral plane. Sketch the planes on which (c) and (d) act w.r.t. the principal planes.

Answer: (a) 1.967, 2.245, -0.441, 0.898, 0.0018; (b) 4, 2, -1; (c) 2.5; (d) 1.667, 2.055 (MPa).
(T.C.D. 1982)

13.70 Complete the stress matrix (MPa) given that the first and second invariants are 6 and -24 respectively.

$$\sigma_{ij} = \begin{bmatrix} 6 & - & 2 \\ - & 0 & 4 \\ 2 & 4 & - \end{bmatrix}.$$

Find the principal stresses then determine graphically the normal and shear stresses on the octahedral and maximum shear planes.

Answer: $\sigma_z = 0$, $\tau_{xy} = 2$; 8, 2, -4; 2, 4.8; 6, 2. (T.C.D. 1980)

13.71 At a point in a material the stress state (kN/m^2) is given by the matrix of components:

$$\sigma_{ij} = \begin{bmatrix} 1860 & 3400 & 1600 \\ 3400 & 3920 & -905 \\ 1600 & -905 & 7445 \end{bmatrix}.$$

Determine the normal and shear stresses on a plane whose normal direction is given by the unit vector; $\mathbf{u}_N = 0.67\mathbf{u}_x + 0.326\mathbf{u}_y + 0.668\mathbf{u}_z$. Also find the unit vector that describes the direction of the shear stress acting on this plane.

Answer: 7097, 1768.7 kN/m^2; $\mathbf{u}_N = -0.746\mathbf{u}_x + 0.362\mathbf{u}_y + 0.573\mathbf{u}_z$.

3D Strain Transformation

13.72 Given a major principal strain of 600μ and strain invariants; $I_1 = 400$ and $I_2 = -4800$, find the remaining principal strains. Also find, in magnitude and direction, the normal and shear strains for the octahedral plane.

Answer: 600, 200, -400, 822, 133.

13.73 Determine, graphically, the normal and shear strains for a plane whose direction cosines are $l = 0.732$, $m = 0.521$ and n (positive) w.r.t. the principal directions $1, 2$ and 3 where the principal strains are 6800×10^{-6}, 1540×10^{-6} and 1130×10^{-6} respectively. Check numerically.
   Answer: 1264×10^{-6}; 662×10^{-6}.

13.74 Determine the principal strains and the corresponding transformation matrix **L** for the principal directions w.r.t. orthogonal axes, in which the microstrain components are $\epsilon_{11} = 10$, $\epsilon_{22} = 10$, $\epsilon_{33} = 60$, $\epsilon_{12} = 30$, $\epsilon_{13} = -20$, $\epsilon_{23} = -20$.
   Answer: $\epsilon_1 = 80$, $\epsilon_2 = 20$, $\epsilon_3 = -20$; $l_{11} = 1/\sqrt{2}$, $l_{12} = -1/\sqrt{2}$, $l_{13} = 0$, $l_{21} = 1/\sqrt{3}$, $l_{22} = 1/\sqrt{3}$, $l_{23} = 1/\sqrt{3}$, $l_{31} = -1/\sqrt{6}$, $l_{32} = -1/\sqrt{6}$, $l_{33} = 2/\sqrt{6}$.

13.75 Determine the principal strains and the deviatoric strain tensor in respect of the absolute microstrain tensor components $\epsilon_x = 400$, $\epsilon_y = 700$, $\epsilon_z = 400$, $\epsilon_{yz} = 200$, $\epsilon_{xy} = \epsilon_{xz} = 0$.
   Answer: $\epsilon_1 = 800$, $\epsilon_2 = 400$, $\epsilon_3 = 300$; $\epsilon_x^D = -100$, $\epsilon_y^D = 200$, $\epsilon_z^D = -100$, $\epsilon_{yz}^D = 200$, $\epsilon_{xy}^D = \epsilon_{xz}^D = 0$.

13.76 The strain state at a point is given by the matrix of its components:

$$\frac{\epsilon_{ij}}{10^{-4}} = \begin{bmatrix} 1 & -3 & \sqrt{2} \\ -3 & 1 & -\sqrt{2} \\ \sqrt{2} & -\sqrt{2} & 4 \end{bmatrix}.$$

Determine the normal strain in the direction defined by the unit vector $\mathbf{u} = (1/2)\mathbf{u}_1 - (1/2)\mathbf{u}_2 + (1/\sqrt{2})\mathbf{u}_3$ and the shear strain between this direction and the perpendicular direction $\mathbf{u} = -(1/2)\mathbf{u}_1 + (1/2)\mathbf{u}_2 + (1/\sqrt{2})\mathbf{u}_3$. Comment on the result.
   Answer: 6×10^{-4}; 0.

13.77 The given microstrain tensor applies to a point in an elastic solid:

$$\mu\epsilon_{ij} = \begin{bmatrix} 600 & 200 & 200 \\ 200 & 0 & 400 \\ 200 & 400 & 0 \end{bmatrix}.$$

Find (a) the principal strains, (b) the magnitudes of the maximum and octahedral shear strains, (c) the normal strain for a direction defined by the unit vector $\mathbf{u}_N = 0.53\mathbf{u}_x + 0.35\mathbf{u}_y + 0.77\mathbf{u}_z$.
   Answer: 800, 200, -400; 600, 9.8; 622.

13.78 Transform the given microstrain tensor

$$\mu\epsilon_{ij} = \begin{bmatrix} 100 & 500 & -500 \\ 500 & 0 & 0 \\ -500 & 0 & -100 \end{bmatrix}$$

from reference axes x_1, x_2 and x_3 to axes x_1', x_2' and x_3', where the unit vectors for directions x_1' and x_2' are $\mathbf{u}_1' = \mathbf{u}_1 + 2\mathbf{u}_2 + 3\mathbf{u}_3$ and $\mathbf{u}_2' = \mathbf{u}_1 + \mathbf{u}_2 - \mathbf{u}_3$ respectively.
   Answer: $\epsilon_{11}' = -128.6$; $\epsilon_{22}' = 666.7$; $\epsilon_{33}' = 538.1$; $\epsilon_{12}' = 138.9$; $\epsilon_{13}' = 198$; $\epsilon_{23}' = -276.2$.

PLANE ELASTICITY THEORY

14.1 CARTESIAN PLANE STRESS AND PLANE STRAIN

Many problems that are plane in nature may be analysed by a simplified form of the general theory of elasticity. In Cartesian coordinates, specifically (see Fig. 11.1), plane stress refers to a class of body with small thickness in the z-direction, typified by a thin plate (Fig. 14.1a) where the stress in the plane of the plate is two-dimensional while the stress through the thickness (σ_z) is zero. Plane strain, on the other hand, refers to a body with a large dimension in the z-direction (Fig. 14.1b) giving, for that direction, either no strain if the ends are constrained or some constant value (ϵ_o).

Stress–Strain (Constitutive) Relations

The non-zero stresses σ_x, σ_y and τ_{xy} may be functions of x and y but not z. In plane stress, $\sigma_z = 0$, when from Eq. (4.12) the strains are

$$\epsilon_x = (\sigma_x - v\sigma_y)/E, \tag{14.1a}$$

$$\epsilon_y = (\sigma_y - v\sigma_x)/E, \tag{14.1b}$$

$$\epsilon_z = -v(\sigma_x + \sigma_y)/E, \tag{14.1c}$$

$$\gamma_{xy} = \tau_{xy}/G = 2(1 + v)\tau_{xy}/E, \tag{14.1d}$$

where ϵ_z, the through-thickness strain, is not zero. In plane strain $\sigma_z \neq 0$ and the strains are

$$\epsilon_x = [\sigma_x - v(\sigma_y + \sigma_z)]/E, \tag{14.2a}$$

585

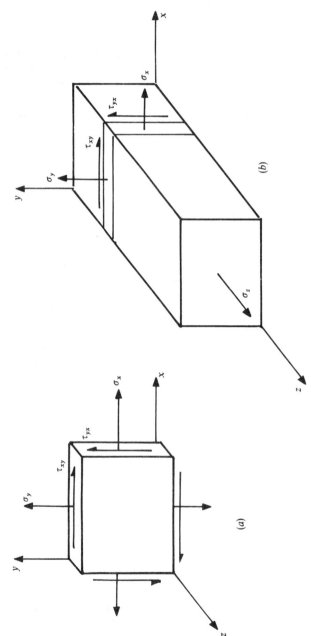

Figure 14.1 Plane stress and plane strain in Cartesian coordinates.

$$\epsilon_y = [\sigma_y - \nu(\sigma_x + \sigma_z)]/E, \qquad (14.2b)$$

$$\epsilon_o = [\sigma_z - \nu(\sigma_x + \sigma_y)]/E, \qquad (14.2c)$$

$$\gamma_{xy} = \tau_{xy}/G = 2(1 + \nu)\tau_{xy}/E. \qquad (14.2d)$$

Since, from Eq. (14.2c), $\sigma_z = \epsilon_o E + \nu(\sigma_x + \sigma_y)$ it follows that plane sections will remain plane under a uniform σ_z when $\nu(\sigma_x + \sigma_y)$ is constant. Moreover, since σ_x, σ_y and τ_{xy} are again functions only of x and y, the following analyses of equilibrium and compatibility will apply to both plane stress and strain Cartesian problems. The theory appropriate to plane polar coordinates will later be derived from transformation.

Equilibrium

Let the stresses vary over the elemental dimensions δx and δy in the manner illustrated in Fig. 14.2(a) and (b). These act together as shown in Fig. 14.1 but, for clarity, are separated here for the x- and y-directions. Body forces due to self-weight are taken to be absent.

In Fig. 14.2(a) (with unit thickness) vertical force equilibrium gives

$$[\sigma_y + (\partial\sigma_y/\partial y)\delta y - \sigma_y]\delta x + [\tau_{yx} + (\partial\tau_{yx}/\partial x)\delta x - \tau_{yx}]\delta y = 0,$$
$$\partial\sigma_y/\partial y + \partial\tau_{yx}/\partial x = 0, \quad (14.3a)$$

and from Fig. 14.2(b), for horizontal force equilibrium,

$$[\sigma_x + (\partial\sigma_x/\partial x)\delta x - \sigma_x]\delta y + [\tau_{xy} + (\partial\tau_{xy}/\partial y)\delta y - \tau_{xy}]\delta x = 0,$$
$$\partial\sigma_x/\partial x + \partial\tau_{xy}/\partial y = 0. \quad (14.3b)$$

Equations (14.3) are the plane equilibrium equations for variations in stress with x and y through the body. Note that $\tau_{xy} = \tau_{yx}$.

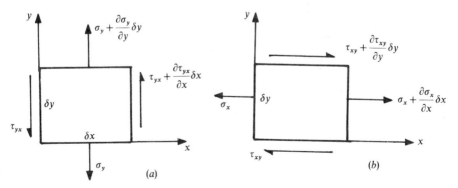

Figure 14.2 Stress variation in the x- and y-directions.

Strain–Displacement Relations

In Chapter 13 (Fig. 13.31(a)) it was shown that initially unstrained perpendicular line elements δx and δy deform and rotate whilst translating into a new position under stress. For plane deformation the u and v displacements that define the shift in the origin from O to O', are different functions of coordinates x and y. Plane sections remain plane as all points in the body displace only within the x–y plane, i.e. u and v do not depend upon z. The displacement gradients in Eqs (13.25), found from the partial differentiation of u and v w.r.t. x and y, again define the normal and shear strain components. Alternatively, in confirmation of this, the normal strain components may be identified with the corresponding length changes in δx and δy:

$$\epsilon_x = \{[\delta x + (\partial u/\partial x)\delta x] - \delta x\}/\delta x = \partial u/\partial x, \tag{14.4a}$$

$$\epsilon_y = \{[\delta y + (\partial v/\partial y)\delta y] - \delta y\}/\delta y = \partial v/\partial y. \tag{14.4b}$$

The small angular distortions are realistically aproximated by

$$e_{xy} = (\partial u/\partial y)\delta y/\delta y = \partial u/\partial y,$$
$$e_{yx} = (\partial v/\partial x)\delta x/\delta x = \partial v/\partial x.$$

Their sum gives the net change in the right angle (Fig. 13.31(b)) that defines the engineering shear strain,

$$\gamma_{xy} = \partial u/\partial y + \partial v/\partial x. \tag{14.4c}$$

Compatibility

It is seen from Eqs (14.4a, b and c) that the three strains ϵ_x, ϵ_y and γ_{xy} depend upon the two displacements u and v. The relationship between the strains expresses the condition of compatibility. That is, by eliminating u and v between Eqs (14.4),

$$\partial^2 \epsilon_x/\partial y^2 + \partial^2 \epsilon_y/\partial x^2 = \partial^2 \gamma_{xy}/\partial x \partial y. \tag{14.5}$$

Equation (14.5) may further be written in terms of the stress components. With plane stress, for example, substituting from Eqs (14.1) leads to

$$\partial^2(\sigma_x - v\sigma_y)/\partial y^2 + \partial^2(\sigma_y - v\sigma_x)/\partial x^2 = 2(1 + v)\partial^2\tau_{xy}/\partial x \partial y. \tag{14.6a}$$

The corresponding plane strain compatibility relationship is, from Eqs (14.2),

$$\partial^2[\sigma_x(1 - v^2) - v\sigma_y(1 + v) - v\epsilon_o E]/\partial y^2,$$
$$+ \partial^2[\sigma_y(1 - v^2) - v\sigma_x(1 + v) - v\epsilon_o E]/\partial x^2 = 2(1 + v)\partial^2\tau_{xy}/\partial x \partial y. \tag{14.6b}$$

The Biharmonic Equation and Component Stresses

The requirement that both equilibrium and compatibility are satisfied is met by combining Eqs (14.3) and (14.6). For plane stress, Eqs (14.3a) and (14.6a) lead to

$$\partial^2\sigma_x/\partial y^2 - v\partial^2\sigma_y/\partial y^2 + \partial^2\sigma_y/\partial x^2 - v\partial^2\sigma_x/\partial x^2 = -2(1+v)\partial^2\sigma_y/\partial y^2,$$

$$(14.7a)$$

and from Eqs (14.3b) and (14.6a),

$$\partial^2\sigma_x/\partial y^2 - v\partial^2\sigma_y/\partial y^2 + \partial^2\sigma_y/\partial x^2 - v\partial^2\sigma_x/\partial x^2 = -2(1+v)\partial^2\sigma_x/\partial x^2.$$

$$(14.7b)$$

Subtracting Eq. (14.7a) from (14.7b),

$$-2(1+v)\partial^2\sigma_x/\partial x^2 + 2(1+v)\partial^2\sigma_y/\partial y^2 = 0,$$
$$\partial^2\sigma_y/\partial y^2 - \partial^2\sigma_x/\partial x^2 = 0. \qquad (14.8a)$$

Adding Eqs (14.7a and b),

$$2[\partial^2\sigma_x/\partial y^2 - v\partial^2\sigma_y/\partial y^2 + \partial^2\sigma_y/\partial x^2 - v\partial^2\sigma_x/\partial x^2]$$
$$= -2(1+v)(\partial^2\sigma_y/\partial y^2 + \partial^2\sigma_x/\partial x^2),$$
$$\partial^2\sigma_x/\partial x^2 + \partial^2\sigma_x/\partial y^2 + \partial^2\sigma_y/\partial y^2 + \partial^2\sigma_y/\partial x^2 = 0,$$
$$\therefore \quad (\partial^2/\partial x^2 + \partial^2/\partial y^2)(\sigma_x + \sigma_y) = 0. \qquad (14.8b)$$

It is seen that Eq. (14.8a) is satisfied when σ_x and σ_y are defined from a stress function $\phi = \phi(x, y)$ (G. B. Airy, 1862) in the following manner:

$$\sigma_x = \partial^2\phi/\partial y^2 \quad \text{and} \quad \sigma_y = \partial^2\phi/\partial x^2 \qquad (14.9a, b)$$

and from either Eq. (14.3a) or Eq. (14.3b) the shear stress is given by

$$\partial\tau_{xy}/\partial x = -\partial\sigma_y/\partial y = -\partial^3\phi/\partial x^2\partial y,$$
$$\therefore \quad \tau_{xy} = -\partial^2\phi/\partial x\partial y. \qquad (14.9c)$$

Combining Eq. (14.8b) with Eqs (14.9a and b) the result may be written as a biharmonic equation in any one of the following three forms:

$$(\partial^2/\partial x^2 + \partial^2/\partial y^2)(\partial^2\phi/\partial y^2 + \partial^2\phi/\partial x^2) = 0, \qquad (14.10a)$$

$$(\partial^2/\partial x^2 + \partial^2/\partial y^2)^2\phi = 0, \qquad (14.10b)$$

$$\nabla^2(\nabla^2\phi) = \nabla^4\phi = 0, \qquad (14.10c)$$

where 'del-squared' is defined as $\nabla^2 = \partial^2/\partial x^2 + \partial^2/\partial y^2$. The reader should check that Eqs (14.9) and (14.10) will also apply to the case of plane strain by combining Eqs (14.3) with Eq. (14.6b).

14.2 CARTESIAN STRESS FUNCTIONS

If a function $\phi(x, y)$ can be found that satisfies $\nabla^4\phi = 0$, it will satisfy both equilibrium and compatibility in plane stress and plane strain expressed in Cartesian coordinates. The stresses are then found from $\phi(x, y)$, through

Eqs (14.9). The constants appearing in the function ϕ must meet the final requirement of satisfying the boundary conditions of a particular problem. That is, the internal stress distribution, which will in general vary throughout the volume of the body, must be in equilibrium with the forces and moments applied around the boundary.

Polynomial Functions

The stress functions of most common use in plate and beam problems, are taken from the polynomial function:

$$\phi(x, y) = ax^2 + bxy + cy^2 + dx^3 + ex^2y + fxy^2 + gy^3 + hx^4$$
$$+ ix^3y + jx^2y^2 + kxy^3 + ly^4 + \cdots.$$

Linear and constant terms are excluded, since these disappear with the second derivative. The following examples illustrate some functions.

Example 14.1 Show that the plane displacement system $u = ax^2y^2, v = byx^3$ is compatible.

The following strain components are found from Eqs (14.4):

$$\epsilon_x = \partial u/\partial x = 2axy^2,$$
$$\epsilon_y = \partial v/\partial y = bx^3,$$
$$\gamma_{xy} = \partial u/\partial y + \partial v/\partial x = 2ax^2y + 3byx^2.$$

Applying the L.H and R.H. sides of Eq. (14.5) separately:

$$\partial^2\epsilon_x/\partial y^2 + \partial^2\epsilon_y/\partial x^2 = \partial(4axy)/\partial y + \partial(3bx^2)/\partial x = 4ax + 6bx,$$
$$\partial^2\gamma_{xy}/\partial x\partial y = \partial(4axy + 6bxy)/\partial y = 4ax + 6bx.$$

Example 14.2 Derive the displacements corresponding to the following stresses in plane strain: $\sigma_x = c(y^2 + 2x)$, $\sigma_y = -cx^2$ and $\tau_{xy} = -2cy$ when $\epsilon_o = 0$. Show that the corresponding strains are compatible.

Clearly, the given stress expressions satisfy the equilibrium equations (14.3). From Eqs (14.2):

$$\sigma_z = v(\sigma_x + \sigma_y),$$
$$\therefore \quad \epsilon_x = \partial u/\partial x = (1 + v)[\sigma_x(1 - v) - v\sigma_y]/E$$
$$= (1 + v)[c(y^2 + 2x)(1 - v) + vcx^2]/E.$$

Integrating:

$$u = (1 + v)[c(y^2x + x^2)(1 - v) + vcx^3/3]/E + f(y), \qquad \text{(i)}$$
$$\epsilon_y = \partial v/\partial y = (1 + v)[\sigma_y(1 - v) - v\sigma_x]/E$$
$$= -(1 + v)[cx^2(1 - v) + vc(y^2 + 2x)]/E.$$

Integrating:

$$v = -(1+v)[cx^2y(1-v) + vc(y^3/3 + 2xy)]/E + g(x), \qquad \text{(ii)}$$

$$\gamma_{xy} = \partial u/\partial y + \partial v/\partial x = 2(1+v)\tau_{xy}/E,$$

$$\therefore \quad \partial u/\partial y + \partial v/\partial x = -4cy(1+v)/E. \qquad \text{(iii)}$$

The strain components ϵ_x, ϵ_y and γ_{xy} satisfy the compatibility condition given in Eq. (14.5). Substituting Eqs (i) and (ii) into (iii) to find the u, v displacements:

$$(1+v)2cyx(1-v)/E + f'(y) - (1+v)[2cyx(1-v) + 2vcy]/E + g'(x)$$
$$= -4cy(1+v)/E,$$
$$f'(y) - 2vcy(1+v)/E + g'(x) = -4cy(1+v)/E,$$

from which it follows that $g'(x) = 0$ and

$$f'(y) = 2vcy(1+v)/E - 4cy(1+v)/E = 2cy(1+v)(v-2)/E,$$
$$\therefore \quad f(y) = cy^2(1+v)(v-2)/E.$$

The displacement expressions are then, from Eqs (i) and (ii),

$$u = (1+v)[c(y^2x + x^2)(1-v) + vcx^3/3]/E + cy^2(1+v)(v-2)/E,$$
$$v = -(1+v)[cx^2y(1-v) + vc(y^3/3 + 2xy)]/E.$$

Constants of integration have been omitted here because no boundary conditions are given. Note that u and v will differ for plane stress where $\sigma_z = 0$.

Example 14.3 The square plate $a \times a$ in Fig. 14.3 is subjected to the plane stresses shown. Determine, for the given boundary conditions, the general expressions for the u, v displacements at any point (x, y) in the plate. What are the displacements for a point originally at the origin? (T.C.D. 1979)

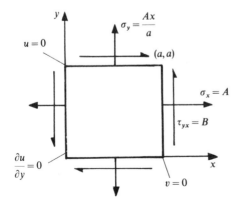

Figure 14.3 Plane stress plate.

The given stresses obviously satisfy the equilibrium equations (14.3). For plane stress, Eqs (14.1 and 14.4) give

$$\epsilon_x = \partial u/\partial x = (\sigma_x - v\sigma_y)/E = A(1 - vx/a)/E,$$
$$u = A(x - vx^2/2a)/E + f(y) + P, \tag{i}$$

$$\epsilon_y = \partial v/\partial y = (\sigma_y - v\sigma_x)/E = A(x/a - v)/E,$$
$$v = A(xy/a - vy)/E + g(x) + Q, \tag{ii}$$

$$\gamma_{xy} = \partial u/\partial y + \partial v/\partial x = 2(1 + v)\tau_{xy}/E = 2B(1 + v)/E. \tag{iii}$$

Substituting (i) and (ii) in (iii) gives

$$f'(y) + Ay/aE + g'(x) = 2B(1 + v)/E.$$

Since the R.H.S. is a constant, say $\alpha + \beta = 2B(1 + v)/E$, it follows that

$$f'(y) + Ay/aE = \alpha, \quad \Rightarrow f(y) = \alpha y - Ay^2/2aE$$
$$g'(x) = \beta, \quad \Rightarrow g(x) = \beta x.$$

Hence Eqs (i) and (ii) become

$$u = A(x - vx^2/2a)/E + \alpha y - Ay^2/2aE + P, \tag{iv}$$
$$v = A(xy/a - vy)/E + \beta x + Q. \tag{v}$$

The following boundary conditions apply to Eqs (iv) and (v):

$u = 0$ at $x = 0$, $y = a$:

$$\therefore \quad 0 = \alpha a - Aa^2/2aE + P, \quad \Rightarrow P = Aa/2E - \alpha a.$$

$v = 0$ at $x = a$, $y = 0$:

$$\therefore \quad 0 = \beta a + Q, \quad \Rightarrow Q = -\beta a.$$

$\partial u/\partial y = 0$ at $x = 0$, $y = 0$

$$\therefore \quad \alpha = 0 \text{ and hence } \beta = 2B(1 + v)/E.$$

Finally, Eqs (iv) and (v) become

$$u = Ax(1 - vx/2a)/E - Ay^2/2aE + Aa/2E,$$
$$v = Ay(x/a - v)/E + 2B(1 + v)(x - a)/E,$$

which give the displacements at the origin ($x = 0$, $y = 0$) of

$$u = Aa/2E \quad \text{and} \quad v = -2Ba(1 + v)/E.$$

Example 14.4 The rectangular section bar in Fig. 14.4 is line loaded as shown. Assuming plane strain conditions, derive the stresses and displacements from the stress function $\phi = Ax^3$. If all the points along the z-axis are fixed, determine the displacements beneath the load. (T.C.D. 1983)

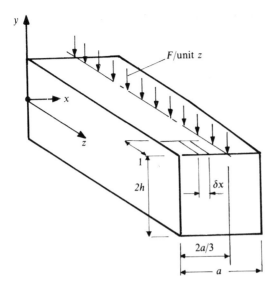

Figure 14.4 Plane strain bar.

Since ϕ satisfies $\nabla^4\phi = 0$, the stress components are, from Eqs (14.9),

$$\sigma_y = \partial^2\phi/\partial x^2 = \partial^2(Ax^3)/\partial x^2 = \partial(3Ax^2)/\partial x = 6Ax,$$

$$\sigma_x = \tau_{xy} = 0.$$

Now, at the boundary, F is the resultant of the internal σ_y distribution. That is, for unit thickness in the z-direction,

$$F = \int_0^a \sigma_y\,dx = 6A\int_0^a x\,dx = 3Aa^2,$$

$$\therefore \quad A = F/3a^2 \quad \text{and} \quad \sigma_y = 2Fx/a^2.$$

Alternatively, A may be found from moment balance:

$$2aF/3 = \int_0^a \sigma_x x\,dx = 6A\int_0^a x^2\,dx, \quad \Rightarrow A = F/3a^2.$$

Since $\epsilon_o = 0$ and $\sigma_x = 0$, Eq. (14.2c) gives $\sigma_z = v\sigma_y$. Then Eqs (14.2a and b) become

$$\epsilon_x = \partial u/\partial x = -v(1 + v)\sigma_y/E = -2Fv(1 + v)x/a^2 E,$$

$$\epsilon_y = \partial v/\partial y = (1 - v^2)\sigma_y/E = 2F(1 - v^2)x/a^2 E.$$

Integrating these leads to the u-, v-displacements:

$$u = -Fv(1 + v)x^2/a^2 E + f(y) + C_1, \tag{i}$$

$$v = 2F(1 - v^2)xy/a^2 E + g(x) + C_2. \tag{ii}$$

Since $\tau_{xy} = 0$, Eqs (14.2d) and (14.4c) become

$$\gamma_{xy} = \partial u/\partial y + \partial v/\partial x = 0. \tag{iii}$$

Substituting Eqs (i) and (ii) into (iii) gives

$$f'(y) + 2F(1 - v^2)y/a^2 E + g'(x) = 0,$$
$$\therefore \quad g'(x) = 0 \qquad \text{and} \qquad f'(y) + 2F(1 - v^2)y/a^2 E = 0,$$
$$f(y) = F(1 - v^2)y^2/a^2 E.$$

Hence Eqs (i) and (ii) become

$$u = -Fv(1 + v)x^2/a^2 E + F(1 - v^2)y^2/a^2 E + C_1, \tag{iv}$$
$$v = 2F(1 - v^2)xy/a^2 E + C_2, \tag{v}$$

where $C_1 = C_2 = 0$ since $u = v = 0$ for $x = y = 0$. At the load point $(x = 2a/3, y = h)$, Eqs (iv) and (v) give

$$u = F[h^2(1 - v^2)/a^2 - 4v(1 + v)/9]/E,$$
$$v = 4Fh(1 - v^2)/3aE.$$

Example 14.5 Determine the stresses and displacements from the plane stress function $\phi = Ax^2 + Bxy + Cy^2$.

Clearly ϕ satisfies $\nabla^4\phi = 0$. The stress components are, from Eqs (14.9),

$$\sigma_x = \partial^2\phi/\partial y^2 = \partial(Bx + 2Cy)/\partial y = 2C,$$
$$\sigma_y = \partial^2\phi/\partial x^2 = \partial(2Ax + By)/\partial x = 2A,$$
$$\tau_{xy} = -\partial^2\phi/\partial x\partial y = -\partial(2Ax + By)/\partial y = -B.$$

That is, the function here provides a solution to a thin plate under the action of uniform stresses along its sides (Fig. 14.5). The displacements

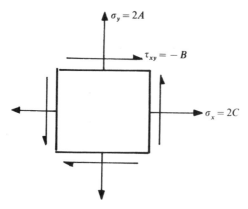

Figure 14.5 Thin plate.

follow from Eqs (14.1) and (14.4):

$$\epsilon_x = \partial u/\partial x = (\sigma_x - v\sigma_y)/E = 2(C - vA)/E,$$
$$u = 2(C - vA)x/E + f(y) + Q_1, \tag{i}$$
$$\epsilon_y = \partial v/\partial y = (\sigma_y - v\sigma_x)/E = 2(A - vC)/E,$$
$$v = 2(A - vC)y/E + g(x) + Q_2, \tag{ii}$$
$$\gamma_{xy} = \partial u/\partial y + \partial v/\partial x = 2(1 + v)\tau_{xy}/E = -2(1 + v)B/E. \tag{iii}$$

Substituting Eqs (i) and (ii) into (iii) leads to

$$f'(y) + g'(x) = -2(1 + v)B/E = \alpha + \beta.$$

Since the derivatives are the respective constants α and β, then $f(y) = \alpha y$ and $g(x) = \beta x$ and Eqs (i) and (ii) become

$$u = 2(C - vA)x/E + \alpha y + Q_1,$$
$$v = 2(A - vC)y/E + \beta x + Q_2.$$

Example 14.6 Show that the stress function $\phi = Dxy^3 + Bxy$ provides the stress distribution for the cantilever, of depth $2h$ and of unit thickness in Fig. 14.6, when it is loaded by a concentrated force P at its free end.

Equation (14.9) gives the stress components:

$$\sigma_x = \partial^2\phi/\partial y^2 = \partial(3Dxy^2 + Bx)/\partial y = 6Dxy, \tag{i}$$
$$\sigma_y = \partial^2\phi/\partial x^2 = \partial(Dy^3 + By)/\partial x = 0,$$
$$\tau_{xy} = -\partial^2\phi/\partial x\partial y = -\partial(3Dxy^2 + Bx)/\partial x = -(3Dy^2 + B). \tag{ii}$$

The boundary conditions are as follows.

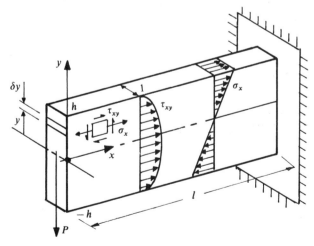

Figure 14.6 Plane cantilever.

(a) $\tau_{xy} = 0$ for $y = \pm h$, since there is no shear stress along the free edges.

$$\therefore \quad -(B + 3Dh^2) = 0,$$
$$B = -3Dh^2.$$

(b) P is the resultant of the internal τ_{xy} distribution acting at the free end. That is,

$$P = \int_{-h}^{h} \tau_{xy} \, dy = 3D \int_{-h}^{h} (h^2 - y^2) \, dy$$

$$= 3D|h^2 y - y^3/3|_{-h}^{h} = 4Dh^3, \quad \Rightarrow D = P/4h^3.$$

Then, from Eqs (i) and (ii):

$$\sigma_x = Pxy/(2h^3/3) = (Px)y/I = My/I,$$
$$\tau_{xy} = 3P(h^2 - y^2)/4h^3 = (P/2I)(h^2 - y^2),$$

which are distributed in the manner shown in Fig. 14.6. Note that the stresses agree with the separate applications of ETB and the flexural shear flow theory (Eq. (10.3)). In the stress function approach x, not z, is used to denote the length of a plane beam within the x, y plane. Then y bounds the depth and z bounds the thickness.

Sinusoidal Function

If a stress function of the form

$$\phi = f(y)\sin(n\pi x/L) = f(y)\sin(\alpha x) \tag{14.11a}$$

is to satisfy $\nabla^4 \phi = 0$, the following condition is found:

$$\frac{d^4 f(y)}{dy^4} - 2\alpha^2 \frac{d^2 f(y)}{dy^2} + \alpha^4 f(y) = 0.$$

The solution is

$$f(y) = (A + By)\exp(\alpha y) + (C + Dy)\exp(-\alpha y)$$
$$= E\cosh\alpha y + F\sinh\alpha y + Hy\cosh\alpha y + Jy\sinh\alpha y, \tag{14.11b}$$

where $\alpha = n\pi/L$. The function will apply directly to sinusoidal edge loadings, where the particular boundary conditions enable the constants E, F, H and J to be found.

Example 14.7 A long thin strip of depth $2h$ is subjected to a normal stress distribution $p\sin(\pi x/L)$ along its longer edges (Fig. 14.7). Determine the constants in Eq. (14.11b) and the stress distribution along the x-axis.

Here $\alpha = \pi/L$ since $n = 1$. The plane stress components are found from

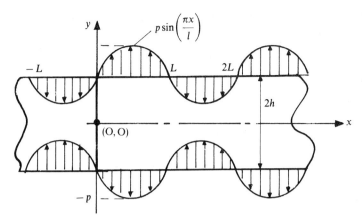

Figure 14.7 Sinusoidal wave.

Eqs (14.9)–(14.11):

$$\sigma_y = \partial^2 \phi/\partial x^2$$
$$= -(E \cosh \alpha y + F \sinh \alpha y + Hy \cosh \alpha y + Jy \sinh \alpha y)\alpha^2 \sin \alpha x, \quad \text{(i)}$$

$$\tau_{xy} = -\partial^2 \phi/\partial x \partial y$$
$$= -[\alpha(E \sinh \alpha y + F \cosh \alpha y) + H(y\alpha \sinh \alpha y + \cosh \alpha y)$$
$$+ J(y\alpha \cosh \alpha y + \sinh \alpha y)]\alpha \cos \alpha x, \quad \text{(ii)}$$

$$\sigma_x = \partial^2 \phi/\partial y^2$$
$$= [\alpha^2(E \cosh \alpha y + F \sinh \alpha y) + H\alpha(y\alpha \cosh \alpha y + 2 \sinh \alpha y)$$
$$+ J\alpha(y\alpha \sinh \alpha y + 2 \cosh \alpha y)] \sin \alpha x. \quad \text{(iii)}$$

In Eq. (i) the boundary condition is $\sigma_y = p \sin \alpha x$ (tensile) for $y = \pm h$ and all x. This gives

$$-\alpha^2(E \cosh \alpha h + F \sinh \alpha h + Hh \cosh \alpha h + Jh \sinh \alpha h) = p,$$
$$-\alpha^2(E \cosh \alpha h - F \sinh \alpha h - Hh \cosh \alpha h + Jh \sinh \alpha h) = p.$$

By addition and subtraction,

$$E = -(p + J\alpha^2 h \sinh \alpha h)/(\alpha^2 \cosh \alpha h), \quad \text{(iv)}$$

$$F = -Hh \cosh \alpha h/\sinh \alpha h. \quad \text{(v)}$$

In Eq. (ii) the boundary condition is $\tau_{xy} = 0$ for $y = \pm h$ and all x. Thus,

$$-[\alpha(E \sinh \alpha h + F \cosh \alpha h) + H(h\alpha \sinh \alpha h + \cosh \alpha h)$$
$$+ J(h\alpha \cosh \alpha h + \sinh \alpha h)] = 0,$$
$$-[\alpha(-E \sinh \alpha h + F \cosh \alpha h) + H(h\alpha \sinh \alpha h + \cosh \alpha h)$$
$$- J(h\alpha \cosh \alpha h + \sinh \alpha h)] = 0.$$

By addition and subtraction,

$$H = -\frac{\alpha F \cosh \alpha h}{h\alpha \sinh \alpha h + \cosh \alpha h}, \tag{vi}$$

$$J = -\frac{\alpha E \sinh \alpha h}{h\alpha \cosh \alpha h + \sinh \alpha h}. \tag{vii}$$

Substituting Eq. (iv) into (vii) leads to

$$J = \frac{\alpha p \sinh \alpha h}{\alpha^2 (\sinh \alpha h \cosh \alpha h + \alpha h)},$$

$$E = -\frac{p(\sinh \alpha h + \alpha h \cosh \alpha h)}{\alpha^2 (\sinh \alpha h \cosh \alpha h + \alpha h)}.$$

Substituting Eq. (v) into (vi) gives $F = H = 0$, which is a consequence of the common amplitude p for each edge distribution. When $y = 0$ the stress state along the x-axis is, from Eqs (i)–(iii),

$$\sigma_y = -E\alpha^2 \sin \alpha x$$

$$= \frac{p \sin \alpha x(\sinh \alpha h + \alpha h \cosh \alpha h)}{\sinh \alpha h \cosh \alpha h + \alpha h},$$

$$\sigma_x = (\alpha^2 E + 2J\alpha) \sin \alpha x$$

$$= p \sin \alpha x \frac{(\sinh \alpha h - \alpha h \cosh \alpha h)}{(\sinh \alpha h \cosh \alpha h + \alpha h)},$$

$$\tau_{xy} = -(F\alpha + H)\alpha \cos \alpha x = 0.$$

Fourier Series

Full range series When the edge loading is not sinusoidal it may, nevertheless, be represented with a sinusoidal series function $g(x)$ of period $2L$ so that the stress function in Eqs (14.11) then becomes

$$\phi(x, y) = [E \cosh \alpha y + F \sinh \alpha y + Hy \cosh \alpha y + Jy \sinh \alpha y]g(x) \tag{14.12}$$

where

$$g(x) = a_o/2 + a_1 \sin(\pi x/L) + a_2 \sin(2\pi x/L) + a_3 \sin(3\pi x/L)$$
$$+ b_1 \cos(\pi x/L) + b_2 \cos(2\pi x/L) + b_3 \cos(3\pi x/L) + \cdots$$

$$= a_o/2 + \sum_{n=1}^{\infty} [a_n \sin(n\pi x/L) + b_n \cos(n\pi x/L)]. \tag{14.13a}$$

If $\psi(x)$ describes the actual edge loading, then the coefficients a_o, a_n and b_n in Eq. (14.13a) are found from identifying $\psi(x)$ with each part of $g(x)$ and integrating over the range $\pm L$ as follows:

$$\int_{-L}^{L} [a_o/2 + a_n \sin(n\pi x/L)\sin(n\pi x/L) + b_n \cos(n\pi x/L)\cos(n\pi x/L)]\,dx$$

$$= \int_{-L}^{L} \psi(x)\,dx + \int_{-L}^{L} \psi(x)\sin(n\pi x/L)\,dx + \int_{-L}^{L} \psi(x)\cos(n\pi x/L)\,dx.$$

This gives

$$\int_{-L}^{L} (a_o/2)\,dx = \int_{-L}^{L} \psi(x)\,dx,$$

$$a_o = (1/L) \int_{-L}^{L} \psi(x)\,dx, \qquad (14.13b)$$

$$\int_{-L}^{L} a_n \sin(n\pi x/L)\sin(n\pi x/L)\,dx = \int_{-L}^{L} \psi(x)\sin(n\pi x/L)\,dx,$$

$$a_n L = \int_{-L}^{L} \psi(x)\sin(n\pi x/L)\,dx,$$

$$a_n = (1/L) \int_{-L}^{L} \psi(x)\sin(n\pi x/L)\,dx, \qquad (14.13c)$$

$$\int_{-L}^{L} b_n \cos(n\pi x/L)\cos(n\pi x/L)\,dx = \int_{-L}^{L} \psi(x)\cos(n\pi x/L)\,dx,$$

$$b_n L = \int_{-L}^{L} \psi(x)\cos(n\pi x/L)\,dx,$$

$$b_n = (1/L) \int_{-L}^{L} \psi(x)\cos(n\pi x/L)\,dx. \qquad (14.13d)$$

Note that outside the range $-L \le x \le L$, $g(x)$ in Eqs (14.13) will not, in general, represent $\psi(x)$ unless $\psi(x)$ is itself periodic.

Example 14.8 A rectangular wave loading of amplitude p and period $2L$ is applied normally to the both longer edges of a thin plate of depth $2h$ as shown in Fig. 14.8. Determine the Fourier series loading function $g(x)$ in Eq. (14.13a). If $L = 4h$ determine σ_x and σ_y at $x = L/2$, $y = 0$ using the result from Example 14.7.

The loading function is;

$$\psi(x) = \begin{cases} +p & \text{for } 0 \le x \le L, \\ -p & \text{for } L \le x \le 2L. \end{cases}$$

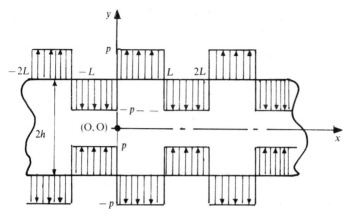

Figure 14.8 Rectangular wave.

From Eq. (14.13b–d)

$$a_o = (1/L)\left[\int_0^L (p)\,dx + \int_L^{2L} (-p)\,dx\right]$$

$$= (1/L)[|px|_0^L - |px|_L^{2L}] = 0,$$

$$a_n = (1/L)\left[\int_0^L p\sin(n\pi x/L)\,dx + \int_L^{2L} (-p)\sin(n\pi x/L)\,dx\right]$$

$$= (1/L)(Lp/n\pi)[|-\cos(n\pi x/L)|_0^L + |\cos(n\pi x/L)|_L^{2L}]$$

$$= (p/n\pi)[(1 - \cos n\pi) + (\cos 2n\pi - \cos n\pi)],$$

$$a_n = (2p/n\pi)(1 - \cos n\pi) \quad \text{for } n = 1, 3, 5, \dots,$$

$$a_n = 0 \quad \text{for } n = 2, 4, 6, \dots,$$

$$b_n = (1/L)\left[\int_0^L p\cos(n\pi x/L)\,dx + \int_L^{2L} (-p)\cos(n\pi x/L)\,dx\right]$$

$$= (1/L)(Lp/n\pi)[|\sin(n\pi x/L)|_0^L - |\sin(n\pi x/L)|_L^{2L}]$$

$$= (p/n\pi)[0 - 0] = 0.$$

Equation (14.13a) becomes

$$g(x) = \sum_{n=1}^{\infty} (2p/n\pi)(1 - \cos n\pi)\sin(n\pi x/L) \quad \text{for } n = 1, 3, 5, \dots$$

$$= (4p/\pi)[\sin(\pi x/L) + (1/3)\sin(3\pi x/L) + (1/5)\sin(5\pi x/L) + \cdots]$$

Since the functions x and y in Eq. (14.12) are separable, the hyperbolic function in y may be taken from the previous example where the same boundary conditions apply. It follows that for $y = 0$ the plane stresses are $\tau_{xy} = 0$ and

$$\sigma_y = (4p/\pi)[\sin(\pi x/L) + (1/3)\sin(3\pi x/L) + (1/5)\sin(5\pi x/L) + \cdots]$$
$$\times [\sinh(\pi h/L) + (\pi h/L)\cosh(\pi h/L)]/[\sinh(\pi h/L)\cosh(\pi h/L) + \pi h/L],$$

$$\sigma_x = (4p/\pi)[\sin(\pi x/L) + (1/3)\sin(3\pi x/L) + (1/5)\sin(5\pi x/L) + \cdots]$$
$$\times [-\sinh(\pi h/L) - (\pi h/L)\cosh(\pi h/L) + 2\sinh(\pi h/L)]/[\sinh(\pi h/L)$$
$$\times \cosh(\pi h/L) + \pi h/L],$$

and, for $x = L/2$, $h/L = 1/4$:

$$\sigma_y = (4p/\pi)(1 - 1/3 + 1/5 - 1/7 + 1/9 - 1/11 \cdots)1.9801/1.835 = 1.01p,$$
$$\sigma_x = (4p/\pi)(0.7609)(-1.9801 + 1.7362)/1.835 = -0.129p.$$

Odd and even loading Note that where $b_n = 0$ for an *odd loading* function, $\psi(-x) = -\psi(x)$, and the Fourier representation $g(x)$ will, in general, contain sines only. For an *even loading* function, $\psi(-x) = \psi(x)$, it will be found that $a_n = 0$, resulting in a Fourier cosine series. Also, when $\psi(x) = -\psi(x + L)$, as in this example, the Fourier series will contain *odd harmonics* only. When $\psi(x) = \psi(x + L)$ the Fourier series will contain only *even harmonics*.

Half-range series Simplified Fourier sine or cosine series may be applied over the half-range $0 \leq x \leq L$. These are

$$g(x) = \sum_{n=1}^{\infty} a_n \sin \alpha x, \qquad (14.14a)$$

where $\alpha = n\pi/L$ and

$$a_n = (2/L) \int_0^L \psi(x) \sin \alpha x \, dx$$

and also

$$g(x) = a_o + \sum_{n=1}^{\infty} b_n \cos \alpha x \qquad (14.14b)$$

where

$$a_o = (1/L) \int_0^L \psi(x) \, dx,$$

$$b_n = (2/L) \int_0^L \psi(x) \cos \alpha x \, dx.$$

Beyond the range of L, the sine series will extend $\psi(x)$ alternately in positive and negative y with period L about the x-axis. The cosine series duplicates $\psi(x)$ with period L along the x-axis. These half-range series then coincide with the full series representation in each case.

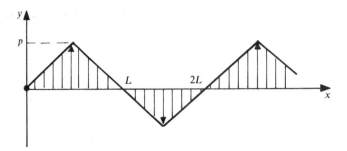

Figure 14.9 Edge loading.

Example 14.9 Find the expression $g(x)$ for the edge loading in Fig. 14.9.

Since $\psi(x)$ alternates about x with period L, then $g(x)$ will follow from the half-range sine series. From the loading function

$$\psi(x) = \begin{cases} 2px/L & \text{for } 0 \le x \le L/2, \\ 2p(1 - x/L) & \text{for } L/2 \le x \le L. \end{cases}$$

Equation (14.14a) becomes

$$a_n = (2/L) \int_0^{L/2} (2px/L) \sin \alpha x \, dx + (2/L) \int_{L/2}^{L} 2p(1 - x/L) \sin \alpha x \, dx \quad \text{(i)}$$

Integrating by parts:

$$\int x \sin \alpha x \, dx = (1/\alpha^2) \sin \alpha x - (x/\alpha) \cos \alpha x$$

and Eq. (i) becomes

$$\begin{aligned}
a_n &= (4p/L^2)|(1/\alpha^2) \sin \alpha x - (x/\alpha) \cos \alpha x|_0^{L/2} - (4p/L\alpha)|\cos \alpha x|_{L/2}^{L} \\
&\quad - (4p/L^2)|(1/\alpha^2) \sin \alpha x - (x/\alpha) \cos \alpha x|_{L/2}^{L} \\
&= (4p/L^2)[(1/\alpha^2) \sin (n\pi/2) - (L/2\alpha) \cos (n\pi/2)] \\
&\quad - (4p/L\alpha)[\cos n\pi - \cos (n\pi/2)] \\
&\quad - (4p/L^2)\{[(1/\alpha^2) \sin n\pi - (L/\alpha) \cos n\pi] \\
&\quad - [(1/\alpha^2) \sin (n\pi/2) - (L/2\alpha) \cos (n\pi/2)]\} \\
&= (8p/L^2\alpha^2) \sin (n\pi/2) = (8p/\pi^2 n^2) \sin (n\pi/2),
\end{aligned}$$

$$\therefore \quad g(x) = \sum_{n=1}^{\infty} a_n \sin (n\pi x/L) = (8p/\pi^2) \sum_{n=1}^{\infty} (1/n^2) \sin (n\pi/2) \sin (n\pi x/L)$$

$$= (8p/\pi^2)[\sin (\pi x/L) - (1/3^2) \sin (3\pi x/L) + (1/5^2) \sin (5\pi x/L) - \cdots].$$

14.3 CYLINDRICAL OR POLAR PLANE STRESS AND STRAIN

Because many plane problems are best described with cylindrical coordinates (r, θ), it becomes necessary to re-express the constitutive relations, the strain–displacement relations, equilibrium, compatibility and the stress functions in terms of r and θ. Fortunately, the number of derivations in the theory may be reduced by suitable transformation from Cartesian coordinates. Firstly, note that the constitutive relations are identical in form to the Cartesian equations (14.1) and (14.2) where r and θ now replace x and y respectively (z remains in each case).

Strain–Displacement Relations

The dependence of the strain components ϵ_r, ϵ_θ and $\gamma_{r\theta}$ upon radial (u) and tangential (v) displacements are found from the distortion of the cylindrical line elements $OP(=\delta r)$ and $OQ = r\delta\theta$ as shown in Fig. 14.10. The radial strain in OP is found from the rate of change of u w.r.t. r as shown. That is,

$$\epsilon_r = [\delta r + (\partial u/\partial r)\delta r - \delta r]/\delta r = \partial u/\partial r. \tag{14.15a}$$

For the tangential or hoop strain in OQ, the increase in length of $r\delta\theta$ is due to both the rate of change of v w.r.t θ and the change in radius from r to $r + u$ as O displaces to O′ with polar coordinates u, v. Then,

$$\epsilon_\theta = [(r + u)\delta\theta + (\partial v/\partial\theta)\delta\theta - r\delta\theta]/r\delta\theta,$$

$$\epsilon_\theta = u/r + (1/r)\partial v/\partial\theta. \tag{14.15b}$$

The true change in the angle POQ, which defines the shear strain, must omit

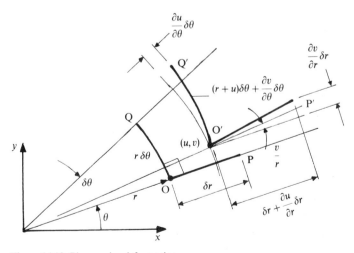

Figure 14.10 Plane polar deformation.

the rigid rotation (v/r radians) that contributes to the new position $O'P'$. This gives

$$\gamma_{r\theta} = (\partial v/\partial r)(\delta r/\delta r) - v/r + (\partial u/\partial\theta)(\delta\theta/r\,\delta\theta),$$
$$\gamma_{r\theta} = \partial v/\partial r - v/r + (1/r)(\partial u/\partial\theta). \tag{14.15c}$$

Equilibrium

Polar equilibrium equations are found from the consideration of radial and tangential force balance for the cylindrical element with unit thickness in Fig. 14.11. Note that the superimposed stress contributions to each direction have been separated in (a) and (b) for clarity. The hoop σ_θ, radial σ_r and shear $\tau_{r\theta}$ stresses vary across the element in the manner shown for the positive r and θ centre-line directions indicated. Since the inclinations of σ_θ and $\tau_{r\theta}$ vary with $\delta\theta$, these have been resolved w.r.t. centre-line directions. The following derivation applies in the absence of body forces. Apart from self-weight, an example of the latter is the centrifugal force produced in a rotating disc or cylinder. This is treated separately in Chapter 15.

For radial, centre-line, equilibrium for Fig. 14.11(a):

$$(\sigma_r + \delta r\,\partial\sigma_r/\partial r)(r + \delta r)\delta\theta - \sigma_r r\,\delta\theta - \tau_{r\theta}\delta r\cos(\delta\theta/2)$$
$$+ (\tau_{r\theta} + \delta\theta\,\partial\tau_{r\theta}/\partial\theta)\delta r\cos(\delta\theta/2)$$
$$- \sigma_\theta\delta r\sin(\delta\theta/2) - (\sigma_\theta + \delta\theta\,\partial\sigma_\theta/\partial\theta)\delta r\sin(\delta\theta/2) = 0.$$

Now when $\delta\theta \to 0$, $\cos(\delta\theta/2) \to 1$ and $\sin(\delta\theta/2) \to \delta\theta/2$. This gives

$$(\sigma_r + \delta r\,\partial\sigma_r/\partial r)(r + \delta r)\delta\theta - \sigma_r r\,\delta\theta - \tau_{r\theta}\delta r + (\tau_{r\theta} + \delta\theta\,\partial\tau_{r\theta}/\partial\theta)\delta r$$
$$- \sigma_\theta\delta r\,\delta\theta/2 - (\sigma_\theta + \delta\theta\,\partial\sigma_\theta/\partial\theta)\delta r\,\delta\theta/2 = 0.$$

Neglecting products of small quantities leads to

$$\sigma_r\delta r\,\delta\theta + r\,\delta r\,\delta\theta\,\partial\sigma_r/\partial r + \delta\theta\,\delta r\,\partial\tau_{r\theta}/\partial\theta - \sigma_\theta\delta r\,\delta\theta/2 - \sigma_\theta\delta r\,\delta\theta/2 = 0.$$

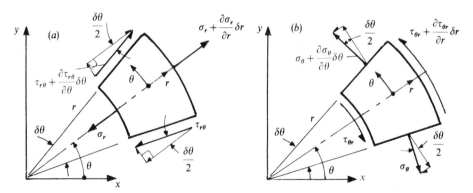

Figure 14.11 Stress variation on a plane cylindrical element.

Dividing through by $r\delta\theta\delta r$, gives the first equilibrium equation:

$$(\sigma_r - \sigma_\theta)/r + (1/r)\partial\tau_{r\theta}/\partial\theta + \partial\sigma_r/\partial r = 0. \tag{14.16a}$$

For tangential centre-line equilibrium (Fig. 14.11b),

$$(\tau_{\theta r} + \delta r\partial\tau_{\theta r}/\partial r)(r + \delta r)\delta\theta - \tau_{\theta r}r\delta\theta + (\tau_{\theta r} + \delta\theta\partial\tau_{\theta r}/\partial\theta)\delta r\sin(\delta\theta/2)$$
$$+ \tau_{\theta r}\delta r\sin(\delta\theta/2) + (\sigma_\theta + \delta\theta\partial\sigma_\theta/\partial\theta)\delta r\cos(\delta\theta/2) - \sigma_\theta\delta r\cos(\delta\theta/2) = 0,$$

and, with similar approximations, this reduces to the second equilibrium equation (noting that $\tau_{\theta r} = \tau_{r\theta}$):

$$(2/r)\tau_{\theta r} + (1/r)\partial\sigma_\theta/\partial\theta + \partial\tau_{\theta r}/\partial r = 0. \tag{14.16b}$$

Stress Component Functions

The reader should check that Eqs (14.16a and b) are satisfied by the following stress functions* where $\phi = \phi(r, \theta)$:

$$\sigma_r = (1/r)\partial\phi/\partial r + (1/r^2)\partial^2\phi/\partial\theta^2, \tag{14.17a}$$

$$\sigma_\theta = \partial^2\phi/\partial r^2, \tag{14.17b}$$

$$\tau_{r\theta} = -\partial[(1/r)(\partial\phi/\partial\theta)]\partial r$$
$$= (1/r^2)\partial\phi/\partial\theta - (1/r)(\partial^2\phi/\partial r\partial\theta). \tag{14.17c}$$

Bi-harmonic Equation

With $r^2 = x^2 + y^2$, $x = r\cos\theta$, $y = r\sin\theta$ (see Fig. 14.11) the following first and second partial derivatives apply:

$$\partial/\partial x = (\partial r/\partial x)\partial/\partial r + (\partial\theta/\partial x)\partial/\partial\theta$$
$$= \cos\theta(\partial/\partial r) - (1/r)\sin\theta(\partial/\partial\theta),$$
$$\partial^2/\partial x^2 = [\cos\theta(\partial/\partial r) - (1/r)\sin\theta(\partial/\partial\theta)][\cos\theta(\partial/\partial r) - (1/r)\sin\theta(\partial/\partial\theta)]$$
$$= \cos\theta(\partial/\partial r)[\cos\theta(\partial/\partial r) - (1/r)\sin\theta(\partial/\partial\theta)]$$
$$\quad - (1/r)\sin\theta(\partial/\partial\theta)[\cos\theta(\partial/\partial r) - (1/r)\sin\theta(\partial/\partial\theta)]$$
$$= \cos^2\theta(\partial^2/\partial r^2) + (1/r^2)\sin\theta\cos\theta(\partial/\partial\theta) + (1/r)\sin^2\theta(\partial/\partial r)$$
$$\quad + (1/r^2)\sin^2\theta(\partial^2/\partial\theta^2) + (1/r^2)\sin\theta\cos\theta(\partial/\partial\theta), \tag{14.18a}$$
$$\partial/\partial y = (\partial r/\partial y)\partial/\partial r + (\partial\theta/\partial y)\partial/\partial\theta,$$
$$= \sin\theta(\partial/\partial r) + (1/r)\cos\theta(\partial/\partial\theta),$$

---

* Equations (14.18) may also be used to transform σ_x and σ_y (Eqs (14.9a and b)) into σ_r and σ_θ (Eqs (14.17a and b)), noting that when $\theta = 0$, $\sigma_x = \sigma_r$ and $\sigma_y = \sigma_\theta$. The shear stress $\tau_{r\theta}$ (Eq. (14.17c)) then follows from either of the equilibrium equations (14.16).

$$\partial^2/\partial y^2 = [\sin\theta(\partial/\partial r) + (1/r)\cos\theta(\partial/\partial\theta)][\sin\theta(\partial/\partial r) + (1/r)\cos\theta(\partial/\partial\theta)]$$
$$= \sin\theta(\partial/\partial r)[\sin\theta(\partial/\partial r) + (1/r)\cos\theta(\partial/\partial\theta)]$$
$$+ (1/r)\cos\theta(\partial/\partial\theta)[\sin\theta(\partial/\partial r) + (1/r)\cos\theta(\partial/\partial\theta)]$$
$$= \sin^2\theta(\partial^2/\partial r^2) - (1/r^2)\sin\theta\cos\theta(\partial/\partial\theta) + (1/r)\cos^2\theta(\partial/\partial r)$$
$$+ (1/r^2)\cos^2\theta(\partial^2/\partial\theta^2) - (1/r^2)\sin\theta\cos\theta(\partial/\partial\theta). \tag{14.18b}$$

From the addition of Eqs (14.18a and b), the biharmonic equation becomes:

$$\nabla^4\phi(x, y) = (\partial^2/\partial x^2 + \partial^2/\partial y^2)^2\phi = 0,$$
$$\nabla^4\phi(r, \theta) = [\partial^2/\partial r^2 + (1/r)(\partial/\partial r) + (1/r^2)(\partial^2/\partial\theta^2)]^2\phi = 0. \tag{14.19}$$

14.4 POLAR STRESS FUNCTIONS

In the search for physically significant functions ϕ that satisfy Eq. (14.19), the following three classes of function are considered.

General function $\phi = \phi(\theta, r)$

A particularly useful function (Boussinesq, 1885) of this type that satisfies Eq. (14.19) has the following form:

$$\phi = Cr\theta\sin\theta.$$

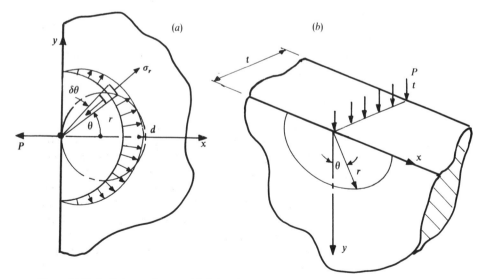

Figure 14.12 Radial stress in a semi-infinite plate under a normal point force.

Applying Eqs. (14.17), the corresponding stresses are $\sigma_\theta = \tau_{r\theta} = 0$ and

$$\sigma_r = (1/r)\partial\phi/\partial r + (1/r^2)\partial^2\phi/\partial\theta^2$$
$$= (C/r)\theta \sin\theta + (1/r^2)(\partial/\partial\theta)(Cr\sin\theta + Cr\theta\cos\theta)$$
$$= (2C/r)\cos\theta. \tag{14.20}$$

This radial stress distribution results from a point force applied normally to the straight edge of a large plate (semi-infinite) as shown in Fig. 14.12(a). For any given radius, σ_r varies with $\cos\theta$ between zero values for $\theta = \pm\pi/2$ and a maximum at $\theta = 0$ as shown.

The constant C in Eq. (14.20) is found from horizontal equilibrium for unit thickness:

$$P = \int_{-\pi 2}^{\pi/2} \sigma_r(r\,d\theta)\cos\theta,$$

$$\therefore \quad P = 2C\int_{-\pi/2}^{\pi/2}\cos^2\theta\,d\theta = C|\theta + (1/2)\sin 2\theta|_{-\pi/2}^{\pi/2} = C\pi.$$

In practice, when the force acts in a compressive line over thickness t (Fig. 14.12b), P is replaced by $-P/t$. Then from equilibrium, $-P/t = C\pi$ and Eq. (14.20) supplies the compressive stress:

$$\sigma_r = (2P/\pi tr)\cos\theta. \tag{14.21}$$

The maximum shear stress (not component $\tau_{r\theta}$) in the plate is given by

$$\tau_{max} = (\sigma_r - \sigma_\theta)/2 = \sigma_r/2. \tag{14.22}$$

For the circle of diameter d, tangential to the force point in Fig. 14.12(a); $r = d\cos\theta$ and Eq. (14.22) becomes

$$\tau_{max} = P/\pi\,dt.$$

That is, both σ_r and τ_{max} are constant around tangent circles. The theory has been confirmed on this basis when, in point loading a plane photoelastic model, the circles appear as isochromatic fringes each of constant shear stress value. The radial stress equation (14.21) also applies to a point tangential (i.e. shear) force applied to the straight edge of a semi-infinite plate provided θ is measured from the force line as shown in Fig. 14.13(a). Moreover, with a different constant C, Eq. (14.20) will also apply to point loading of finite bodies. In particular, normal, horizontal and inclined compressive forces applied to a wedge tip are amenable to this function, again when θ is measured anticlockwise from the force line of action.

In Fig. 14.13(b), for a wedge of thickness t and apex angle 2α, under an

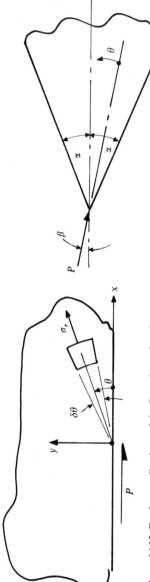

Figure 14.13 Further applications of the Boussinesq function.

inclined compressive tip force P, horizontal equilibrium gives

$$P/t = \int_{-(\alpha-\beta)}^{\alpha+\beta} \sigma_r(r \, d\theta) \cos\theta = 2C \int_{-(\alpha-\beta)}^{\alpha+\beta} \cos^2\theta \, d\theta$$

$$= C|\theta + (1/2)\sin 2\theta|_{-(\alpha-\beta)}^{\alpha+\beta} = C(2\alpha + \sin 2\alpha \cos 2\beta),$$

$$C = P/t(2\alpha + \sin 2\alpha \cos 2\beta),$$

$$\therefore \quad \sigma_r = 2P \cos\theta/rt[2\alpha + (\sin 2\alpha)(\cos 2\beta)] \quad \text{(compressive)}$$

The wedge stresses under horizontal and vertical tip forces follow from substituting $\beta = 0$ and $\beta = \pi/2$ respectively.

Example 14.10 Derive the strain and displacement expressions in the body of a semi-infinite plate of unit thickness under a point compressive normal force P (i.e. P is reversed in Fig. 14.12a).

Combining Eqs (14.1) and (14.15):

$$\epsilon_r = \partial u/\partial r = (\sigma_r - v\sigma_\theta)/E,$$

$$\epsilon_\theta = u/r + (1/r)\partial v/\partial\theta = (\sigma_\theta - v\sigma_r)/E,$$

$$\gamma_{r\theta} = \tau_{r\theta}/G = \partial v/\partial r - v/r + (1/r)\partial u/\partial\theta.$$

Substituting $\sigma_\theta = \tau_{r\theta} = 0$ and $\sigma_r = -(2P/\pi r)\cos\theta$:

$$\partial u/\partial r = -(2P/\pi Er)\cos\theta, \qquad \text{(i)}$$

$$u/r + (1/r)\partial u/\partial\theta = 2vP(\cos\theta)/\pi Er, \qquad \text{(ii)}$$

$$\partial v/\partial r - v/r + (1/r)\partial u/\partial\theta = 0. \qquad \text{(iii)}$$

Integrating Eq. (i):

$$u = -(2P/\pi E)(\cos\theta)\ln r + f(\theta). \qquad \text{(iv)}$$

Substituting (iv) in (ii):

$$-(2P/\pi E)(\cos\theta)\ln r + f(\theta) + \partial v/\partial\theta = (2vP/\pi E)\cos\theta,$$

$$\partial v/\partial\theta = (2vP/\pi E)\cos\theta + (2P/\pi E)(\cos\theta)\ln r - f(\theta),$$

$$v = \int[(2vP/\pi E)\cos\theta + (2P/\pi E)(\cos\theta)\ln r - f(\theta)] \, d\theta$$

$$\hspace{5cm} \text{(v)}$$

$$= (2vP/\pi E)\sin\theta + (2P/\pi E)(\sin\theta)\ln r - \int f(\theta) \, d\theta + g(r).$$

From Eqs (iv) and (v) the partial derivatives are

$$\partial u/\partial\theta = (2P/\pi E)(\sin\theta)\ln r + f'(\theta), \qquad \text{(vi)}$$

$$\partial v/\partial r = (2P/\pi Er)\sin\theta + g'(r). \qquad \text{(vii)}$$

Substituting Eqs (v), (vi) and (vii) into (iii) gives

$$f'(\theta) + (2P/\pi E)(1 - v)\sin\theta + rg'(r) + \int f(\theta)\,d\theta - g(r) = 0.$$

Since the R.H.S. is zero the separate sum of isolated functions in r and θ must equal zero. That is,

$$rg'(r) - g(r) = 0 \quad \text{or} \quad dg/g = dr/r,$$

$$\therefore \quad \ln g = \ln r + \ln A, \quad \Rightarrow g(r) = A_1 r \tag{viii}$$

and

$$f'(\theta) + (2P/\pi E)(1 - v)\sin\theta + \int f(\theta)\,d\theta = 0,$$

$$d^2 f/d\theta^2 + f = -(2P/\pi E)(1 - v)\cos\theta.$$

The solution

$$f(\theta) = A_2 \sin\theta + A_3 \cos\theta - (1 - v)(P/\pi E)\theta\sin\theta \tag{ix}$$

results in the general u, v displacement expressions when Eqs (viii) and (ix) are substituted into (iv) and (v):

$$u = A_2 \sin\theta + A_3 \cos\theta - (2P/\pi E)(\cos\theta)\ln r - (1 - v)(P/\pi E)\theta\sin\theta,$$

$$v = (1 + v)(P/\pi E)\sin\theta + (2P/\pi E)(\sin\theta)\ln r$$
$$- (1 - v)(P/\pi E)\theta\cos\theta + A_1 r + A_2 \cos\theta - A_3 \sin\theta.$$

Since $v = 0$ when $\theta = 0$ along the x-axis, then $A_2 = -A_1 r$. This can only be satisfied by $A_2 = A_1 = 0$ for all r. Because deformation is localized, it is assumed that $u = 0$ at some radius r_o from P along the x-axis ($\theta = 0$). Then

$$u = 0 = A_3 - (2P/\pi E)\ln r_o,$$

$$A_3 = (2P/\pi E)\ln r_o.$$

Finally, the plate displacements at any point θ, $r \le r_o$, may be found from

$$u = (P/\pi E)[2\ln(r_o/r)\cos\theta - (1 - v)\theta\sin\theta],$$

$$v = (P/\pi E)[(1 + v)\sin\theta + 2(\sin\theta)\ln(r/r_o) - (1 - v)\theta\cos\theta].$$

Functions of θ Only, $\phi = \phi(\theta)$

Since this function is independent of r, then $\partial\phi/\partial r = 0$ and the biharmonic Eq. (14.19) becomes

$$[\partial^2/\partial r^2 + (1/r)(\partial/\partial r) + (1/r^2)(\partial^2/\partial\theta^2)][(1/r^2)\partial^2\phi/\partial\theta^2] = 0,$$

$$(\partial^2/\partial r^2)[(1/r^2)\partial^2\phi/\partial\theta^2] + (1/r)(\partial/\partial r)[(1/r^2)\partial^2\phi/\partial\theta^2] + (1/r^4)(\partial^4\phi/\partial\theta^4) = 0,$$

$$(\partial/\partial r)[(-2/r^3)\partial^2\phi/\partial\theta^2] + (1/r)[(-2/r^3)\partial^2\phi/\partial\theta^2] + (1/r^4)\partial^4\phi/\partial\theta^4 = 0,$$

$$(6/r^4)\,d^2\phi/d\theta^2 - (2/r^4)\,d^2\phi/d\theta^2 + (1/r^4)\,d^4\phi/d\theta^4 = 0,$$

$$d^4\phi/d\theta^4 + 4d^2\phi/d\theta^2 = 0.$$

The solution, obtained by putting $\phi = N \exp(m\theta)$, has the final form

$$\phi = A \cos 2\theta + B \sin 2\theta + C\theta. \tag{14.23}$$

The corresponding stress functions Eqs (14.17) become $\sigma_\theta = 0$ and

$$\sigma_r = (1/r^2) \, d^2\phi/d\theta^2, \tag{14.24a}$$

$$\tau_{r\theta} = (1/r^2) \, d\phi/d\theta. \tag{14.24b}$$

Any part of Eq. (14.23) is a valid function, though in practice it has limited applications. Using the function $\phi = B \sin 2\theta + C\theta$ in Eqs (14.24) will prescribe the stresses in the body of a wedge of unit thickness when a moment is applied at its tip. The simple function $\phi = C\theta$ is the most useful, since it represents torsion of a thin disc or rotating wheel, where, from Eqs (14.24), the shear stress is $\tau_{r\theta} = (C/r^2)$ and $\sigma_\theta = \sigma_r = 0$. If the torque is applied to the axis of a solid disc of radius R, then C is found from the equilibrium condition.

$$T = 2\pi \int_0^R \tau_{r\theta} r^2 \, dr = 2\pi C \int_0^R dr = 2\pi CR,$$

$$\therefore \quad C = T/2\pi R.$$

When the torque is applied to the axis of a thin annulus with inner radius

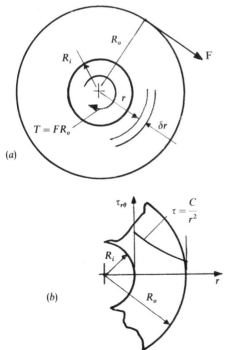

(a)

(b)

Figure 14.14 Disc under torque.

R_i and outer radius R_o (Fig. 14.14a) the equilibrium condition becomes

$$T = 2\pi \int_{R_i}^{R_o} \tau_{r\theta} r^2 \, dr = 2\pi C \int_{R_i}^{R_o} dr = 2\pi C(R_o - R_i),$$

$$C = T/2\pi(R_o - R_i) = FR_o/2\pi(R_o - R_i),$$

and $\tau_{r\theta}$ varies inversely with r^2 from a maximum value at R_i to a minimum at R_o as shown in Fig. 14.14(b).

Functions of r Only, $\phi = \phi(r)$

This function describes symmetry about the z-axis where points in the r, θ plane displace only radially. Thus, $v = \partial v/\partial r = \partial v/\partial \theta = \partial u/\partial \theta = 0$ when (Eqs (14.15) give $\gamma_{r\theta} = 0$ and

$$\epsilon_r = du/dr, \qquad \epsilon_\theta = u/r. \qquad (14.25a, b)$$

Here two strains depend upon a single displacement u. The compatibility condition expressing this dependence is found from

$$\epsilon_r = du/dr = d(r\epsilon_\theta)/dr = r \, d\epsilon_\theta/dr + \epsilon_\theta,$$

$$\therefore \quad \epsilon_r - \epsilon_\theta = r \, d\epsilon_\theta/dr.$$

Since ϕ is independent of θ, then $\partial\phi/\partial\theta = 0$ and Eq. (14.19) reduces to

$$[d^2/dr^2 + (1/r)d/dr][d^2\phi/dr^2 + (1/r)d\phi/dr] = 0,$$
$$d^4\phi/dr^4 + (d^2/dr^2)[(1/r)d\phi/dr] + (1/r)d^3\phi/dr^3 + (1/r)(d/dr)[(1/r)d\phi/dr] = 0,$$
$$d^4\phi/dr^4 + (2/r)d^3\phi/dr^3 - (1/r^2)d^2\phi/dr^2 + (1/r^3)d\phi/dr = 0.$$

The solution, obtained by changing the variable with the substitution $r = \exp(t)$, has the final form

$$\phi = A \ln(r) + Br^2 \ln(r) + Cr^2. \qquad (14.26)$$

Three useful functions ϕ may be identified from Eq. (14.26), where from Eq. (14.17), the associated stress functions become $\tau_{r\theta} = 0$ and

$$\sigma_r = (1/r)d\phi/dr, \qquad (14.27a)$$
$$\sigma_\theta = d^2\phi/dr^2. \qquad (14.27b)$$

(a) $\phi = Cr^2$ Eqs (14.27) give $\sigma_r = \sigma_\theta = 2C$ which describes a plane stress field that is uniform in all directions, i.e., as produced in the wall of a thin-walled spherical vessel under pressure or by the action of coplanar, concurrent forces applied in the plane of a plate.

(b) $\phi = A \ln(r) + Cr^2$ Equation (14.27) supplies the stress components:

$$\sigma_r = 2C + A/r^2, \qquad \sigma_\theta = 2C - A/r^2. \qquad (14.28a, b)$$

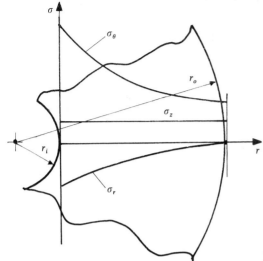

Figure 14.15 Thick-walled cylinder.

This is the familiar Lamé solution (1852) (see page 647) describing the stress distribution in the wall of a long, thick-walled cylinder (Fig. 14.15) subjected to internal and external pressures. The axial stress σ_z, may be found from the plane strain condition that the axial strain ϵ_z, may either be zero or constant. The latter may be identified with a closed-end cylinder where, from horizontal equilibrium between the forces exerted by the axial stress and the internal pressure on the ends,

$$\sigma_z \pi (r_o^2 - r_i^2) = p\pi r_i^2,$$
$$\therefore \quad \sigma_z = p r_i^2 / (r_o^2 - r_i^2). \tag{14.29}$$

Example 14.11 Determine the constants A and C in the Lamé equations given that the cylinder is subjected to internal pressure only and that the axial strain is zero. Plot the distribution of σ_r, σ_θ and σ_z and derive a general expression for the radial displacement.

The boundary conditions are $\sigma_r = -p_i$ for $r = r_i$ and $\sigma_r = 0$ for $r = r_o$. Substituting into Eq. (14.28a) leads to the two simultaneous equations:

$$-p_i = 2C + A/r_i^2, \tag{i}$$
$$0 = 2C + A/r_o^2. \tag{ii}$$

Subtracting (ii) from (i) gives

$$A = -p r_i^2 r_o^2 / (r_o^2 - r_i^2), \qquad 2C = p r_i^2 / (r_o^2 - r_i^2).$$

Substituting from A and $2C$ in Eqs (14.28a and b):

$$\sigma_r = pr_i^2(1 - r_o^2/r^2)/(r_o^2 - r_i^2),$$
$$\sigma_\theta = pr_i^2(1 + r_o^2/r^2)/(r_o^2 - r_i^2).$$

When $\epsilon_z = 0$, Eq. (14.2c) gives, on replacing x and y with r and θ

$$\sigma_z = v(\sigma_r + \sigma_\theta) = 2vpr_i^2/(r_o^2 - r_i^2). \tag{iii}$$

It is apparent from Eqs (iii) and (14.29) that the constant axial stress σ_z depends upon the end condition. The triaxial stresses are distributed as shown in Fig. 14.15. The radial stress σ_r is a maximum in compression and the hoop stress σ_θ is a maximum in tension, both at the inner radius. The radial displacement follows from Eq. (iii) and either of Eqs (14.25):

$$u = r\epsilon_\theta = r[\sigma_\theta - v(\sigma_r + \sigma_z)]/E,$$
$$u = r[\sigma_\theta(1 - v^2) - v\sigma_r(1 + v)]/E$$
$$= prr_i^2[(1 - v^2)(1 + r_o^2/r^2) - v(1 + v)(1 - r_o^2/r^2)]/E(r_o^2 - r_i^2)$$
$$= pr_i^2[(1 - v - 2v^2)r + (1 + v)r_o^2/r]/E(r_o^2 - r_i^2).$$

(c) $\phi = A \ln(r) + Br^2 \ln(r) + Cr^2$ With the complete function ϕ in Eq. (14.26), the radial and tangential stresses, found from Eq. (14.24), are

$$\sigma_r = A/r^2 + B(1 + 2\ln r) + 2C, \tag{14.30a}$$
$$\sigma_\theta = -A/r^2 + B(3 + 2\ln r) + 2C. \tag{14.30b}$$

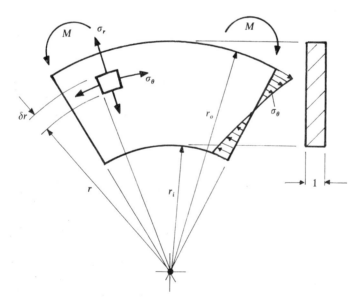

Figure 14.16 Bending of a curved bar.

These provide an exact solution to pure bending of a curved beam (Golovin, 1881). Thus, in Fig. 14.16, for end moments M, applied to a rectangular beam of unit thickness with inner and outer radii r_i and r_o respectively, the boundary conditions are $\sigma_r = 0$ for $r = r_i$ and $r = r_o$, when Eq. (14.30a) gives

$$\sigma_r = [(1/r)\,d\phi/dr]_{r_i} = A/r_i^2 + B(1 + 2\ln r_i) + 2C = 0, \qquad (14.31a)$$

$$\sigma_r = [(1/r)\,d\phi/dr]_{r_o} = A/r_o^2 + B(1 + 2\ln r_o) + 2C = 0, \qquad (14.31b)$$

together with the equilibrium boundary condition

$$M = \int_{r_i}^{r_o} \sigma_\theta r\,dr = \int_{r_i}^{r_o} (d^2\phi/dr^2)r\,dr = |r\,d\phi/dr|_{r_i}^{r_o} - \int_{r_i}^{r_o} (d\phi/dr)\,dr,$$

where, from Eqs (14.31a and b), the first term is zero. Then, substituting from Eq. (14.26):

$$M = -|\phi|_{r_i}^{r_o} = -[A\ln(r_o/r_i) + B(r_o^2\ln r_o - r_i^2\ln r_i) + C(r_o^2 - r_i^2)] \qquad (14.32)$$

The constants A, B and C are found from the simultaneous solution to Eqs (14.31) and (14.32). This results in the final stress expressions

$$\sigma_r = \frac{4M[(r_i^2 r_o^2/r^2)\ln(r_o/r_i) + r_o^2\ln(r/r_o) + r_i^2\ln(r_i/r)]}{(r_o^2 - r_i^2)^2 - 4r_o^2 r_i^2[\ln(r_o/r_i)]^2},$$

$$\sigma_\theta = \frac{4M[-(r_o^2 r_i^2/r^2)\ln(r_o/r_i) + r_o^2\ln(r/r_o) + r_i^2\ln(r_i/r) + (r_o^2 - r_i^2)]}{(r_o^2 - r_i^2)^2 - 4r_o^2 r_i^2[\ln(r_o/r_i)]^2}.$$

These exact distributions satisfy equilibrium, compatibility and the boundary conditions. It follows that the alternative solution to this problem (Eq. (6.14)) differs because it does not satisfy the condition of compatibility. For an extended treatment of the theory of elasticity, in which the Cartesian and polar equilibrium and compatibility conditions are expressed in full stress and strain space, the reader is referred to the following texts: H. Ford and J. M. Alexander, *Advanced Mechanics of Materials*, Longmans, 1963; A. E. H. Love, *A Treatise on the Mathematical Theory of Elasticity*, fourth edition, Cambridge University Press, 1934.

14.5 PLANE FINITE ELEMENTS

It is not the intention to give here a detailed exposition of all that the finite element method entails. For this the reader should refer to one of the many texts that have been devoted to the subject amongst which is *The Finite Element Method in Structural and Continuum Mechanics* by O. C. Zienkiewicz and Y. K. Cheung, McGraw-Hill, 1967. Instead, this introduction will simply reveal how the key elements of the theory of plane elasticity, outlined earlier, are employed with the subdivision of a plane or two-dimensional body into smaller elements interconnected throughout at nodal points. In this technique of discretization, an assemblage of triangular and/or rectangular elements may be chosen to describe the initial unstrained shape of the body. When it is elastically strained, the behaviour of the entire body may be computed from the known elastic behaviour of its elements. The great advantage of the method is that it can provide acceptable numerical solutions to the stresses, strains and displacements suffered by a body under load when these are not readily furnished from classical solutions. This particularly applies where the loading is complex and the geometry is irregular. It should be emphasized, however, that the use of the method on classical bodies cannot better a closed solution that provides for equilibrium and compatibility of its internal stress and strain distributions and matches the external boundary conditions. Indeed, in practice, one should always check the availability of an acceptable alternative before embarking on an expensive computer run.

Triangular Elements

In following the stiffness or displacement finite-element method, the u, v displacements at the nodal points are the unknowns. For the plane triangular elements (e) in Fig. 14.17(a), the nodes are first numbered 1, 2 and 3 in an anticlockwise direction.

Displacements The nodal coordinates are then (x_1, y_1), (x_2, y_2), and (x_3, y_3). The displacements of each node may then be expressed in a function $u(x, y)$ and $v(x, y)$ of these coordinates. The simplest linear functions are assumed:

$$u(x, y) = \alpha_1 + \alpha_2 x + \alpha_3 y, \tag{14.33a}$$

$$v(x, y) = \alpha_4 + \alpha_5 x + \alpha_6 y, \tag{14.33b}$$

where the six unknown coefficients α match the total number of degrees of freedom for the element, i.e. two per node. The linear polynomial will also ensure displacement continuity along the sides of adjacent elements. The constants α_1 and α_4 account for the rigid body translation shown in Fig. 13.31(a). Equation (14.33a and b) are combined in matrix form to give the displacements $\{\delta\} = \{u\,v\}^{\mathrm{T}}$ for any point (x, y) in the element:

$$\{\delta\} = [f]\{\alpha\}, \tag{14.33c}$$

where $\{\alpha\} = \{\alpha_1, \alpha_2, \alpha_3, \alpha_4, \alpha_5, \alpha_6\}^{\mathrm{T}}$ and $[f]$ is the 6 × 2 matrix in x and y:

$$[f] = \begin{bmatrix} 1 & x & y & 0 & 0 & 0 \\ 0 & 0 & 0 & 1 & x & y \end{bmatrix}.$$

In particular, substituting x and y for each node in Eq. (14.33c) gives $\{\delta_1\} = \{u_1 \quad v_1\}^{\mathrm{T}}$, $\{\delta_2\} = \{u_2 \quad v_2\}^{\mathrm{T}}$ and $\{\delta_3\} = \{u_3 \quad v_3\}^{\mathrm{T}}$. These may be combined to give the element nodal point displacements:

$$\{\delta^{\mathrm{e}}\} = [A]\{\alpha\}, \tag{14.34a}$$

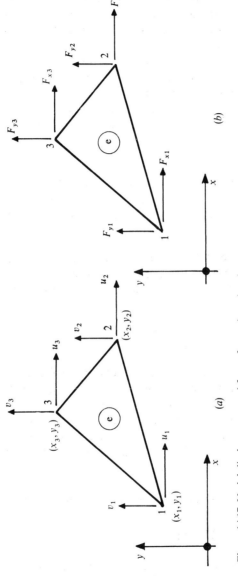

Figure 14.17 Nodal displacements and forces for a triangular element.

where $\{\delta^e\} = \{\{\delta_1\}\{\delta_2\}\{\delta_3\}\}^T$ and $[A]$ is a 6×6 matrix derived from $[f]$. Written in full this becomes

$$\{\delta^e\} = \begin{Bmatrix} \{\delta_1\} \\ \{\delta_2\} \\ \{\delta_3\} \end{Bmatrix} = \begin{Bmatrix} u_1 \\ v_1 \\ u_2 \\ v_2 \\ u_3 \\ v_3 \end{Bmatrix} = \begin{bmatrix} 1 & x_1 & y_1 & 0 & 0 & 0 \\ 0 & 0 & 0 & 1 & x_1 & y_1 \\ 1 & x_2 & y_2 & 0 & 0 & 0 \\ 0 & 0 & 0 & 1 & x_2 & y_2 \\ 1 & x_3 & y_3 & 0 & 0 & 0 \\ 0 & 0 & 0 & 1 & x_3 & y_3 \end{bmatrix} \begin{Bmatrix} \alpha_1 \\ \alpha_2 \\ \alpha_3 \\ \alpha_4 \\ \alpha_5 \\ \alpha_6 \end{Bmatrix}. \qquad (14.34b)$$

The coefficients α then follow from inverting matrix $[A]$ in Eq. (14.34a) to give

$$\{\alpha\} = [A^{-1}]\{\delta^e\}. \qquad (14.34c)$$

The determination of the components of $[A^{-1}]$ amount to solving for α_1, α_2, α_3, α_4, α_5 and α_6 between the six equations that (14.34b) represents. This gives

$$\alpha_1 = \frac{(x_2y_3 - x_3y_2)u_1 + (x_3y_1 - x_1y_3)u_2 + (x_1y_2 - x_2y_1)u_3}{(y_2 - y_3)(x_1 - x_2) - (y_1 - y_2)(x_2 - x_3)},$$

$$\alpha_2 = \frac{(y_2 - y_3)u_1 + (y_3 - y_1)u_2 + (y_1 - y_2)u_3}{(y_2 - y_3)(x_1 - x_2) - (y_1 - y_2)(x_2 - x_3)},$$

$$\alpha_3 = \frac{(x_3 - x_2)u_1 + (x_1 - x_3)u_2 + (x_2 - x_1)u_3}{(y_2 - y_3)(x_1 - x_2) - (y_1 - y_2)(x_2 - x_3)},$$

$$\alpha_4 = \frac{(x_2y_3 - x_3y_2)v_1 + (x_3y_1 - x_1y_3)v_2 + (x_1y_2 - x_2y_1)v_3}{(y_2 - y_3)(x_1 - x_2) - (y_1 - y_2)(x_2 - x_3)},$$

$$\alpha_5 = \frac{(y_2 - y_3)v_1 + (y_3 - y_1)v_2 + (y_1 - y_2)v_3}{(y_2 - y_3)(x_1 - x_2) - (y_1 - y_2)(x_2 - x_3)},$$

$$\alpha_6 = \frac{(x_3 - x_2)v_1 + (x_1 - x_3)v_2 + (x_2 - x_1)v_3}{(y_2 - y_3)(x_1 - x_2) - (y_1 - y_2)(x_2 - x_3)}.$$

Note that as the denominator in these expressions is twice the area (Δ) of the triangular element, it follows that $[A^{-1}]$ in Eq. (14.34c) is

$$[A^{-1}] = \frac{1}{2\Delta} \begin{bmatrix} (x_2y_3 - x_3y_2) & 0 & (x_3y_1 - x_1y_3) & 0 & (x_1y_2 - x_2y_1) & 0 \\ (y_2 - y_3) & 0 & (y_3 - y_1) & 0 & (y_1 - y_2) & 0 \\ (x_3 - x_2) & 0 & (x_1 - x_3) & 0 & (x_2 - x_1) & 0 \\ 0 & (x_2y_3 - x_3y_2) & 0 & (x_3y_1 - x_1y_3) & 0 & (x_1y_2 - x_2y_1) \\ 0 & (y_2 - y_3) & 0 & (y_3 - y_1) & 0 & (y_1 - y_2) \\ 0 & (x_3 - x_2) & 0 & (x_1 - x_3) & 0 & (x_2 - x_1) \end{bmatrix}$$

$$(14.34d)$$

The reader should now check that $[A][A^{-1}] = I$. Combining Eqs (14.33c) and (14.34c) expresses displacements $\{\delta\}$ within the element in terms of those displacements $\{\delta^e\}$ at its nodes. This gives

$$\{\delta\} = [f][A^{-1}]\{\delta^e\}. \tag{14.35}$$

Strains The following strain components now follow from Eqs (14.4) and Eqs (14.33):

$$\epsilon_x = \partial u/\partial x = \alpha_2,$$
$$\epsilon_y = \partial v/\partial y = \alpha_6,$$
$$\gamma_{xy} = \partial u/\partial y + \partial v/\partial x = \alpha_3 + \alpha_5,$$

which obviously satisfy the compatibility equation (14.5). In matrix form, these strains become

$$\{\epsilon\} = [C]\{\alpha\}, \tag{14.36a}$$

where $\{\epsilon\} = \{\epsilon_x \quad \epsilon_y \quad \gamma_{xy}\}^T$, $\{\alpha\} = \{\alpha_1 \quad \alpha_2 \quad \alpha_3 \quad \alpha_4 \quad \alpha_5 \quad \alpha_5 \quad \alpha_6\}^T$ and for which the 6×3 matrix $[C]$ is

$$[C] = \begin{bmatrix} 0 & 1 & 0 & 0 & 0 & 0 \\ 0 & 0 & 0 & 0 & 0 & 1 \\ 0 & 0 & 1 & 0 & 1 & 0 \end{bmatrix}. \tag{14.36b}$$

Substitution of Eq. (14.34c) into Eq. (14.36a) expresses the strains $\{\epsilon\}$ at any point (x, y) within the element in terms of the nodal point displacements $\{\delta^e\}$. That is,

$$\{\epsilon\} = [C][A^{-1}]\{\delta^e\}, \tag{14.37a}$$
$$\{\epsilon\} = [B]\{\delta^e\}, \tag{14.37b}$$

where $[B] = [C][A^{-1}]$ is found from matrix multiplication of Eqs (14.34d) and (14.36b). This gives

$$[B] = \frac{1}{2\Delta} \begin{bmatrix} y_2 - y_3 & 0 & y_3 - y_1 & 0 & y_1 - y_2 & 0 \\ 0 & x_3 - x_2 & 0 & x_1 - x_3 & 0 & x_2 - x_1 \\ x_3 - x_2 & y_2 - y_3 & x_1 - x_3 & y_3 - y_1 & x_2 - x_1 & y_1 - y_2 \end{bmatrix}. \tag{14.37c}$$

Note that, since both $[B]$ in (14.37c) and $\{\delta^e\}$ in Eq. (14.34b) appear only in terms of the nodal point coordinates and their displacements, the strain components throughout the element are constant. This is a consequence of the linear displacement functions assumed in Eqs (14.33).

Constitutive relations The strain components $\{\epsilon\}$ are related to the stress components $\{\sigma\} = \{\sigma_x \quad \sigma_y \quad \tau_{xy}\}^T$, with the following elastic constitutive relation:

$$\{\epsilon\} = [P]\{\sigma\}. \tag{14.38a}$$

However, it is seen from Eqs (14.1) and (14.2) that the 3×3 material flexibility matrix $[P]$ will depend upon whether the problem is one of plane stress or plane strain. For plane stress, Eq. (14.38a) becomes, in full:

$$\left\{ \begin{array}{c} \epsilon_x \\ \epsilon_y \\ \gamma_{xy} \end{array} \right\} = (1/E) \begin{bmatrix} 1 & -v & 0 \\ -v & 1 & 0 \\ 0 & 0 & 2(1+v) \end{bmatrix} \left\{ \begin{array}{c} \sigma_x \\ \sigma_y \\ \tau_{xy} \end{array} \right\}. \tag{14.38b}$$

Note that the remaining 'through thickness' strain ϵ_z in Eqs (14.1c) need not appear, since it depends upon σ_x and σ_y. For plane strain, with $\epsilon_o = 0$ in Eq. (14.2c) the full form of Eq. (14.38a) is

$$\left\{ \begin{array}{c} \epsilon_x \\ \epsilon_y \\ \gamma_{xy} \end{array} \right\} = (1/E) \begin{bmatrix} (1-v^2) & -v(1+v) & 0 \\ -v(1+v) & (1-v^2) & 0 \\ 0 & 0 & 2(1+v) \end{bmatrix} \left\{ \begin{array}{c} \sigma_x \\ \sigma_y \\ \tau_{xy} \end{array} \right\}. \tag{14.38c}$$

Since $[P]$ is a square matrix it may be inverted to find the internal stresses in terms of known strain. The resulting 3×3 material stiffness matrix $[D] = [P^{-1}]$ will differ between plane stress and plane strain. That is,

$$\{\sigma\} = [P^{-1}]\{\epsilon\}, \tag{14.39a}$$

$$\{\sigma\} = [D]\{\epsilon\}. \tag{14.39b}$$

Solving the three simultaneous equations (14.38b) for σ_x, σ_y and τ_{xy} defines $[D]$ for plane stress:

$$[D] = \frac{E}{(1-v^2)} \begin{bmatrix} 1 & v & 0 \\ v & 1 & 0 \\ 0 & 0 & (1-v)/2 \end{bmatrix} \tag{14.39c}$$

and from Eqs (14.38c) for plane strain:

$$[D] = \frac{E(1-v)}{(1+v)(1-2v)} \begin{bmatrix} 1 & v/(1-v) & 0 \\ v/(1-v) & 1 & 0 \\ 0 & 0 & (1-2v)/2(1-v) \end{bmatrix}. \tag{14.39d}$$

We can proceed with matrix $[D]$ provided it is remembered that its components d_{ij} in Eqs (14.39c and d) depend upon the type of plane problem.

Element stiffness matrix An element stiffness matrix $[K^e]$, of dimensions 6×6, relates the nodal point forces $\{F_1\} = \{F_{x1} \quad F_{y1}\}^T, \{F_2\} = \{F_{x2} \quad F_{y2}\}^T$ and $\{F_3\} = \{F_{x3} \quad F_{y3}\}^T$ in Fig. 14.17(b) to the nodal point displacements:

$$\{F^e\} = [K^e]\{\delta^e\}, \tag{14.40a}$$

where the force vector is $\{F^e\} = \{\{F_1\} \ \{F_2\} \ \{F_3\}\}^T = \{F_{x1} \ F_{y1} \ F_{x2} \ F_{y2} \ F_{x3} \ F_{y3}\}^T$ and the displacement vector is $\{\delta^e\} = \{\{\delta_1\} \ \{\delta_2\} \ \{\delta_3\}\}^T = \{u_1 \ v_1 \ u_2 \ v_2 \ u_3 \ v_3\}^T$. The problem is thus resolved into that of defining $[K^e]$ in Eq. (14.40a) in terms of the known nodal point coordinates $(x_i, y_i, i = 1, 2, 3)$

and the elastic constants E and v for the material. For this we can either use the principle of virtual work or stationary potential energy as outlined in Chapter 12. In using virtual work, the principle of virtual displacements becomes the basis for finding $[K^e]$. In the present matrix notation, Eq. (12.16a) is written

$$\{\delta^{ve}\}^T\{F^e\} = \int_V \{\epsilon^v\}^T\{\sigma\}\,dV, \qquad (14.41a)$$

where respectively $\{\delta^{ve}\}$ and $\{\epsilon^v\}$ are the virtual displacements and strains, $\{F^e\}$ and $\{\sigma\}$ are the real forces and stresses all at the nodal points. Equation (14.41a) maintains the necessary equilibrium relationship between the 'real' quantities. In fact $\{F^e\}$ are nodal forces that are statically equivalent to the internal stresses. They become actual forces when identified with those that are applied at nodal points around the boundary. The reversal in the order of the quantities between Eqs (12.16a) and (14.41a), and the choice of which vector to transpose, does not alter the magnitude of the resulting scalar products. In the R.H.S., for example; $\{\sigma\}\{\epsilon\}^T = \{\epsilon\}^T\{\sigma\} = \{\sigma\}^T\{\epsilon\} = \{\epsilon\}\{\sigma\}^T$. Note that $\{\ \}$ denotes a column vector and $\{\ \}^T$ denotes a single-row matrix in these components. Substituting Eqs (14.37b) and (14.39b) into Eq. (14.41a) leads to

$$\{\delta^{ve}\}^T\{F^e\} = \int_V ([B]\{\delta^{ve}\})^T[D][B]\{\delta^e\}\,dV, \qquad (14.41b)$$

where the integration is carried out over the volume V of the element. Since neither $[B]$ nor $[D]$ depends upon x and y and $([B]\{\delta^{ve}\})^T = \{\delta^{ve}\}^T[B]^T$, Eq. (14.41b) becomes

$$\{\delta^{ve}\}^T\{F^e\} = \{\delta^{ve}\}^T[B]^T[D][B]\{\delta^e\} \int dV.$$

Putting $\int dV = V = \Delta t$ where Δ and t are the area and thickness of the triangle and cancelling the virtual displacements gives

$$\{F^e\} = [B]^T[D][B]\{\delta^e\}V. \qquad (14.41c)$$

Comparing Eqs (14.40a) and (14.41c) defines the element stiffness matrix:

$$[K^e] = [B]^T[D][B]V. \qquad (14.42a)$$

Thus, the components of $[K^e]$ may be found explicitly from multiplying the matrices $[B]$ and $[D]$ strictly in the order of the R.H.S. of Eq. (14.42a). The result is the 6×6 symmetrical matrix

$$[K^e] = \begin{bmatrix} k^e_{11} & k^e_{12} & k^e_{13} & k^e_{14} & k^e_{15} & k^e_{16} \\ k^e_{21} & k^e_{22} & k^e_{23} & k^e_{24} & k^e_{25} & k^e_{26} \\ k^e_{31} & k^e_{32} & k^e_{33} & k^e_{34} & k^e_{35} & k^e_{36} \\ k^e_{41} & k^e_{42} & k^e_{43} & k^e_{44} & k^e_{45} & k^e_{46} \\ k^e_{51} & k^e_{52} & k^e_{53} & k^e_{54} & k^e_{55} & k^e_{56} \\ k^e_{61} & k^e_{62} & k^e_{63} & k^e_{64} & k^e_{65} & k^e_{66} \end{bmatrix}, \qquad (14.42b)$$

where the leading diagonal components are

$$k^e_{11} = (t/4\Delta)[d_{11}(y_2 - y_3)^2 + d_{33}(x_3 - x_2)^2]$$
$$k^e_{22} = (t/4\Delta)[d_{22}(x_3 - x_2)^2 + d_{33}(y_2 - y_3)^2]$$
$$k^e_{33} = (t/4\Delta)[d_{11}(y_3 - y_1)^2 + d_{33}(x_1 - x_3)^2] \qquad (14.42\text{c})$$
$$k^e_{44} = (t/4\Delta)[d_{22}(x_1 - x_3)^2 + d_{33}(y_3 - y_1)^2]$$
$$k^e_{55} = (t/4\Delta)[d_{11}(y_1 - y_2)^2 + d_{33}(x_2 - x_1)^2]$$
$$k^e_{66} = (t/4\Delta)[d_{22}(x_2 - x_1)^2 + d_{33}(y_1 - y_2)^2]$$

and the off-diagonal components are

$$k^e_{12} = k^e_{21} = (t/4\Delta)[d_{12}(x_3 - x_2)(y_2 - y_3) + d_{33}(x_3 - x_2)(y_2 - y_3)]$$
$$k^e_{13} = k^e_{31} = (t/4\Delta)[d_{11}(y_3 - y_1)(y_2 - y_3) + d_{33}(x_3 - x_2)(x_1 - x_3)]$$
$$k^e_{14} = k^e_{41} = (t/4\Delta)[d_{12}(x_1 - x_3)(y_2 - y_3) + d_{33}(x_3 - x_2)(y_3 - y_1)]$$
$$k^e_{15} = k^e_{51} = (t/4\Delta)[d_{11}(y_1 - y_2)(y_2 - y_3) + d_{33}(x_3 - x_2)(x_2 - x_1)]$$
$$k^e_{16} = k^e_{61} = (t/4\Delta)[d_{12}(x_2 - x_1)(y_2 - y_3) + d_{33}(x_3 - x_2)(y_1 - y_2)]$$
$$k^e_{23} = k^e_{32} = (t/4\Delta)[d_{12}(x_3 - x_2)(y_3 - y_1) + d_{33}(x_1 - x_3)(y_2 - y_3)]$$
$$k^e_{24} = k^e_{42} = (t/4\Delta)[d_{22}(x_3 - x_2)(x_1 - x_3) + d_{33}(y_3 - y_1)(y_2 - y_3)]$$
$$k^e_{25} = k^e_{52} = (t/4\Delta)[d_{12}(x_3 - x_2)(y_1 - y_2) + d_{33}(x_2 - x_1)(y_2 - y_3)] \qquad (14.42\text{d})$$
$$k^e_{26} = k^e_{62} = (t/4\Delta)[d_{22}(x_2 - x_1)(x_3 - x_2) + d_{33}(y_1 - y_2)(y_2 - y_3)]$$
$$k^e_{34} = k^e_{43} = (t/4\Delta)[d_{12}(x_1 - x_3)(y_3 - y_1) + d_{33}(x_1 - x_3)(y_3 - y_1)]$$
$$k^e_{35} = k^e_{53} = (t/4\Delta)[d_{11}(y_1 - y_2)(y_3 - y_1) + d_{33}(x_1 - x_3)(x_2 - x_1)]$$
$$k^e_{36} = k^e_{63} = (t/4\Delta)[d_{12}(x_2 - x_1)(y_3 - y_1) + d_{33}(x_1 - x_3)(y_1 - y_2)]$$
$$k^e_{45} = k^e_{54} = (t/4\Delta)[d_{12}(x_1 - x_3)(y_1 - y_2) + d_{33}(x_2 - x_1)(y_3 - y_1)]$$
$$k^e_{46} = k^e_{64} = (t/4\Delta)[d_{22}(x_1 - x_3)(x_2 - x_1) + d_{33}(y_1 - y_2)(y_3 - y_1)]$$
$$k^e_{56} = k^e_{65} = (t/4\Delta)[d_{12}(x_2 - x_1)(y_1 - y_2) + d_{33}(x_2 - x_1)(y_1 - y_2)]$$

where $d_{ij} = d_{ji}$ $(i, j = 1, 2, 3)$ as previously defined in Eqs (14.39c and d), depend upon the plane stress and plane strain conditions. For example, $d_{11} = E/(1 - v^2) = d_{22}$ for plane stress, $d_{11} = (1 - v)E/(1 + v)(1 - 2v) = d_{22}$ for plane strain and $d_{13} = d_{23} = 0$ for both conditions. It is seen from Eq (14.40a) that $[K^e]$ will need to be inverted to obtain the node displacements:

$$\{\delta^e\} = [K^{e-1}]\{F^e\}. \qquad (14.40\text{b})$$

The element strains and stresses are then found from Eqs (14.37b) and (14.39). The element stresses may be obtained from their combination:

$$\{\sigma\} = [D][B]\{\delta^e\}, \qquad (14.43\text{a})$$
$$\{\sigma\} = [H^e]\{\delta^e\}, \qquad (14.43\text{b})$$

where the product $[H^e] = [D][B]$ is again explicitly defined in terms of the nodal point coordinates and the elastic constants. That is,

$$[H^e] = (1/2\Delta) \begin{bmatrix} d_{11}(y_2 - y_3) & d_{12}(x_3 - x_2) & d_{11}(y_3 - y_1) & d_{12}(x_1 - x_3) & d_{11}(y_1 - y_2) & d_{12}(x_2 - x_1) \\ d_{21}(y_2 - y_3) & d_{22}(x_3 - x_2) & d_{21}(y_3 - y_1) & d_{22}(x_1 - x_3) & d_{21}(y_1 - y_2) & d_{22}(x_2 - x_1) \\ d_{33}(x_3 - x_2) & d_{33}(y_2 - y_3) & d_{33}(x_1 - x_3) & d_{33}(y_3 - y_1) & d_{33}(x_2 - x_1) & d_{33}(y_1 - y_2) \end{bmatrix}.$$

Overall stiffness matrix Normally, the overall stiffness matrix $[K]$, comprising all contributions from individual element stiffnesses $[K^e]$, is assembled before the inversion process. That is, for all elements:

$$\{F\} = [K]\{\delta\}, \tag{14.44a}$$
$$\{\delta\} = [K^{-1}]\{F\}. \tag{14.44b}$$

The assembly procedure is now illustrated in the following example.

Example 14.12 Identify the necessary input data and assemble the overall stiffness matrix for the 4-element plane stress cantilever shown in Fig. 14.18

With the origin of x, y at point 1 the nodal point coordinates and applied forces are following:

Node	x(mm)	y(mm)	F_x	F_y(kN)
1	0	0	0	0
2	0	50	0	0
3	50	0	0	0
4	50	50	0	0
5	100	0	0	−10
6	100	50	0	0

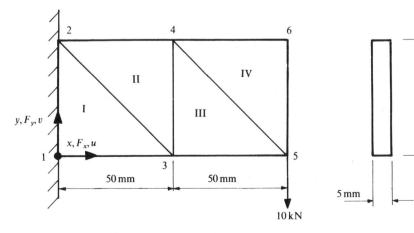

Figure 14.18 Cantilever with triangular elements.

At the fixed nodes corresponding to the reactions, $u_1 = v_1 = u_2 = v_2 = 0$. The element numbers, their associated node connectivities, thickness and elastic properties are

Element number	Connectivity nodes	Material properties		Thickness t (mm)
		E (GPa)	v	
I	1 2 3	200	0.3	5
II	2 3 4	200	0.3	5
III	3 4 5	200	0.3	5
IV	4 5 6	200	0.3	5

In the formulation of $[K]$, first note that $[K^e]$ for each element is 6×6 (see Eq. (14.42b). Since there are two degrees of freedom at each of 6 nodes, the total of 12 degrees of freedom implies that the overall stiffness matrix $[K]$ will have dimensions of 12×12. Consider element I: Eq. (14.40a) becomes $\{F^I\} = [K^I]\{\delta^I\}$. Since the numbering is clockwise, 3 replaces 2 and 2 replaces 3 in the previous analyses to give in Eq. (i).

$$
\begin{Bmatrix} F_{x1} \\ F_{y1} \\ \hline F_{x3} \\ F_{y3} \\ \hline F_{x2} \\ F_{y2} \end{Bmatrix} =
\left[\begin{array}{cc|cc|cc}
k^I_{11} & k^I_{12} & k^I_{13} & k^I_{14} & k^I_{15} & k^I_{16} \\
k^I_{21} & k^I_{22} & k^I_{23} & k^I_{24} & k^I_{25} & k^I_{26} \\
\hline
k^I_{31} & k^I_{32} & k^I_{33} & k^I_{34} & k^I_{35} & k^I_{36} \\
k^I_{41} & k^I_{42} & k^I_{43} & k^I_{44} & k^I_{45} & k^I_{46} \\
\hline
k^I_{51} & k^I_{52} & k^I_{53} & k^I_{54} & k^I_{55} & k^I_{56} \\
k^I_{61} & k^I_{62} & k^I_{63} & k^I_{64} & k^I_{65} & k^I_{66}
\end{array} \right]
\begin{Bmatrix} u_1 \\ v_1 \\ \hline u_3 \\ v_3 \\ \hline u_2 \\ v_2 \end{Bmatrix} , \text{(i)}
$$

with the circled terms (S^I_{11}), (S^I_{13}), (S^I_{12}), (S^I_{31}), (S^I_{33}), (S^I_{32}), (S^I_{21}), (S^I_{23}), (S^I_{22}).

In Eq (i) matrix components k^I_{ij} are again defined by Eqs (14.42c and d) provided subscripts 2 and 3 on x and y are interchanged. These single-element stiffnesses are reassembled within the overall stiffness matrix $[K]$ in Eq. (14.44a) as seen in Eq. (ii).

$$
\begin{Bmatrix} F_{x1} \\ F_{y1} \\ F_{x2} \\ F_{y2} \\ F_{x3} \end{Bmatrix} =
\left[\begin{array}{cccccc|cccccc}
k^I_{11} & k^I_{12} & k^I_{15} & k^I_{16} & k^I_{13} & k^I_{14} & & & & & & \\
k^I_{21} & k^I_{22} & k^I_{25} & k^I_{26} & k^I_{23} & k^I_{24} & & & & & & \\
k^I_{51} & k^I_{52} & k^I_{55} & k^I_{56} & k^I_{53} & k^I_{54} & & & & & & \\
k^I_{61} & k^I_{62} & k^I_{65} & k^I_{66} & k^I_{63} & k^I_{64} & & & & & & \\
k^I_{31} & k^I_{32} & k^I_{35} & k^I_{36} & k^I_{33} & k^I_{34} & & & & & & \\
\end{array} \right]
\begin{Bmatrix} u_1 \\ v_1 \\ u_2 \\ v_2 \\ u_3 \end{Bmatrix} \text{(ii)}
$$

continued

Equation (ii) continued

$$
\begin{Bmatrix} F_{y3} \\ F_{x4} \\ F_{y4} \\ F_{x5} \\ F_{y5} \\ F_{x6} \\ F_{y6} \end{Bmatrix}
\begin{bmatrix} k^{I}_{41} & k^{I}_{42} & k^{I}_{45} & k^{I}_{46} & k^{I}_{43} & k^{I}_{44} & & & & & & \\ & & & & & & & & & & & \\ & & & & & & & & & & & \\ & & & & & & & & & & & \\ & & & & & & & & & & & \\ & & & & & & & & & & & \\ & & & & & & & & & & & \end{bmatrix}
\begin{Bmatrix} v_3 \\ u_4 \\ v_4 \\ u_5 \\ v_5 \\ u_6 \\ v_6 \end{Bmatrix}
\qquad \text{(ii)}
$$

Applying Eq. (14.40a) to the second element II gives $\{F^{II}\} = [K^{II}]\{\delta^{II}\}$. Since the node numbering 234 here is anticlockwise, this gives Eq. (iii),

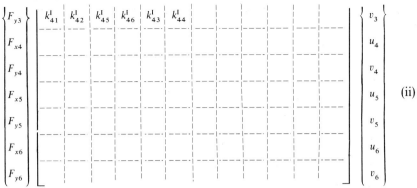

$$
\begin{Bmatrix} F_{x2} \\ F_{y2} \\ F_{x3} \\ F_{y3} \\ F_{x4} \\ F_{y4} \end{Bmatrix}
=
\begin{bmatrix} k^{II}_{11} & k^{II}_{12} & k^{II}_{13} & k^{II}_{14} & k^{II}_{15} & k^{II}_{16} \\ k^{II}_{21} & k^{II}_{22} & k^{II}_{23} & k^{II}_{24} & k^{II}_{25} & k^{II}_{26} \\ k^{II}_{31} & k^{II}_{32} & k^{II}_{33} & k^{II}_{34} & k^{II}_{35} & k^{II}_{35} \\ k^{II}_{41} & k^{II}_{42} & k^{II}_{43} & k^{II}_{44} & k^{II}_{45} & k^{II}_{46} \\ k^{II}_{51} & k^{II}_{52} & k^{II}_{53} & k^{II}_{54} & k^{II}_{55} & k^{II}_{56} \\ k^{II}_{61} & k^{II}_{62} & k^{II}_{63} & k^{II}_{64} & k^{II}_{65} & k^{II}_{66} \end{bmatrix}
\begin{Bmatrix} u_2 \\ v_2 \\ u_3 \\ v_3 \\ u_4 \\ v_4 \end{Bmatrix}, \quad \text{(iii)}
$$

(with circled terms S^{II}_{22}, S^{II}_{23}, S^{II}_{24}, S^{II}_{32}, S^{II}_{33}, S^{II}_{34}, S^{II}_{42}, S^{II}_{43}, S^{II}_{44})

in which k^{II}_{ij} are defined in Eqs (14.42c and d) with subscripts 1, 2 and 3 on x and y, replaced by 2, 3 and 4 respectively. These are assembled within Eq. (14.44a) as can be seen in Eq. (iv).

$$
\begin{Bmatrix} F_{x1} \\ F_{y1} \\ F_{x2} \\ F_{y2} \\ F_{x3} \\ F_{y3} \\ F_{x4} \end{Bmatrix}
\begin{bmatrix} & & & & & & & & & & & \\ & & & & & & & & & & & \\ & k^{II}_{11} & k^{II}_{12} & k^{II}_{13} & k^{II}_{14} & k^{II}_{15} & k^{II}_{16} & & & & & \\ & k^{II}_{21} & k^{II}_{22} & k^{II}_{23} & k^{II}_{24} & k^{II}_{25} & k^{II}_{26} & & & & & \\ & k^{II}_{31} & k^{II}_{32} & k^{II}_{33} & k^{II}_{34} & k^{II}_{35} & k^{II}_{36} & & & & & \\ & k^{II}_{41} & k^{II}_{42} & k^{II}_{43} & k^{II}_{44} & k^{II}_{45} & k^{II}_{46} & & & & & \\ & k^{II}_{51} & k^{II}_{52} & k^{II}_{53} & k^{II}_{54} & k^{II}_{55} & k^{II}_{56} & & & & & \end{bmatrix}
\begin{Bmatrix} u_1 \\ v_1 \\ u_2 \\ v_2 \\ u_3 \\ v_3 \\ u_4 \end{Bmatrix}
\quad \text{(iv)}
$$

continued

Equation (iv) continued

$$
\left\{\begin{array}{c} F_{y4} \\ F_{x5} \\ F_{y5} \\ F_{x6} \\ F_{y6} \end{array}\right\} = \left[\begin{array}{cccccc} & k_{61}^{\text{II}} & k_{62}^{\text{II}} & k_{63}^{\text{II}} & k_{64}^{\text{II}} & k_{65}^{\text{II}} & k_{66}^{\text{II}} \\ & & & & & & \\ & & & & & & \\ & & & & & & \\ & & & & & & \\ & & & & & & \end{array}\right] \left\{\begin{array}{c} v_4 \\ u_5 \\ v_5 \\ u_6 \\ v_6 \end{array}\right\} \quad \text{(iv)}
$$

In applying Eq. (14.40a) to element III, $\{F^{\text{III}}\} = [K^{\text{III}}]\{\delta^{\text{III}}\}$ becomes Eq. (v),

$$
\left\{\begin{array}{c} F_{x3} \\ F_{y3} \\ F_{x5} \\ F_{y5} \\ F_{x4} \\ F_{y4} \end{array}\right\} = \left[\begin{array}{cccccc} k_{11}^{\text{III}}\,(S_{33}^{\text{III}}) & k_{12}^{\text{III}} & k_{13}^{\text{III}} & k_{14}^{\text{III}}\,(S_{35}^{\text{III}}) & k_{15}^{\text{III}} & k_{16}^{\text{III}}\,(S_{34}^{\text{III}}) \\ k_{21}^{\text{III}} & k_{22}^{\text{III}} & k_{23}^{\text{III}} & k_{24}^{\text{III}} & k_{25}^{\text{III}} & k_{26}^{\text{III}} \\ k_{31}^{\text{III}}\,(S_{53}^{\text{III}}) & k_{32}^{\text{III}} & k_{33}^{\text{III}} & k_{34}^{\text{III}}\,(S_{55}^{\text{III}}) & k_{35}^{\text{III}} & k_{35}^{\text{III}}\,(S_{54}^{\text{III}}) \\ k_{41}^{\text{III}} & k_{42}^{\text{III}} & k_{43}^{\text{III}} & k_{44}^{\text{III}} & k_{45}^{\text{III}} & k_{46}^{\text{III}} \\ k_{51}^{\text{III}}\,(S_{43}^{\text{III}}) & k_{52}^{\text{III}} & k_{53}^{\text{III}} & k_{54}^{\text{III}}\,(S_{45}^{\text{III}}) & k_{55}^{\text{III}} & k_{56}^{\text{III}}\,(S_{44}^{\text{III}}) \\ k_{61}^{\text{III}} & k_{62}^{\text{III}} & k_{63}^{\text{III}} & k_{64}^{\text{III}} & k_{65}^{\text{III}} & k_{66}^{\text{III}} \end{array}\right] \left\{\begin{array}{c} u_3 \\ v_3 \\ u_5 \\ v_5 \\ u_4 \\ v_4 \end{array}\right\}, \quad \text{(v)}
$$

and, finally, for element IV, $\{F^{\text{IV}}\} = [K^{\text{IV}}]\{\delta^{\text{IV}}\}$ becomes Eq. (vi).

$$
\left\{\begin{array}{c} F_{x4} \\ F_{y4} \\ F_{x5} \\ F_{y5} \\ F_{x6} \\ F_{y6} \end{array}\right\} = \left[\begin{array}{cccccc} k_{11}^{\text{IV}}\,(S_{44}^{\text{IV}}) & k_{12}^{\text{IV}} & k_{13}^{\text{IV}} & k_{14}^{\text{IV}}\,(S_{45}^{\text{IV}}) & k_{15}^{\text{IV}} & k_{16}^{\text{IV}}\,(S_{46}^{\text{IV}}) \\ k_{21}^{\text{IV}} & k_{22}^{\text{IV}} & k_{23}^{\text{IV}} & k_{24}^{\text{IV}} & k_{25}^{\text{IV}} & k_{26}^{\text{IV}} \\ k_{31}^{\text{IV}}\,(S_{54}^{\text{IV}}) & k_{32}^{\text{IV}} & k_{33}^{\text{IV}} & k_{34}^{\text{IV}}\,(S_{55}^{\text{IV}}) & k_{35}^{\text{IV}} & k_{36}^{\text{IV}}\,(S_{56}^{\text{IV}}) \\ k_{41}^{\text{IV}} & k_{42}^{\text{IV}} & k_{43}^{\text{IV}} & k_{44}^{\text{IV}} & k_{45}^{\text{IV}} & k_{46}^{\text{IV}} \\ k_{51}^{\text{IV}}\,(S_{64}^{\text{IV}}) & k_{52}^{\text{IV}} & k_{53}^{\text{IV}} & k_{54}^{\text{IV}}\,(S_{65}^{\text{IV}}) & k_{55}^{\text{IV}} & k_{56}^{\text{IV}}\,(S_{66}^{\text{IV}}) \\ k_{61}^{\text{IV}} & k_{62}^{\text{IV}} & k_{63}^{\text{IV}} & k_{64}^{\text{IV}} & k_{65}^{\text{IV}} & k_{66}^{\text{IV}} \end{array}\right] \left\{\begin{array}{c} u_4 \\ v_4 \\ u_5 \\ v_5 \\ u_6 \\ v_6 \end{array}\right\}, \quad \text{(vi)}
$$

When the components of $[K^{\text{III}}]$ and $[K^{\text{IV}}]$ are further added to the appropriate locations within the combined Eqs (ii) and (iv), the governing Eq. (14.44a) for the cantilever is as seen in Eq. (vii).

$$
[K]\{\delta\} = \{F\}
$$

$\{F\}$	u_1	v_1	u_2	v_2	u_3	v_3	u_4	v_4	u_5	v_5	u_6	v_6
F_{x1}	k^I_{11}	k^I_{21}	k^I_{51}	k^I_{61}	k^I_{31}	k^I_{41}	0	0	0	0	0	0
F_{y1}	k^I_{12}	k^I_{22}	k^I_{52}	k^I_{62}	k^I_{32}	k^I_{42}	0	0	0	0	0	0
F_{x2}	k^I_{15}	k^I_{25}	$k^I_{55}+k^{II}_{11}$	$k^I_{65}+k^{II}_{21}$	$k^I_{35}+k^{II}_{31}$	$k^I_{45}+k^{II}_{41}$	k^{II}_{51}	k^{II}_{61}	0	0	0	0
F_{y2}	k^I_{16}	k^I_{26}	$k^I_{56}+k^{II}_{12}$	$k^I_{66}+k^{II}_{22}$	$k^I_{36}+k^{II}_{32}$	$k^I_{46}+k^{II}_{42}$	k^{II}_{52}	k^{II}_{62}	0	0	0	0
F_{x3}	k^I_{13}	k^I_{23}	$k^I_{53}+k^{II}_{13}$	$k^I_{63}+k^{II}_{23}$	$k^I_{33}+k^{II}_{33}+k^{III}_{11}$	$k^I_{43}+k^{II}_{43}+k^{III}_{21}$	$k^{II}_{53}+k^{III}_{51}$	$k^{II}_{63}+k^{III}_{61}$	k^{III}_{31}	k^{III}_{41}	0	0
F_{y3}	k^I_{14}	k^I_{24}	$k^I_{54}+k^{II}_{14}$	$k^I_{64}+k^{II}_{24}$	$k^I_{34}+k^{II}_{34}+k^{III}_{12}$	$k^I_{44}+k^{II}_{44}+k^{III}_{22}$	$k^{II}_{54}+k^{III}_{52}$	$k^{II}_{64}+k^{III}_{62}$	k^{III}_{32}	k^{III}_{42}	0	0
F_{x4}	0	0	k^{II}_{15}	k^{II}_{25}	$k^{II}_{35}+k^{III}_{15}$	$k^{II}_{45}+k^{III}_{25}$	$k^{II}_{55}+k^{III}_{55}+k^{IV}_{11}$	$k^{II}_{65}+k^{III}_{65}+k^{IV}_{21}$	$k^{III}_{35}+k^{IV}_{31}$	$k^{III}_{45}+k^{IV}_{41}$	k^{IV}_{51}	k^{IV}_{61}
F_{y4}	0	0	k^{II}_{16}	k^{II}_{26}	$k^{II}_{36}+k^{III}_{16}$	$k^{II}_{46}+k^{III}_{26}$	$k^{II}_{56}+k^{III}_{56}+k^{IV}_{12}$	$k^{II}_{66}+k^{III}_{66}+k^{IV}_{22}$	$k^{III}_{36}+k^{IV}_{32}$	$k^{III}_{46}+k^{IV}_{42}$	k^{IV}_{52}	k^{IV}_{62}
F_{x5}	0	0	0	0	k^{III}_{13}	k^{III}_{23}	$k^{III}_{53}+k^{IV}_{13}$	$k^{III}_{63}+k^{IV}_{23}$	$k^{III}_{33}+k^{IV}_{33}$	$k^{III}_{43}+k^{IV}_{43}$	k^{IV}_{53}	k^{IV}_{63}
F_{y5}	0	0	0	0	k^{III}_{14}	k^{III}_{24}	$k^{III}_{54}+k^{IV}_{14}$	$k^{III}_{64}+k^{IV}_{24}$	$k^{III}_{34}+k^{IV}_{34}$	$k^{III}_{44}+k^{IV}_{44}$	k^{IV}_{54}	k^{IV}_{64}
F_{x6}	0	0	0	0	0	0	k^{IV}_{15}	k^{IV}_{25}	k^{IV}_{35}	k^{IV}_{45}	k^{IV}_{55}	k^{IV}_{65}
F_{y6}	0	0	0	0	0	0	k^{IV}_{16}	k^{IV}_{26}	k^{IV}_{36}	k^{IV}_{46}	k^{IV}_{56}	k^{IV}_{66}

The large number of stiffness calculations that arise with just four elements shows that a computer is an essential requirement in finite element method (FEM) practice involving several hundred elements. Note that $[K]$ like $[K^e]$ is again symmetrical about the leading diagonal and is narrowly banded. The latter depends upon careful node numbering. These two observations are exploited to reduce computer storage space and to increase the speed in solving for $\{\delta\}$ between the simultaneous equations that Eq. (14.44a) contains. Normally the Gaussian elimination and Gauss–Seidal iteration methods are employed. Submatrices are employed to assist both with the assembly of $[K]$ and in the solution for $\{\delta\}$. For example, with the partitioning shown in each of Eqs (i), (iii), (v) and (vi) and the identification of the submatrices S^e_{ij}, shown circled, Eq. (vii) is written

$$
\begin{Bmatrix} \{F_1\} \\ \{F_2\} \\ \{F_3\} \\ \{F_4\} \\ \{F_5\} \\ \{F_6\} \end{Bmatrix}
\begin{bmatrix}
S^I_{11} & S^I_{12} & S^I_{13} & 0 & 0 & 0 \\
S^I_{21} & S^I_{22}+S^{II}_{22} & S^I_{23}+S^{II}_{23} & S^{II}_{24} & 0 & 0 \\
S^I_{31} & S^I_{32}+S^{II}_{32} & S^I_{33}+S^{II}_{33}+S^{III}_{33} & S^{II}_{34}+S^{III}_{34} & S^{III}_{35} & 0 \\
0 & S^{II}_{42} & S^{II}_{43}+S^{III}_{43} & S^{II}_{44}+S^{III}_{44}+S^{IV}_{44} & S^{III}_{45}+S^{IV}_{45} & S^{IV}_{46} \\
0 & 0 & S^{III}_{53} & S^{III}_{54}+S^{IV}_{54} & S^{III}_{55}+S^{IV}_{55} & S^{IV}_{56} \\
0 & 0 & 0 & S^{IV}_{64} & S^{IV}_{65} & S^{IV}_{66}
\end{bmatrix}
\begin{Bmatrix} \{\delta_1\} \\ \{\delta_2\} \\ \{\delta_3\} \\ \{\delta_4\} \\ \{\delta_5\} \\ \{\delta_6\} \end{Bmatrix}
$$

where the individual stiffnesses in Eq. (vii) reappear with the addition of the submatrices. For example, $S^I_{33} + S^{II}_{33} + S^{III}_{33}$ is

$$
\begin{bmatrix} k^I_{33} & k^I_{34} \\ k^I_{43} & k^I_{44} \end{bmatrix}
+ \begin{bmatrix} k^{II}_{33} & k^{II}_{34} \\ k^{II}_{43} & k^{II}_{44} \end{bmatrix}
+ \begin{bmatrix} k^{III}_{11} & k^{III}_{12} \\ k^{III}_{21} & k^{III}_{22} \end{bmatrix}
$$
$$
= \begin{bmatrix} (k^I_{33}+k^{II}_{33}+k^{III}_{11}) & (k^I_{34}+k^{II}_{34}+k^{III}_{12}) \\ (k^I_{43}+k^{II}_{43}+k^{III}_{21}) & (k^I_{44}+k^{II}_{44}+k^{III}_{22}) \end{bmatrix}.
$$

The final step is to ensure that the boundary conditions are met. There are various procedures that can ensure zero displacement at support points. The direct method is to account for $u_1 = v_1 = u_2 = v_2 = 0$ in the assembly of $[K]$. Alternatively, this can be achieved numerically either by increasing the corresponding diagonal stiffnesses k_{ii} in multiplying by a large number, 10^{10} say, or by eliminating the row and column corresponding to the fixed displacement. It is obviously desirable that concentrated loading in this and other examples is made to coincide with a node. Distributed loading must be replaced with statically equivalent forces acting at the element nodes. Where it is not possible to do this by inspection, the principle of virtual displacements is used to equate the work done by the actual loading to the virtual work under the equivalent system (see Zienkiewicz and Cheung *op. cit.*).

2. Rectangular elements

For the rectangular plane element shown in Fig. 14.19(a), the following displacement functions are assumed:

$$u(x, y) = \alpha_1 + \alpha_2 x + \alpha_3 y + \alpha_4 xy, \qquad (14.45a)$$

$$v(x, y) = \alpha_5 + \alpha_6 x + \alpha_7 y + \alpha_8 xy \qquad (14.45b)$$

where the number of coefficients matches the eight degrees of freedom, i.e. four nodes with two degrees per node. These displacements ensure that u and v vary linearly (i) with x along side $y = b$ and (ii) with y along the side $x = a$. Applying Eq. (14.45a and b) to each node again leads to the same form of Eq. (14.35) for the generalized displacement vector, $\{\delta\} = \{f\}[A^{-1}] \cdot \{\delta^e\}$, where $\{\delta\} = \{u \quad v\}^T$, $\{\delta^e\} = \{u_1 \ v_1 \ u_2 \ v_2 \ u_3 \ v_3 \ u_4 \ v_4\}^T$ with

$$f = \begin{bmatrix} 1 & x & y & xy & 0 & 0 & 0 & 0 \\ 0 & 0 & 0 & 0 & 1 & x & y & xy \end{bmatrix}$$

and

$$[A] = \begin{bmatrix} 1 & 0 & 0 & 0 & 0 & 0 & 0 & 0 \\ 0 & 0 & 0 & 0 & 1 & 0 & 0 & 0 \\ 1 & 0 & b & 0 & 0 & 0 & 0 & 0 \\ 0 & 0 & 0 & 0 & 1 & 0 & b & 0 \\ 1 & a & 0 & 0 & 0 & 0 & 0 & 0 \\ 0 & 0 & 0 & 0 & 1 & a & 0 & 0 \\ 1 & a & b & ab & 0 & 0 & 0 & 0 \\ 0 & 0 & 0 & 0 & 1 & a & b & ab \end{bmatrix}. \qquad (14.46)$$

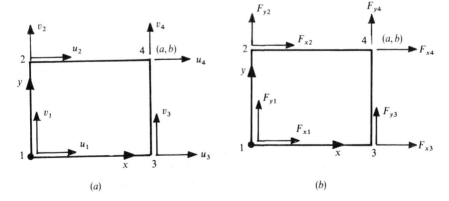

Figure 14.19 Displacements and forces for a rectangular element.

The strain components are, from Eqs (14.4),

$$\epsilon_x = \partial u/\partial x = \alpha_2 + \alpha_4 y,$$
$$\epsilon_y = \partial v/\partial y = \alpha_7 + \alpha_8 x,$$
$$\gamma_{xy} = \partial u/\partial y + \partial v/\partial x = \alpha_3 + \alpha_4 x + \alpha_6 + \alpha_8 y.$$

It is seen that, as the linear variation in these strains satisfies the compatibility equation (14.5), they will ensure strain continuity at the interface between adjacent elements provided the node displacements are matched. The 8×3 matrix $[C]$ appearing in (14.36a) becomes

$$[C] = \begin{bmatrix} 0 & 1 & 0 & y & 0 & 0 & 0 & 0 \\ 0 & 0 & 0 & 0 & 0 & 0 & 1 & x \\ 0 & 0 & 1 & x & 0 & 1 & 0 & y \end{bmatrix} \tag{14.47}$$

where $\{\epsilon\} = \{\epsilon_x \ \epsilon_y \ \gamma_{xy}\}^T$ and $\alpha = \{\alpha_1 \ \alpha_2 \ \alpha_3 \ \alpha_4 \ \alpha_5 \ \alpha_6 \ \alpha_7 \ \alpha_8\}^T$. Again $[B] = [C][A^{-1}]$ in the strain–nodal displacement relation (14.37b). The constitutive relations (14.38a) and (14.39b) remain unaltered since they depend upon the plane condition and not upon the shape of the element. The previous derivation of the element stiffness matrix $[K^e]$ applies except that, as $[B]$ will now contain components in x and y, the matrix product $[B]^T[D][B]$ must be integrated w.r.t. volume V. Writing $\delta V = t\delta x\delta y$ for an elemental rectangle of uniform thickness t, the application of the virtual displacement principal leads to:

$$[K^e] = t \int_x \int_y [B]^T[D][B]\,dx\,dy, \tag{14.48}$$

in which the matrices must first be multiplied before integrating over the element area. In this case, when $[K^e]$ from Eq. (4.48) is substituted into Eq. (14.40a) it relates the following nodal point displacements and forces: $\{\delta^e\} = \{u_1 \ v_1 \ u_2 \ v_2 \ u_3 \ v_3 \ u_4 \ v_4\}^T$ and $\{F^e\} = \{F_{x1} \ F_{y1} \ F_{x2} \ F_{y2} \ F_{x3} \ F_{y3} \ F_{x4} \ F_{y4}\}^T$ as shown in Fig. 14.19(a and b). The overall stiffness matrix $[K]$ may be assembled as before and Eq. (14.44a) solved for the nodal displacements. Finally, Eq. (14.43b) supplies the element stresses $\{\sigma\} = \{\sigma_x \ \sigma_y \ \tau_{xy}\}^T$ where, from Eqs (14.39b), (14.46) and (14.47) it follows that the 3×8 matrix $[H^e]$ is found from

$$[H^*] = [D][B] = [D][C][A^{-1}]$$

$$= \frac{1}{(ab)} \begin{bmatrix} -d_{11}(b-y) & -d_{12}(a-x) & -d_{11}y & d_{12}(a-x) & d_{11}(b-y) & -d_{12}x & d_{11}y & d_{12}x \\ -d_{12}(b-y) & -d_{22}(a-x) & -d_{12}y & d_{22}(a-x) & d_{12}(b-y) & -d_{22}x & d_{12}y & d_{22}x \\ -d_{33}(a-x) & -d_{23}(b-y) & d_{33}(a-x) & -d_{33}y & -d_{33}x & d_{33}(b-y) & d_{33}x & d_{33}y \end{bmatrix}.$$

where x and y are the coordinates of that node for which the three stress components are required. Note that the u, v displacements at any point x, y are supplied by Eq. (14.35). Unlike the case of a triangular element, neither the stress nor the strain is constant within a rectangular element. Consequently, it is found that large-sized rectangular meshes are undesirable if discontinuities in the stresses at common nodes are to be avoided. There are many other types of finite element, including those for beam and plate bending, torsion and axial symmetry.

EXERCISES

Cartesian Equilibrium and Compatibility

14.1 Show that the stresses (σ_x and τ_{xy}) and strains supplied by ETB (Eqs (6.4) and (10.3)) satisfy equilibrium and compatibility for all points in the length and cross-section of an end-loaded cantilever beam.

14.2 Show that the following strains and displacements are all compatible in plane stress: (i) $\epsilon_x = axy; \epsilon_y = by^3; \gamma_{xy} = c - dy^2;$ (ii) $\epsilon_x = q(x^2 + y^2); \epsilon_y = qy^2; \gamma_{xy} = 2qxy;$ (iii) $u = ax^2y^2; v = byx^3$.

14.3 Analysis of plane strain ($\epsilon_z = \gamma_{yz} = \gamma_{xz} = 0$) shows non-zero strains can be represented by
$$\epsilon_x = a + b(x^2 + y^2) + x^4 + y^4; \quad \epsilon_y = c + d(x^2 + y^2) + x^4 + y^4; \quad \gamma_{xy} = e + fxy(x^2 + y^2 - g^2),$$
where $a, b, c, d, e, f,$ and g are constants. Show that these are only possible if $f = 4$ and $b + d + 2g^2 = 0$. Hence derive the displacements from the strains.

14.4 Find the conditions for which the shear strain $\gamma_{xy} = Ax^2y + Bxy + Cx^2 + Dy$ is compatible with the displacement $u = ax^2y^2 + bxy^2 + cx^2y$, $v = ax^2y + bxy$.

14.5 A thin plate is subjected to the following in-plane stress field components; $\sigma_x = ay^3 + bx^2y - cx$, $\sigma_y = dy^3 - e$, $\tau_{xy} = fxy^2 + gx^2y - h$. What are the constraints on the constants a, b, c, d, e, f, g and h such that the stress field satisfies both equilibrium and compatibility?
Answer: $a = 2f/3; b = -f; d = -f/3; c = g = 0; e, f$ and h are unconstrained.

14.6 The stress system in a plate consists of direct tensile stresses $\sigma_x = cyx^2$, $\sigma_y = cy^3/3$ together with a shear stress τ_{xy}. Determine τ_{xy}.
Answer: $\tau_{xy} = -cxy^2$.

14.7 If general expressions for the two direct stress components of a plane stress field have the form $\sigma_x = ax^3 + bx^2y + cxy^2 + dy^3$, $\sigma_y = kx^3 + lx^2y + mxy^2 + ny^3$, where a, b, c, d, k, l, m, n are constants, determine the necessary relationship for these constants if the stress distribution is to be compatible. For the special case where $a = c = d = k = m = 0$, determine expressions for the compatible stress components σ_x, σ_y and τ_{xy}, (C.E.I. Pt.2, 1980).
Answer: $(6a + 3k + c)$ $(3n - b)$ $- (m - 3a)$ $(2b + l + 3d) = 0$; $\sigma_x = 3x^2y$; $\sigma_y = y^3 - 6x^2y$, $\tau_{xy} = 2x^3 - 3xy^2$].

14.8 A square plate of side length a and with sides along the coordinate axes is fixed in position at the origin with the side along the y-axis fixed in direction ($\partial u/\partial y = 0$). If the stresses in the plate are $\sigma_x = Ay/a, \sigma_y = A$, with a consistent value of shear stress, find this value and determine the displacements generally and those existing at the corner (a, a) (I.C. 1967)
Answer: $\tau_{xy} = K; u = (Axy/aE) - vAx/E; v = (Ay/E) - (vAy^2/2aE) + 2Kx(1 + v)/E - Ax^2/2aE;$ and at (a, a) $u = Aa(1 - v)/E, v = Aa(1 - v)/2E + 2Ka(1 + v)/E.$

14.9 Derive general expressions for the displacements at any point x, y in the cantilever of Example 14.6 (Fig. 14.6). Determine the constants and identify the terms in the final displacement expressions for the each of the following boundary conditions:

(i) $u = v = 0$ and $\partial u/\partial y = 0$ at $(1, 0)$;

(ii) $u = v = 0$ and $\partial v/\partial x = 0$ at $(1, 0)$;

Answer:

(i) $u = (P/2EI)(x^2 y - l^2 y + vy^3/3) - Py^3/6GI$;

$\quad v = (P/EI)(l^2 x/2 - vxy^2/2 - x^3/6 - l^3/3) + (Ph^2/2GI)(x - l)$.

(ii) $u = (P/2EI)(x^2 y - l^2 y + vy^3/3) - (Py^3/6GI)(3h^2/y^2 - 1)$;

$\quad v = (P/EI)(l^2 x/2 - vxy^2/2 - x^3/6 - l^3/3) + (Ph^2/2GI)(x - l)$.

Terms with E—deflection due to bending; terms with G—deflection due to shear

Cartesian Stress Functions

14.10 Derive the stress function that supplies the stress components $\sigma_x = cy$, $\sigma_y = cx$, $\tau_{xy} = -k$, where c and k are constants.

Answer: $c(x^3 + y^3)/6 + kxy$.

14.11 Find the condition for which the stress function $\phi = Axy^4 + Bx^3 y^2$ is valid.

Answer: $A + B = 0$.

14.12 The stress function $\phi = Ax^2 + Bxy + Cy^2$ provides the direct and shear stresses in a thin rectangular plate. One corner, at the origin, is fixed in position and direction. Show that the displacements for the diagonally opposite corner (a, b) are $u = 2a(C - vA)/E$, $v = 2b(A - vC)/E$.

14.13 If ϕ in Exercise 14.12 applies to a thick plate in plane strain, where $\epsilon_z = 0$, determine the displacements at the corner (a, b).

Answer: $u = 2x(1 + v)[C(1 - v) - Av]/E$, $v = 2y(1 + v)[A(1 - v) - Cv]/E$.

14.14 Show that the stress function $\phi = Dx^3$ matches (i) in-plane bending of a thin plate under end moments M (Fig. 14.20a) and (ii) linearly varying normal edge forces of total effect F applied to width a (Fig. 14.20b). Taking the thicknesses to be unity, derive the displacement expressions in each case. If, in Fig. 14.20(a) the plate centre is position-fixed, show that the displacements at the point $(a, 0)$ are $u = -Mva^2/2EI$, $v = 0$. If, in Fig. 14.20(b), the plate origin is position-fixed show that displacements at point $(a, 0)$ are $u = -vF/E$, $v = 0$.

Answer: (i) $u = -M(vx^2 + y^2)/2EI + C_1 y + C_2$, $v = Mxy/EI + C_3 x + C_4$ where C are constants; (ii) $u = -F(vx^2 + y^2)/Ea^2 + C_2$, $v = 2Fxy/Ea^2 + C_4$.

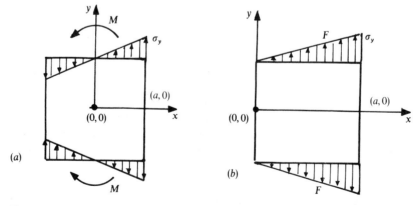

Figure 14.20

14.15 Derive the stress components from the stress function $\phi = Axy^3 + Bxy + Cy^3$ given that the stress resultants are a shear force F and a bending moment M at the section $x = 0$ of a beam with a constant I and depth of $\pm h$ w.r.t the neutral x-axis. How could these stress distributions be made to represent those for an encastre beam carrying a central concentrated load?

Answer: $\sigma_x = Fxy/I + My/I$; $\tau_{xy} = F(h^2 - x^2)/2I$.

14.16 The stress function $\phi = Ax^2 + Bx^2y + Cx^2y^3$ provides the exact solution to the distribution of stress in a beam of depth $2h$ with unit thickness when it is simply supported over a length $2L$ in carrying a uniformly distributed load q/unit length. With the origin at the beam centre, derive the stresses and compare with those supplied by the ETB. What is the vertical displacement at the centre?

Answer: $\sigma_x = (qy/2I)(x^2 - l^2 + 2h^2/5 - 2y^2/3)$; $\qquad \sigma_y = (q/2I)(y^3/3 - yh^2 - 2h^3/3)$; $\tau_{xy} = (qx/2I)(h^2 - y^2)$; $v = (5ql/24EI)[1 + (12h^2/5l^2)(v/2 + 4/5)]$.

14.17 Determine the state of stress represented by the stress function $\phi = Ax^2 - By^3$ when applied to a thin square plate ($a \times a$) with origin at the centre of the left vertical side. Check that both equilibrium and compatibility are satisfied. Show the variation of stress on the plate surfaces. (C.E.I. Pt.2, 1986)

14.18 Show that the stress function $\phi = (q/8c^3)[x^2(y^3 - 3c^2y + 2c^3) - (y^3/5)(y^2 - 2c^2)]$ is valid and determine the problem it solves when applied to the region included in $y \pm c$, $x = 0$ on the side x positive.

14.19 A cantilever beam with rectangular cross-section occupies the region $-a \leq z \leq a$, $-h \leq y \leq h$ and $0 \leq x \leq 1$. The end $x = l$ is built-in and a concentrated force P is applied at the free end ($x = 0$) in the $+y$ direction. Given that the non-zero stress components are $\sigma_x = Cxy$ and $\tau_{xy} = A + By^2$, show that they satisfy equilibrium when $2B + C = 0$ and that $C = 3P/4ah^3$ from vertical force equilibrium at the free end.

14.20 If, for the cantilever in Exercise 14.10, $a = 2$ mm, $h = 4$ mm, $l = 10$ mm and $P = 1$ kN, determine:

  (i) the Airy stress function from which the stresses derive;
  (ii) the principal stresses at the point $Q(5, 2, 0)$;
  (iii) the normal and shear stresses at Q on an oblique plane whose edge is given by $y - x + 3 = 0$.

Hint: see Chapter 13, for the appropriate stress transformation equations.

14.21 Establish the relationship between the constants in the quartic stress function $\phi = Ax^4 + Bx^3y + Cx^2y^2 + Dxy^3 + Ey^4$ if the function is to be valid. Determine further relationships between the constants from a consideration of known resultant normal forces X and Y and shear force S acting on the sides of a square plate of side length a and unit thickness when the plate lies in the positive x, y quadrant with one corner at the origin.

14.22 A rectangular strip of thickness t is simply supported at its ends ($x = \pm l$) and carries on its upper edge ($y = +d$) a uniformly distributed load, w per unit length, acting in the plane of the strip, as shown in Fig. 14.21. The expression

$$\phi = (3w/4t)[(y^3/30d^3)(5x^2 - y^2 - 5l^2 + 2d^2) - (x^2/6d)(3y + 2d)]$$

is proposed as an Airy stress function. Show that ϕ satisfies the compatibility condition and determine expressions for the direct and shear stresses in the strip. Investigate whether the direct stress expression satisfies the boundary equilibrium conditions. (C.E.I. Pt.2, 1971)

14.23 The rectangular plate shown in Fig. 14.22 is of uniform thickness t and is built-in at the end $x = 0$. It is loaded along the upper edge, $y = c$, by a tangential shear flow q. The lower edge of the plate is not loaded. The expression

$$\phi = (q/4t)[xy - xy^2/c - xy^3/c^2 + ly^2/c + ly^3/c^2]$$

is proposed as an appropriate Airy stress function. To what extent does this function satisfy stress boundary conditions and compatibility? (C.E.I. Pt. 2, 1972)

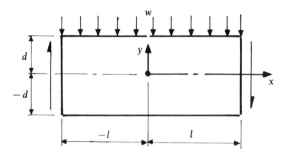

Figure 14.21

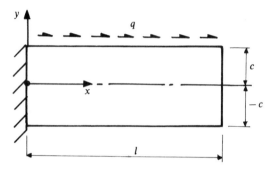

Figure 14.22

14.24 Show that the stress function $\phi = axy/2 - axy^3/2c^2 + aby^3/2c^2$ is biharmonic and that it may be used to supply the stresses in the cantilever of Fig. 14.23. Is the condition that the cantilever is unloaded at its free end met by the function?

14.25 The function $\phi = -xy^3(P/4a^3 + Q/4a^2L) + y^3(PL/4a^3 + Q/4a^2) - xy^2Q/4aL + y^2Q/4a + xy(3P/4a + Q/4L)$ is proposed for the loaded beam in Fig. 14.24. Check that this satisfies $\nabla^4\phi = 0$ and determine the component stresses. To what extent are the boundary conditions satisfied in relation to the ratio L/a?

14.26 A cantilever beam of length l and depth $2h$ carries a downward-acting uniformly distributed load w/unit area along its top and bottom faces under plane strain conditions. Using the stress function $\phi = Px^2y + Qx^2y^3 + Ry^3 + Sy^5$ find the constants from a consideration of

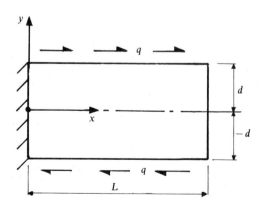

Figure 14.23

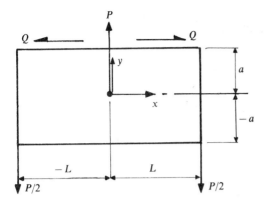

Figure 14.24

the known conditions existing at the beam surface. Take the origin of coordinates at the centre of the free end. (C.E.I. Pt. 2, 1986)

14.27 The stress function $\phi = (3F/2c)(xy - xy^3/3c^2) + Qy^3/4c^3$ is to be used to determine the stress field in a cantilever beam of length L, unit thickness, and depth $2c$, subjected to a given loading. Determine the nature of that loading. (C.E.I. Pt. 2, 1985)

Fourier Series Loading Functions

14.28 Derive the Fourier function $g(x)$ for the edge loadings shown in Fig. 14.25(a)–(d).

Answer: (a) $g(x) = (2p/\pi)\sin \pi x/l - (2p/2\pi)\sin 2\pi x/l + (2p/3\pi)\sin 3\pi x/l - \cdots$;
 (b) $g(x) = (2p/\pi)\{[-\cos \pi l/L + (L/l\pi)\sin \pi l/L]\sin \pi x/l + [(\tfrac{1}{2}\cos 2\pi l/L + (L/4l\pi)\sin 2\pi l/L]$
 $\times \sin 2\pi x/L \cdots\}$;
 (c) $g(x) = p/2 - (4p/\pi^2)[\cos \pi x/l + (1/3^2)\cos 3\pi x/l + (1/5^2)\cos 5\pi x/l \cdots]$;
 (d) $g(x) = 3p/2 - (2p/\pi)[\sin \pi x/l + (1/3)\sin 3\pi x/l + (1/5)\sin 5\pi x/l \cdots]$.

14.29 The stress function in Eqs (14.11) provides the solution to the problem of a long flat strip with its axis in direction x, under a sinusoidal edge loading of wavelength $2L$. Use this stress function to find the normal stress σ_y at the origin (Fig. 14.26) in a strip of width $2h$ loaded with a uniform pressure p under guides that are equally spaced along its edges for the case when $h = L$. (I.C. 1970)

Cylindrical Coordinates

14.30 For a particular plane strain problem the strain–displacement equations in cylindrical coordinates (r, θ, z) are $\epsilon_r = du/dr$, $\epsilon_\theta = u/r$, $\epsilon_z = \gamma_{r\theta} = \gamma_{\theta z} = \gamma_{zr} = 0$. Show that the appropriate compatibility equation in terms of stress is $rv\,d\sigma_r/dr - r(1 - v)d\sigma_\theta/dr + \sigma_r - \sigma_\theta = 0$. State the nature of a problem that the above equations represent. (C.E.I. Pt. 2, 1971)

14.31 A thick-walled cylinder of radii r_i and r_o is pressurized both internally (p_i) and externally (p_o). Derive expressions for the constants A and C in the corresponding stress function: $\phi = A \ln r + Cr^2$.

Answer: $A = -(p_i - p_o)r_o^2/[(r_o/r_i)^2 - 1]$; $2C = [p_i - p_o(r_o/r_i)^2]/[(r_o/r_i)^2 - 1]$.

14.32 A very large plate of thickness t is subjected to a membrane tensile stress σ_o applied around its periphery. If a small circular hole is cut from the middle of the plate, show, from the Lamé equations, that the maximum stress at the edge of the hole is $2\sigma_o$.

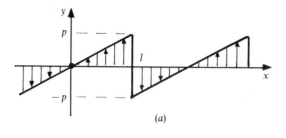

(a)

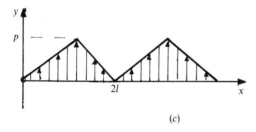

(b)

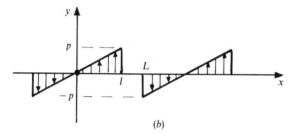

(c)

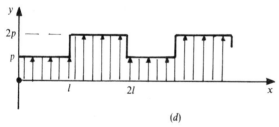

(d)

Figure 14.25

14.33 Determine the hoop stress at the inner and outer radii of a curved bar of 45 mm mean radius with 30 mm square cross-section when it is subjected to a pure moment of 300 N m that induces tension along the top edge. (T.C.D. 1980)

Answer: 54.5 MPa; − 86 MPa.

14.34 The stress function $\phi = C\theta$ prescribes the shear stress distribution in an annular disc of thickness t and inner radius r_i under torsion. If the torque is transmitted to the disc by a keyed shaft of radius r_i, show that there is no discontinuity in shear stress at r_i when $t = r_i/4$.

14.35 The stress function $\phi = B \sin 2\theta + C\theta$, will provide the distribution of radial (σ_r) and shear ($\tau_{r\theta}$) stress in a wedge of apex angle 2α and unit thickness, with a moment M applied in the

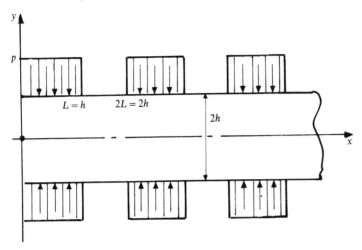

Figure 14.26

plane of the wedge at its tip. Determine the constants and sketch the stress distributions for a given radius r. Measure positive θ anticlockwise from the axis of the wedge with origin at its tip, with M applied in the direction of θ negative.

Answer: $B = M/2(\sin 2\alpha - 2\alpha \cos 2\alpha)$; $C = -M \cos 2\alpha/(\sin 2\alpha - 2\alpha \cos 2\alpha)$.

14.36 Find the maximum radial stress at a radius of 75 mm when the 30° wedge in Fig. 14.13(b) is loaded at its tip with a compressive force of 15 kN across its 5 mm thickness with an inclination of 5° to the wedge axis.

Answer: -78.77 MPa on the load line.

14.37 A wedge of apex angle 30° supports a 10 kN compressive force at right angles to its axis of symmetry. If the wedge thickness is 2 mm, find the maximum stress at a radius of 50 mm and show where this occurs. Show that the maximum radial stress may be approximated by ETB when $2\alpha \leq 5°$.

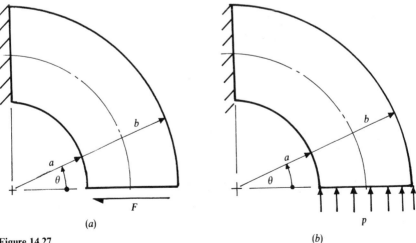

(a)

(b)

Figure 14.27

14.38 The stress function $\phi = (qr^2/2\pi)(\frac{1}{2}\sin 2\theta - \theta)$, prescribes the stresses in a semi-infinite plate. Find the edge loading required to produce the stresses.

Answer: Normal pressure q over radius r in the quadrant $\pi/2 \leq \theta \leq \pi$ with θ measured anticlockwise.

14.39 A small hole of radius a lies at the centre of a large thin plate of thickness t. The plate is subjected to a uniaxial stress σ_o at its boundary. Derive the maximum stress concentration factor from the stress function $\phi = (\sigma_o/4)[r^2 - 2a^2 \ln r - (r - a^2/r)^2 \cos 2\theta]$. If $\sigma_o = 25\,\text{MPa}$ and $t = 5\,\text{mm}$ find the maximum tensile stress in the plate. (T.C.D. 1976)

14.40 The stress components in the body of an infinite plate carrying a remote horizontal tensile stress σ_x, with a hole of radius r_i at the origin of coordinates, are

$$\sigma_r = (\sigma_x/2)[(1 - r_i^2/r^2) + (1 + 3r_i^4/r^4 - 4r_i^2/r^2)\cos 2\theta],$$

$$\tau_{r\theta} = -(\sigma_x/2)(1 - 3r_i^4/r^4 + 2r_i^2/r^2)\sin 2\theta, \quad \sigma_\theta = (\sigma_x/2)[(1 + r_i^2/r^2) - (1 + 3r_i^4/r^4)\cos 2\theta].$$

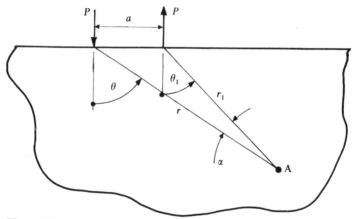

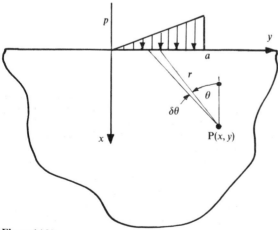

Figure 14.28

Figure 14.29

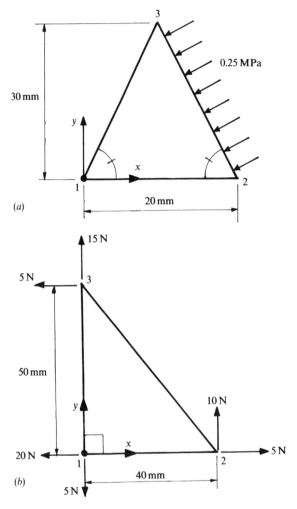

Figure 14.30

Use the principle of superposition to find the stresses when the plate is subjected to pure shear $(\sigma_x, \sigma_y = -\sigma_x)$ at the boundary. Hence show, from the distribution of stress around the boundary, that the greatest stress concentrations are 3 and 4 respectively for each loading. St Venant's principle implies that the effect the hole has in concentrating stress is local. Examine the principle quantitatively in terms of the ratio r/r_i in each case.

14.41 A hole of radius a in an infinite plate is loaded with an internal horizontal pressure p. Taking the origin of coordinates at the hole centre and assuming that the outer boundary is stress free, derive expressions for the polar stress components σ_r, σ_θ and $\tau_{r\theta}$ in the body of the plate and hence find the maximum stress concentration at the hole. (I.C. 1971)

14.42 A window port in a large-diameter, thin-walled pressurized cylinder is idealized as a large flat plate containing a hole with normal boundary stresses $\sigma_x = \sigma$ and $\sigma_y = 2\sigma$ applied to the plate boundary. If the hole is to be free from stress concentration, what thickness of hole reinforcement is necessary?

14.43 Match the stress function $\phi = (Ar^3 + B/r + Cr + Dr \ln r) f(\theta)$ to each of the thin curved beams in Fig. 14.27(a) and (b) with $f(\theta) = \sin \theta$ and $f(\theta) = \cos \theta$ respectively. Examine the local distribution of stress at the fixed and free ends and show their effects on the general stress predictions from the stress function. (I.C. 1966)

Answer: $\phi = F \sin \theta [r^3 - a^2 b^2 / r - 2(a^2 + b^2) r \ln r] / 2[(a^2 - b^2) + (a^2 + b^2) \ln (b/a)]$;
$\phi = -[p(\cos \theta)/8(a + b)][r^3 + a^3(a + 2b)/r + 4abr \ln r]$.

14.44 Employ the Boussinesq function $\phi = Cr\theta \sin \theta$ to obtain separate solutions for the stresses at point A (Fig. 14.28) in a semi-infinite plate of thickness t due to two concentrated forces which constitute a couple Pa. Resolve forces at A to show that the stresses in the θ direction are

$$\sigma_r = (2P/\pi t)[-(1/r)\cos\theta + (1/r_1)\cos\theta_1 \cos^2 \alpha],$$

$$\sigma_\theta = (2P/\pi t r_1)\cos\theta_1 \sin^2 \alpha,$$

$$\tau_{r\theta} = (P/\pi t r_1)\cos\theta_1 \sin 2\alpha.$$

Show that when $a \to 0$ and the couple Pa is replaced by a moment M the stresses become $\sigma_r = (2M/\pi t r^2)\sin 2\theta$, $\sigma_\theta = 0$, $\tau_{r\theta} = (2M/\pi t r^2)\cos^2 \theta$.

14.45 In Fig. 14.29 a linear distribution of loading is applied to the edge of a semi-infinite plate. Employ the Boussinesq function and the plane stress transformation equations for the given coordinate directions to derive integral expressions for the Cartesian stress components at a point $P(x, y)$ in the body of the plate.

(Hint: $\delta\sigma_x = -(2\delta p/\pi r)\cos^3 \theta$, $\delta\sigma_y = -(2\delta p/\pi r)\cos\theta \sin^2 \theta$ and $\delta\tau_{xy} = -(2\delta p/\pi r)\sin\theta \cos^2 \theta$.) (I.C. 1967)

Plane Finite Elements

14.46 Derive the rectangular element stiffness matrices (i) in the case of plane stress and (ii) in the case of plane strain. The dimensions of the element are $a \times b \times t$ (see Fig. 14.19).

14.47 Consider the assembly of the overall stiffness matrix when the plane-stress cantilever beam in Fig. 14.18 is divided into two rectangular elements 1243 and 3465 as shown.

14.48 Determine the explicit forms of the matrix that enables interior element displacements to be found when its nodal displacements are known (i) for a triangular element and (ii) for a rectangular element.

14.49 Determine the stiffness equations $\{F^e\} = [K^e]\{\delta^e\}$ for each of the triangular elements in Fig. 14.30 when (a) is in plane stress and (b) is in plane strain. Take $E = 210\,\text{GPa}$, $v = 0.28$, with uniform thickness 20 mm.

Answer: force vectors (N) are (a) $\{F^e\} = \{0 \ 0 \ -75 \ -25 \ -75 \ -5\}^T$; (b) $\{F\}^e = \{-20 \ -5 \ 5 \ 10 \ -5 \ 15\}^T$.

FIFTEEN

AXIALLY SYMMETRIC STRUCTURES

15.1 INTRODUCTION

A wide variety of load-bearing structures are axially symmetric. These include cylinders, spheres, and plates in which the loading may be due to pressure, direct concentrated forces, distributed loading (including self-weight), the effects of temperature variation and centrifugal forces under rotation. The pressurized thin-walled cylinder and sphere have been treated separately with the application of constitutive relations in Chapter 4 and the solution to a pressurized thick-walled cylinder appeared in a particular stress function $\phi = \phi(r)$ in Chapter 14. This chapter further examines elasticity with axial symmetry. Plasticity behaviour is covered in Chapters 16 and 17.

15.2 MEMBRANE THEORY

Thin-walled pressure vessels, including the cylinder, sphere, ellipsoid and toroid may be treated together with rotating rims in the membrane theory.

Let a common element of membrane in Fig. 15.1, with thickness t, density ρ, wall radii r_θ and r_ϕ, be subjected to the hoop and meridional stresses σ_θ and σ_ϕ due to combined pressure p, self-weight and centrifugal effects. The sides subtend angles $\delta\theta$ and $\delta\phi$ at the axes of symmetry about which the solid may also rotate at radius r with angular velocity ω to induce body forces. Ignoring bending effects and assuming that the radial stress σ_r through the wall is negligibly small compared to σ_θ and σ_ϕ, the following radial equilibrium equation applies to the biaxial stress state in the plane of the

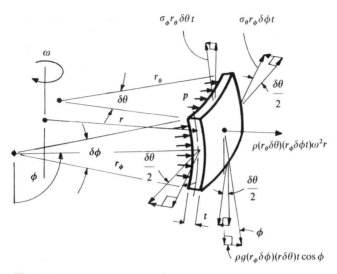

Figure 15.1 Membrane element.

wall:

$$(r_\theta \delta\theta)(r_\phi \delta\phi)(\rho t \omega^2 r + \rho g t \cos\phi + p) = 2\sigma_\theta r_\theta \delta\phi \, t \sin(\delta\theta/2)$$
$$+ 2\sigma_\phi r_\theta \delta\theta \, t \sin(\delta\phi/2).$$

Now, as $\sin(\delta\theta/2) \simeq \delta\theta/2$ and $\sin(\delta\phi/2) \simeq \delta\phi/2$ this equation simplifies to

$$r_\theta r_\phi(\rho\omega^2 r + \rho g \cos\phi + p/t) = \sigma_\theta r_\phi + \sigma_\phi r_\theta,$$
$$\rho\omega^2 r + \rho g \cos\phi + (p/t) = \sigma_\theta/r_\theta + \sigma_\phi/r_\phi. \tag{15.1}$$

A second equilibrium equation that isolates the action of one stress for the particular vessel then enables both stresses to be found from Eq. (15.1). The following simplified analyses apply to some of the more common forms of membrane under typical service loading conditions. For a detailed treatment, the reader is referred to any of the specific texts on shell theory amongst which is J. E. Gibson, *Linear Elastic Theory of Thin Shells*, Pergamon, 1965.

Cylinder and Sphere

For a closed, thin-walled, pressurized cylinder (page 84–87),

$$\sigma_\theta = (r/\cos\alpha)[p/t + \omega^2 \rho r + \rho g \sin\alpha]. \tag{15.3b}$$

Since $r_\theta = r$ and $r_\phi = \infty$, from Eq. (15.1),

$$\sigma_\theta = pr/t + \rho\omega^2 r^2 + \rho g r \cos\phi. \tag{15.2b}$$

In the absence of pressure and self-weight and with $\phi = 90°$, Eqs (15.2) give

$\sigma_\theta = \rho\omega^2 r^2, \sigma_z = 0$, which defines the hoop stress in a rotating rim or cylinder. For a stationary, pressurized sphere $\omega = 0, r_\theta = r_\phi = r$, when Eq. (15.1) confirms the stresses $\sigma_\theta = \sigma_\phi = pr/2t$ that were derived previously (Eq. (4.20)). These stresses will be neither constant nor equal when the sphere rotates, since the radius from the rotation axis varies.

Conical Vessel

For the rotating, pressurized conical vessel in Fig. 15.2, vertical equilibrium for any horizontal section of radius $r = r_\theta \cos\alpha$, expresses the balance between the force due to σ_ϕ and the forces due to pressure and to self-weight ($\rho g\pi rlt$) (centrifugal forces act horizontally). This condition gives

$$\sigma_\phi t(2\pi r)\cos\alpha = p\pi r^2 + \pi r^2 t\rho g/\tan\alpha,$$

$$\sigma_\phi = pr/2t\cos\alpha + \rho gr/2\sin\alpha. \tag{15.3a}$$

Substituting Eq. (15.3a) into Eq. (15.1) with $r_\phi = \infty$ and $\phi = 90 - \alpha$ gives

$$\sigma_\theta = (r/\cos\alpha)[p/t + \omega^2\rho r + \rho g\sin\alpha]. \tag{15.3b}$$

Example 15.1 Determine the speed at which an upturned $30°$ cone $2\,\text{m}$ long and $5\,\text{mm}$ thick may rotate about a vertical axis when it contains a gas at pressure $3\,\text{bar}$ if the maximum allowable stress is $150\,\text{MPa}$. Take the density $\rho = 7100\,\text{kg/m}^3$ for the cone material.

Equation (15.3) show that σ_θ is greatest at the base of the cone where

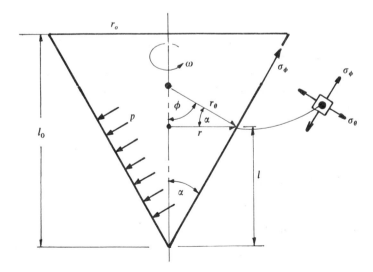

Figure 15.2 Rotating cone.

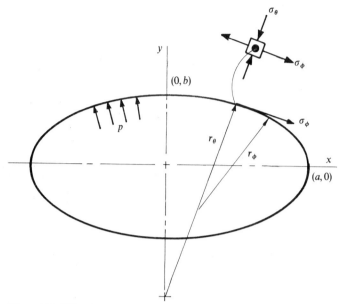

Figure 15.3 Ellipsoidal shell.

$r_o = l_o \tan \alpha = 2 \tan 15° = 0.5358$ m. Substituting into Eq. (15.3b) reveals that the contribution from self-weight is negligible:

$$\omega^2 = [(\sigma_\theta/r)\cos\alpha - p/t - \rho g \sin\alpha]/\rho r$$
$$= [(150 \times 0.9659/535.8) - (3 \times 10^5/10^6 \times 5)$$
$$- (7100 \times 9.81 \times 0.259/10^9)] \times 10^{12}/(7100 \times 535.8)$$
$$= [0.2704 - 0.06 - 0.000\,018](262.87 \times 10^3) = 55303.12,$$
$$\therefore \quad \omega = 235.17 \,\text{rad/s} \quad \text{or} \quad N = 60\omega/2\pi = 2245.67 \,\text{rev/min}.$$

Ellipsoidal Vessel

For the stationary, ellipsoidal, pressurized vessel in Fig. 15.3 the meridional stress σ_ϕ may be isolated from a consideration of equilibrium on a vertical section. This gives, on ignoring self-weight,

$$\sigma_\phi = pr_\theta/2t, \tag{15.4a}$$

where radius $r_\theta = (1/b^2)\sqrt{(a^4 y^2 + b^4 x^2)}$ from ellipse geometry. Substituting Eq. (15.4a) into Eq. (15.1) leads to the hoop stress:

$$\sigma_\theta = (pr_\theta/t)(1 - r_\theta/2r_\phi), \tag{15.4b}$$

in which the meridional radius $r_\phi = (b^2/a^4)r_\theta^3$. Particular stress values of

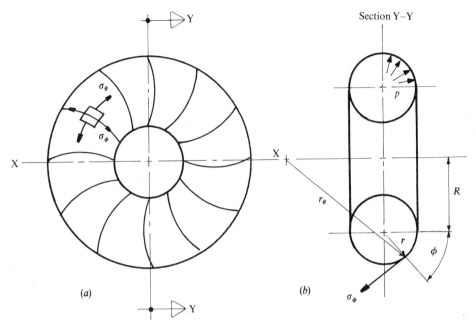

Figure 15.4 Toroidal shell under pressure.

interest, from Eqs (15.4), are

1. at the crown $(0, b)$, where $r_\theta = r_\phi = a^2/b$, giving equal maximum stresses

$$\sigma_\phi = \sigma_\theta = pa^2/2bt;$$

2. at the equator $(a, 0)$, where $r_\theta = a, r_\phi = b^2/a$, giving different streses

$$\sigma_\phi = pa/2t, \qquad \sigma_\theta = (pa/t)(1 - a^2/2b^2).$$

The equatorial hoop stress σ_θ can produce buckling in the knuckle region when it becomes compressive for $a/b > \sqrt{2}$.

Toroid

In the case of a closed, stationary toroidal shell of thickness t and radii r and R (Fig. 15.4) the hoop stress follows directly from the equilibrium equation referred to section X–X.

Ignoring self-weight,

$$2\sigma_\theta(2\pi rt) = 2p\pi r^2,$$
$$\therefore \quad \sigma_\theta = pr/2t. \tag{15.5a}$$

Substituting Eq. (15.5a) in Eq. (15.1) with $r_\phi = r$ and $r_\theta = r + R/\sin\phi$ leads to

$$\sigma_\phi/r = p/t - pr/2tr_\theta = (p/t)[1 - r/2(r + R/\sin\phi)],$$
$$\sigma_\phi = (pr/2t)(2R + r\sin\phi)/(R + r\sin\phi). \tag{15.5b}$$

Example 15.2 Determine the maximum membrane stresses in a stationary toroid of thickness 10 mm with cylindrical and mean radii of 0.25 m and 1.5 m respectively under a pressure of 1 bar. (1 MPa = 10 bar.)

Equation (15.5b) is a maximum when $\phi = 3\pi/2$, giving

$$\sigma_\phi = (pr/2t)(2R - r)/(R - r)$$
$$= (0.1 \times 0.25/2 \times 0.001)(3 - 0.25)/(1.5 - 0.25) = 2.75\,\text{MPa}$$

and from Eq. (15.5a) the constant hoop stress is

$$\sigma_\theta = pr/2t = 0.1 \times 0.25/(2 \times 0.01) = 1.25\,\text{MPa}.$$

Self-weight

Occasionally some account needs to be made of the stresses induced in the presence of self-weight. Take, for example, the element, $\delta\phi$, of suspended spherical dish in Fig. 15.5 with radius r and solid angle 2α under no other external loading. Isolating the self-weight term in Eq. (15.1), the consequent tensile stresses are related, for $r = r_\theta = r_\phi$, in the radial equilibrium equation:

$$\rho g \cos\phi = \sigma_\theta/r + \sigma_\phi/r.$$

The stresses may be separated from the vertical equilibrium equation referred

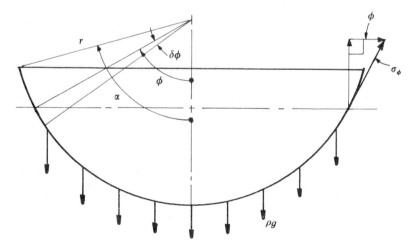

Figure 15.5 Self-weight loading.

to the horizontal section defined by ϕ. Noting that the elemental surface area is $2\pi(r\sin\phi)(r\,\delta\phi)$, then

$$2\pi(r\sin\phi)t\sigma_\phi\sin\phi = 2\pi r^2 t\rho g\int_0^\phi \sin\phi\,\mathrm{d}\phi,$$

$$\sigma_\phi = \rho gr(1-\cos\phi)/\sin^2\phi = \rho gr/(1+\cos\phi), \tag{15.6a}$$

$$\therefore\quad \sigma_\theta = \rho gr\cos\phi - \sigma_\phi = \rho gr[\cos\phi - 1/(1+\cos\phi)]. \tag{15.6b}$$

It is seen from Eqs (15.6) that $\sigma_\phi = \sigma_\theta = \rho gr/2$ for $\phi = 0$ at the crown. The greatest stresses depend upon α, i.e. when $\alpha = 90°$, $\sigma_\phi = \rho gr$ and $\sigma_\theta = -\rho gr$.

15.3   THICK-WALLED CYLINDERS

For the following thick-walled, cylindrical bodies it becomes necessary to consider the variation in radial stress σ_r through the wall.

Monobloc Cylinders under Pressure

In the particular case of a solid cylinder under internal pressure p, a triaxial stress state exists at any point in the wall (Fig. 15.6(a)). The radial stress (σ_r) is no longer negligible compared to the hoop (σ_θ) and axial (σ_z) stresses.

Vertical equilibrium about a horizontal diameter for a cylindrical element of radial thickness δr, over which σ_r varies as shown in Fig. 15.6(b), is expressed by

$$2\sigma_\theta\,\delta r\,L + \sigma_r 2rL = (\sigma_r + \delta\sigma_r)2(r+\delta r)L,$$

in which σ_r (assumed tensile) acts on the inner and outer projected areas. This gives the radial equilibrium equation

$$\sigma_\theta - \sigma_r = r\,\mathrm{d}\sigma_r/\mathrm{d}r. \tag{15.7}$$

The stresses may be separated when it is taken that σ_z is independent of r in maintaining plane cross-sections; i.e. the longitudinal strain ϵ_z is constant in the length for all r. That is, from Eq. (4.12c),

$$\epsilon_z = (1/E)[\sigma_z - v(\sigma_\theta + \sigma_r)] = \text{constant},$$

$$\therefore\quad \sigma_\theta + \sigma_r = (\sigma_z - \epsilon_z E)/v = 2a \quad \text{(constant)}. \tag{15.8}$$

Eliminating σ_θ between Eqs (15.7) and (15.8) and integrating leads to

$$2\int(\mathrm{d}r/r) = \int \mathrm{d}\sigma_r/(a-\sigma_r),$$

$$2\ln r = -\ln(a-\sigma_r) + \ln b,$$

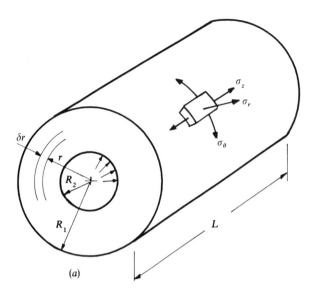

(a)

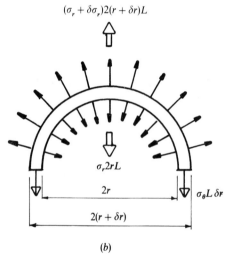

(b)

Figure 15.6 Thick-walled pressurized cylinder.

where b is a constant. This may be combined with Eq. (15.8) to give the radial and hoop stress variations as

$$\sigma_r = a - b/r^2, \tag{15.9a}$$

$$\sigma_\theta = a + b/r^2. \tag{15.9b}$$

Equations (15.9) are the more usual form of the Lamé equations in which the constants a and b are found from the conditions existing at the boundary (where both internal and external pressures may act). They yield the same distributions (see Fig. 14.15) as with the stress function Eqs (14.28). Equation (14.29) again supplies σ_z for a closed cylinder, but where $\epsilon_z = 0$ in plane strain, Eq. (15.8) gives $\sigma_z = v(\sigma_\theta + \sigma_r)$.

Example 15.3 Find the necessary thickness of a hydraulic main, 130 mm internal diameter, required to contain a gauge pressure of 100 bar when the maximum hoop stress is limited to 15 MPa.

Since σ_θ is a maximum (tension) at the inner diameter and the pressure is the compressive radial stress value at the inner diameter, the boundary conditions are (i) $\sigma_\theta = +15$ MPa for $r = 65$ mm and (ii) $\sigma_r = -10$ MPa for $r = 65$ mm. Substituting into Eqs (15.9) gives

$$15 = a + b/(65)^2,$$
$$-10 = a - b/(65)^2,$$

from which $a = 2.5$ and $b = 52\,812.5$. At the outer diameter the pressure is atmospheric, i.e. zero gauge pressure, and Eq. (15.9a) becomes

$$0 = 2.5 - 52\,812.5/r_o^2,$$
$$\therefore \quad r_o = 145.4 \text{ mm} \qquad \text{and} \qquad t = 80.4 \text{ mm}.$$

Example 15.4 A pipe of 150 mm i.d. and 200 mm o.d. fails under hoop tension for an internal pressure of 500 bar. Determine, for a safety factor of 4, the safe internal pressure for a second pipe of the same material and internal diameter but with 40 mm thick walls.

Applying Eqs (15.9a) to the boundary conditions for the first pipe: (i) $\sigma_r = -50$ MPa for $r = 75$ mm and (ii) $\sigma_r = 0$ for $r = 100$ mm gives

$$-50 = a - b(75)^2,$$
$$0 = a - b/(100)^2,$$

from which $a = 64.29$ and $b = 642.86 \times 10^3$. Then, from Eq. (15.9b) with $r = 75$ mm, the hoop stress that caused failure in the bore is

$$\sigma_\theta = a + b/r^2 = 64.29 + (642.86 \times 10^3)/75^2 = 178.6 \text{ MPa}.$$

Hence, the safe working hoop stress for the second cylinder is $\sigma_\theta = 178.6/4 = 44.65$ MPa. Applying Eqs (15.9) with new constants a' and b':

$$44.65 = a' + b'/(75)^2,$$
$$0 = a' - b'/(115)^2,$$

from which $a' = 13.32$ and $b' = 176.16 \times 10^3$. The corresponding internal pressure $\sigma_r = -p$ for $r = 75\,\text{mm}$ is, from Eq. (15.9a),

$$-p = a' - b'/r^2 = 13.32 - (176.16 \times 10^3)/75^2,$$

$$p = 18\,\text{MPa} \quad \text{or} \quad 180\,\text{bar}.$$

Compound Discs and Cylinders with Interference Fits

Shaft and hub When a solid cylindrical shaft or disc is forced into a hub, the external diameter of the shaft and the internal hub diameter are both subjected to compressive radial pressure (p). The magnitude of this interface pressure is controlled by the initial difference (interference) in diameters (Δd). This depends upon the difference between hoop strains at the common radius (r_c), which in turn is a function of each biaxial stress state (σ_θ, σ_r) in the adjacent components (Fig. 15.7). Since the stress in the shaft cannot be infinite at $r = 0$, Eqs (15.9) become

$$\sigma_\theta = \sigma_r = -p = \text{constant.}$$

With an annular hub of outer radius r_o the constants a and b in Eqs (15.9) may be found from the two boundary conditions: (i) $\sigma_r = -p$ for $r = r_c$ and (ii) $\sigma_r = 0$ for $r = r_o$. Hence, the hoop stress σ_θ at $r = r_c$ in the hub may be found. In the absence of axial stress σ_z, for a thin disc (in the z-direction), the relative hoop strain at r_c is given by

$$\Delta\epsilon_\theta = [(\sigma_\theta - v\sigma_r)/E]_{\text{hub}} - [(\sigma_\theta - v\sigma_r)/E]_{\text{shaft}}, \tag{15.10a}$$

$$\Delta u/r_c = [(\sigma_\theta + vp)/E]_{\text{hub}} + p[(1 - v)/E]_{\text{shaft}}. \tag{15.10b}$$

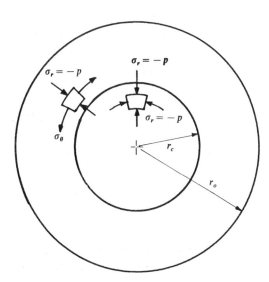

Figure 15.7 Interface stress states.

Then $\Delta d_c = 2\,\Delta u$, in which Δu is the radial displacement of the common radius r_c. Some applications are illustrated in the following examples.

Example 15.5 An oversized steel plug is pressed into a steel hub of 12.5 mm i.d. and 38 mm o.d. If the maximum stress in the hub must not exceed 465 MPa, calculate the necessary interference. Take $E = 208\,\text{GPa}$, $v = 0.3$.

Here $\sigma_\theta = 465\,\text{MPa}$ for $r_c = 6.25\,\text{mm}$. Taken with the second boundary condition for the hub that $\sigma_r = 0$ for $r_o = 19\,\text{mm}$, Eq. (15.9b) supplies

$$465 = a + b/(6.25)^2,$$
$$0 = a - b/(19)^2,$$

from which $a = 45.4$ and $b = 16\,390.5$. The radial stress σ_r at $r_c = 6.25\,\text{mm}$ is, from Eq. (15.9a),

$$\sigma_r = a - b/r^2 = 45.4 - 16\,390.5/(6.25)^2 = -374.2\,\text{MPa}.$$

Thus, $p = 374.2\,\text{MPa}$ and Eq. (15.10b) becomes

$$\Delta u/6.25 = \{[465 + (0.3 \times 374.2)] + 374.2(1 - 0.3)\}/208\,000,$$
$$\Delta u = 6.25(465 + 374.2)/208\,000 = 0.025\,\text{mm},$$
$$\therefore \quad \Delta d_c = 0.050\,\text{mm}.$$

Example 15.6 A steel sleeve of 115 mm o.d. and 75 mm i.d. is shrunk on to a solid shaft to give a circumferential strain at the outer diameter of 0.07 per cent. Determine the interface pressure and the greatest tensile stress in the sleeve. Take $E = 207\,\text{GPa}$, $v = 0.27$.

The sleeve outer boundary conditions are (i) $\sigma_r = 0$ for $r_o = 57.5\,\text{mm}$ and (ii) $\epsilon_\theta = (\sigma_\theta - v\sigma_r)/E = 7 \times 10^{-4}$ for $r_o = 57.5\,\text{mm}$, from which it follows that $\sigma_\theta = 144.9\,\text{MPa}$. Then, from Eqs (15.9),

$$0 = a - b/(57.5)^2,$$
$$144.9 = a + b/(57.5)^2,$$

giving $a = 72.45$ and $b = 239.54 \times 10^3$. At the interface radius, $r_c = 37.5\,\text{mm}$:

$$\sigma_r = a - b/r_c^2 = 72.45 - (239.54 \times 10^3)/(37.5)^2 = -97.89\,\text{MPa}$$
$$(p = 97.89\,\text{MPa}),$$
$$\sigma_\theta = a + b/r_c^2 = 72.45 + (239.54 \times 10^3)/(37.5)^2 = 242.79\,\text{MPa}.$$

Annular discs

Thick-walled discs When one thick-walled disc is shrunk fit with interference on to another, there results a compressive residual hoop stress distribution in the inner disc that effectively reduces the tensile hoop stress resulting from a subsequently applied internal pressure. The effect is to produce a more even hoop stress distribution over the compound thickness. Where internal pressures reach high values, several discs may be compounded to enhance the benefit. In general, for two annular discs in different materials with small thickness in the z-direction, the relative hoop strain at the common interface radius (r_c) is again found from the adjacent plane stress states (σ_θ, σ_r) in Fig. 15.8 and the appropriate constitutive equations. With respective subscripts I and O for the inner and outer discs, the relative interface hoop strain is

$$\Delta\epsilon_\theta = [(1/E)(\sigma_\theta - v\sigma_r)]_O - [(1/E)(\sigma_\theta - v\sigma_r)]_I, \qquad (15.11a)$$

$$\Delta d/d_c = [(1/E)(\sigma_\theta + vp)]_O - [(1/E)[(\sigma_\theta + vp)]_I, \qquad (15.11b)$$

where Δd is the initial interference, p is the common compressive interface pressure and σ_θ is supplied from Eqs (15.9) with the typical boundary conditions $\sigma_r = 0$ for $r = r_i$, $\sigma_r = -p$ for $r = r_c$ in I; $\sigma_r = 0$ for $r = r_o$, $\sigma_r = -p$ for $r = r_c$ in O.

Thin-walled discs When two thin-walled rings are shrunk together with a common compressive pressure p at the interface diameter d_c, the membrane theory gives

$$(\sigma_\theta)_O = pd_c/2t_O \quad \text{(tensile)} \quad \text{and} \quad (\sigma_\theta)_I = pd_c/2t_I \quad \text{(compressive)}.$$

Since $(\sigma_r)_I = (\sigma_r)_O = -p$ and, neglecting the small lateral strain term vp/E, Eq. (15.11b) approximates to

$$\Delta d = (pd_c^2/2)(1/E_I t_I + 1/E_O t_O). \qquad (15.12)$$

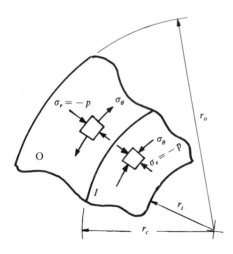

Figure 15.8 Stress state at interface.

Thick-walled cylinders The plane stress analyses in the preceding two subsections ignore any residual axial stress σ_z at the interface following shrinkage. Plane strain conditions may be assumed for compound thick-walled cylinders, e.g. gun barrels, where $\epsilon_z = 0$ gives an axial stress $\sigma_z = v(\sigma_\theta + \sigma_r)$. Since interface triaxial stress states $(\sigma_\theta, \sigma_r, \sigma_z)$ exist, Eqs (15.11) are modified to become

$$\Delta\epsilon_\theta = \{(1/E)[\sigma_\theta - v(\sigma_r + \sigma_z)]\}_O - \{(1/E)[\sigma_\theta - v(\sigma_r + \sigma_z)]\}_I,$$
$$\Delta\epsilon_\theta = \{(1/E)[\sigma_\theta(1 - v^2) - v\sigma_r(1 + v)]\}_O$$
$$- \{(1/E)[\sigma_\theta(1 - v^2) - v\sigma_r(1 + v)]\}_I,$$
$$\Delta d/d_c = \{(1/E)[\sigma_\theta(1 - v^2) + vp(1 + v)]\}_O$$
$$- \{(1/E)[\sigma_\theta(1 - v^2) + vp(1 + v)]\}_I. \tag{15.13}$$

The subsequent pressure distributions in thick-walled compounded discs and cylinders are found from Eqs (15.9) assuming a monobloc cylinder. The net stresses at pressure follow from superposition as illustrated in the following example.

Example 15.7 A brass sleeve (I) of 55 mm (2.2 in) i.d. and 70 mm (2.8 in) o.d., is fitted within a steel hub (O) of 100 mm (4 in) o.d. to give a common interface pressure of 30 MPa (2 tonf/in^2). Determine the initial difference in diameters between sleeve and hub. Plot the residual stresses resulting from this process. Assuming plane stress, what are the net hoop stresses at the inner diameter of each component when a subsequent pressure of 1 kbar is applied? Assume $E_I = 90$ GPa (6000 tonf/in^2), $v_I = 0.33$, $E_O = 210$ GPa (13 400 tonf/in^2) and $v_O = 0.28$.

IN SI UNITS Applying to I the boundary conditions $\sigma_r = 0$ for $r = 27.5$ mm and $\sigma_r = -30$ for $r = 35$ mm,

$$0 = a - b/(27.5)^2,$$
for $$-30 = a - b/(35)^2,$$

which gives $a = -78.5$ and $b = -59.5 \times 10^3$. The hoop stress at the brass interface (I in Fig. 15.8) is then

$$\sigma_\theta = a + b/r_c^2 = -78.5 - (59.5 \times 10^3)/35^2 = -127.1 \text{ MPa (compression).}$$

Applying to O the boundary conditions $\sigma_r = 0$ for $r = 50$ mm and $\sigma_r = -30$ for $r = 35$ mm,

$$0 = a' - b'/(50)^2,$$
$$-30 = a' - b'/(35)^2,$$

$\therefore$ $a' = 28.82$ and $b' = 72.1 \times 10^3$ and the steel interface hoop stress is

$$\sigma_\theta = a' + b'/r_c^2 = 28.82 + (72.1 \times 10^3)/35^2 = 87.62 \text{ MPa} \qquad \text{(tension).}$$

From Eq. (15.11b),

$$\Delta d/d_c = [(1/E)(\sigma_\theta + vp)]_O - [(1/E)(\sigma_\theta + vp)]_I$$
$$= [87.62 + (0.28 \times 30)]/(210 \times 10^3)$$
$$- [-127.1 + (0.33 \times 30)]/(90 \times 10^3)$$
$$= [0.4574 - (-1.302)] \times 10^{-3} = 1.7595 \times 10^{-3},$$
$$\therefore \quad \Delta d = 70 \times 1.7595 \times 10^{-3} = 0.123 \text{ mm}.$$

The residual stress distribution given in Fig. 15.9 is completed from the stress state at $r = r_i$ in the brass where $\sigma_r = 0$ and

$$\sigma_\theta = a + b/r_i^2$$
$$= -78.5 - (59.5 \times 10^3)/(27.5)^2$$
$$= -157.2 \text{ MPa}.$$

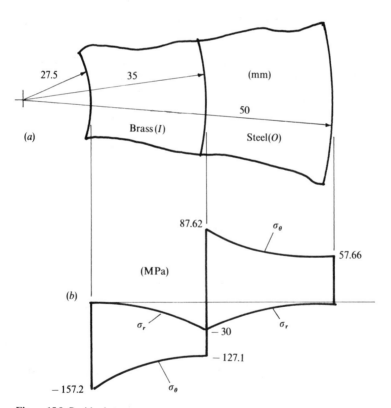

Figure 15.9 Residual stresses.

At $r = r_o$ in the steel, $\sigma_r = 0$ and

$$\sigma_\theta = a' + b'/r_o^2$$
$$= 28.82 + (72.1 \times 10^3)/(50)^2$$
$$= 57.66 \, \text{MPa}.$$

For a subsequent internal pressure of 1 kbar the 'monobloc' boundary conditions become $\sigma_r = -100$ MPa for $r = 27.5$ mm and $\sigma_r = 0$ for $r = 50$ mm. From Eq. (15.9a),

$$-100 = a'' - b''/(27.5)^2,$$
$$0 = a'' - b''/(50)^2.$$

Thus $a'' = 43.37$ and $b'' = 108.43 \times 10^3$, when the applied hoop stresses are

$$\sigma_\theta = a'' + b''/r_i^2$$
$$= 43.37 + (108.43 \times 10^3)/(27.5)^2 = 186.7 \, \text{MPa},$$
$$\sigma_\theta = a'' + b''/r_o^2$$
$$= 43.37 + (108.43 \times 10^3)/(35)^2 = 131.88 \, \text{MPa}.$$

Superimposing these applied values on the residuals in Fig. 15.9, the net hoop stresses become

$$\sigma_\theta = -157.2 + 186.7 = 29.5 \, \text{MPa} \qquad \text{for} \quad r_i = 27.5 \, \text{mm} \quad \text{(brass i.d.)},$$
$$\sigma_\theta = 87.62 + 131.88 = 219.5 \, \text{MPa} \qquad \text{for} \quad r_o = 35 \, \text{mm} \quad \text{(steel i.d.)}.$$

Note that the resulting interference for the above example with long thick cylinders (plane strain) gives $\Delta d = 0.108$ mm from Eq. (15.13).

SOLUTION TO THE INTERFERENCE IN IMPERIAL UNITS For I, $\sigma_r = 0$ for $r = 1.1$ in and $\sigma_r = -2$ for $r = 1.4$ in. That is,

$$0 = a - b/(1.1)^2,$$
$$-2 = a - b/(1.4)^2.$$

This gives $a = -5.22$ and $b = -6.31$, from which the hoop stress at the interface is

$$\sigma_\theta = a + b/r_c^2$$
$$= -5.22 - 6.31/(1.4)^2 = -8.44 \, \text{tonf/in}^2.$$

For O, $\sigma_r = 0$ for $r = 2$ in and $\sigma_r = -2$ for $r = 1.4$ in. That is,

$$0 = a' - b'/(2)^2,$$
$$-2 = a' - b'/(1.4)^2,$$

which gives $a' = 1.923$ and $b' = 7.69$. The interface hoop stress becomes

$$\sigma_\theta = a' + b'/r_c^2$$
$$= 1.923 + 7.69/(1.4)^2 = 5.85 \, \text{tonf/in}^2,$$

and, from Eq. (15.11b),

$$\Delta d/d_c = [5.85 + (0.28 \times 2)]/13\,400 - [-8.44 + (0.33 \times 2)]/6000$$
$$= [0.4784 - (-1.2967)] \times 10^{-3} = 1.7751,$$
$$\therefore \quad \Delta d = 2.8 \times 1.7751 \times 10^{-3} = 0.004\,97 \text{ in.}$$

15.4 ROTATING CYLINDRICAL BODIES

Thin Annular Disc

When a solid or thick-walled annular disc rotates with uniform angular speed ω rad/s, about its z-axis, the centrifugal 'body' force induces radial (σ_r) and hoop (σ_θ) stress distributions in the wall. Provided the z-dimension is small and uniform a plane stress condition $\sigma_z = 0$ will simplify the analysis. Thus, for unit thickness, the radial equilibrium equation for the disc element in Fig. 15.10 is

$$(\sigma_r + \delta\sigma_r)(r + \delta r)\delta\theta - \sigma_r r\,\delta\theta - 2\sigma_\theta\delta r\sin(\delta\theta/2) + \rho\,\delta V\,\omega^2 r = 0.$$

Putting $\delta V = r\,\delta\theta\,\delta r$ and with $\sin(\delta\theta/2) \simeq \delta\theta/2$ in the limit, leads to

$$\sigma_r + r\,\mathrm{d}\sigma_r/\mathrm{d}r - \sigma_\theta + \rho\omega^2 r^2 = 0. \tag{15.14}$$

Now, from Eqs (14.1) and (14.25), the plane stress polar constitutive relations

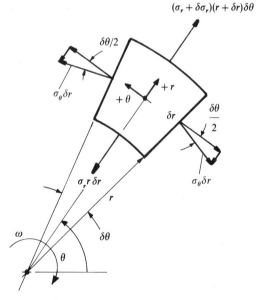

Figure 15.10 Rotating-disc element.

for axial symmetry are

$$\epsilon_\theta = u/r = (1/E)(\sigma_\theta - v\sigma_r),$$
$$\epsilon_r = du/dr = (1/E)(\sigma_r - v\sigma_\theta).$$

Combining these leads to the compatibility condition;

$$r(d\sigma_\theta/dr - v\,d\sigma_r/dr) + (1 + v)(\sigma_\theta - \sigma_r) = 0. \tag{15.15}$$

Then, from Eqs (15.14) and (15.15),

$$r(d\sigma_\theta/dr + d\sigma_r/dr) + (1 + v)\rho\omega^2 r^2 = 0,$$
$$d(\sigma_\theta + \sigma_r)/dr = -(1 + v)\rho\omega^2 r.$$

Integration gives

$$\sigma_\theta + \sigma_r = 2a - (1 + v)\rho\omega^2 r^2/2. \tag{15.16}$$

Eliminating σ_θ between Eqs (15.14) and (15.16) leads to

$$2\sigma_r + r\,d\sigma_r/dr = 2a - (3 + v)\rho\omega^2 r^2/2,$$
$$(1/r)\,d(r^2\sigma_r)/dr = 2a - (3 + v)\rho\omega^2 r^2/2,$$

$$\therefore \quad r^2\sigma_r = \int [2ar - (3 + v)\rho\omega^2 r^3/2]\,dr$$

$$= ar^2 - (3 + v)\rho\omega^2 r^4/8 - b,$$

$$\sigma_r = a - b/r^2 - (3 + v)\rho\omega^2 r^2/8, \tag{15.17a}$$

$$\sigma_\theta = a + b/r^2 - (1 + 3v)\rho\omega^2 r^2/8. \tag{15.17b}$$

The constants a and b are found from the condition that $\sigma_r = 0$ for $r = r_i$ and $r = r_o$. This gives

$$a = \rho\omega^2(r_i^2 + r_o^2)(3 + v)/8, \tag{15.18a}$$

$$b = \rho\omega^2 r_i^2 r_o^2(3 + v)/8. \tag{15.18b}$$

Example 15.8 Determine the hoop stress at the inner $r_i = 50\,\text{mm}$ (2 in) and outer $r_o = 150\,\text{mm}$ (6 in) radii, for a disc when rotating uniformly at 5000 rev/min. Show the σ_θ and σ_r variation with r and the radius for which σ_r is a maximum. Take $v = 0.3$ and $\rho = 7480\,\text{kg/m}^3$ ($0.27\,\text{lb/in}^3$).

SOLUTION TO SI UNITS

$$\omega = 2\pi N/60 = 2\pi \times 5000/60 = 523.6\,\text{rad/s}$$

and, from Eqs (15.18a and b),

$$a = 7480 \times (523.6)^2(50^2 + 150^2)10^{-6} \times 3.3/8$$
$$[\text{kg/m}^3 \times (\text{rad/s})^2 \times \text{m}^2 = \text{kg m/s}^2 \times \text{m}^{-2} \equiv \text{Pa}]$$
$$= 21.15 \times 10^6\,\text{Pa} = 21.15\,\text{MPa},$$

$$b = 7480 \times (523.6)^2 \times 50^2 \times 150^2 \times 10^{-12} \times 3.3/8$$
$$[\text{kg/m}^3 \times (\text{rad/s})^2 \times \text{m}^4 = \text{kg m/s}^2 \equiv \text{N}]$$
$$= 47.583 \times 10^3 \text{ N}.$$

Then, from Eq. (15.17b) with $r_i = 50 \text{ mm}$,

$$\sigma_\theta = [21.15 + (47.583 \times 10^3)/50^2]$$
$$- [1.9 \times 7480(523.6)^2 \times 50^2 \times 10^{-12}/8]$$
$$= 21.15 + 19.033 - 1.218 = 38.97 \text{ MPa}.$$

For $r_o = 150 \text{ mm}$:

$$\sigma_\theta = [21.15 + (47.583 \times 10^3)/150^2]$$
$$- [1.9 \times 7480(523.6)^2 \times 150^2 \times 10^{-12}/8]$$
$$= 21.15 + 2.115 - 10.96 = 12.31 \text{ MPa}.$$

The radial stress is a maximum, when from Eq. (15.17a),

$$d\sigma_r/dr = 2b/r^3 - \rho\omega^2 r(3 + v)/4 = 0,$$
$$r^4 = 8b/\rho\omega^2(3 + v).$$

Substituting for b from Eq. (15.18b) leads to

$$r = \sqrt{(r_i r_o)}. \tag{i}$$

The corresponding maximum radial stress is, from Eqs (15.17a) and (15.18),

$$(\sigma_r)_{max} = a - b/r_i r_o - (3 + v)\rho\omega^2 r_i r_o/8$$
$$= [(3 + v)\omega^2 \rho/8](r_i^2 + r_o^2 - r_i r_o - r_i r_o),$$
$$(\sigma_r)_{max} = (3 + v)(r_o - r_i)^2 \omega^2 \rho/8. \tag{ii}$$

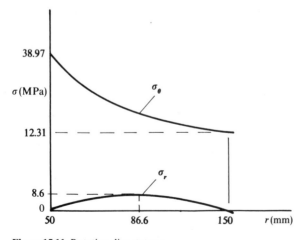

Figure 15.11 Rotating-disc stresses.

In this example, from Eq. (i), $r = \sqrt{(50 \times 150)} = 86.6$ mm and, from Eq. (ii),

$$(\sigma_r)_{max} = 3.3 \times 100^2 \times 523.6^2 \times 7480 \times 10^{-12}/8 = 8.6\,\text{MPa}.$$

The σ_r and σ_θ stress distributions are as shown in Fig. 15.11.

SOLUTION TO MAXIMUM RADIAL AND HOOP STRESSES IN IMPERIAL UNITS From Eq. (i), $r = \sqrt{(r_i r_o)} = \sqrt{(2 \times 6)} = 3.465$ in and again from Eq. (ii),

$$
\begin{aligned}
(\sigma_r)_{max} &= (3 + v)(r_o - r_i)^2 \omega^2 \rho/8g \\
&= (3 + 0.3)(6 - 2)^2 (523.6)^2 \times 0.27/(8 \times 32.2 \times 12) \\
&= 1265\,\text{lbf/in}^2. \tag{iii}
\end{aligned}
$$

The maximum hoop stress occurs at the inside radius. Substituting Eqs (15.18) into Eq. (15.17b) with $r = 2$ in:

$$
\begin{aligned}
\sigma_\theta &= [(3 + v)(r_i^2 + r_o^2 + r_o^2 r_i^2/r^2) - (1 + 3v)r^2]\omega^2 \rho/8g \\
&= [(3 + 0.3)(2^2 + 6^2 + 6^2) - (1 + 0.9)2^2](523.6)^2 \times 0.27/(8 \times 32.2 \times 12) \\
&= 5850\,\text{lbf/in}^2.
\end{aligned}
$$

Note the use of the gravitational acceleration $g = 32.2\,\text{ft/s}^2$ in converting from the pound mass to pound force. For example, the units of Eq. (iii) are

$$
\left[\text{in}^2 \left(\frac{\text{rad}}{\text{s}}\right)^2 \frac{\text{lb}}{\text{in}^3} \times \frac{\text{lbf s}^2}{\text{lb ft}} \times \frac{\text{ft}}{\text{in}}\right] \equiv \frac{\text{lbf}}{\text{in}^2}.
$$

Long Thick-walled Cylinder or Solid Shaft

When the length dimension is not small, a similar analysis to the annular disc but with an assumed plane strain condition, $\epsilon_z = \text{constant}$, leads to the triaxial stress distributions for a rotating cylinder:

$$\sigma_r = a - b/r^2 - (3 - 2v)\rho\omega^2 r^2/8(1 - v), \tag{15.19a}$$

$$\sigma_\theta = a + b/r^2 - (1 + 2v)\rho\omega^2 r^2/8(1 - v), \tag{15.19b}$$

$$\sigma_z = E\epsilon_z + v(\sigma_\theta + \sigma_r). \tag{15.19c}$$

Plane strain modifies only the compatibility condition. The reader should check that Eqs (15.19) satisfy the same equilibrium (Eq. (15.14)) used for plane stress analysis. For a long, rotating solid shaft, $b = 0$ in Eqs (15.19), since both σ_r and σ_θ must be finite at the shaft centre.

Solid Disc

The stresses in a rotating disc are also given by Eqs (15.17) but with $b = 0$, since infinite values are not possible for $r = 0$. That is,

$$\sigma_r = a - \rho\omega^2 r^2(3 + v)/8, \tag{15.20a}$$

$$\sigma_\theta = a - \rho\omega^2 r^2(1 + 3v)/8. \tag{15.20b}$$

These show that $\sigma_r = \sigma_\theta = a$ become the equal maximum stress values at the disc centre when $a = \rho\omega^2 r_o^2(3 + v)/8$ for $\sigma_r = 0$ at the disc o.d. The latter boundary condition will not apply to the inner disc in a rotating compound assembly, since a radial pressure will exist at the interface. This condition is examined for the following disc–hub assembly.

Disc and Hub with Interference Fit

When a compound disc–hub assembly is to rotate at high speed, the initial interference between their diameters may be chosen to prevent slipping and also to limit the maximum hoop stress. The determination of the initial interference and the shrinkage stresses for this assembly have been dealt with previously in Section 15.3 (page 647). When the assembly subsequently rotates, Eqs (15.17) supply the stresses in the hub and Eqs (15.20) supply those for the disc. The net stresses are then found from superimposing the rotational (R) and shrinkage (S) values. For example, at a given radius r in the hub, the net hoop stress is

$$\begin{aligned}
\sigma_\theta &= (\sigma_\theta)_S + (\sigma_\theta)_R \\
&= (a + b/r^2)_S + [a + b/r^2 - \rho\omega^2 r^2(1 + 3v)/8]_R \\
&= (a_S + a_R) + (b_S + b_R)/r^2 - \rho\omega^2 r^2(1 + 3v)/8 \\
&= a + b/r^2 - \rho\omega^2 r^2(1 + 3v)/8,
\end{aligned}$$

where $a = a_S + a_R$ and $b = b_S + b_R$. It follows that Eqs (15.17) and (15.20) may also be used directly for net stresses when the respective constants are combined. This approach can simplify the analysis considerably when net limiting stresses are specified at speed.

Example 15.9 A compound steel assembly comprises a hub with 510 mm i.d. and 710 mm o.d. shrunk on to a solid disc. If an interference of 0.05 mm must be maintained between the hub and disc at a speed of 10 000 rev/min, determine the initial mismatch in the common diameter. Assume plane stress conditions and take $E = 207$ GPa, $v = 0.3$ and $\rho = 7835$ kg/m^3.

Following shrinkage, let the compressive residual pressure $\sigma_r = -p$ exist at r_c as shown in Fig. 15.12(a). With the two boundary conditions, $\sigma_r = 0$ for $r_o = 355$ mm and $\sigma_r = -p$ for $r_c = 255$ mm, a tensile residual hoop stress $\sigma_\theta = 3.132p$ at $r_c = 255$ mm in the hub is found from Eqs (15.9).

  Treating the assembly as a single solid disc, the rotation stresses are found from Eqs (15.20a) with the boundary condition $\sigma_r = 0$ for

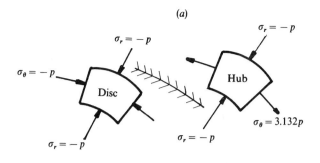

(a)

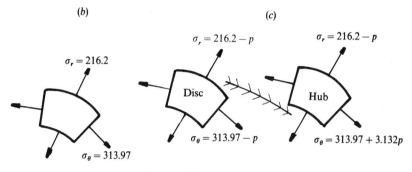

Figure 15.12 Compound cylinder assembly.

$r_o = 355\,\text{mm}$. This gives

$$a = (3 + v)\rho\omega^2 r_o^2/8$$
$$= 3.3 \times 7835 \times (1047.2)^2 \times 355^2/(8 \times 10^{12}) = 446.66.$$

Then, from Eqs (15.20), the rotation stress state at $r_c = 255\,\text{mm}$ is

$$\sigma_r = 446.66 - (255/355)^2 \times 446.66$$
$$= 446.66 - 230.46 = 216.2\,\text{MPa},$$
$$\sigma_\theta = 446.66 - (1.9/3.3) \times 230.46$$
$$= 446.66 - 132.69 = 313.97\,\text{MPa}.$$

The residual and rotational stress states for r_c are shown in Fig. 15.12(a) and (b) respectively. Their sum in (c) gives the net stresses. It follows from Eq. (15.10a) that the required interference in Fig. 15.12(c) is given by

$$0.05/510 = \{[(313.97 + 3.132p) - v(216.2 - p)] - [(313.97 - p) - v(216.2 - p)]\}/207\,000.$$

Hence $p = 4.983\,\text{MPa}$. Then applying Eq. (15.10b) to Fig. 15.12(a), the original interference Δd is found from

$$\Delta d/510 = [(3.132p + vp) - (-p + vp)]/207\,000,$$
$$\Delta d = 4.132 \times 4.983 \times 510/207\,000 = 0.0507 \text{ mm}.$$

Example 15.10 A steel disc of outer radius 250 mm is shrunk on to a hollow axle that has an outer radius of 40 mm and an inner radius of 20 mm. The interference between the disc and axle is 2.5×10^{-3} mm at the common radius. Both the disc and the axle are made of the same material for which $E = 208$ GPa, $v = 0.3$ and $\rho = 7860$ kg/m³. Find the radial stress at the common radius when the system is stationary. Find the angular speed at which the disc becomes loose. (C.E.I. Pt. 2, 1985)

For the axle (Fig. 15.13), the boundary conditions for Eqs (15.9) give

$$\sigma_r = 0 = a - b/20^2,$$
$$\sigma_r = -p = a - b/40^2,$$
$$\therefore \quad a = -1.333p, \qquad b = -533.33p,$$

and at $r_c = 40$ mm,

$$\sigma_\theta = a + b/r_c^2,$$
$$\sigma_\theta = -1.333p - 533.33p/40^2 = -1.666p.$$

For the disc boundary conditions, Eqs (15.9) give

$$\sigma_r = 0 = a' - b'/250^2,$$
$$\sigma_r = -p = a' - b'/40^2,$$
$$\therefore \quad a' = 0.026\,27p, \qquad b' = 1642.04p,$$

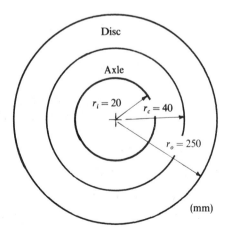

Figure 15.13 Disc–axle assembly.

and for $r_c = 40 \, \text{mm}$,

$$\sigma_\theta = a' + b'/r_c^2 = 0.026\,27p + 1642.04p/40^2 = 1.0525p.$$

Substituting into Eq. (15.11b):

$$E\,\Delta d/d_c = (1.0525p + 0.3p) - (-1.666p + 0.3p),$$
$$208 \times 10^3 \times 2.5 \times 10^{-3}/80 = 2.7185p, \quad \Rightarrow p = 23.91 \, \text{MPa},$$
$$\therefore \quad \sigma_r = -23.91 \, \text{MPa} \quad \text{at} \quad r_c = 40 \, \text{mm}.$$

It is assumed, for rotation, that when the interface pressure is reduced to zero the disc will slip on the axle. Treating the rotating assembly as a solid annular disc the boundary conditions $\sigma_r = 0$ for $r_i = 20$ and $r_o = 250 \, \text{mm}$ in Eq. (15.17a) give

$$\sigma_r = 0 = a'' - b''/20^2 - (7860 \times 3.3 \times 10^{-12} \times 20^2/8)\omega^2,$$
$$0 = a'' - b''/20^2 - (1.297 \times 10^{-6})\omega^2, \tag{i}$$
$$\sigma_r = 0 = a'' - b''/250^2 - (7860 \times 3.3 \times 10^{-12} \times 250^2/8)\omega^2,$$
$$0 = a'' - b''/250^2 - (202.64 \times 10^{-6})\omega^2. \tag{ii}$$

Solving Eqs (i) and (ii) gives

$$a'' = (2.0394 \times 10^{-4})\omega^2, \qquad b'' = 0.08106\omega^2.$$

At $r_c = 40 \, \text{mm}$, the radial stress in Eq. (15.17a) becomes

$$\sigma_r = (2.0394 \times 10^{-4})\omega^2 - 0.08106\omega^2/40^2$$
$$- (7860 \times 3.3 \times 10^{-12} \times 40^2/8)\omega^2,$$
$$\sigma_r = (148.09 \times 10^{-6})\omega^2.$$

Hence, the net radial stress at the interface becomes zero when

$$-p + (148.09 \times 10^{-6})\omega^2 = 0,$$
$$-23.91 + (148.09 \times 10^{-6})\omega^2 = 0,$$
$$\therefore \quad \omega = 401.73 \, \text{rad/s}, \quad \Rightarrow N = 3836.25 \, \text{rev/min}.$$

15.5 THICK-WALLED SPHERE UNDER PRESSURE

Consider the equilibrium condition about a horizontal plane for the hemispherical element of radial thickness δr in Fig. 15.14, where the radial stress σ_r varies by the amount $\delta \sigma_r$.

When both the radial and hoop stress are assumed tensile,

$$\sigma_\theta(2\pi r)\,\delta r + \sigma_r \pi r^2 = (\sigma_r + \delta\sigma_r)\pi(r + \delta r)^2.$$

Neglecting products of small quantities, this reduces to

$$\sigma_\theta - \sigma_r = (r/2)\,d\sigma_r/dr. \tag{15.21}$$

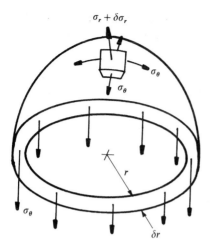

Figure 15.14 Element of sphere.

The constitutive relations for the triaxial stress state are

$$\epsilon_r = du/dr = (1/E)[\sigma_r - v(\sigma_\theta + \sigma_\theta)],$$
$$\epsilon_\theta = u/r = (1/E)[\sigma_\theta - v(\sigma_r + \sigma_\theta)].$$

Since these two strains depend upon the single displacement u, the following compatibility condition expresses their combination:

$$(\sigma_r - 2v\sigma_\theta) = d[r\sigma_\theta(1 - v) - vr\sigma_r]/dr,$$
$$(1 + v)(\sigma_\theta - \sigma_r) = vr \, d\sigma_r/dr - (1 - v)r \, d\sigma_\theta/dr. \tag{15.22}$$

Substituting Eq. (15.21) into Eq. (15.22) and integrating leads to

$$(r/2)(1 - v)d\sigma_r/dr + r(1 - v) \, d\sigma_\theta/dr = 0,$$
$$(1/2) \, d\sigma_r/dr + d\sigma_\theta/dr = 0,$$
$$\therefore \quad (1/2)\sigma_r + \sigma_\theta = A. \tag{15.23}$$

Combining Eqs (15.21) and (15.23):

$$2A - 3\sigma_r = r \, d\sigma_r/dr,$$

$$\int dr/r = \int d\sigma_r/(2A - 3\sigma_r),$$

$$- 3\ln r + \ln B = \ln(2A - 3\sigma_r),$$

$$B/r^3 = (2A - 3\sigma_r),$$

$$\therefore \quad \sigma_r = (1/3)(2A - B/r^3). \tag{15.24}$$

Writing $a = 2A/3, b = B/3$, then from Eqs (15.23) and (15.24), the stresses are

$$\sigma_r = a - b/r^3, \tag{15.25a}$$
$$\sigma_\theta = a + b/2r^3, \tag{15.25b}$$

where a and b follow from the boundary conditions. For example, with internal pressure only, $\sigma_r = -p$ for $r = r_i$ and $\sigma_r = 0$ for $r = r_o$. Then Eqs (15.25) become

$$\sigma_r = pr_i^3(1 - r_o^3/r^3)/(r_o^3 - r_i^3), \qquad (15.26a)$$

$$\sigma_\theta = pr_i^3(1 + r_o^3/2r^3)/(r_o^3 - r_i^3). \qquad (15.26b)$$

Example 15.11 Find the change in the volume capacity for a thick-walled sphere under internal fluid pressure p.

From Eq. (4.19) the volumetric strain is

$$\delta V/V = 3\epsilon_\theta = (3/E)[\sigma_\theta - v(\sigma_\theta + \sigma_r)] = (3/E)[\sigma_\theta(1 - v) - v\sigma_r].$$

Substituting from Eqs (15.26) with $r = r_i$:

$$\delta V/V = (3p/E)[(1 - 2v)r_i^3 + (1 + v)r_o^3/2]/(r_o^3 - r_i^3),$$

where V is the internal volume of the sphere. If a compressible fluid with bulk modulus K transmits the pressure, then the change in its volume is, from Eq. (4.3), $\delta V = -pV/K$.

15.6 RADIAL TEMPERATURE GRADIENT IN CYLINDRICAL BODIES

Short, Thin-walled Mating Rings

Consider an inner (I) and outer (O) ring that fit together initially without interference. The radial and hoop stresses induced at the common diameter d_c, due to a temperature rise T, are found from the compatibility condition

$$\Delta\epsilon_\theta = [(\sigma_\theta - v\sigma_r)/E]_O - [(\sigma_\theta - v\sigma_r)/E]_I = -(\alpha_O - \alpha_I)T \qquad (15.27)$$

provided the linear expansion coefficients obey $\alpha_I > \alpha_O$. Now, at the interface, a common compressive radial stress $(\sigma_r)_O = (\sigma_r)_I = -p$ exists. The membrane hoop stresses are

$$(\sigma_\theta)_O = pd_c/2t_O \qquad \text{and} \qquad (\sigma_\theta)_I = -pd_c/2t_I.$$

Substituting into Eq. (15.27) leads to

$$[(pd_c/2t + vp)/E]_O - [(-pd_c/2t + vp)/E]_I = -(\alpha_O - \alpha_I)T.$$

Neglecting the effect of the small lateral strain vp/E, this leads to

$$p = -\frac{(\alpha_O - \alpha_I)T}{(d_c/2)(1/E_Ot_O + 1/E_It_I)}. \qquad (15.28)$$

Equations (15.12) and (15.28) may be used together to determine the net stresses following a temperature rise in discs with initial interference stresses at ambient temperature.

Example 15.12 Determine the mean hoop stress in an inner copper–outer steel composite annular disc with $d_c = 810 \, \text{mm}$, $t_c = 5 \, \text{mm}$, and $t_s = 7.5 \, \text{mm}$, when the temperature is raised by 25 °C above the unstressed condition. Take $E_s = 210 \, \text{GPa}$, $\alpha_s = 12 \times 10^{-6}/°\text{C}$ and $E_c = 90 \, \text{GPa}$, $\alpha_c = 18 \times 10^{-6}/°\text{C}$.

The interface pressure is found directly from Eq. (15.28):

$$p = \frac{-(12 - 18)10^{-6} \times 25 \times 10^3}{(810/2)[1/(210 \times 7.5) + 1/(90 \times 5)]} = 0.13 \, \text{MPa},$$

$$\therefore \quad (\sigma_\theta)_s = p d_m/2t_c = 0.13 \times 817.5/(2 \times 7.5) = 7.08 \, \text{MPa},$$

$$(\sigma_\theta)_c = -p d_m/2t_c = -0.13 \times 805/(2 \times 5) = -10.46 \, \text{MPa}.$$

Short, Thick-walled Annular Disc

When the inner r_i and outer r_o radii of a thick-walled, stationary, unloaded disc are held at different, steady temperatures, radial and hoop stress variations will exist through the wall. Assuming plane stress conditions, the radial equilibrium Eq. (15.7) again applies. For a given temperature T at radius r in the disc, above the temperature for the stress free condition, the constitutive relations become

$$\epsilon_r = du/dr = (1/E)(\sigma_r - v\sigma_\theta) + \alpha T, \tag{15.29a}$$

$$\epsilon_\theta = u/r = (1/E)(\sigma_\theta - v\sigma_r) + \alpha T. \tag{15.29b}$$

Eliminating u in Eqs (15.29) corresponds to the compatibility condition

$$(1/E)(\sigma_r - v\sigma_\theta) + \alpha T = (1/E)d[r(\sigma_\theta - v\sigma_r)]/dr + d(r\alpha T)/dr,$$

$$(1 + v)(\sigma_r - \sigma_\theta) = r(d\sigma_\theta/dr - vd\sigma_r/dr) + E\alpha r \, dT/dr. \tag{15.30}$$

Substituting Eq. (15.7) into Eq. (15.30) leads to

$$d\sigma_\theta/dr + d\sigma_r/dr = -E\alpha \, dT/dr,$$

$$\therefore \quad \sigma_\theta + \sigma_r = 2a - E\alpha T. \tag{15.31}$$

Subtracting Eq. (15.7) from (15.31) leads to

$$2\sigma_r + r \, d\sigma_r/dr = 2a - E\alpha T,$$

$$(1/r)[d(r^2\sigma_r)/dr] = 2a - E\alpha T,$$

$$r^2\sigma_r = \int (2ar - E\alpha Tr) \, dr - b,$$

$$\sigma_r = a - b/r^2 - (E\alpha/r^2) \int Tr \, dr, \tag{15.32a}$$

$$\sigma_\theta = a + b/r^2 - E\alpha T - (E\alpha/r^2) \int Tr \, dr. \tag{15.32b}$$

Equations (15.32) may be solved for a given function $T = T(r)$ with typical boundary conditions $\sigma_r = 0$ at $r = r_i$ and $r = r_o$. Where external and internal pressure loading exists together with the temperature gradient, Eqs (15.32) remain valid, since they contain the Lamé equations (15.9), but with different constants owing to the change in the boundary conditions. If the disc rotates in the presence of a temperature gradient, then the appropriate rotation terms from Eqs (15.17) may be superimposed upon Eqs (15.32). This gives

$$\sigma_r = a' - b'/r^2 - (E\alpha/r^2) \int Tr \, dr - (3 + v)\rho\omega^2 r^2/8,$$

$$\sigma_\theta = a' + b'/r^2 - E\alpha T - (E\alpha/r^2) \int Tr \, dr - (1 + 3v)\rho\omega^2 r^2/8.$$

Long, Thick-walled Cylinder

Temperature variations arise in the long piping used in processing plant. Though the equilibrium equation (15.7) remains valid, a plane strain condition with $\epsilon_z = $ constant is now more representative. Thus,

$$\epsilon_z = (1/E)[\sigma_z - v(\sigma_r + \sigma_\theta)] + \sigma T,$$

from which the axial stress becomes

$$\sigma_z = E(\epsilon_z - \alpha T) + v(\sigma_r + \sigma_\theta).$$

The radial and circumferential constitutive strains are now

$$\epsilon_r = du/dr = (1/E)[\sigma_r - v(\sigma_\theta + \sigma_z)] + \alpha T,$$
$$\epsilon_\theta = u/r = (1/E)[\sigma_\theta - v(\sigma_r + \sigma_z)] + \alpha T.$$

Eliminating u and σ_z, the compatibility condition becomes

$$(1/E)[\sigma_r - v(\sigma_\theta + \sigma_z)] + \alpha T = (1/E)d\{r[\sigma_\theta - v(\sigma_r + \sigma_z)]\}/dr + d(\alpha rT)/dr,$$
$$(1 + v)(\sigma_r - \sigma_\theta) = r(1 - v^2)d\sigma_\theta/dr - rv(1 + v)d\sigma_r/dr + \alpha Er(1 + v)dT/dr,$$
$$(\sigma_r - \sigma_\theta) = (1 - v)r \, d\sigma_\theta/dr - rv \, d\sigma_r/dr + \alpha Er \, dT/dr,$$
$$(1 - v)(d\sigma_\theta/dr + d\sigma_r/dr) = -E\alpha \, dT/dr,$$
$$\sigma_\theta + \sigma_r = 2a - E\alpha T/(1 - v).$$

Making the comparison with Eq. (15.31), it can be deduced that the factor $(1 - v)$ will appear here in the final stress expressions:

$$\sigma_r = a - b/r^2 - [E\alpha/r^2(1-v)]\int Tr\,dr \qquad (15.33a)$$

$$\sigma_\theta = a + b/r^2 - E\alpha T/(1-v) - [E\alpha/r^2(1-v)]\int Tr\,dr \qquad (15.33b)$$

$$\sigma_z = E(\epsilon_z - \alpha T) + 2v\{a - \alpha ET/2(1-v) - [E\alpha/r^2(1-v)]\int Tr\,dr\}. \qquad (15.33c)$$

Example 15.13 Determine the maximum radial and circumferential thermal stresses at atmospheric pressure for the inner 100 mm and outer 150 mm radii in a cylinder where the temperature are 200 °C and 100 °C respectively. Take $E = 207\,\text{GPa}$, $v = 0.29$, $\alpha = 11 \times 10^{-6}/°C$ and assume that the absolute temperature T varies with radius r according to $T = A + B\ln r$. How are these stresses altered when the internal gauge pressure is raised to 2 kbar when all other conditions prevail?

With the given temperature function, the integral in Eqs (15.33) is

$$\int Tr\,dr = \int r(A + B\ln r)\,dr$$

$$= Ar^2/2 + (Br^2/4)(2\ln r - 1) + \text{constant}.$$

Substituting into Eqs (15.33a and b):

$$\sigma_r = a - b/r^2 - \alpha ET/2(1-v) + \alpha EB/4(1-v), \qquad \text{(i)}$$

$$\sigma_\theta = a + b/r^2 - 3\alpha ET/2(1-v) + \alpha EB/4(1-v) + K, \qquad \text{(ii)}$$

where the constant K arises from integration of the temperature function T (which contains A). A single integration constant is sufficient to ensure that equilibrium is maintained. Substituting Eqs (i) and (ii) into Eq. (15.7) gives $K = E\alpha(T - B/2)/(1-v)$. Then Eqs (i) and (ii) may be written as

$$\sigma_r = a' - b/r^2 - \alpha ET/2(1-v), \qquad \text{(iii)}$$

$$\sigma_\theta = a' + b/r^2 - \alpha ET/2(1-v) - \alpha EB/2(1-v), \qquad \text{(iv)}$$

where $a' = a + \alpha EB/4(1-v)$. The boundary conditions are $\sigma_r = 0$ for $r_i = 100$ mm and $T = 473$ K, and $\sigma_r = 0$ for $r_o = 150$ mm and $T = 373$ K. Substituting into Eq. (iii) gives

$$a' - b/100^2 = 11 \times 10^{-6} \times 207 \times 10^3 \times 473/2(1 - 0.29) = 758.465,$$

$$a' - b/150^2 = 11 \times 10^{-6} \times 207 \times 10^3 \times 373/2(1 - 0.29) = 598.113.$$

Solving these simultaneous equations leads to $a' = 469.78$, $b = -2.8863 \times 10^6$. The constants A and B in the temperature function,

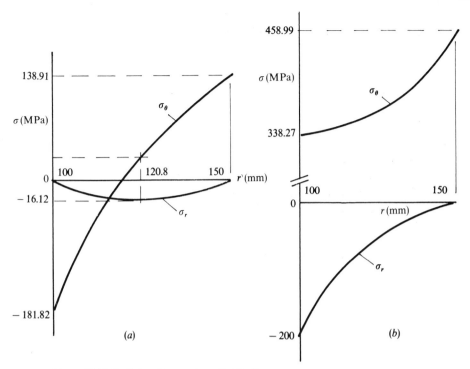

Figure 15.15 Radial and hoop stress distributions.

$T = A + B \ln r$, follow from substituting the boundary values

$$473 = A + B \ln 100, \qquad 373 = A + B \ln 150,$$

$$\therefore \quad A = 1608.87, \qquad B = -246.65.$$

Substituting together with a' and b into Eqs (iii) and (iv), the stress distributions (shown in Fig. 15.15(a)) are finally expressed by

$$\sigma_r = -2110.08 + (2.8863 \times 10^6)/r^2 + 395.5 \ln r, \qquad \text{(v)}$$

$$\sigma_\theta = -1714.57 - (2.8863 \times 10^6)/r^2 + 395.5 \ln r. \qquad \text{(vi)}$$

Particular values are $\sigma_\theta = -181.82\,\text{MPa}$ for $r = 100\,\text{mm}$ and $\sigma_\theta = 138.91\,\text{MPa}$ for $r = 150\,\text{mm}$. The maximum radial stress follows from Eq. (v):

$$d\sigma_r/dr = 395.51/r - (5.7726 \times 10^6)/r^3 = 0,$$

$$\therefore \quad r = 120.8\,\text{mm}, \quad \Rightarrow (\sigma_r)_{\max} = -16.12\,\text{MPa}.$$

A superimposed internally applied pressure of 2 kbar (i.e. $\sigma_r = -200\,\text{MPa}$ for $r = 100\,\text{mm}$) modifies the constants in Eqs (iii) and (iv) to $a' = 629.83$ and $b = 71.366 \times 10^4$. This gives Eqs (v) and (vi) as

$$\sigma_r = -1950 - (71.366 \times 10^4)/r^2 + 395.5 \ln r,$$
$$\sigma_\theta = -1554.49 + (71.366 \times 10^4)/r^2 + 395.5 \ln r.$$

These supply the stress distributions given in Fig. 15.15(b). The stress values at the boundary radii are $\sigma_\theta = 338.27$ MPa for $r = 100$ mm, $\sigma_\theta = 458.99$ MPa for $r = 150$ mm, $\sigma_r = -200$ MPa for $r = 100$ mm and $\sigma_r = 0$ for $r = 150$ mm.

15.7 BENDING OF CIRCULAR PLATES

Consider a thin, initially flat, circular plate of outer radius r_o and uniform thickness t (Fig. 15.16), subjected to a given axisymmetric loading applied normally to its faces. For a given radius r, the radial σ_r and hoop σ_θ stresses will be constant in the same plane, i.e. at a perpendicular distance z from the neutral middle-plane. The latter lies midway between the surfaces of the plate and is unstressed. When the vertical deflection w of the middle-plane is small compared to the plate thickness, the following simplified theory enables w, σ_r and σ_θ to be found. If u is the radial displacement, distance z from the neutral plane, it follows from Fig. 15.16(a) that $u = z\phi$ and $\phi = \delta w/\delta r$, relative to the plate centre where $\phi = 0$ and $w = 0$. Conventionally, the deflection w is positive upwards from the plate centre. For any point (r, z) in the plane of the plate (Fig. 15.16b), the biaxial constitutive relations are

$$\therefore \quad \epsilon_r = du/dr = (\sigma_r - v\sigma_\theta)/E,$$
$$\epsilon_\theta = u/r = (\sigma_\theta - v\sigma_r)/E.$$

Inverting these, the stresses become (see Eqs 4.15)

$$\sigma_r = E(\epsilon_r + v\epsilon_\theta)/(1 - v^2) = E(du/dr + vu/r)/(1 - v^2)$$
$$= Ez(d\phi/dr + v\phi/r)/(1 - v^2), \tag{15.34a}$$

$$\sigma_\theta = E(\epsilon_\theta + v\epsilon_r)/(1 - v^2) = E(u/r + v\,du/dr)/(1 - v^2)$$
$$= Ez(\phi/r + vd\phi/dr)/(1 - v^2). \tag{15.34b}$$

It follows from ETB that the respective moments per unit radial and circumferential length (Fig. 15.16b) that produce these stresses are

$$M_r = \sigma_r I/z \quad \text{and} \quad M_\theta = \sigma_\theta I/z$$

where $I = 1 \times t^3/12$. Substituting from Eqs (15.34):

$$M_r = D(d\phi/dr + v\phi/r), \tag{15.35a}$$

$$M_\theta = D(\phi/r + vd\phi/dr), \tag{15.35b}$$

where $D = Et^3/12(1 - v^2)$ is the flexural stiffness. Now, for small deflections, it follows from Fig. 15.16(a) that the radial curvature R_r in the plane COr

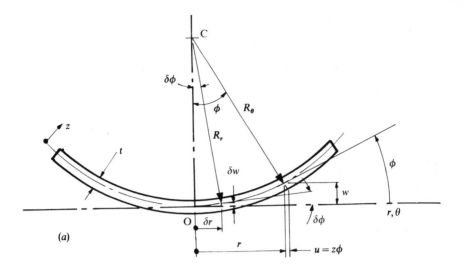

(a)

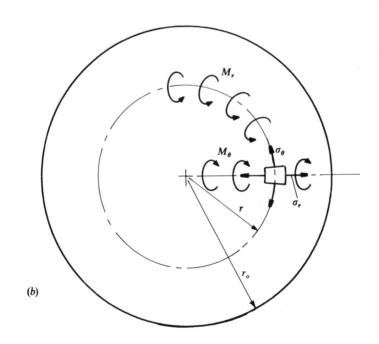

(b)

Figure 15.16 Plate bending.

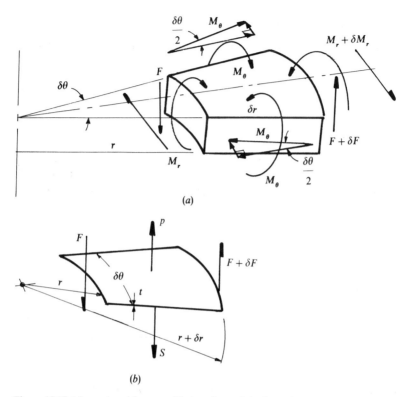

Figure 15.17 Moment and force equilibrium for a plate element.

for elements lying along axis Or is found from $\mathrm{d}r \simeq R_r \, \mathrm{d}\phi$. The circumferential curvature R_θ in the plane $CO\theta$ under M_θ is found from $r \simeq R_\theta \phi$. These lead to the alternative expressions for the bending moments from Eqs (15.35):

$$M_r = D(1/R_r + v/R_\theta), \qquad M_\theta = D(1/R_\theta + v/R_r).$$

Finally, the equilibrium of the corresponding sagging moments acting on central radial plane of the annular element $r \, \delta\theta \times \delta r$ in Fig. 15.17(a) is considered. For the axisymmetric loading shown, the vertical shear force F varies with r and not θ. Thus, when $F = F(r)$, is a downward shear force/unit circumference at the inner annular radius r, moment equilibrium gives

$$(M_r + \delta M_r)(r + \delta r)\delta\theta - M_r r \, \delta\theta - 2M_\theta \, \delta r \sin(\delta\theta/2)$$
$$+ (F + \delta F)(r + \delta r)^2 \, \delta\theta - Fr^2 \delta\theta = 0,$$

in which the right-hand screw rule determines the indicated moment vector directions and the effect of the normal loading over δr is ignored. In the limit, this becomes

$$M_r + r \, \mathrm{d}M_r/\mathrm{d}r - M_\theta + Fr = 0. \tag{15.36}$$

Substituting Eqs (15.35) into Eq. (15.36) leads to

$$(1/r)(\mathrm{d}\phi/\mathrm{d}r + v\phi/r + r\,\mathrm{d}^2\phi/\mathrm{d}r^2 - v\phi/r + v\,\mathrm{d}\phi/\mathrm{d}r - \phi/r - v\,\mathrm{d}\phi/\mathrm{d}r)$$
$$= -F/D,$$

$$\therefore \quad (\mathrm{d}/\mathrm{d}r)[(1/r)\,\mathrm{d}(r\phi)/\mathrm{d}r] = -F/D, \tag{15.37a}$$

or,

$$(\mathrm{d}/\mathrm{d}r)[(1/r)\,\mathrm{d}(r\,\mathrm{d}w/\mathrm{d}r)/\mathrm{d}r] = -F/D, \tag{15.37b}$$

from which the vertical deflection (w) is found from successive integration once the function $F(r)$ has been established from vertical force equilibrium. That is, if $p = p(r)$ is a net upward pressure applied normally to the surface of the elemental ring segment in Fig. 15.17(b), for which S is the self-weight/unit volume, then in general

$$\int_0^{2\pi} (F + \delta F)(r + \delta r)\,\delta\theta + \int_0^{2\pi} pr\,\delta\theta\,\delta r = \int_0^{2\pi} Fr\,\delta\theta + \int_0^{2\pi} \delta r\,\delta\theta\,\delta r\,t,$$

$$2\pi(r + \delta r)(F + \delta F) + p2\pi r\,\delta r = 2\pi rF + 2\pi r\,\delta r\,tS,$$

$$\mathrm{d}(Fr)/\mathrm{d}r + pr - Str = 0 \tag{15.38}$$

where F is found from integration. The following cases illustrate the application of the theory to the derivation of expressions for the maximum deflection and the maximum bending stresses.

Central Concentrated Force P—Edges Simply Supported

In the absence of self-weight and normal pressure, Eq. (15.38) gives $Fr = $ constant. Vertical equilibrium at all radii r in Fig. 15.18, where F now acts upwards around the sides of the disc (opposing P in Fig. 15.18), gives the shear force as

$$2\pi rF = P, \quad \Rightarrow F = P/2\pi r.$$

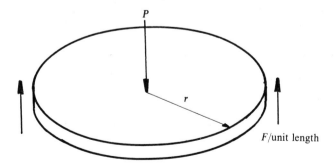

Figure 15.18 Concentrated loading.

Then, from Eq. (15.37b), by successive integration,

$$(d/dr)[(1/r)d(r\,dw/dr)/dr] = -P/2\pi r D,$$
$$(1/r)d(r\,dw/dr)/dr = -(P/2\pi D)\ln r + C_1,$$
$$d(r\,dw/dr)/dr = -(P/2\pi D)r\ln r + C_1 r,$$
$$r\,dw/dr = -(Pr^2/8\pi D)(2\ln r - 1) + C_1 r^2/2 + C_2,$$
$$dw/dr = -(Pr/8\pi D)(2\ln r - 1) + C_1 r/2 + C_2/r,$$
$$w = -(Pr^2/8\pi D)(\ln r - 1) + C_1 r^2/4 + C_2\ln r + C_3. \qquad (15.39)$$

The following boundary conditions determine the constants $C_i (i = 1, 2, 3)$: $dw/dr = 0$ for $r = 0$, $\Rightarrow C_2 = 0$; $w = 0$ for $r = 0$, $\Rightarrow C_3 = 0$. Also $M_r = 0$ for $r = r_0$, when from Eq. (15.35a) and Eq. (15.39),

$$d\phi/dr + v\phi/r = d^2 w/dr^2 + (v/r)\,dw/dr = 0,$$
$$-(Pr/8\pi D)(2/r) - (2\ln r - 1)P/8\pi D + C_1/2$$
$$+ (v/r)[-(Pr/8\pi D)(2\ln r - 1) + C_1 r/2] = 0,$$
$$C_1 = (P/4\pi D)[2\ln r + (1 - v)/(1 + v)].$$

Hence, Eq. (15.39) becomes

$$w = -(Pr^2/8\pi D)(\ln r - 1) + (Pr^2/16\pi D)[2\ln r + (1 - v)/(1 + v)]. \quad (15.40)$$

Equation (15.40) supplies the maximum central deflection for $r = r_o$:

$$w_{max} = (Pr_o^2/16\pi D)[2 + (1 - v)/(1 + v)] = (3Pr_o^2/4\pi E t^3)(1 - v)(3 + v).$$

From Eqs (15.34):

$$\sigma_r = Ez[d^2 w/dr^2 + (v/r)\,dw/dr]/(1 - v^2),$$
$$\sigma_\theta = Ez[(1/r)\,dw/dr + v\,d^2 w/dr^2]/(1 - v^2).$$

Substituting from Eq. (15.40), with $z = t/2$ for the compressive side,

$$\sigma_r = -[3P(1 + v)/2\pi t^2]\ln(r_o/r),$$
$$\sigma_\theta = -[3P(1 + v)/2\pi t^2][\ln(r_o/r) + (1 - v)/(1 + v)].$$

These apply only to finite r-values. The theory avoids the infinite stress values at the plate centre when, in practice, the stresses will be greatest within the small surface area on which P acts.

Central Concentrated Load P—Edges Clamped

Equation (15.39) remains valid but now the following boundary conditions apply:

$$w = 0 \qquad \text{for} \quad r = 0, \quad \Rightarrow C_3 = 0,$$
$$dw/dr = 0 \qquad \text{for} \quad r = 0, \quad \Rightarrow C_2 = 0,$$
$$dw/dr = 0 \qquad \text{for} \quad r = r_o, \quad \Rightarrow C_1 = (P/4\pi D)(2\ln r_o - 1).$$

These give

$$w = (Pr^2/16\pi D)[2\ln(r/r_o) - 1]. \tag{15.41}$$

Equation (15.41) is a maximum at the plate centre for $r = r_o$:

$$w_{\max} = Pr_o^2/16\pi D = (3Pr_o^2/4\pi Et^3)(1 - v^2).$$

Substituting Eq. (15.41) into Eqs (15.34) gives the maximum tensile stresses for the top surface at the fixed edge, i.e. $r = r_o$ and $z = t/2$:

$$\sigma_r = PEt/8\pi D(1 - v^2) = 3P/2\pi t^2,$$
$$\sigma_\theta = PvEt/8\pi D(1 - v^2) = 3Pv/2\pi t^2.$$

Uniformly Distributed Loading W—Edges Simply Supported

When the net loading W, including self-weight, acts downwards, then $p = -W = \text{constant}$ and $S = 0$ in Eq. (15.38). This gives $F = Wr/2$. Alternatively, from Fig. 15.19, the shear force/unit circumferential length is

$$2\pi Fr = W\pi r^2, \quad \Rightarrow F = Wr/2.$$

Substituting into Eq. (15.37b) with successive integration leads to

$$w = -Wr^4/64D + C_1 r^2/4 + C_2\ln r + C_3. \tag{15.42}$$

The following boundary conditions apply: $w = 0$ for $r = 0 \Rightarrow C_3 = 0$; $dw/dr = 0$ for $r = 0 \Rightarrow C_2 = 0$. Also, $M_r = 0$ for $r = r_o$, when, from Eqs (15.35a) and (15.42),

$$Wr_o^2(3 + v)/16D + C_1(1 + v)/2 = 0,$$
$$\therefore \quad C_1 = Wr_o^2(3 + v)/8D(1 + v).$$

Substituting for C_i in Eq. (15.42), the general displacement expression is found:

$$w = (Wr^2/64D)[2r_o^2(3 + v)/(1 + v) - r^2], \tag{15.43}$$

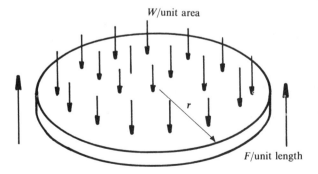

Figure 15.19 Distributed loading.

and for $r = r_o$:

$$w_{max} = (Wr_o^4/64D)(5 + v)/(1 + v) = (3Wr_o^4/16Et^3)(5 + v)(1 - v).$$

Substituting Eq. (15.43) into Eq. (15.34) gives the maximum compressive stresses at the centre top surface, i.e. $r = 0$ and $z = t/2$:

$$\sigma_r = \sigma_\theta = -3Wr_o^2(3 + v)/8t^2.$$

Tensile stresses of equal magnitude act on the bottom surface.

Uniformly Distributed Loading—Edges Fixed

Equation (15.42) remains valid and again $C_2 = C_3 = 0$. The third boundary condition is

$$dw/dr = 0 \quad \text{for} \quad r = r_0, \quad \Rightarrow C_1 = Wr_o^2/8D.$$

The general deflection expression is then

$$w = (Wr^2/64D)(2r_o^2 - r^2), \tag{15.44}$$

and for $r = r_o$:

$$w_{max} = Wr_o^4/64D = (3Wr_o^4/16Et^3)(1 - v^2).$$

Substituting Eq. (15.44) into Eq. (15.34) gives the maximum tensile radial stress for the top surface of the fixed edge ($r = r_o$, $z = t/2$):

$$\sigma_r = 3Wr_o^2/4t^2,$$

and a maximum compressive stress at the centre top surface ($r = 0, z = t/2$):

$$\sigma_\theta = -3Wr_o^2(1 + v)/8t^2.$$

These solutions may be superimposed when a plate is subjected to combined loading. For example, the sum of the stresses and displacements for the previous cases may be applied to simply supported and fixed-edge plates under concentrated and distributed loadings.

Non-standard Cases

Many plate problems require separate treatment for their particular geometry, loading and boundary conditions. The following examples typically illustrate how Eqs (15.34), (15.35) and (15.38) are then applied.

Example 15.14 An annular plate of inner radius $r_i = 38$ mm and outer radius $r_o = 165$ mm is loaded by a uniform pressure p (N/mm^2) on its top surface whilst being clamped around its outer edge and free around its inner edge (Fig. 15.20(a)). Determine the distribution of M_r and M_θ throughout the plate. Take $v = 0.27$.

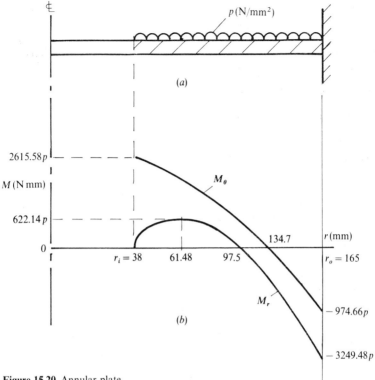

Figure 15.20 Annular plate.

For any radius $38 \le r \le 165$ (Fig. 15.20(a)), vertical equilibrium gives

$$2\pi r F = \pi p(r^2 - r_i^2),$$
$$F = (p/2)(r - r_i^2/r).$$

Substituting into Eq. (15.37b) and integrating leads to

$$(d/dr)[(1/r)\,d(r\phi)/dr] = -(p/2D)(r - r_i^2/r),$$
$$\phi = (p/2D)[(r_i^2 r/4)(\ln r^2 - 1) - r^3/8] + C_1 r/2 + C_2/r,$$

$$\text{(i)}$$

$$d\phi/dr = (p/2D)[(r_i^2/4)(\ln r^2 - 1) - 3r^2/8 + r_i^2/2] + C_1/2 - C_2/r^2.$$

$$\text{(ii)}$$

Substituting Eqs (i) and (ii) into Eq. (15.35a):

$$M_r = D((p/2D)[(r_i^2/4)(\ln r^2 - 1) - 3r^2/8 + r_i^2/2] + C_1/2 - C_2/r^2$$
$$+ (v/r)\{(p/2D)[(r_i^2 r/4)(\ln r^2 - 1) - r^3/8] + C_1 r/2 + C_2/r\}).$$

$$\text{(iii)}$$

The following boundary conditions apply to Eqs (i) and (iii):

$$\phi = dw/dr = 0 \quad \text{for} \quad r_0 = 165\,\text{mm};$$
$$(p/2D)[(38^2 \times 165/4)(\ln 165^2 - 1) - 165^3/8] + 165C_1/2 + C_2/165 = 0,$$
$$-6404.67p/D + 82.5C_1 + C_2/165 = 0. \tag{iv}$$

$$M_r = 0 \quad \text{for} \quad r_i = 38\,\text{mm};$$
$$1222.92p/D + 0.635C_1 - (5.0554 \times 10^{-4})C_2 - 281.45p/D = 0,$$
$$1504.37p/D + 0.635C_1 - (5.0554 \times 10^{-4})C_2 = 0. \tag{v}$$

Solving Eqs (iv) and (v) gives $C_1 = -129.08p/D$ and $C_2 = (2813.64 \times 10^3)p/D$. Substituting C_1 and C_2 into Eq. (iii) leads to the radial moment expression for all radii r:

$$M_r = p[229.24(\ln r^2 - 1) - 0.2044r^2 - (2.054 \times 10^6)/r^2 + 279.03]. \tag{vi}$$

This gives the distribution shown in Fig. 15.20(b), where

$$M_r = 0 \quad \text{for} \quad r_i = 38\,\text{mm (check)},$$
$$M_r = 0 \quad \text{for} \quad r = 97.5\,\text{mm (by trial)}$$

and

$$M_r = -3249.48p \quad \text{for} \quad r_0 = 165\,\text{mm}.$$

A stationary value of M_r in Eq. (vi) occurs for

$$dM_r/dr = 0 = (229.24 \times 2/r) - 0.4088r + 2(2.054 \times 10^6)/r^3,$$
$$0.4088r^4 - 458.48r^2 - (4.108 \times 10^6) = 0,$$

from which $r = 61.48\,\text{mm}$, where, from Eq. (vi), $M_r = 622.14p$.

The circumferential moment expression is found from substituting Eqs (i) and (ii) into Eq. (15.35b):

$$M_\theta = (r_i^2 p/8)(1 + v)(\ln r^2 - 1) - (pr^2/16)(1 + 3v)$$
$$+ D(1 + v)C_1/2 + D(1 - v)C_2 r^2$$
$$= p[229.24(\ln r^2 - 1) - 0.1131r^2 + (2.054 \times 10^6)/r^2 - 81.97].$$

This gives the distribution in Fig. 15.20(b), where

$$M_\theta = 2615.58p \quad \text{for} \quad r_i = 38\,\text{mm},$$
$$M_\theta = -974.66p \quad \text{for} \quad r_0 = 165\,\text{mm}$$

and

$$M_\theta = 0 \quad \text{for} \quad r = 134.7\,\text{mm}.$$

Example 15.15 A solid circular steel plate, 300 mm (12 in) in diameter and 12 mm (1/2 in) thick, is clamped around its outer edge and loaded

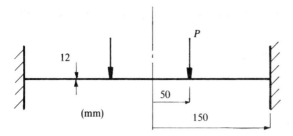

Figure 15.21 Clamped ring.

by a $P = 20\,\text{kN}$ (2 tonf) force ring at a 50 mm (2 in) radius (Fig. 15.21). Determine the central plate deflection. Take $E = 207\,\text{GPa}$ (13 400 tonf/in^2) and $v = 0.27$ for steel.

SOLUTION TO SI UNITS Working in Newtons and millimetres, the plate is considered in two sections:

(I) $0 \le r \le 50$, for which $F = 0$. Integrating Eq. (15.37a) gives

$$\phi = A_1 r/2 + A_2/r. \tag{i}$$

Since $\phi = dw/dr = 0$ for $r = 0$, this gives $A_2 = 0$ in Eq. (i).

(II) $50 \le r \le 150$, for which the vertical equilibrium equation is

$$2\pi r F = P, \quad \Rightarrow F = P/2\pi r.$$

Equation (15.37a) now gives

$$\phi = -(Pr/8\pi D)(2\ln r - 1) + B_1 r/2 + B_2/r. \tag{ii}$$

Now $\phi = 0$ for $r = 150$ mm in Eq. (ii), which gives

$$0 = -(1.0768 \times 10^6)/D + 75B_1 + (6.667 \times 10^{-3})B_2. \tag{iii}$$

Moreover, Eqs (i) and (ii) may be equated for $r = 50$ mm:

$$25A_1 = -(0.2715 \times 10^6)/D + 25B_1 + (20 \times 10^{-3})B_2. \tag{vi}$$

Also, M_r in Eq. (15.35a) must be the same for $r = 50$ mm. This leads to

$$A_1(1 + v)/2 = -(P/8\pi D)[2(1 + v)\ln r + (1 - v)] + B_1(1 + v)/2$$
$$- (B_2/r^2)(1 - v),$$
$$0.635A_1 = -(0.008\,488 \times 10^6)/D + 0.635B_1 - (0.292 \times 10^{-3})B_2. \tag{v}$$

Solving Eqs (iii), (iv) and (v) provides the constants

$$A_1 = (2.082 \times 10^3)/D, \qquad B_1 = (14.534 \times 10^3)/D$$

and

$$B_2 = -(1990.7 \times 10^3)/D.$$

The central deflection is found from matching the deflections at radius $r = 50\,\text{mm}$. Firstly, integrating Eq. (i):

$$w = A_1 r^2/4 + A_3,$$

but, as $w = 0$ for $r = 0$, then $A_3 = 0$. For $r = 50\,\text{mm}$:

$$w = (2.082 \times 10^3)50^2/4D = 1.3013 \times 10^6/D. \qquad \text{(vi)}$$

Integrating Eq. (ii)

$$w = -(Pr^2/8\pi D)(\ln r - 1) + B_1 r^2/4 + B_2 \ln r + B_3, \qquad \text{(vii)}$$

which becomes, for $r = 50\,\text{mm}$,

$$w = -(4.4972 \times 10^6)/D + B_3. \qquad \text{(viii)}$$

Equating (vi) and (viii) gives $B_3 = (5.7985 \times 10^6)/D$. Finally, putting $r = 150\,\text{mm}$ in Eq. (vii), the central deflection becomes

$$w = (5.767 \times 10^6)/D$$
$$= (5.767 \times 10^6) \times 12(1 - 0.27^2)/(207 \times 10^3 \times 12^3) = 0.179\,\text{mm}.$$

SOLUTION TO IMPERIAL UNITS Following a similar procedure, the constants here are

$$A_1 = 0.2084/D, \qquad B_1 = 0.429/D, \qquad A_2 = 0, \qquad B_2 = -0.3183/D.$$

At the radius $r = 2\,\text{in}$, Eq. (i) gives the deflection:

$$w = 0.2084/D, \qquad \text{(ix)}$$

and at the same radius Eq. (ii) gives

$$w = -(\ln 2 - 1)/\pi D + 0.429/D + 0.0796/D + B_3. \qquad \text{(x)}$$

Equating (ix) and (x) gives

$$B_3 = 0.3977/D.$$

Substituting for B_3 and $r = 6\,\text{in}$ in Eq. (vii) leads to the deflection:

$$w = 1.204/D = 1.204 \times 12(1 - v^2)/Et^3$$
$$= 1.204 \times 12(1 - 0.27^2)/[13\,400 \times (1/2)^3] = 0.008\,\text{in}.$$

15.8 NOTE ON GENERAL THEORY OF RECTANGULAR PLATES

It can be shown (see L. G. Jaeger, *Elementary Theory of Elastic Plates*, Pergamon, 1964) that the foregoing theory for circular plates and the theory of elliptical plates are particular cases of the general theory of rectangular plates. This is now summarized from Fig. 15.22 using a general, but different, convention.

When a uniform rectangular plate, lying in the x, y, z coordinate frame in Fig. 15.22(a), is loaded with a normal pressure p over its top surface, the *positive downwards* displacement $w = w(x, y)$ must satisfy the governing equation in any one of the following three forms:

$$D(\partial^4 w/\partial x^4 + 2\partial^4 w/\partial x^2 \partial y^2 + \partial^4 w/\partial y^4) = p, \qquad (15.45a)$$

$$D(\partial^2 w/\partial x^2 + \partial^2 w/\partial y^2)^2 w = p, \qquad (15.45b)$$

$$D(\nabla^2)^2 w = D\nabla^4 w = p, \qquad (15.45c)$$

where $\nabla^2 = \partial^2/\partial x^2 + \partial^2/\partial y^2$ and $D = Et^3/12(1 - v^2)$. The Cartesian stress components in the plane of the plate (Fig. 15.22b) are found from Eqs (14.1)

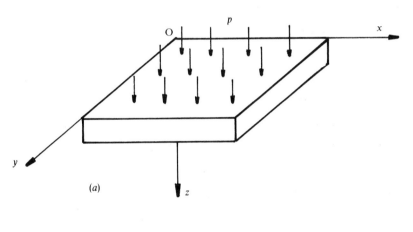

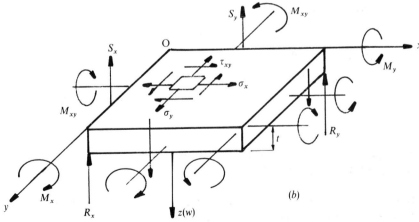

Figure 15.22 Bending and deflection of a rectangular plate.

and (14.4):

$$\sigma_x = - Ez(\partial^2 w/\partial x^2 + v\,\partial^2 w/\partial y^2)/(1 - v^2), \qquad (15.46a)$$
$$\sigma_y = - Ez(\partial^2 w/\partial y^2 + v\,\partial^2 w/\partial x^2)/(1 - v^2), \qquad (15.46b)$$
$$\tau_{xy} = - Ez(\partial^2 w/\partial x\,\partial y)(1 + v). \qquad (15.46c)$$

The corresponding moment components/unit side lengths are, from $M = \sigma I/z$:

$$M_x = - D(\partial^2 w/\partial x^2 + v\,\partial^2 w/\partial y^2), \qquad (15.47a)$$
$$M_y = - D(\partial^2 w/\partial y^2 + v\,\partial^2 w/\partial x^2), \qquad (15.47b)$$
$$M_{xy} = D(1 - v)\partial^2 w/\partial x\,\partial y. \qquad (15.47c)$$

In general, the shear force S_x, acting along y (Fig. 15.22b), is found from moment equilibrium:

$$
\begin{aligned}
S_x &= [\partial M_x/\partial x - \partial M_{xy}/\partial y] \\
&= - D[(\partial^3 w/\partial x^3) + (\partial^3 w/\partial x\,\partial y^2) + (1 - v)(\partial^3 w/\partial x\,\partial y^2)] \\
&= - D[(\partial^3 w/\partial x^3) + (\partial^3 w/\partial x\partial y^2)], \qquad (15.48a)
\end{aligned}
$$

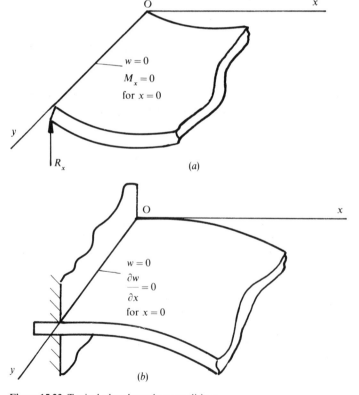

Figure 15.23 Typical plate boundary conditions.

and similarly, the shear force S_y, acting along x, is

$$S_y = [\partial M_y/\partial y - \partial M_{xy}/\partial x]$$
$$= -D[(\partial^3 w/\partial y^3) + (\partial^3 w/\partial y\,\partial x^2) + (1-v)(\partial^3 w/\partial y\,\partial x^2)]$$
$$= -D[(\partial^3 w/\partial y^3) + (\partial^3 w/\partial y\,\partial x^2)]. \tag{15.48b}$$

Equation (15.48) enable the reactions for the edges $x = 0$ and $y = 0$ to be found:

$$R_x = [S_x - \partial M_{xy}/\partial y]$$
$$= -D[(\partial^3 w/\partial x^3) + (2-v)(\partial^3 w/\partial x\,\partial y^2)]_{x=0}, \tag{15.49a}$$
$$R_y = [S_y - \partial M_{xy}/\partial x]$$
$$= -D[(\partial^3 w/\partial y^3) + (2-v)(\partial^3 w/\partial y\,\partial x^2)]_{y=0}. \tag{15.49b}$$

There are three common boundary conditions:

1. When the plate is simply supported along its edge (y-axis in Fig. 15.23a), then $w = 0, \partial^2 w/\partial x^2 = 0$ and $M_x = 0$ for $x = 0$ and all y.
2. When the edge is fixed along its edge (y-axis in Fig. 15.23b), then $w = 0$, and $dw/dx = 0$ for $x = 0$ and all y.
3. When one edge, say $x = 0$, is free, then M_x and R_x are both zero in the respective equations (15.47a) and (15.49a).

Example 15.16 Show that the deflection function $w = Axy(x-a)(y-b)$ is valid for a rectangular plate $a \times b$ under normal pressure p. Determine the loading and moment distribution along the edges and the stresses

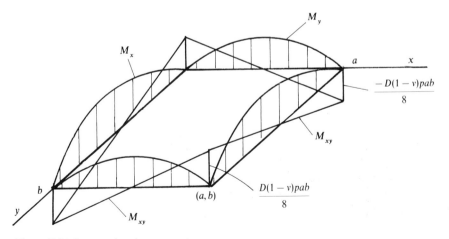

Figure 15.24 Rectangular plate moments.

at the plate bottom centre.

$$w = Axy(x - a)(y - b) = A(x^2y^2 - bx^2y - axy^2 + abxy);$$

$$\partial^2 w/\partial x\,\partial y = A(4xy - 2bx - 2ay + ab), \qquad \partial^2 w/\partial x^2 = A(2y^2 - 2by),$$

$$\partial^2 w/\partial y^2 = A(2x^2 - 2ax), \qquad \partial^3 w/\partial x^2\,\partial y = A(4y - 2b),$$

$$\partial^4 w/\partial x^2\,\partial y^2 = 4A, \qquad \partial^4 w/\partial x^4 = \partial^4 w/\partial y^4 = 0.$$

Substituting these derivatives into Eq. (15.45a) gives $p = 8A =$ constant, i.e. a uniform pressure. Then, from Eqs (15.47):

$$M_x = -2AD[(y^2 - by) + v(x^2 - ax)]$$
$$= -pEt^3[(y^2 - by) + v(x^2 - ax)]/48(1 - v^2),$$

$$M_y = -2AD[(x^2 - ax) + v(y^2 - by)]$$
$$= -pEt^3[(x^2 - ax) + v(y^2 - by)]/48(1 - v^2),$$

$$M_{xy} = D(1 - v)A(4xy - 2bx - 2ay + ab)$$
$$= pEt^3(4xy - 2bx - 2ay + ab)/96(1 + v).$$

For edges $x = 0$ and $x = a$:

$$M_x = -pEt^3y(y - b)/48(1 - v^2),$$
$$M_y = -pEt^3y(y - b)/48(1 - v^2),$$
$$M_{xy} = \pm pEt^3a(b - 2y)/96(1 + v).$$

For edges $y = 0$ and $y = b$:

$$M_x = -pEt^3x(x - a)/48(1 - v)^2,$$
$$M_y = -pEt^3x(x - a)/48(1 - v^2),$$
$$M_{xy} = \pm pEt^3b(a - 2x)/96(1 + v).$$

These distributions are shown in Fig. 15.24.

The following stress components are found from Eqs (15.46):

$$\sigma_x = -2AEz[vx(x - a) + y(y - b)]/(1 - v^2),$$
$$\sigma_y = -2AEz[vy(y - b) + x(x - a)]/(1 - v^2),$$
$$\tau_{xy} = -AEz[4xy - 2bx - 2ay + ab]/(1 + v).$$

Substituting $A = p/8$, $x = a/2$, $y = b/2$ and $z = -t/2$ gives $\tau_{xy} = 0$ and the tensile stress state at the bottom centre:

$$\sigma_x = pEt(a^2v + b^2)/32(1 - v^2), \qquad \sigma_y = pEt(a^2 + vb^2)/32(1 - v^2).$$

Example 15.17 A simply supported square plate $a \times a$ is subjected to a normal pressure distribution of the form $p = p_o \sin(\pi x/a)\sin(\pi y/a)$. Assuming a deflected shape of the form $w = c\sin(\pi x/a)\sin(\pi y/a)$, determine the maximum deflection and bending moment. What are the edge reactions exerted by the simple supports?

The derivatives of the deflection function are

$$\partial w/\partial x = (\pi c/a)\cos(\pi x/a)\sin(\pi y/a),$$
$$\partial^2 w/\partial x^2 = -(\pi 2c/a^2)\sin(\pi x/a)\sin(\pi y/a),$$
$$\partial^3 w/\partial x^3 = -(\pi^3 c/a^3)\cos(\pi x/a)\sin(\pi y/a),$$
$$\partial^2 w/\partial x\,\partial y = (\pi^2 c/a^2)\cos(\pi x/a)\cos(\pi y/a),$$
$$\partial^3 w/\partial x\,\partial y^2 = -(\pi^3 c/a^3)\cos(\pi x/a)\sin(\pi y/a),$$
$$\partial^3 w/\partial y\,\partial x^2 = -(\pi^3 c/a^3)\sin(\pi x/a)\cos(\pi y/a),$$
$$\partial^4 w/\partial x^4 = (\pi^4 c/a^4)\sin(\pi x/a)\sin(\pi y/a) = \partial^4 w/\partial y^4, \qquad \text{(i)}$$

$$\partial^4 w/\partial x^2\,\partial y^2 = (\pi^4 c/a^4)\sin(\pi x/a)\sin(\pi y/a). \qquad \text{(ii)}$$

Substituting Eqs (i) and (ii) into Eq. (15.45a) in order to validate the assumed deflection function:

$$4[(\pi^4 c/a^4)\sin(\pi x/a)\sin(\pi y/a)] = p/D,$$
$$4[(\pi^4 c/a^4)\sin(\pi x/a)\sin(\pi y/a)] = (p_o/D)\sin(\pi x/a)\sin(\pi y/a),$$
$$\therefore \quad 4\pi^4 c/a^4 = p_o/D, \quad \Rightarrow c = a^4 p_o/4\pi^4 D,$$
$$w = (a^4 p_o/4\pi^4 D)\sin(\pi x/a)\sin(\pi y/a). \qquad \text{(iii)}$$

The maximum deflection occurs at the plate centre, where $x = y = a/2$. Substituting into Eq. (iii), this gives

$$w_{\max} = p_o a^4/4D\pi^4.$$

Now from Eq. (15.47a and b),

$$M_x = M_y = (D\pi^2 c/a^2)(1 + v)\sin(\pi x/a)\sin(\pi y/a)$$
$$= (a^2 p_o/4\pi^2)(1 + v)\sin(\pi x/a)\sin(\pi y/a).$$

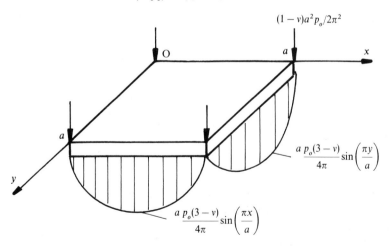

Figure 15.25 Plate reactions.

This is a maximum for $x = a/2$ and $y = a/2$, giving

$$(M_x)_{max} = (M_y)_{max} = p_o a^2 (1 + v)/4\pi^2.$$

Equations (iii) and (15.49) supply the following edge reactions for this plate:

$$R_x = [D\pi^3 c(3 - v)/a^3] \sin(\pi y/a) = [ap_o(3 - v)/4\pi] \sin(\pi y/a),$$
$$R_y = [D\pi^3 c(3 - v)/a^3] \sin(\pi x/a) = [ap_o(3 - v)/4\pi] \sin(\pi x/a),$$

which are shown graphically in Fig. 15.25. Note, from moment equilibrium, that forces of magnitude $2M_{xy} = (1 - v)a^2 p_o/2\pi^2$ will prevent the plate from rising at the corners (a moment/unit length has force units). These are found from Eq. (15.47c), with either the substitution $x = 0$, $y = 0$ or $x = a$, $y = a$.

EXERCISES

Membrane Theory

15.1 What are the axial and hoop stresses induced by a pressure of 10 bar in a long cylinder with 25 mm wall thickness and 1 m diameter when the ends are (a) closed and (b) open, i.e. the end pressure is contained by movable pistons?
Answer: 10, 20; 0, 20 (MPa).

15.2 A spherical balloon 0.2 mm thick and 250 mm diameter has an ultimate tensile strength of 9 MPa. Calculate the bursting pressure.
Answer: 28.8 kPa.

15.3 What is the bursting speed of a thin-rimmed cast iron flywheel of mean diameter 1 m if the ultimate tensile strength of cast iron is 138 MPa and the specific weight is 7200 kg/m³?
Answer: 138 m/s.

15.4 Find the diameter of the largest thin-walled rim that may revolve at 500 rev/min if it is to be made from steel of density 7756 kg/m³, when the maximum tensile stress is limited to 46 MPa. Determine the increase in diameter. Take $E = 207$ GPa.
Answer: 3 m, 0.66 mm.

15.5 A thin steel cylinder with walls 15 mm thick and diameter of 2.1 m is subjected to an internal pressure of 5.5 bar whilst being rotated about its axis at 300 rev/min. Find the maximum value of the hoop stress in the material and the safety factor used in the design. The ultimate tensile strength and density of the steel are 450 MPa and 7760 kg/m³ respectively.
Answer: 45.8 MPa; 9.85.

15.6 A thin steel rim 1.5 m in diameter and 150 mm wide is constructed from 20 mm thick steel plate. Determine, using a factor of safety of 8, the maximum permissible speed in rev/min given that the ultimate tensile strength for the rim material is 463 MPa. The rim is to be constructed in two semi-circular halves with each simple lap joint held together by a single bolt. Determine the bolt diameter when the permissible shear stress of the bolt material is limited to 69 MPa.
Answer: 1080 rev/min; 56 mm.

15.7 A hemispherical concrete shell of mean diameter 15 m and thickness 50 mm rests on its annular rim. Determine the maximum stresses in the shell due to the combined action of self-weight and a normal pressure of 1 bar over its external surface. Take $\rho = 2400$ kg/m³ for concrete.

15.8 Determine the maximum stresses in a hemispherical shell 1.5 m diameter and 10 mm thick when it is filled with water of density 1000 kg/m^3 and simply suspended around its annular rim.

Answer: $\sigma_\theta = -184\,\text{kN/m}^2$; $\sigma_\phi = 184\,\text{kN/m}^2$.

15.9 The vessel in Fig. 15.26 comprises part of a sphere of radius a, subjected to a ring of loading at radii r_1 and r_2 as shown. If the total effect of the load is W, derive the membrane stresses in the vessel.

15.10 Part of a reactor vessel is simplified in Fig. 15.27. The spherical dish, 10.5 m radius, rests on a cylinder, 5.5 m radius, both having a uniform thickness of 75 mm. The vessel supports a central concentrated core load of 200 t and a total rim load of 1800 t. Calculate the membrane stresses in the dish and cylinder when an additional pressure of 6.5 bar is applied as shown (a) when the dish self-weight is negligible and (b) when the dish self-weight is 78 kN/m^3.

15.11 Derive the membrane stresses arising in the tori-spherical ends of long thin cylindrical vessel containing pressure p when the knuckle radius is a and the sphere radius is b.

Answer: sphere, $\sigma_r = \sigma_\theta = pb/2$; knuckle, $\sigma_r = -p(a + b\cos\theta)/2\cos\theta$, $\sigma_\theta = -pr[2 - r/(b\cos\theta)]/2\cos\theta$ (with θ as for torus).

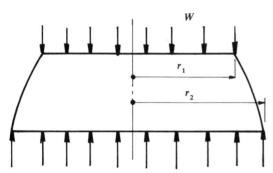

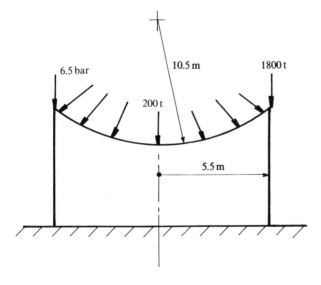

Figure 15.26

Figure 15.27

Monobloc Thick-walled Cylinders and Spheres

15.12 A steel cylinder of 203 mm inside diameter is not to be stressed beyond 123.5 MPa under an internal pressure of 310 bar. Find a suitable external diameter and the magnitude of the strains at the outer surface. Take $E = 207$ GPa, $v = 0.28$.

Answer: 262 mm; 0.039 per cent; 0.0099 per cent.

15.13 A thick-walled cylinder of steel has an i.d. of 100 mm and an o.d. of 200 mm. The internal and external pressures are 550 and 70 bar respectively. Determine the maximum hoop and shear stresses in the cylinder and the change in the outer diameter. Take $E = 210$ GPa, $v = 0.3$.

Answer: 73.5 MPa; 64.25 MPa; 0.1 per cent.

15.14 A cylindrical steel vessel of 127 mm i.d. and 178 mm o.d., with closed ends, contains fluid at a pressure of 350 bar. Determine the axial, hoop and radial stresses and strains at the inner and outer diameters. Take $E = 200$ GPa, $v = 0.28$.

Answer: (36.15, 107, -34.7 MPa); (79, 531, $-373\,\mu$); (36.65, 72.3, 0 MPa); (79, 309.6, $-151.2\,\mu$).

15.15 A cast iron closed cylinder with i.d. of 114 mm and wall thickness 25 mm can safely withstand an internal pressure of 170 bar. What is the safe internal pressure a second cast iron cylinder can withstand for an i.d. of 152 mm and thickness 25 mm when the greatest tensile stress is to remain the same? What is the minimum tensile stress in each cylinder?

Answer: 13.5 MPa; 31.3 MPa; 34.7 MPa.

15.16 An annular plate of 305 mm o.d. and 12.5 mm thick is subjected to a radial compressive force of 2.35 kN/mm of inner circumference, whilst the outer edge remains unloaded. Compare the hoop stress distribution with that when the force is applied around the outer edge whilst the inner edge remains free.

Answer: 46.32; 231.6 MPa; -231.6; -417 MPa.

15.17 A pressure of 400 bar exists within the bore of a pipeline. If the maximum stress is not to exceed 120 MPa, determine the outer diameter.

Answer: 70.71 mm.

15.18 A closed cylinder, of 150 mm i.d. with 75 mm wall thickness, is subjected to an internal gauge pressure of 700 bar. Calculate and plot, for each 25 mm of radius, the distribution of radial and hoop stress through the wall.

Answer: σ_θ, 116.67, 75.83, 56.93, 46.66 MPa; σ_r, -70, -29.17, -10.27, 0 MPa; $\sigma_z = 23.33$ MPa.

15.19 A steel bathysphere of 500 mm i.d. is to withstand an external fluid pressure of 1000 atmospheres. Find the necessary thickness if the maximum compressive stress in the vessel is limited to 230 MPa. Find also the diameter and thickness changes under this pressure.

Compound Cylinders with Interference

15.20 A steel sleeve of 51 mm i.d. and 102 mm o.d. is force-fitted on to a solid shaft. If the external hoop strain is 0.06 per cent, calculate the radial pressure at the common diameter and the maximum tensile stress in the sleeve. Take $E = 207$ GPa.

Answer: 184.8 MPa; 308 MPa.

15.21 A copper bush of 20 mm i.d. and 30 mm o.d., 50 mm long, is pressed firmly into a cast iron hub of 50 mm o.d., with an interference of 0.03 mm. Determine the torque to cause slipping between hub and bush, assuming a coefficient of friction $\mu = 0.4$. $E = 140$ GPa, $v = 0.25$ for cast iron; $E = 100$ GPa, $v = 0.30$ for copper.

Answer: 1.435 kN m.

15.22 A brass sleeve of 50 mm i.d. and 100 mm o.d., 50 mm long, is to be force-fitted on to a solid shaft of diameter 50 mm. The diametral interference of the assembly is 0.04 mm. Calculate the force required to fit the sleeve on the shaft. $E = 200\,\text{GN/m}^2$, $v = 0.3$ for steel; $E = 100\,\text{GN/m}^2$, $v = 0.33$ for brass. Coefficient of friction for the pair, $\mu = 0.2$. (C.E.I. Pt. 2, 1976)

15.23 A circular disc of outer diameter D, containing a central hole of diameter d, is fitted to a solid shaft of nominal diameter d using an interference fit. The disc and shaft are of the same material, with Young's modulus E and Poisson's ratio v. The assembly is required to satisfy the condition that the maximum hoop stress in the disc shall not exceed a value of σ_{limit}. Derive the relationship for the diametral interference δ and show that if D/d is large, then δ is proportional to d. (C.E.I. Pt. 2, 1983)

15.24 A steel wheel ring, 102 mm in i.d., is to be shrunk on to a hollow steel wheel to give a radial pressure of 40 MPa. Calculate the necessary initial difference in the common diameter when the radial thicknesses are both 25 mm. Take $E = 207\,\text{GPa}$.

Answer: 0.087 mm.

15.25 A steel gear blank of 127 mm o.d. is pressed on to a 76 mm diameter hollow steel shaft whose internal diameter is 51 mm. If the hoop strain for the blank o.d. is 0.016 per cent, determine the initial difference in the common diameter and the maximum hoop stress in the blank. What magnitude of torque would initiate slipping for a 50 mm length of contact between blank and shaft? Take $E = 207\,\text{GPa}$, $\mu = 0.2$.

Answer: 0.049 mm; 60.5 MPa; 2.63 kN m.

15.26 A hollow steel tube of 76 mm o.d. with 12.5 mm thick walls has a bronze sleeve of 102 mm o.d. shrunk on to it. If the common radial pressure is 27.5 MPa, calculate the initial interference. Plot the residual stress distributions. $E = 207\,\text{GPa}$, $v = 0.3$ for steel; $E = 117\,\text{GPa}$, $v = 0.33$ for bronze.

Answer: 0.093 mm; σ_θ; -99.3, -72.4, 98.5, 70.7 MPa.

15.27 A compound tube is made by shrinking one steel tube on to another. The diameters of the assembly are 152 mm internal, 305 mm common and 355 mm external. If the shrinkage process results in 0.077 per cent external hoop strain, determine the initial interference and the maximum hoop stress. Take $E = 207\,\text{GPa}$.

Answer: 0.35 mm; 189 MPa.

15.28 A 25 mm thick steel cylinder with an inside diameter of 200 mm has shrunk on to it a second steel cylinder of 300 mm outside diameter. Calculate the interference fit required, prior to assembly, if both tubes are to suffer the same maximum hoop stress when the compound tube is pressurized to $60\,\text{N/mm}^2$. (C.E.I. Pt. 2, 1972)

15.29 The two components of a compound cylinder have inner and outer radii of 175, 200 and 200, 250 mm, and are made to give an interference on the common radius of 1×10^{-1} mm. Find the radial and tangential stresses at the common 200 mm radius due only to the shrinking process. Take $E = 208\,\text{GPa}$. (C.E.I. Pt. 2, 1986)

15.30 A hollow steel tube of 180 mm o.d. and 50 mm thick has a bronze sleeve of 230 mm o.d. shrunk on to it. Find the required interference at the common diameter for a radial shrinkage pressure of 92.5 MPa. Plot the residual stress distributions following shrinkage and find the net hoop stress at the inner diameter of each tube when a pressure of 770 bar is then applied to the assembly. $E = 210\,\text{GPa}$, $v = 0.3$ for steel; $E = 117\,\text{GPa}$, $v = 0.33$ for bronze.

Answer: 0.73 mm; -132 MPa; 416 MPa; -109.4 MPa.

15.31 The nominal dimensions of two cylinders are 40 mm and 80 mm for the inner cylinder and 80 mm and 120 mm for the outer cylinder. If both cylinders are made from the same steel, determine the interference in order that the maximum bore stress for the compound cylinder does not exceed $60\,\text{MN/m}^2$ when it is pressurized to $60\,\text{MN/m}^2$. (C.E.I. Pt. 2, 1979)

15.32 The nominal radii of a compound cylinder are 100, 120 and 140 mm. When an internal pressure of $15\,\text{MN/m}^2$ is applied to the assembly, it is required that the maximum circumferential stress shall be the same in both components. Determine the magnitude of the interaction pressure at this pressure and the initial interference. Take $E = 200\,\text{GN/m}^2$ and $v = 0.3$. (C.E.I. Pt. 2, 1981)

Rotating Cylindrical Bodies

15.33 Find the maximum stress and the stress at the 228 mm o.d. of a solid disc when it is rotating at 12 000 rev/min. Show graphically the radial and hoop stress distributions. Take $v = 0.3$, $\rho = 7480 \, \text{kg/m}^3$.

Answer: 63.6 MPa; 27 MPa.

15.34 A steel disc of 255 mm o.d., 127.5 mm i.d. and 12.5 mm thick rotates at 2000 rev/min. Ignoring the method of grip, determine the stresses due to centrifugal effects. Take $v = 0.3$ and $\rho = 7835 \, \text{kg/m}^3$.

Answer: 2.4 MPa at the i.d.

15.35 At what speed may an annular disc of 30 mm o.d. and 20 mm i.d. rotate if the maximum shear stress is limited to 35 MPa? Take $\rho = 7600 \, \text{kg/m}^3$ and $v = 0.28$.

15.36 A thin uniform steel disc with o.d. 500 mm and i.d. 120 mm, is subjected to a radial tensile stress of 110 MPa at its o.d. due to the blading when rotating at 7000 rev/min. Determine the maximum radial and circumferential stresses and the change in diameter, stating where these occur. Take $E = 200 \, \text{GPa}$, $v = 0.3$ and $\rho = 7850 \, \text{kg/m}^3$.

15.37 Determine the general expression for the circumferential stress in a solid disc of radius b, rotating at an angular frequency ω, when a radial stress component of magnitude σ_b (tensile) is applied to the periphery of the disc. Hence determine the position and magnitude of the maximum circumferential stress in the disc. (C.E.I. Pt. 2, 1977)

15.38 An internal combustion engine has a cast iron flywheel that can be considered to be a uniformly thick disc of 230 mm o.d. and 50 mm i.d. Given that the ultimate tensile strength and density of cast iron are 200 N/mm² and 7180 kg/m³ respectively, calculate the speed at which the flywheel would burst. Take $v = 0.25$. (C.E.I. Pt. 2, 1974)

15.39 Figure 15.28 shows a disc with 36 integrally machined teeth of uniform cross-section arranged symmetrically around the disc. The toothed disc is cut from a single sheet of uniform

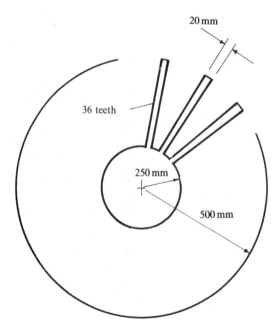

Figure 15.28

thickness. Determine the maximum stress in the disc when it is rotating at 1500 rev/min, assuming that the combined centrifugal forces at the roots of the teeth due to rotation are uniformly distributed around the central 250 mm radius. Take $E = 200$ GPa, $v = 0.3$ and $\rho = 7380$ kg/m^3. (C.E.I. Pt. 2, 1979)

15.40 A thin solid steel disc, 455 mm in diameter, has a steel ring of outer diameter 610 mm and of the same uniform thickness shrunk on to it. If the interference pressure is reduced to zero at a rotational speed of 3000 rev/min calculate, for the common diameter, the diametral interference before assembly and the radial pressure due to shrinkage. Take $E = 207$ GPa, $v = 0.29$ and $\rho = 7756$ kg/m^3. (C.E.I. Pt. 2, 1968)

15.41 A rotating disc and solid shaft steel assembly is designed for a pressure of 34.5 MPa and a hoop stress of 207 MPa at the common 508 mm diameter. If the disc outer diameter is 710 mm, at what maximum speed can the assembly be rotated? Find also the magnitude of the stresses at the centre and at the disc outer diameter. Take $v = 0.28$ and $\rho = 7756$ kg/m^3.

Answer: 3209 rev/min; -11.4, -11.4, 137.6 MPa.

15.42 A solid steel disc of thickness $t = 50$ mm and 760 mm diameter weighs 7844 kg/m^3. A central hole is to be machined in the disc so that, when it is pressed on to a solid steel shaft of diameter $d = 102$ mm, it will not slide on the shaft under an axial force of $W = 30$ kN. Determine the necessary hole size for the disc and the maximum allowable rotational speed if slippage is not to occur. (*Hint*: The common interface pressure p is found from $\mu = W/p\pi dt$.) Take $E = 207$ GPa, $v = 0.3$, $\mu = 0.3$.

Answer: 0.06 mm.

15.43 A thin turbine rotor has inner and outer nominal radii of 40 mm and 125 mm respectively. The effect of the rotating turbine blades is to set up a radial tensile stress of 25 MPa at the outer radius. The rotor is shrunk on to a solid shaft with a radial interference of δ. The fit is to be designed using the assumption that at a speed of 10 000 rev/min the rotor would become loose (i.e. free to slide) on the shaft. Determine the necessary interference fit δ. Take $E = 208$ GPa, $v = 0.3$ and $\rho = 7500$ kg/m^3 for both rotor and shaft. (C.E.I. Pt. 2, 1984)

Effect of Temperature

15.44 If, for the thin annular disc in Example 15.12, the initial mismatch was 0.5 mm, calculate the mean hoop stress when the temperature of the assembly is raised by 25 °C.

15.45 The temperature at the 60 mm i.d. of a turbine disc is 60 °C, while at the 200 mm o.d. it is 160 °C when the disc rotates at 6000 rev/min. If the outer diameter temperature is reduced to 140 °C, determine the increase in rotational speed that is possible when the circumferential stress at the inside diameter remains constant. Take $E = 200$ GPa, $v = 0.25$, $\alpha = 11 \times 10^{-6}/°C$ and $\rho = 7500$ kg/m^3.

15.46 Use the principle of superposition to establish the net stress distribution in a thick-walled cylinder of 100 mm i.d. and 200 mm o.d. when it carries oil at a pressure and temperature of 35 MPa and 800 K. The outer pressure is atmospheric and the temperature is 310 K. Find also the position and magnitude of the greatest hoop stress and the changes in the inner and outer diameters relative to a surrounding temperature of 295 K. Take $E = 200$ GPa, $v = 0.3$ and $\alpha = 12 \times 10^{-6}/K$.

Answer: -259 MPa at i.d.; 0.68 mm i.d.; 1.31 mm o.d.

15.47 A solid steel shaft 0.2 m in diameter has a bronze bush of 0.3 m o.d. shrunk on to it. In order to remove the bush, the whole assembly is raised in temperature uniformly. After a rise of 100 °C the bush can just be moved along the shaft. Neglecting any effect of temperature in the axial direction, calculate the original interface pressure between the bush and shaft. Take $E = 208$ GPa, $v = 0.29$, $\alpha = 12 \times 10^{-6}/°C$ for steel, and $E = 112$ GPa, $v = 0.33$, $\alpha = 18 \times 10^{-6}/°C$ for bronze. (C.E.I. Pt. 2, 1971)

15.48 An aluminium alloy ring of 400 mm o.d. is shrunk on to a solid steel shaft of 200 mm diameter. At 20 °C, the diametral interference fit at the common diameter is 0.2 mm. Find (i) the temperature to which the assembly must be taken in order to just loosen the ring and (ii) the angular speed at which the assembly just loosens when operating at 40 °C. Take $E = 207$ GPa, $v = 0.3$, $\alpha = 11 \times 10^{-6}/°C$, and $\rho = 7850$ kg/m³ for steel, and $E = 69$ GPa, $v = 0.3$, $\alpha = 23 \times 10^{-6}/°C$ and $\rho = 2720$ kg/m³ for aluminium alloy. (C.E.I. Pt. 2, 1982)

Bending of Circular Plates

15.49 An aluminium alloy plate diaphragm is 510 mm in diameter and 6.5 mm thick. It is clamped around its periphery and subjected to a uniform pressure of 0.7 bar. Calculate the values of the maximum bending stress and deflection. Take for aluminium $E = 70$ GPa and $v = 0.3$. (C.E.I. Pt. 2, 1967)

Answer: 2.1 mm; 82.7 MPa; 53.8 MPa.

15.50 A circular plate of 330 mm o.d. and 76 mm i.d. is loaded all around its inner edge by a total vertical force P (kN) while being simply supported around its outer edge. Determine the distribution of radial bending moment and find the position and magnitude of the maximum value. Take $v = 0.27$.

Answer: 67 mm; $1.25P$ (N m).

15.51 The loading for the plate in Exercise 15.50 is replaced by a ring of edge bending moments M_{r1} per unit of outer circumference while the inner edge remains free. Plot the radial moment distribution and repeat for the similar problem when a ring of inner-edge bending moments M_{r2} per unit of circumference is applied in the absence of outer edge loading.

Answer: gradual falls in M_{r1} and M_{r2} with $1/r^2$ to zero.

15.52 The top surface of a simply supported circular steel plate of diameter 1 m and thickness 50 mm is subjected to an increase in pressure of 0.5 bar above atmospheric. Given that density of steel is 7750 kg/m³, calculate the total deflection at the plate centre due to the pressure and self-weight. What is the maximum tensile stress in the plate and the safe pressure when the deflection is restricted to 0.1 mm? Take for steel $E = 207$ GPa, $v = 0.27$.

Answer: 0.094 mm; 6.6 MPa.

15.53 A diaphragm of diameter 220 mm is clamped around its periphery and subjected to a normal gas pressure of 2 bar on one surface. If the central deflection is not to exceed 0.65 mm, determine the thickness and the maximum bending stress. Take $E = 210$ GPa and $v = 0.3$.

15.54 A uniform, circular flat plate is rigidly built-in around the boundary at radius a, and is subjected to a pressure p acting normally to one surface. It is required to double the central deflection by adding a central concentrated normal force W. Find an expression for W. (C.E.I. Pt. 2, 1986)

15.55 The flat end of a 2 m diameter container can be regarded as clamped around its edge. Under operating conditions the plate will be subjected to a uniformly distributed pressure of 0.02 N/mm². Calculate from first principles the maximum deflection and the required thickness of the end plate if the bending stress is not to exceed 150 N/mm². $E = 200$ GPa and $v = 0.3$. (C.E.I. Pt. 2, 1974)

15.56 A thin, horizontal circular flat plate is rigidly built-in around its circumference at radius R. It supports a vertical central concentrated load W. In order to reduce the central deflection under W to one-quarter, a uniform pressure p is applied to the opposite side of the plate. Find an expression for the required value of p in terms of W and the relevant constants. (C.E.I. Pt. 2, 1984)

15.57 A flat circular plate of radius 200 mm and thickness 10 mm is subjected to a uniform pressure $p = 1$ MN/m² acting normally to the surface of one side only. It is semi-rigidly built-in around its outer edge producing an edge moment $m_e = 2$ kN m per metre of circumferential length. Find the largest radial stress in the plate and the central deflection. Take $E = 208\,\text{GPa}$ and $v = 0.3$. (C.E.I. Pt. 2, 1985)

15.58 Figure 15.29 shows the arrangement for an overload pressure warning device. A contact block X is centrally positioned 0.05 mm for the surface of a clamped circular plate, which is then subjected to a uniform pressure p on one side. Electrical contact is made between the block and the plate when the pressure exceeds a set value. The plate is 30 mm in diameter, 0.2 mm thick with $E = 250\,\text{GPa}$, $v = 0.3$, and a tensile yield stress of $200\,\text{MN/m}^2$. Determine the pressure required to activate the warning device and check whether deformation is elastic at this limit. (C.E.I. Pt. 2, 1980)

15.59 Determine an expression relating the radial strain measured by the gauge G in Fig. 15.30 to the applied pressure p for a plate rigidly clamped at a 50 mm radius with thickness 0.2 mm. Take $E = 200\,\text{GPa}$ and $v = 0.3$. (C.E.I. Pt. 2, 1977)

15.60 A cylinder-head valve (Fig. 15.31) of diameter 38 mm is subjected to a gas pressure of $q = 1.4\,\text{MN/m}^2$. Assuming simple supports and that the stem applies a concentrated force P at the centre of the plate, calculate the movement of the stem necessary to lift the plate from its supports. Take $D = 260\,\text{N m}$ and $v = 0.3$. (C.E.I. Pt. 2, 1971)

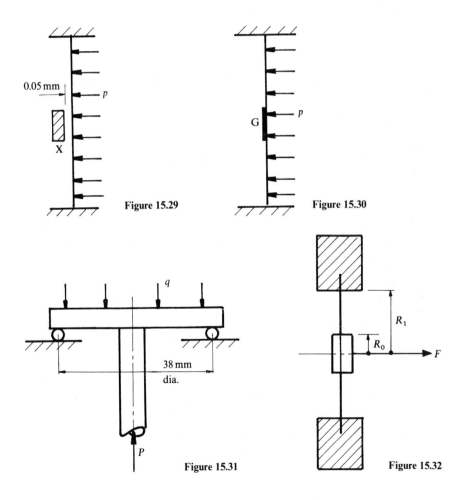

Figure 15.29

Figure 15.30

Figure 15.31

Figure 15.32

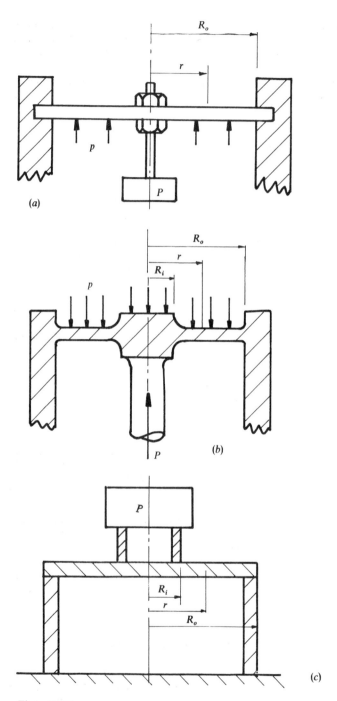

Figure 15.33

15.61 A thin circular plate is rigidly clamped at radius R_1. Two circular metal blocks of radii $R_0 \ll R_1$ are firmly clamped concentrically on either side as shown in Fig. 15.32. Derive an expression for the block deflection when a normal force F is applied at the centre. (C.E.I. Pt. 2, 1982)

15.62 Clearly identify the loading condition for each plate in Fig. 15.33(a–c) by deriving the shear force per unit length of circumference at radius r. State the particular boundary conditions that should be invoked to obtain the bending stresses.

Bending of Rectangular Plates

15.63 Show that for a simply supported circular plate subjected to a normal pressure p, $D\nabla^4 w = p$ reduces to Eq. (15.37b) where $F = pr/2$. (*Hint*: Use the same conversions from Cartesian to polar coordinates as with $\nabla^4\phi = 0$ in Chapter 14, page 605).

15.64 Given that the equilibrium condition $\partial\sigma_x/\partial x + \partial\tau_{xy}/\partial y + \partial\tau_{xz}/\partial z = 0$ applies to a thin rectangular plate, show that the distribution of shear stress τ_{xz} over the thickness t is given by $\tau_{xz} = (2S_{xz}/2t)(1 - 4z^2/t^2)$, where S_{xz} is the shear force/unit length of y.

15.65 The deflection w of a square plate, $a \times a$, lying in the x–y plane, with centre at the origin, is given by $w = w_o \cos(\pi x/a)\cos(3\pi y/a)$, where w_o is the central deflection. Find the surface loading, the edge support reactions and the bending moments at the plate centre. (C.E.I. Pt. 2, 1985)

15.66 A thin elliptical plate with thickness t and semi-axes a and b $(a > b)$ is rigidly clamped around its edges and subjected to a uniform normal pressure p. Show that the deflection is given by the expression; $w = w_o(1 - x^2/a^2 - y^2/b^2)^2$ and find the value for w_o. Determine the top surface stresses, σ_x and σ_y, at the plate centre and at the ends of the major axis.

Answer: $w_o = (pa^4 b^4)/8D(3b^4 + a^2 b^2 + 3a^4)$;

$$\sigma_x = -12pa^2 b^2 (b^2 + va^2)/t^2 (3b^4 + a^2 b^2 + 3a^4);$$

$$\sigma_y = -12pa^2 b^2 (a^2 + vb^2)/t^2 (3b^4 + a^2 b^2 + 3a^4);$$

$$\sigma_x = 6pa^2 b^4/t^2 (3a^4 + a^2 b^2 + 3b^4);$$

$$\sigma_y = 6pa^4 b^2/t^2 (3a^4 + a^2 b^2 + 3b^4).$$

15.67 A simply supported rectangular plate, $a \times 2a$, lies in the x–y plane such that the x-axis bisects the longer sides and the y-axis is coincident with one longer side. If the surface loading of the plate is given by $q = q_o \sin(\pi x/a)$ and the resulting deflection by $w = (qa^4/D\pi^4)[1 + A\cosh(\pi y/a) + B(\pi y/a)\sinh(\pi y/a)]$, find q_o, A and B and calculate the central deflection and the bending moments M_x and M_y. Take $v = 0.30$. (C.E.I. Pt. 2, 1986)

15.68 A simply supported rectangular steel plate with plane dimensions $3\,m \times 4\,m$ and $25\,mm$ thick carries a uniform normal pressure of $15\,kN/m^2$. Find the maximum deflection and the greatest bending stresses. Take $E = 210\,GPa$, $v = 0.28$.

15.69 A simply supported thin plate of edge dimensions $1 \times 1.5\,m$ and thickness $0.025\,m$ is subjected to a sinusoidal loading; $p = p_o \sin(\pi x/a)\sin(\pi y/b)$ with maximum amplitude $2\,kN/m^2$ for the origin at one corner of the plate. Find the maximum values of the bending moment and deflection. Take $E = 200\,GPa$, $v = 0.35$. (T.C.D. 1979)

15.70 A rectangular plate, $3\,m \times 2\,m$ and $30\,mm$ thick, is built-in along one $3\,m$ side and simply supported along the other $3\,m$ side. The remaining edges are free. If the plate carries a uniform normal pressure of $20\,kN/m^2$, determine the maximum deflection and the bending stresses. Take $E = 207\,GPa$ and $v = 0.27$.

15.71 A rectangular steel plate, $1.25\,m \times 1.75\,m$ and $30\,mm$ thick, lies in the x–y plane with the origin of coordinates at one corner. Find the maximum values of deflection and bending moment when the plate carries the pressure distribution $q = 2\sin(\pi x/1.25)\sin(\pi y/1.75)$ (kN/m^2). Take $E = 210\,GPa$ and $v = 0.28$.

YIELD AND STRENGTH CRITERIA

16.1 YIELDING OF DUCTILE METALLIC MATERIALS

Under a uniaxial stress system, a metallic material will begin to deform plastically when the yield stress is reached. It has been seen that this situation can be avoided with a judicious choice of a safety factor in deciding on a safe working elastic stress for the material. However, in practice, the stress state in many loaded structures is often biaxial or triaxial. The question then arises of what magnitudes of these combined stresses will cause the onset of yielding. It is required to find a suitable criterion based upon stress, strain or strain energy for the complex system that can be related to the corresponding quantity at the uniaxial yield point. It is shown that the criteria can all be related to the uniaxial yield stress Y, which is most conveniently measured from a tension test. The following yield criteria have been proposed over the past two centuries, though now it is recognized that those attributed to von Mises and Tresca are the most representative of initial yield behaviour in metallic materials. The criteria are normally expressed in their principal triaxial forms, that is, under the stress system σ_1, σ_2 and σ_3 where $\sigma_1 > \sigma_2 > \sigma_3$ (Fig. 16.1). The stress transformation equations, given in Chapter 13, enable these yield criteria to be expressed in any two- or three-dimensional combination of applied direct and shear stresses.

Maximum Principal Stress Theory (Rankine, 1820–72)

The simplest criterion of yielding states that yielding commences either when the major principal stress σ_1 attains the value of the tensile yield stress Y,

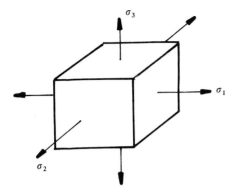

Figure 16.1 Principal stress system.

or when the minor principal stress σ_3, if compressive and of greater numerical magnitude than σ_1, reaches $-Y$. That is,

$$\sigma_1 = Y \quad \text{or} \quad \sigma_3 = -Y. \tag{16.1a, b}$$

If the stress system is a general biaxial one under σ_x, σ_y and τ_{xy}, then from Table 13.1(d), the non-zero principal stresses are

$$\sigma_1 = (\sigma_x + \sigma_y)/2 + (1/2)\sqrt{[(\sigma_x - \sigma_y)^2 + 4\tau_{xy}^2]}, \tag{16.2a}$$

$$\sigma_2 = (\sigma_x + \sigma_y)/2 - (1/2)\sqrt{[(\sigma_x - \sigma_y)^2 + 4\tau_{xy}^2]}. \tag{16.2b}$$

Now, as σ_2 in Eq. (16.2b) is compressive, this must be identified with σ_3 in Eq. (16.1b) according to the convention that $\sigma_1 > \sigma_2 > \sigma_3$ (where $\sigma_2 = 0$). Equations (16.2) are readily reduced to predict initial yielding under other biaxial stress systems. Equations (16.1a, b) ignore the intermediate principal stress (σ_2) and assume, along with all other yield criteria in this section, that the tensile and compressive yield stresses are equal.

Maximum Principal Strain Theory (St Venant, 1797–1886)

This seldom-used theory postulates that yielding under the principal system in Fig. 16.1 commences when the major principal strain attains the magnitude of the uniaxial strain at the elastic limit. When the limiting strain is equated to Y/E it follows from Eqs (4.12a and b) that for tensile and compressive yielding this criterion states

$$\epsilon_1 = Y/E = (1/E)[\sigma_1 - v(\sigma_2 + \sigma_3)],$$
$$\epsilon_3 = -Y/E = (1/E)[\sigma_3 - v(\sigma_1 + \sigma_2)],$$

i.e.,

$$[\sigma_1 - v(\sigma_2 + \sigma_3)] = Y, \tag{16.3a}$$

$$[\sigma_3 - v(\sigma_1 + \sigma_2)] = -Y. \tag{16.3b}$$

In the general biaxial case, substitutions from Eqs (16.2) are made in whichever bracketed quantity [ ] in Eqs (16.3) is numerically the greater.

Maximum Shear Stress Theory (Coulomb, 1773; Tresca, 1868; Guest, 1900)

This often-used theory assumes that yielding, under the principal stress system in Fig. 16.1, begins when the maximum shear stress reaches a critical value. The latter is taken as the maximum shear stress k at the point of yielding under simple tension or compression. Now, from Table 13.1(a), it follows that $k = Y/2$ and acts along planes at $45°$ to the uniaxial stress axis (Fig. 16.2(a)). The greatest shear stress for Fig. 16.1 is given by $\tau_{13} = (\sigma_1 - \sigma_3)/2$ in Eq. (13.20b), and acts along a plane inclined at $45°$ to the 1 and 3 directions (Fig. 16.2b).

Equating the shear stresses for the uniaxial and triaxial cases, $\tau_{13} = k$, leads to the Tresca yield criterion:

$$\sigma_1 - \sigma_3 = Y, \tag{16.4a}$$

where numerical values of σ_1 and σ_3 must be substituted in the L.H.S. with their signs denoting tension or compression, e.g. if $\sigma_1 = 20$, $\sigma_2 = -10$ and $\sigma_3 = -12$ MPa then the L.H.S. value is $20 - (-12) = 32$ MPa, for which the intermediate stress value is irrelevant. The alternative form of Eq. (16.4a), which conveniently avoids symbols, is

Greatest principal stress $-$ least principal stress $=$ tensile yield stress. (16.4b)

In the general biaxial stress case, Eqs (16.2) are substituted into Eq. (16.4b). For the particular case where $\sigma_y = 0$ in Eqs (16.2), the Tresca criterion then appears in its common biaxial form:

$$\sigma_x^2 + 4\tau_{xy}^2 = Y^2. \tag{16.4c}$$

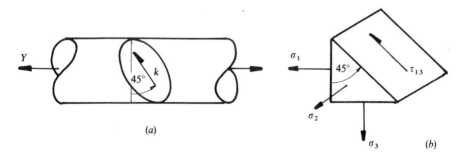

(a)

(b)

Figure 16.2 Maximum shear stresses on a $45°$ plane.

Total Strain Energy Theory (Beltrami, 1885; Haigh, 1920)

The theory assumes that yielding commences for Fig. 16.1 when the total strain energy stored attains the value of the strain energy for uniaxial yielding. For the principal system, the strain energy density in Eq. (11.3c) becomes

$$U = \int_{\epsilon_{ij}} \sigma_{ij} d\epsilon_{ij}$$

$$= \int \sigma_1 d\epsilon_1 + \sigma_2 d\epsilon_2 + \sigma_3 d\epsilon_3$$

$$\equiv (1/2)(\sigma_1 \epsilon_1 + \sigma_2 \epsilon_2 + \sigma_3 \epsilon_3), \qquad (16.5a)$$

which is the sum of the areas enclosed by the linear stress–strain curve for each orthogonal direction (Fig. 16.3).

Substituting Eqs (4.12) into Eq. (16.5a) leads to the equivalent U expression in the principal stresses:

$$U = (1/2E)(\sigma_1^2 + \sigma_2^2 + \sigma_3^2) - (v/E)(\sigma_1 \sigma_2 + \sigma_2 \sigma_3 + \sigma_1 \sigma_3). \qquad (16.5b)$$

Putting $\sigma_1 = Y$ and $\sigma_2 = \sigma_3 = 0$ in Eq. (16.5b) gives $U = Y^2/2E$ at the yield point for simple tension. Equating the U expressions for the uniaxial and triaxial cases (Eq. (16.5b)) gives the total strain energy criterion:

$$(\sigma_1^2 + \sigma_2^2 + \sigma_3^2) - 2v(\sigma_1 \sigma_2 + \sigma_2 \sigma_3 + \sigma_1 \sigma_3) = Y^2. \qquad (16.6a)$$

Note here that the intermediate stress σ_2 contributes to triaxial yielding. Further substitutions for σ_1, σ_2 and σ_3 in Eq. (16.6a) may again be made for the general biaxial stress system. For non-zero principal biaxial stresses σ_1 and σ_2, with $\sigma_3 = 0$, Eq. (16.6a) becomes

$$\sigma_1^2 - 2v\sigma_1 \sigma_2 + \sigma_2^2 = Y^2. \qquad (16.6b)$$

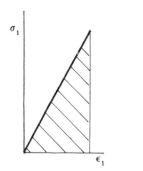
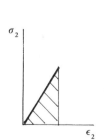
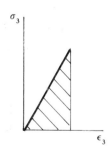

Figure 16.3 Strain energy components.

Shear Strain Energy Theory (Maxwell, 1856; Huber, 1904; von Mises, 1913; Hencky, 1925)

Huber proposed that the total strain energy was composed of dilatational (volumetric) and distortional (shear) components. The former depends upon the mean or hydrostatic component of the applied stress system while the latter is due to the remaining reduced or deviatoric components of stress (see Fig. 4.3). Clark Maxwell implied that hydrostatic stress plays no part in yielding when he was the first to propose that yielding under Fig. 16.1 was due only to the shear strain energy component reaching a critical value (taken as the shear strain energy at the tensile yield point). This was later formalized by von Mises and Hencky in any one of the following three routes leading to the same yield criterion for ductile, initially isotropic, metallic materials, for which there is now considerable experimental support.

Shear strain energy, U_s In Fig. 4.3, the volumetric strain energy density is given by

$$U_v = \int \sigma_m \, d\epsilon + \sigma_m \, d\epsilon + \sigma_m \, d\epsilon$$

$$= (1/2)(\sigma_m \epsilon + \sigma_m \epsilon + \sigma_m \epsilon).$$

Substituting for the strain in any one direction, $\epsilon = \sigma_m/3K$ from Eqs (4.4) and (4.10) leads to

$$U_v = \sigma_m^2/2K = 3(1 - 2v)\sigma_m^2/2E.$$

Substituting from Eq. (4.5):

$$U_v = (1 - 2v)(\sigma_1 + \sigma_2 + \sigma_3)^2/6E. \tag{16.7}$$

Subtracting Eq. (16.7) from Eq. (16.5b) leads to the shear strain energy:

$$U_s = U - U_v$$
$$= (1/2E)(\sigma_1^2 + \sigma_2^2 + \sigma_3^2) - (v/E)(\sigma_1\sigma_2 + \sigma_2\sigma_3 + \sigma_1\sigma_3)$$
$$- (1 - 2v)(\sigma_1 + \sigma_2 + \sigma_3)^2/6E$$
$$= (1 + v)[(\sigma_1 - \sigma_2)^2 + (\sigma_2 - \sigma_3)^2 + (\sigma_1 - \sigma_3)^2]/6E. \tag{16.8a}$$

The value of U_s at the tensile yield point is found from putting $\sigma_1 = Y$, $\sigma_2 = \sigma_3 = 0$ in Eq. (16.8a). This gives

$$U_s = (1 + v)Y^2/3E. \tag{16.8b}$$

Equating (16.8a) and (16.8b) provides the usual form of the von Mises yield criterion:

$$(\sigma_1 - \sigma_2)^2 + (\sigma_2 - \sigma_3)^2 + (\sigma_1 - \sigma_3)^2 = 2Y^2. \tag{16.9a}$$

If one principal stress (say $\sigma_3 = 0$) is zero, the biaxial form of Eq. (16.9a) becomes

$$\sigma_1^2 - \sigma_1\sigma_2 + \sigma_2^2 = Y^2. \tag{16.9b}$$

Substitution from Eqs (16.2) for σ_1 (tensile) and σ_2 (compressive) provides the general biaxial form in terms of σ_x, σ_y and τ_{xy}. When $\sigma_y = 0$, a common biaxial form of the Mises criterion is found:

$$\sigma_x^2 + 3\tau_{xy}^2 = Y^2. \tag{16.9c}$$

Subscripts in this equation and in Eq. (16.4c) are often omitted to account for yielding under all combinations of direct stress and shear.

Octahedral shear stress, τ_o When it is proposed that yielding under the principal stress system in Fig. 16.1 begins when τ_o in Eq. (13.21b) reaches its critical value at the tensile yield point ($\sigma_1 = Y, \sigma_2 = \sigma_3 = 0$), we have

$$(1/3)\sqrt{[(\sigma_1 - \sigma_2)^2 + (\sigma_2 - \sigma_3)^2 + (\sigma_1 - \sigma_3)^2]} = (1/3)\sqrt{(2Y^2)},$$

which again leads to Eq. (16.9a). This is because the normal stress for the octahedral planes, σ_o in Eq. (13.21a), is numerically equal to the mean stress σ_m and causes elastic dilatation only. The distortion and hence yielding is therefore soley due to τ_o. Clearly, energy considerations are unnecessary in formulating the von Mises criterion on this basis.

Deviatoric stress invariants Because the mean or hydrostatic stress plays no part in yielding, it follows that, in a general formulation, the yield criterion is a function of the deviatoric or reduced stress tensor, σ_{ij}^D in Eq. (13.24). Moreover, as yield behaviour is a property of the material itself, the yield function must be independent of the coordinates used to define σ_{ij}^D. That is, yielding is a function of the deviatoric stress invariants, J_1^D, J_2^D and J_3^D. These may be obtained by subtracting σ_m from the absolute stress invariants J_1, J_2 and J_3 in the principal stress cubic equation (13.15b). In principal stress form,

$$J_1^D = (\sigma_1 - \sigma_m) + (\sigma_2 - \sigma_m) + (\sigma_3 - \sigma_m) = (\sigma_1 + \sigma_2 + \sigma_3) - 3\sigma_m = 0,$$

$$\begin{aligned} J_2^D &= (\sigma_1 - \sigma_m)(\sigma_2 - \sigma_m) + (\sigma_2 - \sigma_m)(\sigma_3 - \sigma_m) + (\sigma_1 - \sigma_m)(\sigma_3 - \sigma_m) \\ &= (\sigma_1\sigma_2 + \sigma_2\sigma_3 + \sigma_1\sigma_3) - 3\sigma_m^2 \\ &= (1/3)[(\sigma_1\sigma_2 + \sigma_2\sigma_3 + \sigma_1\sigma_3) - (\sigma_1^2 + \sigma_2^2 + \sigma_3^2)] \\ &= -(1/6)[(\sigma_1 - \sigma_2)^2 + (\sigma_2 - \sigma_3)^2 + (\sigma_1 - \sigma_3)^2], \end{aligned} \tag{16.10a}$$

$$\begin{aligned} J_3^D &= (\sigma_1 - \sigma_m)(\sigma_2 - \sigma_m)(\sigma_3 - \sigma_m) \\ &= \sigma_1\sigma_2\sigma_3 - \sigma_m(\sigma_1\sigma_2 + \sigma_2\sigma_3 + \sigma_1\sigma_3) + 2\sigma_m^3 \\ &= \sigma_1\sigma_2\sigma_3 - (1/3)(\sigma_1 + \sigma_2 + \sigma_3)(\sigma_1\sigma_2 + \sigma_2\sigma_3 + \sigma_1\sigma_3) \\ &\quad + (2/27)(\sigma_1 + \sigma_2 + \sigma_3)^3 \\ &= (1/3)^4[(2\sigma_1 - \sigma_2 - \sigma_3)^3 + (2\sigma_2 - \sigma_1 - \sigma_3)^3 + (2\sigma_3 - \sigma_1 - \sigma_2)^3], \end{aligned} \tag{16.10b}$$

and corresponding to Eq. (13.15b) the deviatoric stress cubic equation is

$$(\sigma^D)^3 - J_1^D(\sigma^D)^2 + J_2^D(\sigma^D) - J_3^D = 0.$$

In order that J_2^D in Eq. (16.10a) becomes positive, this cubic is normally written, with $J_1^D = 0$, as

$$(\sigma^D)^3 - J_2^D(\sigma^D) - J_3^D = 0.$$

Yielding begins when a function of the two non-zero deviatoric invariants attains a critical constant value c:

$$f(J_2^D, J_3^D) = c, \tag{16.11}$$

where c is normally defined from the reduction of the L.H.S. to yielding in simple tension or torsion. With particular functions f, this formulation enables the observed, initial yield behaviour of most metallic materials to be adequately represented (see D.W.A. Rees, *Proc. R. Soc. Lond. A* **383**, 333–357, 1984). When J_3^D is omitted and f is identified with J_2^D only, then Eq. (16.11) simply becomes

$$J_2^D = c, \tag{16.12}$$

where $c = -Y^2/3$ is found from putting $\sigma_1 = Y$ and $\sigma_2 = \sigma_3 = 0$ in Eq. (16.10a). Then, from Eqs (16.10a and 16.12):

$$(1/6)[(\sigma_1 - \sigma_2)^2 + (\sigma_2 - \sigma_3)^2 + (\sigma_1 - \sigma_3)^2] = Y^2/3,$$

which again results in the von Mises yield criterion (Eq. (16.9a)). That is, the second invariant of deviatoric stress reaches a critical, constant value at the yield point of a given material.

Example 16.1 A mild steel tube has a 100 mm mean diameter and 3 mm wall thickness. Determine the torque that can be transmitted by the tube according to the five foregoing yield criteria using a safety factor $S = 2.25$. Take the tensile yield stress $Y = 230$ MPa and $v = 0.3$.

If k is the shear yield stress, then the working shear stress $\tau_w = k/S$. The Batho theory (Eq. (9.26)) gives

$$T = 2At\tau_w = 2(\pi \times 50^2)3\tau_w = (15 \times 10^3)\pi k/S \quad \text{(N mm)} \qquad \text{(i)}$$

where for torsion the principal yield stresses are $\sigma_1 = k$, $\sigma_2 = 0$ and $\sigma_3 = -k$.

(i) MAXIMUM PRINCIPAL STRESS THEORY In this case, Eqs (16.1a and b) give simply $k = Y$ and $-k = -Y$. Hence, from Eq. (i),

$$T = (15 \times 10^3)\pi \times 230/(2.25 \times 10^3) = 4820 \text{ N m}.$$

(ii) MAXIMUM PRINCIPAL STRAIN THEORY Equations (16.3) give identical forms:

$$\sigma_1 - v\sigma_3 = k(1 + v) = Y,$$
$$\sigma_3 - v\sigma_1 = -k(1 + v) = -Y,$$
$$\therefore \quad k = Y/(1 + v),$$

when, with Eq. (i),

$$T = (15 \times 10^3)\pi \times 230/(1.3 \times 2.25 \times 10^3) = 3706 \text{ N m.}$$

(iii) MAXIMUM SHEAR STRESS THEORY Equation (16.4a) here becomes $\sigma_1 - \sigma_3 = 2k = Y$. Hence,

$$\therefore \quad k = Y/2.$$

Substituting into Eq. (i) leads to

$$T = (15 \times 10^3)\pi \times 230/(2 \times 2.25 \times 10^3) = 2408 \text{ N m.}$$

(iv) MAXIMUM STRAIN ENERGY THEORY Equation (16.6b) reduces to $\sigma_1^2 + \sigma_3^2 - 2v\sigma_1\sigma_3 = Y^2$. That is,

$$k^2 + (-k)^2 - 2vk(-k) = Y^2,$$
$$2k^2(1 + v) = Y^2,$$
$$\therefore \quad k = Y/\sqrt{2(1 + v)}.$$

Substituting into Eq. (i):

$$T = (15 \times 10^3)\pi \times 230/[\sqrt{(2 \times 1.3)} \times 2.25 \times 10^3] = 2987.4 \text{ N m.}$$

(v) MAXIMUM SHEAR STRAIN ENERGY THEORY Equation (16.9a) becomes

$$(k - 0)^2 + (0 + k)^2 + (k + k)^2 = 2Y^2,$$
$$\therefore \quad k = Y/\sqrt{3}.$$

Then, from Eq. (i),

$$T = (15 \times 10^3)\pi \times 230/(\sqrt{3} \times 2.25 \times 10^3) = 2778 \text{ N m.}$$

16.2 GRAPHICAL COMPARISON BETWEEN CRITERIA FOR PRINCIPAL BIAXIAL STRESS

The five criteria may be compared graphically (Fig. 16.4) in principal axes σ_1 and σ_2 when the third principal stress $\sigma_3 = 0$. Clearly, the energy criteria (Eqs (16.6b) and (16.9b)) both describe ellipses with a 45° inclination in their major axes. Putting $\sigma_3 = 0$ in Eqs (16.1), (16.3) and (16.4) results in the square, parallelogram and hexagon shown when each respective criterion is applied separately to the stress state existing within each quadrant.

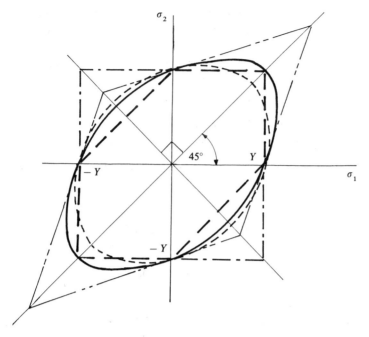

Figure 16.4 Yield loci for the five criteria:
—·— principal stress (Eq. (16.1)); ———··—— Principal strain (Eq. (16.3)); ——— Tresca (Eq. (16.4)); ----- total energy (Eq. (16.6)); —————— von Mises (Eq. (16.9)).

The particular shape is called a yield locus, wherein elastic conditions are predicted to prevail. Plastic behaviour occurs to the exterior of each locus. The Tresca locus is the most conservative of these, while the von Mises ellipse conforms most closely to observed biaxial yield behaviour. Because of the extended elastic ranges exhibited by the principal stress and strain theories in particular quadrants, they lead to unreliable predictions and are therefore not often employed in practice. It is seen from the previous example that the lowest, most reliable, predictions correspond to Tresca and von Mises. Of these two criteria, the following examples further show that, in principal stress space, the simpler Tresca predictions are always the safest but those from von Mises are generally the more realistic. In their principal stress form the yield criteria may be applied directly to predicting yielding in thin- and thick-walled cylinders, rotating discs, plates and blocks under orthogonal forces. It is therefore necessary to call on the appropriate theory for stress calculation from preceding chapters.

Example 16.2 A thin-walled cylinder of 750 mm (30 in) i.d. and 15 mm (1/2 in) wall thickness is subjected to an internal pressures of 2 MPa (300 psi). Calculate the additional axial load the cylinder could carry, with a safety factor of $S = 5$ on yielding under $Y = 300\,\text{MPa}\ (20\,\text{tonf/in}^2)$ from the Tresca and von Mises yield criteria.

SOLUTION TO SI UNITS The stress state for a point in the wall of the cylinder is $\sigma_r = 0$ and, from Eq. (4.17),

$$\sigma_\theta = pd/2t = 2 \times 750/(2 \times 15) = 50\,\text{MPa},$$
$$\sigma_z = pd/4t + W/A = 25 + W/\pi dt$$
$$= 25 + W/(\pi \times 750 \times 15) = 25 + W/(35.34 \times 10^3).$$

Now, in polar form the Tresca equation (16.4b) becomes, for $\sigma_z > \sigma_\theta > \sigma_r$,

$$\sigma_z - \sigma_r = Y/S,$$
$$25 + W/(35.34 \times 10^3) - 0 = 300/5$$
$$W = 1.237\,\text{MN}.$$

The von Mises criterion (Eq. (16.9b)) here becomes

$$\sigma_\theta^2 - \sigma_\theta\sigma_z + \sigma_z^2 = (Y/S)^2,$$
$$50^2 - 50[25 + W/(35.34 \times 10^3)] + [25 + W/(35.34 \times 10^3)]^2 = 60^2,$$
$$[W/(35.34 \times 10^3)]^2 = 1725,$$
$$W = 1.468\,\text{MN}.$$

SOLUTION TO IMPERIAL UNITS

$$\sigma_\theta = pd/2t = 300 \times 30/(2 \times 1/2) = 9000\,\text{lbf/in}^2,$$
$$\sigma_z = pd/4t + W/A = 4500 + W/47.124.$$

From Tresca:

$$\sigma_z = Y/S,$$
$$4500 + W/47.124 = 20 \times 2240/5,$$
$$W = (210.173 \times 10^3)\,\text{lbf} = 93.83\,\text{tonf}.$$

From von Mises:

$$\sigma_\theta^2 - \sigma_\theta\sigma_z + \sigma_z^2 = (Y/S)^2,$$
$$9000^2 - 9000(4500 + W/47.124) + (4500 + W/47.124)^2 = (20 \times 2240/5)^2,$$
$$W = (210.56 \times 10^3)\,\text{lbf} = 94\,\text{tonf}.$$

Example 16.3 Derive expressions for the pressure p to initiate yielding in the bore of a thick-walled closed cylinder with an outside-to-inside diameter ratio of K, according to the Tresca and Mises yield criteria. Determine p for a closed cylinder with 30 mm i.d. and 50 mm o.d., given

$Y = 250 \, \text{MPa}$. What is the effect of 'open ends' where the end force is sustained by bore pistons?

It has previously been shown (page 613) that the principal, triaxial stress state in the wall of a closed cylinder is

$$\sigma_\theta = pr_i^2(1 + r_o^2/r^2)/(r_o^2 - r_i^2), \tag{i}$$

$$\sigma_r = pr_i^2(1 - r_o^2/r^2)/(r_o^2 - r_i^2), \tag{ii}$$

$$\sigma_z = pr_i^2/(r_o^2 - r_i^2), \tag{iii}$$

where $\sigma_\theta > \sigma_z > \sigma_r$. The Tresca equation (16.4b) becomes

$$\sigma_\theta - \sigma_r = Y. \tag{iv}$$

Substituting Eqs (i) and (ii) into Eq. (iv):

$$[pr_i^2/(r_o^2 - r_i^2)][(1 + r_o^2/r^2) - (1 - r_o^2/r^2)] = Y,$$
$$2pr_i^2 r_o^2/r^2(r_o^2 - r_i^2) = Y,$$
$$p = Yr^2(r_o^2 - r_i^2)/2r_i^2 r_o^2.$$

Clearly, the lowest yield pressure occurs in the bore, where, for $r = r_i$,

$$p = (Y/2)(1 - r_i^2/r_o^2) = Y(K^2 - 1)/2K^2. \tag{v}$$

When $K = 50/30 = 1.667$, Eq. (v) gives $p = 80 \, \text{MPa}$. The Mises criterion, Eq. (16.9a), is now written as

$$(\sigma_\theta - \sigma_z)^2 + (\sigma_\theta - \sigma_r)^2 + (\sigma_r - \sigma_z)^2 = 2Y^2. \tag{vi}$$

Substituting Eqs (i)–(iii) into Eq. (vi) with $r = r_i$ leads to

$$6p^2[r_o^2/(r_o^2 - r_i^2)]^2 = 2Y^2,$$
$$p = (Y/\sqrt{3})(1 - r_i^2/r_o^2) = Y(K^2 - 1)/\sqrt{3}K^2. \tag{vii}$$

For $K = 1.667$, Eq. (vii) gives $p = 92.4 \, \text{MPa}$.

With open ends, $\sigma_z = 0$. This does not alter the Tresca prediction, since σ_z remains intermediate. However, the Mises Eq. (vi) becomes

$$\sigma_\theta^2 - \sigma_\theta \sigma_r + \sigma_r^2 = Y^2.$$

Substituting from Eqs (i) and (ii), with $r = r_i$,

$$p^2[(r_o^2 + r_i^2)/(r_o^2 - r_i^2)]^2 + p^2(r_o^2 + r_i^2)/(r_o^2 - r_i^2) + p^2 = Y^2,$$
$$p^2[(K^2 + 1)^2 + (K^2 - 1)(K^2 + 1) + (K^2 - 1)^2]/(K^2 - 1)^2 = Y^2,$$
$$p^2(1 + 3K^4)/(K^2 - 1)^2 = Y^2,$$
$$\therefore \quad p = Y(K^2 - 1)/\sqrt{(3K^4 + 1)}.$$

For $K = 1.667$, the yield pressure here becomes $90.46 \, \text{MPa}$.

16.3 COMPARISON BETWEEN TRESCA AND VON MISES IN σ, τ STRESS SPACE

There are many instances of combined direct and shear stress states where the particular Tresca and Mises criteria in Eqs (16.4c) and (16.9c) are more convenient to apply than their principal stress forms. It is usual to omit the subscripts x and y on σ_x and τ_{xy} in order to embrace both Cartesian and cylindrical coordinates. Figure 16.5 compares the elliptical yield loci described by Mises and Tresca for the positive σ, τ quadrant. The shear yield stress k (i.e. the semi-minor axis) differs, with Tresca being the smaller. In fact, the Tresca predictions of yielding under combined σ, τ are always the more conservative. Experiments conducted on thin-walled tubes of ductile metallic materials under combined tension and torsion have shown that yield points lie between these loci closer to the Mises ellipse (see G. I. Taylor and H. Quinney, *Phil. Trans R. Soc. A* **230**, 323–362, 1931).

In practice, these forms of yield criteria apply to a point in a circular shaft under direct shear or torsion combined with either an axial load or a bending moment. They also apply when such loadings are all applied simultaneously. A similar combined direct-shear stress state exists at points in the length of a loaded beam away from the neutral axis. The following examples illustrate these loadings.

Example 16.4 The loading on the bolt shown in Fig. 16.6 consists of an axial tensile force of $W = 10\,\text{kN}$ together with a transverse shearing force of $F = 5\,\text{kN}$. Estimate, from the Tresca and von Mises criteria, a safe bolt diameter d with a safety factor of 3, given that the tensile yield stress of bolt material is $270\,\text{MPa}$.

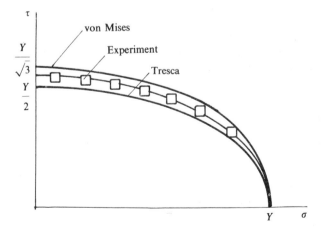

Figure 16.5 Tresca and Mises loci in σ, τ stress space.

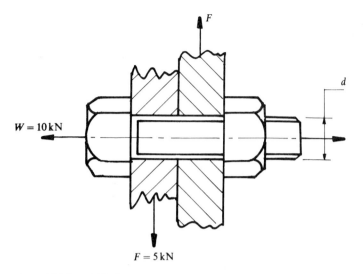

Figure 16.6 Loaded bolt.

The axial stress is

$$\sigma = W/A = 10 \times 10^3/(\pi d^2/4) = 12.73 \times 10^3/d^2 \quad \text{(MPa)}$$

and for a bolt in single shear, the shear stress is

$$\tau = F/A = 5 \times 10^3/(\pi d^2/4) = 6.366 \times 10^3/d^2 \quad \text{(MPa)}.$$

Since $\sigma_y = 0$ in the absence of bearing pressure, the Tresca Equation (16.4c) becomes

$$\sigma^2 + 4\tau^2 = (Y/S)^2,$$
$$(12.73 \times 10^3/d^2)^2 + 4(6.366 \times 10^3/d^2)^2 = (270/3)^2,$$
$$d = \{[(12\,730)^2 + 4(6366)^2]/90^2\}^{1/4} = 14.15\,\text{mm},$$

and, from the von Mises Equation (16.9c),

$$\sigma^2 + 3\tau^2 = (Y/S)^2,$$
$$(12.73 \times 10^3/d^2)^2 + 3(6.366 \times 10^3/d^2)^2 = (270/3)^2,$$
$$d = \{[(12\,730)^2 + 3(6366)^2]/90^2\}^{1/4} = 13.7\,\text{mm}.$$

Example 16.5 A solid steel shaft of 32 mm ($1\frac{1}{4}$ in) diameter transmits 100 kW (135 hp) at 3000 rev/min and simultaneously withstands a bending moment M. Calculate the allowable value of M for a safety factor of 3 according to the Tresca and von Mises yield criteria. Take $Y = 400\,\text{MPa}$ (25 tonf/in^2).

SOLUTION TO SI UNITS Here, $\tau = Tr/J$ (Eq. (9.5)), where $T = 60P/2\pi N$

(Eq. (9.6)). Hence,

$$\tau = (60P/2\pi N)(r/J)$$
$$= (60 \times 100 \times 10^6 \times 16)/[2\pi \times 3000 \times \pi(32)^4/32] = 49.47\,\text{MPa}.$$

Now from ETB (Eq. (6.4):

$$\sigma = My/I = (M \times 16)/[\pi(32)^4/64] = M/3217 \quad (\text{MPa}).$$

Substituting into Eq. (16.4c), Tresca gives,

$$\sigma^2 + 4\tau^2 = (Y/S)^2,$$
$$(M/3217)^2 + 4(49.47)^2 = (400/3)^2,$$
$$M = 287.53 \times 10^3\,\text{N\,mm} = 287.53\,\text{N\,m},$$

while from Eq. (16.9c), von Mises gives

$$\sigma^2 + 3\tau^2 = (Y/S)^2,$$
$$(M/3217)^2 + 3(49.47)^2 = (400/3)^2,$$
$$M = 328.64 \times 10^3\,\text{N\,mm} = 328.64\,\text{N\,m}.$$

SOLUTION TO IMPERIAL UNITS

$$P = 2\pi NT/33\,000 \quad (\text{hp}) \quad (\text{with } T \text{ in lbf ft}),$$
$$\therefore \quad T = 33\,000 \times 135/(2\pi \times 3000) = 236.345\,\text{lbf ft}.$$
$$\tau = Tr/J$$
$$= 236.345 \times 12 \times 0.625/[\pi(1.25)^4/32]$$
$$= 7395.51\,\text{lbf/in}^2 = 3.302\,\text{tonf/in}^2.$$
$$\sigma = My/I$$
$$= M \times 0.625/[\pi(1.25)^4/64] = 5.2152M.$$

From Tresca:

$$\sigma^2 + 4\tau^2 = (Y/S)^2,$$
$$(5.2152M)^2 + 4(3.302)^2 = (25/3)^2,$$
$$M = 0.9746\,\text{tonf in} = 2183\,\text{lbf in}.$$

From von Mises:

$$\sigma^2 + 3\tau^2 = (Y/S)^2,$$
$$(5.2152M)^2 + 3(3.302)^2 = (25/3)^2,$$
$$M = 1.1622\,\text{tonf in} = 2603.3\,\text{lbf in}.$$

Example 16.6 Calculate the maximum uniformly distributed elastic load a 2 m long steel cantilever can bear when its T cross-section in Fig. 16.7 is 500 mm wide and 500 mm deep, with equal flange and web thicknesses

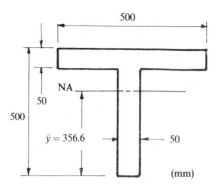

Figure 16.7 T-section.

of 50 mm ($I = 1125 \times 10^6\,\text{mm}^4$ and $\bar{y} = 356.6$ mm from the web bottom). Make the calculation for the flange and web tops and at the neutral axis of bending. Use a von Mises yield criterion where appropriate, with a tensile yield stress of 310 MPa.

At the flange top, $\tau = 0$ and $\sigma = My/I$, where $M = wl^2/2$ and $y = 143.4$ mm. Thus,

$$\sigma = \frac{(w \times 2000^2/2)143.4}{1125 \times 10^6} = 0.255w \quad (\text{MPa}).$$

The flange fibres begin to yield when $0.255w = 310$. Hence, $w = 1216\,\text{N/mm}$.

At the neutral axis, where $\sigma = 0$, the shear stress is (Eq. (10.4))

$$\tau = F(A\bar{y})/Ix,$$

where $F = wl$, $x = 50$ mm and

$$A\bar{y} = (500 \times 50)118.4 + (93.4 \times 50)46.7 = 3.178 \times 10^6\,\text{mm}^3,$$
$$\therefore \quad \tau = (2000w)3.178 \times 10^6/(1125 \times 10^6 \times 50) = 0.113w \quad (\text{MPa}).$$

At the shear yield point, Eq. (16.9c) becomes $3\tau^2 = Y^2$. Hence, $\tau = Y/\sqrt{3}$ and

$$0.113w = 310/\sqrt{3}, \quad \Rightarrow w = 1584\,\text{N/mm}.$$

At the web top ($y = 93.4$ mm, $x = 50$ mm), both direct and shear stresses act:

$$\sigma = My/I = w(2000)^2 \times 93.4/(2 \times 1125 \times 10^6) = 0.166w \quad (\text{MPa}),$$
$$A\bar{y} = 500 \times 50 \times 118.4 = 2.96 \times 10^6\,\text{mm}^3,$$
$$\therefore \quad \tau = F(A\bar{y})/Ix = (2000w)2.96 \times 10^6/(1125 \times 10^6 \times 50)$$
$$= 0.1052w \quad (\text{MPa}).$$

With combined stress yielding, Eq. (16.9c) becomes

$$(0.166w)^2 + 3(0.1052w)^2 = 310^2,$$

$$\therefore \quad w = 1257.4 \, \text{N/mm}.$$

With the least value $w = 1216 \, \text{N/mm}$, yielding starts at the flange top.

Example 16.7 A tubular alloy beam of equilateral triangular section with 30 mm side and 1 mm thick is mounted as a short cantilever 20 mm long. Calculate the magnitude of the end force F_y that may be applied along one vertical side (Fig. 16.8(a)) according to the von Mises and Tresca yield criteria. Use a safety factor of 2 with a tensile yield stress value of 180 MPa.

As with the previous example, this design should consider both bending and shear effects in the beam. The maximum bending stress occurs at points 1 and 3 (where $\tau = 0$) and is given by $\sigma = My/I_x$, where $y = \pm 15 \, \text{mm}$, $M = 20F_y$, and from Example 1.8

$$I_x = a^3 t/4 = 30^3 \times 1/4 = 6750 \, \text{mm}^4,$$

$$\therefore \quad \sigma = (20F_y) \times 15/6750 = 0.0444F_y \quad \text{(MPa)}.$$

Since the design stress is $\sigma = 180/2 = 90 \, \text{MPa}$,

$$\therefore \quad 0.0444F_y = 90, \quad \Rightarrow F_y = 2025 \, \text{N}.$$

The maximum shear stress will follow from the shear flow distribution for this thin-walled closed section. Applying Eq. (10.12) with x a principal axis and F_y positive downwards, the net shear flow is

$$q = q_b + q_o = (F_y/I_x)D_{xb} + q_o.$$

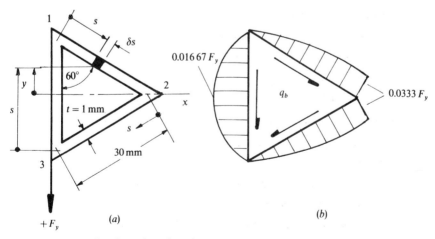

Figure 16.8 Shear flow in a triangular tube.

Working clockwise from point 1, the first moments $D_{xb} = \int_0^s yt \, ds$ are:

$1 \to 2,$ $\quad y = 15 - s/2,$ $\quad t = 1 \, mm:$

$$D_{xb} = \int_0^s (15 - s/2) \, ds = 15s - s^2/4.$$

At 1, $\quad s = 0,$ $\quad D_{xb1} = 0,$

At 2, $\quad s = 30 \, mm,$ $\quad D_{xb2} = 225 \, mm^3.$

$2 \to 3,$ $\quad y = -s/2,$ $\quad t = 1 \, mm:$

$$D_{xb} = -\int_0^s (s/2) \, ds + 225 = 225 - s^2/4.$$

At 2, $\quad s = 0,$ $\quad D_{xb2} = 225 \, mm^3,$

At 3, $\quad s = 30 \, mm,$ $\quad D_{xb3} = 0.$

$3 \to 1,$ $\quad y = s - 15,$ $\quad t = 1 \, mm:$

$$D_{xb} = \int_0^s (s - 15) \, ds = s^2/2 - 15s.$$

At 3, $\quad D_{xb3} = 0,$

At NA, $\quad s = 15 \, mm,$ $\quad D_{xb} = -112.5 \, mm^3,$

At 1, $\quad s = 30 \, mm,$ $\quad D_{xb} = 0.$

Multiplying these D_{xb} values by $F_y/6750$, the q_b distribution is given in Fig. 16.8(b). Taking moments about point 2 with clockwise q_o:

$$(30 \sin 60°)F_y = (0.016\,67 \times 30 \times 2/3)(30 \sin 60°)F_y - 2Aq_o,$$

$$q_o = -0.0222F_y$$

It follows from adding this constant anticlockwise shear flow value to Fig. 16.8(b) that the greatest net shear flow occurs midway between 1 and 3, where $\sigma = 0$. That is,

$$q_{max} = 0.01667F_y + 0.0222F_y = 0.0389F_y,$$

$$\therefore \quad \tau_{max} = q_{max}/t = 0.0389F_y \quad (MPa),$$

and from the von Mises criterion (Eq. (16.9c)), $\sqrt{3}\tau_{max} = (Y/2)$. That is,

$$\sqrt{3} \times 0.0389F_y = 180/2,$$

$$\therefore \quad F_y = 1335.8 \, N.$$

Tresca, Eq. (16.4c), gives $2\tau_{max} = (Y/2)$,

$$\therefore \quad F_y = 1156.8 \, N.$$

In this case shear, not bending, governs the allowable F_y value.

16.4 STRENGTH CRITERIA FOR BRITTLE MATERIALS

The yield criteria considered so far all assume that the yield stresses are of equal magnitude under the forward or reversed application of a given stress. That is, from the symmetry of Figs 16.4 and 16.5, the magnitude of the uniaxial tensile and compressive yield stresses are equal and the positive and negative shear yield stresses are also equal. This assumption is reasonably consistent with the observed initial yield behaviour in ductile metallic materials. However, a similar assumption made for the strengths of most brittle materials would not be consistent with observed behaviour, since these are inherently anisotropic, with different strengths in tension and compression. Moreover, it is worth noting that plastically pre-strained ductile metallic materials exhibit an analogous anisotropy due to a phenomenon known as the Bauschinger effect. This refers to the reduction in reversed yield stress following forward deformation into the plastic range. Initial anisotropy in brittle materials, e.g. rock, ceramic, concrete, cast iron and soils, is often accounted for with the internal friction or Coulomb–Mohr theory. This is an adaption of the major principal stress theory of Rankine with different tensile and compressive yield strengths. Figure 16.9(a) displays each criterion in principal biaxial stress space. A third, modified Mohr theory given in Fig. 16.9(a) is less conservative than Coulomb–Mohr but more conservative than the principal stress theory for the second and third quadrants. Experimental results often lie closest to the intermediate prediction supplied by the modified Mohr criterion.

In Fig. 16.9(b) the Mohr's circles corresponding to combined stress points A, B, C and D in the Couomb–Mohr theory are shown. The respective points A and C, being equi-biaxial tension and compression, appear simply as the extreme points shown. Uniaxial tension or compression, with different ultimate strengths U_t and U_c respectively, are represented by circles passing through A and C tangential to the τ axis. The criterion is that failure under a combined, biaxial stress state is coincident with the point of tangency between the corresponding Mohr's circle and the envelope linking the uniaxial tension and compression circles. Duguet in 1885 proposed the linear envelope in Fig. 16.9(b), where, for example, the strength under pure shear ($\sigma_2 = -\sigma_1$) corresponds to points B in (a) and (b). In the general case, for which tension (say) is combined with shear, the principal stresses at failure become the coordinates of points D in (a) and (b). These are expressed in the equation of the straight line joining U_t and U_c and passing through D. That is,

$$\sigma_2 = (U_c/U_t)\sigma_1 - U_c.$$

When σ_1 is positive and σ_2 and U_c are both negative, this is written as

$$\sigma_1/U_t + \sigma_2/U_c = 1. \tag{16.13a}$$

In the case of compression combined with shear, when σ_2 is positive and σ_1

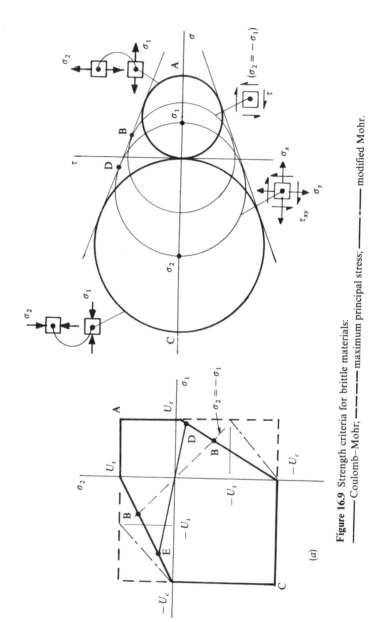

Figure 16.9 Strength criteria for brittle materials:
——— Coulomb–Mohr; ——— maximum principal stress; ——— modified Mohr.

and U_c are both negative, the equation of strength for the line passing through point E is

$$\sigma_2/U_t + \sigma_1/U_c = 1. \qquad (16.13b)$$

For the remaining square envelopes of the strength locus, the stress state is uniaxial and the major principal stress criterion applies. With the usual convention, σ_1(tensile) $> \sigma_2$(compressive) irrespective of signs, it is only necessary to employ Eq. (16.13a), as seen in the following examples. Though these examples specifically employ the Coulomb–Mohr theory, less conservative predictions are available from the two alternative theories, Of these, the modified Mohr is best applied graphically to combined stress states for quadrants two and four in Fig. 16.9(a).

Example 16.8 A brittle material is to be subjected to a direct stress of $\sigma = 75\,\text{MPa}$ and a shear stress τ. What value of τ would cause failure when the direct stress is (i) tensile and (ii) compressive? The tensile and compressive yield strengths for the material are 100 and 200 MPa respectively.

(i) The principal stresses are (Eqs (16.2))

$$\sigma_{1,2} = \sigma/2 \pm (1/2)\sqrt{(\sigma^2 + 4\tau^2)},$$
$$\sigma_1 = 37.5 + (1/2)\sqrt{(75^2 + 4\tau^2)} \quad \text{(tensile)},$$
$$\sigma_2 = 37.5 - (1/2)\sqrt{(75^2 + 4\tau^2)} \quad \text{(compressive)}$$

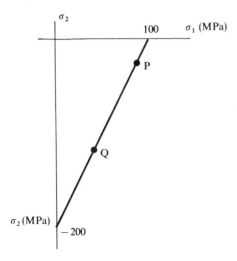

Figure 16.10 Failure locus.

Applying Eq. (16.13a) with $U_t = 100$ and $U_c = -200$ MPa:

$37.5/100 + (1/200)\sqrt{(75^2 + 4\tau^2)} - 37.5/200 + (1/400)\sqrt{(75^2 + 4\tau^2)} = 1,$

$0.1875 + (7.5 \times 10^{-3})\sqrt{(75^2 + 4\tau^2)} = 1,$

$\sqrt{(75^2 + 4\tau^2)} = 108.33,$

$\tau = 39.09$ MPa,

$\therefore \quad \sigma_1 = 91.67$ MPa, $\qquad \sigma_2 = -16.67$ MPa.

These correspond to the intersection point P in Fig. 16.10.
  (ii) The principal stresses now are

$$\sigma_1 = -37.5 + (1/2)\sqrt{(75^2 + 4\tau^2)} \quad \text{(tensile)},$$
$$\sigma_2 = -37.5 - (1/2)\sqrt{(75^2 + 4\tau^2)} \quad \text{(compressive)},$$

and, again applying Eq. (16.13a),

$-37.5/100 + (1/200)\sqrt{(75^2 + 4\tau^2)} + 37.5/200 + (1/400)\sqrt{(75^2 + 4\tau^2)} = 1,$

$\sqrt{(75^2 + 4\tau^2)} = 158.33,$

$\tau = 69.72$ MPa,

$\therefore \quad \sigma_1 = 41.67$ MPa, $\qquad \sigma_2 = -116.67$ MPa.

These correspond to the intersection point Q in Fig. 16.10.

Example 16.9 A brittle material is loaded under the radial combined stress path $R = \sigma/\tau = 2$. Determine these stresses at failure given $U_t = 155$ MPa and $U_c = 675$ MPa.

The corresponding principal stress path is, from Eqs (16.2),

$$\frac{\sigma_1}{\sigma_2} = \frac{\sigma/2 + (1/2)\sqrt{(\sigma^2 + 4\tau^2)}}{\sigma/2 - (1/2)\sqrt{(\sigma^2 + 4\tau^2)}} = \frac{R + \sqrt{(R^2 + 4)}}{R - \sqrt{(R^2 + 4)}},$$

and, for $R = 2$,

$$\sigma_1/\sigma_2 = -5.824. \tag{i}$$

Now with σ_1 tensile and σ_2 compressive, Eq. (16.13a) becomes

$$\sigma_1/155 - \sigma_2/675 = 1 \tag{ii}$$

Solving Eqs (i) and (ii) simultaneously and substituting in Eq. (16.2):

$$\sigma_1 = 149.12 \text{ MPa}, \qquad \sigma_2 = -25.6 \text{ MPa},$$

and

$$\sigma_1/\tau = R/2 + (1/2)\sqrt{(R^2 + 4)} = 2.4142,$$
$$\therefore \quad \tau = 61.77 \text{ MPa}, \qquad \sigma = R\tau = 123.54 \text{ MPa}.$$

If, in Eq. (i), it is assumed that σ_1 is compressive and σ_2 is tensile, then

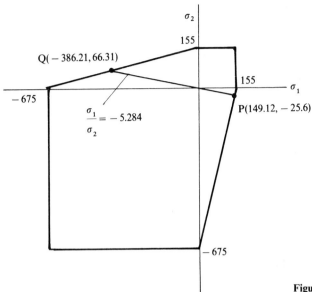

Figure 16.11 Strength locus.

Eq. (16.3b) becomes

$$\sigma_2/155 - \sigma_1/675 = 1. \qquad \text{(iii)}$$

Solving Eqs (i) and (iii) gives

$$\sigma_2 = 66.31\,\text{MPa}, \qquad \sigma_1 = -386.21\,\text{MPa},$$
$$\tau = \sigma_1/2.4142 = -159.98\,\text{MPa},$$
$$\sigma = -319.95\,\text{MPa}.$$

This is also the solution for $R = 2$ with reversal in the direction of τ. Clearly, the material is stronger under the compressive direct stress. Figure 16.11 graphically interprets these stress values as the respective intersections (points P and Q) between the stress path and the strength locus in quadrants two and four.

EXERCISES

Yield Criteria in Principal Stress Form

16.1 At a point in a material the major principal stress is 194 MPa. If the tensile yield stress is 253 MPa, find the minor principal stress that would cause yielding according to the total strain energy ($v = 0.3$), the Tresca and von Mises yield criteria.

Answer: 113.8; 58.67; 91.87 (MPa).

16.2 Given $\sigma_1 = 185\,\text{MPa}$ and $\sigma_3 = 0$, determine the magnitude of the compressive σ_2 that would cause yielding according to the Tresca and von Mises yield criteria. Take $Y = 260\,\text{MPa}$.

16.3 The tensile yield stress of a material is 232 MPa. If the principal biaxial stresses set up in service are 123.5 MPa in tension and 38.5 MPa in compression, calculate the operating safety factors from the Tresca and von Mises yield criteria.

Answer: 1.43; 1.58.

16.4 A steel cube that is subjected to a constant uniaxial tensile stress is lowered to increasing depth in the sea. If the tensile yield stress is 350 MPa and the density of sea water is 1020 kg/m^3, determine the value of the tensile stress at which yielding commences.

Answer: 350 MPa.

16.5 A thin-walled cylinder of diameter d and thickness t is subjected to an internal pressure p. Derive expressions for the pressure required to initiate yielding according to the von Mises and Tresca yield criteria (i) when the radial stress is neglected and (ii) when the magnitude of the radial stress is $-p$. The tensile yield stress for the cylinder material is Y. Compare the yield pressures when $d/t = 25$ and $Y = 300$ MPa.

Answer: (i) $4Yt/\sqrt{3}d$, $2Yt/d$; (ii) $\sqrt{[2Y^2/(3d^2/8t^2 + 3d/2t + 2)]}$, $Y/(1 + d/2t)$; 27.7, 24, 25.63 and 22.2 (N/mm^2).

16.6 A thick-walled cylinder of 25 mm i.d. and 50 mm o.d. is subjected to a constant axial tensile load of 20 kN together with an internal pressure p. If the ends of the cylinder are closed and the external pressure is atmospheric, calculate the value of p that would initiate yielding at the bore according to the Tresca and von Mises yield criteria. Take $Y = 240$ MPa. (T.C.D. 1981)

Answer: 103.8 MPa, 90 MPa.

16.7 A thick-walled cylinder of 100 mm i.d. simultaneously withstands an internal pressure of 10 MPa and a bending moment of 3.5 kN m. Using a factor of safety of 3 for the most highly stressed point in the section, determine the cylinder o.d. based upon the von Mises yield criterion.

16.8 A thick-walled cylinder of 100 mm i.d. and 200 mm o.d. is subjected to an internal/external pressure ratio of 5:1. Calculate the pressure that would initiate yielding in the bore according to the Tresca yield criterion given a tensile yield stress of 250 MPa.

Answer: 117.2 MPa.

16.9 The 2 m internal diameter of a thick cylinder is maintained at atmospheric pressure while the exterior surfaces are subjected to a hydrostatic pressure of 4.5 MPa. Given that the tensile yield stress for the cylinder material is 250 MPa, determine the wall thickness according to the Tresca yield criterion with a safety factor of 3.

16.10 A high-pressure steel pipeline weighing 300 N/m is simply supported at 5 m intervals. If the inner and outer diameters are 30 and 50 mm respectively, estimate the internal pressure and point in bore where yielding begins. Assume a von Mises yield criterion and that the cylinder has open ends over the support length. Take $Y = 250$ MPa. (T.C.D. 1980)

Answer: 84.68 MPa; top of i.d.

16.11 A uniformly thick brass cylinder has an outer radius of 125 mm and an internal radius of 75 mm. It has rigid end closures and is subjected to an internal pressure only. Using the Tresca and Maxwell–von Mises yield criteria with a tensile yield stress of 450 MPa, determine the pressure at which yielding occurs. Assume that the axial stress is uniform. (C.E.I. Pt. 2, 1985)

16.12 A long, thick cylinder of outside diameter $2a$ and inside diameter $2b$ is subjected to an internal pressure p. Determine the maximum pressure permitted by the Tresca criterion if the cylinder is not to deform plastically. Calculate the upper and lower bound values for the longitudinal stress for one closed end containing a pressurized fluid that causes a projectile to be ejected from the open end. Under what conditions will the longitudinal stress affect the onset of yielding? (C.E.I. Pt. 2, 1975)

16.13 A steel rotor disc of uniform thickness 50 mm has an outer rim diameter of 750 mm and a central hole of diameter 150 mm. There are 200 blades each of weight 2.25 N at an effective radius of 425 mm pitched evenly around the periphery. Determine the rotational speed at which yielding first occurs according to the Tresca yield criterion given that the yield stress in simple tension is 690 MPa. Take $E = 207$ GPa, $\rho = 7840$ kg/m^3 and $\nu = 0.29$. (C.E.I. Pt. 2, 1970)

16.14 A steel tube has an internal diameter of 25 mm and an outside diameter of 50 mm. When a second steel tube is shrunk over the outer diameter of the first tube, a condition of yield exists at the inner diameter of each tube. Determine the required interference before shrinkage and the outer diameter of the second tube assuming a Tresca yield criterion. Ignore axial stress. Take $Y = 413.5$ MPa and $E = 207$ GPa.

Answer: 0.13 mm; 100 mm.

16.15 A collar of 200 mm inner diameter slides freely on a hollow column of 160 mm i.d. and 200 mm o.d. When an internal pressure p is introduced into the column to create an interface pressure of 9 MPa, the collar locks and is able to support an axial load of 200 kN. Determine p and check from the Tresca yield criterion that the stress states at each i.d. are elastic within a safety factor of 3 on a tensile yield stress of 300 MPa.

16.16 Show that the von Mises criterion may be written in principal deviatoric stress form as; $(\sigma_1^D)^2 + (\sigma_2^D)^2 + (\sigma_3^D)^2 = 2Y^2/3$.

16.17 Show that the two non-zero (positive) deviatoric stress invariants can be expressed in the general tensor notation form $J_2^D = (1/2)\sigma_{ij}^D\sigma_{ij}^D$ and $J_3^D = (1/3)\sigma_{ij}^D\sigma_{jk}^D\sigma_{ki}^D$.

Yield Criteria in General Biaxial Stress Form

16.18 A specimen of steel has a yield stress of 330 MPa. A 50 mm diameter shaft made from this material has a axial torque of 2.5 kN m applied to it. Assuming that the material yields according to (i) the shear strain energy criterion and (ii) the maximum shear stress criterion, calculate the value of the additional axial tensile load that would cause the shaft material to pass the yield point.

Answer: 547 kN; 508 kN.

16.19 A screwdriver has a steel shank with a tensile yield stress of 300 MPa. Using a safety factor of 10, calculate the minimum shank diameter for a maximum expected torque of 3 N m, based upon the Tresca and von Mises criteria of elastic strength.

Answer: 10 mm; 9.6 mm.

16.20 Find an expression for the stress at the yield point of a solid circular shaft carrying a uniform bending moment M. Find expressions for the same shaft under the action of a torque T according to the five theories of yielding for ductile materials. What are the numerical ratios of M/T in each case?

Answer: 0.5; 0.65; 0.806; 1; 0.865.

16.21 A composite tube of 100 mm external diameter and 50 mm internal diameter, 250 mm long, consists of a brass tube that just fits inside a steel tube. The interface diameter is 75 mm and the tubes are bonded to each other at the interface. Samples of brass and steel tested separately are found to yield in uniaxial tension at 150 MPa and 200 MPa respectively. If yielding is to be avoided in both materials, determine (i) the maximum axial tensile load and (ii) the maximum torque that may be applied to the composite tube. Assume, where necessary, a Tresca yield criterion and take the moduli E for brass and steel as 100 GPa and 200 GPa respectively. (C.E.I. Pt. 2, 1982)

16.22 A solid shaft of 75 mm diameter is subjected to an axial force of 80 kN and a torque of 3.8 kN m. If the yield stress for the shaft material is 278 MPa, calculate the safety factors that have been used in the design according to the Tresca and von Mises yield criteria.

Answer: 3.13; 3.59.

16.23 A 30 mm diameter rod is subjected to a bearing pressure of 120 MPa around its outer surface. Examine the additional loading from torsion and from bending that would produce yielding in the shaft. Take $Y = 310$ MPa.

16.24 A steel shaft carries a bending moment of 2 kN m and a torque of 1.5 kN m. If the tensile yield stress for the shaft material is 280 MPa, determine a suitable shaft diameter based upon a safety factor of 2 and each of the following criteria of elastic strength: (i) maximum shear stress; (ii) total strain energy ($v = 0.25$); (iii) shear strain energy.

16.25 A mild steel tube has a mean diameter of 100 mm and a thickness of 3 mm. Using a safety factor of 2 and a tensile yield stress of 230 MPa, calculate the torque that can be transmitted by this tube if the criterion of failure is (i) the maximum shear stress and (ii) the shear strain energy.
Answer: 2.5 kN m; 2.9 kN m.

16.26 A thin-walled closed tube with internal diameter d and thickness t reaches its yield point under an internal pressure p_y. Working in terms of p_y, determine the torque that would yield the cylinder when (i) p_y is removed and (ii) $p_y/2$ remains.
Answer: $\pi p_y d^3/8$; $\sqrt{3} \pi p_y d^3/16$.

16.27 A 20 mm diameter steel shaft is loaded with an axial torque of 60 N m and an axial tensile force of 15 kN. Given that the tensile yield stress is 300 MPa, determine the safety factors employed with the Tresca and von Mises yield criteria.
Answer: 3.33; 2.61.

16.28 A 32 mm diameter shaft is to transmit 100 kW at 3000 rev/min and simultaneously carry a bending moment of 300 N m. Determine which of the five theories of yielding predicts that the shaft has exceeded its yield point when the tensile yield stress is 365 MPa.
Answer: Tresca and both energy theories.

16.29 A thin circular pipe withstands an axial torque of 1 kN m and an internal pressure of 25 bar. Calculate the thickness according to the Tresca and von Mises yield criteria using a safety factor of 2.5. The tensile yield stress of the pipe material is 200 MPa.
Answer: 2.7 mm; 2 mm.

16.30 A stirrer rotates at 600 rev/min under a thrust of 4 kN inside a chemical vessel where the internal pressure is 120 MPa. Given that the power absorbed by the stirrer is 4 kW, calculate its diameter using a safety factor of 3 and the von Mises yield criterion. The yield stress of the stirrer material is 180 MPa.
Answer: 19 mm.

16.31 A thin-walled 120 mm square tube with 5 mm wall thickness simultaneously withstands a torque of 4.5 kN m and an axial force of 40 kN. Given that the tensile yield stress of the tube material is 310 MPa, find the safety factor that has been employed in the design according to the von Mises yield criterion.

16.32 A box beam section 150 mm wide and 200 mm deep, with respective thicknesses of 20 mm and 15 mm, is riveted lengthwise along one vertical side 60 mm above the neutral axis. Given that the maximum shear force and bending moment in the beam are 200 kN and 65 kN m respectively, find the design safety factors according to the application of the maximum principal strain and total strain energy theories. Take $Y = 300$ MPa, $v = 0.25$ for the beam material.
Answer: 3.34; 3.37.

16.33 State the Tresca and von Mises yield criteria. A particular steel is found to yield at 300 MPa in tension. Determine, for 25 mm diameter samples of the same steel, (i) the torque required to induce yielding according to the Tresca criterion, (ii) the maximum bending moment to initiate yield by both criteria, and (iii) the maximum bending moment required to induce yielding by the von Mises yield criterion, when the cylindrical surface only is subjected to a pressure of 100 MPa. (C.E.I. Pt. 2, 1983)

16.34 A 5 mm diameter rod that exhibits elastic-perfectly plastic behaviour is found to yield in pure tension under a force of 3.5 kN. Determine from the Tresca and von Mises yield criteria the torque required to (i) yield the outer fibres and (ii) produce a fully plastic bar. (C.E.I. Pt. 2, 1981)

16.35 A solid circular section shaft of 125 mm diameter rotates at 30 rad/s. Owing to the configuration of the bearings and the gearing, it is subjected to a maximum bending moment of 9 kN m. If the elastic limit stress in simple tension is 300 MPa, calculate the power which the

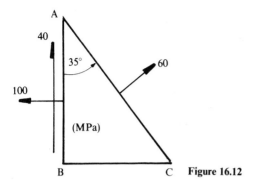

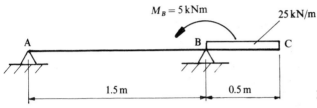

Figure 16.12

Figure 16.13

shaft may transmit using the Tresca yield criterion. When the shaft is transmitting this power, check the safety of the system using the von Mises criterion. (C.E.I. Pt. 2, 1976)

16.36 An element in a loaded structure is subjected to the stress state shown in Fig. 16.12. Determine the complete state of stress of planes AC and BC and find the safety factor that has been used against yielding according to the von Mises yield criterion. Take $Y = 300$ MPa.

16.37 Calculate the safety factors that have been used against yielding at the web–flange interface for the beam in Fig. 16.13 with I-section dimensions of width 50 mm, depth 100 mm, and thickness 5 mm throughout. Take $Y = 250$ MPa.

16.38 The shaft in Fig. 16.14 rotates in a plain bearing at its fixed end and carries a pully belt drive with differing tensions at its free end. Determine a suitable shaft diameter d using a safety

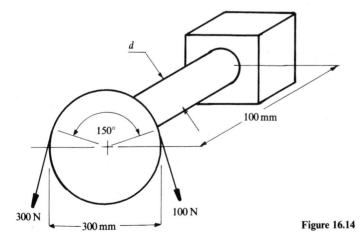

Figure 16.14

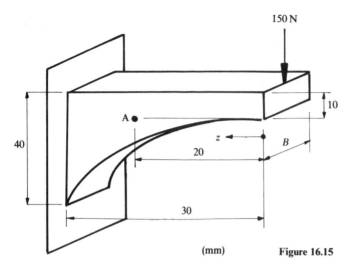

Figure 16.15

factor of 8 with a tensile yield stress of 300 MPa based upon the Tresca and von Mises criteria of elastic strength. (K.P. 1977)

16.39 Given that the depth of the cantilever in Fig. 16.15 varies with length according to $10 + kz^2$, determine the shear and bending stresses at the hole A in terms of the width B. What is the value of B according to the maximum principal stress and total strain energy yield criteria when a safety factor of 5 is used with a yield stress of 100 MPa and $v = 0.3$?

Answer: $\sigma = 4.72/B$, $\tau = 9.45/B$; 0.6 mm; 0.8 mm. (K.P. 1977)

16.40 What are the sectional dimensions of the cheapest tube in Fig. 16.16 to carry a torque of 50 N m given that its cost c (pence/metre length) is $100c = A(60 + y/t)$, where A is the section area in millimetre-squared, t is the thickness and y is the mean side length? The design is to be based upon the total strain energy criterion with $v = 0.3$ and a tensile yield stress of 150 MPa.

Answer: $t = 1.21$ mm; $y = 36.4$ mm. (K.P. 1977)

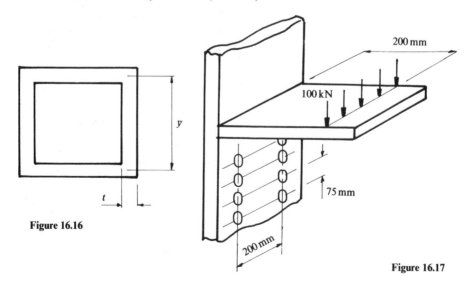

Figure 16.16

Figure 16.17

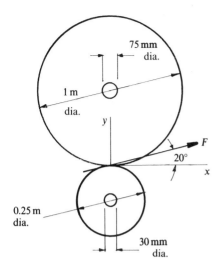

75 mm
dia.

1 m
dia.

y

F

20°

x

0.25 m
dia.

30 mm
dia.

Figure 16.18

16.41 Are 22 mm diameter bolts suitable for the bracket assembly in Fig. 16.17, given allowable stresses values of 120 and 80 MPa in tension and shear?

16.42 A gear wheel 1 m diameter is mounted on a short stub axle 75 mm diameter as shown in Fig. 16.18. It is driven by a small 0.25 m diameter pinion wheel through an axle of 30 mm diameter. The gears mesh on a vertical axis and the line of force between the two gears makes an angle of 20° to a horizontal axis in the plane x, y through the contact point. The gear and pinion axles are constructed from the same material with overhanging cantilever lengths of 200 mm and 100 mm respectively. Determine the maximum power that the unit can transmit when the larger gear wheel rotates at 300 rev/min without yielding of the axles. A sample of the axle material yields at 700 MPa in tension. Calculate the maximum permissible power according to the Tresca and von Mises criteria. (C.E.I. Pt. 2, 1980)

16.43 The belt drive system in Fig. 16.19 transmits power from 300 mm diameter pulleys A to B through a 100 mm diameter shaft. Belts from A are arranged horizontally while those from B are vertical each with tension forces $2P$ and P on their tight and slack sides. The shaft is simply supported in spherical bearings at C and D. Determine the position and magnitude of

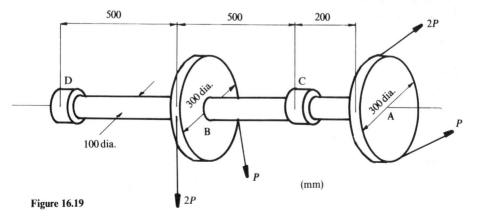

500

500

200

2P

D

300 dia.

C

300 dia.

B

A

100 dia.

P

P

(mm)

Figure 16.19

2P

Figure 16.20

the maximum bending moment on the shaft and also the position and magnitude of the maximum principal stress in the shaft, when transmitting 75 kW at a shaft speed of 500 rev/min. If the yield stress of the shaft material in uniaxial tension is 200 MPa, determine the maximum power that the pulley system can transmit at 500 rev/min without the shaft yielding, assuming a Tresca yield criterion. (C.E.I. Pt. 2, 1977)

16.44 The cross-section of the aluminium alloy composite beam in Fig. 16.20 is constructed by riveting together three similar plates with four identical equal-angle sections with the following local properties of their areas:

Plates: 125×5 mm, $\bar{x} = 62.5$ mm, $\bar{y} = 2.5$ mm, $I_x = 1302$ mm^4, $I_y = 813\,000$ mm^4.
Angles: $50 \times 50 \times 5$ mm, $\bar{x} = \bar{y} = 14.3$ mm, $I_x = I_y = 113\,000$ mm^4.

If the 3 m long, simply supported beam is to carry a central concentrated load W, determine the maximum permissible load W when the rivet holes give rise to an elastic stress concentration (magnification) of 3 and the maximum vertical deflection is not to exceed 2 mm. The Tresca yield criterion is applicable for a maximum uniaxial working stress of 70 MPa. Take $E = 80$ GPa. (C.E.I. Pt. 2, 1979)

Failure of Brittle Materials

16.45 If the maximum principal stress is to be limited to one-third of the 370 MPa tensile strength in cast iron, what is the minor principal stress by Coulomb–Mohr, given a compressive strength of 1390 MPa?
Answer: −926 MPa.

16.46 If the maximum shear stress in a cast iron component is not to exceed 385 MPa, find from the Coulomb–Mohr strength criterion the magnitudes of the major and minor principal stresses given that the ultimate strengths are 1390 and 370 MPa in compression and tension respectively.
Answer: 224.3 MPa; −545.7 MPa.

16.47 A 6 mm diameter cast iron pin is subjected to a torque of 9.8 N m combined with an axial compressive force of 3.5 kN. Compare the safety factors according to the three theories of brittle rupture when the ultimate tensile and compressive strengths are 293 and 965 MPa.
Answer: Coulomb–Mohr, 1.05; Mohr, 1.35; maximum principal stress, 1.6.

SEVENTEEN

INELASTIC DEFORMATION

17.1 INTRODUCTION

The broad title for this chapter actually covers all aspects of deformation behaviour in both metallic and non-metallic materials that is not elastic. Many specialist texts have been devoted to one or other aspect of inelasticity. In this introduction to the subject the study is restricted to the mechanics of plasticity, creep and viscoelasticity. The theory of plasticity is considerably simplified for an elastic-perfectly plastic metallic material, though this idealization will only approximate to real materials that display work hardening. It is shown how uniaxial hardening is represented and incorporated into a flow rule for multiaxial stress states. In creep, the inelastic strain is a function of stress, time and temperature. Particular functional forms are reviewed and applied to metallic structures. Because polymeric and other organic materials behave in a viscoelastic manner, they are normally treated separately. The traditional methods for modelling their simultaneous elastic and time–dependent responses to stress are considered.

17.2 PLASTICITY WITH HARDENING

Simple Tension and Compression

When metallic materials harden with the spread of plasticity, some account needs to be made of the associated increase in flow stress. It is therefore instructive to examine first the hardening behaviour in the stress–strain curve under uniaxial tension or compression. Figure 3.1 is typical of tension test

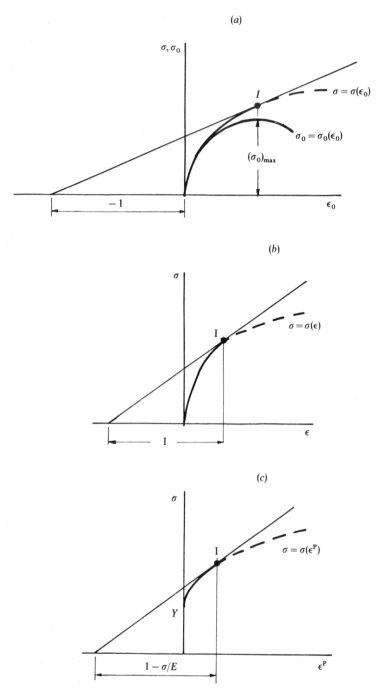

Figure 17.1 Hardening in nominal and true stress–strain axes.

results for ferrous and non-ferrous materials when the engineering stress and strain are conventionally defined from Eqs (3.1) and (3.2). Since these expressions refer to the original test-piece cross-sectional area and length, errors will arise in stress and strain calculations with the large changes in test-piece dimensions that arise within the plastic range. For example, the ultimate tensile strength for a material is not a true stress value, since no account is made of the reduction in cross-sectional area up to the start of necking. Moreover, the stress in a material will actually continue to rise, not fall, as the neck develops before final fracture. It is possible to correct the conventional or nominal tensile stress–strain curve to reflect this behaviour with the alternative true definitions for σ and ϵ:

$$\sigma = W/A, \qquad (17.1)$$

$$\epsilon = \int \mathrm{d}l/l, \qquad (17.2)$$

where A and l are the current area and length. Unlike elastic behaviour, the plastic deformation that is produced by stressing beyond the elastic limit occurs without change in volume. Thus, if A_o and l_o are the original area and length, the nominal engineering stress and strain are;

$$\sigma_o = W/A_o \quad \text{and} \quad \epsilon_o = (l - l_o)/l_o.$$

The incompressibility or constant volume condition gives

$$Al = A_o l_o, \qquad (17.3a)$$

$$\therefore \quad A = A_o l_o/l = A_o/(1 + \epsilon_o). \qquad (17.3b)$$

Substituting Eq. (17.3b) into Eqs (17.1) leads to the true stress:

$$\sigma = (W/A_o)(1 + \epsilon_o) = \sigma_o(1 + \epsilon_o). \qquad (17.4)$$

Integrating Eq. (17.2) provides a measure of the true or logarithmic total strain between the original and current limits of length:

$$\epsilon = \ln(l/l_o) = \ln(1 + \epsilon_o). \qquad (17.5a)$$

The true plastic component of strain ϵ^P is found from subtracting the elastic component, $\epsilon^E = \sigma/E$, from Eq. (17.5a):

$$\epsilon^P = \epsilon - \epsilon^E,$$

$$\epsilon^P = \ln(1 + \epsilon_o) - (\sigma_o/E)(1 + \epsilon_o). \qquad (17.5b)$$

The corresponding corrections afforded to a nominal curve $\sigma_o = \sigma_o(\epsilon_o)$, by Eqs (17.4) and (17.5) are shown in Figs. 17.1(a) and (b).

In Considere's construction (1885)* in Fig. 17.1(a), the true stress at the

---

* References are appended to this chapter on page 777.

point of instability (I) corresponds to the maximum ordinate for σ_o. This condition is expressed in

$$d\sigma_o/d\epsilon_o = 0.$$

Substituting from Eq. (17.4) and differentiating the quotient:

$$d\sigma_o/d\epsilon_o = (1 + \epsilon_o)(d\sigma/d\epsilon_o) - \sigma = 0,$$
$$\therefore \quad d\sigma/d\epsilon_o = \sigma/(1 + \epsilon_o). \tag{17.6a}$$

This shows that the tangent to the true stress curve $\sigma = \sigma(\epsilon_o)$ intersects the engineering strain axis at $\epsilon_o = -1$ (Fig. 17.1(a)). From Eq. (17.5a):

$$d\epsilon/d\epsilon_o = 1/(1 + \epsilon_o).$$

Hence this instability condition, w.r.t. $\sigma = \sigma(\epsilon)$, is given by

$$d\sigma/d\epsilon = (d\epsilon_o/d\epsilon) \times (d\sigma/d\epsilon_o)$$
$$= (1 + \epsilon_o)\sigma/(1 + \epsilon_o),$$
$$\therefore \quad d\sigma/d\epsilon = \sigma, \tag{17.6b}$$

for which a subtangent value of unity appears on a true total strain axis in Fig. 17.1(b). Note that if Eq. (17.5b) is used instead of Eq. (17.5a) in this derivation, then instability on axes of true stress against true plastic strain is expressed in

$$d\sigma/d\epsilon^P = \sigma/(1 - \sigma/E), \tag{17.6c}$$

where here the curve $\sigma = \sigma(\epsilon^P)$ in Fig. 17.1(c) commences at the yield stress Y, not at the origin. Since there is no appreciable difference between Eqs (17.6b and c), a unit subtangent is often employed with this construction. A number of empirical forms of $\sigma = \sigma(\epsilon)$ are available. The simplest of these, is the Hollomon (1945) power function:

$$\sigma = A\epsilon^n,$$

where A is the strength coefficient and $0.1 \leq n \leq 0.55$ is the strain-hardening exponent. Since this equation represents a curve passing through the origin, it will normally apply to a particular range of strain in the plastic region. If it is applied to represent the curve $\sigma = \sigma(\epsilon^P)$ in Fig. 17.1(c), then the greatest errors will arise for low plastic strains $\epsilon < 0.2$ per cent. In order to avoid this, the Ludwik law (1909) represents curves that commence at Y:

$$\sigma = Y + A'(\epsilon^P)^{n'}$$

where A' and $n' < 1$ are similarly named empirical constants.

Example 17.1 Determine the true stress and true strain at the point of tensile instability for the Hollomon function $\sigma = A\epsilon^n$.

$$d\sigma/d\epsilon = (nA)\epsilon^{n-1},$$

where, from Eq. (17.6b), at the point of instability,

$$(nA)\epsilon^{n-1} = \sigma = A\epsilon^n,$$

$$\therefore \quad \epsilon = n \quad \text{and} \quad \sigma = An^n.$$

That is, the strain at which necking begins equals the exponent value for a given material.

Example 17.2 At the point of tensile plastic instability, the engineering stress and strain are 340 MPa (22 tonf/in^2) and 30 per cent respectively. Determine the constants A and n in the Hollomon true stress–true strain law $\sigma = A\epsilon^n$.

IN SI UNITS Substituting $\sigma_o = 340$ MPa and $\epsilon_o = 0.3$ in Eqs (17.4) and (17.5a):

$$\sigma = \sigma_o(1 + \epsilon_o) = 340(1 + 0.3) = 442 \text{ MPa},$$

$$\epsilon = \ln(1 + \epsilon_o) = \ln(1 + 0.3) = 0.2624.$$

Now from Example 17.1, as $n = \epsilon = 0.2624$ the Hollomon law gives

$$442 = A(0.2624)^{0.2624}$$

$$\therefore \quad A = 627.9.$$

IN IMPERIAL UNITS

$$\sigma = \sigma_o(1 + \epsilon_o) = 22(1 + 0.3) = 28.6 \text{ tonf/in}^2.$$

Again, $n = \epsilon = 0.2624$ and the Hollomon law gives

$$28.6 = A(0.2624)^{0.2624}$$

$$\therefore \quad A = 40.63.$$

Tensile necking The use of true stress and strain is adequate to describe the uniform state of plastic deformation in a tensile test-piece up to the start of necking. However, during neck formation in ductile materials, the stress state is neither uniform nor uniaxial within this region. The state of stress in the neck was analysed by Bridgeman (1952), who showed that radial (σ_r) and tangential (σ_θ) stresses are induced under the applied axial stress (σ_z). With the smallest neck radius of a and a radius of curvature R (Fig. 17.2), he proposed the following triaxial stress state for radius $0 \le r \le a$ within the neck:

$$\sigma_r = \sigma_\theta = \bar{\sigma} \ln[(a^2 + 2aR - r^2)/2aR], \qquad (17.7a)$$

$$\sigma_z = \bar{\sigma}[1 + \ln(a^2 + 2aR - r^2)/2aR], \qquad (17.7b)$$

in which $\bar{\sigma}$ is the true equivalent stress for the neck. The distribution of σ_r and σ_z is shown in Fig. 17.2. For an applied axial force W, the mean axial

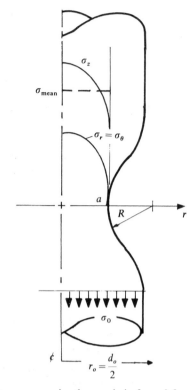

Figure 17.2 Tensile neck.

stress σ_{mean} in the neck is found from

$$W = \pi a^2 \sigma_{\text{mean}} = \int_0^a 2\pi r \sigma_z \, dr.$$

Substituting from Eq. (17.7b) gives

$$\pi a^2 \sigma_{\text{mean}} = 2\pi \bar{\sigma} \int_0^a \{r + r \ln [(a^2 + 2aR - r^2)/2aR]\} \, dr$$
$$= 2\pi \bar{\sigma} \{(aR + a^2/2)[\ln (a^2 + 2aR) - \ln 2aR]\}$$
$$= \pi \bar{\sigma} [(a^2 + 2aR) \ln (1 + a/2R)],$$
$$\therefore \quad \bar{\sigma}/\sigma_{\text{mean}} = [(1 + 2R/a) \ln (1 + a/2R)]^{-1} \qquad (17.8a)$$
$$= [(1 + 1/X) \ln (1 + X)]^{-1}, \qquad (17.8b)$$

where $X = a/2R$ in Eq. (17.8b) and a/r_o was plotted against $\bar{\sigma}/\sigma_{\text{mean}}$ in the form of Bridgeman correction curves for particular materials. Hence, if the smallest diameter of the neck $(2a)$ is measured at intervals during loading (W) beyond the point of instability, the true equivalent stress at those intervals may be calculated. The true strain can also be calculated from the neck diameter $d = 2a$ using the following constant-volume condition from

Eq. (17.3a):

$$(\pi/4)d_o^2 l_o = (\pi/4)d^2 l$$

when, from Eq. (17.2),

$$\epsilon = \ln(l/l_o) = 2\ln(d_o/d). \tag{17.9}$$

The broken lines in Fig. 17.1 shows the corresponding extension to the true stress–strain curves in the presence of necking.

Example 17.3 The minimum radius of a tensile neck is 5.613 mm under an applied force of 59.63 kN. Given that $R = 70.16$ mm for a test-piece of original diameter 12.5 mm, determine the true stress in the neck.

Here $a/2R = 5.613/2 \times 70.16 = 0.04$. Then from Eq. (17.8a):

$$\bar{\sigma}/\sigma_{\text{mean}} = [(1 + 25)\ln(1 + 0.04)]^{-1} = 0.9806,$$

where

$$\sigma_{\text{mean}} = 59.63 \times 10^3/\pi(5.613)^2 = 602.45 \text{ MPa},$$

$$\therefore \quad \bar{\sigma} = 0.9806 \times 602.45 = 590.78 \text{ MPa}.$$

One of the problems with tensile testing is that the range of uniform strain is limited by the formation of neck. Larger strain can be achieved from the compression test on a short cylinder, but unless the ends are well lubricated, non-uniform deformation is manifested with barrelling of the testpiece central region. Here the true strain may be calculated from either expression in Eq. (17.9), depending upon whether the current height (l) or diameter (d) of the cylinder is measured. For perfect lubrication of an isotropic material, the true stress–strain curve for compression will coincide with that for tension.

Torsion

When a cylindrical bar of material hardens with the plastic deformation produced from torsion, the distribution of shear stress through the section (Fig. 9.29) must also be considered. The following Nadai method (1950) allows the shear stress–strain curve (τ_o vs. γ_o) for the outer diameter to be derived from the torque–unit twist curve (T vs. κ in Fig. 17.3). A solid bar, outer radius r_o is in equilibrium with the applied torque T when;

$$T = 2\pi \int_0^{r_o} \tau r^2 \, dr.$$

Substituting from Eq. (9.3), $\gamma = r\kappa$ ($\kappa = \theta/L$ is the twist/unit length):

$$T = (2\pi/\kappa^3) \int_0^{\gamma_o} \tau(\gamma)\gamma^2 \, d\gamma,$$

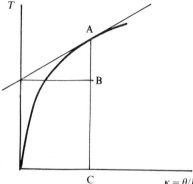

$\kappa = \theta/L$ **Figure 17.3** Nadai's construction.

in which the function $\tau = \tau(\gamma)$ applies to radius r. Differentiating w.r.t. γ_o gives

$$(1/2\pi)\mathrm{d}(T\kappa^3)/\mathrm{d}\gamma_o = \tau_o(r_o\kappa)^2,$$

where $\tau_o = \tau_o(r_o\kappa)$. Since $\mathrm{d}\gamma_o = r_o\,\mathrm{d}\kappa$,

$$(1/2\pi)\mathrm{d}(T\kappa^3)/\mathrm{d}\kappa = \tau_o r_o^3 \kappa^2,$$
$$\therefore \quad \tau_o = (1/2\pi r_o^3 \kappa^2)\mathrm{d}(T\kappa^3)/\mathrm{d}\kappa,$$
$$\tau_o = (1/2\pi r_o^3)(3T + \kappa\,\mathrm{d}T/\mathrm{d}\kappa). \tag{17.10a}$$

The geometrical interpretation of Eq. (17.10a), as given by the Nadai construction in Fig. 17.3, leads to

$$\tau_o = (1/2\pi r_o^3)(3\mathrm{AC} + \mathrm{AB}). \tag{17.10b}$$

The Nadai method usefully reveals the true shear stress–strain behaviour to fracture for a hardening material despite the presence of a stress gradient. Torsion tests conducted on thin-walled tubes avoid a stress gradient but these are prone to buckling well before fracture, so limiting the range of strain.

Example 17.4 If the torque–unit twist diagram for a solid bar can be expressed as $T = B\kappa + A\kappa^n$, derive the corresponding shear stress–strain relationship for the outer radius r_o. Hence determine the torque and shear stress required to produce $\gamma_o = 2$ per cent in a 20 mm diameter alloy bar for which $n = 1/3$, $A = 10^6\,\mathrm{N\,mm}^{4/3}$ and $B = 25 \times 10^6\,\mathrm{N\,mm}^2$.

The linear term in the T expression represents the elastic line. With $n < 1$, the parabolic term dominates in the plastic range. The slope of the T–κ curve is $\mathrm{d}T/\mathrm{d}\kappa = B + nA\kappa^{n-1}$. Substituting into Eq. (17.10a):

$$\tau_o = (1/2\pi r_o^3)[3(B\kappa + A\kappa^n) + \kappa(B + nA\kappa^{n-1})]$$
$$= (1/2\pi r_o^3)[A\kappa^n(3 + n) + 4B\kappa].$$

Putting $\kappa = \gamma_o/r_o$ for the outer radius gives

$$\tau_o = (2B/\pi r_o^4)[\gamma_o + (3 + n)(Ar_o/4B)(\gamma_o/r_o)^n]$$
$$= (B/J)\{\gamma_o + [A(3 + n)r_o^{1-n}/4B]\gamma_o^n\}, \qquad \text{(i)}$$

where $J = \pi r_o^4/2$. For $\gamma_o = 0.02$, $\kappa = 0.02/10 = 0.002$ and the torque is

$$T = (25 \times 10^6 \times 0.002) + 10^6(0.002)^{1/3} = (0.05 + 0.126) \times 10^6$$
$$= 0.176 \times 10^6 \, \text{N mm} = 176 \, \text{N m}.$$

Now $J = \pi(10)^4/2 = 1.571 \times 10^4 \, \text{mm}^4$ and Eq. (i) gives

$$\tau_o = (25 \times 10^2/1.571)\{0.02 + [3.333 \times 10^6 \times 10^{2/3}/(4 \times 25 \times 10^6)](0.02)^{1/3}\}$$
$$= 1591.3(0.02 + 0.042) = 98.66 \, \text{MPa}.$$

Equivalent Stress and Strain

In the analysis of the triaxial stress state in the neck of tensile test-piece, the concept was introduced of a stress equivalent to a true uniaxial stress. In the von Mises criterion (Eq. (16.9a)), yielding under σ_1, σ_2 and σ_3 commences in proportion to the root mean square of these stresses. For a hardening material, this r.m.s., equivalent or effective, value continues to increase beyond the initial yield stress Y within the plastic range. The magnitude of the equivalent uniaxial stress $\bar{\sigma}$ is found from replacing Y by $\bar{\sigma}$ in Eq. (16.9a). This gives

$$\bar{\sigma} = (1/\sqrt{2})\sqrt{[(\sigma_1 - \sigma_2)^2 + (\sigma_2 - \sigma_3)^2 + (\sigma_1 - \sigma_3)^2]}, \qquad (17.11)$$

which implies that the initial yield surface expands uniformly to contain the current stress point within the plastic range (isotropic hardening). Clearly, a uniaxial stress σ_1 becomes the equivalent stress ($\sigma_1 = \bar{\sigma}$) when two of the stresses in Eq. (17.11) are set to zero. A similar expression holds for the plastic strain under a triaxial stress state. Thus if $d\epsilon_1^P$, $d\epsilon_2^P$ and $d\epsilon_3^P$ are the incremental plastic strains following a change $d\sigma_1$, $d\sigma_2$ and $d\sigma_3$ to the stress state, the equivalent plastic strain increment is defined as

$$d\bar{\epsilon}^P = (\sqrt{2}/3)\sqrt{[(d\epsilon_1^P - d\epsilon_2^P)^2 + (d\epsilon_2^P - d\epsilon_3^P)^2 + (d\epsilon_1^P - d\epsilon_3^P)^2]} \quad (17.12)$$

in which the numerical factor $\sqrt{2}/3$ ensures that Eq. (17.12) correctly defines the equivalent strain under uniaxial stress. That is, if the sides of a unit cube permanently change in length by $d\epsilon_1^P$, $d\epsilon_2^P$ and $d\epsilon_3^P$ without changing its volume, then

$$(1 + d\epsilon_1^P)(1 + d\epsilon_2^P)(1 + d\epsilon_3^P) - 1 = 0.$$

Neglecting infinitesimal products, the constant-volume condition is

$$d\epsilon_1^P + d\epsilon_2^P + d\epsilon_3^P = 0. \qquad (17.13)$$

In the case of tension, say, in which the axial strain is $d\epsilon_1^P$, the lateral strains are found from Eq. (17.13) to be $d\epsilon_2^P = d\epsilon_3^P = -d\epsilon_1^P/2$. Substituting into Eq. (17.12) gives $d\bar{\epsilon}^P = d\epsilon_1^P$. Note from pages 555 and 571, that the octahedral shear stress and strain (τ_{oct}, γ_{oct}), which have also been used for equivalence, do not readily reduce to a simple stress state because the numerical factors differ, i.e. for similar substitutions under tension these are $\tau_{oct} = \sqrt{2}\sigma_1/3 = \sqrt{2}\bar{\sigma}/3$ and $d\gamma_{oct}^P = \sqrt{2}d\epsilon_1^P = \sqrt{2}d\bar{\epsilon}^P$. With the equivalent definitions in Eqs (17.11) and (17.12) the plot of $\bar{\sigma}$ vs. $\int d\bar{\epsilon}^P$ will correlate hardening behaviour under all stress states in the single curve $\bar{\sigma} = \bar{\sigma}(\int d\bar{\epsilon}^P) = \bar{\sigma}(\bar{\epsilon}^P)$. The latter is conveniently found from simple tests: tension and compression for a limited range of plastic strain or torsion and balanced biaxial tension (the bulge test) for a larger range of plastic strain. In practice, because materials are initially anisotropic to some degree, it is usually found that points from different tests lie within a narrow scatter band on $\bar{\sigma}$ vs. $\bar{\epsilon}^P$ axes.

The Flow Rule

Levy (1870) and von Mises (1913) proposed a flow rule of plasticity in which the incremental plastic strain tensor $d\epsilon_{ij}^P$ is linearly dependent upon the deviatoric stress tensor σ_{ij}^D in Eq. (13.24). The geometrical interpretation is that during loading the plastic strain increment vector lies in the position of the exterior, normal to a flow surface that expands isotropically in stress space to contain the current stress point. With the surface defined by the von Mises equivalent stress function (Eq. 17.11), the Levy–Mises flow rule is

$$d\epsilon_{ij}^P = d\lambda\sigma_{ij}^D, \qquad (17.14)$$

where $d\lambda$ is scalar factor of proportionality that changes with the progress of deformation under any given stress state. This is defined from simple tension $(\bar{\sigma}, \bar{\epsilon}^P)$ where, with a mean stress $\sigma_m = \bar{\sigma}/3$, Eqs (13.24) and (17.12) give $d\epsilon_1^P = d\bar{\epsilon}^P$, $\sigma_1^D = 2\bar{\sigma}/3$, $\sigma_2^D = \sigma_3^D = -\bar{\sigma}/3$. Hence, from Eq. (17.14),

$$d\lambda = 3d\bar{\epsilon}^P/2\bar{\sigma}, \qquad (17.15)$$

which is inversely proportional to the incremental plastic modulus $\bar{\sigma}/d\bar{\epsilon}^P$ For a principal triaxial stress system σ_1, σ_2 and σ_3, Eq. (17.14) expands to give, with Eq. (17.15), the principal plastic strain increments:

$$d\epsilon_1^P = (d\bar{\epsilon}^P/\bar{\sigma})[\sigma_1 - (\sigma_2 + \sigma_3)/2], \qquad (17.16a)$$
$$d\epsilon_2^P = (d\bar{\epsilon}^P/\bar{\sigma})[\sigma_2 - (\sigma_1 + \sigma_3)/2], \qquad (17.16b)$$
$$d\epsilon_3^P = (d\bar{\epsilon}^P/\bar{\sigma})[\sigma_3 - (\sigma_1 + \sigma_2)/2], \qquad (17.16c)$$

in which Eq. (17.13) holds. Often the analogy is made with the elastic constitutive Eqs (4.12), where $v = 1/2$ for incompressible plasticity and $1/E$ is replaced by the incremental compliance $d\bar{\epsilon}^P/\bar{\sigma}$. For non-linear hardening, where the uniaxial function is $\bar{\sigma} = \bar{\sigma}(\int d\bar{\epsilon}^P)$, the compliance is $d\bar{\epsilon}^P/\bar{\sigma} = d\bar{\sigma}/\bar{\sigma}\bar{\sigma}'$ where $\bar{\sigma}' = d\bar{\sigma}/d\bar{\epsilon}^P$. The R.H.S. of Eqs (17.16) can then be expressed in stress and integrated. Numerical methods are required to solve Eqs (17.11)–(17.16) except in the case of the simplest uniform biaxial and triaxial stress states, where the stress components increase proportionately or follow a stepped path during loading.

Example 17.5 A pressure-vessel steel has a stress–strain curve represented by $\bar{\sigma} = \sigma_o(1 + B\bar{\epsilon}^P)^n$. At what internal pressure would tensile plastic instability occur in (a) a thin-walled sphere and (b) a closed cylinder? Given $B = 200$ and $n = 0.1$, determine the equivalent and maximum strain reached at the point of instability in each case. (I.C. 1972)

(a) The hoop and meridional stresses are (Eq. 4.20) $\sigma_\theta = \sigma_\phi = pr/2t$:

$$\therefore \quad \ln \sigma_\theta = \ln p + \ln r - \ln t - \ln 2,$$
$$d\sigma_\theta/\sigma_\theta = dp/p + dr/r - dt/t.$$

Since $dp/p = 0$ at the point of instability and $d\epsilon_\theta^P = dr/r$, $d\epsilon_r^P = dt/t$, then

$$d\sigma_\theta/\sigma_\theta = d\epsilon_\theta^P - d\epsilon_r^P. \tag{i}$$

From Eq. (17.11), with $\sigma_1 = \sigma_\theta$, $\sigma_2 = \sigma_\phi = \sigma_\theta$ and $\sigma_3 = \sigma_r = 0$, the equivalent stress is

$$\bar{\sigma} = (1/\sqrt{2})\sqrt{[0 + \sigma_\theta^2 + \sigma_\theta^2]} = \sigma_\theta. \tag{ii}$$

Replacing subscripts 1, 2 and 3 by θ, ϕ and r in Eqs (17.13) and (17.14):

$$d\epsilon_\theta^P = (2d\lambda/3)\sigma_\theta/2,$$
$$d\epsilon_\phi^P = (2d\lambda/3)\sigma_\theta/2 = d\epsilon_\theta^P,$$
$$d\epsilon_r^P = -(d\epsilon_\theta^P + d\epsilon_\phi^P) = -2d\epsilon_\theta^P,$$
$$\therefore \quad d\epsilon_\theta^P = d\epsilon_\phi^P = -d\epsilon_r^P/2, \tag{iii}$$

when from Eq. (17.12) the equivalent plastic strain is

$$d\bar{\epsilon}^P = \sqrt{2/3}\sqrt{[(3d\epsilon_\theta^P)^2 + (3d\epsilon_\theta^P)^2]} = 2d\epsilon_\theta^P. \tag{iv}$$

From Eqs (ii) and (iv), $d\bar{\sigma}/d\bar{\epsilon}^P = d\sigma_\theta/2d\epsilon_\theta^P$. Then from Eqs (i) and (iii):

$$d\bar{\sigma}/d\bar{\epsilon}^P = \sigma_\theta(1 - d\epsilon_r^P/d\epsilon_\theta^P)/2 = 3\sigma_\theta/2,$$
$$d\bar{\sigma}/d\bar{\epsilon}^P = 3\bar{\sigma}/2.$$

Thus, in a similar plot to Fig. 17.1(b), the subtangent value is $2/3$. The equivalent plastic instability strain is found from

$$d\bar{\sigma}/d\bar{\epsilon}^P = Bn\bar{\sigma}/(1 + B\bar{\epsilon}^P) = 3\bar{\sigma}/2,$$
$$\therefore \quad \bar{\epsilon}^P = (1/B)(2Bn/3 - 1),$$
$$= (1/200)(2 \times 200 \times 0.1/3 - 1) = 0.062,$$
$$\therefore \quad \epsilon_\theta^P = \bar{\epsilon}^P/2 = 0.062/2 = 0.031.$$

The equivalent stress at instability is then $\bar{\sigma} = \sigma_o(2nB/3)^n$, which gives the critical pressure as

$$p = (2t\sigma_o/r)(2nB/3)^n.$$

(b) For a thin cylinder the stresses are $\sigma_r = 0$ and, from Eq. (4.17), $\sigma_\theta = pr/t = 2\sigma_z$ and $\sigma_r = 0$. Hence Eq. (i) again applies to the point of instability. The equivalent stress (Eq. (17.11)) is now

$$\bar{\sigma}_\theta = (1/\sqrt{2})\sqrt{[(\sigma_\theta - \sigma_z)^2 + \sigma_\theta^2 + \sigma_z^2]}$$
$$= (1/\sqrt{2})\sqrt{[(\sigma_\theta/2)^2 + \sigma_\theta^2 + (\sigma_\theta/2)^2]},$$
$$\bar{\sigma} = \sqrt{3}\sigma_\theta/2, \tag{v}$$

and, from Eqs (17.13) and (17.14), $d\epsilon_z^P = 0$ and

$$d\epsilon_\theta^P = (2d\lambda/3)[\sigma_\theta - (1/2)(0 + \sigma_\theta/2)] = (d\lambda/2)\sigma_\theta,$$
$$d\epsilon_r^P = (2d\lambda/3)[0 - (1/2)(\sigma_\theta + \sigma_\theta/2)] = -(d\lambda/2)\sigma_\theta,$$
$$\therefore \quad d\epsilon_r^P = -d\epsilon_\theta^P. \tag{vi}$$

The equivalent plastic strain is, from Eq. (17.12),

$$d\bar{\epsilon}^P = (\sqrt{2}/3)\sqrt{[(d\epsilon_\theta^P)^2 + (2d\epsilon_\theta^P)^2 + (d\epsilon_\theta^P)^2]} = (2/\sqrt{3})d\epsilon_\theta^P. \tag{vii}$$

Combining Eqs (v) and (vii):

$$d\bar{\sigma}/d\bar{\epsilon}^P = 3d\sigma_\theta/4d\epsilon_\theta^P,$$

where, from Eqs (i) and (vi),

$$d\bar{\sigma}/d\bar{\epsilon}^P = 3(d\epsilon_\theta^P - d\epsilon_r^P)\sigma_\theta/4d\epsilon_\theta^P = 3\sigma_\theta/2$$
$$\therefore \quad d\bar{\sigma}/d\bar{\epsilon}^P = \sqrt{3}\bar{\sigma},$$

i.e. a subtangent of $1/\sqrt{3}$ in this case. Hence, from the given law,

$$d\bar{\sigma}/d\bar{\epsilon}^P = Bn\bar{\sigma}/(1 + B\bar{\epsilon}^P) = \sqrt{3}\,\bar{\sigma},$$
$$\therefore \quad \bar{\epsilon}^P = (1/B)(Bn/\sqrt{3} - 1)$$
$$= (1/200)(200 \times 0.1/\sqrt{3} - 1) = 0.0527,$$
$$\therefore \quad \epsilon_\theta^P = (\sqrt{3}/2) \times 0.0527 = 0.0457.$$

The corresponding equivalent stress is $\bar{\sigma} = \sigma_o(Bn/\sqrt{3})^n$, from which the critical pressure is

$$p = (2t\sigma_o/\sqrt{3}r)(Bn/\sqrt{3})^n.$$

17.3 NON-HARDENING MULTIAXIAL PLASTICITY

In the case of a non-hardening material (elastic-perfectly plastic) where $\bar{\sigma} = Y = $ constant, Eqs (17.16) may be written in total plastic strain form. Normally, however, the non-hardening condition enables the plastic stresses to be found from combining the equilibrium condition with the yield criterion without recourse to the plastic strain behaviour. That is, the problem is statically determinate. Moreover, the yield criterion is only necessary when the stress state is complex. The non-hardening plastic behaviour of beams in bending and solid shafts under torsion, considered in Chapters 6 and 9, were solved from the application of equilibrium alone. The following examples illustrate some further applications of this simplified approach to determine the 'principal plastic stress' distributions in structures with axial symmetry. These include a thick-walled pressurized cylinder, with different end conditions, and a rotating disc.

Example 17.6 Derive expressions for the internal pressure required to penetrate a plastic zone to a radial depth c in a non-hardening thick-walled cylinder with inner and outer radii r_i and r_o and closed ends, assuming a von Mises yield criterion. Show graphically the distributions of radial, hoop and axial stress when $c = (r_i + r_o)/2$ and the corresponding residual stress distributions when the pressure is released.

Since the outer annulus $c \leq r \leq r_o$ is elastic, the radial and hoop stresses are supplied by Lamé (see page 613). The boundary conditions are, for the interface radius c, $\sigma_r = -p_c$ and for outer radius r_o, $\sigma_r = 0$. These give

$$\sigma_\theta = p_c c^2 (1 + r_o^2/r^2)/(r_o^2 - c^2), \qquad \text{(i)}$$

$$\sigma_r = p_c c^2 (1 - r_o^2/r^2)/(r_o^2 - c^2), \qquad \text{(ii)}$$

$$\sigma_z = p_c r_i^2/(r_o^2 - r_i^2) = (\sigma_\theta + \sigma_r)/2. \qquad \text{(iii)}$$

At the interface a state of yield exists. Substituting Eqs (i)–(iii) into Eq. (16.9a) with $r = c$ and θ, r and z replacing 1, 2 and 3 leads to

$$p_c = (Y/\sqrt{3})(1 - c^2/r_o^2)$$

and, therefore, from Eqs (i)–(iii) the stresses in the elastic zone are

$$\sigma_\theta = (Yc^2/\sqrt{3}r_o^2)(1 + r_o^2/r^2), \qquad \text{(iv)}$$

$$\sigma_r = (Yc^2/\sqrt{3}r_o^2)(1 - r_o^2/r^2), \qquad \text{(v)}$$

$$\sigma_z = Yc^2/\sqrt{3}r_o^2. \qquad \text{(vi)}$$

The radial equilibrium Eq. (15.7) applies to the plastic annulus $r_i \leq r \leq c$:

$$\sigma_\theta - \sigma_r = r\, d\sigma_r/dr \qquad \text{(vii)}$$

and again assuming, with Eq. (iii), that $\sigma_z = (\sigma_r + \sigma_\theta)/2$ for closed ends (i.e. $d\epsilon_z^P = 0$ from Eq. (17.14)), the Mises criterion becomes simply

$$\sigma_\theta - \sigma_r = 2Y/\sqrt{3}. \qquad \text{(viii)}$$

Combining Eqs (vii) and (viii) and integrating gives

$$\sigma_r = (2Y/\sqrt{3})\ln r + K,$$

in which K is found from the condition that $\sigma_r = -p_c$ for $r = c$. This gives

$$K = -p_c - (2Y/\sqrt{3})\ln c = -(Y/\sqrt{3})(1 - c^2/r_o^2) - (2Y/\sqrt{3})\ln c$$

and therefore the stresses in the plastic zone are

$$\sigma_r = (2Y/\sqrt{3})\ln(r/c) - (Y/\sqrt{3})(1 - c^2/r_o^2), \qquad \text{(ix)}$$

$$\sigma_\theta = (2Y/\sqrt{3})\ln(r/c) + (Y/\sqrt{3})(1 + c^2/r_o^2), \qquad \text{(x)}$$

$$\sigma_z = (2Y/\sqrt{3})\ln(r/c) + (Y/\sqrt{3})(c^2/r_o^2). \qquad \text{(xi)}$$

Now as $\sigma_r = -p_i$ for $r = r_i$, this gives the required internal pressure:

$$p_i = (2Y/\sqrt{3})\ln(c/r_i) + (Y/\sqrt{3})(1 - c^2/r_o^2). \qquad \text{(xii)}$$

The pressure p_{ult} required to produce a fully plastic cylinder is found from putting $c = r_o$ in Eq. (xii):

$$p_{\text{ult}} = (2Y/\sqrt{3})\ln(r_o/r_i).$$

The elastic (Eqs (iv)–(vi)) and plastic (Eqs (ix)–(xi)) stress distributions under pressure p_i with $c = (r_i + r_o)/2$ are shown in Fig. 17.4(a).

When the pressure p_i is released, the elastic stresses σ_E under p_i recover, leaving a residual stress σ_R. That is,

$$\sigma_R = \sigma - \sigma_E,$$

where σ_E take the usual Lamé form with p_i replacing p_c in Eqs (i)–(iii) for $r_i \le r \le r_o$. Subtracting these from the corresponding stresses σ under pressure in each zone leads to the residuals:

For the outer zone, $c \le r \le r_o$:

$$\sigma_{\theta R} = \sigma_\theta - \sigma_{\theta E}$$
$$= (Yc^2/\sqrt{3}r_o^2)(1 + r_o^2/r^2) - p_i r_i^2(1 + r_o^2/r^2)/(r_o^2 - r_i^2),$$

$$\sigma_{rR} = \sigma_r - \sigma_{rE}$$
$$= (Yc^2/\sqrt{3}r_o^2)(1 - r_o^2/r^2) - p_i r_i^2(1 - r_o^2/r^2)/(r_o^2 - r_i^2),$$

$$\sigma_{zR} = (\sigma_{\theta R} + \sigma_{rR})/2$$
$$= Yc^2/\sqrt{3}r_o^2 - p_i r_i^2/(r_o^2 - r_i^2).$$

For the inner zone $r_i \le r \le r_o$:

$$\sigma_{\theta R} = \sigma_\theta - \sigma_{\theta E}$$
$$= (2Y/\sqrt{3})\ln(r/c) + (Y/\sqrt{3})(1 + c^2/r_o^2) - p_i r_i^2(1 + r_o^2/r^2)/(r_o^2 - r_i^2),$$

$$\sigma_{rR} = \sigma_r - \sigma_{rE}$$
$$= (2Y/\sqrt{3})\ln(r/c) - (Y/\sqrt{3})(1 - c^2/r_o^2) - p_i r_i^2(1 - r_o^2/r^2)/(r_o^2 - r_i^2),$$

$$\sigma_{zR} = \sigma_z - \sigma_{zE}$$
$$= (2Y/\sqrt{3})\ln(r/c) + Yc^2/\sqrt{3}r_o^2 - p_i r_i^2/(r_o^2 - r_i^2).$$

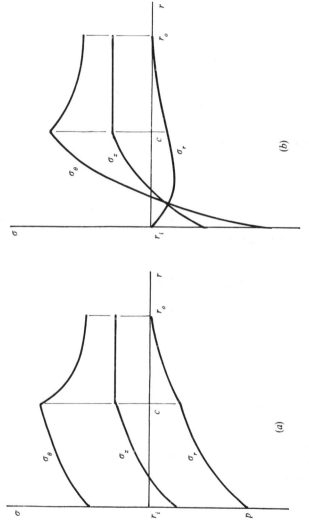

Figure 17.4 Elastic and plastic stress distributions.

These are distributed as shown in Fig. 17.4(b). The compressive stresses left within the inner core of the cylinder are particularly beneficial in improving fatigue strength when they oppose and reduce subsequently applied cyclic tensile stresses (σ_θ and σ_z) due to fluctuating internal pressure. The process of pre-pressurizing thick cylinders in this way is known as autofrettage. If the Tresca yield criterion $\sigma_\theta - \sigma_r = Y$ is preferred instead of the von Mises, it follows from Eq. (viii) that all the forgoing Mises stresses including the residuals σ_R need only be multiplied by $\sqrt{3}/2$. The resulting Tresca σ_θ and σ_r expressions will also apply to an open-ended cylinder and a thin disc, since with $\sigma_z = 0$ in each case it remains the intermediate stress. The Mises criterion is more difficult to combine with the equilibrium condition for an open cylinder. The following example illustrates Nadai's approach (1950) in this case.

Example 17.7 Determine the internal pressure required to (a) initiate yielding, (b) produce full plasticity, and (c) yield to the mean radius c in a thin disc of non-hardening von Mises material with yield stress $Y = 500\,\text{MPa}$ and radii $r_i = 25\,\text{mm}$, $r_o = 62.5\,\text{mm}$. What is the maximum residual hoop stress for (c)?

(a) With reference to page 706, when $\sigma_z = 0$, the initial yield pressure is given by

$$p = Y(K^2 - 1)/\sqrt{(3K^4 + 1)}. \qquad (i)$$

For $K = 62.5/25 = 2.5$, Eq. (i) gives $p = 500(6.25 - 1)/\sqrt{(117.19 + 1)} = 241.46\,\text{MPa}$.

(b) Equations (15.7) and (16.9b) again apply to this case:

$$\sigma_\theta^2 - \sigma_\theta\sigma_r + \sigma_r^2 = Y^2, \qquad (ii)$$

$$\sigma_\theta - \sigma_r = r\,d\sigma_r/dr. \qquad (iii)$$

Putting $\sigma = (\sigma_\theta + \sigma_r)/\sqrt{2}$ and $\sigma' = (\sigma_\theta - \sigma_r)/\sqrt{2}$ in Eq. (ii) leads to the ellipse

$$\sigma^2/a^2 + \sigma'^2/b^2 = 1$$

where $a = \sqrt{2}Y$ and $b = \sqrt{(2/3)}Y$ appear with the parameters (Fig. 17.5(a)):

$$\sigma = a\sin\theta = \sqrt{2}Y\sin\theta, \qquad \sigma' = b\cos\theta = \sqrt{(2/3)}\,Y\cos\theta.$$

Solving for σ_θ and σ_r:

$$\sigma_\theta = (1/\sqrt{2})(\sigma + \sigma') = Y[\sin\theta + (1/\sqrt{3})\cos\theta] = (2Y/\sqrt{3})\sin(\theta + \pi/6), \qquad (iv)$$

$$\sigma_r = (1/\sqrt{2})(\sigma - \sigma') = Y[\sin\theta - (1/\sqrt{3})\cos\theta] = (2Y/\sqrt{3})\sin(\theta - \pi/6). \qquad (v)$$

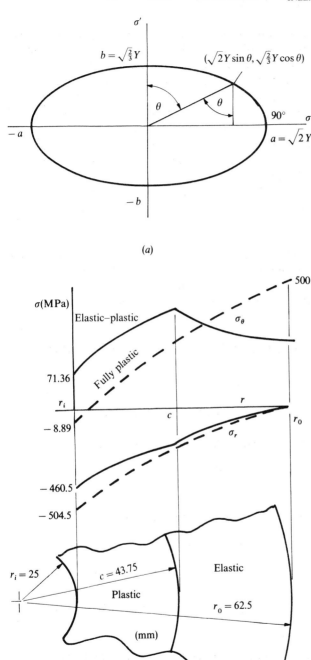

Figure 17.5 Yielding in a thin disc.

Substituting Eqs (iv) and (v) into Eq. (iii) leads to

$$r\,d[\sin(\theta - \pi/6)]/dr = [\sin(\theta + \pi/6) - \sin(\theta - \pi/6)] = \cos\theta.$$

Putting $y = \cos\theta$ enables the variables to be separated into

$$dy/\sqrt{(1 - y^2)} + dy/\sqrt{3} = -(2/\sqrt{3})\,dr/r$$

and integrating gives

$$A^2/r^2 = \cos\theta\exp(-\sqrt{3}\theta), \tag{vi}$$

where A is an integration constant, found from the boundary condition that $\sigma_r = 0$ for $r = r_o$. Hence, from Eq. (v), $\theta_o = \pi/6$ and Eq. (vi) gives

$$A^2 = (\sqrt{3}/2)r_o^2\exp(-\sqrt{3}\pi/6),$$
$$\therefore\quad r_o^2/r^2 = (2/\sqrt{3})\cos\theta\exp[\sqrt{3}(\pi/6 - \theta)]. \tag{vii}$$

This equation enables θ to be found for any radius $r_i \le r \le r_o$. Then σ_θ and σ_r follow from Eqs (iv) and (v). In particular, as $\sigma_r = -p$ for $r = r_i$ in Eq. (v), the fully plastic pressure is

$$p_i = (2Y/\sqrt{3})\sin(\pi/6 - \theta_i), \tag{viii}$$

where, correspondingly, θ_i is found from Eq. (vii):

$$(r_o/r_i)^2 = (2/\sqrt{3})\cos\theta_i\exp[\sqrt{3}(\pi/6 - \theta_i)]. \tag{ix}$$

The radial stress remains compressive under pressure for the valid range $-\pi/3 < \theta < \pi/6$. Thus, for $r_o/r_i = 2.5$ the disc becomes fully plastic when, from Eq. (ix),

$$(62.5/25)^2 = (2/\sqrt{3})\cos\theta_i\exp[\sqrt{3}(\pi/6 - \theta_i)].$$

A trial solution gives $\theta_i = -30.9° = -0.539\,\text{rad}$. Then, from Eq. (viii),

$$p_i = (2 \times 500/\sqrt{3})\sin(\pi/6 + 0.539) = 504.5\,\text{MPa},$$

and from Eq. (iv), $\sigma_\theta = -8.89\,\text{MPa}$.

(c) From the previous example (Eqs (i) and (ii)), the stresses in the elastic zone (Fig. 17.5(b)) $c \le r \le r_o$ are $\sigma_z = 0$ and

$$\sigma_\theta = p_cc^2(1 + r_o^2/r^2)/(r_o^2 - c^2),$$
$$\sigma_r = p_cc^2(1 - r_o^2/r^2)/(r_o^2 - c^2) \tag{x}$$

where the interface pressure p_c is given from Eq. (i) for $K = r_o/c$. The stresses in the plastic zone $r_i \le r \le c$ are again supplied by Eqs (iv) and (v). However, the constant A in Eq. (vi) is now found from the condition that σ_r is common to both zones at $r = c$. That is, from Eqs (v) and (x),

$$Y(1 - r_o^2/c^2)/\sqrt{(3r_o^4/c^4 + 1)} = (2Y/\sqrt{3})\sin(\theta_c - \pi/6),$$
$$(1 - r_o^2/c^2)/\sqrt{(r_o^4/c^4 + 1/3)} = 2\sin(\theta_c - \pi/6). \tag{xi}$$

Having found θ_c, to satisfy Eq. (xi), it follows from Eq. (vi) that

$$A^2 = c^2 \cos \theta_c \exp(-\sqrt{3}\theta_c),$$
$$c^2/r^2 = (\cos \theta / \cos \theta_c) \exp[\sqrt{3}(\theta_c - \theta)]. \qquad \text{(xii)}$$

This equation enables θ to be found for any radius $r_i \le r \le c$. In particular, when $r = r_i$, $\theta = \theta_i$, from which the internal pressure p_i required to produce an elastic-plastic cylinder is again found from Eq. (viii). Now,

$$c/r_i = (25 + 62.5)/2 \times 25 = 1.75,$$
$$\therefore \quad r_o/c = (r_o/r_i) \times (r_i/c) = 2.5/1.75 = 1.4268.$$

Substituting into Eq. (xi):

$$-0.4910 = 2 \sin (\theta_c - \pi/6).$$

A trial solution gives $\theta_c = 15.6° = 0.2723$ rad. Then, from Eq. (xii) with $r = r_i$ and $\theta = \theta_i$,

$$1.75^2 = 1.038\,25 \cos \theta_i \exp[\sqrt{3}(0.2723 - \theta_i)].$$

The trial solution yields $\theta_i = -22.9° = -0.3997$ rad. Hence, from Eq. (viii),

$$p_i = (2Y/\sqrt{3}) \sin (\pi/6 + 0.3997) = 460.5 \,\text{MPa}.$$

The maximum bore hoop stress is found from Eq. (iv) for $\theta = \theta_i$:

$$\sigma_\theta = (2Y/\sqrt{3}) \sin (\pi/6 - 0.3997) = 71.36 \,\text{MPa}.$$

Upon release of p_i the elastic hoop stress (page 613) recovers:

$$\sigma_{\theta E} = p_i(1 + r_o^2/r_i^2)/(r_o^2/r_i^2 - 1)$$
$$= 460.5(1 + 2.5^2)/(2.5^2 - 1) = 635.93 \,\text{MPa}$$

and, therefore, the residual compressive hoop stress is

$$\sigma_{\theta R} = \sigma_\theta - \sigma_{\theta E} = 71.36 - 635.93 = -564.57 \,\text{MPa},$$

indicating that reversed yielding occurs at the bore, since Y is exceeded. Figure 17.5(b) shows the σ_r and σ_θ distributions for the elastic-plastic (e.p.) and fully plastic (f.p.) disc.

Example 17.8 Determine the angular velocity ratio ω_2/ω_1 for a uniformly thin solid disc of radius r_o where ω_1 is the speed to initiate yielding and ω_2 is the speed for which the elastic-plastic interface reaches the mean radius $c = r_o/2$. What is this ratio when the disc becomes fully plastic? Assume a Tresca yield criterion and take $v = 1/3$. Plot the distribution of σ_r and σ_θ under ω_1 and ω_2.

The elastic stresses are given by Eqs (15.20) with constant a:

$$\sigma_r = \rho\omega^2(3+v)(r_o^2 - r^2)/8, \tag{i}$$

$$\sigma_\theta = \rho\omega^2[(3+v)r_o^2 - (1+3v)r^2]/8. \tag{ii}$$

These are distributed as shown in Fig. 17.6(a). Since both are tensile, the Tresca criterion (Eq. (16.4b)) employs the greater value together with the zero axial stress σ_z. This gives simply

$$\sigma_\theta = Y. \tag{iii}$$

Clearly, yielding will commence at the centre where σ_θ is a maximum. Substituting Eq. (ii) into (iii) for $r = 0$ supplies the speed ω_1 to initiate yielding:

$$\rho\omega_1^2(3+v)r_o^2/8 - 0 = Y,$$

$$\therefore \quad \omega_1^2 = 8Y/\rho(3+v)r_o^2 = 2.4Y/\rho r_o^2. \tag{iv}$$

In the elastic-plastic case, the equilibrium equation (15.14) holds for the

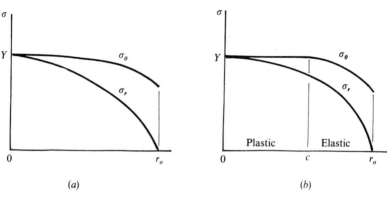

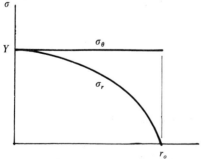

(c)

Figure 17.6 Radial and hoop stress distributions.

inner plastic core $0 \le r \le c$. Combining this with Eq. (iii) leads to

$$d(r\sigma_r)/dr = Y - \rho\omega^2 r^2,$$
$$r\sigma_r = Yr - \rho\omega^2 r^3/3 + A,$$
$$\sigma_r = Y - \rho\omega^2 r^2/3 + A/r. \qquad (v)$$

As $\sigma_\theta = \sigma_r = Y$ for $r = 0$, then $A = 0$ in Eq. (v). For the outer elastic annulus $(c \le r \le r_o)$, Eqs (15.17) supply the stresses. The following boundary conditions enable the constants a and b to be found: $\sigma_r = 0$ for $r = r_o$ in Eq. (15.17a), giving

$$0 = a - b/r_o^2 - (3 + v)\rho\omega^2 r_o^2/8,$$

and $\sigma_\theta = Y$ for $r = c$ in Eq. (15.17b), giving

$$Y = a + b/c^2 - (1 + 3v)\rho\omega^2 c^2/8.$$

Solving these simultaneously leads to

$$a = \{Y - (\rho\omega^2/8)[(3 + v)r_o^2 - (1 + 3v)c^2]\}c^2/(r_o^2 + c^2) + (3 + v)\rho\omega^2 r_o^2/8,$$
$$b = \{Y - (\rho\omega^2/8)[(3 + v)r_o^2 - (1 + 3v)c^2]\}r_o^2 c^2/(r_o^2 + c^2).$$

The angular speed ω_2^2 is found from the condition that σ_r is common to both zones at the interface radius $c = r_o/2$. In the elastic zone, Eq. (15.17a) gives

$$\sigma_r = a - b/c^2 - (3 + v)\rho\omega_2^2 c^2/8.$$

Substituting for a and b with $c = r_o/2$ leads to

$$\sigma_r = \{Y - (\rho\omega_2^2/8)[(3 + v)r_o^2 - (1 + 3v)c^2]\}(c^2 - r_o^2)/(c^2 + r_o^2)$$
$$+ (\rho\omega_2^2/8)(3 + v)(r_o^2 - c^2),$$
$$\sigma_r = 3(3 + v)\rho\omega_2^2 r_o^2/32 - [Y - \rho\omega_2^2 r_o^2(11 + v)/32]. \qquad (vi)$$

In the plastic zone, Eq. (v) gives, for $r = c = r_o/2$,

$$\sigma_r = Y - \rho\omega_2^2 r_o^2/12. \qquad (vii)$$

Equating (vi) and (vii):

$$8\dot{Y}/5 = \rho\omega_2^2 r_o^2[1/12 + 3(3 + v)/32 + 3(11 + v)/160] = 0.6083\rho\omega_2^2 r_o^2,$$
$$\therefore \quad \omega_2^2 = 2.63Y/\rho r_o^2 \qquad (viii)$$

Then from Eqs (iv) and (viii), the speed ratio is

$$\omega_2/\omega_1 = \sqrt{(2.63/2.4)} = 1.047.$$

When the disc becomes fully plastic, Eq. (v) applies for $0 \le r \le r_o$. Again, $\sigma_r = Y$ for $r = 0$, giving $A = 0$. Also, as $\sigma_r = 0$ for $r = r_o$,

$$0 = Y - \rho\omega^2 r_o^2/3.$$

The fully plastic speed ω^P is therefore

$$(\omega^P)^2 = 3Y/\rho r_o^2$$

when from Eq. (iv) the speed ratio is

$$\omega^P/\omega_1 = \sqrt{(3/2.4)} = 1.118.$$

The distributions in σ_r and σ_θ for the elastic, elastic-plastic and fully plastic discs are shown in Fig. 17.6(a), (b) and (c) respectively.

17.4 CREEP

Creep is the time-dependent deformation that occurs in a material subjected to constant loading over a prolonged period. While some metallic materials such as lead, copper and mild steel do creep at room temperature under sufficiently high loads, the phenomenon is normally associated with the load capacity at higher temperatures, e.g. stainless steel in steam and chemical plant operating in excess of 450 °C, and nickel-base alloys in gas turbines at 750°C and above. The understanding of the mechanics of creep begins with a macroscopic representation of the accumulation of creep strain ϵ_c with time t at a given stress σ and temperature T.

When σ and T are constant under uniaxial stress conditions, the three-stage creep curve in Fig. 17.7(a) is normally observed. Though differences in creep response of a material may arise when one stage dominates under a particular σ and T combination, the basic shape of the creep curve remains unaltered (Fig. 17.7(b)). Following the initial application of the stress, in which the instantaneous strain ϵ_0 may be elastic ϵ^E or elastic plus plastic $\epsilon^E + \epsilon^P$, three stages of creep ensue as follows.

Primary creep (OA) is a period of work-hardening in which the creep rate $d\epsilon_c/dt$ decreases with time (also known as transient creep). As a result, the material becomes harder to deform as the internal stress increases with the dislocation density.

Secondary creep (AB) in which there is a balance between work-hardening and thermal softening. The latter is a recovery process activated by the energy within the dislocation structure. The result is that in the region AB the creep rate is constant and the material becomes neither harder nor softer. When AB is absent, the balance is achieved in a single point of inflection at which the creep rate is the minimum value for the curve. The secondary region normally forms the basis for engineering design, where a given creep strain can be tolerated, e.g. 0.01 per cent in 10^4 h ($1\frac{1}{7}$ years) or 0.1 per cent in 10^5 h ($11\frac{1}{2}$ years).

Tertiary creep (BC) results from necking, cracking and metallurgical instability. It is characterized by an increasing creep rate culminating in fracture at point C.

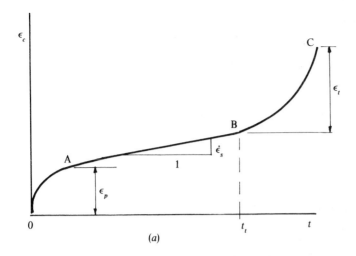

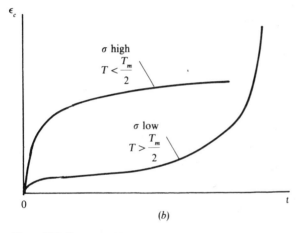

Figure 17.7 Creep curves.

Time Dependence of Creep Strain

A number of empirical functions $\epsilon_c = \epsilon_c(t)$ have been proposed to describe one or all of these stages within a specified σ and T range. In the lower range of homologous temperatures ($T/T_m < 0.5$, where T_m is the absolute melt temperature) primary creep dominates (Fig. 17.7(b)). When the stress is low, a limited amount of logarithmic primary (or transient) creep is observed:

$$\epsilon_c = \alpha \ln t$$

where $\alpha = \alpha(\sigma, T)$. As σ increases, primary creep is better described by the

Bailey power law (1935):

$$\epsilon_c = \alpha t^m \qquad (17.17)$$

where $0 < m < 1$. For $m = 1/3$, Eq. (17.17) appears within Andrade's law (1910):

$$\epsilon_c = (1 + \alpha t^{1/3}) \exp(\beta t) - 1 \simeq \alpha t^{1/3},$$

since $\exp(kt) \to 1$ as $t \to 0$ in primary creep. With increasing time, Andrade's β component of flow appears. This is now identified with the component of secondary creep, particularly for the higher temperature range $T/T_m > 0.5$ (Fig. 17.7b). An equivalent description of high-temperature creep is simply achieved with the addition of a linear term to Eq. (17.17):

$$\epsilon_c = \alpha t^m + \beta t, \qquad (17.18)$$

where $\beta = \beta(\sigma, T)$. Graham and Walles (1955, 1958) extended Eq. (17.18) for $m = 1/3$, with the addition of a tertiary cubic time term for a wider range of σ and T in which the material will fail:

$$\epsilon_c = \alpha t^{1/3} + \beta t + \gamma t^3 \qquad (17.19)$$

where $\gamma = \gamma(\sigma, T)$ and for which good agreement was shown between the complete creep curves of many high-temperature engineering alloys. An alternative to Eq. (17.19) is the three-part time function of Davis et al. (1969). This employs Garofalo's (1965) exponential-primary term added to linear-secondary and exponential-tertiary terms:

$$\epsilon_c = \epsilon_P[1 - \exp(-mt)] + \dot{\epsilon}_s t + \epsilon_L \exp[p(t - t_t) - 1], \qquad (17.20)$$

where, with reference to Fig. 17.7(a), ϵ_P in Eq. (17.20) is the maximum amount of primary creep strain, $\dot{\epsilon}_s$ is the secondary creep rate, and t_t is the time at the start of tertiary creep. The constants α, β, γ, m, n, ϵ_L and p in Eqs (17.17)–(17.20) are determined empirically for the given σ and T conditions. Note that $\epsilon_o + \epsilon_c$ is the net strain under σ.

Stress Dependence of Secondary (Minimum) Creep Rate

For a constant temperature T, the family of creep curves in Fig. 17.8 applies to stress levels $\sigma_1 > \sigma_2 > \sigma_3 > \sigma_4$. As the curves display different amounts and rates of primary, secondary and tertiary creep, it follows that the strain function $\epsilon_c = \epsilon_c(t, \sigma)$ is complex. However, in designing for creep, the function is normally restricted to the secondary region where the creep rate $\dot{\epsilon}_s$ depends solely upon σ. For lower stresses, the linear law applies:

$$\dot{\epsilon}_s = A\sigma,$$

where $A = A(T)$. For a medium stress range, the widely used Norton power law (1929) applies:

$$\dot{\epsilon}_s = A\sigma^n \qquad (17.21)$$

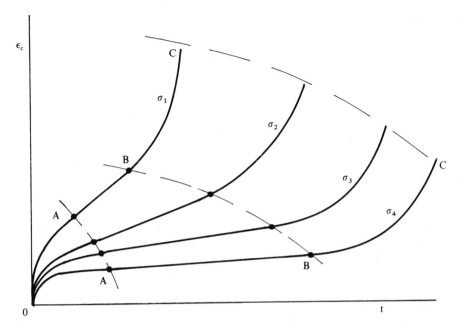

Figure 17.8 Constant-T creep curves.

where $3 \le n \le 8$ is a material constant found from a double-log plot between $\dot{\epsilon}_s$ and σ. However, as the stress is increased, n becomes dependent upon σ. The exponential law may then be preferable:

$$\dot{\epsilon}_s = A \exp(B\sigma),$$

where B is a material constant. Garofalo (1965) proposed a hyperbolic sine law over the complete range of stress:

$$\dot{\epsilon}_s = A[\sinh(C\sigma)]^p \qquad (17.22)$$

where C and p are experimentally determined constants.

Temperature Dependence of Secondary Creep Rate

For a constant stress σ, a further family of creep curves (Fig. 17.9) applies to the absolute temperatures $T_1 > T_2 > T_3 > T_4$. As the temperature increases, the relative proportion of the lifetime expended in secondary (t_{AB}) and tertiary (t_{BC}) alter in a complex manner.

These trends would become cumbersome if expressed in a single function $\epsilon_c = \epsilon_c(t, T)$. However, the dependence of the secondary creep rate upon temperature $\dot{\epsilon}_s = \dot{\epsilon}_s(T)$ is uniquely defined by the thermal activation energy necessary for the dislocation recovery processes of climb and cross-slip (Dorn,

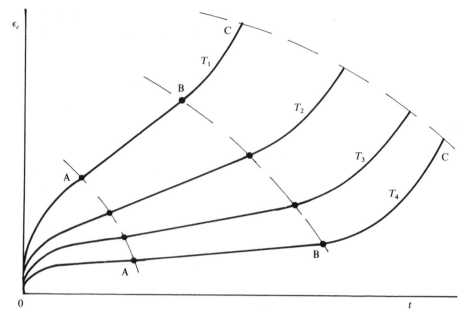

Figure 17.9 Constant-σ creep curves.

1957). This leads to the Arrhenius expression,

$$\dot{\epsilon}_s = D \exp(-Q/RT), \tag{17.23}$$

where D is a constant, $R = 1.98$ cal/mol (8.29 J/mol) is the characteristic gas constant and Q is the creep activation energy: for example, the value for aluminium is $Q = 30$ kcal/mol (125.6 kJ/mol) and for copper $Q = 20$ kcal/mol (83.7 kJ/mol). These are found from an $\dot{\epsilon}_s$ vs. $1/T$ plot (Fig. 7.10) illustrated in the following example.

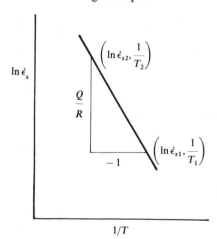

Figure 17.10 Arrhenius plot.

Example 17.9 The secondary creep rates given were obtained from a series of tensile creep tests at different temperatures all under the same load:

$T(K)$	290	300	310	320	330
$\dot{\epsilon}_s(h^{-1})$	4.8×10^{-6}	2.74×10^{-5}	1.4×10^{-4}	6.44×10^{-4}	2.7×10^{-3}

Determine the activation energy in Eq. (17.23) and the constant for secondary creep and so identify the material. (T.C.D. 1984)

Note first that equating constant load to constant stress implies that the change in section area is negligible up to and during secondary creep. A true constant stress test reduces the load to account for loss in section. Taking the logarithm of Eq. (17.23) leads to:

$$\ln \dot{\epsilon}_s = -Q/RT + \ln D.$$

Hence the plot of $\ln \dot{\epsilon}_s$ vs. $1/T$ in Fig. 17.10 results in a straight line of slope $-Q/R$. That is, from the 2 points chosen to lie on this line:

$$-Q/R = (\ln \dot{\epsilon}_{s2} - \ln \dot{\epsilon}_{s1})/(1/T_2 - 1/T_1)$$
$$= [\ln(1.5 \times 10^{-3}) - \ln(3 \times 10^{-6})]/(3.07 - 3.48)10^{-3}$$
$$= -15\,158,$$
$$\therefore \quad Q = 1.98 \times 15\,158 = 30\,012 \text{ cal/mol } (125.6 \text{ kJ/mol}),$$

so identifying aluminium or one of its alloys. The constant D is found from

$$\ln D = \ln \dot{\epsilon}_s + Q/RT = \ln(1.5 \times 10^{-3}) + 30\,012 \times /(1.98 \times 325.73)$$
$$= -6.502 + 46.534 = 40.03,$$
$$\therefore \quad D = 243 \times 10^{15}.$$

Stress and Temperature Dependence of the Secondary Creep Rate

The separate equations describing the stress and temperature dependence of $\dot{\epsilon}_s$ may be combined in a single function $\dot{\epsilon}_s = (\sigma, T)$. For example, when $A = A(T)$ in Eq. (17.21) is identified with Eq. (17.23), this gives $\dot{\epsilon}_s$ for the medium σ range:

$$\dot{\epsilon}_s = D\sigma^n \exp(-Q/RT). \tag{17.24}$$

Alternatively, when the function $A = A(T)$ in Eq. (17.22) is identified with Eq. (17.23) over the whole σ range:

$$\dot{\epsilon}_s = D[\sinh(C\sigma)]^p \exp(-Q/RT). \tag{17.25}$$

Stress, Temperature and Time Dependence of Creep Strain

Depending upon the time dependence, there are various forms for the general creep strain function $\epsilon_c = \epsilon_c(\sigma, T, t)$. When primary creep strain is ignored and only secondary creep strain is required, then $\dot{\epsilon}_s = \dot{\epsilon}_s t$, with $\dot{\epsilon}_s$ given in Eq. (17.24) or Eq. (17.25). A commonly used function for primary creep strain combines the forms of Eqs (17.17) and (17.24) to give

$$\epsilon_c = D\sigma^n t^m \exp(-Q/RT). \tag{17.26a}$$

If both primary and secondary creep are required, Eq. (17.18) indicates that a linear time term may be added to Eq. (17.26) provided the stress and temperature dependence of creep strain for each stage remains unaltered. This gives

$$\epsilon_c = D\sigma^n t^m \exp(-Q/RT) + D'\sigma^n t \exp(-Q/RT),$$

where D and D' are material constants.

The Equation of State

The state equation assumes that for a given creep time or strain, the creep strain rate $\dot{\epsilon}_c$ is a function of only the current stress and temperature (Ludwik, 1909). The implication is that the history of strain does not inflence $\dot{\epsilon}_c$. Two state equations—time and strain hardening—result from this. In their general forms these are respectively

$$\dot{\epsilon}_c = \dot{\epsilon}_c(\sigma, T, t), \tag{17.27a}$$

$$\dot{\epsilon}_c = \dot{\epsilon}_c(\sigma, T, \epsilon_c) \tag{17.27b}$$

which are related through the creep curve function $\epsilon_c = \epsilon_c(\sigma, T, t)$. More usually, the state equations appear with separate functions of σ, T, t and ϵ_c (see Eqs (17.24)–(17.26). The differences between Eqs (17.27a and b) are illustrated in Fig. 17.11 under uniaxial stepped stress conditions. When, after primary creep time t_1, the stress is increased from σ_1 to σ_2 at a constant

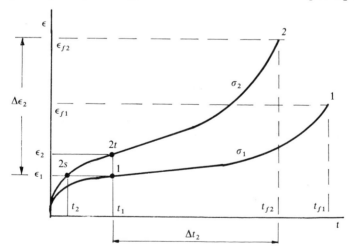

Figure 17.11 Time and strain hardening.

temperature, then according to Eqs (17.27), $\dot{\epsilon}_c$ depends either upon the elapsed time or accumulated strain. With time hardening (see for example, Eq. (17.30a), the new strain rate $\dot{\epsilon}_c$ depends upon σ_2 at t_1 (path 1–2t), while for strain hardening $\dot{\epsilon}_c$ depends upon σ_2 at ϵ_1 (path 1 − 2s). For $T = $ constant, Eq. (17.26a) describes the primary strain in the following form

$$\epsilon_c = A\sigma^n t^m. \tag{17.26b}$$

The respective time hardening law (Eq. (17.27a)) is found from differentiating Eq. (17.26b) w.r.t. time

$$\dot{\epsilon}_c = mA\sigma^n t^{m-1}. \tag{17.28a}$$

Eliminating t between Eqs (17.26b) and (17.28a) gives the corresponding strain hardening prediction

$$\dot{\epsilon}_c = (mA^{1/m})\sigma^{n/m}\epsilon_c^{(m-1)/m}. \tag{17.28b}$$

Experiment shows that strain hardening is the more realistic, although time hardening may be used reliably for gradual stress changes. If the respective time and strain to fracture under σ_1 and σ_2 are t_{f1}, ϵ_{f1} and t_{f2}, ϵ_{f2} as shown, Eqs (17.27a and b) supply the remaining life Δt_2 and strain $\Delta\epsilon_2$ according to the time and strain fraction rules:

$$\sum(t/t_f) = t_1/t_{f1} + \Delta t_2/t_{f2} = 1, \tag{17.29a}$$

$$\sum(\epsilon/\epsilon_f) = \epsilon_1/\epsilon_{f1} + \Delta\epsilon_2/\epsilon_{f2} = 1. \tag{17.29b}$$

Of these, experiment has shown that the time fraction equation (17.28a) due to Robinson (1938) is more reliable than a fracture strain criterion. However, creep strain prior to fracture appears implicitly in the popular Monkman–Grant (1956) criterion for rupture:

$$\dot{\epsilon}_s t_f^k = \text{constant}. \tag{17.29c}$$

Normally, as $k \simeq 1$, the constant is equivalent to the secondary component of the fracture strain. This is often referred to as the true creep strain, which remains constant over a wide range of stress and temperature. One useful application of the time hardening equation (17.27a) is for the prediction of stress relaxation behaviour. Relaxation refers to the stress response to a stepped strain input. This is the opposite effect to creep, that is, the strain response to a stepped stress input. Consider, for example, the strain input to a bolt on initial tightening. As the initial stress σ_o will relax with time, it becomes necessary to know when the bolts should be retightened to prevent the stress from falling below a critical level. If σ is the stress after time t, then for constant net strain,

$$\sigma/E + \epsilon_c = \text{constant}. \tag{17.30a}$$

Differentiating Eq. (17.30a) w.r.t. time:

$$(1/E)\,d\sigma/dt + \dot{\epsilon}_c = 0, \tag{17.30b}$$

Substituting Eq. (17.28a) into Eq. (17.30b) leads to

$$(1/E)\,d\sigma/dt + mA\sigma^n t^{m-1} = 0,$$

$$\int_0^t t^{m-1}\,dt = -(1/AEm)\int_{\sigma_o}^{\sigma} d\sigma/\sigma^n,$$

$$(1/m)|t^m|_0^t = -[1/AEm(1-n)]|\sigma^{1-n}|_{\sigma_o}^{\sigma},$$

$$t^m = [1/AE(n-1)](1/\sigma^{n-1} - 1/\sigma_o^{n-1}). \tag{17.31a}$$

If Eq. (17.27b) is preferred, then from Eq. (17.30a) the stress after strain ϵ_c is found from

$$(1/E)\,d\sigma/d\epsilon_c + 1 = 0,$$

$$\therefore \quad \sigma = -E\epsilon_c + C,$$

where $C = \sigma_o$, since $\sigma = \sigma_o$ for $\epsilon_c = 0$. Substituting from Eq. (17.26b) gives the relaxation time.

$$t = [(\sigma_o - \sigma)/EA\sigma^n]^{1/m}. \tag{17.31b}$$

Example 17.10 In a cylindrical vessel a pressure of 1 MPa (150 lbf/in²) acts on a circular cover plate of area 0.13 m² (16 in diameter). The plate is held in position by twenty 25 mm (1 in) diameter steel bolts, equispaced around its rigid rim. What should be the initial tightening stress in the bolts in order that a safety factor of 2 is maintained after 10 000 h of creep relaxation at 455°C (850°F)? After what time should the bolts be re-tightened to prevent leakage around the plate? Ignore primary creep but account for elasticity with $E = 172\,\text{GPa}$ (25×10^6 lbf/in²) and a secondary creep rate of $\dot\epsilon_s = 44.3 \times 10^{-16}\sigma^4$ ($\dot\epsilon_s = 10 \times 10^{-24}\sigma^4$) ($\dot\epsilon_s$ in h⁻¹ and σ in MPa (lbf/in²)). (C.E.I. Pt.2, 1969)

SOLUTION TO SI UNITS The force on the plate is $1 \times 0.13 = 0.13$ MN. Hence, the minimum working stress in the bolts is given by

$$\sigma_w = 0.13 \times 4 \times 10^6/(20 \times \pi \times 25^2) = 13.24\,\text{MPa}.$$

The allowable relaxed stress is $\sigma = 2 \times \sigma_w = 26.48$ MPa. Then, from Eq. (17.31a), with $m = 1$ for secondary creep, the initial tightening stress σ_o is

$$\sigma_o = 1/[1/\sigma^{n-1} - AEt(n-1)]^{1/(n-1)}$$
$$= 1/[1/26.48^3 - 44.3 \times 10^{-16} \times 172 \times 10^3 \times 10000 \times 3]^{1/3}$$
$$= 1/[(5.385 - 2.2)10^{-5}]^{1/3},$$
$$\sigma_o = 31.55\,\text{MPa}.$$

Note that Eq. (17.31b) gives $\sigma_o = 30.23$ MPa.

Leakage begins when the stress relaxes to the minimum working value $\sigma = 13.24\,\mathrm{MPa}$. Again from Eq. (17.31a), the corresponding time is

$$t = (1/\sigma^{n-1} - 1/\sigma_o^{n-1})/AE(n-1)$$
$$= (1/13.24^3 - 1/31.55^3)/(3 \times 172 \times 10^3 \times 44.3 \times 10^{-16})$$
$$= 174\,550\,\mathrm{h}.$$

Equation (17.31b) here gives a less reliable prediction of $72\,5620\,\mathrm{h}$.

SOLUTION TO IMPERIAL UNITS Here the minimum working stress per bolt is

$$\sigma_w = \frac{150 \times (\pi/4)(16)^2}{20 \times (\pi/4)(1)^2} = 1920\,\mathrm{lbf/in}^2,$$

and the allowable relaxed stress after $10\,000\,\mathrm{h}$ of relaxation is $\sigma = 3840\,\mathrm{lbf/in}^2$. The initial tightening stress σ_o is again found from

$$\sigma_o = 1/[1/\sigma^{n-1} - AEt(n-1)]^{1/n-1}$$
$$= 1/(1/3840^3 - 10 \times 10^{-24} \times 25 \times 10^6 \times 10\,000 \times 3)^{1/3}$$
$$= 4617\,\mathrm{lbf/in}^2.$$

The leakage time is given by

$$t = (1/\sigma^{n-1} - 1/\sigma_o^{n-1})/AE(n-1)$$
$$= (1/1920^3 - 1/4617^3)/(10 \times 10^{-24} \times 25 \times 10^6 \times 3)$$
$$= 174\,833\,\mathrm{h}.$$

Damage in Tertiary Creep

In equations (17.29a and b) the fractions of ductility (ϵ/ϵ_f) and life (t/t_f) expended in creep may be taken as measures of creep damage. The concept of creep damage was extended by Katchanov (1959) and Rabotnov (1969) within a more general formulation of brittle tensile creep rupture for low stress and long life in which a dominant tertiary stage results from early grain-boundary cavitation and cracking. According to the theory, the loss of continuity is expressed in a structural damage parameter ω, which takes boundary values of $\omega = 0$ at the start of tertiary time ($t = 0$ for simplicity) and $\omega = 1$ at fracture, $t = t_f$. During tertiary creep, the strain rate ϵ_c and damage rate $\dot{\omega}$ are assumed to depend upon σ, T and ω. For a given temperature, the following functions, consistent with these boundary conditions, were assumed:

$$\dot{\epsilon} = A\sigma^n (1 - \omega)^{-q}, \tag{17.32a}$$

$$\dot{\omega} = B\sigma^k (1 - \omega)^{-r}, \tag{17.32b}$$

where A, B, k, n, q and r are positive material constants. The coupled equations (17.32) may be integrated together to give the equation to the tertiary creep curve. From Eq. (17.32b),

$$\int_0^{\omega} (1 - \omega)^r \, d\omega = B\sigma^k \int_0^t dt, \tag{17.33a}$$

which gives the time $t = t_f$ to rupture for $\omega = 1$:

$$t_f = 1/B(1 + r)\sigma^k. \tag{17.33b}$$

With the upper limit of Eq. (17.33a) as $0 < \omega < 1$, it integrates to

$$1 - (1 - \omega)^{1+r} = B(1 + r)\sigma^k t,$$

where, from Eq. (17.33b),

$$(1 - \omega)^{1+r} = 1 - t/t_f. \tag{17.33c}$$

Substituting Eq. (17.33c) into (17.32a) gives the tertiary creep strain:

$$\epsilon = A\sigma^n \int_0^t (1 - t/t_f)^{-q/(1+r)} \, dt$$
$$= A\sigma^n t_f \lambda [1 - (1 - t/t_f)^{1/\lambda}],$$
$$\epsilon/\epsilon_f = 1 - (1 - t/t_f)^{1/\lambda}, \tag{17.34}$$

where $\lambda = (1 + r)/(1 - q + r)$ and the rupture strain is $\epsilon_f = A\sigma^n t_f \lambda$. Equation (17.34) supplies a creep curve that is concave upwards (Fig. 17.12), consistent with an increasing creep rate in the tertiary region. The ductility ratio λ is also given as

$$\lambda = \epsilon_f/A\sigma^n t_f = \epsilon_f/\dot{\epsilon}_s t_f,$$

where, from Eq. (17.32a), $\dot{\epsilon}_s = A\sigma^n$ for $\omega = 0$ at the start of tertiary creep (the

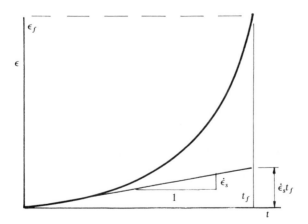

Figure 17.12 Tertiary creep curve.

end of secondary creep). Thus, λ in Eq. (17.34) may be calculated from the measured quantities: minimum (or secondary) creep rate $\dot{\epsilon}_s$, the rupture strain ϵ_f and the rupture time t_f as shown in Fig. 17.12.

Correlation of Uniaxial Creep Data

A number of state-equation parameters have been proposed to correlate both the creep deformation and rupture life of a material under given steady stress and temperature conditions. In the Dorn (1957) approach, the parameter $\theta = \theta(t, T)$ appears in the strain function, $\epsilon_c = \epsilon_c(\sigma, \theta)$. When the independent variables σ and θ in this function are separated, ϵ_c may be plotted against θ, for a constant σ, to produce a single curve over a range of T (Fig. 17.13a).

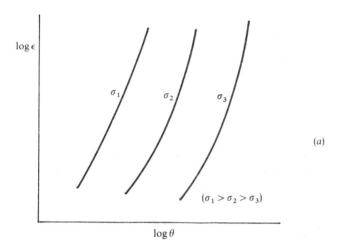

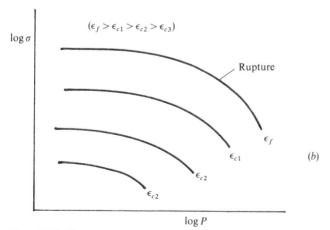

Figure 17.13 Creep parameters.

With the following Dorn parameter,

$$\theta = t \exp(-Q/RT),$$

the method was found to work better for pure metals than for alloys.

In the Larson–Miller (1952), Manson–Haferd (1953) and Graham–Walles (1958) approaches, parameters $P = P(t, T)$ appear with creep strain ϵ_c in the stress function $\sigma = \sigma(\epsilon_c, P)$. Thus, for a constant ϵ_c, the plot of σ vs. P produces a single curve for a range of T (Fig. 17.13b). The parameters are

$$P_{LM} = T(C + \log_{10} t), \tag{17.35a}$$

$$P_{MH} = (T - T_a)/(\log_{10} t - \log_{10} t_a), \tag{17.35b}$$

$$P_{GW} = t(T' - T)^{-20} \quad \text{for } T' > T \quad \text{and} \quad P_{GW} = t(T - T')^{20} \quad \text{for } T > T', \tag{17.35c}$$

where, $C \simeq 20$, t_a, T_a and T' are constants. Each of Eqs (17.35) was found to represent the creep behaviour of engineering alloys within an acceptable scatter band.

Example 17.11 Two tensile creep rupture tests on austenitic stainless steel showed that $t_f = 10^4$ h under $\sigma = 121$ MPa at 650°C and that $t_f = 5 \times 10^4$ h under $\sigma = 28$ MPa at 815°C. Estimate the life under 11 MPa at 925°C using the Larson–Miller parameter with $C = 20$.

Substituting the two test conditions in Eq. (17.35a) gives

$$P_{LM} = (650 + 273)(20 + \log 10^4) = 923(20 + 4) = 22\,152,$$
$$P_{LM} = (815 + 273)[20 + \log(5 \times 10^4)] = 1088(20 + 4.699) = 26\,872.5.$$

If it is assumed that P_{LM} and σ plot linearly on double-log axes within one logarithmic decade of stress, then $\sigma = N(P_{LM})^m$ so that

$$\log \sigma = m \log P_{LM} + \log N,$$
$$\log 121 = m \log 22\,152 + \log N,$$
$$\log 28 = m \log 26\,872.5 + \log N,$$

from which $m = -7.5757$, $\log N = 35$. Then for $\sigma = 11$ MPa, $P_{LM} = 30\,379$,

$$\log t_f = P_{LM}/T - 20 = 30\,379/(925 + 273) - 20 = 5.358,$$
$$\therefore \quad t_f = 227\,986 \text{ h}.$$

Since the actual life is closer to 30 000 h, the example illustrates the danger in extrapolating data without the reference to the curve embracing all the test conditions.

Creep of Structures

In practice it becomes necessary to apply the function $\epsilon_c = \epsilon_c(\sigma, T, t)$ to structures that creep. Where the problem involves multiaxial stress states that vary throughout the volume, numerical analyses are employed. However, creep under simpler stress systems in specific structures, e.g. beams, torsion bars and thin-walled pressure vessels, may be treated analytically. These are examined here.

Beams in bending Assuming that the power law $\epsilon_c = A\sigma^n f(t)$ (see Eq. (7.26b)) applies to the beam material at constant temperature, the moment equilibrium relation to be satisfied is (Eq. (6.2))

$$M = \int_A \sigma \, dA \, y. \tag{17.36}$$

The condition of zero axial force; $F = \int_A \sigma \, dA = 0$ (page 128) again implies that the neutral axis passes through the centroid of the section. In addition, plane sections remain plane when, from Eq. (6.1), $\epsilon = y/R$. Thus, when instantaneous elastic and plastic strains are ignored, a simplified analysis gives the creep strain as

$$\epsilon_c = y/R = A\sigma^n f(t). \tag{17.37}$$

Substituting Eq. (17.37) into (17.36):

$$M = [ARf(t)]^{-1/n} \int_A y^{(1 + 1/n)} \, dA = I_c [ARf(t)]^{-1/n}, \tag{17.38a}$$

where the moment of 'inelastic' area is

$$I_c = \int_A y^{(1 + 1/n)} \, dA. \tag{17.38b}$$

For example, with a rectangular section $b \times d$ and $\delta A = b \, \delta y$, Eq. (17.38b) gives

$$I_c = 2b \int_0^{d/2} y^{1 + 1/n} \, dy = [2bn/(1 + 2n)](d/2)^{(1 + 2n)/n}. \tag{17.38c}$$

Combining Eqs (17.37) and (17.38a) completes the comparison with ETB (Eq. (6.41)):

$$M/I_c = 1/[ARf(t)]^{1/n} = \sigma/y^{1/n}. \tag{17.39}$$

The skeletal position y_{sk} in the section, for which the elastic and creep stresses would be the same under the given M, is found from equating (6.4) and (17.39):

$$\sigma = My_{sk}/I = My_{sk}^{1/n}/I_c. \tag{17.40a}$$

This gives

$$y_{sk} = (I/I_c)^{n/(n-1)}. \tag{17.40b}$$

Thus, as an elastic stress calculation applies to this point, it is often used to simplify design procedures involving creep.

Provided creep deflections are small, the flexure equation (7.3) may be combined with Eq. (17.39) to give

$$d^2v/dz^2 = 1/R = Af(t)[M(z)/I_c]^n. \tag{17.41}$$

When elastic $\epsilon^E = \sigma/E$, plastic ϵ^P and thermal strain $\epsilon^T = \alpha T$ also make a significant contribution to the net strain, then Eq. (17.37) becomes

$$y/R = \sigma/E + \epsilon^P + \alpha T + \epsilon_c,$$
$$\therefore \quad \sigma = E(y/R - \epsilon^P - \alpha T - \epsilon_c),$$

which must be substituted into Eq. (17.36) and integrated numerically.

Example 17.12 A 0.5 m long cantilever with rectangular section 50 mm wide $\times$ 25 mm deep carries an end load of 45 N at 300°C. Neglecting elasticity and self-weight, determine the deflection at the centre and the end due to the effect of secondary creep after 10 h if its rate is given by $\dot{\epsilon}_s = 3.5 \times 10^{-6}\sigma^2$ with $\dot{\epsilon}_s$ in h^{-1} and σ in MPa. What is the maximum stress in the beam and where is the skeletal point? (I.C. 1966)

Here $M(z) = 45z$ (N mm) with the origin of z at the free end. The secondary creep strain is simply $\epsilon_c = 3.5 \times 10^{-6}\sigma^2 t$. Equation (17.38c) gives

$$I_c = (2 \times 50 \times 2/5)(25/2)^{2.5} = 22\,097.09.$$

Hence, Eq. (17.41) becomes

$$d^2v/dz^2 = (3.5 \times 10^{-6})t(45z/22\,097.09)^2 = (1.452 \times 10^{-11}t)z^2,$$
$$dv/dz = (1.452 \times 10^{-11}t)z^3/3 + C_1,$$
$$v = (1.452 \times 10^{-11}t)z^4/12 + C_1 z + C_2.$$

The boundary conditions are

$$dv/dz = 0 \text{ when } z = 500 \text{ mm}, \quad \Rightarrow C_1 = -0.605 \times 10^{-3},$$
$$v = 0 \text{ when } z = 500 \text{ mm}, \quad \Rightarrow C_2 = -0.075\,63t + 0.3025,$$
$$\therefore \quad v = (1.452 \times 10^{-11}t)z^4/12 - (0.605 \times 10^{-3})z - 0.075\,63t + 0.3025.$$

After $t = 10$ h at $z = 0$ mm:

$$v = -0.7563 + 0.3025 = -0.4538 \text{ mm}.$$

After $t = 10$ h at $z = 250$ mm.

$$v = 0.4726 - 0.1513 - 0.7563 + 0.3025 = -0.1324 \text{ mm}.$$

The maximum stress occurs at the fixed end, where $M = 45 \times 500 = 22\,500$ N mm and $y = 12.5$ mm. Then, from Eq. (17.39),

$$\sigma = My^{1/n}/I_c$$
$$= 22\,500 \times (12.5)^{0.5}/22\,097.09 = 3.6 \text{ MPa},$$

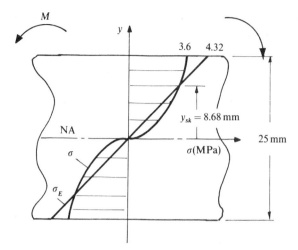

Figure 17.14 Skeletal point.

which is independent of secondary time. Stress is distributed throughout the section as shown in Fig. 17.14. The position of the skeletal point is also shown at the intersection between the creep and elastic stress distributions. From Eq. (17.40b),

$$y_{sk} = [50 \times 25^3/(12 \times 22\,097.09)]^2 = 8.68 \text{ mm}.$$

Note that the initial elastic stress ($\sigma = My/I$) redistributes under M during creep to a more uniform condition through the depth.

Circular shaft in torsion For a thick-walled tube of inner and outer radii r_i and r_o under torque T, the creep shear strain γ at radius r after time t is given by

$$\gamma = r\theta/l = A[\tau(r)]^n f(t), \tag{17.42a}$$

in which the elastic and plastic shear strain under T are ignored. Equation (17.42a) is readily derived from $\epsilon_c = A\sigma^n f(t)$, when for torsional equivalence, $\epsilon_c = \gamma/\sqrt{3}$ and $\sigma = \sqrt{3}\tau$ from Eqs (17.11) and (17.12). The shear stress is found from Eq. (17.42a):

$$\tau(r) = [\theta/Alf(t)]^{1/n} r^{1/n} = Kr^{1/n}. \tag{17.42b}$$

Now, the equilibrium condition is (see page 287)

$$T = 2\pi \int_{r_i}^{r_o} \tau(r)r^2 \, dr. \tag{17.43a}$$

Substituting Eq. (17.42b) into (17.43a) and integrating leads to

$$T = 2\pi Kn(r_o^{3+1/n} - r_i^{3+1/n})/(1+3n). \tag{17.43b}$$

Eliminating K between Eqs (17.42b) and (17.43b) gives

$$\tau(r) = \frac{(1 + 3n)Tr^{1/n}}{2\pi n(r_o^{3+1/n} - r_i^{3+1/n})} \qquad (17.44)$$

and combining Eqs (17.42a) and (17.44) to complete the comparison with ETT (for $n = 1$, see Eq. (9.5)), we have for creep,

$$T/J_c = [\theta/Af(t)]^{1/n} = \tau/r^{1/n}, \qquad (17.45)$$

where $J_c = 2\pi n(r_o^{3+1/n} - r_i^{3+1/n})/(1 + 3n)$ for a hollow shaft. With a solid shaft, $r_i = 0$ and $J_c = 2\pi n r_o^{3+1/n}/(1 + 3n)$. The skeletal radius r_{sk} at which the elastic and creep shear stresses are the same under a given T is found from

$$\tau = Tr_{sk}/J = Tr_{sk}^{1/n}/J_c. \qquad (17.46a)$$

For a tube, Eq. (17.46a) gives

$$r_{sk} = \left\{ \frac{(1 + 3n)(r_o^4 - r_i^4)}{4n[r_o^{(3n+1)/n} - r_i^{(3n+1)/n}]} \right\}^{n/(n-1)}. \qquad (17.46b)$$

Example 17.13 Determine the position of the skeletal radius for a tube of inner radius 5 mm (1/4 in) and outer radius 25 mm (1 in), when it is subjected to a constant torque of 1 kN m (750 lbf ft) under creep conditions for a stress index of $n = 5$ in secondary creep. Compare the elastic and steady-state creep shear stress distributions.

IN SI UNITS Substituting $r_i = 5$ mm, $r_o = 25$ mm and $n = 5$ into Eq. (17.46b):

$$r_{sk} = \left\{ \frac{16(25^4 - 5^4)}{20(25^{3.2} - 5^{3.2})} \right\}^{1.25}$$

$$r_{sk} = (10.55)^{1.25} = 19.015 \text{ mm},$$

and from Eq. (17.44) the creep shear stress distribution is given by

$$\tau_c = 16 \times 1 \times 10^6 \times r^{0.2}/(10\pi \times 29\,572.13) = 17.22r^{0.2} \quad \text{(MPa)}. \quad \text{(i)}$$

The elastic shear stress is found from ETT (Eq. (9.5)):

$$\tau_E = Tr/J = 2 \times 1 \times 10^6 \times r/\pi(25^4 - 5^4) = 1.632r \quad \text{(MPa)}. \qquad \text{(ii)}$$

The elastic and creep stress distributions from Eqs (i) and (ii) are shown in Fig. 17.15. It is evident that the initial, instantaneous elastic stress under T redistributes during primary creep to attain the more uniform steady condition under secondary creep.

SOLUTION IN IMPERIAL UNITS From Eq. (17.46b), the skeletal radius is

$$r_{sk} = \left\{ \frac{16[(1)^4 - (1/4)^4]}{20[(1)^{3.2} - (1/4)^{3.2}]} \right\}^{1.25} = 0.764 \text{ in},$$

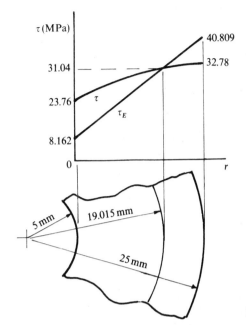

Figure 17.15 Stress distributions.

and from Eq. (17.44) the creep stress distribution is

$$\tau_c = \frac{16 \times 750 \times 12 r^{0.2}}{2\pi \times 5[(1)^{3.2} - (1/4)^{3.2}]} = (4.639 \times 10^3) r^{0.2} \, \text{lbf/in}^2.$$

Multiaxial stress When creep deformation is predominantly the result of the same dislocation mechanisms associated with low-temperature plasticity, then the time derivatives of Eqs (17.16) will supply the creep rates in the case of multiaxial principal stress states. The principal secondary creep rates corresponding to $\dot{\epsilon}_s = d\epsilon^P/dt$ are, from Eq. (17.16),

$$\dot{\epsilon}_{1s} = (\dot{\bar{\epsilon}}_s/\bar{\sigma})[\sigma_1 - (\sigma_2 + \sigma_3)/2], \tag{17.47a}$$

$$\dot{\epsilon}_{2s} = (\dot{\bar{\epsilon}}_s/\bar{\sigma})[\sigma_2 - (\sigma_1 + \sigma_3)/2], \tag{17.47b}$$

$$\dot{\epsilon}_{3s} = (\dot{\bar{\epsilon}}_s/\bar{\sigma})[\sigma_3 - (\sigma_1 + \sigma_2)/2], \tag{17.47c}$$

where the equivalent stress $\bar{\sigma}$ is defined in Eq. (17.11) and the equivalent secondary creep rate $\dot{\bar{\epsilon}}_s$ is the time derivative of Eq. (17.12). Combining Eqs (17.47) with the Norton uniaxial law (Eq. (17.21)) establishes the coefficient in Eqs (17.47) as

$$\dot{\bar{\epsilon}}_s/\bar{\sigma} = A\bar{\sigma}^{n-1}.$$

This time-dependent plasticity approach cannot account for the increase in creep rates that arise throughout life in certain cavity–forming alloys, e.g. stainless steels and nimonics, without a Katchanov–Rabotnov damage modification. That is, from Eq. (17.32a), the coefficient in Eqs (17.47) becomes

$$\dot{\bar{\epsilon}}/\bar{\sigma} = A\bar{\sigma}^{n-1}(1 - \omega)^{-q},$$

in which ω is supplied from Eq. (17.32b).

Example 17.14 A thin-walled pressure vessel of diameter d and wall thickness t is to be fitted with hemispherical ends. In addition, the vessel is to be designed to withstand an internal pressure p when operating under creep conditions. Assuming that the most satisfactory design life is obtained when the circumferential strain rates at the cylinder–hemisphere interface are matched, calculate the thickness of the ends if the material obeys the uniaxial Norton law $\bar{\epsilon}_s = A\bar{\sigma}^4$. Neglect elastic strains. (I.C. 1970).

With zero radial stress everywhere, the circumferential creep rate for the cylinder (subscript c) is, from the polar form of Eq. (17.47a),

$$\dot{\epsilon}_\theta = A\bar{\sigma}^{n-1}(\sigma_\theta - \sigma_z/2), \qquad (i)$$

where, for biaxial stresses, the equivalent stress (Eq. 17.11) becomes

$$\bar{\sigma}_c = \sqrt{(\sigma_\theta^2 - \sigma_\theta\sigma_z + \sigma_z^2)}. \qquad (ii)$$

Substituting Eq. (ii) into (i) with $\sigma_\theta = pd/2t_c$, $\sigma_z = pd/4t_c$ leads to

$$\dot{\epsilon}_\theta = A\bar{\sigma}_c^{n-1}(3pd/8t_c)$$
$$= A(3pd/16t_c)^{(n-1)/2}(3pd/8t_c). \qquad (iii)$$

For the ends (subscript s), Eqs (17.11) and (17.47a) reduce to

$$\dot{\epsilon}_\theta = A\bar{\sigma}_s^{n-1}(\sigma_\theta - \sigma_\phi/2), \qquad (iv)$$
$$\bar{\sigma}_s = \sqrt{(\sigma_\theta^2 - \sigma_\theta\sigma_\phi + \sigma_\phi^2)}. \qquad (v)$$

Substituting Eq. (v) into (iv) with $\sigma_\theta = \sigma_\phi = pd/4t_s$ gives

$$\dot{\epsilon}_\theta = A\bar{\sigma}_s^{n-1}(pd/8t_s)$$
$$= A(pd/16t_s)^{(n-1)/2}(pd/8t_s). \qquad (vi)$$

Equating (iii) and (vi) gives the thickness ratio

$$t_c/t_s = [3^{(n+1)/2}]^{1/n} = 3^{(n+1)/2n} = 3^{5/8} = 1.987.$$

17.5 VISCOELASTICITY

The traditional method for representing the viscoelastic strain response of non-metallic materials to stress is through the use of rheological models. Firstly, an ideal elastic solid is modelled with a Hookean spring, that is associated with the proportionality that exists between stress and strain:

$$\sigma = E\epsilon_s, \tag{17.48a}$$

where E is the elastic modulus. For a shear mode of deformation, Eq. (17.48a) would be written $\gamma = G\tau_s$. Secondly, for a perfectly viscous fluid, a Newtonian dashpot is associated with the linear relationship between stress and rate of strain:

$$\sigma = \mu\dot{\epsilon}_d \tag{17.48b}$$

where μ is the coefficient of viscosity. Where the observed behaviour is neither ideally elastic nor perfectly viscous, a spring–dashpot combination is employed to match particular viscoelastic effects. In the case of the representation of creep, relaxation and recovery phenomena, for example, the usefulness of series and parallel combinations of a single spring and dashpot are first examined.

Two-element Maxwell Model

In this series combination (Fig. 17.16(a)) the net strain ϵ under stress σ is the sum of the strains in the spring and dashpot. That is

$$\epsilon = \epsilon_s + \epsilon_d. \tag{17.49a}$$

Differentiating Eq. (17.49a) w.r.t. time t and substituting from Eqs (17.48) provide the governing constitutive equation:

$$d\epsilon/dt = (1/E)\,d\sigma/dt + \sigma/\mu. \tag{17.49b}$$

For *creep*, where $\sigma = \sigma_o = $ constant, Eq. (17.49b) becomes

$$d\epsilon/dt = \sigma_o/\mu,$$
$$\therefore \quad \epsilon = (\sigma_o/\mu)t + A_1. \tag{17.50a}$$

For $t = 0$, $\epsilon = \sigma_o/E$, indicating that the instantaneous strain is elastic only. This gives the constant $A_1 = \sigma_o/E$, whereupon Eq. (17.50a) becomes

$$\epsilon = \sigma_o(t/\mu + 1/E). \tag{17.50b}$$

This linear relationship between ϵ and t is clearly unsatisfactory when creep accumulates at a decreasing or increasing rate (Fig. 17.16(b)).

When, in Fig. 17.16(b) the stress σ_o is completely removed after time t_1 an immediate *recovery* of the elastic spring component of strain $\epsilon_s = \sigma_o/E$ occurs. However, as no viscous dashpot strain is recovered, regardless of the elapsed time, the model does not reproduce observed behaviour. In the case of *relaxation*, $\epsilon = \epsilon_o = $ constant and Eq. (17.49b) becomes

$$d\sigma/\sigma = -(E/\mu)\,dt,$$

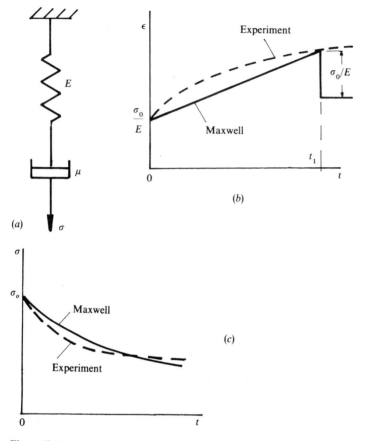

Figure 17.16 Creep and relaxation for the Maxwell model.

which integrates to

$$\ln \sigma = -(E/\mu)t + A_2. \qquad (17.51a)$$

For $t = 0$, $\sigma = \sigma_o$ giving $A_2 = \ln \sigma_o$. Then Eq. (17.51a) gives the relaxed stress:

$$\sigma = \sigma_o \exp(-Et/\mu), \qquad (17.51b)$$

in which $\sigma \to 0$ as $t \to \infty$. This exponential stress decay (Fig. 17.16(c)) is consistent with observed behaviour.

Two-element Voigt or Kelvin Model

For the parallel combination in Fig. 17.17(a), the net stress σ is the sum of the stresses for the spring and dashpot. That is,

$$\sigma = \sigma_s + \sigma_d. \qquad (17.52a)$$

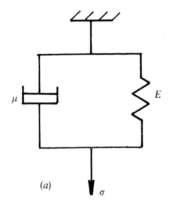

(a)

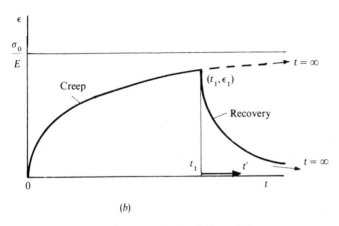

(b)

Figure 17.17 Creep and recovery for the Voigt model.

Substituting Eqs (17.48a and b) into Eq. (17.52a) leads to the governing constitutive equation:

$$\sigma = E\epsilon + \mu\, d\epsilon/dt. \qquad (17.52b)$$

For *creep*, $\sigma = \sigma_o$ and Eq. (17.52b) results in the differential equation

$$d\epsilon/dt + (E/\mu)\epsilon = \sigma_o/\mu. \qquad (17.53)$$

Integrating Eq. (17.53) after separating the variables leads to

$$\int d\epsilon/(\sigma_o - E\epsilon) = (1/\mu)\int dt$$

$$-(1/E)\ln(\sigma_o - E\epsilon) = t/\mu + A_1. \qquad (17.54a)$$

Here $\epsilon = 0$ for $t = 0$, since the dashpot prevents instantaneous strain. Hence,

$A_1 = -(1/E)\ln\sigma_o$ and Eq. (17.54a) gives the creep strain as

$$\epsilon = (\sigma_o/E)[1 - \exp(-Et/\mu)]. \tag{17.54b}$$

It is seen from Fig. 17.17(b) that Eq. (17.54b) gives a primary creep curve in which $\epsilon \to \sigma_o/E$ as $t = \infty$. Here the strain magnitude is clearly incorrect. With the removal of σ_o at time t_1, Eq. (17.53) predicts *recovery* of strain ϵ after time $t' = t - t_1$ as

$$d\epsilon/dt' + (E/\mu)\epsilon = 0. \tag{17.55a}$$

Since $\epsilon = \epsilon_1$ (the creep strain at t_1) for $t' = 0$, Eq. (17.55a) integrates to

$$\epsilon = \epsilon_1 \exp(-Et'/\mu) = \epsilon_1 \exp[-E(t - t_1)/\mu], \tag{17.55b}$$

in which ϵ_1 is given by Eq. (17.54b) for $t = t_1$. Then Eq. (17.55b) becomes

$$\epsilon = (\sigma_o/E)[1 - \exp(-Et_1/\mu)]\exp[-E(t - t_1)/\mu],$$
$$\epsilon = (\sigma_o/E)\exp(-Et/\mu)[\exp(Et_1/\mu) - 1]. \tag{17.55c}$$

It is seen from Fig. 17.17(b) that $\epsilon = \epsilon_1$ for $t = t_1$ and $\epsilon \to 0$ as $t \to \infty$. Apart from the absence of instantaneous elastic recovery, the model is consistent with observed behaviour.

Under *relaxation* conditions where $\epsilon = \epsilon_o = $ constant Eq. (17.52b) gives $\sigma = E\epsilon$. The model therefore fails to admit relaxation of the initial stress.

The Standard Linear Solid (SLS)

When the Voigt model is connected in series with a second Hookean spring of different stiffness, it can be expected that the resulting SLS (Fig. 17.18(a)) will provide a better model of viscoelastic behaviour. Here there is equality in stress between the Voigt element (subscripts v and 2) and the single spring (subscript 1). This gives

$$\sigma = E_1\epsilon_1 = E_2\epsilon_v + \mu\, d\epsilon_v/dt. \tag{17.56}$$

Moreover, the total strain ϵ under an applied stress σ is

$$\epsilon = \epsilon_1 + \epsilon_v,$$
$$\therefore \quad \epsilon_v = \epsilon - \sigma/E_1. \tag{17.57}$$

Substituting Eq. (17.57) into (17.56) leads to the constitutive equation

$$d\epsilon/dt + (E_2/\mu)\epsilon = (\sigma/\mu)(1 + E_2/E_1) + (1/E_1)\,d\sigma/dt. \tag{17.58}$$

For *creep*, $\sigma = \sigma_o$, $d\sigma_o/dt = 0$ and Eq. (17.58) becomes

$$d\epsilon/dt + (E_2/\mu)\epsilon = (\sigma_o/\mu)(1 + E_2/E_1). \tag{17.59a}$$

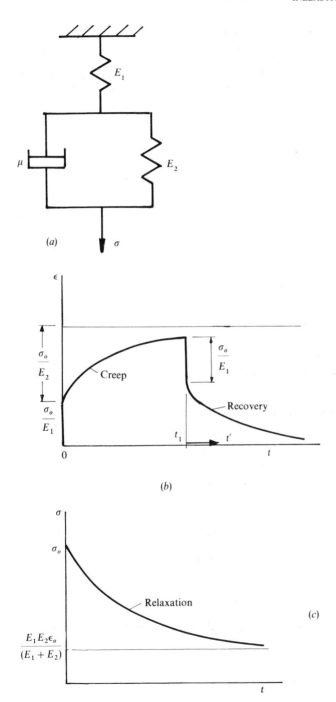

Figure 17.18 Creep, recovery and relaxation from the SLS.

Separating the variables leads to

$$\mu \int \frac{d\epsilon}{[\sigma_o + E_2(\sigma_o/E_1 - \epsilon)]} = \int dt$$

and integration, with $\epsilon = \sigma_o/E$ for $t = 0$, gives the net strain as

$$\epsilon = \sigma_o/E_1 + (\sigma_o/E_2)[1 - \exp(-E_2 t/\mu)]. \tag{17.59b}$$

Equation (17.59b) displays an instantaneous strain σ_o/E_1 followed by time-dependent creep strain that approaches σ_o/E_2 with decreasing rate as $t \to \infty$ (Fig. 17.18(b)). On removal of σ_o after time t_1, the elastic strain σ_o/E_1 is immediately recovered. Thereafter, for $t > t_1$ in Fig. 17.18b, it follows from Eq. (17.55c) that the *recovery* of creep strain conforms to

$$\epsilon = (\sigma_o/E_2)\exp(-E_2 t/\mu)[\exp(E_2 t_1/\mu) - 1]. \tag{17.59c}$$

That is, $\epsilon \to 0$ as $t \to \infty$. The SLS model thus provides a good representation of observed creep and recovery behaviour in polymers.

For *relaxation*, where $\epsilon = \epsilon_o$ and $d\epsilon_o/dt = 0$, Eq. (17.58) becomes

$$d\sigma/dt + (E_1 + E_2)\sigma/\mu = E_1 E_2 \epsilon_o/\mu. \tag{17.60a}$$

With the condition that $\sigma = \sigma_o = E_1 \epsilon_o$ for $t = 0$, the solution to Eq. (17.60a) is

$$\sigma = E_1 \epsilon_o\{E_1 \exp[-(E_1 + E_2)t/\mu] + E_2\}/(E_1 + E_2). \tag{17.60b}$$

The additional elastic constant E_1 appearing in Eq. (17.60b) enables observed relaxation behaviour to be followed closely. Note from Fig. 17.18(c) that the stress does not relax to zero as $t \to \infty$, but to the asymptotic stress value $\sigma = E_1 E_2 \epsilon_o/(E_1 + E_2)$.

Other Models

There are numerous other combinations of springs and dashpots. The SLS is the simplest possible model to exhibit both creep and relaxation. Parallel combinations of the Maxwell model and series combinations of the Voigt model have been extensively used each with different relaxation $(\mu/E)_m$ and retardation $(\mu/E)_v$ times. By increasing the number of such models without limit, a discrete spectrum of such times enables observed relaxation and creep behaviour to be fitted to any required degree of accuracy (Ferry, 1970). Specific forms of known creep and relaxation behaviour may be modelled more conveniently within a limited combination. For example, a suitable rheological model for concrete employs a Maxwell element (E_1, μ_1) and two Voigt elements $(E_2, \mu_2$ and $E_3, \mu_3)$ all connected in series. This gives, for creep

$$\epsilon = \sigma_o\{1/E_1 + t/\mu_1 + (1/E_2)[1 - \exp(-E_2 t/\mu_2)] + (1/E_3)[1 - \exp(-E_3 t/\mu_3)]\}.$$

A more simplified form of this model was usefully employed by Burger for a representation of the observed creep behaviour in fibre-reinforced resins and bituminous materials. This appears in the following example.

Example 17.15 Deduce the creep response to a stress σ_o when a Maxwell model (E_1 and μ_1) is connected in series with a Voigt model (E_2 and μ_2) as shown in Fig. 17.19(a). Show that as $t \to \infty$ then the creep rate $\epsilon \to \sigma_o/\mu_1$ and that when the dashpot μ_1 is absent the SLS prediction applies.

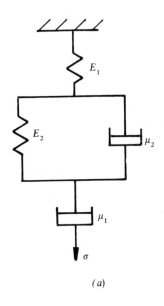

(a)

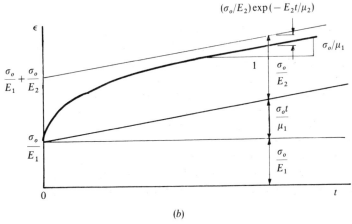

(b)

Figure 17.19 Burger's four-parameter model.

The creep strain is simply found from superimposing the separate Voigt and Maxwell predictions. That is, from the addition of Eqs (17.50b and 17.54b)

$$\epsilon = \epsilon_v + \epsilon_m$$
$$= (\sigma_o/E_2)[1 - \exp(-E_2 t/\mu_2)] + \sigma_o(t/\mu_1 + 1/E_1). \qquad (i)$$

It is seen that the net creep curve in Fig. 17.19(b) is the sum of those in Figs 17.16 and 17.17. The curve shows an instantaneous elastic response σ_o/E_1, followed by primary creep at a decreasing rate. The magnitude of the latter is found from differentiating Eq. (i):

$$\dot{\epsilon} = (\sigma_o/\mu_1)[1 + (\mu_1/\mu_2)\exp(-E_2 t/\mu_2)]. \qquad (ii)$$

It is apparent from Eq. (ii) that a steady rate $\dot{\epsilon} = \sigma_o/\mu_1$ is approached as $t \to \infty$. Clearly, when the dashpot (μ_1) is absent, Eq. (i) reduces to SLS prediction in Eq. (17.59b).

Boltzmann's Superposition Principle (Boltzmann, 1876)

This linear superposition or hereditary principle is useful when dealing with stress or strain inputs that vary with time. In the case of creep under a stepped stress input (Fig. 17.20(a)), the principle asserts that the corresponding net strain response is simply the linear addition of the individual strain responses to each stress level. For a linear viscoelastic material obeying $\epsilon(t) = \sigma C(t)$, the creep strain is linear in stress for a given time. It follows that the time function or compliance $C(t) = \epsilon(t)/\sigma$ is then a constant for that particular time.

When the stress changes from σ_1 to $\sigma_1 + \sigma_2$ at time τ (Fig. 17.20(a)), the

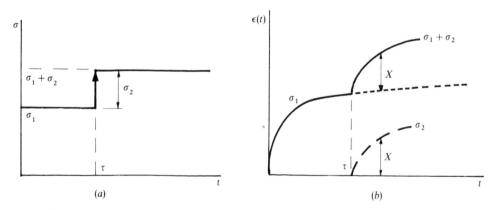

Figure 17.20 Creep superposition under a stress step.

net strain for time $t > \tau$ is increased by the amount X to give:

$$\epsilon(t) = \sigma_1 C(t) + \sigma_2 C(t - \tau), \tag{17.61a}$$

as illustrated in Fig. 17.20(b). In the more general case of a non-linear viscoelastic material, $\epsilon(t) = \phi(\sigma, t)$, the net strain response to this step becomes

$$\epsilon(t) = \phi(\sigma_1, t) + \phi(\sigma_2, t - \tau). \tag{17.61b}$$

When the step corresponds to the removal of σ_1 after time τ (Fig. 17.21(a)), the recovery phenomenon (Fig. 17.21b) will appear for $t > \tau$ with the substitution of $\sigma_2 = -\sigma_1$ in Eqs (17.61). For a linear viscoelastic material, this substitution gives the net strain at $t > \tau$

$$\epsilon(t) = \sigma_1 C(t) - \sigma_1 C(t - \tau). \tag{17.61c}$$

If, however, the creep strain reaches its asymptotic or equilibrium value $\epsilon(\infty)$ (as with $t \to \infty$ in the Voigt model of Fig. 17.17(b)), then after the stress is altered the net strain at $t > \infty$ becomes

$$\epsilon(t) = \sigma_1 C(\infty) + \sigma_2 C(t - \tau).$$

When σ varies with time, this may be treated as a series of steps in an integral formulation. Let an increase in stress $d\sigma$ apply to time τ. According to the principle, the increase in strain after time $t > \tau$ for a linear material is given by

$$d\epsilon(t) = d\sigma C(t - \tau). \tag{17.62a}$$

However, $d\sigma = d\tau(d\sigma/d\tau)$, since σ continuously varies with τ. The net strain at time t then follows from integrating Eq. (17.62a):

$$\epsilon(t) = \int_{-\infty}^{t} C(t - \tau)(d\sigma/d\tau)\,d\tau, \tag{17.62b}$$

in which the lower limit will account for any stress history prior to $t = 0$ as illustrated in Fig. 17.22(a) and (b).

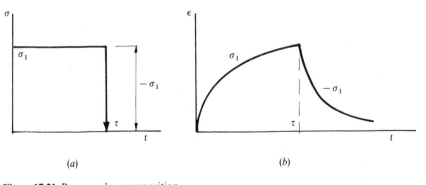

Figure 17.21 Recovery by superposition.

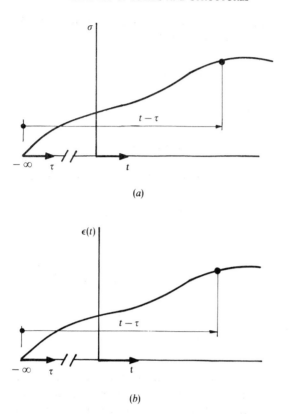

(a)

(b)

Figure 17.22 Continuously varying stress history.

If there has been no prior history of stress, then the lower limit may be taken as zero. In this linear form, Eq. (17.62b) is known as the convolution integral. For creep in non-linear viscoelastic materials, Eqs (17.62a and b) are replaced by the general formulation:

$$d\epsilon(t) = (\partial\phi/\partial\sigma)\,d\sigma, \tag{17.63a}$$

$$\epsilon(t) = \int_{-\infty}^{t} [\partial\phi(\sigma, t - \tau)/\partial\sigma](d\sigma/d\tau)\,d\tau. \tag{17.63b}$$

One particular form of Eq. (17.63b), due to Leaderman (1955), is that for which the variables are separable: $\phi(\sigma, t) \equiv C(t)f(\sigma)$. Equation (17.63b) gives

$$\epsilon(t) = \int_{0}^{t} C(t - \tau)(\partial f/\partial\tau)\,d\tau.$$

Under stress relaxation conditions, where the strain input varies continuously with time, it is possible to deduce the $\sigma(t)$ expressions from a comparison with Eqs (17.62b) and (17.63b). They are, for linear $[\sigma = \epsilon M(t)]$ and for non-linear $[\sigma = \theta(\epsilon, t)]$ viscoelasticity, respectively,

$$\sigma(t) = \int_{-\infty}^{t} M(t - \tau)(d\epsilon/d\tau)\,d\tau,$$

$$\sigma(t) = \int_{-\infty}^{t} [\partial\theta(\epsilon, t - \tau)/\partial\epsilon](d\epsilon/d\tau)\,d\tau.$$

These are known as the Duhamel integrals. It is important to recognize that the relaxation modulus $M(t)$ for a linear viscoelastic material, and the function $\theta(\epsilon, t)$ for non-linear viscoelasticity, are not, in general, the reciprocals of the corresponding creep functions. That is,

$$M(t) \neq 1/C(t) \qquad \text{and} \qquad \theta(\epsilon, t) \neq 1/\phi(\epsilon, t).$$

Example 17.16 Isochronous data for the torsional creep of a linear viscoelastic polymer has established the following time dependence for the rigidity modulus:

t (min)	10	10^2	10^3	10^4	10^5
$G(t)$ (GPa)	11.25	10.62	10	9.1	7.75

A torque of 65 N m is applied rapidly to a tube of this material of 50 mm mean diameter, 1.5 mm thick and 175 mm long. If the torque is held constant for 10^3 min and then removed, determine the maximum angle of twist at the end of this period and 10^2 min after unloading. (I.C. 1969)

Isothermal creep data for polymers is often presented in isochronous form. That is, stress–strain curves, each associated with a given creep time, are obtained from a family of creep curves. Since for a linear viscoelastic material, $\gamma(t) = \tau C(t)$, a single modulus for each time may be identified with the slope of the stress–strain line. In the case of torsion, that is

$$G(t) = \tau/\gamma(t) = 1/C(t),$$

or,

$$\gamma(t) = \tau/G(t) = r\theta(t)/l. \qquad \text{(i)}$$

From the Bato theory (Eq. (9.26)) the constant shear stress is

$$\tau = T/2At = 65 \times 10^3/(2 \times \pi(25)^2 \times 1.5) = 11\,\text{MPa},$$

and from Eq. (i), for $t = 10^3$ min,

$$\theta(10^3) = l\tau/rG(10^3) = 175 \times 11/(25 \times 10 \times 10^3) = 77 \times 10^{-4}\,\text{rad} = 0.44°.$$

Applying the superposition principle to unloading (see Fig. 17.21), the torsional equivalent of Eq. (17.61c) is

$$\gamma(t) = \tau C(t) - \tau C(t - \tau)$$
$$= \tau/G(t) - \tau/G(t - \tau). \qquad \text{(ii)}$$

Here $t = 10^3 + 10^2 = 1100\,\text{min}$, $\tau = 10^3\,\text{min}$, $t - \tau = 10^2\,\text{min}$. Interpolating from the table above, $G(t) = 9.99\,\text{GPa}$ and $G(t - \tau) = 10.62\,\text{GPa}$. Equation (ii) gives

$$\gamma(t) = 11[(1/9.99) - (1/10.62)] \times 10^{-3} = 0.0653 \times 10^{-3}$$

and, from Eq. (i),

$$\theta(t) = l\gamma(t)/r = 175 \times 0.0653 \times 10^{-3}/25$$
$$= 0.457 \times 10^{-3}\,\text{rad} = 0.0262°.$$

Example 17.17 A linear viscoelastic material obeys the creep law $\epsilon(t) = A\sigma t^m$. Determine the creep response to a stress input that increases linearly according to $\sigma = kt$ (k is a constant) for $0 \le t \le t_1$ and whereafter the stress remains constant at kt_1 for $t_1 \le t \le 2t_1$ (Fig. 17.23(a)). If at time $2t_1$ the stress is abruptly removed, determine the creep strain at t_1 and $2t_1$ and the net strain at time $4t_1$ using Boltzmann's superposition principle.

During the linear increase in stress, where $d\sigma/d\tau = k$, Eq. (17.62b) becomes

$$\epsilon(t) = kA \int_0^t (t - \tau)^m \, d\tau$$
$$= (kA)t^{m+1}/(m+1). \qquad \text{(i)}$$

Equation (i) supplies the creep strain at time t_1 as

$$\epsilon(t_1) = (kA)t_1^{m+1}/(m+1). \qquad \text{(ii)}$$

During the constant stress period the creep strain is

$$\epsilon(t) = A\sigma(t - t_1)^m = (kA)t_1(t - t_1)^m. \qquad \text{(iii)}$$

Adding Eqs (i) and (iii) gives the net strain for $t_1 \le t \le 2t_1$.

$$\epsilon(t) = (kA)t^{m+1}/(m+1) + (kA)t_1(t - t_1)^m. \qquad \text{(iv)}$$

Equation (iv) supplies the creep strain at time $t = 2t_1$ as

$$\epsilon(2t_1) = (kA)t_1^{m+1}[1 + 2^{m+1}/(m+1)].$$

Applying Eq. (17.61c) for the net strain in the interval $2t_1 \le t \le 4t_1$:

$$\epsilon(t) = (kA)t^{m+1}/(m+1) + (kA)t_1(t - t_1)^m - (kA)t_1(t - 2t_1)^m, \qquad \text{(v)}$$

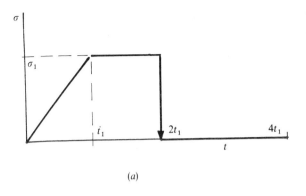

(a)

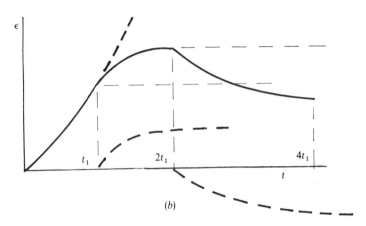

(b)

Figure 17.23 Stress and strain responses for linear viscoelasticity.

and for $t = 4t_1$, Eq. (v) gives the net strain:

$$\epsilon(4t_1) = (kA)t_1^{m+1}[4^{m+1}/(m+1) + 3^m - 2^m].$$

The separate strain responses are illustrated in Fig. 17.23(b). They are

$$0 \leq t \leq t_1, \quad \text{creep under a linear stress increase,}$$
$$t_1 \leq t \leq 2t_1, \quad \text{creep under a constant stress,}$$
$$t \geq 2t_1, \quad \text{recovery following stress removal.}$$

The net strain response (heavy line) is the sum of these.

REFERENCES

Andrade, N. da C. (1910). *Proc. R. Soc.* A **84**, 1.
Bailey, R. W. (1935). *Proc. Inst. Mech. Eng.* **131**, 131.

Boltzmann, L. (1876). *Pog. Ann. Physik* **7**, 624.

Bridgeman, P. W. (1952). *Studies in Large Plastic Flow and Fracture*, p. 9. McGraw-Hill, New York.

Considere, A. (1885). *Ann. Ponts et Chaussees* **9**(6), 574.

Davis, P. W, Evans, W. J., Williams, K. R. and Wilshire, B. (1969). *Scripta Metall.* **3**, 671.

Dorn, J. E. (1957). *Creep and Recovery*, American Society for Metals, Cleveland, Ohio.

Evans, W. J. (1982). *Recent Advances in the Creep and Fracture of Engineering Materials and Structures* (ed. B. Wilshire and D. R. J. Owen), p. 135. Pineridge, Swansea.

Ferry, J. D. (1970). *Viscoelastic Properties of Polymers*, Wiley, New York.

Garofalo, F. (1965). *Fundamentals of Creep and Creep Rupture in Metals*. Macmillan, New York.

Graham, A. and Walles, K. F. A. (1955). *J. Iron Steel Inst.* **179**, 105.

Graham, A. and Walles, K. F. A. (1958). *Aero. Res. Cl. Rpt. on Regularities in Creep and Hot Fatigue Data, Pt. I*, No. 379. HMSO, London.

Hollomon, J. H. (1945). *Am. Inst. Mech. Eng.* **162**, 268.

Katchanov, L. M. (1959). *Izv. Akad. Nauk. SSSR (Mech. Eng.)* **3**, 84.

Larson, F. R. and Miller, J. (1952). *Trans. Am. Soc. Mech. Eng.* **74**, 765.

Leaderman, H. (1955) *J. Polymer. Sci.*, **16**, 261.

Levy, M. (1870). *C.r. hebd. Seanc. Acad. Sci., Paris* **70**, 1323.

Ludwik, P. (1909). *Elements der Technologischen Mechanik*. Verlag Julius Springer, Berlin.

Manson, S. S. and Haferd, A. M. (1953). N.A.C.A. Tech. Note 2890.

Mises, R. von. (1913). *Nader Ges. Wiss. Gottingen*, **4**, p. 582.

Monkman, F. C. and Grant, N. J. (1956). *Proc. Am. Soc. Test. Mater.* **56**, 593.

Nadai, A. (1950). *Theory of Flow and Fracture of Solids*, McGraw-Hill, New York.

Norton, F. H. (1929). *Creep of Steel at High Temperatures*. McGraw-Hill, New York.

Rabotnov, Y. N. (1969). *Creep Problems in Structural Members*. North Holland, Amsterdam.

Robinson, E. L. (1938). *Trans. Am. Soc. Mech. Eng.* **100**, 253.

EXERCISES

Plasticity and Hardening Behaviour

17.1 Convert the nominal load–extension data in Exercise 3.12 to true stress and natural strain up to the instability point and thus determine the strain hardening coefficient and exponent in the law $\sigma = A\epsilon^n$.

17.2 If the true stress–strain curve for a material is defined by $\sigma = A\epsilon^n$, determine the nominal tensile strength. Given $A = 800 \, \text{MPa}$ and $n = 0.2$, find the ultimate tensile strength and the true stress at instability.

Answer: An^n/e^n; 475 MPa; 580 MPa.

17.3 Show that when the Vocé exponential law $\sigma = \sigma_\infty - (\sigma_\infty - \sigma_o)\exp(-\epsilon/\mu)$ describes true stress–strain behaviour then the stress and strain at the point of instability are $\sigma_u = \sigma_\infty/(1 + \mu)$ and $\epsilon_u = \mu \ln[(1 + \mu)(\sigma_\infty - \sigma_o)/\mu\sigma_\infty]$, where μ, σ_o and σ_∞ are constants.

17.4 Given that the stress and strain at the point of instability are $\sigma_u = Kn^n$ and $\epsilon_u = n - \epsilon_o$, show that these derive from the Swift law $\sigma = K(\epsilon_o + \epsilon)^n$, where K, ϵ_o and n are constants.

17.5 Use the Bridgeman correction factor to determine the true stress–strain behaviour beyond the point of tensile instability for the results referred to in Exercise 17.1.

17.6 Given that $T = T_o(1 + A\kappa)^n$ describes the non-linear portion of the torque vs. unit-twist curve for a solid bar, determine the corresponding τ vs. γ relationship for the bar material. If $n = 1/4$ and $A = 4$, determine the torque required to produce a shear stress of 400 MPa at the 15 mm outer radius of a solid bar. What is the corresponding shear strain?

17.7 The following torque–twist/unit length data applies to a 25 mm diameter, 75 mm parallel length solid bar. Establish the shear stress–shear strain behaviour for the bar material.

T(Nm)	514	582	729	797	910	983.4	1068	1124.7
$\theta°$/mm	0.131	0.197	0.394	0.525	0.787	1.05	1.58	2.10

T(Nm)	1158.6	1192.5	1221.9	1245.6	1265.9	1284	1295	1402
$\theta°$/mm	2.49	2.89	3.28	3.68	4.07	4.46	4.72	8.79

17.8 Determine the equivalent stress and strain at the point of instability for a thin-walled, open-ended pressurized cylinder of mean diameter d and thickness t, if the material hardens according to the law $\bar{\sigma} = \sigma_o(1 + B\bar{\epsilon})^n$. What is the pressure that causes this instability?

Answer: $\bar{\sigma} = \sigma_o(2nB/3)^n$; $\bar{\epsilon}_u = (1/B)(2nB/3 - 1)$; $p = (4t\sigma_o/d)(2nB/3)^n$.

17.9 Why does fracture in the parallel length of a circular section tensile test piece initiate from a point on its axis? How does a superimposed hydrostatic pressure influence the nominal axial strain required to cause tensile necking? Explain with reference to the Bridgeman analysis.

17.10 Determine the true fracture strain of a ductile material in terms of (i) the percentage elongation at fracture and (ii) the percentage reduction of area at fracture in a tensile test.

17.11 Find the condition for which the hoop stress at the bore of a closed pressurized thick cylinder is zero when it is fully plastic. Assume a non-hardening von Mises material.

17.12 Show that the pressure required to penetrate a plastic zone to a radial depth c in a non-hardening thick-walled cylinder, of inner and outer radii r_i and r_o respectively, is independent of the end condition according to the Tresca yield criterion. If $r_o/r_i = 3.5$ and $Y = 375$ MPa, determine the pressure to produce full plasticity and the maximum hoop and radial stresses when the ends are open.

17.13 Determine the internal pressure required to cause the plastic zone to penetrate to the mean radius in a thick-walled closed cylinder with respective inner and outer diameters of 25 and 100 mm. Plot the corresponding distribution of σ_θ, σ_r and σ_z throughout the wall when this pressure is held and when it is removed. Assume a von Mises yield criterion and take $Y = 350$ MPa, $E = 207$ GPa and $v = 0.27$.

17.14 Derive the expression for the internal pressure required to produce a fully plastic thick-walled spherical shell. Plot the distribution of radial and hoop stress through the wall at this pressure.

17.15 Show, from the Nadai approach (page 740), that it is not possible to fully yield a thin annular disc of non-hardening von Mises material under internal pressure when the radius ratio exceeds 2.963 and that the pressure then remains constant at $2Y/\sqrt{3}$.

17.16 A thin solid disc of uniform thickness and outside radius R, rotates at an angular velocity ω (rad/s). Determine the ratio between the speed necessary to initiate plastic flow and that for the flow to extend to 90 per cent and 100 per cent of the volume of the disc. Assume a Tresca yield criterion for which the constant flow stress is Y. The density is ρ and Poisson's ratio $v = 0.3$. (I.C. 1971).

Answer: 1.112; 1.114.

17.17 Establish the stationary residual stress distributions corresponding to each flow speed in Exercise 17.16.

17.18. Determine the speeds necessary to (i) initiate yielding (ii) produce plastic flow to the mean radius and (iii) produce full plasticity, in a thin annulus of inner and outer diameters of 50 and 300 mm respectively. Sketch the distributions of radial and hoop stresses in each case. Take $Y = 310$ MPa = constant; $\rho = 7750$ kg/m^3; $v = 0.28$.

17.19 What internal pressure will result in an equivalent residual compressive stress of magnitude equal to Y (the yield stress) in a closed, thick-walled, non-hardening, von Mises cylinder of inner and outer radii a and b respectively?

Answer: $(2Y/\sqrt{3})(1 - a^2/b^2)$.

Creep

17.20 The following short-time transient creep behaviour of aluminium applies to two stress levels σ_1 and σ_2 at $100\,^\circ$C:

t(min)	$\frac{1}{6}$	$\frac{1}{2}$	1	$1\frac{1}{2}$	2	3	4	5	6
$\epsilon_1(\%)$	0.31	0.34	0.36	0.37	0.38	0.39	0.40	0.41	0.42
$\epsilon_2(\%)$	0.9	1.47	2.03	2.48	2.84	3.48	4.0	4.5	5.0

t(min)	7	8	9	10	11	12	13	14	15
$\epsilon_1(\%)$	0.426	0.435	0.44	0.445	0.45	0.455	0.46	0.465	0.47
$\epsilon_2(\%)$	5.36	5.75	6.2	6.5	6.8	7.1	7.4	7.7	8

Show that the laws relating ϵ to t ($\epsilon_1 = \epsilon_o + b \ln t$ and $\epsilon_2 = \epsilon_o + at^m$) depend upon stress, where ϵ_o is the instantaneous strain and a, b and m are constants.

17.21 A series of $100\,^\circ$C tensile creep tests on an aluminium alloy gave the following secondary creep rates:

σ(MPa)	100	150	200	250
$\dot{\epsilon}_s(\text{h}^{-1})$	9.2×10^{-6}	5.1×10^{-5}	1.55×10^{-4}	4.2×10^{-4}

Determine the law relating $\dot{\epsilon}_s$ to σ.

Answer: $\dot{\epsilon}_s = 10^{-13}\sigma^4$.

17.22 Isochronous creep data at 400 h obey a linear stress–strain law $\epsilon = M\sigma$ over a low range of stress for the following temperatures:

$T(^\circ C)$	650	550	500	450
$M(10^{-6})$	53.3	28	16	9.8

Determine the function $M = M(T)$ and hence find the stress that may be applied to give a creep strain of 0.4 percent in 400 h.

17.23 The following results apply to the short-time creep of lead under $\sigma = 11.5$ MPa at room temperature. Examine the applicability of the Graham–Walles and Davis representations to the full creep curve.

t(min)	0.25	0.5	1	2	3	4	6	8	10	15	20	25	30	35	40	42	44	46
$\epsilon(\%)$	1.25	1.6	2.06	2.77	3.26	3.6	4.23	4.77	5.2	6.17	7.03	7.89	8.77	9.94	12.49	14.57	17.94	26.3

17.24 The following data apply to the creep deformation in a Ni–Cr alloy at 820°C.

t (h)	20	100	200	1000	2000	3000	4000	5000
ϵ (%)	1	1.6	2	3.2	4	4.8	6	8.5

Given that the minimum creep rate occurs at 200 h, estimate the constants θ_1, θ_2, θ_3 and θ_4 in the prediction of Evans (1982):

$$\epsilon_c = \theta_1[1 - \exp(-\theta_2 t)] + \theta_3[\exp(\theta_4 t) - 1].$$

17.25 The following data apply to the tensile creep of lead at room temperature. Confirm that the respective Norton, Monkman–Grant, and Katchanov–Rabotnov laws relate $\dot{\epsilon}_s$ to σ, $\dot{\epsilon}_s$ to t_f, and σ to t_f. Determine the constants in each law.

σ (MPa)	11.52	10.4	9.86	9.25	8.91	8.80
t_f (h)	3.05	10	18.5	37	52	67.1
$\dot{\epsilon}_s$ (h^{-1})	0.0265	0.0075	0.0052	0.00263	0.00225	0.0019

Answer: $\dot{\epsilon}_s = 37 \times 10^{-4} \sigma^{10}$; $\dot{\epsilon}_s t_f = 0.1$; $t_f = 10^{13} \sigma^{-11.85}$.

17.26 In a constant-temperature creep test, the stress is increased in equal increments $\Delta\sigma$ at regular intervals Δt. Given that the creep law for the material is $\epsilon_c = A\sigma^n t^m$, show, from the time and strain-hardening laws, that the respective creep strain after N steps is

$$\epsilon_c = A(\Delta t)^m \sum [N^m - (N-1)^m](N\,\Delta\sigma)^n \quad \text{and} \quad \epsilon_c = A(\Delta t)^m [\sum (N\,\Delta\sigma)^{n/m}]^m.$$

17.27 A stainless steel pipeline of 40 mm bore and 2 mm wall thickness is to carry fluid at pressure of 0.6 bar and temperature of 510°C for 9 years. Given a limiting creep strain of 1 per cent and a secondary creep law $\dot{\epsilon}_s = A\sigma^5 \exp(-Q/RT)$, comment upon the safety of the design with the following information supplied for a stress of 200 MPa:

T (°C)	618	640	660	683	707
$\dot{\epsilon}_s/10^{-8}$ (s^{-1})	10	17	43	77	200

17.28 At stress levels of 230 MPa and 155 MPa, the Larson–Miller parameter for a nickel-base alloy has values of 25 and 26 respectively. Estimate the maximum operating temperature at a stress of 155 MPa for a life of 10^4 h. If a component of this alloy spent 10 per cent of its life at 230 MPa and the remaining life at 155 MPa, determine the time to failure from the life fraction rule. (I.C. 1969)

Answer: 810°C; 5880 h.

17.29 The uniaxial creep life of an alloy is 10^5 h at 900 K and 120 MPa, while at 300 MPa and 800 K the life is 10^4 h. Using the Larson–Miller and Manson–Haferd parameters (P), interpolate linearly on $\log P - \log \sigma$ axes to compare temperature predictions for which the life is 10^5 h at 200 MPa.

17.30 Find the width of a cantilever beam 750 mm long × 75 mm deep carrying a 2.5 kN force at its free end, when the deflection there is restricted to 10 mm in 10^4 h at 500°C. Take the secondary creep law $\dot{\epsilon}_s = 3 \times 10^{-13} \sigma$ ($\dot{\epsilon}_s$ in h^{-1} and σ in MPa). Neglecting self-weight, what will be the maximum stress and strain in the beam after this time? (I.C. 1971)

17.31 An alloy steel shaft of 50 mm outer diameter and 38 mm inner diameter, 2.75 m long carries a steady torque of 500 N m at 450°C. Given that the secondary creep rate

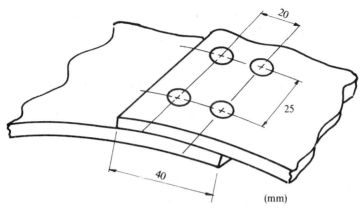

(mm)   **Figure 17.24**

$\dot{\epsilon}_s = (30.5 \times 10^{-4})\sigma^3$, with $\dot{\epsilon}_s$ in h^{-1} and σ in MPa, applies to uniaxial tension at $450°C$, determine the angular twist after 10^4 h. (I.C. 1965)

17.32 Compare the times taken for the stress in a bolt between two rigid flanges to relax to one half of its initial value (σ_o) according to the time- and strain-hardening equations of state. Assume the creep law $\epsilon_c = A\sigma^3 t^{1/2}$. (I.C. 1968)

Answer: $t_{SH}/t_{TH} = 2.18$.

17.33 A thin-walled cylindrical vessel of inside diameter d with hemispherical ends operates under secondary creep conditions where $\dot{\epsilon}_s = A\sigma^{4.8}$. Find the ratio between the hemisphere and cylinder thicknesses when each of the following conditions exists at the joint:

    (i) the circumferential strain rates are the same;
    (ii) the longitudinal strain rates are the same;
    (iii) the hoop stresses are the same.

17.34 A cylindrical vessel is riveted with the single lap circumferential joint shown in Fig. 17.24. The 2 mm diameter rivets are formed to produce an initial axial stress of 160 MPa. If a pressure of 0.35 MPa must be exerted between the lapping surfaces to prevent leakage, determine the life of the joint at 500°C. Ignore primary creep, take $E = 70$ GPa and employ the following secondary creep data with a Norton law (K.P. 1975):

σ (MPa)	80	100	120	140	160
$\dot{\epsilon}_s/10^{-9}$ (h^{-1})	50	150	380	800	1570

(K.P. 1975)

Viscoelasticity

17.35 Explain why the principal mechanical model representations of viscoelastic behaviour give only a qualitative picture.

17.36 State the creep compliance $C(t) = \epsilon(t)/\sigma_o$ and relaxation moduli $M(t) = \sigma(t)/\epsilon_o$ for the Maxwell, Voigt and SLS models.

17.37 Determine the strain function $\epsilon(t)$ when the Maxwell, Voigt and SLS models are loaded at a constant stress rate $R = d\sigma/dt$. If after time t_1, the stress is abruptly removed, find the recovery response $\epsilon(t > t_1)$ for the Voigt and SLS models.

Answer: $\epsilon_m(t) = R(1/E + t/\mu)$, $\epsilon_v(t) = (R/E)[t - \tau_o(1 - \exp(-t/\tau_o))]$.

17.38 Determine the stress function $\sigma(t)$ when the Maxwell, Voigt and SLS models are loaded under a constant strain rate $T = d\epsilon/dt$.

Answer: $\sigma_m(t) = \mu T[1 - \exp(-t/\tau_o)]$; $\sigma_v(t) = T(\mu + Et)$.

17.39 A spring of stiffness E_1 is connected in series with a dashpot of viscosity μ_1. The two are then connected in parallel with a second spring of stiffness E_2. Derive and sketch the stress relaxation response $\sigma(t)$ to a strain input ϵ_o. Given that $\sigma(t) = \epsilon_o M(t)$, determine the ratio $M(t)/E_1$ for $t = 20$ s when $E_1 = 5E_2$.

17.40 Show that when the Maxwell (E_1, μ_1) and Voigt models (E_2, μ_2) are connected in parallel the creep strain is given by

$$\epsilon(t) = \frac{(\sigma_o/E_1)[1/E_2 + t/\mu_2 - (1/E_2 + t/\mu_2)\exp(-E_1t/\mu_1)]}{(1/E_1 + 1/E_2) + t/\mu_2 - (1/E_1)\exp(-E_1t/\mu_1)}.$$

17.41 Sketch the strain–time curve represented by the following equation:

$$\epsilon(t) = \sigma/E_1 + \sigma t/\mu_1 + (\sigma/E_2)[1 - \exp(-E_2t/\mu_2)].$$

Determine the model from which this is derived and discuss whether this is realistic for creep and relaxation in polymeric materials. (C.E.I. Pt.2, 1971)

17.42 A Voigt and Maxwell model are connected in series. Identify those elements of the combined model that represent the following components of strain: (i) instant elasticity, (ii) irrecoverable flow, and (iii) retarded elasticity. Show where these appear in the corresponding creep curve. A certain polymer is modelled by this system where $E = 50\,\text{kN/m}^2$, $\mu = 10^6\,\text{N s/m}^2$ for Voigt and $E = 10^6\,\text{kN/m}^2$, $\mu = 100 \times 10^6\,\text{N s/m}^2$ for Maxwell. The polymer is subjected to a direct stress of $6\,\text{kN/m}^2$ for only 30 s. Determine the strain after times of 30 s, 60 s and 2000 s. (C.E.I. Pt.2, 1974)

17.43 A viscous dashpot $\dot{\epsilon} = \mu\sigma$ is connected in series with a non-linear spring: $\sigma = E\epsilon - (a\epsilon)^2$ where μ, a and E are constants. Determine the constitutive relation and deduce the relaxation behaviour for the model. (I.C. 1969)

Answer: $\epsilon = \sigma/\mu + (\sigma/E)[1 - (2a/E)^2\sigma]^{-1/2}$.

17.44 Derive and sketch the creep and relaxation responses for the model in Fig. 17.25 given that the spring stiffnesses are E_1 and E_2 and that the dashpot compliance is $C(t)$. If $E_1 = 4E_2$, find the relaxation modulus $M(t)/E_1$ for $t = 10$ s, given that the relaxation time is 100 s. (I.C. 1968)

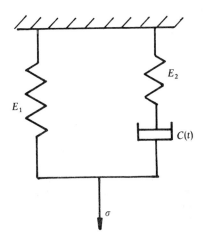

Figure 17.25

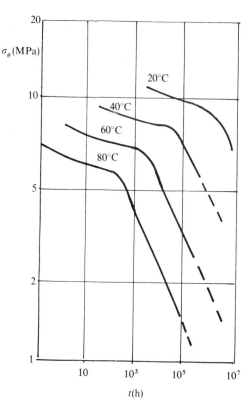

$t(h)$ **Figure 17.26**

17.45 Stress rupture data for high-density polyethylene are given in Fig. 17.26. Ignoring axial stress, estimate a suitable pipe thickness that is to sustain an internal pressure of 0.5 MPa at a temperature of 80°C for 12×10^6 h.

17.46 If for Example 17.16 the torque is increased from 65 N m to 100 N m after a time of 10^3 min, find the angular twist after 10^5 min. If the total torque is then removed, what angular twist remains after an equal recovery period of 10^5 min?

17.47 A linear viscoelastic material has a creep compliance function of the form $C(t) = t^m/E_o$. Derive the expression for constant stress-rate loading and outline how the solution could be employed to describe saw-tooth loading. (I.C. 1972)

17.48 A uniform, light cantilever with second moment of area I supports a uniformly distributed load w_1 from its centre to its free end. This load is applied suddenly at time t_1 and held constant until time t_2 when an additional uniformly distributed load w_2 is suddenly superimposed upon w_1, occupying the same length. If the cantilever material obeys the Voigt model creep behaviour, derive the deflection of the free end at time $t > t_2$, using the Boltzman superposition principle. (I.C. 1969)

EIGHTEEN
THE MECHANICS OF FATIGUE AND FRACTURE

18.1 HIGH-CYCLE FATIGUE

So far in the analyses of elastic forces, moments and torques it has been assumed that these remain constantly applied to load-bearing components. In practice, however, loads are often cyclic in nature, e.g. random wind loading on an aircraft, reversed torsion of a watch spring, repeated bending of an axle and crankshaft in a road vehicle, and pressure rippling in piping. High-cycle fatigue is the name given to describe the failure resulting from cyclic elastic stresses in which the peak levels are less than the static yield stress. This is a common cause of failure, indicating that there is normally a limit to the number of elastic cycles a given material can withstand. There are many factors that influence this limit. Those concerned specifically with the cycle are the stress range, mean stress, changes in the stress amplitude, stress state and stress concentration; those accounting for the effect of the environment include temperature and corrosion.

All theoretical predictions of their influence have been examined experimentally from the standard fatigue test. This test, for example, may impose a constant-amplitude reversed stress (Fig. 18.1(a)) under cyclic bending, torsion or tension–compression to a suitable test piece until it fails in a number (N) of fatigue cycles. By testing geometrically identical test pieces to failure, each at a different stress amplitude (S), the usual S–N curve in Fig. 18.1(b) may be constructed. Alternatively, an S–$\log N$ plot may be preferred, to distinguish more clearly the existence or otherwise of a fatigue limit. This is the asymptotic S_L value that the curve approaches in carbon steels, so indicating an infinite life in this material for $S < S_L$. However, for

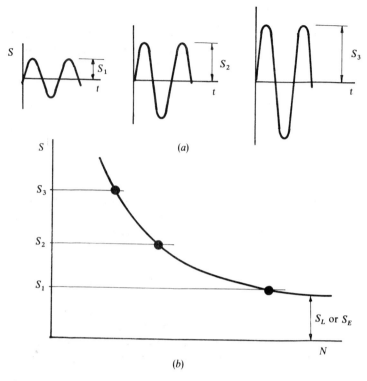

Figure 18.1 Reversed stress cycle and associated S–N curve.

most other materials a fatigue limit does not occur, since the curve falls continuously with increasing cycles. It then becomes necessary to employ an endurance limit S_E for a specified life, e.g. 10^9 or 10^{10} cycles. In general, the S–N curve reveals the fatigue life, that is, the number of cycles to failure under a given stress amplitude (or semi-range of stress). Note that N indicates the probability of failure from an assumed distribution of life at each stress level. This is to account for the normally wide scatter observed in fatigue life. When S_L or S_E is regarded as material property under specified test conditions, it may be expressed as a fraction of the ultimate tensile strength (UTS). That is, an endurance ratio E_R is defined as

$$E_R = S_L(\text{or } S_E)/\text{UTS},\qquad(18.1)$$

where, $0.3 < E_R < 0.7$ for materials under the fully reversed stress cycle in Fig. 18.1(a). Equation (18.1) provides a useful method for estimating fatigue strength, e.g. for mild steel with a UTS of 455 MPa, an average S_L value is $0.5 \times 455 = \pm 227.5$ MPa. The following factors within the stress cycle will, however, influence the reversed fatigue strength.

Mean Stress and Stress Range

For stress cycles that are not symmetric about the time base, then a mean stress σ_M for the cycle is defined as the average of the maximum and minimum peak stresses. The stress range is the difference between these peak values. They are

$$\sigma_M = (\sigma_{max} + \sigma_{min})/2, \qquad (18.2a)$$

$$\Delta\sigma = \sigma_{max} - \sigma_{min}. \qquad (18.2b)$$

Four types of cycle, given in Fig. 18.2 (a)–(d), may then be classified as follows.

(a) Reversed cycle: zero mean stress with equal σ_{max} and σ_{min} magnitudes.
(b) Alternating cycle: σ_{max} is positive, σ_{min} is negative with a different magnitude so that σ_M may be either positive or negative.
(c) Repeated cycle: $\sigma_{min} = 0$, σ_{max} and $\sigma_M = \sigma_{max}/2$ are either both tensile or compressive.
(d) Fluctuating cycle: σ_{max}, σ_{min} and σ_M are either all positive or all negative.

Experiment has shown that as σ_M in (b), (c) and (d) increases so the safe range of cyclic stress $\Delta\sigma$, defining the fatigue or endurance limit, decreases. A number of design rules predict the reducing effect that σ_M has on the safe range of stress ($\Delta\sigma_o$) under the reversed stress cycle (Fig. 18.2a). The limiting conditions are then:

(i) when $\sigma_M = 0$, $\Delta\sigma_o = 2S_L$ or $\Delta\sigma_o = 2S_E$ for the reversed cycle; and
(ii) when $\Delta\sigma = 0$, $\sigma_M = Y$ or $\sigma_M = UTS$ for a static tension test.

These three limiting points, lying on the normalized axes of Fig. 18.3, have been connected with a parabola or the straight lines as shown. The Gerber

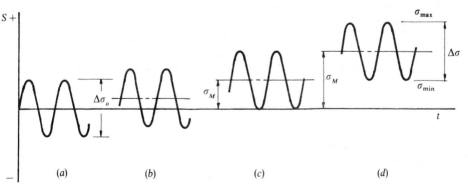

(a)   (b)   (c)   (d)

Figure 18.2 Types of stress cycle.

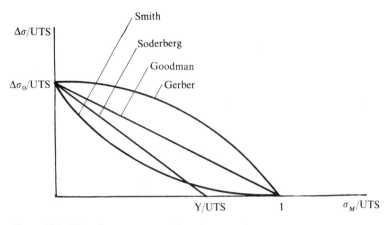

Figure 18.3 Effect of mean stress on the safe range of stress.

parabola, the Goodman line (1899)* and Smith's empirical equation (Smith and Collins, 1942) all employ $\sigma_M = $ UTS, while Soderberg (1930) employed the more conservative limiting condition $\sigma_M = Y$. These are respectively:

Gerber: $\quad \Delta\sigma = \Delta\sigma_o[1 - (\sigma_M/\text{UTS})^2]$ $\qquad$ (18.3a)

Goodman: $\quad \Delta\sigma = \Delta\sigma_o[1 - \sigma_M/\text{UTS}]$ $\qquad$ (18.3b)

Smith: $\quad \Delta\sigma = \Delta\sigma_o(1 - \sigma_M/\text{UTS})/(1 + \sigma_M/\text{UTS})$ $\quad$ (18.3c)

Soderberg: $\quad \Delta\sigma = \Delta\sigma_o[1 - \sigma_M/Y]$. $\qquad$ (18.3d)

Equations (18.3) enable the safe $\Delta\sigma$ to be estimated for a given σ_M. Test results for ductile materials are normally found to lie in the area bounded by the Soderberg and Gerber predictions. Results for brittle materials lie closest to the Smith prediction.

Example 18.1 The reversed stress fatigue limit for a low-carbon steel is $\pm 230\,\text{MPa}$ ($\pm 15\,\text{tonf/in}^2$) its UTS is $870\,\text{MPa}$ ($56.7\,\text{tonf/in}^2$) and its yield stress is $380\,\text{MPa}$ ($25\,\text{tonf/in}^2$). Estimate the safe range of stress for (i) a repeated cycle based upon the Gerber and Goodman predictions and (ii) an alternating tensile cycle with mean stress $185\,\text{MPa}$ ($12\,\text{tonf/in}^2$) based upon the Soderberg prediction. What are the peak stresses for (ii)?

SOLUTION TO SI UNITS

(i) Here $\Delta\sigma_o = 2S_L = 460\,\text{MPa}$ and $\sigma_M = \Delta\sigma/2$ for a repeated cycle.

---

* References are appended to this chapter on page 830.

Substituting into Eq. (18.3a), the safe Gerber range is found from

$$\Delta\sigma = 460[1 - (\Delta\sigma/2 \times 870)^2],$$
$$(\Delta\sigma)^2 + 6581.74(\Delta\sigma) - 3027.6 \times 10^3 = 0,$$
$$\therefore \quad \Delta\sigma = 431 \text{ MPa}.$$

Substituting into Eq. (18.3b), the Goodman prediction becomes

$$\Delta\sigma = 460[1 - (\Delta\sigma/2 \times 870)],$$
$$\Delta\sigma = 363 \text{ MPa}.$$

(ii) Here $\sigma_M = 185 \text{ MPa}$, $\Delta\sigma_o = 460 \text{ MPa}$ and $Y = 380 \text{ MPa}$. Equation (18.3d) gives the safe stress range for the alternating cycle as

$$\Delta\sigma = 460[1 - 185/380] = 236 \text{ MPa}.$$

Combining Eqs (18.2a and b), the peak stresses are

$$\sigma_{\min} = \sigma_M - \Delta\sigma/2 = 185 - 236/2 = 67 \text{ MPa},$$
$$\sigma_{\max} = \sigma_M + \Delta\sigma/2 = 185 + 236/2 = 303 \text{ MPa}.$$

SOLUTION TO IMPERIAL UNITS
(i) $\sigma_M = \Delta\sigma/2$, $\Delta\sigma_o = 2S_L = 30 \text{ tonf/in}^2$ and the UTS $= 56.7 \text{ tonf/in}^2$.
Gerber:

$$\Delta\sigma = 30\{1 - [\Delta\sigma/(2 \times 56.7)^2]\},$$
$$(\Delta\sigma)^2 + 428.67\,\Delta\sigma - 12860 = 0,$$
$$\Delta\sigma = 28.16 \text{ tonf/in}^2.$$

Goodman:

$$\Delta\sigma = 30[1 - (\Delta\sigma/2 \times 56.7)],$$
$$\Delta\sigma = 23.7 \text{ tonf/in}^2.$$

(ii) $\sigma_M = 12 \text{ tonf/in}^2$, $\Delta\sigma_o = 30 \text{ tonf/in}^2$ and $Y = 25 \text{ tonf/in}^2$.
Soderberg:

$$\Delta\sigma = 30[1 - (12/25)] = 15.6 \text{ tonf/in}^2,$$
$$\therefore \quad \sigma_{\min} = 4.2 \text{ tonf/in}^2, \quad \sigma_{\max} = 19.8 \text{ tonf/in}^2.$$

Cumulative Damage

In practice the stress amplitude may vary, so complicating the prediction of fatigue life. A large number of investigations have examined the effect of block pattern and random changes to the stress amplitude. These are often compared with predictions from a hypothesis due to Miner (1945).

Consider the three blocks of reversed stress cycles in Fig. 18.4(a), where n_1, n_2 and n_3 cycles are applied at respective semi-stress ranges S_1, S_2 and

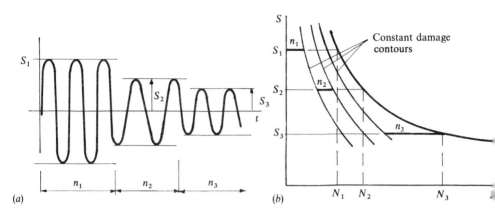

Figure 18.4 Damage under a variable amplitude cycle.

S_3. If N_1, N_2 and N_3 are the corresponding fatigue lives (Fig. 18.4(b)) then the life expended, or the damage incurred, at each S, is given by the cycle ratios n_1/N_1, n_2/N_2 and n_3/N_3. According to Miner's linear damage rule, failure occurs when the sum of the cycle ratios equals unity. In general, that is,

$$\sum (n/N) = 1. \tag{18.4}$$

The right-hand side of Eq. (18.4) has often been replaced with another constant that accounts for the observed stress dependence of the cumulative cycle ratio. Normally, with a high-to-low block stress sequence, $\sum (n/N) < 1$; with a low-to-high sequence, $\sum (n/N) > 1$. Alternatively, if D is the damage under cycle ratio $R = n/N$, these observations on stress dependence will appear when a given non-linear function $D = D(R)$ is assigned to each semi-stress range. A simple function proposed by Marco and Starkey (1954) is the power law

$$D = (R)^x, \tag{18.5a}$$

where $x = x(S)$ is a positive constant for a given S. Corten and Dolan (1956) proposed the alternative function:

$$D = an^b, \tag{18.5b}$$

where $a = a(S)$ and b is a constant independent of S. In the non-linear Eqs (18.5), a different D–R curve may be identified with each S level as illustrated in Fig. 18.5. For the high-to-low stress sequence in Fig. 18.4(a) the damage D_1 incurred by the cycle ratio $R = n_1/N_1$ is either governed by curve $D_1 = (R_1)^{x_1}$ or $D_1 = a_1 n_1^b$ as defined by Eqs (18.5a and b). The change to stress level S_2, where the curve $D_2 = (R_2)^{x_2}$ or $D_2 = a_2 n_2^b$ applies, occurs at a constant damage value. Similarly for the change to S_3, where another D–R curve describes the accumulation of damage. Although failure again occurs

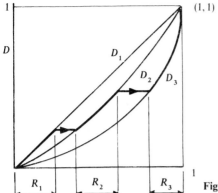

Figure 18.5 Non-linear damage rule.

when D attains a value of unity, it is seen from the relative disposition of the D–R curves in this approach that the cumulative cycle ratio $R_1 + R_2 + R_3$ is less than 1, in accordance with observation. Moreover, for a low-to-high stress sequence S_3–S_2–S_1 these curves will also accord with $\sum R > 1$. Miner's rule (Eq. (18.4)) implies that a single D–R curve applies to all stress levels. This simplistic, stress-independent approach accounts for its widespread use, particularly in the case of random cyclic loading where the non-linear concept becomes impracticable.

Example 18.2 A steel component in a machine tool is subjected to reversed cyclic loading at 100 cycles/day in a continuous sequence involving three stages; 200 cycles at $\pm$ 160 MPa, 200 cycles at $\pm$ 140 MPa and 600 cycles at $\pm$ 100 MPa. If the respective fatigue lives at these stress levels are 10^4, 10^5 and 2×10^5 cycles, estimate the life of the component in days according to (i) Miner's rule and (ii) a non-linear approach where damage curves $D = \sqrt{R}, D = R$ and $D = R^2$ apply to the semi-stress ranges (S) of 160, 140 and 100 MPa respectively.

(i) Let the component fail under P sequences of the three stages. Then, from Eq. (18.4),

$$P(n_1/N_1 + n_2/N_2 + n_3/N_3) = 1,$$
$$P[200/10^4 + 200/10^5 + 600/(2 \times 10^5)] = 1,$$
$$\therefore \quad P = 40$$

and for 1000 cycles/sequence at a rate of 100 cycles/day, the life is $40 \times 1000/100 = 400$ days.

(ii) With reference to Fig. 18.6(a), the proper application of the non-linear approach clearly requires the laborious calculation of the damage value for every change in stress. However, by assuming,

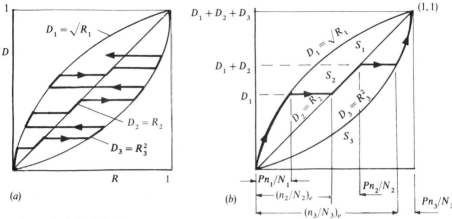

Figure 18.6 Cyclic loading sequence.

conservatively, that stresses of the same peak level are all applied before the stress is increased, then Fig. 18.6(b) may be applied to this problem. Thus, for P_{n_1}/N_1, cycles at S_1 the damage is

$$D_1 = \sqrt{R_1} = \sqrt{(P \times 200/10^4)} = 0.1414\sqrt{P}.$$

The equivalent cycle ratio $(n_2/N_2)_e$ that would cause the same damage at S_2 is found from

$$D_1 = 0.1414\sqrt{P} = (n_2/N_2)_e \tag{i}$$

and, for n_2/N_2 cycles at this stress level,

$$\therefore \quad D_1 + D_2 = R_2 = (n_2/N_2)_e + Pn_2/N_2,$$

where, from Eq. (i),

$$D_1 + D_2 = 0.1414\sqrt{P} + 200P/10^5.$$

This is equivalent to the cycle ratio $(n_3/N_3)_e$ at S_3:

$$D_1 + D_2 = 0.1414\sqrt{P} + 200P/10^5 = (n_3/N_3)_e^2$$
$$\therefore \quad (n_3/N_3)_e = \sqrt{(0.1414\sqrt{P} + 200P/10^5)}, \tag{ii}$$

and for Pn_3/N_3 cycles to failure:

$$D_1 + D_2 + D_3 = R_3^2 = 1,$$
$$[(n_3/N_3)_e + Pn_3/N_3]^2 = 1.$$

Substituting from Eq. (ii):

$$[\sqrt{(0.1414\sqrt{P} + 200P/10^5)} + 600P/(2 \times 10^5)]^2 = 1,$$
$$\sqrt{(0.1414\sqrt{P} + 200P/10^5)} + 300P/10^5 = 1.$$

The solution is $P = 30$ sequences, for which the life is

$$30 \times 1000/100 = 300 \, \text{days},$$

which indicates that Miner's prediction is probably on overestimation for the high-to-low order of stressing.

Stress State

Fatigue under cyclic biaxial and triaxial stress states may be treated in a similar manner to yielding (Chapter 16). Here the uniaxial yield stress (Y) is replaced with a uniaxial fatigue strength (S_o) typically obtained under reversed stress conditions from a push–pull or a reversed bending test. Since fatigue cracks initiate at a free surface, where one principal stress is zero, it is often only necessary to consider biaxial fatigue criteria. For example, when principal biaxial stress ranges $\Delta\sigma_1$ and $\Delta\sigma_2$ are applied in a reversed cycle, it follows from a von Mises type criterion (Eq. (16.9b)) that the combination of stress ranges required for fatigue failure in a ductile material is given by

$$\Delta\sigma_1^2 - \Delta\sigma_1 \Delta\sigma_2 + \Delta\sigma_2^2 = \Delta\sigma_o^2, \tag{18.6a}$$

where $\Delta\sigma_o = 2S_o$. In the presence of a mean stress σ_M, Eq. (18.6a) becomes

$$\Delta\sigma_1^2 - \Delta\sigma_1 \Delta\sigma_2 + \Delta\sigma_2^2 = \Delta\sigma^2, \tag{18.6b}$$

where $\Delta\sigma$ applies here to the given σ_M. When only the reversed fatigue strength is available, $\Delta\sigma$ is found from one of Eqs (18.3). A conservative Tresca-type criterion (Eq. (16.4b)) is often preferred.

$$\text{Greatest principal } \Delta\sigma - \text{least principal } \Delta\sigma = \Delta\sigma_o, \tag{18.7}$$

in which the least $\Delta\sigma$ may be negative for a wholly compressive cycle but would normally be taken as zero (the third principal stress range) when biaxial stress ranges $\Delta\sigma_1$ and $\Delta\sigma_2$ are both tensile. In the case of combined cyclic tension and torsion, where respective ranges of reversed stress $\Delta\sigma$ and $\Delta\tau$ are applied, Gough and Pollard (1935) expressed Eq. (16.9c) in a fatigue criterion for ductile steels:

$$(\Delta\sigma/\Delta\sigma_o)^2 + (\Delta\tau/\Delta\tau_o)^2 = 1, \tag{18.8}$$

where $\Delta\sigma_o$ and $\Delta\tau_o$ are the critical stress ranges under the separate application of reversed bending and torsion respectively, i.e. twice their reversed fatigue strengths. Equation (18.8) was experimentally verified for 12 ductile steels. Brittle cast iron did not conform to Eq. (18.8). The following modified form of a maximum principal strain type criterion (Eq. (16.3)) was found to be more appropriate:

$$(\Delta\sigma_o/\Delta\tau_o - 1)(\Delta\sigma/\Delta\sigma_o)^2 + (2 - \Delta\sigma_o/\Delta\tau_o)(\Delta\sigma/\Delta\sigma_o) + (\Delta\tau/\Delta\tau_o)^2 = 1.$$

Findlay and Mathur (1956) modified Eq. (18.8) to account for fatigue failure in both ductile and brittle materials:

$$(\Delta\sigma/\Delta\sigma_o)^a + (\Delta\tau/\Delta\tau_o)^2 = 1,$$

where $a = \Delta\sigma_o/\Delta\tau_o$ is unity for brittle fatigue and is equal to 2 for ductile fatigue. It should be noted that where safe ranges of stress need to be applied in practice, then $\Delta\sigma_o$ and $\Delta\tau_o$ in these equations incorporate suitable safety factors.

Example 18.3 A thin-walled steel cylinder ($d/t = 20$) with closed ends is subjected to a range of internal pressure $p/3$ about a mean value of p. Given that the reversed fatigue limit is ± 175 MPa, and that the yield stress is 320 MPa, find p for an infinite number of pressure cycles based upon the von Mises and Tresca criteria of equivalent fatigue strength. Use a Soderberg law and neglect radial stress.

Applying Eq. (18.3d) to find the uniaxial stress range in the presence of a mean stress $\sigma_M = p$:

$$\Delta\sigma = \Delta\sigma_o(1 - \sigma_M/Y)$$
$$= 350(1 - p/320) = 350 - 1.0938p. \qquad \text{(i)}$$

The principal stress ranges are

$$\Delta\sigma_\theta = \Delta p\, d/2t = (p/3)20/2 = 10p/3, \qquad \text{(ii)}$$
$$\Delta\sigma_z = \Delta p\, d/4t = (p/3)20/4 = 5p/3. \qquad \text{(iii)}$$

Substituting Eqs. (i)–(iii) into Eq. (18.6b) gives

$$(10p/3)^2 - (5p/3)(10p/3) + (5p/3)^2 = (350 - 1.0938p)^2.$$

This leads to the following quadratic in p:

$$p^2 + 107.28p - 17164 = 0,$$

for which the positive root is

$$p = 87.93 \text{ MPa}.$$

The Tresca criterion (Eq. (18.7)) simply becomes

$$10p/3 - 0 = 350 - 1.0938p,$$
$$p = 79.06 \text{ MPa}.$$

Stress Concentrations

The fatigue strength of a material depends strongly upon concentrations in stress that arise from abrupt changes in section at holes and fillets and from poor surface finish. The stress concentration for these 'notches' is defined as

$$K_t = \frac{\text{maximum stress at notch}}{\text{average stress at the notch section}}. \tag{18.9}$$

For example, in Fig. 18.7, the following K_t values apply to the stress concentration in the fillet radius with $D = 50 \, \text{mm}$ and $d = 25 \, \text{mm}$:

$$K_t = 2.6 \quad \text{for } r = 1.5 \, \text{mm},$$
$$K_t = 1.9 \quad \text{for } r = 3 \, \text{mm},$$
$$K_t = 1.4 \quad \text{for } r = 6 \, \text{mm}.$$

The effect that a particular notch has on reducing fatigue strength may be assessed from the strength reduction factor $(K_f \geq 1)$ defined as

$$K_f = \frac{\text{plain fatigue strength } (S)}{\text{notched fatigue strength } (S_n)}. \tag{18.10}$$

Normally, K_f is greater for long endurances but, because different materials display widely different K_f values for the same K_t, a notch sensitivity index $0 \leq q \leq 1$ relates K_t to K_f in the following form:

$$q = (K_f - 1)/(K_t - 1), \tag{18.11}$$

in which $q \rightarrow 1$ as $K_f \rightarrow K_t$ for a fully notch-sensitive material, and $q \rightarrow 0$ as $K_f \rightarrow 1$ in a material that is insensitive to notches. Note that where K_t is very high for sharp notches, q for the material will be unrealistically low. Moreover, K_f, and hence q in Eq. (18.11), will further increase with (i) geometry, since the likelihood of an internal defect will increase with physical size; (ii) static strength of the material; (iii) decreasing grain size; and (iv) type of cylic loading in the presence of stress gradients. The application of Eq. (18.11) depends upon knowing K_t and K_f for a given geometry and material. Elastic stress concentrations have now been well documented in graphical, tabular and empirical forms (Peterson, 1974). The notched uniaxial fatigue strength S_n (in megapascals) for ductile materials with the geometries given in Fig. 18.8(a)–(c) are available from Heywood's (1956) empirical formula:

$$S_n = (S/K_t)\{1 + 2[(K_t - 1)/K_t]\sqrt{(a/R)}\}. \tag{18.12a}$$

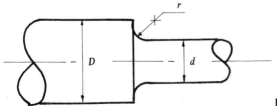

Figure 18.7 Fillet concentration.

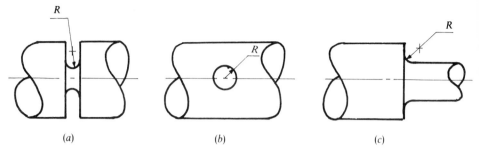

Figure 18.8 Representative geometries with stress concentration.

Combining Eqs (18.10) and (18.12a), they appear in the K_f form:

$$K_f = \frac{K_t}{1 + 2[(K_t - 1)/K_t]\sqrt{(a/R)}}, \qquad (18.12b)$$

where R is the indicated radius (millimeteres) in Fig. 18.8(a–c) and a values are the following respective notch alleviation factors: (a) $a = (104.3/\text{UTS})^2$, (b) $a = (173.8/\text{UTS})^2$, (c) $a = (139/\text{UTS})^2$, with UTS in megapascals. An alternative strength reduction factor, K_f in Eq. (18.10), is the widely used expression proposed by Neuber (1946):

$$K_f = 1 + (K_t - 1)/[1 + \sqrt{(A/R)}], \qquad (18.13)$$

where A is an experimentally determined material constant which was again related (graphically) to the ultimate tensile strength.

Example 18.4 A steel crossbeam containing a 75 mm (3 in) diameter hole ($K_t = 3$) is subjected to a repeated bending cycle. Using the Goodman relationship, compare the maximum bending stress predictions of Heywood and Neuber (with $\sqrt{A} = 0.16$, $\sqrt{A} = 0.035$ in imperial units) for infinite life given that the UTS = 1100 MPa (70 tonf/in²) and that the reversed bending fatigue strength is ± 460 MPa (± 30 tonf/in²) (I.C. 1969)

SOLUTION TO SI UNITS The repeated bending fatigue strength is found from the Goodman equation (18.3b) with $\Delta\sigma_o = 920$ MPa and $\sigma_M = \Delta\sigma/2$. That is,

$$\Delta\sigma = 920[1 - \Delta\sigma/(2 \times 1100)]$$
$$\Delta\sigma = 920 - 0.4182\Delta\sigma, \quad \Rightarrow \Delta\sigma = 648.72 \text{ MPa}.$$

Now, for the stress concentration in Fig. 18.8(b),

$$\sqrt{(a/R)} = 173.8/(\text{UTS} \times \sqrt{R}) = 173.8/(1100 \times \sqrt{37.5}) = 0.0258$$

and substituting into the Heywood equation (18.12b):

$$K_f = \frac{3}{1 + 2[(3 - 1)/3]0.0258} = 2.9.$$

Then from Eq. (18.10), $S_n = S/K_f$ and, since $S = \Delta\sigma$ expresses both the maximum stress and the range for the unnotched repeated cycle, the corresponding notched values are

$$S_n = \Delta\sigma/K_f = 648.72/2.9 = 227.74\,\text{MPa}.$$

For the Neuber prediction in Eq. (18.13),

$$K_f = 1 + (3 - 1)/[1 + 0.16/\sqrt{37.5}] = 2.95,$$
$$\therefore \quad S_n = \Delta\sigma/K_f = 648.72/2.95 = 210\,\text{MPa}.$$

which shows that the Neuber prediction is more conservative.

SOLUTION TO IMPERIAL UNITS With $\Delta\sigma_o = 60\,\text{tonf/in}^2$, $\sigma_M = \Delta\sigma/2$ and the UTS $= 70\,\text{tonf/in}^2$, Eq. (18.3b) gives

$$\Delta\sigma = 60[1 - \Delta\sigma/(2 \times 70)],$$
$$\Delta\sigma = 42\,\text{tonf in}^2.$$

With UTS in ksi, the stress concentration in Fig. 18.81(b) becomes $a = (5/\text{UTS})^2$

$$\therefore \quad \sqrt{a/R} = 5/(\text{UTS} \times \sqrt{R}),$$
$$= 5 \times 1000/(70 \times 2240 \times \sqrt{1.5}) = 0.026$$

From Heywood's equation (18.12b),

$$K_f = \frac{3}{1 + (2 \times 2 \times 0.026/3)} = 2.899,$$
$$\therefore \quad S_n = \Delta\sigma/K_f = 42/2.899 = 14.49\,\text{tonf/in}^2.$$

In Neuber's equations (18.13), $\sqrt{A} = 0.035$ for UTS $= 70\,\text{tonf/in}^2$ (Forrest, 1970).

$$\therefore \quad K_f = 1 + \frac{(3 - 1)}{1 + 0.035/\sqrt{1.5}} = 2.945,$$
$$\therefore \quad S_n = \Delta\sigma/K_f = 42/2.945 = 14.2\,\text{tonf/in}^2.$$

Residual Stresses

The presence of residual surface stresses can drastically alter the fatigue strength of a material. The general rule is that when the sense of the residual stress opposes that within a fatigue cycle, an improvement in fatigue strength

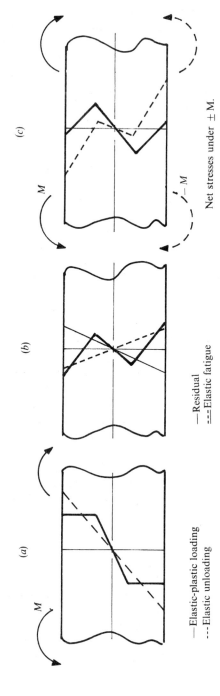

Figure 18.9 Interaction between residual and cyclic bending stresses.

— Elastic-plastic loading
--- Elastic unloading

— Residual
--- Elastic fatigue

Net stresses under ± M.

will be found. A residual compressive stress, for example, would be beneficial to the fatigue strengths under repeated or fluctuating tensile cycles in Figs (18.2(c) and (d)). This is because their effect is to reduce the mean stress, which in turn raises the $S-N$ curve. On the other hand, tensile residual stresses will effectively lower the $S-N$ curve to the detriment of the fatigue strengths under tensile cycling. It has long been known that favourable compressive residual stresses may be induced by mechanical deformation and thermal treatment processes. These include cold rolling, shot-and-hammer peening, honing, carburizing and nitriding. In contrast, heavy grinding and welding can produce surface cracks in addition to tensile residual stresses, both of which will impair the fatigue strength under tensile cycling. It has been shown in Chapters 6, 9 and 17 respectively that plastic pre-straining of beams, torsion bars and thick cylinders (autofrettage) results in a partially compressive residual stress distribution in each case. This condition can be used advantageously when these structures are to operate under tensile fatigue conditions. Figure 18.9(a) illustrates the elastic-plastic bending stress distribution through the depth of an elastic-perfectly plastic beam under bending (see also Fig. 6.34, page 168). When the bending moment is removed, only elastic stress recovers, leaving the residual stress distribution given in Fig. 18.9(b). Superimposed on this in a finer continuous line is the elastic peak stress distribution for a subsequently applied cyclic hogging bending moment. It is seen from the net stress distribution in Fig. 18.9(c) that the surface bending stresses are advantageously reduced. However, this will only apply for the given sense of the cyclic bending stress. Should the direction of the cyclic bending moment be changed to sagging, then the additive effect of the surface stresses become detrimental to fatigue strength (broken lines in Figs. 18.9(a) and (b)). An analogous explanation will hold for the interaction between cyclic elastic and residual stresses in torsion (Fig. 9.30). In a thick cylinder, the interaction between the residual stresses (Fig. 17.4) and repeated elastic pressure stresses is triaxial. Normally the range of bore shear stress is used to assess the considerable benefit that autofrettage has on the fatigue strength of a cylinder under pulsating pressure (Rees, 1987).

Temperature

Environmental effects are best understood from a qualitative understanding of the fatigue process itself. At low–medium temperatures the high-cycle fatigue failure is characterized by its absence of gross macroscopic plastic deformation, though the initiation and propagation of a crack does involve plasticity at sites of high stress concentration on a microscopic scale. That is, the long period of stage I crack initiation is associated with coarsening during slip-band formation at a free surface along planes of maximum shear stress, while the shorter stage II crack propagation period is the result of progressive plastic sharpening and blunting of a crack front that advances in a direction normal to the major principal tensile stress.

The characteristic appearance of fatigue failures consists of two regions. In the stage II propagation area the rate of advance is related to the spacing of characteristic fatigue striations or beach markings. One such marking is associated with a single application of cyclic stress. The fracture surface may be smooth initially as it emanates from the initiation site but often becomes rougher as the increased rate of crack propagation results in a wider spacing of fatigue markings. The featureless stage I initiation site spreads across only a few grains in polycrystalline materials and is therefore not visible though it would normally lie within a narrow 45° shear lip on the surface. Finally, when the remaining uncracked area is unable to withstand peak cyclic stresses it fails catastrophically in either a brittle or ductile manner leaving a respective crystalline or fibrous appearance.

This situation will prevail with increasing temperature up to $T < T_m/2$, where most materials show only a slight fall in fatigue strength. The exception is for steels, titanium and aluminium alloys, which actually show improved fatigue and static strength within this range owing to the strain-ageing phenomenon. However, for increasing temperatures beyond $T > T_m/2$, all materials display a reduction in fatigue strength owing to the simultaneous presence of creep under the peak and mean stress in the cycle. The result is that the fracture path changes from the normal transcrystalline fatigue failure to intercrystalline creep failure.

Corrosion

A corrosive environment, which assists with the formation of an initiation site, through pitting and scaling, will drastically reduce the normal fatigue strength for air. This is most marked for higher strength materials. In salt water, for example, there is a 50–80 per cent loss in fatigue strength for steels and a 60 per cent loss in aluminium alloys, while copper remains unaffected. Moreover, for high-strength steels in particular, there is no evidence of a fatigue limit. The experimental data show simply that the resistance of a given material to corrosion fatigue depends upon its resistance to corrosion in the absence of stress. However, when corrosion and fatigue occur simultaneously the chemical attack greatly increases the crack propagation rate. Moreover, corrosive environments, e.g. oxidizing and sulphidizing, may become even more aggressive with increasing temperature. The initiation phase too is reduced as many more pits form in the presence of stress. The application of cyclic stress impedes the formation of anti-corrosion products in the active zone allowing pits to penetrate inward and sharpen into the cracks that subsequently propagate. There are a number of methods available for corrosion-fatigue protection. The surface may be coated with a metal that must remain continuous if it is to be fully effective but will also protect the base metal by electrochemical action when an interruption occurs. This is achieved by selecting a coating that is anodic to the base. For example, zinc, being anodic to steel, provides good protection. Nickel, cadmium, chromium, copper and aluminium coatings may also be used, provided that they are not cathodic to the base, e.g. as with copper on steel, since this can actually reduce the resistance to corrosion fatigue. Non-metallic coatings formed by anodizing, varnish, paints, and epoxy resins have all been found to be effective in protecting certain metals from some corrosive environments (Forrest, 1970). Mechanical surface treatments such as shot peening, rolling, and nitriding are also effective because the compressive residual stresses imparted by each process will act to close cracks, so preventing the penetration of the corrosive medium. Chemical inhibitors such as sodium carbonate and emulsifying oils, when employed as adherent films, give protection to steels in particular. A sacrificial anode of aluminium or magnesium is often used to provide protection for steel immersed in sea water. This method is often referred to as cathodic protection because the steel is then the cathode in an electrolyte of brine. Care must be taken with this method to avoid over-protection where an excess of atomic hydrogen can penetrate and embrittle the steel.

18.2 LOW-CYCLE FATIGUE

Low endurances ($N \leq 10^4$ cycles in Fig. 18.1(b)) are not of great practical interest when long life is required under stress cycling. Within the low-cycle fatigue regime the inherent ductility of a material becomes a controlling factor in withstanding the plastic deformation that occurs in every cycle.

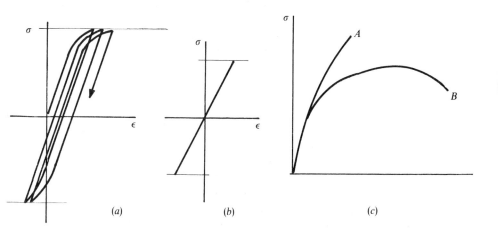

Figure 18.10 Stress cycling in low- and high-cycle fatigue.

Figure 18.10(a) illustrates typical stress–strain hysteresis loops corresponding to stress, or load, cycling between positive and negative limits of equal magnitude. It is seen that the loops progress in the direction of positive (or negative) strain, indicating an accumulation of strain with each cycle. In contrast, under a dominant elastic strain range (Fig. 18.10(b)) for the high-cycle fatigue regime, the cyclic life is dependent upon the material's strength.

Figure 18.10(c) presents a comparative approach to the understanding of material behaviour in ductility-dependent low-cycle fatigue and strength-dependent high-cycle fatigue. This is simply that the more ductile material B has a greater resistance to low-cycle fatigue than A, while the stronger material A has a greater high-cycle fatigue resistance than B. The low-cycle fatigue behaviour of a material becomes more important to consider in practice when plasticity occurs under alternating limits of cyclic strain. Since the latter may be induced either by stress or temperature, they are treated separately.

Strain-induced Stress Cycling

The abbreviation LCF normally refers to this condition. Figures 18.11(a) and (b) illustrate typical stress–strain hysterisis loops corresponding to symmetrical strain cycles bounded by fixed limits of positive and negative strain of equal magnitude. Under the constant total strain range $\Delta\epsilon^T$, the range of stress $\Delta\sigma$, during the first 10 per cent of cyclic life, will initially increase (Fig. 18.11(a)) for a strain-hardening material or will fall (Fig. 18.11b) for a strain-softening material.

As a rule, annealed materials harden while prior-worked materials soften when $Y/UTS > 0.8$. Thereafter, a cyclically stable state is normally attained in which $\Delta\sigma$ remains constant. Cyclic stability accounts for 80–90 per cent of life before a further continuous fall in $\Delta\sigma$ leads to rapid failure in N_f cycles as shown. The resistance of the material to fatigue failure under plastic strain cycling is also governed by its ductility. For lower endurances in LCF, where plastic strains dominate, the Coffin–Manson (Coffin, 1954; Manson, 1953) law expresses this in the form

$$\Delta\epsilon^P = CN_f^{-\beta}, \tag{18.14}$$

where C and β ($0.4 \leq \beta \leq 0.7$) are positive material constants and $\Delta\epsilon^P$ is the range of plastic strain identified with the width of a stable hysteresis loop. Note that C may be found directly from the true plastic fracture strain $\Delta\epsilon^P = \epsilon_f^P$ in a tension test for which $N_f = 1/4$ in Eq. (18.14). Normally, however, C and β are identified with the slope and intercept in a $\log(\Delta\epsilon^P)$ vs. $\log(N_f)$ plot (Fig. 18.12(a)) obtained from a number of LCF tests each with a different $\Delta\epsilon^P$.

When, for greater LCF life, the range of cyclic elastic ($\Delta\epsilon^E$) and plastic strains are of comparable magnitude, then the life N_f is better related to the

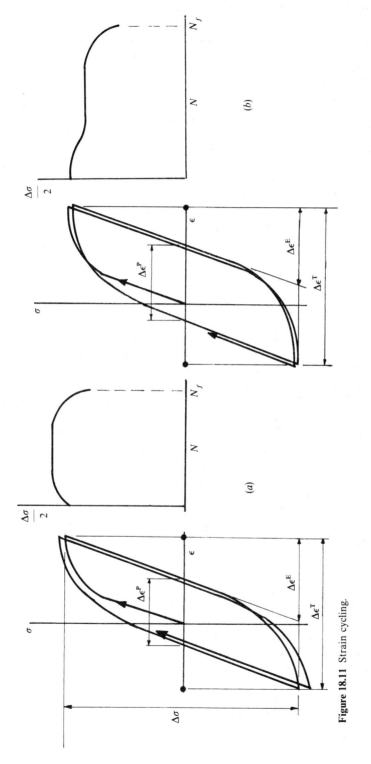

Figure 18.11 Strain cycling.

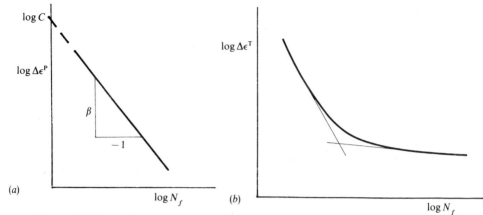

Figure 18.12 Dependence of strain range upon cycle life.

total strain range through the addition of a further term to account for the influence of cyclic elasticity on LCF life (Basquin, 1910). Here the strength of the material controls the resistance to elastic strain cycling. This leads to Manson's (1965) universal slopes relationship, which describes the $\log \Delta \epsilon^T$ vs. $\log N_f$ curve in Fig. 18.12(b):

$$\Delta \epsilon^T = \Delta \epsilon^E + \Delta \epsilon^P,$$
$$\Delta \epsilon^T = 3.5(\text{UTS}/E)N_f^{-0.12} + D^{0.6}N_f^{-0.6}, \tag{18.15}$$

in which $D = \ln[100/(100 - R_A)]$, where R_A is the percentage reduction in area in static tension failure for which E and UTS take their usual meaning. Equation (18.15) applies to the transcrystalline LCF failure for many ductile materials at room temperature under high cyclic frequencies. Further modifications become necessary to account for the reduced life owing to creep at higher temperature and lower frequencies of cycling.

Example 18.5 A material conforms to the Coffin–Manson law $\Delta \epsilon^P N_f^{0.4} = 0.25$. If it is subjected to $\Delta \epsilon^P = 0.015$ for 250 cycles then $\Delta \epsilon^P = 0.0075$ for the remainder of its life, how many cycles can it endure by Miner's rule?

The Coffin–Manson law supplies the life N_f for a given plastic strain range in this material as

$$N_f = (0.25/\Delta \epsilon^P)^{2.5}. \tag{i}$$

Substituting $\Delta \epsilon_1^P = 0.015$ and $\Delta \epsilon_2^P = 0.0075$ in Eq. (i) gives, $N_{f1} = 1134$ cycles and $N_{f2} = 6415$ cycles respectively. Then, from Miner's rule

(Eq. (18.4):

$$n_1/N_{f1} + n_2/N_{f2} = 1,$$
$$250/1134 + n_2/6415 = 1$$
$$\therefore \quad n_2 = 5000 \, \text{cycles}.$$

Strain-induced Temperature Cycling

A fixed range of plastic strain will also be induced when a material is prevented from natural expansion under temperature cycling. The general term 'thermal fatigue' describes the failure resulting from both constrained and unconstrained non-uniform expansion, sometimes under a single cycle (thermal shock), in a brittle material or from repeated cycles in a ductile material.

When the encastre beam in Fig. 18.13(a) is rapidly temperature-cycled in the range ΔT, the total strain induced from restraining the free expansion is

$$\Delta \epsilon^T = \alpha \Delta T = \Delta \epsilon^P + \Delta \epsilon^E,$$

where α is the thermal expansion coefficient. It follows that the range of plastic strain (Fig. 18.13(b)) is given by

$$\Delta \epsilon^P = \alpha \Delta T - \Delta \sigma / E. \tag{18.16}$$

Equations (18.14) and (18.16) may be combined to estimate N_f under these conditions, despite the asymmetry of the cycle. It follows from Eq. (18.16) that, for a given ΔT, the parameters governing thermal fatigue strength are α, $\Delta \epsilon^P$, $\Delta \sigma$ and E. A further property of the material, controlling its resistance to $\Delta \epsilon^P$, is the thermal conductivity k. High k-values are generally

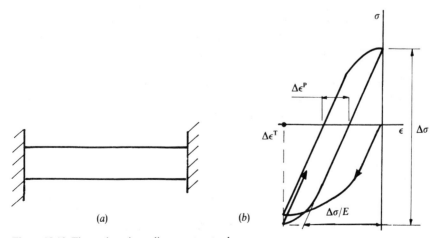

(a)   (b)

Figure 18.13 Thermal strain cycling an encastre beam.

beneficial to reducing temperature gradients and induced strains. The following two thermal fatigue factors have been proposed to quantify the ability to a material to withstand thermal strain cycling: $\Delta\sigma k/\alpha E \equiv Yk/\alpha E$ ($\Delta\sigma = 2Y$ in the absence of hardening) and $\Delta\epsilon^P k/\alpha$. In each case high values are desirable.

18.3 LINEAR ELASTIC FRACTURE MECHANICS

The recent attention paid to the behaviour of cracks under static and cyclic loading has enabled realistic predictions of safe working stresses to be made for components operating in various environments. This introduction first examines the mechanics of purely brittle fracture, then incorporates some corrections that have been made for crack-tip plasticity and finally considers

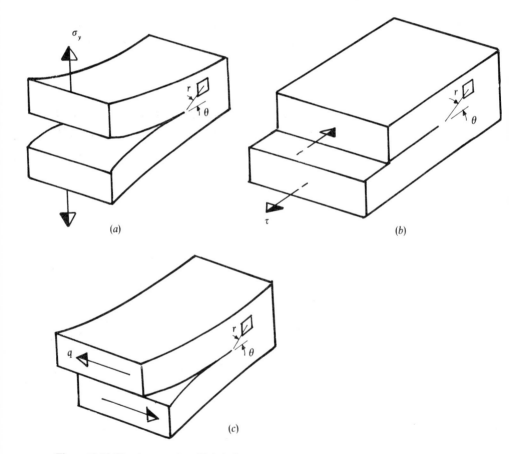

Figure 18.14 The three modes of brittle fracture.

some applications to fatigue where corrosion may also influence behaviour. The reader is referred to specialist texts for the application of fracture mechanics to both ductile fracture and creep crack growth.

A crack may propagate in one of three modes (Fig. 18.14(a)–(c)). The tension opening mode I in (a) is most often referred to, but the two shear modes II and III in (b) and (c) describe respective brittle fractures from sliding and antiplane tearing. In order to determine the critical loading that would cause failure under each mode, a failure criterion must be specified. There are two criteria; one is based upon the stress state around the crack tip and the other upon the release of elastic strain energy produced by crack extension. It is instructive to show first that the two approaches are compatible in the case of mode I opening of a crack of length $2a$ in a large (infinite) plate subjected to a normal boundary stress σ.

Stress Function Approach

Westergard (1939) examined the stress state in the vicinity of the crack tip. With the origin of x–y axes at the crack centre and x aligned with the crack length (Fig. 18.15(a)), the following Cartesian stress components σ_x, σ_y and τ_{xy} were derived for a point (r, θ) in the infinite plate when, for radius r from the crack tip, and $r \ll a$,

$$\sigma_y = \sigma(a/2r)^{1/2} \cos(\theta/2)[1 + \sin(\theta/2)\sin(3\theta/2)], \qquad (18.17a)$$

$$\sigma_x = \sigma(a/2r)^{1/2} \cos(\theta/2)[1 - \sin(\theta/2)\sin(3\theta/2)], \qquad (18.17b)$$

$$\tau_{xy} = \sigma(a/2r)^{1/2} \sin(\theta/2)\cos(\theta/2)\cos(3\theta/2). \qquad (18.17c)$$

For a thin plate, plane stress prevails with $\sigma_z = 0$. In the case of a thick plate,

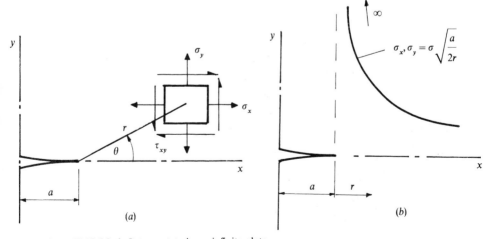

Figure 18.15 Mode I stress state in an infinite plate.

when plane strain is assumed for $\epsilon_o = 0$ in Eq. (14.2c), this gives the through-thickness stress as

$$\sigma_z = v(\sigma_x + \sigma_y) = 2v\sigma(a/2r)^{1/2}\cos(\theta/2). \tag{18.17d}$$

When $\theta = 0$ for a point on the crack x-axis, close to the crack tip, Eqs (18.17) give $\tau_{xy} = 0$ and

$$\sigma_y = \sigma_x = \sigma(a/2r)^{1/2}. \tag{18.18}$$

If the crack length is $2a = 1$ mm, Eq. (18.18) shows that a boundary stress $\sigma = 20$ MPa is magnified fivefold in the components of σ_x and σ_y when $r = 1/100$ mm, as shown in Fig. 18.15(b). Of these stress components, the magnitude of opening stress σ_y governs crack stability. Irwin (1957) examined the elastic opening stress expressions for a number of loaded cracked plates with different geometries. He discovered that σ_y had the common form

$$\sigma_y = [K_{\mathrm{I}}/(2\pi r)^{1/2}]\cos(\theta/2)[1 + \sin(\theta/2)\sin(3\theta/2)], \tag{18.19}$$

in which the mode I stress intensity factor K_{I}, distinguished one plate from another. For example, the infinite-plate K_{I} is found from comparing Eqs (18.17a) and (18.19):

$$K_{\mathrm{I}} = \sigma\sqrt{(a\pi)}. \tag{18.20}$$

The other plates considered were similarly characterized with their particular stress intensity factors. They have a similar form

$$K_{\mathrm{I}} = Q\sigma\sqrt{a}, \tag{18.21}$$

where Q is either a constant, for infinite plates, or a function of finite-plate geometry. Some of the more common crack geometries are given in Fig. 18.16. The particular forms of Eq. (18.21) are as follows.

(a) A semi-infinite plate containing an edge crack of length a:

$$K_{\mathrm{I}} = 1.12\sigma\sqrt{(a\pi)}.$$

(b) An infinite plate containing an embedded circular crack or a semi-circular surface crack, each of radius a, lying in a plane normal to σ:

$$K_{\mathrm{I}} = (2/\pi)\sigma\sqrt{(a\pi)}.$$

(c) An infinite plate containing an embedded elliptical flaw $2a \times 2b$ or a semi-elliptical surface crack of width $2b$ in which the depth a is less than half the plate thickness, each lying normal to σ:

$$K_{\mathrm{I}} = (1.12/\phi)\sigma\sqrt{(a\pi)},$$

in which ϕ varies with the ratio a/b as follows:

a/b	0	0.2	0.4	0.6	0.8	1.0
ϕ	1.0	1.05	1.15	1.28	1.42	$\simeq \pi/2$

Clearly when $a/b = 1$ this closely approximates to the case in (b).

(d) An infinite plate containing a through-thickness crack of length $2a$ with a concentrated opening force P applied normal to the crack surfaces, distance b from the crack centre:

$$K_1 = P\sqrt{[a/\pi(a-b)(a+b)]}. \qquad (18.22)$$

When the force P is aligned with y, then $b = 0$ and Eq. (18.22) becomes

$$K_1 = P/\sqrt{(\pi a)}.$$

(e) A plate of finite width w, containing a central crack of length $2a$, where $a \le 0.3w$, with a normally applied remote stress σ:

$$K_1 = \sigma\sqrt{[w\tan(a\pi/w)]} \simeq \sigma\sqrt{(a\pi)}(1 + \pi^2 a^2/6w^2 + \cdots),$$

or in the alternative form

$$K_1 = \sigma\sqrt{[(a\pi)\sec(a\pi/w)]}.$$

(f) A plate of finite width w, containing two symmetrical edge cracks each of depth a:

$$K_1 = \sigma\sqrt{[w\tan(\pi a/w) + (0.1w)\sin(2\pi a/w)]} \qquad \text{(approximately)}.$$

It is seen that as the stress intensity ahead of the crack can be expressed in the single parameter K_1, then failure will occur when K_1 reaches a critical value K_{1c}. This material property is known as the fracture toughness (units: $\text{MN/m}^{3/2} \equiv \text{MPa}\sqrt{m}$ or $\text{N/mm}^{3/2}$). In a similar manner, the stress intensity factors K_{II} and K_{III} may be derived for various geometries under the shear loading modes II and III. In an infinite plate, for example, containing a

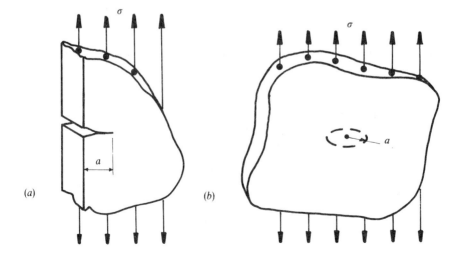

(a)   (b)

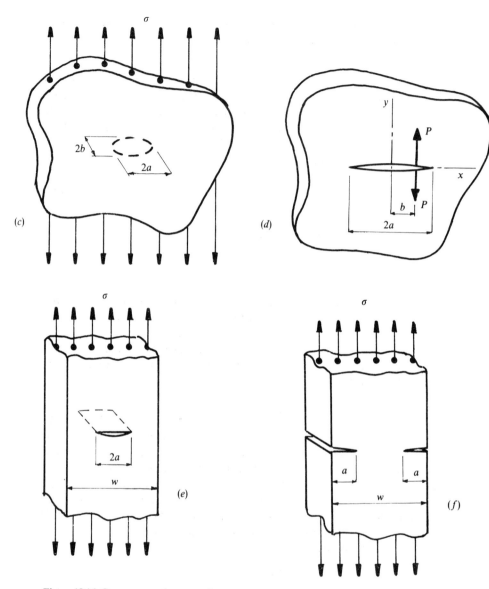

Figure 18.16 Common crack geometries.

through-thickness crack of length $2a$ under the II and III modes, the respective stress intensities are

$$K_{\mathrm{II}} = \sqrt{\tau(a\pi)} \qquad \text{and} \qquad K_{\mathrm{III}} = q\sqrt{(a\pi)},$$

where the operative shear stresses in each case are shown in Figs. 18.14(b) and (c). Critical values of K_{II} and K_{III} then define the fracture toughness of the material. Note that for a given geometry the stress intensities under combined modes of loading are additive.

Example 18.6 A 5 mm (0.2 in) long crack exists in an infinite steel plate of fracture toughness $K_{Ic} = 105\,\text{MN/m}^{3/2}$ (43 tonf/in$^{3/2}$). Calculate the maximum allowable design stress that could be applied around the boundary. If the yield stress for the plate material is 1500 MPa (97 tonf/in^2), comment on the type of failure that would occur if this design stress were exceeded.

IN SI UNITS   Replacing K_I in Eq. (18.20) with K_{Ic}, the maximum allowable stress for $a = 2.5$ mm

$$\sigma = K_{Ic}/\sqrt{(a\pi)} = 105/\sqrt{(\pi \times 2.5 \times 10^{-3})} = 1185\,\text{MPa}.$$

IN IMPERIAL UNITS

$$\sigma = K_{Ic}/\sqrt{(a\pi)} = 43/\sqrt{(\pi \times 0.1)} = 76.72\,\text{tonf/in}^2.$$

Since σ is less than the yield stress, i.e. elastic, then failure in this high-strength, low-ductility material would be brittle. In the low-to-medium temperature range $(T/T_m < 0.5)$ the fracture path is transcrystalline (transgranular). That is, the crack path proceeds from grain to grain along crystallographic cleavage planes characterized by a smooth, bright, granular appearance.

Irwin's (1964) estimation of crack tip plasticity   In medium strength materials, the local stresses in the vicinity of the crack tip will normally exceed the yield stress under plane stress conditions. The following correction to linear elastic fracture mechanics provides an equivalent crack length in the presence of localized crack tip plasticity. Thus, when the σ_y distribution in Fig. 18.17(a) is cut off at an ordinate equal to the yield stress value Y, the elastic stress distribution does not extend within radius r_y from the crack tip, although the area A_1 is assumed to be available for further yielding.

For an infinite plate, from Eqs (18.18) and (18.20):

$$\sigma_y = \sigma(a/2r)^{1/2} = K_I/(2\pi r)^{1/2},$$
$$\therefore\quad r_y = K_I^2/2\pi Y^2. \tag{18.23a}$$

The area beneath the curve for $r_y \leq r \leq 0$ is given by

$$A_1 + A_2 = \int_0^{r_y} K_I\,dr/(2\pi r)^{1/2} = (2/\pi)^{1/2}K_I r_y^{1/2},$$

and, substituting from Eq. (18.23a),

$$A_1 + A_2 = K_I^2/\pi Y = 2Y r_y.$$

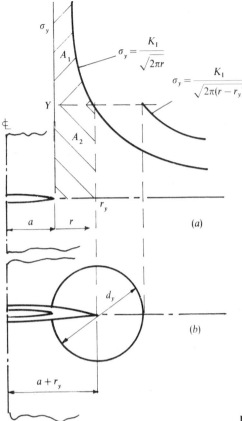

Figure 18.17 Crack-tip plasticity.

But as $A_2 = Yr_y$, it follows that $A_1 = A_2$. Hence, the plastic zone (Fig. 18.17(b)) is taken to be a circle of diameter $d_y = 2r_y$. This modifies the σ_y distribution as shown so that the fictitious crack tip of half-length $a + r_y$ is assumed to lie at the centre of the plastic zone. Differences between the stresses σ_x, σ_y and σ_z (Eqs (18.17)) are necessary for plasticity to occur. The presence of a through-thickness stress σ_z in thick plates under plane strain conditions creates a central triaxial stress state close to the root of the crack. At mid-thickness the plastic zone size is estimated to be three times smaller than under plane stress, where $\sigma_z = 0$. Thus, for plane strain, Eq. (18.23a) becomes

$$r_y = K_I^2/6\pi Y^2. \qquad (18.23b)$$

This radius increases from the plate centre to its edges where, with $\sigma_z = 0$, Eq. (18.23a) will again apply. The effect that a plastic zone has on the fracture toughness in each case may be assessed from the equivalent crack length.

For an infinite plate, Eq. (18.20) is written as

$$K^*_{Ic} = \sigma\sqrt{[\pi(a + r_y)]} = \sigma\sqrt{(a\pi)}\sqrt{(1 + r_y/a)},$$
$$K^*_{Ic} = K_{Ic}\sqrt{[1 + (1/\beta)(\sigma/Y)^2]}, \tag{18.24}$$

where $\beta = 2$ for plane stress and $\beta = 6$ for plane strain.

Dugdale's (1960) model of crack tip plasticity This is an alternative analysis for the plastic zone size under a plane stress condition. The method employs the known solution to the stress intensity under a normal force to the crack boundary (Eq. (18.22)) in order to express the closing force in a non-hardening plastic zone when the crack is assumed to extend elastically to length c (Fig. 18.18). That is, for a plate of unit thickness with a closing force $\delta P = Y \, db$ applied at a distance b ($a \le b \le c$) from the origin in a crack of semi-length c, as shown, Eq. (18.22) becomes

$$K_I = 2Y \int_a^c \sqrt{[c/\pi(c + b)(c - b)]} \, db$$
$$= 2Y(c/\pi)^{1/2} \cos^{-1}(a/c).$$

To obtain the same stress intensity for an elastic crack of length $2c$, it follows that

$$\sigma\sqrt{(c\pi)} = 2Y(c/\pi)^{1/2} \cos^{-1}(a/c),$$
$$\therefore \quad a/c = \cos(\pi\sigma/2Y).$$

Putting $d_y = c - a$ the cosine expansion reveals that the diameter of the zone under a limited spread of plasticity is

$$d_y = 2r_y = \pi K_I^2/8Y^2. \tag{18.25a}$$

In this case, Dugdale's correction to Eq. (18.20) for the fracture toughness

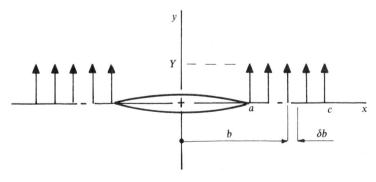

Figure 18.18 Dugdale's solution.

of an infinite plate becomes

$$K_{Ic}^* = K_{Ic}\sqrt{[1 + (\pi^2/8\beta)(\sigma/Y)^2]}, \tag{18.25b}$$

where again $\beta = 2$ and $\beta = 6$ for plane stress and strain respectively so that Eq. (18.25a) is written as

$$r_y = \pi K_I^2/8\beta Y^2 \tag{18.25c}$$

Dependence of K_{Ic} upon plate thickness Clearly, Eqs (18.24) and (18.25b) show that K_{Ic}^* is strongly dependent upon plate thickness. In general, since the toughness of a material decreases with decreasing plasticity, it follows that the true fracture toughness (K_{Ic}) for a material is the lower limiting value that a plane strain condition will ensure (Fig. 18.19). The limiting toughness is particularly important to consider in high-strength alloys and structural steels at lower temperatures, since these are all at risk from brittle failure. When the plastic zone size is small compared to the thickness (B), the brittle transcrystalline cleavage fracture surface will be flat except for small shear lips indicative of plane stress conditions at its free edges. With decreasing plate thickness, the spread of plasticity increases and a full plane stress condition is approached with a 45° ductile transcrystalline shear failure exhibiting a dull fibrous appearance. Figure 18.19 further shows that mixed 45°-shear and 0°-cleavage fractures will appear for plate thicknesses lying within the transition from plane stress to plane strain.

The assumption that the plastic zone is circular is more representative in the case of plane stress. With the restricted spread of plasticity under plane strain conditions, etch-pit and finite-element investigations have shown that the plastic zone extends as two lobes each inclined at θ to the crack axis (Fig. 18.20). Rice and Johnson (1970) showed, for a small amount of plane

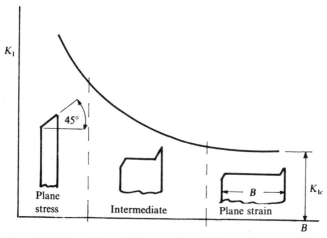

Figure 18.19 Dependence of K_{Ic} upon plate thickness.

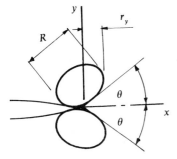

Figure 18.20 Plane strain plasticity.

strain yielding, that $\theta = 70°$, where the greatest extent (R) and forward penetration (r_y for $\theta = 0$) of plasticity were

$$R = 0.155(K_{\mathrm{I}}/Y)^2,$$
$$r_y = 0.04(K_{\mathrm{I}}/Y)^2.$$

As the latter compares with respective coefficients of 0.053 and 0.066 from Eqs (18.23b) and (18.25c), the simplified equivalent crack-length method will not lead to misleading results.

Example 18.7 The heat-affected zone in a welded infinite steel plate contains a through-thickness crack aligned with the weld direction, as shown in Fig. 18.21. If the boundary stress σ is to be 2/3 of the yield stress (Y), compare the tolerable crack lengths, assuming brittle failure when the plate is (a) stress relieved after welding and (b) as-welded, containing tensile residual stresses of the order Y. If the plate is thin, what is the effect of crack tip plasticity on this comparison? (T.C.D. 1979)

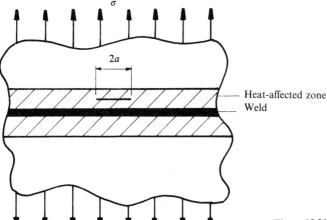

Figure 18.21 Welded plate.

(a) STRESS RELIEVED The tolerable crack length ($2a$) is found from Eq. (18.20) as

$$2a = (2/\pi)(K_{1c}/\sigma)^2, \tag{i}$$

and with $\sigma = 2Y/3$, Eq. (i) becomes

$$2a = (9/2\pi)(K_{1c}/Y)^2 \tag{ii}$$

(b) AS-WELDED Here the tensile residual stress Y assists the boundary stress $2Y/3$ to produce an effective applied stress $\sigma = 5Y/3$. Substituting into Eq. (i) leads to a tolerable crack length:

$$2a = (18/25\pi)(K_{1c}/Y)^2. \tag{iii}$$

Dividing (iii) by (ii) shows that stress-relieving increases the permissible crack length by a factor of 25/4.

Using Eq. (18.24) for the presence of plane stress plasticity, Eq. (i) becomes

$$2a = 2K_{1c}^{2*}/\pi\sigma^2(1 + \sigma^2/2Y^2). \tag{iv}$$

Substituting $\sigma = 2Y/3$ and $\sigma = 5Y/3$ into Eq. (iv), the ratio between the tolerable cracks lengths for stress relieved and welded conditions is

$$\frac{a_{sr}}{a_w} = \frac{25(1 + 25/18)}{4(1 + 2/9)} = 12.2.$$

This shows that there is a further twofold increase in the crack ratio compared with that for a brittle material.

Energy Balance Approach

Another characterization of brittle fracture employs the conservation of energy principle for the derivation of an alternative toughness parameter. Griffith (1921) considered the energy balance at the point of fracture between the loss in strain (or potential) energy for a loaded cracked plate and the surface energy required to form new crack surfaces. It was shown that the elastic work done in deforming the upper crack surface in an infinite plate may be found more readily from applying a stress, with magnitude equal to the boundary stress, normal to the crack surfaces (Fig. 18.22). The crack-opening displacement v is found from integrating the following constitutive relation for both conditions of plane stress ($\sigma_z = 0$) and plane strain ($\epsilon_z = 0$):

$$\epsilon_y = \partial v/\partial y = (1/E)[\sigma_y - v(\sigma_x + \sigma_z)]. \tag{18.26a}$$

Griffith showed that when $x \leq a$, the integration of Eq. (18.26a) led to

$$v = (\sigma/2E)(1 + v)(1 + \alpha)(a^2 - x^2)^{1/2}, \tag{18.26b}$$

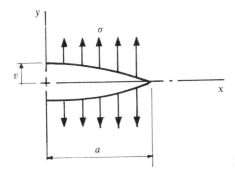

Figure 18.22 Crack-opening displacement.

where $\alpha = (3 - v)/(1 + v)$ for plane stress (a thin plate) and $\alpha = 3 - 4v$ for plane strain (a thick plate). Since half the amount U of strain energy stored is equal to the work done in deforming the upper crack surface, it follows from Eq. (11.3c) that U for this uniaxial stress system is given by

$$U = \int_V \int_\epsilon \sigma \, d\epsilon \, dV,$$

$$U = (1/2) \int_V \sigma \epsilon_y \, dV.$$

Hence for the upper crack surface $-a \leq x \leq a$ the work done in a plate of thickness B is

$$U = (1/2) \int_x \int_y \int_z \sigma (\partial v / \partial y)(dx \, dy \, dz)$$

$$= (2 \times 1/2) \int_{z=0}^B \int_{x=0}^a \sigma v \, dx \, dz$$

$$= B \int_0^a \sigma v \, dx, \tag{18.27}$$

where B is the plate thickness. Substituting Eq. (18.26b) into (18.27):

$$U = (B\sigma^2/2E)(1 + v)(1 + \alpha) \int_0^a (a^2 - x^2)^{1/2} \, dx$$

$$= a^2 B\pi (1 + v)(1 + \alpha)\sigma^2/8E. \tag{18.28}$$

If S_o (units: J/m^2 or N mm/mm^2) is the surface tension per unit area of crack surface (the specific surface energy), then the total surface energy (U_s) of the upper crack is given by

$$U_s = 2aS_o B. \tag{18.29}$$

Griffith postulated that unstable crack growth δa occurs when the strain energy released equals or exceeds the energy required to extend the crack by this amount. That is,

$$\delta U/\delta a \geq \delta U_s/\delta a,$$

where, from Eqs (18.28) and (18.29),

$$aB\pi(1 + v)(1 + \alpha)\sigma^2/4E \geq 2S_oB,$$

$$\sigma \geq \sqrt{[8S_oE/a\pi(1 + v)(1 + \alpha)]}. \tag{18.30a}$$

For the plane stress condition $1 + \alpha = 4/(1 + v)$, when Eq. (18.30a) becomes

$$\sigma \geq \sqrt{(2S_oE/a\pi)}, \tag{18.30b}$$

while for plane strain $1 + \alpha = 4(1 - v)$, when Eq. (18.30a) becomes

$$\sigma \geq \sqrt{[2S_oE/a\pi(1 - v^2)]}. \tag{18.30c}$$

Note from Eqs (18.29) that the critical rate of strain energy released (w.r.t. the crack length) is given as

$$\delta U_s/\delta a = 2S_oB.$$

It follows from this that G_{Ic}, the critical energy release rate for a plate of unit thickness under mode I opening, is

$$G_{Ic} = (1/B)(\delta U/\delta a) = 2S_o. \tag{18.31a}$$

This is Griffith's linear elastic fracture mechanics fracture parameter, which may also be interpreted as a critical crack extension force per unit width. The factor of 2 appears with the creation of a crack of length $2a$. In

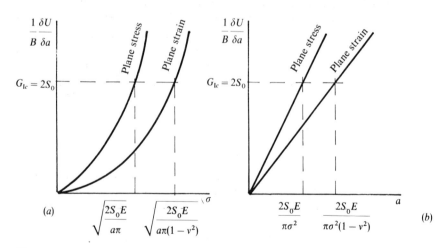

Figure 18.23 Energy for plane stress and plane strain fracture.

Figs 18.23(a) and (b) the energy balances appropriate to plane stress and plane strain fracture are represented graphically to give (a) the critical fracture stress for a given crack length and (b) the critical crack length for a given applied stress. Completely brittle materials containing sharp cracks, such as glass, conform to Eq. (18.31a), i.e. the fracture energy is equal to twice the surface energy (S_o). For most metallic materials the effect that crack-tip plasticity has in blunting the crack tip invalidates Eq. (18.31a), since the fracture energy is several orders or magnitude higher than the surface energy. However, with limited crack-tip plasticity, both Orowan (1955) and Irwin (1957) recognized that fracture still occurred at a critical release rate of elastic energy from the plate. Equation (18.31a) is then written as

$$G_{Ic} = (1/B)(\delta U/\delta a) = 2(S_o + S_p) \qquad (18.31b)$$

in which an irreversible plastic work term $S_p \gg S_o$ is added to the surface energy S_o. It follows from Eqs (18.30b and c) that the fracture stress for an infinite plate under conditions of plane stress and plane strain are respectively

$$\sigma = \sqrt{(EG_{Ic}/\pi a)}, \qquad (18.32a)$$

$$\sigma = \sqrt{[EG_{Ic}/\pi a(1 - v^2)]}. \qquad (18.32b)$$

Combining Eqs (18.20) and Eq. (18.32), the following general relationships between K_{Ic} and G_{Ic} are found for plane stress and plane strain linear elastic fracture mechanics:

$$G_{Ic} = K_{Ic}^2/E, \qquad (18.33a)$$

$$G_{Ic} = K_{Ic}^2(1 - v^2)/E. \qquad (18.33b)$$

Example 18.8 A large thin-walled alloy steel tube has a uniaxial yield stress $Y = 1200\,\text{MPa}$ and an elastic modulus $E = 200\,\text{GPa}$. A separate test on a cracked infinite plate showed that the critical rate of energy released at fracture was $24\,\text{kJ/m}^2$. The allowable circumferential stress in the tube is based upon Y with a factor of safety of 1.5. Calculate the maximum permissible size of longitudinal defect assuming (i) a purely brittle fracture and (ii) the presence of crack-tip plasticity. (T.C.D. 1980)

(i) The allowable circumferential stress is simply

$$\sigma = Y/1.5 = 1200/1.5 = 800\,\text{MPa}.$$

In the brittle case, Eq. (18.20) gives the allowable defect size. Combining this with Eq. (18.33a) for plane stress, in units of newtons and millimetres,

$$a = EG_{Ic}/\pi\sigma^2$$
$$= 200 \times 10^3 \times 24/\pi \times (800)^2 = 2.4\,\text{mm}.$$
$$\therefore \quad 2a = 4.8\,\text{mm}.$$

(ii) Crack tip plasticity. Under plane stress conditions, the radius of the plastic zone is estimated from combining Eqs (18.23a) and (18.33a):

$$r_Y = (1/2\pi)(K_{1c}^2/Y^2) = (1/2\pi)(EG_{1c}/Y^2).$$

Then, from Eq. (18.24), assuming that the given G_{1c} is the critical rate of energy release for non-brittle fracture:

$$a + r_y = K_{1c}^{2*}/\pi\sigma^2 = EG_{1c}/\pi\sigma^2,$$
$$\therefore \quad a = EG_{1c}/\pi\sigma^2 - r_y,$$
$$a = EG_{1c}/\pi\sigma^2 - (1/2\pi)(EG_{1c}/Y^2)$$
$$= (EG_{1c}/\pi\sigma^2)(1 - \sigma^2/2Y^2)$$
$$= 2.4(1 - 2/9) = 1.8666 \, \text{mm},$$
$$\therefore \quad 2a = 3.73 \, \text{mm}.$$

Determination of Fracture Toughness Parameters K_{1c} and G_{1c}

In order to avoid the prohibitive cost involved in the determination of K_{1c} and G_{1c} from testing infinite plates to fracture, a number of smaller test pieces with finite dimensions have evolved, which ensure a plane strain condition. The effect of the nearness of the boundary to the crack surface has been overcome by an analytical boundary collocation method for obtaining the K_{1c} expression. Alternatively, an experimental compliance calibration may be employed for the determination of G_{1c}.

K_{1c} **determination** The British Standard 5447:1977 outlines the procedures for testing to failure (a) three-point bend, (b) compact tension and (c) single-edge notch test pieces given in Figs. 18.24. When the dimensions of these standard test pieces are chosen to satisfy the given conditions, the plane strain fracture toughness K_{1c} is directly calculated from

$$K_{1c} = P_c f(a/w)/Bw^{1/2}, \tag{18.34}$$

where B and w are the respective thicknesses and widths shown, P_c is the load at fracture and the compliance function $f(a/w)$ accounts for the influence that the finite boundaries have on the stress intensity in each case. For Figs 18.24 (a)–(c), these are respectively

(a) $f(a/w) = 6[1.93(a/w)^{1/2} - 3.07(a/w)^{3/2} + 14.53(a/w)^{5/2} - 25.11(a/w)^{7/2}$
$\qquad + 25.8(a/w)^{9/2}]$;

(b) $f(a/w) = 29.6(a/w)^{1/2} - 185.5(a/w)^{3/2} + 655.7(a/w)^{5/2} - 1017(a/w)^{7/2}$
$\qquad + 638.9(a/w)^{9/2}$;

(c) $f(a/w) = 1.99(a/w)^{1/2} - 0.41(a/w)^{3/2} + 18.7(a/w)^{5/2} - 38.48(a/w)^{7/2}$
$\qquad + 53.85(a/w)^{9/2}.$

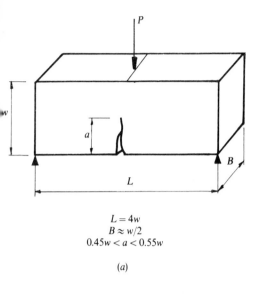

$L = 4w$
$B \approx w/2$
$0.45w < a < 0.55w$

(a)

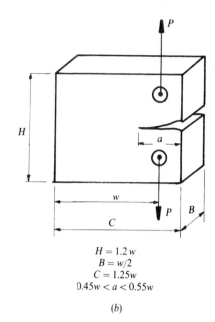

$H = 1.2\,w$
$B = w/2$
$C = 1.25w$
$0.45w < a < 0.55w$

(b)

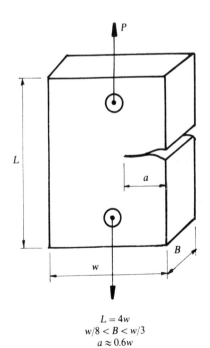

$L = 4w$
$w/8 < B < w/3$
$a \approx 0.6w$

(c)

Figure 18.24 Standard K_{1c} test pieces.

The standard provides a further check that K_{1c}, calculated from Eq. (18.34), is the lower limiting plane strain value when

$$B \simeq a \simeq (w - a) \geq 2.5(K_{1c}/Y)^2. \tag{18.35}$$

Example 18.9 A compact tension test is conducted on a steel sample of thickness $B = 50$ mm with a crack length $a = 50$ mm and a w-dimension of 100 mm (Fig. 18.24b). If this test provides a fracture toughness value of $K_{1c} = 110$ MN/m$^{3/2}$, estimate the fracture load and find the minimum yield strength for which the test is valid. Determine also the critical rate of strain energy released and the energy required to create the crack surfaces. Take $E = 220$ GPa and $v = 0.26$ (T.C.D. 1982)

Here $a/w = 50/100 = 0.5$. Substituting into the compliance function $f(a/w)$ corresponding to Fig. 18.24(b).

$$f(a/w) = 20.93 - 65.58 + 115.91 - 89.9 + 28.24 = 9.596.$$

Then, from Eq. (18.34),

$$P_c = K_{1c}Bw^{1/2}/f(a/w)$$
$$= 110 \times 50 \times 10^{-3} \times (0.1)^{1/2}/9.596 = 0.181 \text{ MN} = 181 \text{ kN}.$$

Equation (18.35) confirms the validity of this test, since $B \geq 2.5(K_{1c}/Y)^2$. The minimum yield strength is therefore

$$Y = \sqrt{(2.5K_{1c}^2/B)}$$
$$= \sqrt{(2.5 \times 110^2/0.05)} = 777.8 \text{ MPa}.$$

Under these plane strain conditions Eq. (18.33b) applies. Therefore, the critical rate of strain energy released per unit thickness is

$$G_{1c} = K_{1c}^2(1 - v^2)/E$$
$$= (110 \times 10^3)^2(1 - 0.26^2)/(220 \times 10^6) = 51.28 \text{ kJ/m}^2.$$

It follows from Eq. (18.31a) that this is also the specific surface energy required to create the two crack surfaces. The total energy consumed is

$$U = G_{1c}(Ba)$$
$$= 51.28 \times 10^3(50 \times 50 \times 10^{-6}) = 128.2 \text{ J}.$$

G_{1c} determination The previous example illustrates how G_{1c} may be estimated indirectly from K_{1c}. Alternatively, a useful compliance technique enables G_{1c} to be obtained from direct measurements of the applied force P and the deflection Δ beneath the force, up to the point of brittle fracture. Figure 18.25 shows the linear elastic P vs. Δ response for plates with initial crack lengths of a and $a + \delta a$. For the longer crack there is a reduction in stiffness as seen in the shallower slope P/Δ, i.e. an increase in compliance $\lambda = \Delta/P$. When

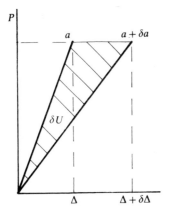

Figure 18.25 Energy released from crack extension.

the crack extends by the amount δa under a constant force, it follows from Eq. (18.31a) that the energy absorbed in the crack formation is

$$\delta U = BG_{Ic}\delta a. \tag{18.36}$$

Now, under a constant force, the increase in compliance is $\delta\lambda = \delta\Delta/P$ and the release of strain or potential energy stored the plate is the shaded area in Fig. 18.25. That is,

$$\delta U = P\,\delta\Delta/2 = P^2\,\delta\lambda/2. \tag{18.37}$$

Equating (18.36) and (18.37) for the fracture condition gives G_{Ic} as

$$G_{Ic} = (P^2/2B)(\delta\lambda/\delta a). \tag{18.38}$$

Since Eq. (18.38) is independent of the loading configuration, it would still apply had the crack extended without load-point displacement (fixed grips). The compliance calibration involves testing a number of identical test piece geometries with different crack lengths to establish plots typified by Fig. 18.25. In this way the plot of λ vs. a may be found. Then, in applying Eq. (18.38), $\delta\lambda/\delta a$ is the slope of this plot at the particular value of fracture force P. It follows from Eqs (18.33b) and (18.38) that G_{Ic} is readily converted to K_{Ic} for plane strain:

$$K_{Ic} = \sqrt{\{[P^2 E/2B(1 - v^2)](\delta\lambda/\delta a)\}}. \tag{18.39}$$

The experimental compliance method only requires the measurement of P and Δ in a test piece of given geometry. This method is often preferred to a theoretical derivation, the complexity of which is typified by the $f(a/w)$ expressions given with the standard test pieces in Fig. 18.24.

Example 18.10 Tests on an edge-cracked, high-tensile alloy steel plate model gave the following displacements beneath an applied force of 1 MN:

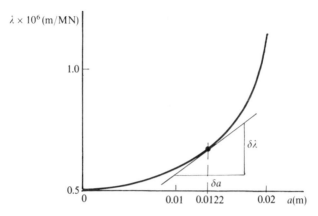

Figure 18.26 Compliance calibration.

a (mm)	0	5	10	15	20
$\Delta/10^{-3}$ (mm)	0.503	0.539	0.612	0.793	1.194

The crack was machined to the new length after each test. If a tensile member 12.5 mm thick, made from the same steel with the properties $K_{\text{Ic}} = 123 \text{ MN/m}^{3/2}$, $Y = 1540 \text{ MPa}$, is to support an axial force of 5 MN, estimate, from the model results, the maximum permissible crack length if brittle fracture is to be avoided. Take $E = 207 \text{ GPa}$ and $v = 0.27$.

First check the plane strain condition from Eq. (18.35):

$$B \geq 2.5(K_{\text{Ic}}/Y)^2 = 2.5(123/1540)^2 = 0.016 \text{ m}.$$

Since $B = 0.0125$ m, it appears that a mixed-mode fracture is likely here. For linear elastic behaviour the direct measurement of load point displacement under a constant load of 1 MN gives a compliance conversion factor, $\lambda = \Delta/P = \Delta \times 10^{-6} \text{ (m/MN)}$, corresponding to each crack length a. This enables the plot of λ vs. a in Fig. 18.26 to be constructed.

Now it follows from Eq. (18.39) that the critical value of $\delta\lambda/\delta a$, under a force of 5 MN, is

$$\delta\lambda/\delta a = 2 \times 0.0125(1 - 0.27^2)123^2/5^2 \times 207\,000,$$
$$= 32.576 \times 10^{-6} \text{ MN}^{-1}.$$

The plane stress condition, i.e. without the $(1 - v^2)$ factor in Eq. (18.39), gives a gradient $\delta\lambda/\delta a = 35.14 \times 10^{-6} \text{ MN}^{-1}$. The lower plane strain gradient provides the safer allowable crack length of $a = 12.2$ mm (Fig. 18.26).

Application of Linear Elastic Fracture Mechanics to Fatigue

The brittle nature of fatigue failure implies that fatigue strength, like fracture toughness, also depends upon the crack-tip parameter K. According to Eq. (18.21) during crack propagation under a repeated or fluctuating tensile stress cycle of constant stress range $\Delta\sigma$ (Fig. 18.2), the range of stress intensity ΔK continuously changes. Thus for all the plate geometries shown in Fig. 18.16, this gives

$$\Delta K = Q\,\Delta\sigma\sqrt{a}, \qquad (18.40)$$

where $\Delta\sigma = \sigma_{max} - \sigma_{min}$. It is now well established that, following initiation, the cyclic rate of crack propagation da/dN in Fig. 18.27(a) is a function of

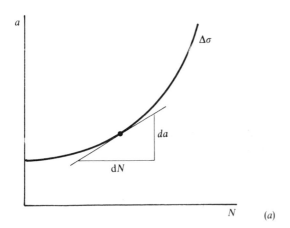

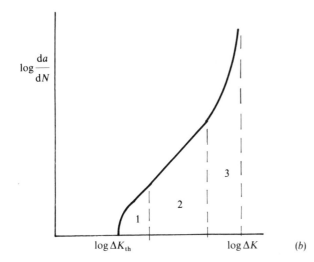

Figure 18.27 Fatigue crack growth.

the range of stress intensity ΔK in modes I, II and III (Paris, 1964). That is,

$$da/dN = f(\Delta K) = g(\Delta\sigma, a). \qquad (18.41)$$

Equation (18.41) will account for the life when the initiation phase is absent in the presence of pre-existing defects.

Many experiments have shown that the logarithmic plot of da/dN vs. ΔK in Fig. 18.27(b) correlates the growth rates for separate tests each with a different $\Delta\sigma$-value. The sigmoidal shape of this curve for non-aggresive environments indicates three regions of fatigue crack growth. Region 1 is associated with low and non-propagating rates of crack growth as ΔK lies near its threshold value (ΔK_{th}). For the mid-range growth in region 2, a power-law form of Eq. (18.41) applies:

$$da/dN = C(\Delta K)^n, \qquad (18.42)$$

where C and n are material constants. The exponent n normally lies in the range 2–7, depending upon the environment, cyclic frequency, mean stress and temperature. In region 3, unstable, accelerating crack growth precedes fracture as K_{max} approaches K_{lc}. Most estimates of mode I fatigue life in structures are based upon combining Eq. (18.40) with Eq. (18.42) for region 2. This gives

$$da/dN = C(Q\,\Delta\sigma\sqrt{a})^n = CQ^n(\Delta\sigma)^n a^{n/2}. \qquad (18.43a)$$

Then, for a crack of initial length a_o that extends to a length a_f when the structure fails in N_f cycles, Eq. (18.43a) becomes

$$\int_{a_o}^{a_f} da/a^{n/2} = CQ^n(\Delta\sigma)^n \int_0^{N_f} dN$$

and provided Q does not vary with a, this integrates directly to give the cyclic life as

$$N_f = \frac{2}{(n-2)CQ^n(\Delta\sigma)^n}\left[\frac{1}{a_o^{(n-2)/2}} - \frac{1}{a_f^{(n-2)/2}}\right]. \qquad (18.43b)$$

It is seen that Eq. (18.43b) will not apply to those plates in Fig. 18.15 for which Q is trigonometric in a. In this case a numerical integration of Eq. (18.43a) will supply N_f.

Example 18.11 Calculate the number of cycles to failure when a cyclic repeated stress of range 175 MPa is applied to an infinite plate containing a 0.2 mm long crack. Take $K_{lc} = 54\,\text{MN/m}^{3/2}$ and the rate of crack growth as $da/dN = 40 \times 10^{-12}(\Delta K)^4$ m/cycle.

Substituting $Q = \sqrt{\pi}$ and $n = 4$ into Eq. (18.43b) gives

$$N_f = (1/a_o - 1/a_f)/C\pi^2(\Delta\sigma)^4. \qquad (i)$$

Now, since $(\Delta K_o)^2 = (\Delta\sigma)^2(a_o\pi)$ and $(\Delta K_f)^2 = (\Delta\sigma)^2(a_f\pi)$, Eq. (i) becomes

$$N_f = \frac{\pi(\Delta\sigma)^2/(\Delta K_o)^2 - \pi(\Delta\sigma)/(\Delta K_f)^2}{C\pi^2(\Delta\sigma)^4}$$

$$= \frac{(\Delta K_f/\Delta K_o)^2 - 1}{C\pi(\Delta\sigma\,\Delta K_f)^2}. \tag{ii}$$

This equation shows that $\Delta K_f, \Delta K_o$ and $\Delta\sigma$ govern the life. Under a repeated cycle $\sigma_{max} = \Delta\sigma$, $K_o = \Delta K_o$ and $\Delta K_f = K_{Ic}$. Substituting $K_o = 175\sqrt{(0.1 \times 10^{-3} \times \pi)} = 3.102\,\text{MN/m}^{3/2}$ with the given values for $K_{Ic}, \Delta\sigma$ and C in Eq. (ii):

$$N_f = \frac{(54/3.102)^2 - 1}{40 \times 10^{-12} \times \pi(175 \times 54)^2}$$

$$N_f = 26\,919 \text{ cycles.}$$

Effect of mean stress Since the compressive part of the reversed and alternating cycles in Fig. 18.2 do not contribute to crack growth, then ΔK in Eq. (18.42) applies to the tensile part of these cycles. For wholly tensile cycles, the question arises of how fracture mechanics can account for the effect that mean stress has in reducing cycle life (Fig. 18.3). Composite plots similar to Fig. 18.27, with different ratios of $\sigma_{max}/\sigma_{min}$, have shown that crack growth rates increase, mostly in region 3, following an increase in this stress ratio. Because the growth rate varies inversely with the toughness K_c (under modes I, II and III), Forman *et al.* (1967) modified Eq. (18.42) to include these effects:

$$da/dN = C\,\Delta K^n/[(1 - R)K_c - \Delta K],$$

where now $R = K_{max}/K_{min}$. Other investigators have re-defined ΔK in Eq. (18.42) with an effective stress intensity range $\Delta K_{eff} = K_{max} - K_{op}$ in Fig. 18.28, where ΔK_{op} accounts for that part of the tensile cycle for which the crack may remain partially closed in the presence of a high mean stress.

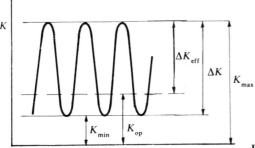

Figure 18.28 Effective opening range.

Effect of environment Under static load conditions, the effect of an aggressive environment may cause failure when stress levels are less than those calculated from K_{Ic}. This is because the effects of stress corrosion cracking, hydrogen and liquid-metal embrittlement enable existing flaws to grow to the necessary critical size. With such environmentally assisted cracking (EAC) phenomena acting under mode I, it becomes necessary to re-define a lower critical stress intensity K_{IEAC} if failure is not to occur by this means over long periods. Normally K_{IEAC} is found for the most common material and environment combinations in laboratory tests lasting at least 100 h on constant bolt-loaded compact tension test pieces (Fig. 18.24(b)). Ratios K_{IEAC}/K_{Ic} are approximately 1/3 but have been found to be as low as 1/10 for alloy steels in an NaCl solution. Intercrystalline failure occurs for stress intensities in the region $K_{IEAC} < K_I < K_{Ic}$. Here the logarithm of the crack growth rate da/dt varies with K_I for most materials in the manner shown in Fig. 18.29. Normally the two regions of strong dependence of growth rate on K_I are interrupted by one of relative insensitivity. The threshold K_I level is identified with K_{IEAC} in this plot.

Fatigue crack growth rates are also sensitive to both the cycle frequency and the temperature. In general, for a given ΔK, the growth rates increase as the environmental attack is enhanced with increasing temperature and decreasing cyclic frequency. This particularly applies to square waveforms incorporating periods of dwell under the maximum and minimum stresses in the cycle. Equation (18.41) will account for this provided the function $f(\Delta K)$ is determined experimentally under simulated conditions. In simplifying life prediction, the effects of the environment on fatigue need only be considered in terms of a change in C and n for the region 2 in Fig. 18.27(b). Alternatively, the separate effects of environment and frequency on fatigue

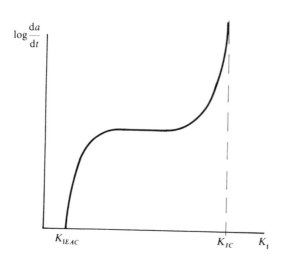

Figure 18.29 Environmental cracking.

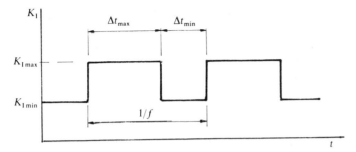

Figure 18.30 Square wave pattern.

crack growth may be superimposed when it is assumed that fatigue and environmentally assisted cracking act independently. The net crack growth rate is then the sum of the individual growth rates, as shown in the following example. In applying this method, the baseline fatigue crack growth data should be established in an inert atmosphere where cyclic frequency and temperature are known to have little effect on the growth rate.

Example 18.12 An aluminium alloy is subjected to cyclic tensile stressing at a frequency $f = 1\,\text{Hz}$ with short, constant dwell periods at the maximum and minimum stresses (Fig. 18.30) where the corresponding stress intensities are $K_{I\max} = 15\,\text{MPa}\sqrt{\text{m}}$ and $K_{I\min} = 8\,\text{MPa}\sqrt{\text{m}}$ respectively. The crack growth rate in argon is given by the power law $da/dN = 13.2 \times 10^{-12}\,\Delta K^{2.9}$ metres/cycle for this fluctuating cycle. Determine the crack growth rate in metres/cycle for an aggressive environment where cracking occurs under the given maximum and minimum stress intensities at growth rates of $da/dt = 1.5 \times 10^{-8}\,\text{m/s}$ and $da/dt = 9 \times 10^{-10}\,\text{m/s}$ respectively. How does the net growth rate depend upon frequency? (T.C.D. 1984)

In the fatigue cycle, $\Delta K = 7\,\text{MPa}\sqrt{\text{m}}$, and the Paris law gives the corresponding crack growth rate as

$$da/dN = 13.2 \times 10^{-12}(7)^{2.9}$$
$$= 3.73 \times 10^{-9}\,\text{m/cycle}.$$

For the square wave in Fig. 18.30, let $\Delta t_{\max}$ and $\Delta t_{\min}$ be the times spent under $K_{I\max}$ and $K_{I\min}$ where the cycle period is

$$\Delta t_{\max} + \Delta t_{\min} = 1/f. \qquad (\text{i})$$

The corresponding environmental crack rates per cycle are

$$(da/dN)_{\max} = (da/dt)_{\max}\,\Delta t_{\max}, \qquad (\text{ii})$$

$$(da/dN)_{\min} = (da/dt)_{\min}\,\Delta t_{\min}, \qquad (\text{iii})$$

Since $\Delta t_{max} = \Delta t_{min}$ and $f = 1$ cycle/s, then from Eq. (i), $\Delta t_{max} = \Delta t_{min} = 0.5$ s. Substituting into Eqs (ii) and (iii):

$$(da/dN)_{max} = 1.5 \times 10^{-8} \times 0.5 = 7.5 \times 10^{-9} \, m/cycle,$$
$$(da/dN)_{min} = 9 \times 10^{-10} \times 0.5 = 0.45 \times 10^{-9} \, m/cycle.$$

The net crack growth rate per cycle is then

$$(da/dN)_{net} = (3.73 + 7.5 + 0.45)10^{-9} = 11.68 \times 10^{-9} \, m/cycle.$$

It is seen from Eq. (i) that the growth rates in Eqs (ii) and (iii) depend upon frequency. In general, when $r = \Delta t_{max}/\Delta t_{min} = $ constant, Eq. (i) gives

$$\Delta t_{max} = r/f(1 + r), \qquad \text{(iv)}$$
$$\Delta t_{min} = 1/f(1 + r). \qquad \text{(v)}$$

Substituting Eqs (iv) and (v) into (ii) and (iii):

$$(da/dN)_{max} = (da/dt)_{max} r/f(1 + r),$$
$$(da/dN)_{min} = (da/dt)_{min}/f(1 + r).$$

The net growth rate per cycle then takes the general form:

$$(da/dN)_{net} = C(\Delta K)^n + (da/dt)_{max} r/f(1 + r) + (da/dt)_{min}/f(1 + r),$$

in which the contribution from fatigue (the first term) is independent of cycle frequency.

REFERENCES

Basquin, D. H. (1910). *Proc. Am. Soc. Test. Mater.* **10**, 625.
Coffin, L. F. (1954). *Trans Am. Soc. Mech. Eng.* **76**, 931.
Corten, H. T. and Dolan, T. J. (1956). *Proceedings of International Conference on Fatigue*, p. 235. Institution of Mechanical Engineers, London.
Dugdale, D. S. (1960). *J. Mech. Phys. Solids* **8**, 100.
Findlay, W. N. and Mathur, P. N. (1956). *Proc. Soc. Exp. Stress Anal.* **XIV**(1), 35.
Forman, R. G., Kearney, V. E. and Engle, R. M. (1967). *J. Basic Eng., Trans. Am. Soc. Mech. Eng.* **89**, 459.
Forrest, P. G. (1970). *Fatigue of Metals*. Pergamon Press, Oxford.
Goodman, J. (1899). *Mechanics Applied to Engineering*. Longmans, London.
Gough, H. J. and Pollard, H. V. (1935). *Proc. Inst. Mech. Eng.* **13**, 3.
Griffith, A. A. (1921). *Phil. Trans. R. Soc. A* **221**, 163.
Heywood, R. B. (1956). *Colloquium on Fatigue, Stockholm (1955)*, p. 92, Springer-Verlag, Berlin.
Irwin, G. R. (1957). *J. Appl. Mech.* **24**, 109.
Irwin, G. R. (1964). *Appl. Mater. Res.* **3**, 65.
Manson, S. S. (1953). N.A.C.A., Tech. Note. 2933.
Manson, S. S. (1965). *Exp. Mech.* **5**, 193.
Marco, S. M. and Starkey, W. L. (1954). *Trans Am. Soc. Mech. Eng.* **76**, 627.
Miner, A. M. (1945). *J. Appl. Mech.* **12**, A159.
Neuber, H. (1946). *Theory of Notch Stresses*. J. W. Edwards, Michigan.

Orowan, E. (1955). *Weld. J. Res. Suppl.* **20**, 157s.

Paris, P. C. (1964). *Fatigue—An Interdisciplinary Approach*, p. 107. Syracuse University Press.

Peterson, R. E. (1974). *Stress Concentration Factors*, Wiley, New York.

Rees, D. W. A. (1987). *Int. J. Pressure Vessels Piping* **30**, 57.

Rice, J. R. and Johnson, M. A. (1970). *Inelastic Behavior of Solids*, p. 641. McGraw-Hill, New York.

Smith, J. O. and Collins, W. L. (1942). *Proc. Am. Soc. Test. Mater.* **42**, 639.

Soderberg, C. R. (1930). *Trans Am. Soc. Mech. Eng.* **52**, 13.

Westergard, H. M. (1939). *J. Appl. Mech.* **61**, A49.

EXERCISES

Fatigue

18.1 A steel has an ultimate tensile strength of 475 MPa and a yield strength of 350 MPa with a fatigue limit under a fully reversed stress cycle of ± 180 MPa. Estimate the maximum stress that would give infinite life for a minimum cyclic stress limit of 110 MPa according to the Gerber, Goodman and Soderberg predictions.

18.2 For a particular steel the endurance limit is expressed as $0.6 \times$ UTS for reversed bending at 10^5 cycles. Construct a Goodman diagram and determine the permissible stress range to avoid failure at 10^5 cycles when the mean stress is increased to $0.25 \times$ UTS. (C.E.I. Pt. 2, 1980)

18.3 It is required to design a tie bolt. Fatigue data, obtained from reversed bending tests, are available for the particular steel to be used. Describe how such data can be used to determine a safe working stress, taking into account that: (i) the bolt will sustain a constant tensile load P on which will be superimposed a cyclic load of amplitude $0.25P$; (ii) the bolt has a constant cross-sectional area A except at a small circumferential V-shaped groove; and (iii) the amplitude is constant except for a 5 per cent increase on two occasions in the bolt life. (C.E.I. Pt. 2, 1977)

18.4 A titanium alloy is to be subjected to a fluctuating tensile cycle in which the mean stress is 450 MPa. If an endurance of 10^8 cycles corresponds to a fully reversed stress cycle of amplitude 650 MPa, calculate the allowable stress amplitude to give the same endurance for the fluctuating cycle. The yield and ultimate tensile stresses for the alloy are 800 and 1400 MPa respectively.

18.5 The fatigue behaviour of an aluminium alloy, under reversed stress conditions, is given by $\Delta \sigma \, bN_f = C$, where $\Delta \sigma$ is the stress range, N_f is the cycle life, and b and C are constants. Given that $N_f = 10^8$ cycles when $\Delta \sigma = 340$ MPa and $N_f = 10^5$ cycles when $\Delta \sigma = 520$ MPa, determine the values of b and C. It is required that a component made from the same material should not fail in less than 10^7 cycles. Calculate the permissible amplitude of the cyclic tensile stress if the component suffers a tensile mean stress of 20 MPa and the yield stress of the material is 400 MPa. (C.E.I. Pt. 2, 1973)

18.6 Cyclic bending stresses of 75, 60 and 40 MPa are applied to a beam in a lifting machine. If that part of the life spent at each stress level is 30, 50 and 20 per cent, estimate the working life in days when the machine operates continuously at 10 cycles/day. Take the respective fatigue lives as 10^3, 10^4 and 10^5 cycles respectively.

Answer: 284 days.

18.7 Estimate the working life in days of a torsional spring that is subjected to reversed cyclic shear stresses of 60, 50, 40 and 35 MPa for 10, 50, 20 and 20 per cent respectively of its life, when the spring operates continuously at 40 cycles/day. The fatigue lives at the given stresses are 10^3, 10^4, 10^5 and 10^6 cycles respectively.

Answer: 164 days.

18.8 A centre-zero moving coil has a torsion spring of circular section 0.15 mm diameter with effective length of 50 mm. It is used for inspection at a rate of 5 components/minute. A satisfactory component twists the wire through $\pm 20°$, but rejected components twist the wire through $30°$. The latter comprise 10 per cent of the total. Estimate the working life of the spring in years for a working year of 50 weeks of 5 days/week and 8 h/day. If the deflection for rejects is limited to $\pm 25°$, determine the percentage increase in life. Take $G = 48.25$ GPa and the following fatigue data under reversed torsion:

Life (cycles)	10^5	10^6	10^7	10^8
Shear stress (MPa)	42.7	29	22.8	20.7

18.9 An aircraft wing experiences the block loading spectrum below repeatedly in service. Calculate from Miner's rule the number of such spectra that may be applied, given the following fatigue lives N_f at each stress amplitude:

$\Delta\sigma$ (MPa)	345	290	240	215	195
$n \times 10^3$ (cycles)	20	43	36	30	27
$N_f \times 10^3$ (cycles)	65	300	950	3100	42 000

18.10 The loading spectrum on an aluminium alloy aircraft component is defined by the number of cycles (n) occurring at various reversed stress levels in every 100×10^3 cycles as given. Also tabulated is the 90 per cent survival fatigue life for each stress level. Determine an expected fatigue life for the component based upon Miner's hypothesis of damage. (C.E.I. Pt. 2, 1970)

$\Delta\sigma$ (MPa)	344.5	289.4	241.2	213.6	192.9
$n \times 10^3$ (cycles)	2	8	17	28	45
$N_f \times 10^3$ (cycles)	70	290	960	3000	40 000

18.11 In two-step fatigue tests, repeated stress amplitudes of 250 and 300 MPa are applied in a low–high and high–low sequence. If, for each sequence, the initial stress is applied for one-half the cycle life, calculate the remaining life under the final stress according to (i) Miner's rule and (ii) when the following non-linear damage laws are assumed for each stress level:

$$D = (n/N)^2 \text{ for 300 MPa}, \qquad N_f = 140\,000$$

and

$$D = (n/N)^3 \text{ for 250 MPa}, \qquad N_f = 260\,000.$$

Taking failure to occur for $D = 1$, establish the effect of the stress sequence according to the non-linear theory.

18.12 A cylindrical vessel of 760 mm internal diameter and 6.5 mm wall thickness is subjected to internal pressure fluctuating fron $P/2$ to P. The fatigue limit of the material at 10^7 cycles under reversed direct stress is ± 230 MPa and the static yield stress in simple tension is 310 MPa. Assuming a shear strain energy criterion of equivalent fatigue strength under biaxial cyclic stress and using the Soderberg diagram, determine a nominal value for P to give an endurance of 10^7 cycles. Neglect the radial stress. (C.E.I. Pt. 2, 1967)

18.13 How does Heywood's method of estimating fatigue strength of a notched component overcome some of the limitations of the notch sensitivity index method? Describe its use for an alloy steel having a plain fatigue strength in rotating bending of ± 770 MPa made into a stepped shaft with diameters 150 and 75 mm, joined by a 4.75 mm fillet radius. (I.C. 1970)

18.14 Show that the following low-cycle fatigue results for steel obey the Coffin–Manson law $\Delta\epsilon^P = CN_f^\beta$, where C and β are constants to be determined:

$\Delta\epsilon^P$	0.04	0.021	0.016	0.0084
N_f	100	500	1000	5000

If a component made from the same steel is subjected to $\Delta\epsilon^P = 2$ per cent for 300 cycles, then $\Delta\epsilon^P = 1$ per cent for 500 cycles, find the cycle life remaining corresponding to $\Delta\epsilon^P = 0.75$ per cent, assuming that Miner's rule applies.

18.15 At a temperature of 800 °C the ends of a steel bar are clamped rigidly. Thereafter, the bar is subjected to a temperature cycle of repeated cooling and heating between the limits of 100 °C and 800 °C until failure occurs by thermally induced fatigue. Calculate the endurance, ignoring work-hardening, when a Coffin–Manson law $N^{1/2}\Delta\epsilon^P = 0.04$ applies to the steel. Assuming that Miner's rule applies, how many thermal cycles between 800 °C and 400 °C remain if the bar had previously been cycled from 800 °C to 500 °C for 1000 cycles? Take $Y = 310$ MPa, $E = 207$ GPa and $\alpha = 10 \times 10^{-6}/°C$. (I.C. 1971)

18.16 A fatigue crack is assumed initiated when a surface crack of length 10^{-6} m is created. The corresponding percentage of cycle life consumed is given by $1000N_i = \sqrt{(2.02)N_f^{2.02}}$, where N_i and N_f are the number of cycles to initiation and failure respectively. Determine the cyclic lifetime of two specimens, one having an N_i/N_f ratio of 0.01, corresponding to a stress range of $\Delta\sigma_1$, and the other having an N_i/N_f ratio of 0.99, corresponding to a stress range of $\Delta\sigma_2$. If the crack at failure is 1 mm deep, determine the mean crack propagation rate at $\Delta\sigma_1$ and the mean crack nucleation rate at $\Delta\sigma_2$. (C.E.I. Pt. 2, 1975)

Fracture

18.17 A glass sheet of uniform thickness fails under a stress of 60 MPa. Estimate the size of the inherent defect in glass given that $E = 72$ GPa and $S_o = 1$ J/m^2. (C.E.I. Pt. 2, 1972)

18.18 Determine the stress concentration factor at a point 0.01 mm away from the tip of a sharp 0.5 mm edge crack in a semi-infinite plate (see Fig. 18.16(a)) when a stress of 20 MPa is applied around the boundary.
   Answer: 5.

18.19 A large, thick steel plate contains a single edge crack of length 1 mm (see Fig. 18.16(a)). If the steel has a fracture toughness of $K_{1c} = 53$ MN/m$^{3/2}$ and a yield strength of 950 MPa, find the stress at which fast fracture would occur and determine whether the fracture would be brittle or ductile.

18.20 A thin steel sheet 3.5 mm thick by 305 mm wide contains a semi-elliptical surface flaw 25 mm long $\times$ 0.5 mm deep (see Fig. 18.16(c)). If the plane strain fracture toughness is $K_{1c} = 66$ MN/m$^{3/2}$ and the yield strength is 1930 MPa, estimate the remote stress that would cause failure when applied normal to the crack area. Do not neglect crack-tip plasticity.
   Answer: 1492 MPa.

18.21 A weld in an infinite plate contains residual stresses approximately equal to the parent plate yield stress $Y = 800$ MPa. The allowable applied stress is to be based upon Y with a safety factor of 2. If the fracture toughness value for the plate material is $K_{1c} = 60$ MN/m$^{3/2}$, find the maximum permissible size of crack in the weld.

18.22 Find the critical defect size for a large-diameter thick-walled steel cylinder of 500 mm o.d. and 375 mm i.d. that is to sustain an internal pressure of 350 MPa if $K_{1c} = 100$ MN/m$^{3/2}$ and $Y = 1500$ MPa. Check on the validity of an assumed plane strain condition.

18.23 A single edge notch tension test is conducted on a steel plate 15 mm thick by 100 mm wide with a sharp notch 50 mm deep (see Fig. 18.24(c)). If the fracture load is 400 kN, evaluate the fracture toughness and the critical rate of strain energy release. Is K_{1c} valid? Take $E = 210$ GPa, $v = 0.27$ and $Y = 1235$ MPa.
   Answer: $K_{1c} = 809.5$ MN/m$^{3/2}$; $\delta U/\delta a = 0.043$ MJ/m.

18.24 A light alloy three-point bend specimen (Fig. 18.24(a)) is 50 mm thick and spans 200 mm between the supports with $w = 100$ mm and $a = 53$ mm. If the test piece fails under a force of 53.2 kN, calculate

    (i) the fracture toughness with a check on its validity for a yield stress of 450 MPa;
    (ii) the minimum value of yield stress for the calculated K_{Ic} that would ensure plane strain,
    (iii) the size of the plastic zone,
    (iv) the critical rate of strain energy release and
    (v) the specific fracture energy.

Take $v = 0.31$ and $E = 75$ MPa.
    Answer: 39.42 MN/m$^{3/2}$; 278.7 MPa; 407 μm; 0.937 kJ/m; 9.365 kJ/m^2.

18.25 A plate 100 mm wide, 5 mm thick containing a central crack 8 mm long (see Fig. 18.16(e)) fails under an applied tensile force of 200 kN. If the yield stress of the plate material is 500 MPa, evaluate the fracture toughness, assuming both brittle and ductile failure. (T.C.D. 1979)
    Answer: 45 MN/m$^{3/2}$; 51.87 MN/m$^{3/2}$.

18.26 The stress function $\phi = Ar^{3/2}\cos^3(\theta/2)$ prescribes the stress in the vicinity of a sharp crack when the origin for r and θ lies at its tip. Express A in terms of K when, for a remotely applied stress, there is no stress applied normally to the crack surfaces. Hence show that the stress components are $\sigma_{x,y} = [K/(2\pi r)^{1/2}]\cos(\theta/2)[1 \pm \sin(\theta/2)]$. (I.C. 1971)

18.27 A function of the form $C(\theta)r^q$, where $q > 1$, prescribes the stress state around the tip of an edge notch for a 90° included angle with the origin for r, θ at the notch tip. Determine the value of the constant q. (I.C. 1971)

18.28 A tie bar, 150 mm wide and 12.5 mm thick, is to withstand a tensile stress of 100 MPa. Given that edge cracks 2.5 mm deep and semi-elliptical surface cracks 10 mm long by 1 mm deep (see Fig. 18.16) are present, select the most suitable steel from two that are available with fracture toughnesses and yield strength values of 100 MN/m$^{3/2}$, 1350 MPa and 54.5 MN/m$^{3/2}$, 1700 MPa respectively. If the quoted K_{Ic} for the chosen steel is to be checked from three-point bending a beam of width 50 mm, thickness 12.5 mm, span 200 mm and crack length 15 mm, what should be the fracture load of the selected steel? (I.C. 1969)

18.29 Show that the radius r_y of the plastic zone ahead of a blunt crack of tip radius ρ is expressed by $r_y = (2/\pi)(K_1/Y)^2[1 + (\rho/r_y) + \sqrt{(2\rho/r_y)}]$ given that the opening stress distribution is $\sigma_y = K_1(1 + \rho/2r)/(2\pi r)^{1/2}$. Take the origin of coordinates at $\rho/2$ from the notch tip. (T.C.D. 1983)

18.30 Two large steel plates, each 3 m wide, contain 2.5 mm central cracks through their thickness. If, for one plate, $K_{Ic} = 115$ MN/m$^{3/2}$, $Y = 910$ MPa, and for the other, $K_{Ic} = 55$ MN/m$^{3/2}$, $Y = 1035$ MPa, which plate could support a tensile force of 4.5 MN with the lesser weight?

18.31 Inspection of a large bolt, with outer and root diameters of 100 mm and 92 mm respectively, reveals a circumferential crack 1 mm deep around the root of one thread. If, for the bolt material, $K_{Ic} = 50$ MN/m$^{3/2}$ and $Y = 300$ MPa, estimate the axial tensile force P to cause failure given that the stress intensity factor for a bar of radius R with a single circumferential crack of depth a is $K = Pf(a/R)/(\pi R^3)^{1/2}$, where the compliance function is given as

$$f(a/R) = [1.12(a/R)^{1/2} - 1.18(a/R)^{3/2} + 0.74(a/R)^{5/2}]/(1 - a/R)^{3/2}.$$

(T.C.D. 1984)

Fatigue Crack Growth

18.32 Calculate the pressure that would cause a thin cylinder of 7.5 m i.d. and 7.5 mm thick, to fail when it contains a through-thickness crack 12.5 mm long aligned with the longitudinal direction. If this vessel is to be subjected to 4000 cycles of repeated internal pressure, determine the maximum cyclic pressure, based upon a safety factor of 2, given the crack growth rate $da/dN = 50 \times 10^{-12}(\Delta K)^4$ m/cycle for ΔK in MN/m$^{3/2}$.

18.33 A steel plate is 350 mm wide and 35 mm thick. It contains an edge crack 9 mm deep. If $K_{Ic} = 82$ MN/m$^{3/2}$, find the cyclic life when a remotely applied tensile force varies between limits of 1.75 MN and 2.75 MN. Take $da/dN = 4.62 \times 10^{-12}(\Delta K)^{3.3}$ m/cycle.

18.34 A thin-walled steel cylinder 8 m i.d. with 50 mm wall thickness, is to withstand 3000 repeated cycles at a maximum pressure of 5.5 MPa. If the rate of growth of pre-existing cracks follows the law $da/dN = 24 \times 10^{-15}(\Delta K)^4$ (m/cycle) and the fracture toughness is $K_{Ic} = 210$ MN/m$^{3/2}$, find the test pressure for 3000 safe cycles.

18.35 An infinite aluminium alloy plate, containing a 10 mm crack, is subjected to a remotely applied stress that fluctuates between 5 and 60 MPa. How many cycles must be applied to double the crack length? Take $da/dN = 45.55 \times 10^{-12}(\Delta K)^3$ m/cycle.

18.36 An aluminium alloy panel is subjected to a constant-amplitude repeated stress cycle in which the maximum value is $\sigma_{max} = 175$ MPa. If the panel contains a central crack 1.5 mm long, find the number of cycles to failure given that the critical crack length is 35 mm and the growth rate is $da/dN = 15.2 \times 10^{-4}(\Delta K)^{2.77}$ mm/cycle, where $\Delta K = 11.1\sigma_{max}\sqrt{a}$.

INDEX